LONDON MATHEMATICAL SOCIETY LECTURE NOTE SERIES

The titles below are available from booksellers, or from Cambridge University Press at
http://www.cambridge.org/mathematics

366 Highly oscillatory problems, B. ENGQUIST, A. FOKAS, E. HAIRER & A. ISERLES (eds)
367 Random matrices: High dimensional phenomena, G. BLOWER
368 Geometry of Riemann surfaces, F.P. GARDINER, G. GONZÁLEZ-DIEZ & C. KOUROUNIOTIS (eds)
369 Epidemics and rumours in complex networks, M. DRAIEF & L. MASSOULIÉ
370 Theory of p-adic distributions, S. ALBEVERIO, A.YU. KHRENNIKOV & V.M. SHELKOVICH
371 Conformal fractals, F. PRZYTYCKI & M. URBAŃSKI
372 Moonshine: The first quarter century and beyond, J. LEPOWSKY, J. MCKAY & M.P. TUITE (eds)
373 Smoothness, regularity and complete intersection, J. MAJADAS & A. G. RODICIO
374 Geometric analysis of hyperbolic differential equations: An introduction, S. ALINHAC
375 Triangulated categories, T. HOLM, P. JØRGENSEN & R. ROUQUIER (eds)
376 Permutation patterns, S. LINTON, N. RUŠKUC & V. VATTER (eds)
377 An introduction to Galois cohomology and its applications, G. BERHUY
378 Probability and mathematical genetics, N. H. BINGHAM & C. M. GOLDIE (eds)
379 Finite and algorithmic model theory, J. ESPARZA, C. MICHAUX & C. STEINHORN (eds)
380 Real and complex singularities, M. MANOEL, M.C. ROMERO FUSTER & C.T.C WALL (eds)
381 Symmetries and integrability of difference equations, D. LEVI, P. OLVER, Z. THOMOVA & P. WINTERNITZ (eds)
382 Forcing with random variables and proof complexity, J. KRAJÍČEK
383 Motivic integration and its interactions with model theory and non-Archimedean geometry I, R. CLUCKERS, J. NICAISE & J. SEBAG (eds)
384 Motivic integration and its interactions with model theory and non-Archimedean geometry II, R. CLUCKERS, J. NICAISE & J. SEBAG (eds)
385 Entropy of hidden Markov processes and connections to dynamical systems, B. MARCUS, K. PETERSEN & T. WEISSMAN (eds)
386 Independence-friendly logic, A.L. MANN, G. SANDU & M. SEVENSTER
387 Groups St Andrews 2009 in Bath I, C.M. CAMPBELL *et al* (eds)
388 Groups St Andrews 2009 in Bath II, C.M. CAMPBELL *et al* (eds)
389 Random fields on the sphere, D. MARINUCCI & G. PECCATI
390 Localization in periodic potentials, D.E. PELINOVSKY
391 Fusion systems in algebra and topology, M. ASCHBACHER, R. KESSAR & B. OLIVER
392 Surveys in combinatorics 2011, R. CHAPMAN (ed)
393 Non-abelian fundamental groups and Iwasawa theory, J. COATES *et al* (eds)
394 Variational problems in differential geometry, R. BIELAWSKI, K. HOUSTON & M. SPEIGHT (eds)
395 How groups grow, A. MANN
396 Arithmetic differential operators over the p-adic integers, C.C. RALPH & S.R. SIMANCA
397 Hyperbolic geometry and applications in quantum chaos and cosmology, J. BOLTE & F. STEINER (eds)
398 Mathematical models in contact mechanics, M. SOFONEA & A. MATEI
399 Circuit double cover of graphs, C.-Q. ZHANG
400 Dense sphere packings: a blueprint for formal proofs, T. HALES
401 A double Hall algebra approach to affine quantum Schur–Weyl theory, B. DENG, J. DU & Q. FU
402 Mathematical aspects of fluid mechanics, J.C. ROBINSON, J.L. RODRIGO & W. SADOWSKI (eds)
403 Foundations of computational mathematics, Budapest 2011, F. CUCKER, T. KRICK, A. PINKUS & A. SZANTO (eds)
404 Operator methods for boundary value problems, S. HASSI, H.S.V. DE SNOO & F.H. SZAFRANIEC (eds)
405 Torsors, étale homotopy and applications to rational points, A.N. SKOROBOGATOV (ed)
406 Appalachian set theory, J. CUMMINGS & E. SCHIMMERLING (eds)
407 The maximal subgroups of the low-dimensional finite classical groups, J.N. BRAY, D.F. HOLT & C.M. RONEY-DOUGAL
408 Complexity science: the Warwick master's course, R. BALL, V. KOLOKOLTSOV & R.S. MACKAY (eds)
409 Surveys in combinatorics 2013, S.R. BLACKBURN, S. GERKE & M. WILDON (eds)
410 Representation theory and harmonic analysis of wreath products of finite groups, T. CECCHERINI-SILBERSTEIN, F. SCARABOTTI & F. TOLLI
411 Moduli spaces, L. BRAMBILA-PAZ, O. GARCÍA-PRADA, P. NEWSTEAD & R.P. THOMAS (eds)
412 Automorphisms and equivalence relations in topological dynamics, D.B. ELLIS & R. ELLIS
413 Optimal transportation, Y. OLLIVIER, H. PAJOT & C. VILLANI (eds)
414 Automorphic forms and Galois representations I, F. DIAMOND, P.L. KASSAEI & M. KIM (eds)
415 Automorphic forms and Galois representations II, F. DIAMOND, P.L. KASSAEI & M. KIM (eds)
416 Reversibility in dynamics and group theory, A.G. O'FARRELL & I. SHORT
417 Recent advances in algebraic geometry, C.D. HACON, M. MUSTAŢĂ & M. POPA (eds)
418 The Bloch–Kato conjecture for the Riemann zeta function, J. COATES, A. RAGHURAM, A. SAIKIA & R. SUJATHA (eds)
419 The Cauchy problem for non-Lipschitz semi-linear parabolic partial differential equations, J.C. MEYER & D.J. NEEDHAM
420 Arithmetic and geometry, L. DIEULEFAIT *et al* (eds)
421 O-minimality and Diophantine geometry, G.O. JONES & A.J. WILKIE (eds)
422 Groups St Andrews 2013, C.M. CAMPBELL *et al* (eds)
423 Inequalities for graph eigenvalues, Z. STANIĆ
424 Surveys in combinatorics 2015, A. CZUMAJ *et al* (eds)
425 Geometry, topology and dynamics in negative curvature, C.S. ARAVINDA, F.T. FARRELL & J.-F. LAFONT (eds)
426 Lectures on the theory of water waves, T. BRIDGES, M. GROVES & D. NICHOLLS (eds)
427 Recent advances in Hodge theory, M. KERR & G. PEARLSTEIN (eds)
428 Geometry in a Fréchet context, C.T.J. DODSON, G. GALANIS & E. VASSILIOU
429 Sheaves and functions modulo p, L. TAELMAN
430 Recent progress in the theory of the Euler and Navier–Stokes equations, J.C. ROBINSON, J.L. RODRIGO, W. SADOWSKI & A. VIDAL-LÓPEZ (eds)

London Mathematical Society Lecture Note Series: 427

Recent Advances in Hodge Theory

Period Domains, Algebraic Cycles, and Arithmetic

Edited by

MATT KERR
Washington University in St. Louis

GREGORY PEARLSTEIN
Texas A & M University

Shaftesbury Road, Cambridge CB2 8EA, United Kingdom

One Liberty Plaza, 20th Floor, New York, NY 10006, USA

477 Williamstown Road, Port Melbourne, VIC 3207, Australia

314–321, 3rd Floor, Plot 3, Splendor Forum, Jasola District Centre, New Delhi – 110025, India

103 Penang Road, #05–06/07, Visioncrest Commercial, Singapore 238467

Cambridge University Press is part of Cambridge University Press & Assessment,
a department of the University of Cambridge.

We share the University's mission to contribute to society through the pursuit of
education, learning and research at the highest international levels of excellence.

www.cambridge.org
Information on this title: www.cambridge.org/9781107546295

First published 2016

A catalogue record for this publication is available from the British Library

Library of Congress Cataloging-in-Publication data
Recent advances in Hodge theory : period domains, algebraic cycles,
and arithmetic / edited by Matt Kerr, Washington University in St Louis,
Gregory Pearlstein, Texas A & M University.
pages cm. – (London Mathematical Society lecture note series; 427)
Includes bibliographical references.
ISBN 978-1-107-54629-5 (pbk.)
1. Algebraic cycles–Congresses. 2. Hodge theory–Congresses. 3. Geometry,
Algebraic–Congresses. 4. Differential-algebraic equations–Congresses.
I. Kerr, Matthew D., 1975– editor. II. Pearlstein, Gregory, 1970, editor.
QA564.R426 2013
514′.74–dc23 2015035249

ISBN 978-1-107-54629-5 Paperback

Contents

PART (II) ALGEBRAIC CYCLES AND NORMAL FUNCTIONS

PART (III) THE ARITHMETIC OF PERIODS

(III.A) MOTIVES, GALOIS REPRESENTATIONS, AND AUTOMORPHIC FORMS

(III.B) MODULAR FORMS AND ITERATED INTEGRALS

Preface

The summer school and conference on "Recent advances in Hodge theory: Period domains, algebraic cycles, and arithmetic" was held from June 10-20, 2013 at the Pacific Institute of Mathematical Sciences, on the beautiful campus of the University of British Columbia in Vancouver, Canada. It was organized by Matt Kerr (Washington University in St. Louis), James D. Lewis (University of Alberta), and Gregory Pearlstein (Texas A&M University).

The conference received generous financial and administrative assistance from PIMS, and we are grateful to Ruth Situma and the rest of the PIMS staff for their hard work. In addition to PIMS, it was supported by the NSF (Conference Grant DMS-1259024), NSA, and Foundation Compositio Mathematica. We are grateful to all these sponsor organizations, whose support allowed for a large number of junior participants, as well as a significant number of participants from outside North America. This led to a vibrant and memorable event.

This proceedings volume for the 2013 Hodge theory conference is edited by Matt Kerr and Gregory Pearlstein. All papers in this volume have been rigorously peer-reviewed, in several instances by more than one referee. The editors wish to express their appreciation to all the referees for their efforts, especially given the short time-frame. In addition, we want to thank Sam Harrison and the staff at Cambridge University Press for their assistance in preparing the volume.

A couple of the articles contained herein congratulate our co-organizer James Lewis on his 60th birthday, which was celebrated on the last day of the conference. We wanted to add our congratulations and thanks, both for his work as co-organizer and for his friendship, support, and inspiring mathematical presence over the years.

St. Louis, MO — Matt Kerr
College Station, TX — Gregory Pearlstein
May 2015

James D. Lewis

Introduction

In its simplest form, Hodge theory is the study of *periods* – integrals of algebraic differential forms which arise in the study of complex geometry, number theory and physics. Its difficulty and richness arise in part from the non-algebraicity of these integrals. According to the beautiful conjectures of Hodge, Bloch and Beilinson, what algebraic structure they do have should be explained by "algebraic cycles." There has been much recent progress on these conjectures and on classifying spaces for periods, as well as their asymptotics and arithmetic.

The main goal of the Vancouver conference was to bring together a diverse community of world-leading mathematicians – in Hodge theory, arithmetic geometry, algebraic cycles, complex geometry, and representation theory – around the common theme of *period maps* (considered broadly). With an intensive summer school followed by 32 research talks, it attracted over 85 participants from the US, Canada, Mexico, Europe and Japan, and sparked several new collaborations. With this book, we hope to draw an even larger audience into this area, and to cement the impact of the conference. In particular, this volume includes careful write-ups of expository talks by Wushi Goldring, Radu Laza and Richard Hain, as well as papers presenting key recent developments in each of several focus areas. We hope that it is useful for graduate students and seasoned researchers alike.

Overview of this volume

The birth of modern Hodge theory began with the work of P. Griffiths, who devised an extension of Lefschetz's original proof of the Hodge conjecture using normal functions and variations of Hodge structure. Despite the fact that this program did not lead to the desired fruition in higher codimension, it has had a lasting impact on the subject. Alongside the study of *algebraic cycles*, the last half-century has seen the development of rich theories of *Hodge theory at the boundary* and *symmetries of Hodge structures*. In particular, by the recent work of Griffiths and others, the Hodge Conjecture itself can now be stated in terms of the asymptotic behavior of normal functions; while Mumford-Tate (symmetry) groups of Hodge structures have led to proofs of the Hodge and Beilinson-Hodge conjectures in special cases.

Algebraic cycles (and normal functions), Mumford-Tate groups (and representation theory), and the Hodge-theoretic boundary (degenerations, compactifications) were major themes of the conference, and remain so for this volume, with sections organized around asymptotics and cycles, and Mumford-Tate groups appearing throughout in papers by Green-Griffiths, Laza-Zheng, Kaplan, Arapura, and Abdulali. Another closely related theme of the volume is the *arithmetic of periods*, which is older than Hodge theory proper – dating back to Euler's work on relations between multiple zeta values, and later work on modular forms and L-functions. Again, Hodge theory at the boundary plays a critical role, as can be seen in Hain's contribution.

We now turn to an outline of the volume's contents.

(I) Hodge theory at the boundary
Period domains and their compactifications

The volume begins with a detailed version of R. Laza's expository lectures on *Classical period domains* (i.e., the unconstrained Hermitian symmetric case of Mumford-Tate domains), written with Z. Zhang. This article reviews the classification of Hermitian symmetric domains and the associated locally symmetric (i.e. connected Shimura) varieties, and treats the representation-theoretic description of variations of Hodge structure over the latter, including the classification of VHS of abelian variety type and Calabi-Yau type. It concludes with a discussion of the Hodge-theoretic meaning of the boundary components in the Baily-Borel and toroidal compactifications.

This is followed by two articles studying the behavior of invariant metrics on boundary components of M-T domains, in the classical (i.e. mixed Shimura variety) resp. period domain cases. The first article, by Burgos, Kramer, and Kühn, considers the behavior at infinity of the (metrized) Jacobi line bundle on the universal elliptic curve $E^0(N)$. The authors show that its "naive" extension as a $\mathbb{Q}$-line bundle encounters singularities at the double points of Néron N-gons, which are directly connected to the "height jumping" phenomenon studied by Brosnan, Hain, and Pearlstein. To obtain an extension functorial with respect to pullback, they instead consider a limit over all possible toroidal compactifications, and show that Chern-Weil theory is compatible with intersections on the resulting object.

The second article is by A. Kaplan. By the work of Griffiths, Schmid, and others, the classifying spaces of pure Hodge structures are homogeneous spaces which admit an invariant hermitian metric with negative holomorphic sectional curvature along horizontal directions. In contrast, due to the presence of a unipotent radical, there is no such statement in general for classifying spaces of graded-polarized mixed Hodge structures. Nonetheless, Kaplan shows that at each point F of such a classifying space $\mathcal{M}$, there exists a pair of subgroups $G_{\mathbb{R}}$ and e^{Λ} such that (i) $\mathcal{M} = G_{\mathbb{R}}e^{\Lambda}F$ and (ii) there is a natural Hermitian structure on $\mathcal{M}$ such that for any $g \in G_{\mathbb{R}} \cup e^{\Lambda}$, left translation by g is an isometry from F to gF.

Period maps and algebraic geometry

Classically, the differential of the period mapping sends the tangent space of the Kuranishi space of a smooth projective variety to the tangent space of the period domain. This section begins with an article by Green and Griffiths on *deformation theory and limiting mixed Hodge structures* (LMHS), which took shape out of the authors' desire to extend this differential "to the boundary," by considering first-order deformations of a singular variety. Making use of the work of Friedman, Fujisawa, and Steenbrink, they treat both the map from semistable smoothing deformations (in several variables) to LMHS, and from non-smoothing deformations to the tangent space of the boundary component classifying LMHS.

The relation between mirror symmetry and variational Hodge theory comes into view in the next two papers. In a series of joint works with Kato and Nakayama, S. Usui has defined logarithmic partial compactifications of classifying spaces for mixed Hodge structures. The purpose of his contribution to this volume is to study open mirror symmetry by combining this framework with Iritani's $\hat{\Gamma}$-integral structure and recent work of Morrison and Walcher. In particular, the truncated normal function arising from the domainwall tension is shown to extend to a log normal function, i.e. a section of the log Néron model.

The second article, by 5 authors (Clingher, Doran, Harder, Novoseltsev, and Thompson), is motivated by a 2006 paper of Doran and Morgan which classified the Calabi-Yau VHS with $h^{2,1} = 1$ over $\mathbb{P}^1 \backslash \{0, 1, \infty\}$ (and monodromies of prescribed forms) into fourteen cases, finding CY-threefold realizations for all but the "14th case." This article is a study of two families of $K3$-fibered $h^{2,1} = 3$ CY 3-folds, and the main result is that a 1-parameter singular subfamily of one of them (which is in fact the mirror of a bidegree (2,12) complete intersection in $\mathbb{WP}(1, 1, 1, 1, 4, 6)$) realizes the 14th case VHS. The authors also obtain a counterexample to a conjecture of Morrison, by showing that this 1-parameter family has no CY resolution, while its mirror admits a CY smoothing.

(II) Algebraic cycles and normal functions

Higher Chow cycles were introduced by S. Bloch to "geometrize" higher algebraic K-theory, and these days are also (in the smooth case) known as motivic cohomology classes. This section begins with two papers on these "generalized" algebraic cycles. In his article, Asakura constructs real-regulator indecomposable K_1-cycles (more precisely, $CH^2(X,1)$) on families of elliptic surfaces with "colliding semistable singular fibers." This gives a simple, general approach to constructions in the literature that have required difficult computations. The article of de Jeu, Lewis and Patel reports new evidence for the "relative Beilinson-Hodge conjecture" formulated in work of Lewis with S.-J. Kang, especially in the case of $CH^r(X,1)$. Unlike the original B-H conjecture, the relative version is expected to be valid even when the varieties are not defined over a number field.

At the heart of current thought on the Hodge conjecture in the last decade, two intertwined programs have emerged. The approach referred to above, proposed by Griffiths and Green (and influenced by work of Clemens and Thomas) reduces the conjecture to the existence of singularities for certain several-variable admissible normal functions obtained from Hodge classes. While this criterion pertains *a priori* to degenerations of normal functions, a result of Schnell reveals the importance of estimates on the *dimension* of its zero loci, which have recently been proven to be algebraic by three different groups of researchers. Another approach, championed by Voisin (building on work of Cattani, Deligne and Kaplan), is to break the Hodge conjecture into two pieces: first, to show that the locus of Hodge classes in a variation is defined over a number field; and second, to prove the Hodge conjecture on arithmetic varieties. Key to this approach is showing that a given family of Hodge classes is *absolute*, extending Deligne's theorem for abelian varieties.

The papers of M. Saito and Saito-Schnell in this volume contain an important advance in this direction. Saito's paper considers an admissible normal function ν of an admissible variation of MHS of strictly negative weights, and an irreducible component Z of its zero locus (known to be algebraic by work of Brosnan-Pearlstein, Schnell, and Kato-Nakayama-Usui). The results of his paper imply that if Z contains a k-rational point ($k \subset \mathbb{C}$ of finite transcendence degree), then Z is defined over k; here ν need *not* come from an algebraic cycle. The joint paper with Schnell proves the comparable result for the Hodge loci of a VHS of weight zero, which is closely related to the theory of absolute Hodge classes when $k = \bar{\mathbb{Q}}$.

The remaining papers in this section concern two of the basic conjectures on (usual) algebraic cycles: the generalized Hodge conjecture (GHC) and Bloch conjecture. An observation of Grothendieck is that, for a variety X, one can deduce the GHC from the Hodge conjecture for certain products $X \times Y_i$. (The idea is that all Hodge substructures of $H^*(X)$ must be Tate twists of Hodge substructures with $V^{n,0} \neq 0$ that are contained in H^* of some Y_i.) Abdulali's article collects together several known cases of the GHC for abelian varieties, many due to him, which can be proved in this way (and variants thereof). The paper that follows it, by Pedrini and Weibel, provides many examples of surfaces of general type with $p_g(S) = 0$ for which Bloch's conjecture holds, for all values of K_S^2 less than 9. The authors' method, which proceeds by using involutions to show the vanishing of the transcendental motive $t_2(S)$, leads to a more uniform treatment of known examples and produces several new ones.

(III) Arithmetic of periods

Motives, Galois representations, and automorphic forms

Another writeup of an expository minicourse from the summer school begins this section: an introduction to the Langlands program by W. Goldring, focusing on the history and current status of results associating Galois representations to automorphic representations. This story begins with modular forms on the upper half plane, specifically Hecke eigenforms of weight at least 2, whose realization in the étale cohomology of the modular curve $X_0(N)$ produces a 2-dimensional representation of $Gal(\bar{\mathbb{Q}}/\mathbb{Q})$. The article then proceeds to a thorough explanation of Langlands's parametrization of the (infinite-dimensional) archimedean components of automorphic representations via L-packets, with particular attention to discrete series representations and limits thereof, and concludes with clear statements of the conjectures and recent results on the Langlands correspondence itself.

Turning to the contribution of S. Patrikis, a classical construction in complex algebraic geometry (due to Kuga and Satake) associates to any $K3$ surface an abelian variety, with a precise relationship between their Hodge structures. A major innovation of Patrikis is his recasting of the Kuga-Satake construction in terms of the existence of a geometric lift of a motivic Galois representation, and his proposal that this is merely the first instance of a general phenomenon. In this paper he demonstrates, using Katz's theory of middle convolutions, the existence of "generalized Kuga-Satake lifts" of motives whose ℓ-adic realizations are isomorphic to some fiber of an SO_5 or SO_6-cohomologically rigid local system. This gives the first non-trivial examples of his proposed lift beyond the Tannakian subcategory of motives generated by abelian varieties and Artin motives.

The subsection concludes with an article of Yoshida on the *fundamental periods of a motive M* over $\mathbb{Q}$. Recall that the Betti-de Rham comparison isomorphism produces a period matrix, whose determinant was refined by Deligne to (conjecturally) describe the critical values of the L-function $L(M,s)$. The fundamental periods are these Deligne periods together with further period invariants, previously introduced by the author to study the change of Deligne periods under linear-algebra operations. In this paper, he gives applications to Siegel modular forms, Harris's quadratic periods, a formula of Blasius (for periods of a tensor product of motives), and also generalizes fundamental periods to motives over number fields.

Modular forms and iterated integrals

Arapura's note points out a simple construction of a *motive* with H^k having any prescribed list of Hodge numbers. (This is distinct from Schreider's results on projective varieties with prescribed Hodge numbers, and Abdulali's theorem that every CM Hodge structure is motivic.) He also makes the interesting conjecture that all Hodge structures of rank 2 and odd level that arise from a motive over $\mathbb{Q}$, in fact arise from cusp forms in the natural way (via higher Kuga varieties), and offers some supporting evidence.

The volume concludes with a substantial article by Hain which builds on his expository and research talks from the conference. This is a study of the *completion of a modular group* with respect to its inclusion into $SL_2(\mathbb{Q})$, which is an object which unifies the theory of iterated integrals of modular forms (as in the work of Manin) with that of admissible variations of MHS over the modular curve. One Hodge-theoretic application is the explicit construction of normal functions associated to Hecke eigenforms. Another focus of the article is on relations in the "maximal Eisenstein quotient" of the completion of $SL_2(\mathbb{Z})$. Using a computation of Terasoma, Hain proves that Pollack's quadratic relations in the derivation Lie algebra of a rank two free Lie algebra lift to this quotient, and are thus motivic.

Conference Participants

Abdulali, Salman	East Carolina University
Arapura, Donu	Purdue University
Asakura, Masanori	Hokkaido University
Banu, Bogdan	Vrije University, Amsterdam
Brosnan, Patrick	University of Maryland
Brunebarbe, Yohan	Insitut Mathématique de Jussieu
Burgos Gil, Jose	ICMAT
Caibar, Mirel	Ohio State University
Carayol, Henri	Université de Strasbourg
Carlson, James	University of Utah
Carter, Brandon	University of Michigan
Chen, Huachen	Ohio State University
Dan, Ananyo	Humboldt Universität zu Berlin
Daniel, Jeremy	Université Paris Diderot
del Angel, Pedro	CIMAT
Elizondo, E. Javier	UNAM
Eskandari, Payman	University of Toronto
da Silva Jr., Genival	Washington University in St. Louis
de Jeu, Rob	Vrije University, Amsterdam
Gangl, Herbert	Durham University
Goldring, Wushi	Princeton University
Goswami, Souvik	University of Alberta
Green, Mark	Univ. of California - Los Angeles
Griffiths, Phillip	Institute for Advanced Study
Hain, Richard	Duke University
Hayama, Tatsuki	Tsinghua University
Howell, Nick	University of Oregon
Izadi, Elham	Univ. of California - San Diego
Kang, Su-Jeong	Providence College
Kappes, Andre	Goethe-Universität Frankfurt

Katzarkov, Ludmil	Univ. of Miami/Univ. Wien
Kaye, Adam	University of Michigan
Keast, Ryan	Washington University in St. Louis
Kerr, Matt	Washington University in St. Louis
Klingler, Bruno	Université Paris Diderot
Kobyzev, Ivan	University of Western Ontario
Kordek, Kevin	Duke University
Laza, Radu	Stony Brook University
le Bras, Arthur-Cesar	Ecole Normale Superieure
Lee, Pak-Hin	Columbia University
Lefebvre, Jerome	University of British Columbia
Lewis, James	University of Alberta
Li, Zhiyuan	Stanford University
Liu, Hang	Vrije University, Amsterdam
Martel, Justin	University of British Columbia
McKernan, James	Massachusetts Inst. of Technology
Mendez, Hector	University of Alberta
Moonen, Ben	University of Amsterdam
Nguyen, Dong Quan	University of British Columbia
Nie, Zhaohu	Utah State University
Pal, Suchandan	University of Michigan
Patel, Deepam	Vrije University, Amsterdam
Patrikis, Stefan	Institute for Advanced Study
Pearlstein, Gregory	Texas A&M University
Pedrini, Claudio	Universita di Genova
Puttick, Alexandre	Paris Sud 11 - Orsay
Quadri, Jadesola Lateefah	University of Ibadan
Rahmati, Mohammad Reza	CIMAT
Raskind, Wayne	Wayne State University
Robles, Colleen	Texas A&M University
Ronagh, Pooya	University of British Columbia
Sadoughianzadeh, Reza	McMaster University
Saito, Morihiko	RIMS Kyoto University
Sarkar, Soumen	University of Regina
Scarfy, Justin	University of British Columbia
Schmid, Wilfried	Harvard University
Schnell, Christian	Stony Brook University
She, Yiwei	University of Chicago
Shinder, Evgeny	MPIM Bonn
Shnidman, Ariel	University of Michigan
Singh, Vijaykumar	Simon Fraser University
Sreekantan, Ramesh	Indian Statistical Institute

Taheri, Heidar	University of British Columbia
Terasoma, Tomohide	University of Tokyo
Usui, Sampei	Osaka University
Venkatesh, Anil	Duke University
Vologodsky, Vadim	University of Oregon
Walcher, Johannes	McGill University
Xia, Jie	Columbia University
Yang, Sen	Louisiana State University
Yin, Qizheng	Univ. Amsterdam/Paris VI
Yoshida, Hiroyuki	Kyoto University
Zhang, Zheng	Stony Brook University
Zhao, Xiaolei	University of Michigan
Zucker, Steven	Johns Hopkins University

PART I

Hodge theory at the boundary

I.A Period domains and their compactifications

1

Classical Period Domains

Radu Laza and Zheng Zhang

We survey the role played by Hermitian symmetric domains in the study of variations of Hodge Structure. These are extended notes based on the lectures given by the first author in Vancouver at the "Advances in Hodge Theory" school (June 2013).

Introduction

There are two classical situations where the period map plays an essential role for the study of moduli spaces, namely the moduli of principally polarized abelian varieties and the moduli of polarized K3 surfaces. What is common for these two situations is the fact that the period domain is in fact a Hermitian symmetric domain. It is well known that the only cases when a period domain is Hermitian symmetric are weight 1 Hodge structures and weight 2 Hodge structures with $h^{2,0} = 1$.

In general, it is difficult to study moduli spaces via period maps. A major difficulty in this direction comes from the Griffiths' transversality relations. Typically, the image Z of the period map in a period domain $\mathbf{D}$ will be a transcendental analytic subvariety of high condimension. The only cases when Z can be described algebraically are when Z is a Hermitian symmetric subdomain of $\mathbf{D}$ with a totally geodesic embedding (and satisfying the horizontality relation). This is closely related to the geometric aspect of the theory of Shimura varieties of Deligne. It is also the case of unconstrained period subdomains in the sense of [GGK12]. We call this case classical, in contrast to the "non-classical" case when the Griffiths' transversality relations are non-trivial.

The purpose of this survey is to review the role of Hermitian symmetric domains in the study of variations of Hodge structure. Let us give a brief

The first author was partially supported by NSF grants DMS-1200875 and DMS-125481. The second author was partially supported by NSF grant DMS-125481.

overview of the content of the paper. In Section 1, we review the basic definitions and properties of Hermitian symmetric domains (Section 1.1) and their classification (Section 1.2) following [Mil04]. The classification is done by reconstructing Hermitian symmetric domains from the associated (semisimple) Shimura data, which are also convenient for the purpose of constructing variations of Hodge structure over Hermitian symmetric domains (Section 1.3). As a digression, we also include the discussion that if the universal family of Hodge structures over a period subdomain satisfies Griffiths transversality then the subdomain must be Hermitian symmetric (i.e. unconstrained $\Rightarrow$ Hermitian symmetric). Section 2 concerns locally symmetric varieties which are quotients of Hermitian symmetric domains. We first review the basic theory of locally symmetric domains and provide some examples of algebraic varieties whose moduli spaces are birational to locally symmetric domains (Section 2.1), and then give a representation theoretic description of variations of Hodge structure on locally symmetric domains (Section 2.2) following [Mil13]. Using the description, we discuss the classification of variations of Hodge structure of abelian variety type and Calabi-Yau type following [Del79] and [FL13] respectively. Baily-Borel and toroidal compactifications of locally symmetric varieties and their Hodge theoretic meanings are reviewed in Section 3.

1 Hermitian Symmetric Domains

In this section, we review the basic concepts and properties related to Hermitian Symmetric domains with an eye towards the theory of Shimura varieties and Hodge theory. The standard (differential geometric) reference for the material in this section is Helgason [Hel78] (see also the recent survey [Viv13]). For the Hodge theoretic point of view, we refer to the original paper of Deligne [Del79] and the surveys of Milne [Mil04] [Mil13].

1.1. Hermitian symmetric spaces and their automorphisms

1.1.1. Hermitian symmetric spaces

We start by recalling the definition of Hermitian symmetric spaces.

Definition 1.1. A *Hermitian manifold* is a pair (M,g) consisting of a complex manifold M together with a Hermitian metric g on M. A Hermitian manifold (M,g) is *symmetric* if additionally

(1) (M,g) is homogeneous, i.e. the holomorphic isometry group $\mathrm{Is}(M,g)$ acts transitively on M;

(2) for any point $p \in M$, there exists an involution s_p (i.e. s_p is a holomorphic isometry and $s_p^2 = \mathrm{Id}$) such that p is an isolated fixed point of s_p (such an involution s_p is called a *symmetry* at p).

A connected symmetric Hermitian manifold is called a *Hermitian symmetric space*. (If there is no ambiguity, we will use M to denote the Hermitian manifold (M,g).)

Note that if (M,g) is homogeneous, it suffices to check Condition (2) at a point (i.e. it suffices to construct a symmetry s_p at some point $p \in M$). Also, the automorphism group $\mathrm{Is}(M,g)$ consists of holomorphic isometries of M:

$$\mathrm{Is}(M,g) = \mathrm{Is}(M^\infty, g) \cap \mathrm{Hol}(M),$$

where M^∞ denotes the underlying C^∞ manifold, $\mathrm{Is}(M^\infty,g)$ is the group of isometries of (M^∞,g) as a Riemannian manifold, and $\mathrm{Hol}(M)$ is the group of automorphisms of M as a complex manifold (i.e. the group of holomorphic automorphisms).

Example 1.2. There are three basic examples of Hermitian symmetric spaces:

(a) the upper half plane $\mathfrak{H}$;
(b) the projective line $\mathbb{P}^1$ (or the Riemann sphere endowed with the restriction of the standard metric on $\mathbb{R}^3$);
(c) any quotient $\mathbb{C}/\Lambda$ of $\mathbb{C}$ by a discrete additive subgroup $\Lambda \subset \mathbb{C}$ (with the natural complex structure and Hermitian metric inherited from $\mathbb{C}$).

To illustrate the definition, we discuss the example of the upper half plane $\mathfrak{H}$. First, it is easy to see that $\mathfrak{H}$, endowed with the metric $\frac{dz d\bar{z}}{y^2}$, is a Hermitian manifold. Clearly, $\mathfrak{H}$ is homogeneous with respect to the natural action of $\mathrm{SL}_2(\mathbb{R})$, given by

$$\begin{pmatrix} a & b \\ c & d \end{pmatrix} z := \frac{az+b}{cz+d}, \quad \text{for } z \in \mathfrak{H}.$$

In fact, $\mathrm{Is}(\mathfrak{H}) \cong \mathrm{SL}_2(\mathbb{R})/\{\pm I\}$. Finally, the isomorphism $z \mapsto -\frac{1}{z}$ is an involution at the point $i \in \mathfrak{H}$. Since $\mathfrak{H}$ is connected, we conclude that the upper half space $\mathfrak{H}$ is a Hermitian symmetric space.

The three examples above represent the three basic classes of Hermitian symmetric spaces. Specifically, we recall the following:

Definition 1.3. Let M be a Hermitian symmetric space.

(1) M is said to be of *Euclidean type* if it is isomorphic to $\mathbb{C}^n/\Lambda$ for some discrete additive subgroup $\Lambda \subset \mathbb{C}^n$.

(2) M is said to be *irreducible* if it is not of Euclidean type and can not be written as a product of two Hermitian symmetric spaces of lower dimensions.
(3) M is said to be of *compact type* (resp. *noncompact type*) if it is the product of compact (resp. noncompact) irreducible Hermitian symmetric spaces. Moreover, Hermitian symmetric spaces of noncompact type are also called *Hermitian symmetric domains*.

Every Hermitian symmetric space can be decomposed uniquely into a product of Hermitian symmetric spaces of these three types:

Theorem 1.4 (Decomposition Theorem). *Every Hermitian symmetric space M decomposes uniquely as*

$$M = M_0 \times M_- \times M_+,$$

where M_0 is a Euclidean Hermitian symmetric space and M_- (resp. M_+) is a Hermitian symmetric space of compact type (resp. of noncompact type). Moreover, M_- (resp. M_+) is simply connected and decomposes uniquely as a product of compact (resp. noncompact) irreducible Hermitian symmetric spaces.

Proof. See [Hel78, Ch. VIII], especially Proposition 4.4, Theorem 4.6 and Proposition 5.5. □

In this survey, we are mostly interested in Hermitian symmetric domains (or, equivalently, Hermitian symmetric spaces of noncompact type). Note that the terminology is justified by the Harish-Chandra embedding theorem: *every Hermitian symmetric space of noncompact type can be embedded into some $\mathbb{C}^n$ as a bounded domain.* Conversely, every bounded symmetric domain $D \subset \mathbb{C}^n$ has a canonical Hermitian metric (called the Bergman metric) which makes D a Hermitian symmetric domain. For instance, the bounded realization of the upper half plane $\mathfrak{H}$ is the unit ball $\mathcal{B}_1 \subset \mathbb{C}$.

1.1.2. Automorphism groups of Hermitian symmetric domains

Let (D, g) be a Hermitian symmetric domain. Endowed with the compact-open topology, the group $\mathrm{Is}(D^\infty, g)$ of isometries has a natural structure of (real) Lie group. As a closed subgroup of $\mathrm{Is}(D^\infty, g)$, the group $\mathrm{Is}(D, g)$ inherits the structure of a Lie group. Let us denote by $\mathrm{Is}(D, g)^+$ (resp. $\mathrm{Is}(D^\infty, g)^+$, $\mathrm{Hol}(D)^+$) the connected component of $\mathrm{Is}(D, g)$ (resp. $\mathrm{Is}(D^\infty, g)$, $\mathrm{Hol}(D)$) containing the identity.

Proposition 1.5. *Let (D, g) be a Hermitian symmetric domain. The inclusions*

$$\mathrm{Is}(D^\infty, g) \supset \mathrm{Is}(D, g) \subset \mathrm{Hol}(D)$$

induce identities

$$\mathrm{Is}(D^\infty, g)^+ = \mathrm{Is}(D, g)^+ = \mathrm{Hol}(D)^+.$$

Proof. See [Hel78, Lemma 4.3]. □

Since D is homogeneous, one can recover the smooth structure of D as a quotient Lie group of $\mathrm{Is}(D,g)^+$ by the stabilizer of a point. Specifically,

Theorem 1.6. *Notations as above.*

(1) $\mathrm{Is}(D,g)^+$ *is an adjoint (i.e. semisimple with trivial center) Lie group.*
(2) *For any point* $p \in D$, *the subgroup* K_p *of* $\mathrm{Is}(D,g)^+$ *fixing* p *is compact.*
(3) *The map*
$$\mathrm{Is}(D,g)^+/K_p \to D, \quad gK_p \mapsto g \cdot p$$
is an $\mathrm{Is}(D,g)^+$*-equivariant diffeomorphism. In particular,* $\mathrm{Is}(D,g)^+$ *(hence* $\mathrm{Hol}(D)^+$ *and* $\mathrm{Is}(D^\infty,g)^+$*) acts transitively on* D.

Proof. See [Hel78, Ch. IV], especially Theorem 2.5 and Theorem 3.3. □

In particular, every irreducible Hermitian symmetric domain is diffeomorphic to H/K for a unique pair (H,K) (obtained as above) with H a connected noncompact simple adjoint Lie group and K a maximal connected compact Lie group (cf. [Hel78, Ch. VIII, §6]). Conversely, given such a pair (H,K), we obtain a smooth homogenous manifold H/K. The natural question is how to endow H/K with a complex structure and a compatible Hermitian metric so that it is a Hermitian symmetric domain. This can be done in terms of standard Lie theory (see [Viv13, §2.1] and the references therein). However, we shall answer this question from the viewpoint of Shimura data. Specifically, we shall replace the Lie group H by an algebraic group G, replace cosets of K by certain homomorphisms $u : U_1 \to G$ from the circle group U_1 to G, and then answer the question in terms of the pairs (G,u).

To conclude this subsection (and as an initial step to produce a Shimura datum), we discuss how to associate a $\mathbb{R}$-algebraic group G to the real Lie group $\mathrm{Hol}(D)^+$ in such a way that $G(\mathbb{R})^+ = \mathrm{Hol}(D)^+$. The superscript $^+$ in $G(\mathbb{R})^+$denotes the neutral connected component relative to the real topology (vs. the Zariski topology). We shall follow [Mil11] for the terminologies on algebraic groups, and also refer the readers to it for the related background materials. For example, we say an algebraic group is *simple* if it is non-commutative and has no proper normal algebraic subgroups, while *almost simple* if it is non-commutative and has no proper normal connected algebraic subgroup (N.B. an almost simple algebraic group can have finite center).

Proposition 1.7. *Let (D,g) be a Hermitian symmetric domain, and let* $\mathfrak{h} = \mathrm{Lie}(\mathrm{Hol}(D)^+)$. *There is a unique connected adjoint real algebraic subgroup G of* $\mathrm{GL}(\mathfrak{h})$ *such that (inside* $\mathrm{GL}(\mathfrak{h})$*)*

$$G(\mathbb{R})^+ = \mathrm{Hol}(D)^+.$$

Moreover, $G(\mathbb{R})^+ = G(\mathbb{R}) \cap \mathrm{Hol}(D)$ *(inside* $\mathrm{GL}(\mathfrak{h})$*); therefore $G(\mathbb{R})^+$ is the stabilizer in $G(\mathbb{R})$ of D.*

Proof. We sketch the proof of the first statement here, and refer the readers to [Mil04, Prop. 1.7] and references therein for details and the second statement.

Since $\mathrm{Hol}(D)^+$ is adjoint, its adjoint representation on the Lie algebra $\mathfrak{h}$ is faithful, and thus there exists an algebraic group $G \subset \mathrm{GL}(\mathfrak{h})$ such that $\mathrm{Lie}(G) = [\mathfrak{h},\mathfrak{h}]$ (inside $\mathfrak{gl}(\mathfrak{h})$). Because $\mathrm{Hol}(D)^+$ is semisimple, $[\mathfrak{h},\mathfrak{h}] = \mathfrak{h}$ and so $G(\mathbb{R})^+ = \mathrm{Hol}(D)^+$ (inside $\mathrm{GL}(\mathfrak{h})$). □

1.2. Classification of Hermitian symmetric domains

Consider the circle group $U_1 = \{z \in \mathbb{C} \mid |z| = 1\}$. Motivated by the following fact, one can think of a point of D as a homomorphism $U_1 \to G$.

Theorem 1.8. *Let D be Hermitian symmetric domain. For each $p \in D$, there exists a unique homomorphism $u_p : U_1 \to \mathrm{Hol}(D)^+$ such that $u_p(z)$ fixes p and acts on T_pD as multiplication by z.*

Proof. See [Mil04, Thm. 1.9]. □

Remark 1.9. Using the uniqueness of u_p one can easily see that $\mathrm{Hol}(D)^+$ acts on the set of u_p's via conjugation. Clearly, given two different points $p \neq p'$ we choose $f \in \mathrm{Hol}(D)^+$ with $f(p) = p'$, then $f \circ u_p(z) \circ f^{-1}$ $(z \in U_1)$ satisfies the conditions in Theorem 1.8 for p', and thus $u_{p'} = f \circ u_p \circ f^{-1}$.

Example 1.10. Let $p = i \in \mathfrak{H}$. As previously noted, we have $\mathrm{Hol}(\mathfrak{H}) = \mathrm{PSL}_2(\mathbb{R})$. The associated real algebraic group (compare Proposition 1.7) is $(\mathrm{PGL}_2)_{\mathbb{R}}$, and it holds: $\mathrm{PGL}_2(\mathbb{R})^+ = \mathrm{PSL}_2(\mathbb{R})$ (N.B. the group PSL_2 is not an algebraic group). To define $u_i : U_1 \to \mathrm{PSL}_2(\mathbb{R})$ we first consider the homomorphism

$$h_i : U_1 \to \mathrm{SL}_2(\mathbb{R}), z = a + ib \mapsto \begin{pmatrix} a & b \\ -b & a \end{pmatrix}.$$

It is easy to verify that $h_i(z)$ fixes i. Since

$$\frac{d}{dw}\left(\frac{aw+b}{-bw+a}\right)\Big|_{w=i} = \frac{a^2+b^2}{(a-ib)^2} = \frac{z}{\bar{z}} = z^2,$$

$h_i(z)$ acts on the tangent space $T_i\mathfrak{H}$ as multiplication by z^2. Thus, for $z \in U_1$, we choose a square root $\sqrt{z} \in U_1$ and set

$$u_i(z) := h_i(\sqrt{z}).$$

The homomorphism $u_i : U_1 \to \mathrm{PSL}_2(\mathbb{R}) = \mathrm{SL}_2(\mathbb{R})/\pm I$ is independent of the choice of $\sqrt{z}$ (since $h_i(-1) = -I$). Thus, u_i satisfies the conditions of Theorem 1.8 at the point $i \in \mathfrak{H}$.

Since $G(\mathbb{R})^+(= \mathrm{Hol}(D)^+)$ acts transitively on D, set-theoretically we can view D as the $G(\mathbb{R})^+$-conjugacy class of $u_p : U_1 \to G(\mathbb{R})$. (Later, we will see that u_p is an algebraic homomorphism). This viewpoint suggests a connection between Hermitian symmetric domains and variations of Hodge structure. Namely, recall that one can view a Hodge structure as a representation of the Deligne torus $\mathbb{S} := \mathrm{Res}_{\mathbb{C}/\mathbb{R}}\mathbb{G}_m$. Then, if we define $h_p : \mathbb{S} \to G$ by $h_p(z) = u_p(z/\bar{z})$, any representation $G \to \mathrm{GL}(V)$ of G (e.g. $\mathrm{Ad} : G \to \mathrm{GL}(\mathrm{Lie}(G))$), composed with h_p for all $p \in D$, will produce a variation of Hodge structure on D.

Conversely, given an abstract pair $(G, u : U_1 \to G)$ with G a real adjoint algebraic group and u an algebraic homomorphism it is natural to ask the following questions:

Question 1.11. For a pair (G,u) as above, we let D be the $G(\mathbb{R})^+$-conjugacy class of u. Denote by K_u the subgroup of $G(\mathbb{R})^+$ fixing u. There is a bijection $G(\mathbb{R})^+/K_u \to D$ and so the space D has a natural smooth structure.

(1) Under what conditions can D be given a nice complex structure (or a Hermitian structure)? Under what additional conditions is D a Hermitian symmetric space?
(2) Under what conditions is K_u compact?
(3) Under what conditions is D be a Hermitian symmetric domain (i.e. of noncompact type)?

1.2.1. Representations of U_1

Let T be an algebraic torus defined over a field k, and let K be a Galois extension of k splitting T. The character group $X^*(T)$ is defined by $X^*(T) = \mathrm{Hom}(T_K, \mathbb{G}_m)$. If r is the rank of T, then $X^*(T)$ is a free abelian group of rank r which comes equipped with an action of $\mathrm{Gal}(K/k)$. In general, to give a representation ρ of T on a k-vector space V amounts to giving an $X^*(T)$-grading $V_K = \bigoplus_{\chi \in X^*(T)} V_\chi$ on $V_K := V \otimes_k K$ with the property that

$$\sigma(V_\chi) = V_{\sigma\chi}, \quad \text{all } \sigma \in \mathrm{Gal}(K/k), \quad \chi \in X^*(T).$$

Here V_χ is the K-subspace of V_K on which $T(K)$ acts through χ:

$$V_\chi = \{v \in V_K \mid \rho(t)(v) = \chi(t) \cdot v, \ \forall\, t \in T(K)\}.$$

For instance, we can regard U_1 as a real algebraic torus. As an $\mathbb{R}$-algebraic group, the K-valued points (with K an $\mathbb{R}$-algebra) of U_1 are

$$U_1(K) = \left\{ \begin{pmatrix} a & b \\ -b & a \end{pmatrix} \in M_{2\times 2}(K) \,\middle|\, a^2+b^2 = 1 \right\}.$$

In particular, $U_1(\mathbb{R})$ is the circle group and $U_1(\mathbb{C})$ can be identified with $\mathbb{C}^*$ through

$$\begin{pmatrix} a & b \\ -b & a \end{pmatrix} \mapsto a+ib, \text{ conversely } z \mapsto \begin{pmatrix} \frac{1}{2}(z+\frac{1}{z}) & \frac{1}{2i}(z-\frac{1}{z}) \\ -\frac{1}{2i}(z-\frac{1}{z}) & \frac{1}{2}(z+\frac{1}{z}) \end{pmatrix}.$$

Noting that $X^*(U_1) \cong \mathbb{Z}$ and complex conjugation acts on $X^*(U_1)$ as multiplication by -1, we obtain the following proposition.

Proposition 1.12. *Consider a representation ρ of U_1 on a $\mathbb{R}$-vector space V. Then $V_{\mathbb{C}} = \bigoplus_{n\in\mathbb{Z}} V_{\mathbb{C}}^n$ with the property that $\overline{V_{\mathbb{C}}^n} = V_{\mathbb{C}}^{-n}$, where $V_{\mathbb{C}}^n = \{v \in V_{\mathbb{C}} \mid \rho(z)(v) = z^n \cdot v, \quad \forall z \in \mathbb{C}^*\}$. Moreover, if V is irreducible, then it must be isomorphic to one of the following types.*

(a) *$V \cong \mathbb{R}$ with U_1 acting trivially (so $V_{\mathbb{C}} = V_{\mathbb{C}}^0$).*

(b) *$V \cong \mathbb{R}^2$ with $z = x+iy$ acting as $\begin{pmatrix} x & -y \\ y & x \end{pmatrix}^n$ for some $n > 0$ (so $V_{\mathbb{C}} = V_{\mathbb{C}}^n \oplus V_{\mathbb{C}}^{-n}$).*

In particular, every real representation of U_1 is a direct sum of representations of these types.

Remark 1.13. Let V be a $\mathbb{R}$-representation of U_1 and write $V_{\mathbb{C}} = \bigoplus_{n\in\mathbb{Z}} V_{\mathbb{C}}^n$ as above. Because $\overline{V_{\mathbb{C}}^0} = V_{\mathbb{C}}^0$, the weight space $V_{\mathbb{C}}^0$ is defined over $\mathbb{R}$; in other words, it is the complexification of the real subspace V^0 of V defined by $V \cap V_{\mathbb{C}}^0$: $V^0 \otimes_{\mathbb{R}} \mathbb{C} = V_{\mathbb{C}}^0$. The natural homomorphism $V/V^0 \to V_{\mathbb{C}}/\bigoplus_{n\leq 0} V_{\mathbb{C}}^n \cong \bigoplus_{n>0} V_{\mathbb{C}}^n$ is a $\mathbb{R}$-linear isomorphism.

The representations of U_1 have the same description no matter if we regard it as a Lie group or an algebraic group, and so every homomorphism $U_1 \to \mathrm{GL}(V)$ of Lie groups is algebraic. In particular, the homomorphism $u_p : U_1 \to \mathrm{Hol}(D)^+ \cong G(\mathbb{R})^+$ is algebraic for any $p \in D$. Let K_p be the subgroup of $G(\mathbb{R})^+$ fixing p. By Theorem 1.8, $u_p(z)$ acts on the $\mathbb{R}$-vector space

$$\mathrm{Lie}(G)/\mathrm{Lie}(K_p) \cong T_pD$$

as multiplication by z, and it acts on $\mathrm{Lie}(K_p)$ trivially. Suppose $T_pD \cong \mathbb{C}^k$ and identify it with $\mathbb{R}^{2k}$ by $(a_1+ib_1,\ldots,a_k+ib_k) \mapsto (a_1,b_1,\ldots,a_k,b_k)$, then it is easy to write down the matrix of multiplication by $z = x+iy$ and conclude that

T_pD (as a real representation of U_1) splits into a direct sum of $\mathbb{R}^2$'s as in the Part (*b*) of the previous proposition with $n = 1$.

Accordingly, we can determine the representation $\mathrm{Ad} \circ u_p : U_1 \to G \to \mathrm{GL}(\mathrm{Lie}(G))$ (because $u_p(z)$ is contained in the stabilizer K_p of p, the action of $u_p(z)$ on T_pD is induced by the adjoint representation). It splits into a direct sum of 1-dimensional real vector spaces (as in Part (*a*) of Proposition 1.12) and 2-dimensional spaces (as in Part (*b*) with $n = 1$). Taking the complexification of the representation Lie(G), we obtain the following proposition.

Proposition 1.14. *Notations as above. Only the characters z, 1 and z^{-1} occur in the representation of U_1 on $\mathrm{Lie}(G)_{\mathbb{C}}$ defined through u_p.*

1.2.2. Cartan involutions

Let G be a connected algebraic group defined over $\mathbb{R}$, and let $g \mapsto \bar{g}$ denote complex conjugation on $G(\mathbb{C})$.

Definition 1.15. An involution θ of G is said to be *Cartan* if the group

$$G^{(\theta)}(\mathbb{R}) := \{g \in G(\mathbb{C}) \mid g = \theta(\bar{g})\}$$

is compact.

Example 1.16. Let $G = \mathrm{SL}_2$, and let θ be the conjugation by $\begin{pmatrix} 0 & 1 \\ -1 & 0 \end{pmatrix}$. Since

$$\theta\left(\overline{\begin{pmatrix} a & b \\ c & d \end{pmatrix}}\right) = \begin{pmatrix} 0 & 1 \\ -1 & 0 \end{pmatrix} \cdot \overline{\begin{pmatrix} a & b \\ c & d \end{pmatrix}} \cdot \begin{pmatrix} 0 & 1 \\ -1 & 0 \end{pmatrix}^{-1} = \begin{pmatrix} \bar{d} & -\bar{c} \\ -\bar{b} & \bar{a} \end{pmatrix},$$

we have

$$\mathrm{SL}_2^{(\theta)}(\mathbb{R}) = \left\{ \begin{pmatrix} a & b \\ c & d \end{pmatrix} \in \mathrm{SL}_2(\mathbb{C}) \middle| d = \bar{a},\ c = -\bar{b} \right\} = \mathrm{SU}_2(\mathbb{R}).$$

Clearly, the group $\mathrm{SU}_2(\mathbb{R})$ is compact, and hence θ is a Cartan involution for SL_2.

Theorem 1.17. *A connected $\mathbb{R}$-algebraic group G has a Cartan involution if and only if it is reductive, in which case any two Cartan involutions are conjugate by an element of $G(\mathbb{R})$.*

Proof. See [Sat80, Ch. I, Thm. 4.2 and Cor. 4.3]. □

Example 1.18. Let G be a connected $\mathbb{R}$-algebraic group.

(a) The identity map is a Cartan involution if and only if $G(\mathbb{R})$ is compact. Moreover, it is the only Cartan involution of G.

(b) Let $G = \mathrm{GL}(V)$ with V a real vector space of dimension n. Fix a basis of V, then G has an involution given by $\theta : M \mapsto (M^t)^{-1}$. On $G(\mathbb{C}) = \mathrm{GL}_n(\mathbb{C})$, $M = \theta(\bar{M})$ if and only if $M\bar{M}^t = I$ (i.e. $M \in \mathrm{U}(n)$). Thus θ is a Cartan involution. Note that different choices of bases give different Cartan involutions, and the previous theorem says that all Cartan involutions of G arise in this way.

(c) ([Sat80, Ch. I, Cor. 4.4]) Let $G \hookrightarrow \mathrm{GL}(V)$ be a faithful representation. Then G is reductive if and only if it is stable under $g \mapsto g^t$ for a suitable choice of a basis for V, in which case the restriction of $g \mapsto (g^t)^{-1}$ to G is a Cartan involution. Furthermore, all Cartan involutions of G arise in this way from the choice of a basis of V.

(d) Let θ be an involution of G. Then there is a unique real form $G^{(\theta)}$ of $G_{\mathbb{C}}$ such that complex conjugation on $G^{(\theta)}(\mathbb{C})$ is $g \mapsto \theta(\bar{g})$. So the Cartan involutions of G correspond to the compact forms of $G_{\mathbb{C}}$.

Now let us go back to Hermitian symmetric domains. Let D be a Hermitian symmetric domain. As before, G is the associated real adjoint algebraic group (cf. Proposition 1.7), and $u_p : U_1 \to G$ is an algebraic homomorphism attached to a point $p \in D$.

Proposition 1.19. *The conjugation by $u_p(-1)$ is a Cartan involution of G.*

Proof. Let s_p be a symmetry at p. Denote by $\mathrm{Inn}(s_p)$ the conjugation of G by s_p. $\mathrm{Inn}(s_p)$ is an involution because $s_p^2 = \mathrm{Id}$. According to Section [Hel78, §V.2], the real form of $G_{\mathbb{C}}$ defined by the involution $\mathrm{Inn}(s_p)$ (cf. Example 1.18 (*d*)) is that associated to the compact dual of the symmetric space. As a result, a symmetry at a point of a symmetric space gives a Cartan involution of G if and only if the space is of noncompact type. In particular, $\mathrm{Inn}(s_p)$ is Cartan. On the other hand, both $u_p(-1)$ and s_p fix p and acts as multiplication by (-1) on T_pD, and hence $u_p(-1) = s_p$ (cf. [Mil04, Prop. 1.14], which also implies the uniqueness of symmetries at a point of a Hermitian symmetric domain.) □

Note that Example 1.16 is cooked up in this way.

1.2.3. Classification of Hermitian symmetric domains in terms of real groups

We will classify (pointed) Hermitian symmetric domain in this section. Let D be a Hermitian symmetric domain. We have already discussed the first two statements of the following theorem.

Theorem 1.20. *Let G be the associated adjoint real algebraic group of D. The homomorphism $u_p : U_1 \to G$ attached to a point $p \in D$ satisfies the following properties:*

(a) *only the character* z, 1 *and* z^{-1} *occur in the representation of* U_1 *on* $\mathrm{Lie}(G)_{\mathbb{C}}$ *defined by* u_p;
(b) *The conjugation of* G *by* $u_p(-1)$ *is a Cartan involution;*
(c) $u_p(-1)$ *does not project to* 1 *in any simple factor of* G.

Proof. See Proposition 1.14 and 1.19 for Part (a) and (b). Suppose $u_p(-1)$ projects to 1 for some simple factor G_1 (which corresponds to a noncompact irreducible factor of D, see Theorem 1.4), then the Cartan involution $\mathrm{Inn}(u_p(-1))$ is the identity map on G_1. But by Example 1.18 (a), this implies that $G_1(\mathbb{R})$ is compact, which is a contradiction. □

The properties (a), (b), (c) in Theorem 1.20 turn out to be an answer to Question 1.11.

Theorem 1.21. *Let* G *be a real adjoint algebraic group, and let* $u: U_1 \to G$ *be a homomorphism satisfying* (a), (b) *and* (c) *of Theorem 1.20. Then the set* D *of conjugates of* u *by elements of* $G(\mathbb{R})^+$ *has a natural structure of a Hermitian symmetric domain, such that* $G(\mathbb{R})^+ = \mathrm{Hol}(D)^+$ *and* $u(-1)$ *is the symmetry at* u *(regarded as a point of* D*).*

Proof. (Sketch) Let K_u be the subgroup of $G(\mathbb{R})^+$ fixing u (i.e. the centralizer of u). By (b), $\theta := \mathrm{Ad}(u(-1))$ is a Cartan involution for G. So $G^{(\theta)}(\mathbb{R}) = \{g \in G(\mathbb{C}) \mid g = u(-1)\cdot \bar{g} \cdot u(-1)^{-1}\}$ is compact. Since $K_u \subset G(\mathbb{R})^+$, $\bar{g} = g$ for any $g \in K_u$, and so $K_u \subset G^{(\theta)}(\mathbb{R})$. As K_u is closed, it is also compact. The natural bijection $D \cong (G(\mathbb{R})^+/K_u)\cdot u$ endows D with the structure of a smooth (homogeneous) manifold.

With this structure, the (real) tangent space at u is $T_uD = \mathrm{Lie}(G)/\mathrm{Lie}(K_u)$. Note that $\mathrm{Lie}(G)$ a real representation of U_1 via $\mathrm{Ad} \circ u$. Using the notations in Proposition 1.12, (a) gives that $\mathrm{Lie}(G)_{\mathbb{C}} = \mathrm{Lie}(G)_{\mathbb{C}}^{-1} \oplus \mathrm{Lie}(G)_{\mathbb{C}}^{0} \oplus \mathrm{Lie}(G)_{\mathbb{C}}^{1}$. Clearly, $K_u = \mathrm{Lie}(G)_{\mathbb{C}}^{0} \cap \mathrm{Lie}(G)$. Using the natural isomorphism

$$\mathrm{Lie}(G)/\mathrm{Lie}(K_u) \to \mathrm{Lie}(G)_{\mathbb{C}}/\mathrm{Lie}(G)_{\mathbb{C}}^{0} \oplus \mathrm{Lie}(G)_{\mathbb{C}}^{-1} \cong \mathrm{Lie}(G)_{\mathbb{C}}^{1},$$

the tangent space T_uD can be identified with $\mathrm{Lie}(G)_{\mathbb{C}}^{1}$, the complex vector space of $\mathrm{Lie}(G)_{\mathbb{C}}$ on which $u(z)$ acts as multiplication by z. This endows T_pD with a $\mathbb{C}$-vector space structure. (In particular, the corresponding almost complex structure J is $u(i)$.) Since D is homogenous, this induces a structure of an almost complex manifold on D, which is integrable (cf. [Wol84, §8.7.9]).

The action of K_u on D induces an action of it on T_uD. As K_u is compact, there is a K_u-invariant positive definite form on T_uD (cf. Proposition 1.18 of [Mil04]), which is compatible with the complex structure J of T_uD because $J = u(i) \in K_u$. Now use the homogeneity of D to move the bilinear form to each tangent space, which will make D into a Hermitian manifold. It is not difficult to see that $u(-1)$ is the symmetric at u and D is a Hermitian symmetric space.

Finally, D is a Hermitian symmetric domain because of (b) and (c). The proof is quite similar with the one of Theorem 1.20. □

Remark 1.22. As we saw in the proof of Theorem 1.21, the condition (b) guarantees that K_u is compact. If further assuming (a) holds, then one can endow D with the structure of a Hermitian symmetric space. The space D is a Hermitian symmetric domain because of (b) and (c).

As a corollary, we can classify Hermitian symmetric domains in terms of such pairs.

Corollary 1.23. *There is a natural one-to-one correspondence between isomorphism classes of pointed Hermitian symmetric domains and pairs (G,u) consisting of a real adjoint algebraic group G and a non-trivial homomorphism $u : U_1 \to G$ satisfying* (a), (b), (c) *in Theorem 1.20.*

1.2.4. Classification of Hermitian symmetric domains in terms of Dynkin diagrams

Let us now focus on irreducible Hermitian symmetric domains. The irreducibility of the domain implies that the associated adjoint algebraic group is a simple algebraic group. Let G be a simple adjoint group over $\mathbb{R}$, and let u be a homomorphism $U_1 \to G$ satisfying (a) and (b) of Theorem 1.20. (N.B. the condition (c) then holds trivially.)

Lemma 1.24. *Let $G_{\mathbb{C}}$ be the scalar extension of G from $\mathbb{R}$ to $\mathbb{C}$, and $\mu = u_{\mathbb{C}} : \mathbb{G}_m \to G_{\mathbb{C}}$. Then*

(1) *$G_{\mathbb{C}}$ is also simple;*
(2) *Only the characters z, 1, z^{-1} occur in the action of* $\mathrm{Ad} \circ \mu : \mathbb{G}_m \to \mathrm{Lie}(G_{\mathbb{C}})$.

Proof. See Page 21 of [Mil04] or Page 478 of [Mil13] for Part (1). Part (2) follows from Theorem 1.20 (a). □

Proposition 1.25. *The map $(G,u) \mapsto (G_{\mathbb{C}}, u_{\mathbb{C}})$ defines a bijection between the sets of isomorphism classes of pairs consisting of*

(1) *a simple adjoint algebraic group over $\mathbb{R}$ and a conjugacy class of $u : U_1 \to G$ satisfying* (a) *and* (b) *in Theorem 1.20, and*
(2) *a simple adjoint algebraic group over $\mathbb{C}$ and a conjugacy class of cocharacters satisfying* (2) *of Lemma 1.24.*

Proof. See [Mil04, Prop. 1.24]. □

Example 1.26. Let μ be a cocharacter of $(\mathrm{PGL}_2)_{\mathbb{C}}$. Let θ be the conjugation of $\mathrm{PGL}_2(\mathbb{C})$ by

$$\begin{pmatrix} 0 & 1 \\ -1 & 0 \end{pmatrix}.$$

The same computation as in Example 1.16 shows that the involution $\bar{\theta}$, defined by $g \mapsto \theta(\bar{g})$, is the identity on the compact form PGU_2 of $(\mathrm{PGL}_2)_{\mathbb{C}}$. Consider another involution on $\mathrm{PGL}_2(\mathbb{C})$ given by $g \mapsto \mu(-1) \circ \bar{\theta}(g) \circ \mu(-1)^{-1}$. By Example 1.18 (*d*), there is a real form H of $(\mathrm{PGL}_2)_{\mathbb{C}}$ such that complex conjugation on $H(\mathbb{C}) = \mathrm{PGL}_2(\mathbb{C})$ is the involution as above. Also define $u := \mu|_{U_1}$ which takes value in $H(\mathbb{R})$. As $\mu(-1)^2 = \mathrm{Id}$, the conjugation by $u(-1)$ is an involution of H. By construction, it is a Cartan involution. In this way, we obtain a pair (H, u) as in (1) of Proposition 1.25.

In particular, if μ is the scalar extension of u_i which is defined in Example 1.10, then

$$\mu(-1) = \begin{pmatrix} 0 & 1 \\ -1 & 0 \end{pmatrix}.$$

As $\mu(-1) \circ \bar{\theta}(g) \circ \mu(-1)^{-1} = \bar{g}$, the corresponding real form H of $(\mathrm{PGL}_2)_{\mathbb{C}}$ is $(\mathrm{PGL}_2)_{\mathbb{R}}$. Also, it is clear that $u = u_i$.

Let $G_{\mathbb{C}}$ be a simple algebraic group. We choose a maximal torus T, and let $X^*(T) = \mathrm{Hom}(T, \mathbb{G}_m)$ (resp. $X_*(T) = \mathrm{Hom}(\mathbb{G}_m, T)$) be the character (resp. cocharacter) group. Note that there is a natural pairing $\langle -,- \rangle : X^*(T) \times X_*(T) \to \mathrm{End}(\mathbb{G}_m) \cong \mathbb{Z}$ between $X^*(T)$ and $X_*(T)$ (see Page 335 of [Mil11]). Choose a set of simple roots $(\alpha_i)_{i \in I}$. The nodes of the Dynkin diagram of $(G_{\mathbb{C}}, T)$ are also indexed by I. Recall that highest root is the unique root $\tilde{\alpha} = \sum_{i \in I} n_i \alpha_i$ such that, for any other root $\sum_{i \in I} m_i \alpha_i$, $n_i \geq m_i$. We say that an root α_i (or the corresponding node) is *special* if $n_i = 1$ in the expression of $\tilde{\alpha}$.

Theorem 1.27. *The isomorphism classes of irreducible Hermitian symmetric domains are classified by the special nodes on connected Dynkin diagrams.*

Proof. See [Mil04, Thm. 1.25]. For completeness, we include the proof here. Notations as above. By Theorem 1.21 and Proposition 1.25, it suffices to construct a bijection between the conjugacy classes of $\mu : \mathbb{G}_m \to G_{\mathbb{C}}$ satisfying (2) of Lemma 1.24, and special nodes of the Dynkin diagram of $G_{\mathbb{C}}(\mathbb{C})$. Since all maximal tori are conjugate, we can assume that μ is in the cocharacter group $X_*(T) \subset X_*(G_{\mathbb{C}})$ of T. Moreover, there is a unique representative μ such that $\langle \alpha_i, \mu \rangle \geq 0$ for all $i \in I$ because the Weyl group acts transitively and freely on the Weyl chambers. Now (2) of Lemma 1.24 is equivalent to $\langle \alpha, \mu \rangle \in \{-1, 0, 1\}$ for all roots α. Since μ is non-trivial, not all values can be 0, so there must be a (unique) simple root α_i such that $\langle \alpha_i, \mu \rangle = 1$, which is in fact a special root (otherwise $\langle \tilde{\alpha}, \mu \rangle > 1$). The other direction easily follows from the fact that $\langle -,- \rangle : X^*(T) \times X_*(T) \to \mathbb{Z}$ is a perfect pairing. □

The special roots of connected Dynkin diagrams are listed in the following table.

Table 1. *Special roots of connected Dynkin diagrams.*

Type	$\widetilde{\alpha}$	Special root
A_n	$\alpha_1+\cdots+\alpha_n$	$\alpha_1,\cdots,\alpha_n$
B_n	$\alpha_1+2\alpha_2+\cdots+2\alpha_n$	α_1
C_n	$2\alpha_1+\cdots+2\alpha_{n-1}+\alpha_n$	α_n
D_n	$\alpha_1+2\alpha_2+\cdots+2\alpha_{n-2}+\alpha_{n-1}+\alpha_n$	$\alpha_1, \alpha_{n-1}, \alpha_n$
E_6	$\alpha_1+2\alpha_2+2\alpha_3+3\alpha_4+2\alpha_5+\alpha_6$	α_1,α_6
E_7	$2\alpha_1+2\alpha_2+3\alpha_3+4\alpha_4+3\alpha_5+2\alpha_6+\alpha_7$	α_7
E_8,F_4,G_2		none

1.3. Hermitian symmetric domains and Hodge structures

The goal of this section is twofold: on one hand, given a Hermitian symmetric domain D, we show how to use the associated pair (G,u) (cf. Proposition 1.7 and Theorem 1.8) to construct variations of Hodge structure over D; on the other hand, given a real vector space V, we consider certain sets of Hodge structures on V and show that they can be endowed with structures of Hermitian symmetric domains (in fact, every irreducible Hermitian symmetric domain can be obtained in this way). We refer the readers to [GGK12, Chap. 1,2] for the background of Hodge structures and variations of Hodge structure.

1.3.1. A closer look at Condition (a) of Theorem 1.20

Given a pair (G,u) as in Theorem 1.20, we have seen in Theorem 1.21 that the $G(\mathbb{R})^+$-conjugacy class of u has a natural structure of a Hermitian symmetric domain. In this subsection, we would like to consider a more general situation.

Let $\mathbb{S} = \mathrm{Res}_{\mathbb{C}/\mathbb{R}}\mathbb{G}_m$ be the Deligne torus. We consider the following pairs $(\mathbf{G},h)$ where $\mathbf{G}$ is a reductive (see for example Page 16 of [Mil11]) algebraic group over $\mathbb{R}$ and $h:\mathbb{S}\to\mathbf{G}$ is an algebraic homomorphism. We denote by X the conjugacy class of h by elements of $\mathbf{G}(\mathbb{R})$ (not $\mathbf{G}(\mathbb{R})^+$). Note that one can produce such a pair from a Hermitian symmetric domain D. Specifically, we set $\mathbf{G}$ to be the adjoint algebraic group G in Proposition 1.7 and define h by $h(z)=u(z/\bar{z})$ with $u=u_p$ (cf. Theorem 1.8) for some $p\in D$ (see also Example 1.10). In this case, D will be a connected component of X, the $G(\mathbb{R})$-conjugacy class of h (cf. [Mil04, Prop. 4.9]).

Let Z_h be the centralizer of h in $\mathbf{G}(\mathbb{R})$. Then the orbit map identifies X with $\mathbf{G}(\mathbb{R})/Z_h$. We view X as a homogenous manifold via this identification.

For any real representation $\rho:\mathbf{G}\to\mathrm{GL}(V)$ and any $h'\in X$, the composition $\rho\circ h':\mathbb{S}\to\mathbf{G}\to\mathrm{GL}(V)$ defines a real Hodge structure on V (e.g. Page 26 of [Mil04]). In other words,

$$V\otimes_{\mathbb{R}}\mathbb{C}=\bigoplus_{p,q}V_{h'}^{p,q},$$

where $V^{p,q}_{h'} = \{v \in V_{\mathbb{C}} \mid (\rho(h'(z)))(v) = z^{-p}\bar{z}^{-q} \cdot v,\ \forall z \in \mathbb{S}(\mathbb{R}) = \mathbb{C}^*\}$. In particular, over $\mathbb{R}$ we have the weight space decomposition:

$$V = \bigoplus_{n\in\mathbb{Z}} V_{n,h'},\quad V_{n,h'} \otimes_{\mathbb{R}} \mathbb{C} = \bigoplus_{p+q=n} V^{p,q}_{h'}$$

(i.e. $v \in V_{n,h'}$ if and only if $(\rho(h'(r)))(v) = r^n \cdot v$ for all $r \in \mathbb{G}_m(\mathbb{R}) = \mathbb{R}^*$, here $\mathbb{G}_m$ is mapped into $\mathbb{S}$ via $\mathbb{G}_m \xrightarrow{w} \mathbb{S}$, $r \mapsto r^{-1}$).

Remark 1.28. For a Hodge structure $\varphi : \mathbb{S} \to \mathrm{GL}(V)$, the standard convention in the theory of Shimura variety is $\varphi_{\mathbb{C}}(z_1,z_2)(v^{p,q}) = z_1^{-p} z_2^{-q} \cdot v^{p,q}$ (cf. [Del79, (1.1.1.1)]). Meanwhile, a different convention $\varphi_{\mathbb{C}}(z_1,z_2)(v^{p,q}) = z_1^p z_2^q \cdot v^{p,q}$ is largely used in Hodge theory (e.g. Page 31 of [GGK12]). We shall use different conventions in different contexts.

Lemma 1.29. *The following statements are equivalent.*

(1) *For all representations (V,ρ) of $\mathbf{G}$, the weight space decomposition of V induced by $h' \in X$ is independent of the h'.*
(2) *For any $h' \in X$, the real Hodge structure on* $\mathrm{Lie}(\mathbf{G})$ *defined by* $\mathrm{Ad} \circ h'$ *is pure of weight* 0.

Proof. See [Del79, 1.1.13(α)]. See also [Con, Lemma 5.1]. □

Assume X satisfies one of the properties in the previous lemma, then for any representation (V,ρ) the weight spaces $V_{n,h'}$ are independent of $h' \in X$, and so we have a trivial vector bundle $X \times (V_n)_{\mathbb{C}} \to X$ for every weight n. Furthermore, the Hodge filtration $F^{\bullet}_{n,h'}$ on $(V_n)_{\mathbb{C}}$ induced by $\rho \circ h'$ defines (as h' varies) a filtration on $X \times (V_n)_{\mathbb{C}}$ by subbundles $\mathcal{F}^{\bullet}_n$.

We want to put a complex structure on X such that (1) $\mathcal{F}^p_n$ will be a holomorphic subbundle ($0 \le p \le n$); (2) $\mathcal{F}^p_n$'s satisfy Griffiths transversality for the natural connection on $X \times (V_n)_{\mathbb{C}}$. To do this, we need the following axiom. Recall that the *type* of a Hodge structure $V_{\mathbb{C}} = \bigoplus_{p,q} V^{p,q}$ is the set of (p,q) such that $V^{p,q}$ is non-empty.

(**Axiom I**) The Hodge structure on $\mathrm{Lie}(\mathbf{G})$ given by $\mathrm{Ad} \circ h'$ for any $h' \in X$ is of type $\{(-1,1),(0,0),(1,-1)\}$.

Note that pairs $(\mathbf{G},h)$ coming from Hermitian symmetric domains clearly satisfy Axiom I (cf. Theorem 1.20 (*a*)). Also, if Axiom I is satisfied, then Lemma 1.29 (2) automatically holds.

Assuming Axiom I, we can endow X with a complex structure as follows (compare to Theorem 1.21). Let $\mathfrak{g} = \mathrm{Lie}(\mathbf{G})$. Also fix $h' \in X$ and denote the Hodge structure on $\mathfrak{g}$ induced by $h' \in X$ by $\{\mathfrak{g}^{p,q}_{\mathbb{C}}\}$. Then there is a natural isomorphism $T_{h'}X \cong \mathfrak{g}/\mathfrak{g}^{0,0}$, where $\mathfrak{g}^{0,0}$ is the real descent of $\mathfrak{g}^{0,0}_{\mathbb{C}}$. Because $T_{h'}X = \mathfrak{g}/\mathfrak{g}^{0,0} \subset \mathfrak{g}^{1,-1}_{\mathbb{C}} \oplus \mathfrak{g}^{-1,1}_{\mathbb{C}}$, $\mathrm{Ad}(h'(i))$ acts on $T_{h'}X$ as multiplication by -1.

Define $J_{h'} = \mathrm{Ad}(h'(\mathrm{e}^{\frac{\pi \mathrm{i}}{4}}))$. Since $J_{h'}^2 = -\mathrm{Id}$, this defines a complex structure on $T_{h'}X$. Moving $J_{h'}$ around using the homogeneity of X, we obtain an almost complex structure on X.

Theorem 1.30. *Let* $\mathbf{G}$ *be a reductive group over* $\mathbb{R}$ *and let* X *be the* $\mathbf{G}(\mathbb{R})$*-conjugacy class of an algebraic homomorphism* $h : \mathbb{S} \to \mathbf{G}$*. If* $(\mathbf{G}, X)$ *satisfies Axiom I, then the almost complex structure defined by* $\{J_{h'}\}$ *is integrable.*

For any representation V *of* $\mathbf{G}$ *and any integers* n *and* p*,* $\mathcal{F}_n^p$ *is a holomorphic vector bundle on* X *with respect to this complex structure. Moreover,* $\mathcal{F}_n^\bullet$ *satisfies Griffiths transversality for the connection* $\nabla = 1 \otimes \mathrm{d} : (V_n)_\mathbb{C} \otimes \mathcal{O}_X \to (V_n)_\mathbb{C} \otimes \Omega_X^1$*. (In other words,* $(X \times (V_n)_\mathbb{R}, \nabla, \mathcal{F}_n^\bullet)$ *forms a real variation of Hodge structure of weight* n *over* X*.)*

Proof. See [Del79, Prop. 1.1.14] or [Mil04, Prop. 5.9]. See also [Con, Prop. 5.3] and [Lev, Thm. 3.7]. □

Remark 1.31. Axiom I is one of Deligne's axioms in the definition of a Shimura datum, for which we refer the readers to [Mil04, Def. 5.5]. Also, see [Mil04, Def. 4.22] for the definition of a connected Shimura datum. For the connections between (connected) Shimura data and Hermitian symmetric domains, see Proposition 4.8, Proposition 5.7 and Corollary 5.8 of op. cit.

1.3.2. Parameter spaces for certain Hodge structures

Fix a real vector space V and an integer n. Let T be a set of tensors (i.e. multilinear maps $V \otimes \cdots \otimes V \to \mathbb{R}$) including a nondegenerate bilinear form t_0, and let $d : \mathbb{Z} \times \mathbb{Z} \to \mathbb{N}$ be a function with the property that (1) $d(p,q) = 0$ unless $p + q = n$; (2) $d(p,q) = d(q,p)$.

For a real Hodge structure V of weight m, a tensor $V^{\otimes 2r} \to \mathbb{R}(-mr)$ of V is said to be a Hodge tensor if it is a morphism of Hodge structures (of type (0,0)). Define $S(d,T)$ be the set of all Hodge structures φ of weight n on V such that

(1) $\dim V_\varphi^{p,q} = d(p,q)$;
(2) each $t \in T$ is a Hodge tensor for φ;
(3) t_0 is a polarization (e.g. Page 32 of [GGK12]) for φ.

For every $\varphi \in S(d,T)$, we consider the corresponding Hodge filtrations $F_\varphi^\bullet$. Denote by $\mathbf{d}$ the sequence of dimensions of F_φ^i for $0 \leq i \leq n$. Because $\dim V_\varphi^{p,q} = d(p,q)$, $\mathbf{d}$ is independent of φ. Consider the flag variety $\mathrm{Fl}(\mathbf{d}, V_\mathbb{C})$ of type $\mathbf{d}$, then $S(d,T)$ acquires a topology as a subspace of $\mathrm{Fl}(\mathbf{d}, V_\mathbb{C})$.

Note that by construction there is an universal family of Hodge structures $\{\varphi\}_{\varphi \in S(d,T)}$ over $S(d,T)$.

Theorem 1.32. *Let S^+ be a connected component of $S(d,T)$.*

(1) *If nonempty, then S^+ has a unique complex structure for which $\{\varphi\}_{\varphi\in S^+}$ is a holomorphic family of Hodge structures. (Alternatively, the map $S^+ \to \mathrm{Fl}(\mathbf{d}, V_{\mathbb{C}})$, $\varphi \mapsto F^\bullet_\varphi$ is holomorphic.)*
(2) *With this complex structure, S^+ is a Hermitian symmetric domain if $\{\varphi\}_{\varphi\in S^+}$ is a variation of Hodge structure.*
(3) *Every irreducible Hermitian symmetric domain is of the form S^+ for a suitable V, n, d and T.*

Proof. (Sketch) See [Mil04, Thm. 2.14]. We sketch the proof of the statement (2) here and refer the readers to op. cit. for the proof of the statements (1) and (3). For simplicity, we further assume that $T = \{t_0\}$. Choose a point $\varphi \in S^+$, and denote by h_0 the corresponding homomorphism $h_0 : \mathbb{S} \to \mathrm{GL}(V)$. Let $\mathbf{G}$ be the algebraic subgroup of $\mathrm{GL}(V)$ whose elements fix t_0 up to scalar. Then h_0 factors through $\mathbf{G}$: $h_0 : \mathbb{S} \to \mathbf{G} \hookrightarrow \mathrm{GL}(V)$. Let $w : \mathbb{G}_m \to \mathbb{S}$ be the homomorphism defined by $r \mapsto r^{-1}$. Because V has a single weight n, $h_0 \circ w$ maps $\mathbb{G}_m$ into the center of $\mathbf{G}$. As a result, there exists a homomorphism $u_0 : U_1 \to \mathbf{G}^{\mathrm{ad}}$ such that $h_0(z) = u_0(z/\bar{z})$ modulo the center $Z(\mathbf{G})(\mathbb{R})$. (Note that there is an exact sequence $0 \to \mathbb{G}_m \overset{w}{\to} \mathbb{S} \to U_1 \to 0$, where w is defined as above and the other homomorphism $\mathbb{S} \to U_1$ is defined by $z \mapsto z/\bar{z}$.)

We now show that $(\mathbf{G}^{\mathrm{ad}}, u_0)$ satisfies Theorem 1.20 (a), (b) and (c). Let $\mathfrak{g}$ be $\mathrm{Lie}(\mathbf{G})$ with the Hodge structure provided by $\mathrm{Ad} \circ h_0$. One can verify that $\mathfrak{g}$ is a sub-Hodge structure of $\mathrm{End}(V)$ (with the natural Hodge structure of weight 0 induced from (V, φ)). Note that we have

$$\mathfrak{g}/\mathfrak{g}^{0,0} \cong T_\varphi S^+ \subset T_\varphi(\mathrm{Fl}(\mathbf{d}, V_{\mathbb{C}})) \cong \mathrm{End}(V_{\mathbb{C}})/F^0\mathrm{End}(V_{\mathbb{C}}),$$

where $\mathfrak{g}^{0,0}$ is the real descent of $\mathfrak{g}^{0,0}_{\mathbb{C}} \subset \mathfrak{g}_{\mathbb{C}}$. If the universal family $\{\varphi\}_{\varphi\in S^+}$ satisfies Griffiths transversality, then $\mathfrak{g}/\mathfrak{g}^{0,0} \subset F^{-1}\mathrm{End}(V_{\mathbb{C}})/F^0\mathrm{End}(V_{\mathbb{C}})$. This implies that $\mathfrak{g}$ is of type $\{(1,-1),(0,0),(-1,1)\}$, and so u_0 satisfies (a).

Let $\mathbf{G}^1$ be the subgroup of $\mathbf{G}$ whose elements fix t_0 ($(\mathbf{G}^1)^{\mathrm{ad}} \cong \mathbf{G}^{\mathrm{ad}}$). Let $C = h_0(i) = u_0(-1)$ be the Weil operator. As t_0 is a polarization of the Hodge structure (V, φ), t_0 is $\mathbf{G}^1$-invariant and $(2\pi i)^n t_0(-, C-)$ is symmetric and positive definite. By [Mil04, Prop. 1.20], this implies that u_0 satisfies (b).

The set S^+ can be viewed as a connected component of the space of homomorphisms $U_1 \to (\mathbf{G}^1)^{\mathrm{ad}}$. By [Del79, 1.1.12], it is equal to the set of conjugates of u_0 by elements of $(\mathbf{G}^1)^{\mathrm{ad}}(\mathbb{R})^+$. Now discard any compact factors of $(\mathbf{G}^1)^{\mathrm{ad}}$ and apply Theorem 1.21. □

When $T = \{t_0\}$, $S(d, t_0)$ is the Griffiths period domain. In general, $S(d,T)$ is a subdomain of $S(d, t_0)$.

As a result of the theorem, if the universal family of Hodge structures on a period subdomain satisfies Griffiths transversality, then the domain must be a Hermitian symmetric domain.

2 Locally symmetric varieties and Hodge theory

In this section, we review the basic theory of locally symmetric varieties. We also explore the role of locally symmetric varieties in studying moduli spaces and variations of Hodge structure.

2.1. Locally symmetric varieties

2.1.1. Motivations from algebraic geometry

One of the most important applications of Hodge theory in algebraic geometry is to study moduli spaces via *period maps*. Let $\mathcal{M}$ be a moduli space of certain smooth complex algebraic varieties X, choose an integer $0 \leq k \leq \dim X$ (typically $k = \dim X$) and let $\mathbf{D}$ be a period domain parametrizing polarized Hodge structure of weight k which has the same Hodge numbers as $H^k_{\text{prim}}(X,\mathbb{Q})$. A period map

$$\mathcal{P} : \mathcal{M} \to \Gamma\backslash\mathbf{D}$$

is defined by associating to X the polarized Hodge structure of weight k on the primitive cohomology group $H^k_{\text{prim}}(X,\mathbb{Q})$ (Γ is a suitable discrete group acting properly and discontinuously on $\mathbf{D}$). The ideal situation is when $\mathcal{P}$ is birational. For that, one needs to prove that $\mathcal{P}$ is injective (Torelli type results) and $\mathcal{P}$ is dominant. However, due to Griffiths transversality, except for principally polarized abelian varieties and K3 type situations, $\mathcal{P}$ is never dominant. Because of this, we consider subdomains of period domains $\mathbf{D}$. In general, the periods lie in Mumford-Tate subdomains of $\mathbf{D}$, and the image Z of $\mathcal{P}$ is typically highly transcendental. If the periods satisfy enough algebraic relations, then they belong to a Hermitian symmetric domain.

Theorem 2.1. *Let Z be a closed horizontal subvariety of a classifying space $\mathbf{D} = \mathbf{G}(\mathbb{R})/\mathbf{K}$ for Hodge structures and let $\Gamma = \mathrm{Stab}_Z \cap \mathbf{G}(\mathbb{Z})$. Assume that*

(i) *$\Gamma\backslash Z$ is strongly quasi-projective;*
(ii) *Z is semi-algebraic in $\mathbf{D}$ (i.e. open in its Zariski closure in the compact dual $\check{\mathbf{D}}$).*

Then Z is a Hermitian symmetric domain $G(\mathbb{R})/K$, whose embedding in $\mathbf{D}$ is an equivariant, holomorphic, horizontal embedding.

Proof. See [FL13, Thm. 1.4]. □

In other words, the only case when the image of a period map can be described purely algebraically is when it is a locally symmetric domain. They are slight generalizations of the classical cases of principally polarized abelian varieties and K3 surfaces. A number of such semi-classical examples have been constructed, including n points on $\mathbb{P}^1$ with $n \leq 12$ [DM86], algebraic curves of genus 3 or 4 [Kon00], cubic surfaces [ACT02], cubic threefolds [ACT11] and some examples of Calabi-Yau varieties [Voi93], [Bor97], [Roh09], [Roh10], [GvG10] (See also [DK07]). In such situations, one can use the rich structures of arithmetic locally symmetric domains (e.g. the existence of natural compactifications, the theory of automorphic forms) to study the moduli spaces, see for example [Laz14, §2.2] and [Loo14].

Example 2.2. We briefly review the example of Kondō [Kon00] realizing the moduli space of smooth genus 3 curves as a ball quotient. A canonical genus 3 curve C is a plane quartic. To it one can associate a quartic K3 surface S by taking the cyclic μ_4-cover of $\mathbb{P}^2$ branched along C. Specifically, if $C = V(f_4)$, then $S = V(f_4(x_0,x_1,x_2)+x_3^4) \subset \mathbb{P}^3$. In this way, one gets a period map

$$\mathcal{P} : \mathcal{M}_3^{nh} \to \mathcal{F}_4 \cong \mathcal{D}/\Gamma$$

from the moduli of non-hyperelliptic genus 3 curves to the period domain of degree 4 K3 surfaces. Since the resulting K3 Hodge structures are special (they have multiplication by μ_4), the image of $\mathcal{P}$ will lie in a Mumford-Tate subdomain. In this situation, the subdomain will be a 6-dimensional complex ball $\mathcal{B}$ embedded geodesically into the 19-dimensional Type IV domain $\mathcal{D}$. In conclusion, one gets

$$\mathcal{P} : \mathcal{M}_3^{nh} \to \mathcal{B}/\Gamma' \subset \mathcal{D}/\Gamma,$$

which turns out to be birational.

2.1.2. Quotients of Hermitian symmetric domains

We discuss locally symmetric domains in this section. Let us start by defining some special (discrete) subgroups of an algebraic group or a Lie group. Let G be an algebraic group over $\mathbb{Q}$. For an injective homomorphism $r : G \to \mathrm{GL}_n$, we let

$$G(\mathbb{Z})_r = \{g \in G(\mathbb{Q}) \mid r(g) \in \mathrm{GL}_n(\mathbb{Z})\}.$$

Note that $G(\mathbb{Z})_r$ is independent of r up to commensurability (cf. [Bor69, Cor. 7.13]), so r can sometimes be omitted from the notation. A subgroup Γ of $G(\mathbb{Q})$ is *arithmetic* if it is commensurable with $G(\mathbb{Z})_r$ for some r. (In other words, $\Gamma \cap G(\mathbb{Z})_r$ has finite index in both Γ and $G(\mathbb{Z})_r$.) Note that every arithmetic subgroup Γ contains a torsion free subgroup of $G(\mathbb{Q})$ of finite index (cf. [Bor69, Prop. 17.4]).

As an example, let us consider

$$\Gamma(N) := r(G(\mathbb{Q})) \cap \{A \in \mathrm{GL}_n(\mathbb{Z}) \mid A \equiv I \bmod N\},$$

and define a congruence subgroup of $G(\mathbb{Q})$ to be any subgroup containing $\Gamma(N)$ as a subgroup of finite index. Although $\Gamma(N)$ depends on the choice of the embedding r, congruence subgroups do not. Every congruence subgroup is an arithmetic subgroup.

Recall that a *lattice* of a Lie group is a discrete subgroup of finite covolume with respect to an equivariant measure. Consider a connected adjoint Lie group H with no compact factors (e.g. $\mathrm{Hol}(D)^+$ for a Hermitian symmetric domain D), and let Γ be a subgroup of H. If there exists a simply connected (cf. Page 199 of [Mil11]) algebraic group G over $\mathbb{Q}$ and a surjective homomorphism $\varphi : G(\mathbb{R}) \to H$ with compact kernel such that Γ is commensurable with $\varphi(G(\mathbb{Z}))$, then we also say that Γ is an arithmetic subgroup of H. In fact, such a subgroup is a lattice of H (cf. Page 484 of [Mil13]), and so Γ is an *arithmetic lattice*.

We now discuss the quotient of a Hermitian symmetric domain D by a certain discrete subgroup Γ of $\mathrm{Hol}(D)^+$ (e.g. a lattice or an arithmetic lattice). If Γ is torsion free, then it acts freely on D, and there is a unique complex structure on $\Gamma\backslash D$ such that the natural quotient map $D \to \Gamma\backslash D$ is holomorphic. In this case, D is also the universal covering space of $\Gamma\backslash D$ with Γ the group of deck transformations; the choice of a point of D determines an isomorphism of Γ with the fundamental group of $\Gamma\backslash D$. Moreover, it is easy to see that for each $p \in \Gamma\backslash D$, there is an involution s_p defined in a neighborhood of p having p as an isolated point. (In other words, $\Gamma\backslash D$ is "locally symmetric.")

Note that a discrete group Γ of $\mathrm{Hol}(D)^+$ is a lattice (i.e. $\Gamma\backslash\mathrm{Hol}(D)^+$ has finite volume) if and only if $\Gamma\backslash D$ has finite volume.

Let H be a connected semisimple Lie group with finite center. We say that a lattice Γ in H is *irreducible* if $\Gamma \cdot N$ is dense in H for every noncompact closed normal subgroup N of H. If we further assume that H has trivial center and no compact factor, then any lattice in H decomposes into irreducible lattices as in [Mil13, Thm. 3.1]. In particular, one can decompose locally symmetric domains as follows.

Theorem 2.3. *Let D be a Hermitian symmetric domain with $H = \mathrm{Hol}(D)^+$. Let Γ be a lattice in H. Then D can be written uniquely as a product $D = D_1 \times \cdots \times D_r$ of Hermitian symmetric domains such that $\Gamma_i := \Gamma \cap \mathrm{Hol}(D_i)^+$ is an irreducible lattice in $\mathrm{Hol}(D_i)^+$ and $\Gamma_1\backslash D_1 \times \cdots \times \Gamma_r\backslash D_r$ is a finite covering of $\Gamma\backslash D$.*

Proof. See [Mil13, Thm 3.2]. □

Recall that a connected semisimple algebraic group can be written as an almost direct product of its almost simple subgroups (called almost simple factors) (cf. [Mil11, Thm. 17.16]). We say a simply connected or adjoint algebraic group G over $\mathbb{Q}$ is *of compact type* (resp. *of noncompact type*) if $G_i(\mathbb{R})$ is compact (resp. noncompact) for every almost simple factor G_i of G (see also [Mil13, Def. 3.7]). Recall also that the rank of a semisimple algebraic group G over $\mathbb{R}$ is the dimension of a maximal split torus in G.

Theorem 2.4. *Let D be a Hermitian symmetric domain with $H = \mathrm{Hol}(D)^+$. Let Γ be a lattice in H. If $\mathrm{rank}(\mathrm{Hol}(D_i)^+) \geq 2$ in Theorem 2.3, then there exists a simply connected algebraic group G of noncompact type over $\mathbb{Q}$ and a surjective homomorphism $\varphi : G(\mathbb{R}) \to H$ with compact kernel such that Γ is commensurable with $\varphi(G(\mathbb{Z}))$. (In particular, Γ is an arithmetic lattice of H.) Moreover, such a pair (G, φ) is unique up to a unique isomorphism.*

Proof. See [Mil13, Thm. 3.13]. We also include the proof here. By Theorem 2.3 we can assume that the lattice Γ is irreducible. The existence of (G, φ) just means that Γ is arithmetic. By Margulis arithmeticity theorem (see for example [Mil13, Thm. 3.12]), the only possibility one has to rule out is that H is isogenous to $\mathrm{SU}(1,n)$ or $\mathrm{SO}(1,n)$, which can not happen because of the assumption that $\mathrm{rank}(H) \geq 2$. See Page 485 of op. cit. for the proof that G is of noncompact type.

Because Γ is irreducible, G is almost simple (cf. Theorem 3.9 of op. cit.). Let (G_1, φ_1) be a second pair. Because the kernel of φ_1 is compact, its intersection with $G_1(\mathbb{Z})$ is finite, and so there exists an arithmetic subgroup Γ_1 of $G_1(\mathbb{Q})$ such that $\varphi_1|_{\Gamma_1}$ is injective. Because $\varphi(G(\mathbb{Z}))$ and $\varphi_1(\Gamma_1)$ are both commensurable with Γ, they are commensurable, which implies that there exists an arithmetic subgroup Γ' of $G(\mathbb{Q})$ such that $\varphi(\Gamma') \subset \varphi_1(\Gamma_1)$. By Margulis superrigidity theorem (e.g. Theorem 3.10 of op. cit.), the homomorphism $\Gamma' \xrightarrow{\varphi} \varphi_1(\Gamma_1) \cong \Gamma_1$ can be lifted uniquely to a homomorphism $\alpha : G \to G_1$ such that $\varphi_1(\alpha(\gamma)) = \varphi(\gamma)$ for all γ in a subgroup $\Gamma'' \subset \Gamma'$ of finite index. By Borel density theorem (e.g. Theorem 3.8 of op. cit.), the subgroup $\Gamma' \subset G(\mathbb{Q})$ is Zariski dense in G, and so $\varphi_1 \circ \alpha(\mathbb{R}) = \varphi$. Since G and G_1 are almost simple, α is an isogeny. Because G_1 is simply connected, α is an isomorphism. It is unique because it is uniquely determined on an arithmetic subgroup of G. □

A few remarks on the algebraic structure of locally symmetric domains. Recall that there is a functor $X \mapsto X^{\mathrm{an}}$ associating to a smooth complex algebraic variety X a complex manifold X^{an}. This functor is faithful, but far from surjective both on objects and on arrows. However, if we restrict the functor to closed subvarieties of the projective spaces $\mathbb{P}^n_{\mathbb{C}}$, then it produces an equivalence of categories between smooth projective complex varieties and closed submanifolds of $(\mathbb{P}^n_{\mathbb{C}})^{\mathrm{an}}$ (Chow's theorem). By the Baily-Borel theorem, every quotient $\Gamma \backslash D$ of a Hermitian symmetric domain D by a torsion free arithmetic subgroup Γ of $\mathrm{Hol}(D)^+$ can be realized (canonically) as a Zariski open subvariety of a projective variety and hence has a canonical structure of an algebraic variety. Now by a *locally symmetric variety* we mean a smooth complex algebraic variety X such that X^{an} is isomorphic to $\Gamma\backslash D$ for a Hermitian symmetric domain D and a torsion free subgroup $\Gamma \subset \mathrm{Hol}(D)^+$ (see also Footnote 15 on Page 488 of [Mil13]).

To obtain an interesting arithmetic theory, one needs to put further restrictions on a locally symmetric variety X. When $X^{\mathrm{an}} \cong \Gamma\backslash D$ for an arithmetic subgroup Γ of $\mathrm{Hol}(D)^+$, we call X an *arithmetic locally symmetric variety*. The group Γ is usually a lattice, so by Margulis arithmeticity theorem nonarithmetic locally symmetric varieties can only occur in very few cases. For an arithmetic locally symmetric variety X with $X^{\mathrm{an}} \cong \Gamma\backslash D$, we let (G,φ) be the pair associated to $\Gamma\backslash D$ as in Theorem 2.4. If there exists a congruence subgroup Γ_0 of $G(\mathbb{Z})$ such that Γ contains $\varphi(\Gamma_0)$ as a subgroup of finite index, then X will have very rich arithmetic structures; such arithmetic locally symmetric varieties are called *connected Shimura varieties*.

We refer the readers to [Mil04, Chap. 4, 5] for the formal definitions of connected Shimura varieties and Shimura varieties.

2.2. Variations of Hodge structure on locally symmetric domains

We first describe general variations of Hodge structure over locally symmetric domains following [Del79] and [Mil13, Chap. 8], and then turn to the discussion of two special types of variations of Hodge structure, namely those of abelian variety type and Calabi-Yau type. In what follows, we shall always let D be a Hermitian symmetric domain and let Γ be an torsion free arithmetic lattice of $\mathrm{Hol}(D)^+$, and use $D(\Gamma)$ to denote the arithmetic locally symmetric variety.

2.2.1. Description of the variations of Hodge structure on $D(\Gamma)$

According to Theorem 2.3, D decomposes uniquely into a product $D = D_1 \times \cdots \times D_r$ such that $\Gamma_i = \Gamma \cap \mathrm{Hol}(D_i)^+$ is an irreducible lattice of $\mathrm{Hol}(D_i)^+$ and the map $D(\Gamma_1) \times \cdots \times D(\Gamma_r) \to D(\Gamma)$ is a finite covering. We further assume that

$$\mathrm{rank}(\mathrm{Hol}(D_i)^+) \geq 2 \tag{2.5}$$

for each $1 \leq i \leq r$. According to Margulis arithmeticity theorem, there exists a pair (G,φ) where G is a simply connected $\mathbb{Q}$-algebraic group and $\varphi : G(\mathbb{R}) \to \mathrm{Hol}(D)^+$ is a surjective homomorphism with compact kernel such that $\varphi(G(\mathbb{Z}))$ is commensurable with Γ; moreover, such a pair is unique up to a unique isomorphism. (cf. Theorem 2.4.)

We also fix a point $o \in D$. By Theorem 1.8, there exists a unique homomorphism $u : U_1 \to \mathrm{Hol}(D)^+$ such that $u(z)$ fixes o and acts on T_oD as multiplication by z.

Let

$$G_{\mathbb{R}}^{\mathrm{ad}} = G_{\mathrm{c}} \times G_{\mathrm{nc}},$$

where $G_{\mathbb{R}}^{\mathrm{ad}}$ is the quotient of $G_{\mathbb{R}}$ by its center and G_{c} (resp. G_{nc}) is the product of the compact (resp. noncompact) simple factors of $G_{\mathbb{R}}^{\mathrm{ad}}$. The homomorphism $\varphi : G(\mathbb{R}) \to \mathrm{Hol}(D)^+$ factors through G_{nc} and defines an isomorphism of Lie groups $G_{\mathrm{nc}}(\mathbb{R})^+ \to \mathrm{Hol}(D)^+$. Now we define $\bar{h} : \mathbb{S} \to G_{\mathbb{R}}^{\mathrm{ad}}$ by

$$\bar{h}(z) = (h_{\mathrm{c}}(z), h_{\mathrm{nc}}(z)) \in G_{\mathrm{c}}(\mathbb{R}) \times G_{\mathrm{nc}}(\mathbb{R}), \tag{2.6}$$

where $h_{\mathrm{c}}(z) = 1$ and $h_{\mathrm{nc}}(z) = u(z/\bar{z})$ in $G_{\mathrm{nc}}(\mathbb{R})^+ \cong \mathrm{Hol}(D)^+$. Note that $\mathbb{G}_m$ can be embedded into $\mathbb{S}$ via the exact sequence

$$0 \to \mathbb{G}_m \xrightarrow{w} \mathbb{S} \to U_1 \to 0,$$

which is defined on the real valued points by $r \overset{w}{\mapsto} r^{-1}$and $z \mapsto z/\bar{z}$ respectively. It is clear that $\bar{h}$ factors through $\mathbb{S}/\mathbb{G}_m$. Moreover, the $G^{\mathrm{ad}}(\mathbb{R})^+$-conjugates of $\bar{h}$ can be identified with D through $g\bar{h}g^{-1} \mapsto g \cdot o$.

Proposition 2.7. *Notations as above. The pair $(G, \bar{h})$ associated to the arithmetic locally symmetric domain $D(\Gamma)$ and a point $o \in D$ satisfies the following properties.*

(1) *The Hodge structure on* $\mathrm{Lie}(G_{\mathbb{R}}^{\mathrm{ad}})$ *defined by* $\mathbb{S} \xrightarrow{\bar{h}} G_{\mathbb{R}}^{\mathrm{ad}} \xrightarrow{\mathrm{Ad}} \mathrm{GL}(\mathrm{Lie}(G_{\mathbb{R}}^{\mathrm{ad}}))$ *is of type* $\{(1,-1),(0,0),(-1,1)\}$*;*
(2) *The conjugation by $\bar{h}(i)$ is a Cartan involution of $G_{\mathbb{R}}^{\mathrm{ad}}$.*

Proof. By definition, $h_{\mathrm{nc}}(z) = u(z/\bar{z})$ under the identification $G_{\mathrm{nc}}(\mathbb{R})^+ \cong \mathrm{Hol}(D)^+$. Because $G_{\mathbb{R}}^{\mathrm{ad}}$ has trivial center, $\bar{h}$ satisfies (1) and (2) if and only if u satisfies (a) and (b) of Theorem 1.20. □

Let $\mathbf{G}$ be a reductive group over $\mathbb{Q}$ and let $h : \mathbb{S} \to \mathbf{G}_{\mathbb{R}}$ be a homomorphism. To state the main results of this subsection, we define the weight homomorphism

$$w_h := h \circ w$$

where $w : \mathbb{G}_m \to \mathbb{S}$ is given as above by $r \mapsto r^{-1}$ (N.B. to give a Hodge structure on a $\mathbb{Q}$-vector space V amounts to giving a homomorphism $\mathbb{S} \to \mathrm{GL}(V_{\mathbb{R}})$ such that w_h is defined over $\mathbb{Q}$). We also consider the following condition on h.

(**Axiom II***) The conjugation by $h(i)$ is a Cartan involution of $\mathbf{G}_{\mathbb{R}}/w_h(\mathbb{G}_m)$.

Axiom II* can be motivated from the following fact. Let V be a faithful representation of $\mathbf{G}$, if w_h is defined over $\mathbb{Q}$, then the homomorphism

$h : \mathbb{S} \to \mathbf{G}_{\mathbb{R}}$ defines a rational Hodge structure on V; assume that $\mathbf{G}$ is the Mumford-Tate group of V, then V is polarizable if and only if $(\mathbf{G}, h)$ satisfies Axiom II*. (cf. [Del79, 1.1.18(a)] and [Mil13, Prop. 6.4]. Roughly speaking, a Cartan involution produces a bilinear form invariant under the group action, but the Mumford-Tate group $\mathbf{G}$ only preserves a polarization up to scalar, so we consider a Cartan involution on the quotient of the Mumford-Tate group by $w_h(\mathbb{G}_m)$.)

A Hodge structure is said *of CM type* if it is polarizable and its Mumford-Tate group is a torus. Also, by a variation of integral Hodge structure we mean a variation of rational Hodge structure that admits an integral structure (i.e. the local system of $\mathbb{Q}$-vector spaces comes from a local system of free $\mathbb{Z}$-modules). The underlying local system of a variation of Hodge structure is determined by the monodromy representation. See for example [Voi03, §3.1] for the definition and properties of the monodromy representation. Finally, we denote by $\mathbf{G}^{\mathrm{der}}$ the derived subgroup (cf. Page 187 of [Mil11]) of $\mathbf{G}$.

Theorem 2.8. *Let $D(\Gamma)$ be an arithmetic locally symmetric domain satisfying (2.5). Let G be the simply connected $\mathbb{Q}$-algebraic group associated to $D(\Gamma)$ as in Theorem 2.4. Choose a point $o \in D$ and define $\bar{h}$ as in (2.6). To give a polarizable variation of integral Hodge structure on $D(\Gamma)$ such that some fiber is of CM type and the monodromy representation has finite kernel is the same as giving a triple $(\mathbf{G}, h : \mathbb{S} \to \mathbf{G}_{\mathbb{R}}, \rho : G \to \mathrm{GL}(V))$, where V is a $\mathbb{Q}$-representation of G and $\mathbf{G} \subset \mathrm{GL}(V)$ is a reductive algebraic group defined over $\mathbb{Q}$, such that*

(1) *The homomorphism h satisfies Axiom II* and w_h is defined over $\mathbb{Q}$;*
(2) *The representation ρ factors through $\mathbf{G}$ and $\rho(G) = \mathbf{G}^{\mathrm{der}}$;*
(3) *The composition $\mathrm{Ad} \circ h : \mathbb{S} \to \mathbf{G}_{\mathbb{R}} \to \mathbf{G}_{\mathbb{R}}^{\mathrm{ad}} \cong G_{\mathbb{R}}^{\mathrm{ad}}$ is equal to $\bar{h}$.*

Proof. See [Mil13, Summary 8.6]. □

Remark 2.9. The reductive group $\mathbf{G}$ should be thought of as the generic Mumford-Tate group of a polarizable variation of Hodge structure on $D(\Gamma)$. Also, we need to assume the variation of Hodge structure is polarizable and integral so that $\rho(G) = \mathbf{G}^{\mathrm{der}}$ (cf. [Mil13, Thm. 6.22]).

For every arithmetic locally symmetric variety, there exists a triple $(\mathbf{G}, h, \rho)$ satisfying the conditions in Theorem 2.8, and hence there is a polarizable variation of integral Hodge structure on the variety. See Pages 512 – 514 of [Mil13] for details.

2.2.2. Symplectic representations

In this subsection, we show how to construct a family of abelian varieties (equivalently, polarizable variations of integral Hodge structure of length 1) on an arithmetic locally symmetric variety $D(\Gamma)$ following [Del79, §1.3] and

[Mil13, Chap. 10, 11]. For simplicity, we assume that D is irreducible. Also, we assume that $\mathrm{rank}(\mathrm{Hol}(D)^+) \geq 2$ as in (2.5).

According to Theorem 2.4, there is a unique simply connected $\mathbb{Q}$-algebraic group G of non-compact type and a surjective homomorphism $\varphi : G(\mathbb{R}) \to \mathrm{Hol}(D)^+$ with compact kernel such that $\varphi(G(\mathbb{Z}))$ is commensurable with Γ. Note that φ factors through $G^{\mathrm{ad}}_{\mathbb{R}}(\mathbb{R})$ and induces an isomorphism of Lie groups $G^{\mathrm{ad}}_{\mathbb{R}}(\mathbb{R})^+ \to \mathrm{Hol}(D)^+$. Fix a point $o \in D$ and let $\bar{h} : \mathbb{S} \to G^{\mathrm{ad}}_{\mathbb{R}}$ be as defined in (2.6), then $\varphi(\bar{h}(z))$ fixes $o \in D$ and acts on T_oD as multiplication by $z/\bar{z}$.

By Theorem 2.8, variations of Hodge structure on $D(\Gamma)$ corresponds to certain representations of G. We now define symplectic representations and show that they corresponds to families of abelian varieties on $D(\Gamma)$.

Let V be a rational vector space and ψ be a nondegenerate alternating form on V. Denote by $\mathrm{GSp}(V,\psi)$ the group of symplectic similitudes (the algebraic subgroup of $\mathrm{GL}(V)$ whose elements preserves ψ up to scalar). The derived subgroup of $\mathrm{GSp}(V,\psi)$ is the symplectic group $\mathrm{Sp}(V,\psi)$. Also, let $D(\psi)$ be the set of Hodge structures which are of type $\{(-1,0),(0,-1)\}$ and are polarized by $2\pi i\psi$.

Definition 2.10. A homomorphism $G \to \mathrm{GL}(V)$ with finite kernel is a *symplectic representation* of $(G, \bar{h} : \mathbb{S} \to G^{\mathrm{ad}}_{\mathbb{R}})$ if there exists a pair $(\mathbf{G}, h : \mathbb{S} \to \mathbf{G}_{\mathbb{R}})$ consisting of a reductive $\mathbb{Q}$-algebraic group $\mathbf{G}$ and a homomorphism h, a nondegenerate alternating form ψ on V, and a factorization of $G \to \mathrm{GL}(V)$ through $\mathbf{G}$:

$$G \xrightarrow{\phi} \mathbf{G} \xrightarrow{\xi} \mathrm{GL}(V)$$

such that

(1) $\xi \circ h \in D(\psi)$;
(2) $\phi(G) = \mathbf{G}^{\mathrm{der}}$ and $\xi(\mathbf{G}) \subset \mathrm{GSp}(V,\psi)$;
(3) The composition $\mathrm{Ad} \circ h : \mathbb{S} \to \mathbf{G}_{\mathbb{R}} \to \mathbf{G}^{\mathrm{ad}}_{\mathbb{R}} \cong G^{\mathrm{ad}}_{\mathbb{R}}$ is equal to $\bar{h}$.

Recall that a family on a connected complex manifold is said to be faithful if the monodromy representation is injective.

Theorem 2.11. *Let $D(\Gamma)$ be an arithmetic locally symmetric variety with D irreducible (for simplicity only) and* $\mathrm{rank}(\mathrm{Hol}(D)^+) \geq 2$, *and let $(G,\bar{h})$ be the pair associated to $D(\Gamma)$ and a point $o \in D$ as above. There exists a faithful family of abelian varieties on $D(\Gamma)$ having a fiber of CM type if and only if $(G,\bar{h})$ admits a symplectic representation.*

Proof. (Sketch) This is [Mil13, Thm. 11.8]. Let $g : \mathcal{A} \to D(\Gamma)$ be a faithful family of abelian varieties. By Theorem 2.8, the polarizable variation of integral Hodge structure $R^1 g_* \mathbb{Q}$ produces a certain triple $(\mathbf{G}, h : \mathbb{S} \to \mathbf{G}_{\mathbb{R}}, \rho : G \to \mathrm{GL}(V))$ with $\mathbf{G}$ its generic Mumford-Tate group. Because the family is faithful, ρ has finite kernel. Also, by [Mil13, Lemma 10.15] there exists an alternating form ψ on V such that ρ induces a homomorphism $G \to \mathbf{G} \hookrightarrow \mathrm{GSp}(V, \psi)$. Then one can easily check that ρ is a symplectic representation of $(G, \bar{h})$.

Conversely, given a symplectic representation $\rho : G \to \mathrm{GL}(V)$ of $(G, \bar{h})$ as in Definition 2.10, using the notations there, the triple $(\xi(\mathbf{G}), \xi_{\mathbb{R}} \circ h, \xi \circ \phi)$ satisfies the conditions (1) (cf. also [Mil13, Prop. 6.4]), (2) and (3) in Theorem 2.8. In this way, we get a polarizable variation of integral Hodge structure of type $\{(-1,0),(0,-1)\}$ which must come from a family of abelian varieties (see for example [Moo99, Thm. 2.2]). □

In the rest of this subsection we study symplectic representations. Because D is irreducible, there is no harm to study symplectic representations over $\mathbb{R}$. Let $\rho : G \to \mathrm{GL}(V)$ be a symplectic representation of $(G, \bar{h})$. After scalar extension to $\mathbb{R}$, we assume that G is an almost simple and simply connected $\mathbb{R}$-algebraic group without compact factors and view V as a real representation of G. If V is irreducible, then $\mathrm{End}_G(V)$ is a division algebra over $\mathbb{R}$ (Shur's Lemma), and so there are three possibilities:

$$\mathrm{End}_G(V) = \begin{cases} \mathbb{R} \text{ (real type)}, \\ \mathbb{C} \text{ (complex type)}, \\ \mathbb{H} \text{ (quaternionic type)}. \end{cases}$$

Accordingly, we have ($V_{\mathbb{C}} := V_{\mathbb{R}} \otimes_{\mathbb{R}} \mathbb{C}$)

$$V_{\mathbb{C}} = \begin{cases} V_+ & \text{(real type)}, \\ V_+ \oplus V_-, V_+ \ncong V_- & \text{(complex type)}, \\ V_+ \oplus V_-, V_+ \cong V_- & \text{(quaternionic type)}, \end{cases}$$

where $V_\pm$ are irreducible complex $G(\mathbb{C})$-representations and $V_+^\vee \cong V_-$. In practice, one can use [GGK12, Theorem (IV.E.4)] to distinguish these cases.

We now classify the irreducible real symplectic representations of the pairs $(G, \bar{h})$. Define $\bar{\mu} : \mathbb{G}_m \to G^{\mathrm{ad}}_{\mathbb{C}}$ by $\bar{\mu}(z) = \bar{h}_{\mathbb{C}}(z, 1)$, where $\bar{h}_{\mathbb{C}} : \mathbb{G}_m \times \mathbb{G}_m \to G^{\mathrm{ad}}_{\mathbb{C}}$ is the complexification of $\bar{h}$. Let $u : U_1 \to \mathrm{Hol}(D)^+ \cong G^{\mathrm{ad}}(\mathbb{R})^+$ be the homomorphism associated to the point $o \in D$ as in Theorem 1.8. Because $\bar{h}_{\mathbb{C}}(z_1, z_2) = u_{\mathbb{C}}(z_1/z_2)$ as in (2.6), the homomorphism $\bar{\mu}$ is the scalar extension of u: $\bar{\mu}(z) = \bar{h}_{\mathbb{C}}(z, 1) = u_{\mathbb{C}}(z)$.

Fix a maximal torus T of $G^{\mathrm{ad}}_{\mathbb{C}}$, and let $X^*(T) = \mathrm{Hom}(T, \mathbb{G}_m)$ (resp. $X_*(T) = \mathrm{Hom}(\mathbb{G}_m, T)$) be the character (resp. cocharacter) group. There is a natural pairing $\langle -, - \rangle : X^*(T) \times X_*(T) \to \mathrm{End}(\mathbb{G}_m) \cong \mathbb{Z}$ between $X^*(T)$ and $X_*(T)$. Let $R \subset X^*(T)$ (resp. $R^\vee \subset X_*(T)$) be the corresponding root system (resp. coroot system). We also denote by $Q(R)$ the lattice generated by R. (In this case $Q(R) = X^*(T)$, but we will not use this.)

Recall that the lattice of weights is $P(R) = \{\varpi \in X^*(T)_{\mathbb{Q}} \mid \langle \varpi, \alpha^\vee \rangle \in \mathbb{Z} \text{ all } \alpha^\vee \in R^\vee\}$. Choose a set $B = \{\alpha_1, \cdots, \alpha_n\}$ of simple roots such that $\langle \alpha, \bar{\mu} \rangle \geq 0$ for all $\alpha \in B$, then the fundamental weights are the dual basis $\{\varpi_1 \cdots, \varpi_n\}$ of $\{\alpha_1^\vee, \cdots, \alpha_n^\vee\}$, and the dominant weights are the elements $\sum n_i \varpi_i$ with $n_i \in \mathbb{N}$.

Also, there is a unique permutation τ of simple roots (or the corresponding Dynkin diagram or the fundamental weights) such that $\tau^2 = \mathrm{Id}$ and the map $\alpha \mapsto -\tau(\alpha)$ extends to the action of the Weyl group. Usually τ is called the opposition involution. Explicitly, τ acts nontrivially on the root systems of type A_n ($\alpha_i \leftrightarrow \alpha_{n+1-i}$), D_n with n odd ($\alpha_{n-1} \leftrightarrow \alpha_n$) and E_6 ($\alpha_1 \leftrightarrow \alpha_6$), and trivially on the other root systems.

Theorem 2.12. *Notations as above. Let V be an irreducible real representation of G, and ϖ be the highest weight of an irreducible summand W (e.g. V_+ or V_-) of $V_{\mathbb{C}}$. The representation V is a symplectic representation of $(G, \bar{h})$ if and only if*

$$\langle \varpi + \tau(\varpi), \bar{\mu} \rangle = 1. \tag{2.13}$$

Proof. (Sketch) (Step 1) By [Del79, Lemma 1.3.3] or [Mil13, Prop. 10.4], a representation $\rho : G \to \mathrm{GL}(V)$ is a symplectic representation if there exist a pair $(\mathbf{G}, h : \mathbb{S} \to \mathbf{G}_{\mathbb{R}})$ and a factorization $\rho = \xi \circ \phi$ of ρ as in Definition 2.10, such that (1) $\xi \circ h$ is of type $\{(-1,0),(0,-1)\}$; (2) $\phi(G) = \mathbf{G}^{\mathrm{der}}$; and (3) $\mathrm{Ad} \circ h = \bar{h}$. In other words, the nondegenerate alternating form ψ is not needed in the first place.

(Step 2) Consider the projective system $(T_n, T_{nd} \to T_n)$, where the index set is $\mathbb{N} - \{0\}$ (ordered by divisibility), $T_n = \mathbb{G}_m$, and $T_{nd} \to T_n$ is given by $z \mapsto z^d$. Denote by $\tilde{\mathbb{G}}_m$ its inverse limit.

By [Del79, 1.3.4], $\tilde{\mathbb{G}}_m$ is the algebraic universal covering of $\mathbb{G}_m$, so we can lift $\bar{\mu} : \mathbb{G}_m \to G_{\mathbb{C}}^{\mathrm{ad}}$ to $\tilde{\mu} : \tilde{\mathbb{G}}_m \to G_{\mathbb{C}}$. Then $W \subset V_{\mathbb{C}}$ is a representation of $\tilde{\mathbb{G}}_m$. According to [Mil13, §10.2], such a representation $\tilde{\mathbb{G}}_m \to \mathrm{GL}(W)$ can be represented by a homomorphism $f : T_n \to \mathrm{GL}(W)$ and defines a gradation $W = \oplus W_r$ ($r \in (1/n)\mathbb{Z}$) with $f(z)$ acting on W_r by multiplication by z^{nr}. We call the r for which $W_r \neq 0$ the weights of the representation of $\tilde{\mathbb{G}}_m$ on W. One can check that the weights do not depend on the representative f.

The most important observation here is as follows: the nontrivial irreducible representation W occurs in a symplectic representation if and only if $\tilde{\mu}$ has exactly two weights a and $a+1$ on W (cf. [Del79, Lemma 1.3.5]).

We show the "only if" direction here. For $h : \mathbb{S} \to \mathbf{G}_{\mathbb{R}}$, we define $\mu_h : \mathbb{G}_m \to \mathbf{G}_{\mathbb{C}}$ by $\mu_h(z) = h_{\mathbb{C}}(z, 1)$ as in [Del79, 1.1.1, 1.1.11]. Because $\mathrm{Ad} \circ h = \bar{h}$ as in Definition 2.10 (3), we have $\phi_{\mathbb{C}} \circ \tilde{\mu} = \mu_h \cdot \nu$ with ν in the center of $\mathbf{G}_{\mathbb{C}}$. On W, μ_h has weights 0 and 1 (see Definition 2.10 (1)). If a is the unique weight of ν on W, then the only weights of $\tilde{\mu}$ on W is a and $a+1$. We need the observation in Step 1 for the other direction, see [Mil13, Lemma 10.6].

(Step 3) Note that the differential of $\tilde{\mu}$ equals the differential of $\bar{\mu}$. The conclusion in Step 2 can be rephrased as follows: if ϖ is the highest weight of W, then the representation W occurs in a symplectic representation if and only if $\langle \varpi + \tau(\varpi), \bar{\mu} \rangle = 1$. This is [Del79, (1.3.6.1)]. In fact, the lowest weight of W is $-\tau(\varpi)$, and the weights β of W are of the form ϖ + (a $\mathbb{Z}$-linear combination of roots $\alpha \in R$). Because $\langle \alpha, \bar{\mu} \rangle \in \mathbb{Z}$ for all roots α, $\langle \beta, \bar{\mu} \rangle$ takes values a and $a+1$ if and only if $\langle -\tau(\varpi), \bar{\mu} \rangle = \langle \varpi, \bar{\mu} \rangle - 1$, which is clearly equivalent to (2.13). □

To apply (2.13), we make the following two observations. Because $\varpi + \tau(\varpi) \in Q(R)$, $\langle \varpi + \tau(\varpi), \bar{\mu} \rangle \in \mathbb{Z}$ for every dominant weight ϖ. Moreover, $\langle \varpi + \tau(\varpi), \bar{\mu} \rangle > 0$. So only the fundamental weights $\{\varpi_1, \cdots, \varpi_n\}$ can satisfy (2.13) ([Del79, Lemma 1.3.7]).

Also, by the proof of Theorem 1.27 there exists a special node α_s (determined by the irreducible Hermitian symmetric domain D) such that, for simple roots $\alpha \in B = \{\alpha_1, \cdots, \alpha_n\}$,

$$\langle \alpha, \bar{\mu} \rangle = \begin{cases} 1 & \text{if } \alpha = \alpha_s, \\ 0 & \text{if } \alpha \neq \alpha_s. \end{cases}$$

Express a weight ϖ as a $\mathbb{Q}$-linear combination of the simple roots $\{\alpha_i\}$ (cf. [Bou02]), then $\langle \varpi + \tau(\varpi), \bar{\mu} \rangle = 1$ if and only if the coefficient of α_s in $\varpi + \tau(\varpi)$ equals 1.

Example 2.14. (Type A_{n-1}) In this case,

$$\varpi_i = \frac{n-i}{n}\alpha_1 + \frac{2(n-i)}{n}\alpha_2 + \cdots + \frac{i(n-i)}{n}\alpha_i + \frac{i(n-i-1)}{n}\alpha_{i+1} + \cdots + \frac{i}{n}\alpha_{n-1},$$

for $1 \leq i \leq n-1$. The opposition involution τ switches the nodes i and $n-i$: $\tau(\varpi_i) = \varpi_{n-i}$, and so

$$\tau(\varpi_i) = \frac{i}{n}\alpha_1 + \frac{2i}{n}\alpha_2 + \cdots + \frac{(n-i)i}{n}\alpha_{n-i} + \frac{(n-i)(i-1)}{n}\alpha_{n+1-i} + \cdots + \frac{n-i}{n}\alpha_{n-1}.$$

If $i \leq n-i$, one can easily compute the coefficient of a simple root α_j in $\varpi_i + \tau(\varpi_i)$:

$$\text{the coefficient of } \alpha_j \text{ in } \varpi_i + \tau(\varpi_i) = \begin{cases} j & \text{if } 1 \leq j \leq i, \\ i & \text{if } i \leq j \leq n-i, \\ n-j & \text{if } n-i \leq j \leq n-1. \end{cases}$$

The special root α_s can be any α_j with $1 \leq j \leq n-1$. We drop the cases that $\alpha_s = \alpha_1$ and $\alpha_s = \alpha_{n-1}$ so that the assumption (2.5) holds. Choose a special root $\alpha_s = \alpha_j$ for $2 \leq j \leq n-2$. It is easy to see that for the coefficient to be 1, ϖ_i must be ϖ_1. Similarly, if $i > n-i$, only the fundamental weight ϖ_{n-1} satisfies (2.13).

Example 2.15. (Type E_6 and E_7) In the E_6 case, the special root $\alpha_s = \alpha_1$ or α_6, and the opposite involution switches α_1 and α_6. We seek a fundamental weight ϖ such that $\varpi = a\alpha_1 + \cdots + b\alpha_6$ with $a + b = 1$. But there is no such a fundamental weight for the root system E_6, and hence there is no corresponding symplectic representation.

Similarly, there is no symplectic representation associated to the Hermitian symmetric domains of type E_7. In fact, $\alpha_s = \alpha_7$ and the opposite involution is trivial in the E_7 case. Therefore, a fundamental weight ϖ satisfies (2.13) if and only if $\varpi = \cdots + \frac{1}{2}\alpha_7$, but no fundamental of E_7 is of this form.

If a fundamental weight satisfies (2.13), then we call the corresponding node a *symplectic node*. They are listed as follows.

Table 2. *List of symplectic nodes.*

Type	Symplectic node
(A_n, α_1)	$\varpi_1, \cdots, \varpi_n$
(A_n, α_i), $1 < i < n$	ϖ_1, ϖ_n
(B_n, α_1), $n \geq 2$	ϖ_n
(C_n, α_n)	ϖ_1
(D_n, α_1), $n \geq 4$	ϖ_{n-1}, ϖ_n
(D_4, α_4)	ϖ_1, ϖ_3
(D_n, α_n), $n \geq 5$	ϖ_1
(E_6, α_1)	none
(E_7, α_7)	none

Remark 2.16. As discussed in [Mil13, Chap. 9], one needs to take motives and Hodge classes (v.s. algebraic varieties and algebraic classes) into consideration in order to realize all but a small number of Shimura varieties as moduli varieties. Let us also mention the celebrated theorem of Deligne saying that Hodge classes are the same as absolute Hodge classes for abelian varieties. The readers can find further discussions in [Del82] and [CS11].

Example 2.17. Consider the special root α_n of the root system A_{2n-1} with $n \geq 2$. By Theorem 1.27, it corresponds to an irreducible Hermitian symmetric domain D.

Let $K = \mathbb{Q}(\sqrt{-d})$ ($d \in \mathbb{Z}^+$) be an imaginary quadratic field extension of $\mathbb{Q}$. Denote by V a K-vector space of dimension $2n$, and set $H : V \times V \to K$ to be a K-Hermitian form on V whose signature is (n, n). There exists a K-basis on V, on which H is given by

$$H(z, w) = az_1\bar{w}_1 + \cdots + z_n\bar{w}_n - z_{n+1}\bar{w}_{n+1} - \cdots - z_{2n}\bar{w}_{2n},$$

where $a \in \mathbb{Q}^+$. Let us assume $a = 1$ for simplicity. With respect to such a basis, we define a $\mathbb{Q}$-algebraic group G by

$$G(R) = \left\{ A \in \mathrm{GL}_{2n}(K \otimes_{\mathbb{Q}} R) \,\middle|\, \overline{A}^t \begin{pmatrix} I_n & 0 \\ 0 & -I_n \end{pmatrix} A = \begin{pmatrix} I_n & 0 \\ 0 & -I_n \end{pmatrix} \right\},$$

where R is a $\mathbb{Q}$-algebra (the matrix $\overline{A}$ is obtained by taking the conjugate of every entry in A by $\overline{k \otimes r} := \bar{k} \otimes r$). The algebraic group G is simply connected (cf. Page 232 of [Mil11]), and $G(\mathbb{R})$ is isomorphic to the special unitary group $\mathrm{SU}(n,n)$.

Since $\mathrm{Hol}(D)^+ \cong \mathrm{PSU}(n,n)$, there exists a natural homomorphism $\varphi : G(\mathbb{R}) \twoheadrightarrow \mathrm{Hol}(D)^+$. We choose an irreducible lattice Γ in $\mathrm{PSU}(n,n) \cong \mathrm{Hol}(D)^+$ which is commensurable with $\varphi(G(\mathbb{Q}) \cap \mathrm{GL}_{2n}(\mathbb{Z}))$. The pair (G, φ) then satisfies the conditions in Theorem 2.4 for the arithmetic locally symmetric variety $D(\Gamma)$.

Consider the natural representation $\rho : G \to \mathrm{GL}(V)$. Let $V_{\mathbb{R}} = V \otimes_{\mathbb{Q}} \mathbb{R}$, then $\rho_{\mathbb{R}} : G_{\mathbb{R}} \to \mathrm{GL}(V_{\mathbb{R}})$ is isomorphic to the standard representation of $\mathrm{SU}(n,n)$. As representations of $G(\mathbb{C}) \cong \mathrm{SL}_{2n}(\mathbb{C})$, we have $V_{\mathbb{C}} = V \otimes_{\mathbb{R}} \mathbb{C} \cong V_+ \oplus V_-$ where V_+ has highest weight ϖ_1 and $V_- \cong V_+^\vee$ has highest weight $\tau(\varpi_1) = \varpi_{2n-1}$. In this case, both ϖ_1 and ϖ_{2n-1} correspond to symplectic nodes, and hence ρ is a symplectic representation. So by Theorem 2.11 we obtain a family of abelian varieties on $D(\Gamma)$ (or on D by pulling back via $D \to D(\Gamma)$).

This is a family of abelian varieties of Weil type over D. An abelian variety of Weil type consists of an abelian variety X of dimension $2n$ and an imaginary quadratic field extension $K \hookrightarrow \mathrm{End}(X) \otimes \mathbb{Q}$, such that for all $k \in K$ the action of k on T_0X has n eigenvalues $\sigma(k)$ and n eigenvalues $\overline{\sigma(k)}$ (here we fix an embedding $\sigma : K \subset \mathbb{C}$). We refer the readers to [vG94, §5.3 – §5.12] for the explicit constructions of families of abelian varieties of Weil type over D.

2.2.3. Hermitian variations of Hodge structure of Calabi-Yau type

In this subsection, we consider Hodge structures of Calabi-Yau type.

Definition 2.18. A rational (resp. real) Hodge structure V of Calabi-Yau (CY) type is an effective rational (resp. real) Hodge structure[1] of weight n such that $V^{n,0}$ is 1-dimensional.

Based on earlier work of Gross ([Gro94]) and Sheng-Zuo ([SZ10]), Friedman and Laza classified $\mathbb{R}$-variations of Hodge structures of CY type over irreducible Hermitian symmetric domains in [FL13] and [FL14]. In this subsection, we only discuss Friedman-Laza's classification for irreducible Hermitian symmetric domains of tube type. All the irreducible tube domains

[1] In this subsection, we switch to the other convention: if $\varphi : \mathbb{S} \to \mathrm{GL}(V_{\mathbb{R}})$ defines a Hodge structure of weight n, then $h(z)$ ($z \in \mathbb{S}(\mathbb{R})$) acts on $V^{p,q}$ as multiplication by $z^p \bar{z}^q$.

Table 3. *Hermitian symmetric domains of tube type.*

Label	(R,α_s)	$G(\mathbb{R})$	K	$\mathbb{R}$-rank
$\mathrm{I}_{n,n}$	(A_{2n-1},α_n)	$\mathrm{SU}(n,n)$	$\mathrm{S}(\mathrm{U}(n)\times\mathrm{U}(n))$	n
II_{2n}	(D_{2n},α_{2n})	$\mathrm{Spin}^*(4n)$	$U_1\times_{\mu_n}\mathrm{SU}(2n)$	n
III_n	(C_n,α_n)	$\mathrm{Sp}(2n,\mathbb{R})$	$\mathrm{U}(n)$	n
IV_{2n-1}	(B_n,α_1)	$\mathrm{Spin}(2,2n-1)$	$\mathrm{Spin}(2)\times_{\mu_2}\mathrm{Spin}(2n-1)$	2
IV_{2n-2}	(D_n,α_1)	$\mathrm{Spin}(2,2n-2)$	$\mathrm{Spin}(2)\times_{\mu_2}\mathrm{Spin}(2n-2)$	2
EVII	(E_7,α_7)	$\mathrm{E}_{7,3}$	$\mathrm{U}(1)\times_{\mu_3}\mathrm{E}_6$	3

are tabulated as follows (the first column is standard Siegel's notation; the second column lists the corresponding Dynkin diagrams and special roots (cf. Theorem 1.27); the third column gives the real simply connected algebraic groups for the unique simple adjoint algebraic groups associated to Hermitian symmetric domains (cf. Proposition 1.7); the fourth column lists the corresponding maximal compact subgroup; the last column gives the real ranks of tube domains).

Let $\mathcal{D}$ be an irreducible Hermitian symmetric domain, and let $\mathbf{D}$ be a classifying space of polarized rational Hodge structures with fixed Hodge numbers. Following Definition 2.1 of [FL13], we call the variations of Hodge structure induced by an equivariant, holomorphic and horizontal embedding of $\mathcal{D}\hookrightarrow\mathbf{D}$ a *Hermitian variation of Hodge structure*. They are the variations of Hodge structure parameterized by Hermitian symmetric domains considered by Deligne [Del79]. In the terminology of [GGK12], $\mathcal{D}\subset\mathbf{D}$ is an unconstrained Mumford-Tate domain (and hence also a Hermitian symmetric domain).

By Proposition 1.7, there is a unique simple adjoint real algebraic group associated to $\mathcal{D}$; we denote by G its algebraic universal covering (N.B. these simply connected algebraic groups are listed in Table 3, see also [Gro94, §1]). Fix a reference point $o\in\mathcal{D}$. According to Theorem 1.8, there is a homomorphism $u: U_1\to G^{\mathrm{ad}}$. We define $\bar{h}:\mathbb{S}\to G^{\mathrm{ad}}$ by $\bar{h}(z)=u(z/\bar{z})$ as in (2.6).

Choosing a suitable arithmetic subgroup of $\mathrm{Hol}(\mathcal{D})^+$, we assume that there is an algebraic group $G_{\mathbb{Q}}$ is defined over $\mathbb{Q}$ with $G_{\mathbb{Q}}\otimes_{\mathbb{Q}}\mathbb{R}\cong G$. To give a Hermitian rational variation of Hodge structure over $\mathcal{D}$, one must give a $\mathbb{Q}$-representation $\rho_{\mathbb{Q}}: G_{\mathbb{Q}}\to\mathrm{GL}(V)$ satisfying the conditions in Theorem 2.8. Following [FL13, §2.1], we assume that the induced real representation $\rho: G\to\mathrm{GL}(V_{\mathbb{R}})$ is irreducible. As variations of real Hodge structure are mainly concerned in this section, we shall focus on the representation ρ.

The question is which irreducible representations of G correspond to Hermitian variations of Hodge structure of CY type over D. As in Theorem 1.27, the Hermitian symmetric domain D determines a root system R together with a special root α_i. We call the corresponding fundamental weight ϖ_i (i.e. $\varpi_i(\alpha_j^\vee) = \delta_{ij}$) a *cominuscule weight*, and call the irreducible representation V_{ϖ_i} of $G(\mathbb{C})$ with highest weight ϖ_i a *cominuscule representation*.

Let $V_\mathbb{R}$ be an irreducible G-representation. Recall that $V_\mathbb{R}$ may be of real type, complex type or quaternionic type. Specifically, $V_\mathbb{C}$ may be irreducible (real type) or reducible (complex type or quaternionic type); if $V_\mathbb{C}$ is reducible, then we can write $V_\mathbb{C} = V_+ \oplus V_-$, where V_+ and V_- are irreducible representations of $G(\mathbb{C})$ and $V_+^\vee \cong V_-$. We distinguish the complex case from the quaternionic case depending on whether $V_+ \cong V_-$ (quaternionic type) or not (complex type). We now show that if $V_\mathbb{R}$ induces a CY Hermitian variation of Hodge structure over D, then the highest weight of V_+ or V_- ($V_+ = V_\mathbb{C}$ in the real case) must be a multiple of the corresponding cominuscule weight.

In what follows, let us focus on tube domains D.

Lemma 2.19. *Let D be an irreducible Hermitian symmetric domain of tube type, and let G be defined as above. Suppose D corresponds to (R, α_i) (so ϖ_i is the corresponding cominuscule weight), and let $V_{n\varpi_i}$ be the irreducible representation of $G(\mathbb{C})$ with highest weight $n\varpi_i$ ($n \in \mathbb{N}^+$). Then there exists a real G-representation $V_\mathbb{R}$ such that $V_\mathbb{R} \otimes_\mathbb{R} \mathbb{C} = V_{n\varpi_i}$.*

Proof. The condition that D is of tube type is equivalent to that $\tau(\alpha_i) = \alpha_i$ where τ is the opposition involution. Let V_{ϖ_i} be a cominuscule representation. Because the dual representation has highest weight $\tau(\varpi_i)$, we have $V_{\varpi_i} \cong V_{\varpi_i}^\vee$. Now one can verify the reality of the representation using [GGK12, Thm. (IV.E.4)]. The same argument works for $V_{n\varpi_i}$. □

Example 2.20. Let D be an tube domain corresponding to (A_5, α_3). Then $G = \mathrm{SU}(3,3)$ and the cominuscule weight is ϖ_3. Because $\tau(\varpi_3) = \varpi_3$, the cominuscule representation can not be of complex type. We now determine whether it is of real type or quaternionic type using [GGK12, Thm. (IV.E.4)].

In the root system A_5, we have

$$2\varpi_3 = \alpha_1 + 2\alpha_2 + 3\alpha_3 + 2\alpha_4 + \alpha_5.$$

(Denote the coefficients of α_i by m_i.) The only noncompact root in this case is α_3 (cf. Page 335 of [Kna02]). Because

$$\sum_{\alpha_i \text{ compact}} m_i = 1 + 2 + 2 + 1 = 6$$

is even, the cominuscule representation is of real type.

In the proof of Theorem 1.27 we see that there is a $\mu \in X_*(G^{\mathrm{ad}}_{\mathbb{C}})$ ($\mu = u_{\mathbb{C}}$) such that

$$\langle \alpha, \mu \rangle = \begin{cases} 1 & \text{if } \alpha = \alpha_i, \\ 0 & \text{if } \alpha \neq \alpha_i, \end{cases}$$

where α_i is the special root associated to the domain D. Following [FL13], we shall use H_0 to denote μ, and use $\varpi(H_0)$ to denote the pairing $\langle \varpi, H_0 \rangle$.

Proposition 2.21. *Notations as above. Let $\rho : G \to \mathrm{GL}(V_{\mathbb{R}})$ be an irreducible representation and λ be the highest weight of an irreducible factor V_+ of $V_{\mathbb{C}}$. Possibly replace V_+ with V_-, we can assume that $\tau\lambda(H_0) \leq \lambda(H_0)$. Then a necessary condition for ρ to arise from a CY Hermitian variation of Hodge structure over D is*

$$\varpi(H_0) < \lambda(H_0) \text{ for all weights } \varpi \neq \lambda \text{ of } V_+.$$

Furthermore, this condition implies that λ is a multiple of the fundamental cominuscule weight ϖ_i associated to the domain D. In particular, if D is a tube domain, then such representations $V_{\mathbb{R}}$ are of real type.

Proof. Consider the following commutative diagram:

$$\begin{array}{ccccccc} U_1 & \xrightarrow{i} & \mathbb{S} & \xrightarrow{h} & \mathbf{G}_{\mathbb{R}} & \longrightarrow & \mathrm{GL}(V_{\mathbb{R}}) \\ {\scriptstyle 2:1}\downarrow & & {\scriptstyle p}\downarrow & & {\scriptstyle \mathrm{Ad}}\downarrow & & \\ U_1 & = & U_1 & \xrightarrow{u} & G^{\mathrm{ad}} & & \end{array}$$

where $i : U_1 \hookrightarrow \mathbb{S}$ is the kernel of the norm map $\mathrm{Nm} : \mathbb{S} \to \mathbb{G}_m$ ($\mathrm{Nm}(z) = z\bar{z}$), and $p : \mathbb{S} \to U_1$ is defined by $z \mapsto z/\bar{z}$. (Note also that $\mathbf{G}^{\mathrm{ad}}_{\mathbb{R}} \cong G^{\mathrm{ad}}$.) In the situation considered here, the Hodge decomposition on $V_{\mathbb{C}}$ is the weight decomposition of $V_{\mathbb{C}}$ with respective to $U_1(\mathbb{C}) \cong \mathbb{C}^*$ via $h_{\mathbb{C}} \circ i_{\mathbb{C}}$: $V^{p,q}$ corresponds to the eigenspace for the character z^{p-q}. If $V_{\mathbb{R}}$ is of real type, then by the above diagram the weights of $\mathbb{G}_m$ on $V_{\mathbb{C}}$ via $h_{\mathbb{C}} \circ i_{\mathbb{C}}$ are $\{2\varpi(H_0) \mid \varpi \in \mathcal{X}(V_+)\}$, where $\mathcal{X}(V_+)$ denotes the weights of the irreducible $G(\mathbb{C})$-representation V_+ (note that $V_+ = V_{\mathbb{C}}$ in this case). If $V_{\mathbb{R}}$ is of complex or quaternionic case, then $V_{\mathbb{C}} = V_+ \oplus V_-$, and the weights of $h \circ i$ on $V_{\mathbb{C}}$ are $\{\pm 2(\varpi(H_0) - c) \mid \varpi \in \mathcal{X}(V_+)\}$, where the constant c comes from the action of the center of $\mathbf{G}_{\mathbb{R}}$ on V_+ (cf. [FL13, §2.1.2]).

Since all the other weights of V_+ are obtained from λ by subtracting positive roots, it follows that

$$\max_{\varpi \in \mathcal{X}(V_+)} \varpi(H_0) = \lambda(H_0).$$

Using the description of the weights of $h_{\mathbb{C}} \circ i_{\mathbb{C}}$ on $V_{\mathbb{C}}$, we see that the CY condition ($\dim_{\mathbb{C}} V^{n,0} = 1$) implies that the above maximal is attained only for the highest weight λ. In other words, for other weights $\varpi \neq \lambda$ of V_+, $\varpi(H_0) < \lambda(H_0)$.

Let α_i be the special root associated to D. By applying the reflection in another simple root $\alpha_j \neq \alpha_i$, we get

$$s_{\alpha_j}(\lambda)(H_0) = (\lambda - \lambda(\alpha_j^\vee) \cdot \alpha_j)(H_0) = \lambda(H_0) - \lambda(\alpha_j^\vee) \cdot \alpha_j(H_0) = \lambda(H_0).$$

Because $s_{\alpha_j}(\lambda) \in \mathcal{X}(V_+)$, $s_{\alpha_j}(\lambda) = \lambda$, which is equivalent to $\lambda(\alpha_j^\vee) = 0$. Now we can conclude that $\lambda = n\varpi_i$. The last assertion follows from the previous lemma. □

Let D be an irreducible Hermitian symmetric domain of tube type, and let α_i (resp. ϖ_i) be the corresponding special root (resp. cominuscule weight). As shown in [Gro94] and [FL13], if $V_\mathbb{R}$ is an irreducible representation of G such that $V_\mathbb{C}$ has highest weight $n\varpi_i$ ($n \in \mathbb{N}^+$) as a $G(\mathbb{C})$-representation, then $V_\mathbb{R}$ induces a CY $\mathbb{R}$-variation of Hodge structures over D. If $V_\mathbb{C}$ is the cominuscule representation (i.e. the highest weight is ϖ_i), then we call the induced CY $\mathbb{R}$-variation of Hodge structure the *canonical* $\mathbb{R}$*-variation of Hodge structure of CY type* over D.

Theorem 2.22. *For every irreducible Hermitian symmetric domain D of tube type, there exists a canonical* $\mathbb{R}$*-variation of Hodge structure* $\mathcal{V}$ *of CY type parameterized by D. Any other irreducible CY* $\mathbb{R}$*-variation of Hodge structure can be obtained from the canonical one by taking the unique irreducible factor of* $\mathrm{Sym}^\bullet \mathcal{V}$ *of CY type.*

Proof. See [Gro94, §3] and [FL13, Thm. 2.22]. □

Remark 2.23. For an irreducible tube domain D, the weight of the canonical $\mathbb{R}$-variation of Hodge structure is also equal to the real rank of D which can be found in Table 3.

Remark 2.24. One may wonder what happens if the irreducible domain D is not of tube type. Let V_+ be a cominuscule representation of $G(\mathbb{C})$. Sheng and Zuo [SZ10] have noted V_+ carries a $\mathbb{C}$-Hodge structure of CY type, and so $V_+ \oplus V_+^\vee$ will carry a $\mathbb{R}$-Hodge structure. However, this Hodge structure is typically not of CY type. To fix this, one needs to apply the operation "half twist" defined by van Geemen [vG01]. See [FL13, §2.1.3] for details.

3 Compactifications of locally symmetric varieties and their Hodge theoretic meanings

Let $\Gamma\backslash D$ be a quotient of a Hermitian symmetric domain D by an arithmetic subgroup Γ of $\mathrm{Hol}(D)^+$. In this section, we shall review the Baily-Borel compactification and toroidal compactification of $\Gamma\backslash D$ and discuss their Hodge theoretic meanings. The emphasis will be on examples, especially when D is a classical period domain.

3.1. The Baily-Borel compactifications

The Baily-Borel compactification $(\Gamma\backslash D)^*$ of $\Gamma\backslash D$ is defined by

$$(\Gamma\backslash D)^* := \mathrm{Proj}(\bigoplus_{n\geq 0} A_n),$$

where A_n is the vector space of automorphic forms on D for the n-th power of the canonical automorphy factor. The graded $\mathbb{C}$-algebra $A = \bigoplus_{n\geq 0} A_n$ is finitely generated and hence we are allowed to take the associated projective scheme. The canonical map

$$\Gamma\backslash D \to (\Gamma\backslash D)^*$$

realizes $\Gamma\backslash D$ as a Zariski-open subvariety of the normal projective algebraic variety $(\Gamma\backslash D)^*$. The readers can find the details in [BB66].

We briefly recall the (set-theoretical) construction of the Baily-Borel compactifications for irreducible locally symmetric domains. Suppose D is irreducible and write $D = H/K$, where $H = \mathrm{Hol}(D)^+$ and K is a maximal compact subgroup of H. Then the boundary components (a.k.a. *cusps*) of $(\Gamma\backslash D)^*$ are determined by rational maximal parabolic subgroups of H, which can usually be described in combinatorial terms. More specifically, let F_P be a boundary component of the closure of D in the Harish-Chandra embedding (it is itself a Hermitian symmetric space associated with a certain parabolic subgroup P of H), then the normalizer $\mathcal{N}(F_P)$, defined by $\mathcal{N}(F_P) := \{g \in H \mid g(F_P) = F_P\}$, is a maximal parabolic subgroup of H. We say such a boundary component F_P is *rational* if its normalizer $\mathcal{N}(F_P)$ can be defined over $\mathbb{Q}$. The arithmetic subgroup Γ preserves the rational boundary components, and so we obtain the Baily-Borel compactification by taking the quotient of $D \bigsqcup \partial D$, where ∂D is the disjoint union of all rational boundary components, by the action of Γ. Moreover, every boundary component of $(\Gamma\backslash D)^*$ has a structure of a locally symmetric variety of lower dimension.

Here are some properties of the Baily-Borel compactification.

(1) The Baily-Borel compactification is canonical. In other words, the construction does not depend on any choice.
(2) The boundary components usually have high codimension (≥ 2 except for some low dimensional locally symmetric domains).
(3) The Baily-Borel compactification is minimal in the following sense. Let S be a smooth variety and $\overline{S}$ a smooth simple normal crossing (partial)

compactification of S. Then any locally liftable map $S \to \Gamma\backslash D$ extends to a regular map $\overline{S} \to (\Gamma\backslash D)^*$ (see [Bor72]).

3.1.1. The Satake-Baily-Borel compactification of $\mathcal{A}_g$

Fix a free $\mathbb{Z}$-module $V_{\mathbb{Z}}$ of rank $2g$, and a nondegenerate skew-symmetric bilinear form Q on $V_{\mathbb{Z}}$. We let D be the classifying space of polarized weight 1 Hodge structures on $V_{\mathbb{Z}}$. To be specific, we have

$$D = \{F \in \mathrm{Gr}(g, V_{\mathbb{C}}) \mid Q(F,F) = 0,\ iQ(F,\overline{F}) > 0\}.$$

This is a Hermitian symmetric domain of type III and we have that

$$D \cong \mathrm{Sp}(2g,\mathbb{R})/\mathrm{U}(g).$$

Taking Q to be the standard symplectic form, the domain D can be identified with the Siegel upper half-space $\mathfrak{H}_g$, the space of symmetric $g \times g$ complex matrices with positive definite imaginary part. The group $\mathrm{Sp}(2g,\mathbb{Z})$ acts on $D \cong \mathfrak{H}_g$ via the fractional linear transformations, and we set

$$A_g := \mathrm{Sp}(2g,\mathbb{Z})\backslash\mathfrak{H}_g.$$

The rational boundary component F_{W_0} corresponds to the choice of a totally isotropic subspace $W_0 \subset V_{\mathbb{Q}}$ ($V_{\mathbb{Q}} := V_{\mathbb{Z}} \otimes_{\mathbb{Z}} \mathbb{Q}$). Since the group $\mathrm{Sp}(2g,\mathbb{Z})$ acts transitively on the set of isotropic subspaces W_0 of fixed dimension, the boundary component of $\mathcal{A}_g^*$ is indexed by the dimension $\nu \in \{0,\cdots,g\}$ of the isotropic subspaces W_0. Furthermore, the choice of W_0 defines a weight filtration $W_\bullet$ on $V_{\mathbb{Q}}$:

$$W_{-1} := \{0\} \subset W_0 \subset W_1 := (W_0)^\perp \subset W_2 := V_{\mathbb{Q}}.$$

The polarization Q induces a nondegenerate symplectic form $\overline{Q}$ on the graded piece $\mathrm{Gr}_1^W := W_1/W_0$. The boundary component F_{W_0} can then be described as the classifying space $D_{g'}$ ($g' = g - \nu$) of $\overline{Q}$-polarized Hodge structure of weight 1 on Gr_1^W (see [Cat84] Page 84), and so $A_{g'} = \mathrm{Sp}(2g',\mathbb{Z})\backslash\mathfrak{H}_{g'} \cong (\mathrm{Sp}(2g,\mathbb{Z}) \cap \mathcal{N}(F_{W_0}))\backslash D_{g'}$ gives a boundary component of $\mathcal{A}_g^*$. The compactification $\mathcal{A}_g^*$ then admits the following stratification

$$\mathcal{A}_g^* = \mathcal{A}_g \sqcup A_{g-1} \sqcup \cdots \sqcup A_0.$$

3.1.2. The Baily-Borel compactifications for modular varieties of orthogonal type

We let L be an integral lattice (i.e. a free $\mathbb{Z}$-module equipped with a symmetric bilinear form $(-,-)$) of signature $(2,n)$ ($n > 3$). Consider the Hermitian symmetric domain of type IV

$$D_L := \{x \in \mathbb{P}(L \otimes_{\mathbb{Z}} \mathbb{C}) \mid (x,x) = 0,\ (x,\bar{x}) > 0\}^+,$$

where the superscript $+$ denotes a choice of one of the two connected components. Let $\mathrm{O}^+(L)$ be the subgroup of $\mathrm{O}(L)$ of real spinor norm 1 (in other words, the subgroup fixing the connected components). A *modular variety of orthogonal type* is a quotient

$$\mathcal{F}_L(\Gamma) = \Gamma \backslash D_L$$

of D_L by an arithmetic subgroup Γ of $\mathrm{O}^+(L \otimes \mathbb{Q})$ (that is, $\Gamma \subset \mathrm{O}(L \otimes \mathbb{Q})$ and $\Gamma \cap \mathrm{O}^+(L)$ is of finite index in both Γ and $\mathrm{O}^+(L)$). Some important examples of orthogonal type modular varieties are the period spaces of (lattice) polarized K3 surfaces. For instance, if one takes the lattice L_{2d} of signature $(2, 19)$ defined by

$$L_{2d} := \mathrm{U}^{\oplus 2} \oplus \mathrm{E}_8^{\oplus 2} \oplus \langle -2d \rangle,$$

and sets $\Gamma := \mathrm{O}^+(L_{2d}) \cap \widetilde{\mathrm{O}}(L_{2d})$ (where the group $\widetilde{\mathrm{O}}(L_{2d})$ is the kernel of the natural homomorphism $\mathrm{O}(L) \to \mathrm{O}(L^\vee/L)$), then the quotient $\Gamma \backslash D_{L_{2d}}$ is the period space of polarized K3 surfaces of degree $2d$.

The rational maximal parabolic subgroups are the stabilizers of isotropic subspaces of $L_\mathbb{Q} := L \otimes_\mathbb{Z} \mathbb{Q}$. Because of the signature, such spaces have dimensions 2 or 1. The Baily-Borel compactification $\mathcal{F}_L(\Gamma)^*$ decomposes into the disjoint union (see for example [GHS13, Thm. 5.11])

$$\mathcal{F}_L(\Gamma)^* = \mathcal{F}_L(\Gamma) \sqcup \coprod_\Pi X_\Pi \sqcup \coprod_l Q_l,$$

where l and Π run though representatives of the finitely many Γ-orbits of isotropic lines and isotropic planes in $L_\mathbb{Q}$ respectively. Each X_Π is a modular curve, each Q_l is a point. X_Π and Q_l are usually referred to as 1-dimensional and 0-dimensional boundary components. For the period space $\Gamma \backslash D_{L_{2d}}$ of polarized K3 surfaces of degree $2d$, the 1-dimensional and 0-dimensional boundary components correspond to the degenerate fibers in Kulikov degenerations of type II and type III respectively.

It is also an interesting problem to compare $(\Gamma \backslash D_{L_{2d}})^*$ with certain compactifications of the moduli space using GIT via the period map, see [Laz14, §2.3] for a survey.

3.2. Toroidal compactifications

We discuss toroidal compactifications $\overline{\Gamma \backslash D}^\Sigma$ (or simply $\overline{\Gamma \backslash D}$) in this section. The general reference is the book [AMRT75]. Roughly speaking, $\overline{\Gamma \backslash D}$ is obtained by adding a divisor at each cusp of $(\Gamma \backslash D)^*$. Locally in the analytic topology near a cusp, the toroidal compactification is a quotient of an open part of a toric variety over the cusp, which depends on a choice of admissible fan in a suitable cone. Let us also note that

(1) The construction of toroidal compactifications depends on certain suitable choices. The compactification $\overline{\Gamma\backslash D}$ can be chosen to be projective, and to have at worst finite quotient singularities.
(2) For every suitable choice, there is a natural morphism $\overline{\Gamma\backslash D} \to (\Gamma\backslash D)^*$, inducing the identity morphism on $\Gamma\backslash D$.

The readers can find the details of the construction of $\overline{\Gamma\backslash D}$ in [AMRT75, §3.5]. We now describe the toroidal compactification of $\mathcal{A}_g$ following [Cat84] (see also Section 1 of [CMGHL14]). We will not discuss their modular meanings here, instead we refer the readers to [Laz14, §2.2.1] and references therein.

Notations as in Section 3.1.1. The construction of $\overline{\mathcal{A}_g}$ is relative over $\mathcal{A}_g^*$. Let W_0 be a totally isotropic subspace of $V_{\mathbb{Q}}$ of dimension $\nu \leq g$, which corresponds to a boundary component F_{W_0} of $\mathcal{A}_g^*$. Consider then the real Lie subalgebra $\mathfrak{n}(W_0)$ of $\mathfrak{sp}(V_{\mathbb{R}}, Q)$:

$$\mathfrak{n}(W_0) := \{N \in \mathfrak{sp}(V_{\mathbb{R}}, Q) \mid Im(N) \subset W_0 \otimes_{\mathbb{Q}} \mathbb{R}\}.$$

For all elements $N \in \mathfrak{n}(W_0)$, we have $Q(N^2(v), w) = -Q(N(v), N(w)) = 0$ where $v, w \in V_{\mathbb{R}}$, and so $N^2 = 0$. For a rational element N, it defines a weight filtration $W_\bullet(N)$ on $V_{\mathbb{Q}}$:

$$W_0(N) := Im(N) \subset W_1(N) := \mathrm{Ker}(N) = Im(N)^{\perp} \subset W_2(N) := V_{\mathbb{Q}},$$

which is compatible with the weight filtration $W_\bullet$ induced by W_0:

$$W_0(N) \subset W_0 \subset W_1 (= W_0^{\perp}) \subset W_1(N).$$

In particular, we obtain a natural surjection

$$\mathrm{Gr}_2^W (= V_{\mathbb{Q}}/W_1) \twoheadrightarrow \mathrm{Gr}_2(N) (= V_{\mathbb{Q}}/W_1(N)).$$

Because N is a nilpotent symplectic endomorphism, we have a natural isomorphism

$$\mathrm{Gr}_2(N) \to \mathrm{Gr}_0(N) \to \mathrm{Gr}_2(N)^{\vee}, \;\; v \mapsto N(v) \mapsto Q(N(\cdot), v),$$

which gives a non-degenerate bilinear form Q_N on $\mathrm{Gr}_2(N)$. It is easy to see that Q_N is symmetric. Pulling it back via the natural surjection $\mathrm{Gr}_2^W \to \mathrm{Gr}_2(N)$, we get a symmetric bilinear form $\tilde{Q}_N$ on Gr_2^W. So there is a natural map (defined over $\mathbb{Q}$)

$$\mathfrak{n}(W_0) \xrightarrow{\cong} \mathrm{Hom}(\mathrm{Sym}^2 \mathrm{Gr}_{2,\mathbb{R}}^W, \mathbb{R})$$

which is an isomorphism.

We now define

$$\mathfrak{n}^+(W_0) = \{N \in \mathfrak{n}(W_0) \mid \tilde{Q}_N \text{ is positive definite}\},$$

and denote by $Cl(\mathfrak{n}^+(W_0))$ the closure of $\mathfrak{n}^+(W_0)$ in $\mathfrak{n}(W_0)$. The weight filtration $W_\bullet$ defined by W_0 is the weight filtration $W_\bullet(N)$ of any element $N \in \mathfrak{n}^+(W_0)$, as well as any cone

$$\sigma = \left\{ \sum_{i=1}^{r} \lambda_i N_i \;\middle|\; \lambda_i \in \mathbb{R}, \lambda_i > 0, N_i \in Cl(\mathfrak{n}^+(W_0)) \right\}$$

which contains an element of $\mathfrak{n}^+(W_0)$ (see Page 78 of [Cat84]). For any such cone σ, we define $B(\sigma) \subset \check{D}$ (note that $\check{D} = \{F \in \mathrm{Gr}(g, V_\mathbb{C}) \mid Q(F,F) = 0\}$ is the compact dual of D) to be the set of all those filtrations in $\check{D}$ which, together with $W_\bullet$, define a mixed Hodge structure polarized by every $N \in \mathrm{Int}(\sigma)$. It is not difficult to see that $B(\sigma) = \exp(\sigma_\mathbb{C}) \cdot D$, where $\sigma_\mathbb{C} = \{\sum_{i=1}^{r} \lambda_i N_i \mid \lambda_i \in \mathbb{C}, N_i \in Cl(\mathfrak{n}^+(W_0))\}$.

The boundary component associated to the cone σ is then defined as

$$\mathcal{B}(\sigma) := B(\sigma)/\exp(\sigma_\mathbb{C}).$$

Since $\exp(\sigma_\mathbb{C})$ acts trivially on Gr_1^W, $\mathcal{B}(\sigma)$ factors over the boundary component F_{W_0} of the Baily-Borel compactification by considering the Hodge structure on the graded piece Gr_1^W. Moreover, if τ is a face of σ, then there is a natural morphism $\mathcal{B}(\tau) \to \mathcal{B}(\sigma)$ which respects the projections to F_{W_0}.

In order to attach the boundary components to D so that the extended space (i.e. the disjoint union of D with the boundary components) is compatible with the action of the arithmetic group, we need a suitable choice of a collection Σ of rational polyhedral cones in $Cl(\mathfrak{n}^+(W_0))$, which is called an *admissible rational polyhedral decomposition* of $\mathfrak{n}^+(W_0)$. See for example [CMGHL14, §1.3, 1.4] for further discussions.

One then sets

$$D^\Sigma = \bigcup_{W_0} \bigcup_{\sigma \in \Sigma} \mathcal{B}(\sigma).$$

The action of the arithmetic group $\mathrm{Sp}(2g,\mathbb{Z})$ extends to D^Σ, and so (set-theoretically) we have

$$\overline{\mathcal{A}_g}^\Sigma = \mathrm{Sp}(2g,\mathbb{Z}) \backslash D^\Sigma.$$

References

[ACT02] D. Allcock, J. A. Carlson, and D. Toledo. The complex hyperbolic geometry of the moduli space of cubic surfaces. *J. Algebraic Geom.*, 11(4):659–724, 2002.

[ACT11] D. Allcock, J. A. Carlson, and D. Toledo. The moduli space of cubic threefolds as a ball quotient. *Mem. Amer. Math. Soc.*, 209(985):xii+70, 2011.

[AMRT75] A. Ash, D. Mumford, M. Rapoport, and Y. Tai. *Smooth compactification of locally symmetric varieties*. Math. Sci. Press, Brookline, Mass., 1975.

[BB66] W. L. Baily, Jr. and A. Borel. Compactification of arithmetic quotients of bounded symmetric domains. *Ann. of Math. (2)*, 84:442–528, 1966.

[Bor69] A. Borel. *Introduction aux groupes arithmétiques*. Publications de l'Institut de Mathématique de l'Université de Strasbourg, XV. Actualités Scientifiques et Industrielles, No. 1341. Hermann, Paris, 1969.

[Bor72] A. Borel. Some metric properties of arithmetic quotients of symmetric spaces and an extension theorem. *J. Differential Geometry*, 6:543–560, 1972. Collection of articles dedicated to S. S. Chern and D. C. Spencer on their sixtieth birthdays.

[Bor97] C. Borcea. *K*3 surfaces with involution and mirror pairs of Calabi-Yau manifolds. In *Mirror symmetry, II*, volume 1 of *AMS/IP Stud. Adv. Math.*, pages 717–743. Amer. Math. Soc., Providence, RI, 1997.

[Bou02] N. Bourbaki. *Lie groups and Lie algebras. Chapters 4–6*. Elements of Mathematics (Berlin). Springer-Verlag, Berlin, 2002.

[Cat84] E. H. Cattani. Mixed Hodge structures, compactifications and monodromy weight filtration. In *Topics in transcendental algebraic geometry (Princeton, N.J., 1981/1982)*, volume 106 of *Ann. of Math. Stud.*, pages 75–100. Princeton Univ. Press, Princeton, NJ, 1984.

[CMGHL14] S. Casalaina-Martin, S. Grushevsky, K. Hulek, and R. Laza. Extending the Prym map to toroidal compactifications of the moduli space of abelian varieties. arXiv:1403.1938, 2014.

[Con] B. Conrad. Handout on Shimura Data. Electronic notes available at http://math.stanford.edu/~conrad/shimsem.

[CS11] F. Charles and C. Schnell. Notes on absolute Hodge classes. Notes based on lectures given at the ICTP summer school on Hodge theory and related topics (June 2010, Trieste), available at arXiv:arXiv:1101.3647, 2011.

[Del79] P. Deligne. Variétés de Shimura: interprétation modulaire, et techniques de construction de modèles canoniques. In *Automorphic forms, representations and L-functions*, Proc. Sympos. Pure Math., XXXIII, pages 247–289. AMS, Providence, R.I., 1979.

[Del82] P. Deligne. Hodge cycles on abelian varieties. In *Hodge cycles, motives, and Shimura varieties*, pages 9–100. Springer, 1982.

[DK07] I. V. Dolgachev and S. Kondō. Moduli of *K*3 surfaces and complex ball quotients. In *Arithmetic and geometry around hypergeometric functions*, volume 260 of *Progr. Math.*, pages 43–100. Birkhäuser, Basel, 2007.

[DM86] P. Deligne and G. D. Mostow. Monodromy of hypergeometric functions and nonlattice integral monodromy. *Inst. Hautes Études Sci. Publ. Math.*, 63:5–89, 1986.

[FL13] R. Friedman and R. Laza. Semialgebraic horizontal subvarieties of Calabi-Yau type. *Duke Math. J.*, 162(12):2077–2148, 2013.

[FL14] R. Friedman and R. Laza. On some hermitian variations of hodge structure of calabi-yau type with real multiplication. *The Michigan Mathematical Journal*, 63(1):83–99, 2014.

[GGK12] M. Green, P. A. Griffiths, and M. Kerr. *Mumford-Tate groups and domains: Their geometry and arithmetic*, volume 183 of *Annals of Mathematics Studies*. Princeton University Press, Princeton, NJ, 2012.

[GHS13] V. Gritsenko, K. Hulek, and G. K. Sankaran. Moduli of K3 surfaces and irreducible symplectic manifolds. In *Handbook of moduli. Vol. I*, volume 24 of *Adv. Lect. Math. (ALM)*, pages 459–526. Int. Press, Somerville, MA, 2013.

[Gro94] B. H. Gross. A remark on tube domains. *Math. Res. Lett.*, 1(1):1–9, 1994.

[GvG10] A. Garbagnati and B. van Geemen. Examples of Calabi-Yau threefolds parametrised by Shimura varieties. *Rend. Semin. Mat. Univ. Politec. Torino*, 68(3):271–287, 2010.

[Hel78] S. Helgason. *Differential geometry, Lie groups, and symmetric spaces*, volume 80 of *Pure and Applied Mathematics*. Academic Press, Inc. [Harcourt Brace Jovanovich, Publishers], New York-London, 1978.

[Kna02] A. W. Knapp. *Lie groups beyond an introduction*, volume 140 of *Progress in Mathematics*. Birkhäuser Boston Inc., Boston, MA, second edition, 2002.

[Kon00] S. Kondō. A complex hyperbolic structure for the moduli space of curves of genus three. *J. Reine Angew. Math.*, 525:219–232, 2000.

[Laz14] R. Laza. Perspectives on the construction and compactification of moduli spaces. arXiv:1403.2105, to appear in CRM Lecture Electronic notes, 2014.

[Lev] B. Levin. Hermitian symmetric domains. Electronic notes available at http://math.stanford.edu/~conrad/shimsem.

[Loo14] E. Looijenga. Moduli spaces and locally symmetric varieties. Notes based on three lectures given 6th MSJ-SI conference "Development of Moduli Theory" (June 2013, Kyoto), available at arXiv:1404.3854, 2014.

[Mil04] J. S. Milne. Introduction to Shimura Varieties. Electronic notes available at http://www.jmilne.org/math/xnotes/svi.pdf, 2004.

[Mil11] J. S. Milne. Algebraic Groups, Lie Groups, and their Arithmetic Subgroups. Electronic notes available at www.jmilne.org/math/Course Notes/ALA.pdf, 2011.

[Mil13] J. S. Milne. Shimura varieties and moduli. In *Handbook of Moduli II*, volume 25 of *Advanced Lectures in Mathematics*, pages 467–548. Int. Press, Somerville, MA, 2013.

[Moo99] B. Moonen. Electronic notes on Mumford-Tate groups. Electronic notes available at http://www.math.ru.nl/personal/bmoonen, 1999.

[Roh09] J. C. Rohde. *Cyclic coverings, Calabi-Yau manifolds and complex multiplication*, volume 1975 of *Lecture Electronic notes in Mathematics*. Springer-Verlag, Berlin, 2009.

[Roh10] J. C. Rohde. Maximal automorphisms of Calabi-Yau manifolds versus maximally unipotent monodromy. *Manuscripta Math.*, 131(3-4):459–474, 2010.

[Sat80] I. Satake. *Algebraic structures of symmetric domains*, volume 4 of *Kanô Memorial Lectures*. Iwanami Shoten, Tokyo; Princeton University Press, Princeton, N.J., 1980.

[SZ10] M. Sheng and K. Zuo. Polarized variation of Hodge structures of Calabi–Yau type and characteristic subvarieties over bounded symmetric domains. *Math. Ann.*, 348(1):211–236, 2010.

[vG94] B. van Geemen. An introduction to the Hodge conjecture for abelian varieties. In *Algebraic cycles and Hodge theory (Torino, 1993)*, volume 1594 of *Lecture Electronic notes in Math.*, pages 233–252. Springer, Berlin, 1994.

[vG01] B. van Geemen. Half twists of Hodge structures of CM-type. *J. Math. Soc. Japan*, 53(4):813–833, 2001.

[Viv13] F. Viviani. A tour on Hermitian symmetric manifolds. arXiv: 1310.3665, 2013.

[Voi93] C. Voisin. Miroirs et involutions sur les surfaces $K3$. *Astérisque*, (218):273–323, 1993. Journées de Géométrie Algébrique d'Orsay (Orsay, 1992).

[Voi03] C. Voisin. *Hodge theory and complex algebraic geometry. II*, volume 77 of *Cambridge Studies in Advanced Mathematics*. Cambridge University Press, Cambridge, 2003. Translated from the French by Leila Schneps.

[Wol84] J. A. Wolf. *Spaces of constant curvature*. Publish or Perish, Inc., Houston, TX, fifth edition, 1984.

STONY BROOK UNIVERSITY, DEPARTMENT OF MATHEMATICS, STONY BROOK, NY 11794

E-mail address: rlaza@math.sunysb.edu

STONY BROOK UNIVERSITY, DEPARTMENT OF MATHEMATICS, STONY BROOK, NY 11794

E-mail address: zzhang@math.sunysb.edu

2

The singularities of the invariant metric on the Jacobi line bundle

José Ignacio Burgos Gil, Jürg Kramer and Ulf Kühn

ABSTRACT. A theorem by Mumford implies that every automorphic line bundle on a pure open Shimura variety, equipped with an invariant smooth metric, can be uniquely extended as a line bundle on a toroidal compactification of the variety, in such a way that the metric acquires only logarithmic singularities. This result is the key of being able to compute arithmetic intersection numbers from these line bundles. Hence, it is natural to ask whether Mumford's result remains valid for line bundles on mixed Shimura varieties.

In this paper we examine the simplest case, namely the Jacobi line bundle on the universal elliptic curve, whose sections are the Jacobi forms. We will show that Mumford's result cannot be extended directly to this case and that a new type of singularity appears.

By using the theory of b-divisors, we show that an analogue of Mumford's extension theorem can be obtained. We also show that this extension is meaningful because it satisfies Chern-Weil theory and a Hilbert-Samuel type formula.

Contents

1 Introduction

In [10], [11], Faltings introduced the notion of logarithmically singular metrics on a projective variety defined over a number field and proved that they satisfy

Burgos Gil was partially supported by the MICINN research projects MTM2010-17389 and MTM2013-42135. Kramer acknowledges support from the DFG Graduate School *Berlin Mathematical School* and from the DFG International Research Training Group *Moduli and Automorphic Forms.*

a Northcott type property, namely that the set of algebraic points not lying on the singular set of the metric with bounded height and degree, is finite. A prominent example of logarithmically singular metric is the Hodge bundle ω on a toroidal compactification of the moduli space of principally polarized abelian varieties of dimension g (with level structure if you do not want to work with stacks) $\overline{\mathscr{A}}_g$ equipped with the Petersson metric.

On the other hand, Mumford [22] introduced the concept of a good metric on a vector bundle, which is a class of singular metrics. He showed that, even being singular, Chern-Weil theory carries over to good metrics. He also proved that the invariant metric on a fully decomposable automorphic vector bundle on a toroidal compactification of the quotient of a hermitian symmetric domain by an arithmetic group is a good metric. This fact allowed him to extend Hirzebruch's proportionality principle to non-compact varieties.

The conclusion of the above facts is that the natural metrics that appear when studying vector bundles on toroidal compactifications of pure Shimura varieties are singular, but the singularities are mild enough so we can use the metrics to study geometric and arithmetic problems.

In [7] and [8], the authors developed a general theory of arithmetic intersections with logarithmically singular metrics that has been extensively used to compute arithmetic intersection numbers [5, 20, 6, 15, 16, 4, 1, 2, 12].

It is natural to ask whether this theory of logarithmically singular metrics can be extended to mixed Shimura varieties, to obtain geometric and arithmetic information of them.

In this paper we examine the first example of a mixed Shimura variety, namely the universal elliptic curve of full level N over the modular curve $E^0(N) \to Y(N)$. On it we consider the Jacobi line bundle, whose sections are the Jacobi forms, equipped with the translation invariant metric.

It turns out that, in this case, a new kind of singularity appears. These singularities are concentrated in codimension two. Therefore, if we remove a set of codimension two, we can extend the Jacobi line bundle to a line bundle with a good hermitian metric on a partial compactification of $E^0(N)$. Since algebraic line bundles can be uniquely extended along codimension two subsets, we obtain a line bundle with a singular metric on a compactification $E(N)$ of the universal elliptic curve.

It turns out that this naive approach is not a good idea. First, it is not functorial. If we consider different toroidal compactifications of $E^0(N)$, then the resulting extensions are not compatible. Second, even if the characteristic forms associated with the metric are locally integrable and define cohomology classes, they fail to satisfy a Chern-Weil theory. The cohomology class of the characteristic form does not agree with the characteristic class of the extended line bundle.

In this paper we propose a different approach to understand the extension of the Jacobi line bundle to a compactification of the universal elliptic curve. The "right" extension is not a line bundle, but a b-$\mathbb{Q}$-Cartier divisor. That is, a limit of different Cartier divisors with rational coefficients on all possible toroidal compactifications of $E^0(N)$. Defined in this way, the extension is obviously functorial because we are taking into account all possible toroidal compactifications. What is remarkable is that, with this interpretation, Chern-Weil theory allows us to interpret intersection products in terms of integrals of singular differential forms (see Theorems 5.2 and 5.6). Moreover, there is a Hilbert-Samuel type formula relating the asymptotic of the dimension of the space of Jacobi forms with the self-intersection of the b-divisor (Theorem 5.1).

The non-functoriality of the naive extension is exactly the height jumping introduced by Hain (see [13] and [23]).

Acknowledgments We have benefited from many discussions with colleagues on the subject of this paper. We want to thank S. Boucksom, R. de Jong, B. Edixhoven, D. Holmes, G. Freixas, A. von Pippich, and M. Sombra for many useful discussions. We thank specially A. von Pippich for pointing to us the Tornheim zeta function that is computed in [25] and R. de Jong that has computed independently the self-intersection product in Theorem 4.11, for sharing with us his work on the asymptotics of the Néron height pairing [14], that gives a complementary point of view on the results of this paper. We also would like to thank the anonymous referees for their careful reading of the manuscript and their comments and suggestions.

This research has been conducted during visits of the authors to the Humboldt University of Berlin, the ICMAT at Madrid and the University of Barcelona. Our thanks go to these institutions for their hospitality.

2 The universal elliptic surface

In the whole paper we fix an integer $N \geq 3$. In this section we will revisit the definition of the universal elliptic surface of level N lying over the modular curve of level N. In particular, we will recall the construction of its smooth toroidal compactification. For further details and references the reader is referred to [17].

THE MODULAR CURVE OF LEVEL N. Let $\mathbb{H}$ denote the upper half-plane given by

$$\mathbb{H} := \{\tau \in \mathbb{C} \mid \tau = \xi + i\eta,\ \eta > 0\}$$

and $\mathbb{H}^* := \mathbb{H} \cup \mathbb{P}^1(\mathbb{Q})$ the extended upper half-plane. The principal congruence subgroup

$$\Gamma(N) := \left\{ \begin{pmatrix} a & b \\ c & d \end{pmatrix} \in \mathrm{SL}_2(\mathbb{Z}) \,\middle|\, a \equiv d \equiv 1 \bmod N, b \equiv c \equiv 0 \bmod N \right\}$$

of level N acts in the usual way by fractional linear transformations on $\mathbb{H}$; this action naturally extends to $\mathbb{H}^*$. The quotient space $X(N) := \Gamma(N)\backslash\mathbb{H}^*$ is called the modular curve of level N; it is the compactification of $Y(N) := \Gamma(N)\backslash\mathbb{H}$ by adding the so-called cusps.

The modular curve $X(N)$ is a compact Riemann surface of genus

$$g_N = 1 + \frac{N-6}{12} \frac{[\mathrm{SL}_2(\mathbb{Z}) : \Gamma(N)]}{2N},$$

where the index of $\Gamma(N)$ in $\mathrm{SL}_2(\mathbb{Z})$ is given as

$$[\mathrm{SL}_2(\mathbb{Z}) : \Gamma(N)] = N^3 \prod_{p|N} \left(1 - \frac{1}{p^2}\right).$$

The number p_N of cusps of $X(N)$ is given by

$$p_N = \frac{[\mathrm{SL}_2(\mathbb{Z}) : \Gamma(N)]}{2N};$$

we denote the cusps by $P_1 := [\infty], P_2, \ldots, P_{p_N}$. We recall that $\Gamma(N)$ is a normal subgroup of $\mathrm{SL}_2(\mathbb{Z})$ and that the quotient group $\mathrm{SL}_2(\mathbb{Z})/\Gamma(N)$ acts transitively on the set of cusps (with stabilizers of order N). Therefore, it suffices in the sequel to consider the cusp $P_1 = [\infty]$. Since $N \geq 3$ the group $\Gamma(N)$ is torsion-free. Therefore, $X(N)$ has no elliptic points.

We recall that the modular curve $X(N)$ is the moduli space of elliptic curves with a full level N-structure. A point $[\tau] \in X(N)$ corresponds to the isomorphism class of elliptic curves determined by $\mathbb{C}/(\mathbb{Z}\tau \oplus \mathbb{Z})$ with N-torsion given by $(\mathbb{Z}\frac{\tau}{N} \oplus \mathbb{Z}\frac{1}{N})/(\mathbb{Z}\tau \oplus \mathbb{Z})$.

THE UNIVERSAL ELLIPTIC SURFACE OF LEVEL N. We consider the product $\mathbb{H} \times \mathbb{C}$ consisting of elements (τ, z) with $\tau \in \mathbb{H}$ and $z = x + iy \in \mathbb{C}$. On $\mathbb{H} \times \mathbb{C}$ the semi-direct product $\Gamma(N) \ltimes \mathbb{Z}^2$ acts by the assignment

$$\left[\begin{pmatrix} a & b \\ c & d \end{pmatrix}, (\lambda, \mu) \right] (\tau, z) := \left(\frac{a\tau + b}{c\tau + d}, \frac{z + \lambda\tau + \mu}{c\tau + d} \right),$$

where $\left(\begin{smallmatrix} a & b \\ c & d \end{smallmatrix}\right) \in \Gamma(N)$ and $(\lambda, \mu) \in \mathbb{Z}^2$. Since $N \geq 3$, the group $\Gamma(N)$ is torsion-free. Hence, the action of $\Gamma(N) \ltimes \mathbb{Z}^2$ on $\mathbb{H} \times \mathbb{C}$ is free and the quotient space

$$E^0(N) := (\Gamma(N) \ltimes \mathbb{Z}^2)\backslash(\mathbb{H} \times \mathbb{C})$$

is a smooth complex surface together with a smooth surjective morphism

$$\pi^0 \colon E^0(N) \longrightarrow Y(N)$$

with fiber $(\pi^0)^{-1}([\tau]) = \mathbb{C}/(\mathbb{Z}\tau \oplus \mathbb{Z})$.

The surface $E^0(N)$ is known to extend to a compact complex surface $E(N)$ together with a surjective morphism

$$\pi \colon E(N) \longrightarrow X(N),$$

the so-called universal elliptic surface of level N. To describe this extension, it suffices to describe the fibers $\pi^{-1}(P_j)$ above the cusps $P_j \in X(N)$ ($j = 1,\dots,p_N$). These are given as N-gons, more precisely as

$$\pi^{-1}(P_j) = \bigcup_{\nu=0}^{N-1} \Theta_{j,\nu},$$

where $\Theta_{j,\nu} \cong \mathbb{P}^1(\mathbb{C})$ is embedded into $E(N)$ with self-intersection number -2, while otherwise

$$\Theta_{j,\nu} \cdot \Theta_{j,\nu'} = \begin{cases} 1 & \nu' = \nu \pm 1, \\ 0 & |\nu - \nu'| \geq 2; \end{cases}$$

here and subsequently, the indices ν, ν' have to be read modulo N.

In terms of local coordinates the situation above the cusp $P_1 = [\infty]$ can be described as follows: The irreducible fiber $\Theta_\nu := \Theta_{1,\nu} \subset E(N)$ can be covered by two affine charts $W_\nu^0, W_\nu^1 \subset E(N)$, where W_ν^0 contains the point $\Theta_\nu \cap \Theta_{\nu+1}$ and W_ν^1 contains the point $\Theta_\nu \cap \Theta_{\nu-1}$. Since Θ_ν and $\Theta_{\nu+1}$ intersect transversally, we can choose coordinates u_ν, v_ν on the chart W_ν^0 in such a way that $\Theta_\nu|_{W_\nu^0}$ is given by the equation $v_\nu = 0$ and $\Theta_{\nu+1}|_{W_\nu^0}$ by the equation $u_\nu = 0$. Using that $\Theta_\nu \cdot \Theta_\nu = -2$ we obtain that the coordinates of W_ν^1 are given by $u_\nu^{-1}, u_\nu^2 v_\nu$. The open subset $W_{\nu+1}^1$ agrees with W_ν^0. Hence, we deduce

$$u_{\nu+1} = v_\nu^{-1}, \quad v_{\nu+1} = u_\nu v_\nu^2.$$

We finally note the relations

$$u_\nu v_\nu = q_N := \mathrm{e}^{2\pi i\tau/N}, \quad u_\nu^{\nu+1} v_\nu^{\nu} = \zeta := \mathrm{e}^{2\pi i z}. \tag{2.1}$$

If we want to work with different cusps we will denote by $W_{j,\nu}^0$ and $W_{j,\nu}^1$ the analogous affine charts around points over the cusp P_j.

We conclude by introducing the zero section

$$\varepsilon \colon X(N) \longrightarrow E(N)$$

and by recalling that the arithmetic genus of $E(N)$ is given by

$$p_{a,N} = \frac{[\mathrm{SL}_2(\mathbb{Z}) : \Gamma(N)]}{24} - 1 = \frac{N p_N}{12} - 1.$$

JACOBI FORMS. Modular forms can be interpreted as global sections of line bundles on the modular curve. The Jacobi forms play a similar role for the universal elliptic curve.

Definition 2.2. Let k,m be non-negative integers. A holomorphic function $f\colon \mathbb{H}\times\mathbb{C}\to\mathbb{C}$ is called *Jacobi form of weight k, index m for* $\Gamma(N)$, if it satisfies the following properties:

(i) The function f satisfies the functional equations

$$f\left(\frac{a\tau+b}{c\tau+d},\frac{z+\lambda\tau+\mu}{c\tau+d}\right)(c\tau+d)^{-k}$$

$$\times\exp\left(2\pi im\left(\lambda^2\tau+2\lambda z-\frac{c(z+\lambda\tau+\mu)^2}{c\tau+d}\right)\right)=f(\tau,z) \qquad (2.3)$$

for all $\left[\left(\begin{smallmatrix}a&b\\c&d\end{smallmatrix}\right),(\lambda,\mu)\right]\in\Gamma(N)\ltimes\mathbb{Z}^2$.

(ii) At the cusp $P_1=[\infty]$, the function f has a Fourier expansion of the form

$$f(\tau,z)=\sum_{\substack{n\in\mathbb{N},r\in\mathbb{Z}\\4mn-Nr^2\geq 0}}c(n,r)q_N^n\zeta^r, \qquad (2.4)$$

and similar Fourier expansions at the other cusps.

We denote the vector space of Jacobi forms of weight k, index m for $\Gamma(N)$ by $J_{k,m}(\Gamma(N))$.

If condition (ii) on the Fourier expansions is restricted to the summation over $n\in\mathbb{N}_{>0}$ and $r\in\mathbb{Z}$ such that $4mn-Nr^2>0$, the function f is called *Jacobi cusp form of weight k, index m for* $\Gamma(N)$ and the span of these functions is denoted by $J_{k,m}^{\mathrm{cusp}}(\Gamma(N))$.

If condition (ii) on the Fourier expansions is dropped, the function f is called *weak Jacobi form of weight k, index m for* $\Gamma(N)$. The span of these functions is denoted by $J_{k,m}^{\mathrm{weak}}(\Gamma(N))$.

Definition 2.5. Given integers k, m and $N>0$, the condition (2.3) is a cocycle condition that defines a line bundle $L_{k,m,N}$ on $E^0(N)$. This line bundle is called the *Jacobi line bundle*. Its space of global sections, $H^0(E^0(N),L_{k,m,N})$, agrees with the space of weak Jacobi forms of weight k and index m for $\Gamma(N)$.

RIEMANN THETA FUNCTIONS. The Riemann theta function $\theta_{1,1}\colon\mathbb{H}\times\mathbb{C}\to\mathbb{C}$ is defined by the convergent power series

$$\theta_{1,1}(\tau,z):=\sum_{n\in\mathbb{Z}}\exp\left(\pi i\tau\left(n+\frac{1}{2}\right)^2+2\pi i\left(z+\frac{1}{2}\right)\left(n+\frac{1}{2}\right)\right) \qquad (2.6)$$

and satisfies the functional equation

$$\theta_{1,1}\left(\frac{a\tau+b}{c\tau+d}, \frac{z+\lambda\tau+\mu}{c\tau+d}\right)(c\tau+d)^{-1/2}$$

$$\times \exp\left(\pi i\left(\lambda^2\tau+2\lambda z-\frac{c(z+\lambda\tau+\mu)^2}{c\tau+d}\right)\right)=\chi\begin{pmatrix} a & b \\ c & d\end{pmatrix}\theta_{1,1}(\tau,z)$$

for all $\left[\left(\begin{smallmatrix} a & b \\ c & d\end{smallmatrix}\right),(\lambda,\mu)\right] \in \mathrm{SL}_2(\mathbb{Z}) \ltimes \mathbb{Z}^2$ with a character $\chi(\cdot)$, which is an 8-th root of unity. Therefore, $\theta_{1,1}^8$ is a weak Jacobi form of weight 4, index 4 for $\Gamma(1)=\mathrm{SL}_2(\mathbb{Z})$. Moreover, from the definition of power series (2.6) it follows that $\theta_{1,1}^8$ is a Jacobi form.

DIMENSION FORMULAE. We recall the dimension formulae for the space of Jacobi forms. For simplicity, we restrict ourselves to the case $m=k=4\ell$. We denote by $j\colon E^0(N)\to E(N)$ the open immersion. From [17], we cite the following result.

Proposition 2.7. *There is a distinguished subsheaf $\mathcal{F}_\ell$ of the sheaf $j_*L_{4\ell,4\ell,N}$ such that we have an isomorphism*

$$J^{\mathrm{cusp}}_{4\ell,4\ell}(\Gamma(N)) \cong H^0(E(N),\mathcal{F}_\ell).$$

In particular, the dimension of $J^{\mathrm{cusp}}_{4\ell,4\ell}(\Gamma(N))$ is given, when N divides 4ℓ, by

$$\begin{aligned}\dim J^{\mathrm{cusp}}_{4\ell,4\ell}&(\Gamma(N))\\ &= p_N\left(\frac{8N\ell^2}{3}-N\ell-\frac{N}{4}Q\Big(\frac{16\ell}{N}\Big)-\frac{N}{2}\sum_{\substack{\Delta\mid 16\ell/N,\,\Delta<0\\ 16\ell/(N\Delta)\ \text{squarefree}}} H(\Delta)\right)\\ &= \frac{8Np_N}{3}\ell^2+o(\ell^2),\end{aligned}$$

where $Q(n)$ denotes the largest integer whose square divides n and $H(\Delta)$ is the Hurwitz class number.

Proof. The first statement is [17, Theorem 2.6], the second statement is [17, Theorem 3.8], noting that

$$[\mathrm{SL}_2(\mathbb{Z}):\Gamma(N)]=2Np_N.$$

To prove the asymptotic estimate one uses that $Q(n)$ is at most $\sqrt{n}$, that the number of divisors of an integer n is $o(n^\varepsilon)$ for any $\varepsilon>0$ and that, by the Brauer-Siegel theorem, the Hurwitz class number $H(\Delta)$ is $o(|\Delta|^{1/2+\varepsilon})$ for any $\varepsilon>0$. □

Remark 2.8. The asymptotic growth in ℓ of the dimension of the space of Jacobi forms and of the space of Jacobi cusp forms is the same. To see this we have to estimate the number of conditions on the vanishing of the Fourier coefficients $c(n,r)$ for $4mn - Nr^2 = 0$ with $m = 4\ell$.

Using the transformation given by $a = d = \lambda = 1$ and $b = c = \mu = 0$, we see that the coefficients of the Fourier expansion (2.4) satisfy the periodicity relation

$$c(n+Nr+Nm, r+2m) = c(n,r). \tag{2.9}$$

Given a Jacobi form of weight k and index m, to impose that it is a Jacobi cusp form is equivalent to impose, at each cusp, the conditions

$$c(n,r) = 0, \text{ for } 4mn - Nr^2 = 0.$$

In view of the periodicity condition (2.9), this is a finite number of conditions that grows linearly with the index. Therefore, the difference $\dim J_{4\ell,4\ell}(\Gamma(N)) - \dim J^{\mathrm{cusp}}_{4\ell,4\ell}(\Gamma(N))$ grows linearly with ℓ and we deduce the asymptotic formula

$$\dim J_{4\ell,4\ell}\big(\Gamma(N)\big) = \frac{8Np_N}{3}\ell^2 + o(\ell^2).$$

Translation invariant metric. Here we recall the translation invariant metric on the line bundle $L_{k,m,N}$.

Definition 2.10. For $f \in J^{\mathrm{weak}}_{k,m}(\Gamma(N))$, we define

$$\|f(\tau,z)\|^2 := |f(\tau,z)|^2 \exp(-4\pi m y^2/\eta)\eta^k,$$

where we recall that $\eta = \mathrm{Im}(\tau)$ and $y = \mathrm{Im}(z)$.

Lemma 2.11. *For $f \in J^{\mathrm{weak}}_{k,m}(\Gamma(N))$, we have*

$$\left\| f\left(\frac{a\tau+b}{c\tau+d}, \frac{z+\lambda\tau+\mu}{c\tau+d}\right)\right\|^2 = \|f(\tau,z)\|^2$$

for all $\big[\big(\begin{smallmatrix} a & b \\ c & d\end{smallmatrix}\big), (\lambda,\mu)\big] \in \Gamma(N) \ltimes \mathbb{Z}^2$. In particular, this shows that $\|\cdot\|$ induces a hermitian metric on the line bundle $L_{k,m,N}$.

Proof. This is a straightforward calculation. □

Lemma 2.12. *Locally, in the affine chart W^0_ν over the cusp $P_1 = [\infty]$, the hermitian metric $\|\cdot\|$ is described by the formula*

$$\begin{aligned}\log\left(\|f(\tau,z)\|^2\right)\Big|_{W^0_\nu} &= \log\left(|f(\tau,z)|^2\right)\Big|_{W^0_\nu} \\ &+ \frac{m}{N}\left((\nu+1)^2\log(u_\nu\bar{u}_\nu) + \nu^2\log(v_\nu\bar{v}_\nu) - \frac{\log(u_\nu\bar{u}_\nu)\log(v_\nu\bar{v}_\nu)}{\log(u_\nu\bar{u}_\nu)+\log(v_\nu\bar{v}_\nu)}\right) \\ &+ k\log\left(-\frac{N}{4\pi}\big(\log(u_\nu\bar{u}_\nu)+\log(v_\nu\bar{v}_\nu)\big)\right).\end{aligned}$$

Proof. Taking absolute values, we derive from (2.1)

$$\begin{aligned}\eta &= -\frac{N}{2\pi}\log|q_N| = -\frac{N}{2\pi}\log|u_\nu v_\nu| \\ &= -\frac{N}{4\pi}\left(\log(u_\nu\bar{u}_\nu) + \log(v_\nu\bar{v}_\nu)\right),\end{aligned}$$

$$\begin{aligned}y &= -\frac{1}{2\pi}\log|\zeta| = -\frac{1}{2\pi}\log|u_\nu^{\nu+1}v_\nu^{\nu}| \\ &= -\frac{1}{4\pi}\left((\nu+1)\log(u_\nu\bar{u}_\nu) + \nu\log(v_\nu\bar{v}_\nu)\right).\end{aligned}$$

With these formulae we compute

$$\begin{aligned}-\frac{4\pi m y^2}{\eta} &= \frac{m}{N}\frac{\left((\nu+1)\log(u_\nu\bar{u}_\nu) + \nu\log(v_\nu\bar{v}_\nu)\right)^2}{\log(u_\nu\bar{u}_\nu) + \log(v_\nu\bar{v}_\nu)} \\ &= \frac{m}{N}\left((\nu+1)^2\log(u_\nu\bar{u}_\nu) + \nu^2\log(v_\nu\bar{v}_\nu) - \frac{\log(u_\nu\bar{u}_\nu)\log(v_\nu\bar{v}_\nu)}{\log(u_\nu\bar{u}_\nu) + \log(v_\nu\bar{v}_\nu)}\right).\end{aligned}$$

From this the proof follows immediately from the definition of the hermitian metric $\|\cdot\|$. □

3 Mumford-Lear extensions and b-divisors

In this section we will introduce Mumford-Lear extensions of a line bundle and relate them with b-divisors. We first recall the different notions of growth for metrics and differential forms that will be useful in the sequel.

NOTATIONS. Let X be a complex algebraic manifold of dimension d and D a normal crossing divisor of X. Write $U = X \setminus D$, and let $j\colon U \longrightarrow X$ be the inclusion. We will denote by $\mathscr{E}_X^*$ the sheaf of algebras of smooth complex differential forms on X and by $\mathscr{E}_U^*$ the restriction of this sheaf to U.

Let V be an open coordinate subset of X with coordinates $z_1,\dots,z_d$; we put $r_i = |z_i|$. We say that V *is adapted to* D, if the divisor D is locally given by the equation $z_1\cdots z_k = 0$. We assume that the coordinate neighborhood V is small enough; more precisely, we will assume that all the coordinates satisfy $r_i < 1/e^e$, which implies that $\log 1/r_i > e$ and $\log(\log 1/r_i) > 1$.

If f and g are two functions with non-negative real values, we will write $f \prec g$, if there exists a constant $C > 0$ such that $f(x) \leq C\cdot g(x)$ for all x in the domain of definition under consideration.

LOG-LOG GROWTH FORMS.

Definition 3.1. We say that a smooth complex function f on $X \setminus D$ has *log-log growth along D*, if we have

$$|f(z_1,\dots,z_d)| \prec \prod_{i=1}^{k} \log(\log(1/r_i))^M \tag{3.2}$$

for every coordinate subset V adapted to D and some positive integer M. The *sheaf of differential forms on X with log-log growth along D* is the subalgebra of $j_*\mathscr{E}_U^*$ generated, in each coordinate neighborhood V adapted to D, by the functions with log-log growth along D and the differentials

$$\begin{aligned} &\frac{\mathrm{d}z_i}{z_i \log(1/r_i)}, \frac{\mathrm{d}\bar{z}_i}{\bar{z}_i \log(1/r_i)}, && \text{for } i = 1,\dots,k, \\ &\mathrm{d}z_i, \mathrm{d}\bar{z}_i, && \text{for } i = k+1,\dots,d. \end{aligned}$$

If D is clear from the context, a differential form with log-log growth along D will be called a *log-log growth form.*

Note that, to check that a function or a form has log-log growth it is enough to check the defining condition at the elements of an open covering of X made of coordinate subsets adapted to D.

DOLBEAULT ALGEBRA OF PRE-LOG-LOG FORMS. Clearly, the forms with log-log growth form an algebra but not a differential algebra. To remedy this we impose conditions on the derivatives as well.

Definition 3.3. A log-log growth form ω such that $\partial\omega$, $\bar{\partial}\omega$ and $\partial\bar{\partial}\omega$ are also log-log growth forms is called a *pre-log-log form (along D).* The sheaf of pre-log-log forms is the subalgebra of $j_*\mathscr{E}_U^*$ generated by the pre-log-log forms. We will denote this complex by $\mathscr{E}_X^*\langle\langle D\rangle\rangle_{\mathrm{pre}}$. The pre-log-log forms of degree zero are called *pre-log-log* functions.

The sheaf $\mathscr{E}_X^*\langle\langle D\rangle\rangle_{\mathrm{pre}}$, together with its real structure, its bigrading, and the usual differential operators ∂, $\bar{\partial}$ is easily shown to be a sheaf of Dolbeault algebras. Moreover, it is the maximal subsheaf of Dolbeault algebras of the sheaf of differential forms with log-log growth.

METRICS WITH LOGARITHMIC GROWTH AND PRE-LOG METRICS. Let L be a line bundle on X and let $\|\cdot\|$ be a smooth hermitian metric on $L|_U$.

Definition 3.4. We will say that the metric $\|\cdot\|$ has *logarithmic growth (along D)* if, for every point $x \in X$, there is a coordinate neighbourhood V of x adapted to D, a nowhere zero regular section s of L on V, and an integer $M \geq 0$ such that

$$\prod_{i=1}^{k} \log(1/r_i)^{-M} \prec \|s(z_1,\dots,z_d)\| \prec \prod_{i=1}^{k} \log(1/r_i)^M \tag{3.5}$$

Definition 3.6. We will say that the metric $\|\cdot\|$ is a *pre-log metric (along D),* if it has logarithmic growth and, for every rational section s of L, the function $\log\|s\|$ is a pre-log-log form along $D\setminus\operatorname{div}(s)$ on $X\setminus\operatorname{div}(s)$.

MUMFORD-LEAR EXTENSIONS. We are now able to define Mumford-Lear extensions. For the remainder of the section we fix a complex algebraic manifold X of dimension d, D and U as before, and we also fix a hermitian line bundle $\overline{L}=(L,\|\cdot\|)$ on U.

Definition 3.7. We say that $\overline{L}$ *admits a Mumford-Lear extension to X*, if there is an integer $e\geq 1$, a line bundle $\mathcal{L}$ on X, an algebraic subset $S\subset X$ of codimension at least 2 that is contained in D, a smooth hermitian metric $\|\cdot\|$ on $\mathcal{L}|_U$ that is a pre-log metric along $D\setminus S$ and an isometry $\alpha\colon (L,\|\cdot\|)^{\otimes e}\to(\mathcal{L}|_U,\|\cdot\|)$. The 5-tuple $(e,\mathcal{L},S,\|\cdot\|,\alpha)$ is called a *Mumford-Lear extension* of $\overline{L}$. When the isomorphism α, the metric and the set S can be deduced from the context, we will denote the Mumford-Lear extension by $(e,\mathcal{L})$. If $e=1$, we will denote it by the line bundle $\mathcal{L}$.

Remark 3.8. The name Mumford-Lear extension arises because they generalize (in the case of line bundles) the extensions of hermitian vector bundles considered by Mumford in [22] and the extensions of line bundles considered by Lear in his thesis [21].

The Mumford-Lear extensions satisfy the following unicity property.

Proposition 3.9. *Assume that $\overline{L}$ admits a Mumford-Lear extension to X. Let $(e_1,\mathcal{L}_1,S_1,\|\cdot\|_1,\alpha_1)$ and $(e_2,\mathcal{L}_2,S_2,\|\cdot\|_2,\alpha_2)$ be two Mumford-Lear extensions of $\overline{L}$ to X. Then, there is a unique isomorphism*

$$\beta\colon\mathcal{L}_1^{\otimes e_2}\to\mathcal{L}_2^{\otimes e_1}$$

such that the diagram

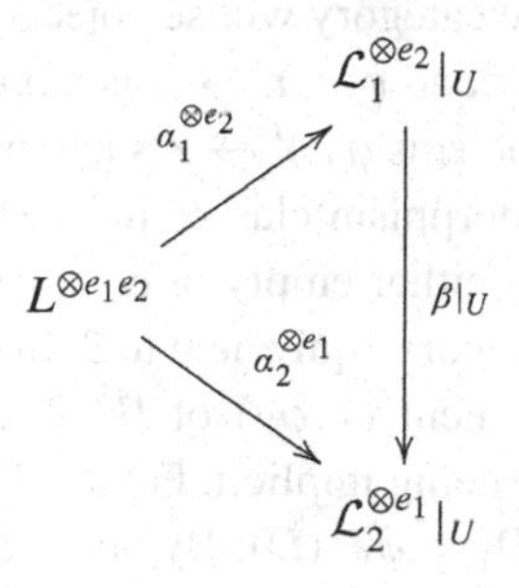

is commutative.

Proof. The composition $\alpha_2^{\otimes e_1}\circ(\alpha_1^{-1})^{\otimes e_2}$ defines an isomophism between the line bundles $\mathcal{L}_1^{\otimes e_2}|_U$ and $\mathcal{L}_2^{\otimes e_1}|_U$ that is the only one that makes the diagram in the theorem commutative. Put $S = S_1 \cup S_2$. The proof of [22, Proposition 1.3] shows that this isomorphism extends uniquely to an isomorphism $\beta_1 \colon \mathcal{L}_1^{\otimes e_2}|_{X\setminus S} \to \mathcal{L}_2^{\otimes e_1}|_{X\setminus S}$. Since X is smooth and S has codimension 2, the isomorphism β_1 extends to a unique isomorphism $\beta \colon \mathcal{L}_1^{\otimes e_2} \to \mathcal{L}_2^{\otimes e_1}$ satisfying the condition of the proposition. □

The next result is an immediate consequence of Proposition 3.9.

Corollary 3.10. *Assume the hypothesis of the previous proposition. Let s be a rational section of L, so $\alpha_1(s^{\otimes e_1})^{\otimes e_2}$ and $\alpha_2(s^{\otimes e_2})^{\otimes e_1}$ are rational sections of $\mathcal{L}_1^{\otimes e_2}$ and $\mathcal{L}_2^{\otimes e_1}$, respectively. Then,*

$$\operatorname{div}(\alpha_1(s^{\otimes e_1})^{\otimes e_2}) = \operatorname{div}(\alpha_2(s^{\otimes e_2})^{\otimes e_1})$$

as Cartier divisors on X. □

This result allows us to associate to each rational section of L a $\mathbb{Q}$-Cartier divisor on X. We will denote by $\mathbb{Q}\text{-Ca}(X)$ the group of $\mathbb{Q}$-Cartier divisors of X.

Definition 3.11. Assume that $\overline{L}$ admits a Mumford-Lear extension to X and let $(e, \mathcal{L}, S, \|\cdot\|, \alpha)$ be one such extension. Let s be a rational section of L. Then, we define *the divisor of s on X* as the $\mathbb{Q}$-Cartier divisor

$$\operatorname{div}_X(s) = \frac{1}{e}\operatorname{div}(\alpha(s^{\otimes e})) \in \mathbb{Q}\text{-Ca}(X),$$

where $\operatorname{div}(\alpha(s^{\otimes e}))$ is the divisor of $\alpha(s^{\otimes e})$ viewed as a rational section of $\mathcal{L}$ on the whole of X.

MUMFORD-LEAR EXTENSIONS AND BIRATIONAL MORPHISMS. We now consider Mumford-Lear extensions on different birational models of X.

Notation 3.12. Let $\mathcal{C}$ be the category whose objects are pairs (Y, φ_Y), where Y is a smooth complex variety and $\varphi_Y \colon Y \to X$ is a proper birational morphism, and whose arrows are morphisms $\varphi \colon Y \to Z$ such that $\varphi_Z \circ \varphi = \varphi_Y$. We denote by $\mathcal{BIR}(X)$ the set of isomorphism classes in $\mathcal{C}$. Since the set of morphisms between two objects of $\mathcal{C}$ is either empty or contains a single element, the set $\mathcal{BIR}(X)$ is itself a small category equivalent to $\mathcal{C}$. In fact, $\mathcal{BIR}(X)$ is a directed set. As a shorthand, an element (Y, φ_Y) of $\mathcal{BIR}(X)$ will be denoted by the variety Y, the morphism φ_Y being implicit. For an object Y of $\mathcal{BIR}(X)$ we will denote $U_Y = \varphi_Y^{-1}(U)$ and $D_Y = \varphi_Y^{-1}(D)$. By abuse of notation we will denote also by φ_Y the induced proper birational morphism from U_Y to U. Finally, we will denote by $\mathcal{BIR}'(X)$ the subset consisting of the elements Y with D_Y a normal crossing divisor. This is a cofinal subset.

Definition 3.13. We say that $\overline{L}$ *admits all Mumford-Lear extensions over* X if, for every object Y of $\mathcal{BIR}'(X)$, the hermitian line bundle $\varphi_Y^*\overline{L}$ on U_Y admits a Mumford-Lear extension to Y.

Definition 3.14. Assume that $\overline{L}$ admits all Mumford-Lear extensions over X. For every $Y \in \mathcal{BIR}'(X)$, choose a Mumford-Lear extension $(e', \mathcal{L}', S', \|\cdot\|', \alpha')$ of $\varphi_Y^*\overline{L}$ to Y. Then, the *divisor of* s *on* Y is defined as

$$\operatorname{div}_Y(s) = \frac{1}{e'} \operatorname{div}(\alpha'(s^{\otimes e'})) \in \mathbb{Q}\text{-Ca}(Y).$$

The $\mathbb{Q}$-Cartier divisors of Definition 3.14 do not need to be compatible with inverse images. As we will see in concrete examples, it may happen that there are morphisms $\varphi\colon Y \to Z$ in $\mathcal{BIR}'(X)$ such that

$$\varphi^* \operatorname{div}_Z(s) \neq \operatorname{div}_Y(s).$$

This lack of compatibility with inverse images is related with the phenomenon of height jumping (see [13] and [23] for a discussion of height jumping).

In contrast, the divisors associated with Mumford-Lear extensions are compatible with direct images.

Proposition 3.15. *Assume that* $\overline{L}$ *admits all Mumford-Lear extensions over* X. *Let* $\varphi\colon Y \to Z$ *be a morphism in* $\mathcal{BIR}'(X)$ *and* s *a section of* L. *Then,*

$$\varphi_* \operatorname{div}_Y(s) = \operatorname{div}_Z(s).$$

Proof. Let T be the subset of Z, where φ^{-1} is not defined. Since Z is smooth, hence normal, by Zariski's main theorem, T has codimension at least 2. Write $W = Z \setminus T$ and let $U' = U_Z \cap W$. Then, $\overline{L}$ induces a line bundle on U' that admits a Mumford-Lear extension to W.

Since T has codimension 2, the restriction map

$$\mathbb{Q}\text{-Ca}(Z) \to \mathbb{Q}\text{-Ca}(W)$$

is an isomorphism. Moreover, using the definition is easy to see that

$$\operatorname{div}_Y(s)|_W = \operatorname{div}_W(s) = \operatorname{div}_Z(s)|_W.$$

Thus, the proposition follows from the commutativity of the diagram

$$\begin{array}{ccc} \mathbb{Q}\text{-Ca}(Y) & \longrightarrow & \mathbb{Q}\text{-Ca}(W) \\ \downarrow{\scriptstyle \varphi_*} & & \| \\ \mathbb{Q}\text{-Ca}(Z) & \xrightarrow{\cong} & \mathbb{Q}\text{-Ca}(W). \end{array}$$

□

B-DIVISORS. Recall that the Zariski-Riemann space of X is the projective limit

$$\mathfrak{X} = \varprojlim_{\mathcal{BIR}(X)} Y.$$

We are not going to use the structure of this space, which is introduced here merely in order to make later definitions more suggestive.

For the definition of b-divisors, we will follow the point of view of [3]. The groups $\mathbb{Q}\text{-Ca}(Y)$, $Y \in \mathcal{BIR}(X)$, form a projective system with respect to the push-forward morphisms and an inductive system with respect to the pull-back morphisms. We define the group of $\mathbb{Q}$-Cartier divisors on $\mathfrak{X}$ as the inductive limit

$$\mathbb{Q}\text{-Ca}(\mathfrak{X}) = \varinjlim_{\mathcal{BIR}(X)^{\text{op}}} \mathbb{Q}\text{-Ca}(Y)$$

and the group of $\mathbb{Q}$-Weil divisors on $\mathfrak{X}$ as the projective limit

$$\mathbb{Q}\text{-We}(\mathfrak{X}) = \varprojlim_{\mathcal{BIR}(X)} \mathbb{Q}\text{-Ca}(Y).$$

Since, for any morphism φ in $\mathcal{BIR}(X)$ the composition $\varphi_* \circ \varphi^*$ is the identity, it is easy to see that there is a map $\mathbb{Q}\text{-Ca}(\mathfrak{X}) \to \mathbb{Q}\text{-We}(\mathfrak{X})$. Note also that, since $\mathcal{BIR}'(X)$ is cofinal, the above projective and inductive limit can be taken over $\mathcal{BIR}'(X)$.

The group of $\mathbb{Q}$-Weil divisors of $\mathfrak{X}$ is closely related to the group of b-divisors of X defined in [24]. Thus, a $\mathbb{Q}$-Weil divisor of $\mathfrak{X}$ will be called a b-divisor of X.

The following definition makes sense thanks to Proposition 3.9.

Definition 3.16. Assume that $\overline{L}$ admits all Mumford-Lear extensions over X. Let s be rational section of L. Then, the *b-divisor associated* to s is

$$\text{b-div}(s) = (\text{div}_Y(s))_{Y \in \mathcal{BIR}'(X)} \in \mathbb{Q}\text{-We}(\mathfrak{X}).$$

When it is needed to specify with respect to which metric we are compactifying the divisor, we will write $\text{b-div}(s, \|\cdot\|)$.

INTEGRABLE B-DIVISORS. From now on we restrict ourselves to the case when X is a surface. We want to extend the intersection product of divisors as much as possible to b-divisors.

It is clear that there is an intersection pairing

$$\mathbb{Q}\text{-Ca}(\mathfrak{X}) \times \mathbb{Q}\text{-We}(\mathfrak{X}) \to \mathbb{Q}$$

defined as follows. Let $C \in \mathbb{Q}\text{-Ca}(\mathfrak{X})$ and $E \in \mathbb{Q}\text{-We}(\mathfrak{X})$. Then, there is an object $Y \in \mathcal{BIR}'(X)$ and a divisor $C_Y \in \mathbb{Q}\text{-Ca}(Y)$ such that C is the image of C_Y. Let E_Y be the component of E on Y. Then, by the projection formula, the intersection product $C_Y \cdot E_Y$ does not depend on the choice of Y. Thus, we define

$$C \cdot E = C_Y \cdot E_Y.$$

But, in general, we can not define the intersection product of two elements of $\mathbb{Q}\text{-We}(\mathfrak{X})$. The following definition is the analogue for b-divisors of the concept of an L^2-function. Recall that, since $\mathcal{BIR}'(X)$ is a directed set, it is in particular a net.

Definition 3.17. A divisor $C = (C_Y)_{Y \in \mathcal{BIR}'(X)} \in \mathbb{Q}\text{-We}(\mathfrak{X})$ is called integrable if the limit

$$\lim_{\overrightarrow{\mathcal{BIR}'(X)}} C_Y \cdot C_Y$$

exists and is finite.

Proposition 3.18. *Let $C_1, C_2 \in \mathbb{Q}\text{-We}(\mathfrak{X})$. If C_1 and C_2 are integrable, then*

$$\lim_{\overrightarrow{\mathcal{BIR}'(X)}} C_{1,Y} \cdot C_{2,Y}$$

exists and is finite.

Proof. Let $C = (C_Y) \in \mathbb{Q}\text{-We}(\mathfrak{X})$ and let $\varphi \colon Y \to Z$ be an arrow in $\mathcal{BIR}'(X)$. Since $\varphi_* C_Y = C_Z$, we deduce that

$$C_Y = \varphi^* C_Z + E,$$

where E is an exceptional divisor for the morphism φ. Hence,

$$C_Y \cdot C_Y = (\varphi^* C_Z + E) \cdot (\varphi^* C_Z + E) = C_Z \cdot C_Z + E \cdot E.$$

Thus, by the Hodge index theorem,

$$C_Y \cdot C_Y - C_Z \cdot C_Z = E \cdot E \leq 0.$$

Hence,

$$\begin{aligned} 0 &\geq (C_{1,Y} \pm C_{2,Y})^2 - (C_{1,Z} \pm C_{2,Z})^2 \\ &= C_{1,Y}^2 - C_{1,Z}^2 + C_{2,Y}^2 - C_{2,Z}^2 \pm 2(C_{1,Y} \cdot C_{2,Y} - C_{1,Z} \cdot C_{2,Z}). \end{aligned}$$

Therefore,

$$\begin{aligned} C_{1,Y} \cdot C_{2,Y} - C_{1,Z} \cdot C_{2,Z} &\leq -\frac{1}{2}(C_{1,Y}^2 - C_{1,Z}^2 + C_{2,Y}^2 - C_{2,Z}^2) \\ &= \frac{1}{2}(|C_{1,Y}^2 - C_{1,Z}^2| + |C_{2,Y}^2 - C_{2,Z}^2|) \end{aligned}$$

and

$$C_{1,Y}\cdot C_{2,Y}-C_{1,Z}\cdot C_{2,Z}\geq \frac{1}{2}(C_{1,Y}^2-C_{1,Z}^2+C_{2,Y}^2-C_{2,Z}^2)$$
$$=-\frac{1}{2}(|C_{1,Y}^2-C_{1,Z}^2|+|C_{2,Y}^2-C_{2,Z}^2|).$$

Thus,

$$|C_{1,Y}\cdot C_{2,Y}-C_{1,Z}\cdot C_{2,Z}|\leq \frac{1}{2}(|C_{1,Y}^2-C_{1,Z}^2|+|C_{2,Y}^2-C_{2,Z}^2|).$$

Thus, the convergence of $(C_{1,Y}^2)_Y$ and $(C_{2,Y}^2)_Y$ implies the convergence of $(C_{1,Y}\cdot C_{2,Y})_Y$. □

4 The Mumford-Lear extension of the Jacobi line bundle

In this section we will study the Mumford-Lear extensions of the Jacobi line bundle.

THE FUNCTIONS $f_{n,m}$. We first study a family of functions that will be useful later. Let (n,m) be a pair of coprime positive integers. Let u,v be coordinates of $\mathbb{C}^2$ and denote $U\subset\mathbb{C}^2$ the open subset defined by $|uv|<1$. Let $D\subset U$ be the normal crossing divisor given by the equation $uv=0$.

Proposition 4.1. *Let $f_{n,m}$ be the function on U given by*

$$f_{n,m}(u,v)=\frac{1}{nm}\frac{\log(u\overline{u})\log(v\overline{v})}{n\log(u\overline{u})+m\log(v\overline{v})}.$$

This function satisfies the following properties.

(i) *The function $f_{n,m}$ is a pre-log-log function along $D\setminus\{(0,0)\}$.*
(ii) *The equality $\partial\overline{\partial}f_{n,m}\wedge\partial\overline{\partial}f_{n,m}=0$ holds. The differential forms $f_{n,m}$, $\partial f_{n,m}$, $\overline{\partial}f_{n,m}$, and $\partial\overline{\partial}f_{n,m}$ and all the products between them are locally integrable. Moreover, any product of $\partial\overline{\partial}f_{n,m}$ with a pre-log-log form along D is also locally integrable.*
(iii) *Let $\varphi\colon U\to U$ be the map given by $(s,t)\mapsto(st,t)$. Note that this map identifies U with a chart of the blow-up of U along $(0,0)$. Then,*

$$\varphi^*f_{n,m}(s,t)=\frac{1}{nm(n+m)}\log(t\overline{t})+f_{n,n+m}(s,t).$$

Proof. Put

$$P_{n,m}(u,v)=n\log(u\overline{u})+m\log(v\overline{v}),$$
$$a=\log(v\overline{v})\frac{\mathrm{d}u}{u},\qquad b=\log(u\overline{u})\frac{\mathrm{d}v}{v}.$$

With these notations, we have

$$\partial f_{n,m} = \frac{1}{nmP_{n,m}^2}(n\log(u\bar{u})b + m\log(v\bar{v})a), \tag{4.2}$$

$$\partial\bar{\partial} f_{n,m} = \frac{2}{P_{n,m}^3}(b-a)\wedge(\bar{a}-\bar{b}), \tag{4.3}$$

$$\partial f_{n,m}\wedge\partial\bar{\partial} f_{n,m} = \frac{2}{nmP_{n,m}^4}a\wedge b\wedge(\bar{a}-\bar{b}). \tag{4.4}$$

From equation (4.3), it follows that $\partial\bar{\partial} f_{n,m}\wedge\partial\bar{\partial} f_{n,m} = 0$.

We now prove (i). Let $p = (0,v_0) \in D\setminus\{(0,0)\}$. Let V be a neighborhood of p such that $|\log(v\bar{v})| \leq K$, $u\bar{u} < 1$ and

$$n|\log(u\bar{u})| \geq 2mK \geq 2m|\log(v\bar{v})|,$$

for some positive constant K. Therefore, on all the points of V, the estimate

$$|P_{n,m}| \geq \frac{n}{2}|\log(u\bar{u})|$$

holds. Then, for $(u,v) \in V$,

$$|f_{n,m}(u,v)| \leq \frac{2}{n^2m}\frac{|\log(u\bar{u})\log(v\bar{v})|}{|\log(u\bar{u})|} \leq \frac{2K}{n^2m}. \tag{4.5}$$

Similarly, if t_1 and t_2 are smooth tangent vectors on V with bounded coefficients, from (4.2) and (4.3), we derive

$$|\partial f_{n,m}(t_1)| \leq \frac{C_1}{|\log(u\bar{u})|^2|u|} \tag{4.6}$$

$$|\partial\bar{\partial} f_{n,m}(t_1,t_2)| \leq \frac{C_2}{|\log(u\bar{u})|^3|u|^2} \tag{4.7}$$

for suitable positive constants C_1 and C_2. The estimates (4.5), (4.6) and (4.7) show that $f_{n,m}$ is a pre-log-log function. Thus, we have proved (i).

Since pre-log-log forms are always locally integrable (cf. [8, Proposition 7.6]), in order to check (ii), it is only necessary to study a neighborhood of the point $(0,0)$. Thus, we restrict ourselves to the open W defined by $|u| < 1/e$ and $|v| < 1/e$.

We show the local integrability of a form of the type $\partial\bar{\partial} f_{n,m}\wedge\psi$ for a pre-log-log form ψ, the other cases being analogous.

By the definition of pre-log-log forms, equation (4.3) shows that $\partial\bar{\partial} f_{n,m}\wedge\psi = g(u,v)\,\mathrm{d}u\wedge\mathrm{d}\bar{u}\wedge\mathrm{d}v\wedge\mathrm{d}\bar{v}$, with g a function satisfying

$$|g(u,v)| \leq C_1\frac{|\log(\log(u\bar{u}))\log(\log(v\bar{v}))|^M}{|P_{n,m}^3 u\bar{u}v\bar{v}|}$$

for certain positive constants C_1 and M. Using the geometric vs. arithmetic mean inequality, and the fact that the logarithm grows slower than any polynomial, we see that g can be bounded as

$$|g(u,v)| \leq \frac{C_2}{|\log(u\overline{u})\log(v\overline{v})|^{1+\varepsilon}u\overline{u}v\overline{v}},$$

with C_2 and ε positive. Since the differential form

$$\frac{du \wedge d\overline{u} \wedge dv \wedge d\overline{v}}{|\log(u\overline{u})\log(v\overline{v})|^{1+\varepsilon}u\overline{u}v\overline{v}}$$

is locally integrable, we deduce that $\partial\overline{\partial} f_{n,m} \wedge \psi$ is locally integrable.

Every product between a smooth form and any of the forms $f_{n,m}$, $\partial f_{n,m}$, $\overline{\partial} f_{n,m}$, and $\partial\overline{\partial} f_{n,m}$, will satisfy growth estimates not worse than the one satisfied by $\partial\overline{\partial} f_{n,m} \wedge \psi$, except the product $\partial\overline{\partial} f_{n,m} \wedge \partial\overline{\partial} f_{n,m}$. Since this last product is zero we conclude (ii).

The statement (iii) follows from a direct computation. □

THE MUMFORD-LEAR EXTENSION OF THE JACOBI LINE BUNDLE TO $E(N)$. We now denote by $D = E(N) \setminus E^0(N)$. This is a normal crossings divisor. Let Σ be the set of double points of D and put $D^0 = D \setminus \Sigma$ for the smooth part of D. Let H be the divisor of $E(N)$ defined as the image of the zero section $X(N) \to E(N)$.

Consider the divisor on $E(N)$ given by

$$C = 8H + \sum_{j=1}^{p_N} \sum_{\nu=0}^{N-1} \left(N - 4\nu + \frac{4\nu^2}{N} \right) \Theta_{j,\nu}, \tag{4.8}$$

Choose a smooth hermitian metric $\|\cdot\|'$ on $\mathcal{O}(C)$ and let s be a section of $\mathcal{O}(C)$ with $\operatorname{div} s = C$.

Proposition 4.9. *The hermitian line bundle $\overline{L} = (L_{4,4,N}, \|\cdot\|)$ satisfies the following properties.*

(i) *The restriction of the metric $\|\cdot\|$ to $E^0(N)$ is smooth. Moreover, the divisor of the restriction of $\theta_{1,1}^8$ to $E^0(N)$ is $8H$. Therefore, there is a unique isomorphism $\alpha : L_{4,4,N} \to \mathcal{O}(C)\,|_{E^0(N)}$ that sends $\theta_{1,1}^8$ to s.*

(ii) *Each point p belonging to only one component $\Theta_{j,\nu}$ has a neighborhood V on which*

$$\log \|\theta_{1,1}^8\|^2 = \log \|s\|'^2 + \psi_1,$$

where ψ_1 is pre-log-log along D^0.

(iii) *On the affine coordinate chart $W_{j,\nu}^0$ defined in Section 2, we can write*

$$\log \|\theta_{1,1}^8\|^2 = \log \|s\|'^2 + \psi_2 - \frac{4}{N} \frac{\log(u_\nu\overline{u}_\nu)\log(v_\nu\overline{v}_\nu)}{\log(u_\nu\overline{u}_\nu) + \log(v_\nu\overline{v}_\nu)},$$

where ψ_2 is pre-log-log along D.

Consequently, if we denote also by $\|\cdot\|$ *the singular metric on* $\mathcal{O}(C)$ *induced by* α *and* $\|\cdot\|$, *then the 5-tuple* $(1, \mathcal{O}(C), \Sigma, \|\cdot\|, \alpha)$ *is a Mumford-Lear extension of the hermitian line bundle* $\overline{L}$ *to* $E(N)$ *and the divisor of* $\theta_{1,1}^8$ *on the universal elliptic surface* $E(N)$ *is given by*

$$\mathrm{div}_{E(N)}(\theta_{1,1}^8) = C.$$

Proof. The metric $\|\cdot\|$ on $L_{4,4,N}$ over the open subset $E^0(N)$ is induced by a smooth metric on the trivial line bundle over $\mathbb{H} \times \mathbb{C}$. Since $N \geq 3$, the map $\mathbb{H} \times \mathbb{C} \to E^0(N)$ is étale. Hence, the metric $\|\cdot\|$ is smooth on $E^0(N)$. Therefore, the components of $\mathrm{div}_{E(N)}(\theta_{1,1}^8)$ that meet the open subset $E^0(N)$ come from the theta function. Is well known that, for fixed $\tau \in \mathbb{H}$, the zeros of the Riemann theta function $\theta_{1,1}(\tau, z)$ are located at $z \in \mathbb{Z}\tau \oplus \mathbb{Z}$; all the zeros are simple. This proves that the restriction of $\mathrm{div}_{E(N)}(\theta_{1,1}^8)$ to $E^0(N)$ is given by $8H$. This finishes the proof of (i).

By the normality of the group $\Gamma(N)$ in $\mathrm{SL}_2(\mathbb{Z})$, in order to prove that $\overline{L}$ admits a Mumford-Lear extension and compute the divisor $\mathrm{div}_{E(N)}(\theta_{1,1}^8)$, it suffices to work over the cusp $P_1 = [\infty]$.

Consider the open affine chart $W = W_{1,\nu}^0$. By Lemma 2.12,

$$\begin{aligned}
\log\left(\|\theta_{1,1}^8(\tau,z)\|^2\right)\Big|_W &= \log\left(|\theta_{1,1}^8(\tau,z)|^2\right)\Big|_W \\
&+ \frac{4}{N}\left((\nu+1)^2\log(u_\nu\bar{u}_\nu) + \nu^2\log(v_\nu\bar{v}_\nu) - \frac{\log(u_\nu\bar{u}_\nu)\log(v_\nu\bar{v}_\nu)}{\log(u_\nu\bar{u}_\nu)+\log(v_\nu\bar{v}_\nu)}\right) \\
&+ 4\log\left(-\frac{N}{4\pi}\left(\log(u_\nu\bar{u}_\nu)+\log(v_\nu\bar{v}_\nu)\right)\right).
\end{aligned}$$

We first study the term $\log|\theta_{1,1}^8|^2$. For this, we rewrite expression (2.6) defining $\theta_{1,1}$ in terms of the local coordinates u_ν, v_ν. Using formulas (2.1), we obtain

$$\begin{aligned}
\theta_{1,1}(\tau,z) &= \sum_{n\in\mathbb{Z}} \mathrm{e}^{\pi i(n+1/2)} q_N^{N/2(n+1/2)^2} \zeta^{(n+1/2)} \\
&= \sum_{n\in\mathbb{Z}} \mathrm{e}^{\pi(n+1/2)} u_\nu^{N/2(n+1/2)^2+(\nu+1)(n+1/2)} v_\nu^{N/2(n+1/2)^2+\nu(n+1/2)}.
\end{aligned}$$

Since the vertical component $\Theta_{1,\nu}$ is characterized by the equation $v_\nu = 0$, the multiplicity of $\theta_{1,1}$ along $\Theta_{1,\nu}$ is given by

$$\min_{n\in\mathbb{Z}}\left(\frac{N}{2}n^2 + \left(\frac{N}{2}+\nu\right)n + \frac{N}{8} + \frac{\nu}{2}\right).$$

For a real number x we write $\lfloor x \rfloor$ for the biggest integer smaller or equal to x and $\epsilon(x) = x - \lfloor x \rfloor$. Then, one easily checks that

$$\min_{n\in\mathbb{Z}}\left(\frac{N}{2}n^2 + \left(\frac{N}{2}+\nu\right)n + \frac{N}{8} + \frac{\nu}{2}\right) = \frac{N}{2}\left(\epsilon^2\left(-\frac{\nu}{N}\right) - \epsilon\left(-\frac{\nu}{N}\right)\right) + \frac{N}{8} - \frac{\nu^2}{2N}.$$

Note that this quantity depends on the value of ν and not just on the residue class of ν modulo N. This is because $\theta_{1,1}(\tau,z)$ is a multi-valued function on $E^0(N)$.

Similarly, the multiplicity of $\theta_{1,1}$ along $\Theta_{1,\nu+1}$ is given by

$$\frac{N}{2}\left(\epsilon^2\left(-\frac{\nu+1}{N}\right)-\epsilon\left(-\frac{\nu+1}{N}\right)\right)+\frac{N}{8}-\frac{(\nu+1)^2}{2N}.$$

Therefore, on $W\setminus H$, we can write

$$\begin{aligned}\log|\theta_{1,1}^8|^2 &= \left(4N\left(\epsilon^2\left(-\frac{\nu+1}{N}\right)-\epsilon\left(-\frac{\nu+1}{N}\right)\right)+N-\frac{4(\nu+1)^2}{N}\right)\log(u_\nu\bar{u}_\nu)\\ &\quad+\left(4N\left(\epsilon^2\left(-\frac{\nu}{N}\right)-\epsilon\left(-\frac{\nu}{N}\right)\right)+N-\frac{4\nu^2}{N}\right)\log(v_\nu\bar{v}_\nu)+\psi_3,\end{aligned}$$

where ψ_3 is a smooth function.

We next consider the remaining terms of the expression of $\log\|\theta_{1,1}^8\|^2$. The term

$$4\log\left(-\frac{N}{4\pi}(\log(u_\nu\bar{u}_\nu)+\log(v_\nu\bar{v}_\nu))\right)$$

is pre-log-log along D.

The terms $\frac{4\nu^2}{N}\log(v_\nu\bar{v}_\nu)$ and $\frac{4(\nu+1)^2}{N}\log(u_\nu\bar{u}_\nu)$ add $4\nu^2/N$ and $4(\nu+1)^2/N$ to the multiplicity of the components $\Theta_{1,\nu}$ and $\Theta_{1,\nu+1}$ respectively. Summing up, we obtain that

$$\begin{aligned}\log\left(\|\theta_{1,1}^8(\tau,z)\|^2\right)\Big|_{W\setminus H} &= \left(4N\left(\epsilon^2\left(-\frac{\nu}{N}\right)-\epsilon\left(-\frac{\nu}{N}\right)\right)+N\right)\log(v_\nu\bar{v}_\nu)\\ &\quad+\left(4N\left(\epsilon^2\left(-\frac{\nu+1}{N}\right)-\epsilon\left(-\frac{\nu+1}{N}\right)\right)+N\right)\log(u_\nu\bar{u}_\nu)\\ &\quad-\frac{4}{N}\left(\frac{\log(u_\nu\bar{u}_\nu)\log(v_\nu\bar{v}_\nu)}{\log(u_\nu\bar{u}_\nu)+\log(v_\nu\bar{v}_\nu)}\right)+\psi_2,\end{aligned}$$

where ψ_2 is pre-log-log along D.

In order to finish the proof of (iii), it only remains to observe that, for $\nu = 0,\dots,N$,

$$4N\left(\epsilon^2\left(-\frac{\nu}{N}\right)-\epsilon\left(-\frac{\nu}{N}\right)\right)=\frac{4\nu^2}{N}-4\nu.$$

Statement (ii) follows from (iii) and the fact that, by Proposition 4.1 (i), the term

$$\frac{4}{N}\left(-\frac{\log(u_\nu\bar{u}_\nu)\log(v_\nu\bar{v}_\nu)}{\log(u_\nu\bar{u}_\nu)+\log(v_\nu\bar{v}_\nu)}\right)$$

is pre-log-log along D^0. □

THE SELF-INTERSECTION OF C. By Proposition 4.9, the Mumford-Lear extension of $\overline{L}$ to $E(N)$ is isomorphic to $\mathcal{O}(C)$. We next compute the self-intersection $C \cdot C$.

Proposition 4.10. *The self-intersection product $C \cdot C$ is given by*

$$C \cdot C = \frac{16(N^2+1)p_N}{3N}.$$

In particular, for $N = 4$, we have $p_4 = 6$, hence $C \cdot C = 136$.

Proof. Using the adjunction formula (see for instance [17], proof of Proposition 3.2), we obtain

$$H \cdot H = -\frac{Np_N}{12}.$$

Moreover,

$$H \cdot \Theta_{j,\nu} = \begin{cases} 1, & \text{if } \nu = 0, \\ 0, & \text{otherwise,} \end{cases}$$

and

$$\Theta_{j,\nu} \cdot \Theta_{j',\nu'} = \begin{cases} -2, & \text{if } j = j',\ \nu = \nu' \\ 1, & \text{if } j = j',\ \nu \equiv \nu' \pm 1 \mod N \\ 0, & \text{otherwise.} \end{cases}$$

From these intersection products and the explicit description of C in (4.8), we derive the result. □

THE B-DIVISOR OF THE JACOBI LINE BUNDLE.

Theorem 4.11. *The line bundle $\overline{L} = (L_{4,4,N}, \|\cdot\|)$ admits all Mumford-Lear extensions over $E(N)$. Moreover, the associated b-divisor is integrable and the equality*

$$\text{b-div}(\theta_{1,1}^8) \cdot \text{b-div}(\theta_{1,1}^8) = \frac{16Np_N}{3}$$

holds.

Proof. Recall that $\Sigma \subset D$ denotes the set of double points of D. By Proposition 4.9 (ii), the restriction $\overline{L}|_{E(N)\setminus\Sigma}$ has a pre-log metric along $D \setminus \Sigma$. Therefore, if $p \notin \Sigma$ and $\varphi\colon X \to E(N)$ is the blow-up of $E(N)$ at p, we deduce that $\varphi^*\mathcal{O}(C)$ is a Mumford-Lear extension of $\overline{L}$ and that

$$\operatorname{div}_X(\theta_{1,1}^8) = \varphi^* \operatorname{div}_{E(N)}(\theta_{1,1}^8).$$

Assume now that $p \in \Sigma$ and $\varphi\colon X \to E(N)$ is the blow-up of $E(N)$ at p. Write Σ_X for the set of double points of the total transform of D. Then, $\#\Sigma_X = \#\Sigma + 1$, because we can write $\Sigma_X = (\Sigma \setminus \{p\}) \cup \{p_1, p_2\}$, where $\{p_1, p_2\}$ is the intersection of the exceptional divisor E of the blow-up with the strict transform of D.

Proposition 4.1 (iii) and Proposition 4.9 (iii) imply that

$$(N, \varphi^*\mathcal{O}(NC) \otimes \mathcal{O}(-2E))$$

is a Mumford-Lear extension of $\overline{L}$ to X (in this case the 2-codimensional set is Σ_X, and the isomorphism and the metric are the ones induced by α and $\|\cdot\|$). Moreover,

$$\operatorname{div}_X(\theta_{1,1}^8) = \varphi^* \operatorname{div}_{E(N)}(\theta_{1,1}^8) - \frac{2}{N} E.$$

A similar phenomenon occurs on any smooth surface birational to $E(N)$. To describe it we need a little of terminology. Let $\varphi\colon X \to E(N)$ be a proper birational morphism with X smooth. A point $p \in X$ will be called mild if the metric of $\varphi^*\overline{L}$ is smooth or pre-log in a neighborhood of p. Put $\Sigma_X \subset X$ for the set of non-mild points. Let n, m be positive integers with $\gcd(n,m) = 1$. We will say that a point p has type (n,m) and multiplicity μ if there is a coordinate neighborhood centered at p, with coordinates (u, v) such that

$$\log \|\theta_{1,1}^8\| = \log \|s\|' + \psi - \frac{\mu}{nm} \frac{\log(u\bar{u})\log(v\bar{v})}{n\log(u\bar{u}) + m\log(v\bar{v})},$$

where ψ is a smooth function on the pre-image of $E^0(N)$ with log-log growth along the pre-image of $E(N) \setminus E^0(N)$. Observe that $E(N)$ has Np_N non-mild points, all of type $(1,1)$ and multiplicity $4/N$.

Assume that Σ_X is finite and that $\overline{L}$ admits a Mumford-Lear extension $(e_X, \mathcal{O}(C_X), \Sigma_X, \|\cdot\|, \alpha)$ to X. Let D_X be the total transform of D to X. If $\varphi : X' \to X$ is the blow-up at a mild point $p \notin \Sigma_X$, then $\Sigma_{X'} = \varphi^{-1}\Sigma_X$ is finite and $(e_X, \varphi^*\mathcal{O}(C_X))$ is a Mumford-Lear extension of $\overline{L}$ to X'. In particular,

$$\operatorname{div}_{X'}(\theta_{1,1}^8) = \varphi^* \operatorname{div}_X(\theta_{1,1}^8).$$

Let now $\varphi\colon X' \to X$ be the blow-up of X at a point $p \in \Sigma_X$, with type (n,m) and multiplicity a/b, with a, b integers. Then, by Proposition 4.1(iii), $\Sigma_{X'} = (\Sigma_X \setminus \{p\}) \cup \{p_1, p_2\}$, where $\{p_1, p_2\}$ is the intersection of the exceptional divisor $E_{X'}$ of the blow-up with the strict transform of D_X. Moreover,

$$(bnm(n+m)e_X, \varphi^*\mathcal{O}(bnm(n+m)C_X) \otimes (-aE_{X'}))$$

is a Mumford-Lear extension of $\overline{L}$ to X'. Hence,

$$\operatorname{div}_{X'}(\theta_{1,1}^8) = \varphi^* \operatorname{div}_X(\theta_{1,1}^8) - \frac{a}{bnm(n+m)} E_{X'}.$$

Note also that the singular point p gives rise to two points in $\Sigma_{X'}$, both of multiplicity a/b, one of type $(n+m, m)$ and the other of type $(n, n+m)$. Since the self-intersection of the exceptional divisor $E_{X'}$ is -1, we deduce that

$$\operatorname{div}_{X'}(\theta_{1,1}^8)^2 = \varphi^* \operatorname{div}_X(\theta_{1,1}^8)^2 - \frac{a^2}{b^2n^2m^2(n+m)^2}. \tag{4.12}$$

Since the elements of $\mathcal{BIR}'(E(N))$ can be obtained by successive blow-ups at points, for all $X \in \mathcal{BIR}'(E(N))$, the set Σ_X is finite and $\overline{L}$ admits a Mumford-Lear extensions to X. Hence, $\overline{L}$ admits all Mumford-Lear extensions over $E(N)$.

From the previous discussion, it is clear that, to study the b-divisor b-div$(\theta^8_{1,1})$, we can forget the blow-ups at mild points and concentrate on blow-ups along non-mild points.

Consider the labeled binary tree with root labeled by $(1,1)$ and such that, if a node is labeled (n,m), the two child nodes are labeled $(n+m,m)$ and $(n,n+m)$. Then, the labels of the tree are in bijection with the set of ordered pairs of coprime positive integers. This tree also describes the type of the non-mild points that appear by successive blow-ups starting with a point of type $(1,1)$.

By equation (4.12) and this description of the singular points that appear in the tower of blow-ups, we deduce that the b-divisor b-div$(\theta^8_{1,1})$ is integrable if and only if the series

$$\sum_{\substack{n>0,m>0\\(n,m)=1}} \frac{1}{n^2m^2(n+m)^2}$$

is absolutely convergent. Since this is the case, we conclude that the b-divisor b-div$(\theta^8_{1,1})$ is integrable. Moreover, since $X(N)$ has p_N cusps and over each cusp $E(N)$ has N points of type $(1,1)$ and multiplicity $4/N$, we deduce from equation (4.12)

$$\text{b-div}(\theta^8_{1,1})^2 = C\cdot C - \frac{4^2Np_N}{N^2}\sum_{\substack{n>0,m>0\\(n,m)=1}} \frac{1}{n^2m^2(n+m)^2}.$$

Now we compute

$$\sum_{\substack{n>0,m>0\\(n,m)=1}} \frac{1}{n^2m^2(n+m)^2} = \frac{\sum_{n>0,m>0}\frac{1}{n^2m^2(n+m)^2}}{\sum_{k>0}\frac{1}{k^6}} = \frac{\zeta(2,2;2)}{\zeta(6)} = \frac{\frac{1}{3}\zeta(6)}{\zeta(6)} = \frac{1}{3},$$

where $\zeta(2,2;2)$ is the special value of the Tornheim zeta function that is computed in [25].

Therefore,

$$\text{b-div}(\theta^8_{1,1})^2 = C\cdot C - \frac{16p_N}{3N} = \frac{16(N^2+1)}{3N}p_N - \frac{16}{3N}p_N = \frac{16Np_N}{3}$$

concluding the proof of the theorem. □

Remark 4.13. We can rewrite the formula in Theorem 4.11 as

$$\text{b-div}(\theta^8_{1,1})^2 = 4\cdot 4\cdot \frac{[\mathrm{SL}_2(\mathbb{Z}):\Gamma(N)]}{2}\frac{\zeta(2,2;2)}{\zeta(6)}.$$

Thus, this degree can be interpreted as the product of the weight of the Jacobi form, its index, the degree of the map $X(N) \to X(1)$ and a quotient of zeta values.

5 Interpretation and open questions

In the previous section we have seen that, when taking into account the invariant metric, the natural way to extend the Cartier divisor $\operatorname{div}(\theta_{1,1}^8)$ associated to a section of the Jacobi line bundle, from the universal family of elliptic curves to a compactification of it, is not as a Cartier divisor, but as a $\mathbb{Q}$-b-divisor. In particular, this implies that we can not restrict ourselves to a single toroidal compactification, but we have to consider the whole tower of toroidal compactifications. Considering purely the arithmetic definition of Jacobi forms, this fact was already observed by the third author [19, Remark 2.19].

In this section we will give further evidence that $\text{b-div}(\theta_{1,1}^8)$ is the natural extension of $\operatorname{div}(\theta_{1,1}^8)$ by showing that it satisfies direct generalizations of classical theorems on hermitial line bundles. We will also state some open problems and future lines of research.

A HILBERT-SAMUEL FORMULA. First, we observe that $\text{b-div}(\theta_{1,1}^8)^2$ satisfies a Hilbert-Samuel type formula.

Theorem 5.1. *For each $N \geq 3$, the equality*

$$\text{b-div}(\theta_{1,1}^8)^2 = \lim_{\ell\to\infty} \frac{\dim J_{4\ell,4\ell}(\Gamma(N))}{\ell^2/2!}$$

holds.

Proof. By Remark 2.8 and Theorem 4.11 we have

$$\lim_{\ell\to\infty} \frac{\dim J_{4\ell,4\ell}\big(\Gamma(N)\big)}{\ell^2/2!} = \lim_{\ell\to\infty} \frac{\frac{8Np_N}{3}\ell^2 + o\big(\ell^2\big)}{\ell^2/2!} = \frac{16Np_N}{3} = \text{b-div}(\theta_{1,1}^8)^2.$$

□

CHERN-WEIL THEORY. The second task is to show that the self-intersection product in the sense of b-divisors is compatible with Chern-Weil theory. Let X be a proper complex variety, $U \subset X$ a dense open subset, L a line bundle on U, and $\|\cdot\|$ a smooth metric on L. Then, we set

$$c_1(L,\|\cdot\|) = \frac{1}{2\pi i}\partial\bar{\partial}\log\|s\|^2,$$

where $s \neq 0$ is a rational section of L. Note that $c_1(L,\|\cdot\|)$ does not depend on the section s. Moreover, it is a smooth form on U, but does not necessarily extend to the whole of X. Assume that L is the restriction to U of a line bundle L' on X such that the metric $\|\cdot\|$ is singular along $X \setminus U$. Due to the presence of singularities, even if the form $c_1(L, \|\cdot\|)$ is locally integrable on X, the cohomolgy class of the current $[c_1(L, \|\cdot\|)]$ does not need to represent the class of the line bundle L'.

In particular, let us write

$$c_1(L_{4,4,N}, \|\cdot\|) = \frac{1}{2\pi i}\partial\bar{\partial}\log\|\theta^8_{1,1}\|^2.$$

The objective of this paragraph is to make precise the idea that the form $c_1(L_{4,4,N}, \|\cdot\|)$ does not represent the first Chern class of a particular Mumford-Lear extension of $L_{4,4,N}$, but rather the class of the b-divisor b-div$(\theta^8_{1,1})$.

Theorem 5.2. *For each $N \geq 3$, the equality*

$$\text{b-div}(\theta^8_{1,1})^2 = \int_{E^0(N)} c_1(L_{4,4,N}, \|\cdot\|)^{\wedge 2} \tag{5.3}$$

holds.

Proof. First, observe that the integral on the right-hand side is an improper integral, but, by propositions 4.1 and 4.9, we know that it is well defined and finite.

We now give two proofs that both terms of equation (5.3) agree. Both methods consist in computing the integral on the right-hand side. In the first method, we replace the singular metric with another singular metric for which classical Chern-Weil theory applies and hence the corresponding integral is given by intersection theory and then we use the theory of residues to compare both integrals. The second method is based on computing the integral on the right-hand side explicitly on a fundamental domain.

Let C be the divisor of Proposition 4.9. Then, $\mathcal{O}(C)$ is a Mumford-Lear extension of $L_{4,4,N}$ to $E(N)$. By Proposition 4.9 (iii) there exists a pre-log metric $\|\cdot\|'$ on the line bundle $\mathcal{O}(C)$ such that each double point $p_{j,\nu} = \Theta_{j,\nu} \cap \Theta_{j,\nu+1} \in W^0_{j,\nu}$ has a neighborhood in which

$$\log\|\theta^8_{1,1}\|^2 = \log\|\theta^8_{1,1}\|'^2 - \frac{4}{N}\frac{\log(u_\nu\bar{u}_\nu)\log(v_\nu\bar{v}_\nu)}{\log(u_\nu\bar{u}_\nu)+\log(v_\nu\bar{v}_\nu)}. \tag{5.4}$$

Write $\omega = c_1(L_{4,4,N}, \|\cdot\|)$, $\omega' = c_1(\mathcal{O}(C), \|\cdot\|')$ and

$$f = \log\|\theta^8_{1,1}\|^2 - \log\|\theta^8_{1,1}\|'^2.$$

Thus,

$$\omega = \omega' + \frac{1}{2\pi i}\partial\bar{\partial}f.$$

Since Chern-Weil theory can be extended to pre-log singularities ([22], [7]), the equality

$$\int_{E(N)} \omega'^{\wedge 2} = C \cdot C$$

holds. Since

$$\int_{E(N)} \omega^{\wedge 2} = \int_{E(N)} \omega'^{\wedge 2} - \int_{E(N)} \mathrm{d}\left(\frac{2}{2\pi i}\partial f \wedge \omega' + \frac{1}{(2\pi i)^2}\partial f \wedge \partial\bar{\partial} f\right),$$

we are led to compute the second integral of the right-hand side of the previous equation. Note that the minus sign in the above formula comes from the fact that $\mathrm{d}\partial = -\partial\bar{\partial}$. Since pre-log-log forms have no residues, in order to compute this integral we can focus on the double points $p_{j,\nu}$, $j = 1,\dots,p_N$, $\nu = 0,\dots,N-1$ of D. For each point $p_{j,\nu}$ and $0 < \varepsilon < 1/e$, let $V_{j,\nu,\varepsilon}$ be the polycylinder

$$V_{j,\nu,\varepsilon} = \{(u_\nu, v_\nu) \in W^0_{j,\nu} \mid |u_\nu| \leq \varepsilon,\ |v_\nu| \leq \varepsilon\}.$$

Then, by Stokes theorem,

$$\begin{aligned} &-\int_{E(N)} \mathrm{d}\left(\frac{2}{2\pi i}\partial f \wedge \omega' + \frac{1}{(2\pi i)^2}\partial f \wedge \partial\bar{\partial} f\right) \\ &= \sum_{j=1}^{p_N}\sum_{\nu=0}^{N-1} \lim_{\varepsilon\to 0}\int_{\partial V_{j,\nu,\varepsilon}} \frac{2}{2\pi i}\partial f \wedge \omega' + \frac{1}{(2\pi i)^2}\partial f \wedge \partial\bar{\partial} f. \end{aligned}$$

Using that ω' is a pre-log-log form and equation (4.2), it is easy to see that

$$\lim_{\varepsilon\to 0}\int_{\partial V_{j,\nu,\varepsilon}} \frac{2}{2\pi i}\partial f \wedge \omega' = 0.$$

As a shorthand, write (u, v) for the coordinates (u_ν, v_ν) of $W^0_{j,\nu}$. We decompose the boundary of $V_{j,\nu,\varepsilon}$ as $\partial V_{j,\nu,\varepsilon} = A_\varepsilon \cup B_\varepsilon$, where

$$A_\varepsilon = \{(u,v) \in W^0_{j,\nu} \mid |u| \leq \varepsilon,\ |v| = \varepsilon\},$$
$$B_\varepsilon = \{(u,v) \in W^0_{j,\nu} \mid |u| = \varepsilon,\ |v| \leq \varepsilon\}.$$

Using equations (5.4) and (4.4) and taking care of the canonical orientation of a complex manifold, we see that

$$\int_{A_\varepsilon} \frac{1}{(2\pi i)^2}\partial f \wedge \partial\bar{\partial} f = \frac{16}{N^2}\int_0^\varepsilon \frac{2(\log(\varepsilon^2))^2\log(r^2)2r\,\mathrm{d}r}{(\log(r^2)+\log(\varepsilon^2))^4 r^2} = -\frac{16}{6N^2}.$$

Similarly,

$$\int_{B_\varepsilon} \frac{1}{(2\pi i)^2}\partial f \wedge \partial\bar{\partial} f = \frac{16}{N^2}\int_0^\varepsilon \frac{2(\log(\varepsilon^2))^2\log(r^2)2r\,\mathrm{d}r}{(\log(r^2)+\log(\varepsilon^2))^4 r^2} = -\frac{16}{6N^2}.$$

Hence,

$$\lim_{\varepsilon\to 0}\int_{\partial V_{j,\nu,\varepsilon}}\left(\frac{2}{2\pi i}\partial f\wedge\omega'+\frac{1}{(2\pi i)^2}\partial f\wedge\partial\bar{\partial}f\right)=-\frac{16}{3N^2}.$$

Therefore,

$$\int_{E^0(N)} c_1(L_{4,4,N},\|\cdot\|)^{\wedge 2}=C\cdot C-\frac{16p_N}{3N}=\text{b-div}(\theta_{1,1}^8)^2.$$

We next give a second proof of this equality. Let (τ,z) be the usual coordinate of $\mathbb{H}\times\mathbb{C}$, and write $\tau=\xi+i\eta$ and $z=x+iy$. Let $p\colon\mathbb{H}\times\mathbb{C}\to E^0(N)$ be the quotient map. By the explicit description of the translation invariant metric of Definition 2.10, a direct computation yields

$$p^*c_1(L_{4,4,N},\|\cdot\|)^{\wedge 2}=\frac{16}{\pi}\frac{\mathrm{d}x\wedge\mathrm{d}y}{\eta}\wedge\frac{\mathrm{d}\xi\wedge\mathrm{d}\eta}{\eta^2}.$$

Thus,

$$\int_{E^0(N)} c_1(L_{4,4,N},\|\cdot\|)^{\wedge 2}=\frac{[\mathrm{SL}_2(\mathbb{Z}):\Gamma(N)]}{2}\int_K\frac{16}{\pi}\frac{\mathrm{d}x\wedge\mathrm{d}y}{\eta}\wedge\frac{\mathrm{d}\xi\wedge\mathrm{d}\eta}{\eta^2},$$

where $K\subset\mathbb{H}\times\mathbb{C}$ is the subset containing the pairs (τ,z) with $|\mathrm{Re}(\tau)|\leq 1/2$, $|\tau|>1$ and z in the parallelogram with vertices $0,1,\tau,1+\tau$. Observe that the pre-factor $[\mathrm{SL}_2(\mathbb{Z}):\Gamma(N)]/2$ is the degree of the map $X(N)\to X(1)$.

The above integral is easily done, in fact it is a classical integral, and gives

$$\int_{E^0(N)} c_1(L_{4,4,N},\|\cdot\|)^{\wedge 2}=\frac{[\mathrm{SL}_2(\mathbb{Z}):\Gamma(N)]}{2}\frac{16}{\pi}\frac{\pi}{3}=\frac{16Np_N}{3}.$$

□

Remark 5.5. Recall the function

$$f_{1,1}(x,y)=\frac{\log(x\bar{x})\log(y\bar{y})}{\log(x\bar{x})+\log(y\bar{y})}.$$

The heart of the first proof of Theorem 5.2 is the relation

$$-\mathrm{Res}_{(0,0)}\left(\frac{1}{(2\pi i)^2}\partial f_{1,1}\wedge\partial\bar{\partial}f_{1,1}\right)=\frac{1}{3}=\sum_{\substack{n>0,m>0\\(n,m)=1}}\frac{1}{n^2m^2(n+m)^2}$$

between the residue at $(0,0)$ of the differential form $\frac{1}{(2\pi i)^2}\partial f_{1,1}\wedge\partial\bar{\partial}f_{1,1}$ and the harmonic double value $\zeta(2,2;2)/\zeta(6)$. This gives us a geometric interpretation of this harmonic double value.

Note also that the second proof relates the value of $\text{b-div}(\theta_{1,1}^8)^2$ with the volume of the modular curve and hence with the zeta value

$$\zeta(-1)=-\frac{1}{4\pi}\int_{\mathcal{F}}\frac{\mathrm{d}\xi\wedge\mathrm{d}\eta}{\eta^2},$$

where $\mathcal{F}$ is the standard fundamental domain for the action of $\mathrm{SL}_2(\mathbb{Z})$ on $\mathbb{H}$. In fact, the formula in Theorem 4.11 can be rewritten as

$$\text{b-div}(\theta_{1,1}^8)^2 = 4 \cdot 4 \cdot \frac{[\mathrm{SL}_2(\mathbb{Z}) : \Gamma(N)]}{2}(-4\zeta(-1)),$$

yielding the obvious formula

$$\zeta(2,2;2) = -4 \cdot \zeta(6) \cdot \zeta(-1).$$

INTERSECTIONS WITH CURVES. Similarly, we also note that the intersection of $\text{b-div}(\theta_{1,1}^8)$ with a complete curve can also be computed using the differential form $c_1(L_{4,4,N}, \|\cdot\|)$.

To a proper curve C contained in $E(N)$, we associate the b-divisor that, on each $X \in \mathcal{BIR}'(E(N))$ consists of the strict transform of C on X. We will denote this divisor by $\text{b-div}(C)$. Note that this b-divisor is not integrable because, by taking successive blow-ups in points of C, the self-intersection of the strict transform of C can be made arbitrarily negative. Assume that C is irreducible and is not contained in $D = E(N) \setminus E^0(N)$. Then, the product $\text{b-div}(\theta_{1,1}^8) \cdot \text{b-div}(C)$ is well defined because after a finite number of blow-ups on the double points of D and of its total transforms, the strict transform of C will not meet any double point of the total transform of D.

Theorem 5.6. *The equality*

$$\text{b-div}(\theta_{1,1}^8) \cdot \text{b-div}(C) = \int_C c_1(L_{4,4,N}, \|\cdot\|)$$

holds.

Proof. Let $X \to E(N)$ be a birational morphism obtained by successive blow-ups on double points of D and of its total transforms and such that the strict transform of C in X, that we denote by C_X, does not meet any double point of the total transform of D to X. Then,

$$\text{b-div}(\theta_{1,1}^8) \cdot \text{b-div}(C) = \mathrm{div}_X(\theta_{1,1}^8) \cdot C_X.$$

Let $(e, \mathcal{L}, S, \alpha, \|\cdot\|)$ be a Mumford-Lear extension of $\overline{L} = (L_{4,4,N}, \|\cdot\|)$ to X. Denote by $s = \alpha(\theta_{1,1}^{8e})$ the rational section of $\mathcal{L}$ determined by $\theta_{1,1}^8$. Since the metric $\|\cdot\|$ is pre-log on $X \setminus S$, we deduce that

$$\mathrm{div}_X(\theta_{1,1}^8) \cdot C_X = \frac{1}{e}\mathrm{div}(s) \cdot C_X = \frac{1}{e}\int_C c_1(\mathcal{L}, \|\cdot\|) = \int_C c_1(L_{4,4,N}, \|\cdot\|).$$

□

A TORIC ANALOGUE OF THE SINGULAR METRIC. We now give an interpretation of the harmonic double value $\zeta(2,2;2)/\zeta(6)$ in terms of toric varieties and the volume of a convex surface.

Consider the projective plane $\mathbb{P}^2$ with projective coordinates $(x_0 : x_1 : x_2)$ and the canonical line bundle $\mathcal{O}(1)$. On this line bundle we can put the canonical metric given by

$$\|x_0\|_{\text{can}} = \frac{|x_0|}{\max(|x_0|, |x_1|, |x_2|)}.$$

This metric is continuous. We have an open immersion $(\mathbb{C}^*)^2 \hookrightarrow \mathbb{P}^2$ that sends the point (z_1, z_2) to $(1 : z_1 : z_2)$. We define the valuation map $\text{val}\colon (\mathbb{C}^*)^2 \to \mathbb{R}^2$ by

$$\text{val}(z_1, z_2) = (-\log|z_1|, -\log|z_2|).$$

The function $\log(\|x_0\|_{\text{can}})$ is constant along the fibers of val. Thus, there exists a function $\Psi_{\text{can}}\colon \mathbb{R}^2 \to \mathbb{R}$ such that

$$\log\|x_0(p)\|_{\text{can}} = \Psi_{\text{can}}(\text{val}(p)).$$

This function is explicitly given by

$$\Psi_{\text{can}}(u, v) = \min(0, u, v).$$

The projective plane $\mathbb{P}^2$ is a toric variety with the action of $(\mathbb{C}^*)^2$ given by

$$(\lambda, \mu)(x_0 : x_1 : x_2) = (x_0 : \lambda x_1 : \mu x_2).$$

The theory of toric varieties tells us that the polytope associated to $\text{div}(x_0)$ is the stability set of Ψ_{can}:

$$\begin{aligned}\Delta &= \{x \in (\mathbb{R}^2)^\vee \mid x(u,v) \geq \Psi_{\text{can}}(u,v),\ \forall (u,v) \in \mathbb{R}^2\}\\ &= \text{conv}((0,0),(1,0),(0,1)).\end{aligned}$$

Moreover,

$$\text{div}(x_0)^2 = 2\,\text{Vol}(\Delta) = 1,$$

where the volume is computed with respect to the Haar measure that gives $\mathbb{Z}^2$ covolume 1.

Now we want to modify the canonical metric to introduce a singularity of the same type as the singularity of the translation invariant metric on the Jacobi line bundle at the double points. We define the metric $\|\cdot\|_{\text{sing}}$ by

$$\log\|x_0\|_{\text{sing}} = \begin{cases} -\frac{\log(|x_1/x_0|)\log(|x_2/x_0|)}{\log(|x_1/x_0|)+\log(|x_2/x_0|)} & \text{if } |x_0| \geq \max(|x_1|, |x_2|),\\ -\max(\log(|x_1/x_0|), \log(|x_2/x_0|)) & \text{otherwise.}\end{cases}$$

As before, the function $\log\|x_0\|_{\mathrm{sing}}$ is constant along the fibers of val and defines a function $\Psi_{\mathrm{sing}}\colon \mathbb{R}^2 \to \mathbb{R}$ that is given explicitly by

$$\Psi_{\mathrm{sing}}(u,v) = \begin{cases} \frac{uv}{u+v}, & \text{if } u,v \geq 0,\\ u, & \text{if } u \leq \min(0,v),\\ v, & \text{if } v \leq \min(0,u). \end{cases}$$

The function Ψ_{sing} is conic but is not piecewise linear. Assume that we can extend the theory of toric varieties to toric b-divisors. Then, to Ψ_{sing} we would associate the convex figure

$$\Delta_{\mathrm{sing}} = \{x \in (\mathbb{R}^2)^\vee \mid x(u,v) \geq \Psi_{\mathrm{sing}}(u,v),\ \forall (u,v) \in \mathbb{R}^2\},$$

and we should obtain

$$\text{b-div}(x_0, \|\cdot\|_{\mathrm{sing}})^2 = 2\,\mathrm{Vol}(\Delta_{\mathrm{sing}}). \tag{5.7}$$

We see that this is indeed the case.

Theorem 5.8. *The equation* (5.7) *holds.*

Proof. Arguing as in the proof of Theorem 4.11, we see that

$$\mathrm{div}(x_0)^2 - \text{b-div}(x_0, \|\cdot\|_{\mathrm{sing}})^2 = \sum_{\substack{n>0,m>0\\(n,m)=1}} \frac{1}{n^2m^2(n+m)^2} = \frac{1}{3}.$$

The stability set Δ_{sing} can be explicitly computed, and is given by

$$\Delta_{\mathrm{sing}} = \{(x,y) \in (\mathbb{R}^2)^\vee \mid x,y \geq 0,\ x+y \leq 1,\ \sqrt{x}+\sqrt{y} \geq 1\}.$$

Thus,

$$2\,\mathrm{Vol}(\Delta) - 2\,\mathrm{Vol}(\Delta_{\mathrm{sing}}) = 2\int_0^1 (1-\sqrt{x})^2\,\mathrm{d}x = \frac{1}{3}.$$

□

Remark 5.9. In fact, since in the theory of toric varieties, the blow-ups have an explicit description in terms of fans, it is possible to interpret the equation $\zeta(2,2;2) = 1/3 \cdot \zeta(6)$ is terms of an infinite triangulation of $\Delta \setminus \Delta_{\mathrm{sing}}$.

OPEN QUESTIONS. In this paper we have examined a particular example and observed, just by comparing numbers, that several classical results should be extendable to b-divisors and singular metrics with a shape similar to the one of the translation invariant metrics. We are in the process of investigating the following questions.

(1) Theorem 5.1 shows that the translation invariant metric encodes the asymptotic behavior of the space of Jacobi forms. It is possible to define global sections of a b-divisor. We can ask what is the exact relationship between the space of Jacobi forms and the global sections of the b-divisor $\mathrm{div}(\theta_{1,1}^8)$. Moreover, once this is settled, we can ask whether there is a Riemann-Roch theorem or a Hilbert-Samuel theorem for b-divisors that imply directly Theorem 5.1.
(2) By Theorem 5.8, it is clear that much of the theory of toric varieties could be extended to toric b-divisors and singular metrics on toric varieties.
(3) Theorem 5.2 shows that Chern-Weil theory of singular metrics can be useful to study b-divisors. It would be interesting to generalize this theorem to higher dimensions. In this direction, with R. de Jong and D. Holmes, we have shown that the local integrability property extends, at least, to the case of toroidal compactifications of families of abelian varieties.
(4) The original motivation of this paper is to be able to define and study the height of cycles on the universal elliptic curve with respect to the Jacobi line bundle equipped with the translation invariant metric, extending the work in [18]. First, it is clear how to define the height of an algebraic point of $E^0(N)$ and one may wonder whether the new singularities are mild enough so that the Northcott property is still true. We can also define the height of an algebraic curve not contained in the divisor D. But it is not clear how to define the height of $E(N)$. The naive definition of that height would give the value $-\infty$, but it should be possible to extract a meaningful finite number. To this end, the study of toric varieties might be useful, because the techniques developed in [9] can be extended to the singular metrics of Theorem 5.8. In this case, we obtain that the stability set of the function associated to the metric is no longer a polytope but a convex set. In this case the regularized height should be defined from the integral along this convex set of the roof function, in analogy with [9, Theorem 5.2.5].

References

1. R. Berndt and U. Kühn, *On Kudla Green function for signature (2,2) I*, arXiv:1205.6417, 2012.
2. ———, *On Kudla Green function for signature (2,2) II*, arXiv:1209.3949, 2012.
3. S. Boucksom, C Favre, and M. Jonsson, *Differentiability of volumes of divisors and a problem of Teissier*, J. Algebraic Geom. **18** (2009), 279–308.
4. J. H. Bruinier, B. Howard, and T. Yang, *Heights of Kudla-Rapoport divisors and derivatives of L-functions*, arXiv:1303.0549, 2013.
5. J. H. Bruinier, J. I. Burgos Gil, and U. Kühn, *Borcherds products and arithmetic intersection theory on Hilbert modular surfaces*, Duke Math. J. **139** (2007), no. 1, 1–88.

6. J. H. Bruinier and T. Yang, *Faltings heights of CM cycles and derivatives of L-functions*, Invent. Math. **177** (2009), 631–681.
7. J. I. Burgos Gil, J. Kramer, and U. Kühn, *Arithmetic characteristic classes of automorphic vector bundles*, Documenta Math. **10** (2005), 619–716.
8. ———, *Cohomological arithmetic Chow rings*, J. Inst. Math. Jussieu **6** (2007), no. 1, 1–172.
9. J. I. Burgos Gil, P. Philippon, and M. Sombra, *Arithmetic geometry of toric varieties. Metrics, measures and heights*, e-print arXiv:1105.5584v2, 2012.
10. G. Faltings, *Endlichkeitssätze für abelsche Varietäten über Zahlkörpern*, Invent. Math. **73** (1983), 349–366.
11. ———, *Finiteness theorems for abelian varieties over number fields*, Arithmetic Geometry (G. Cornell and J.H. Silverman, eds.), Springer-Verlag, 1986, pp. 9–27.
12. G. Freixas i Montplet, *Heights and metrics with logarithmic singularities*, J. Reine Angew. Math. **627** (2009), 97–153.
13. R. Hain, *Normal functions and the geometry of moduli spaces of curves*, Handbook of Moduli, Volume I (G. Farkas and I. Morrison, eds.), Advanced Lectures in Mathematics, vol. XXIV, International Press, Boston, 2013.
14. D. Holmes and R. de Jong, *Asymptotics of the Néron height pairing*, arXiv: 1304.4768 [math.AG], 2013.
15. B. Howard, *Complex multiplication cycles and Kudla-Rapoport divisors*, Ann. Math. **176** (2012), 1097–1171.
16. ———, *Complex multiplication cycles and Kudla-Rapoport divisors II*, arXiv: 1303.0547, 2013.
17. J. Kramer, *A geometrical approach to the theory of Jacobi forms*, Compositio Math. **79** (1991), 1–19.
18. ———, *An arithmetic theory of Siegel-Jacobi forms*, Habilitationsschrift, ETH Zürich, 1992.
19. ———, *An arithmetic theory of Jacobi forms in higher dimensions*, J. Reine Angew. Math. **458** (1995), 157–182.
20. S. Kudla, M. Rappoport, and T. Yang, *Modular forms and special cycles on shimura curves*, Annals of Mathematics Studies, vol. 161, Princeton University Press, 2006.
21. D. Lear, *Extensions of normal functions and asymptotics of the height pairing*, Ph.D. thesis, University of Washington, 1990.
22. D. Mumford, *Hirzebruch's proportionality theorem in the non-compact case*, Invent. Math. **42** (1977), 239–272.
23. G. Pearlstein, SL_2*-orbits and degenerations of mixed Hodge structure*, J. Differential Geom. **74** (2006), 1–67.
24. V.V. Shokurov, *Prelimiting flips*, Proc. Steklov Inst. Math. **240** (2003), 75–213.
25. Leonard Tornheim, *Harmonic double series*, Amer. J. Math. **72** (1950), 303–314.

INSTITUTO DE CIENCIAS MATEMÁTICAS (CSIC-UAM-UCM-UCM3). CALLE NICOLÁS CABRERA 15, CAMPUS UAB, CANTOBLANCO, 28049 MADRID, SPAIN.

E-mail address: burgos@icmat.es

URL: http://www.icmat.es/miembros/burgos/

HUMBOLDT-UNIVERSITÄT ZU BERLIN, INSTITUT FÜR MATHEMATIK, UNTER DEN LINDEN 6, D-10099 BERLIN, GERMANY.
E-mail address: kramer@math.hu-berlin.de
URL: http://www.math.hu-berlin.de/~kramer/

FACHBEREICH MATHEMATIK (AZ), UNIVERSITÄT HAMBURG, BUNDESSTRASSE 55, D-20146 HAMBURG, GERMANY.
E-mail address: kuehn@math.uni-hamburg.de
URL: http://www.math.uni-hamburg.de/home/kuehn/

3

Symmetries of Graded Polarized Mixed Hodge Structures

Aroldo Kaplan

1 Introduction

A decomposition theorem is proved for morphisms of graded polarized mixed Hodge structures of given Hodge numbers. It implies that certain metric in the corresponding Hodge domain is maximally symmetric among the hermitian mixed metrics that have been proposed [HP] [P][PP][U].

The results are related to those of Hertling on classifying spaces of polarized mixed Hodge structures and of Brieskorn lattices [H][HS], and also to that of Kato, Nakayama and Usui on SL_2-orbits in period domains [KNU][U]. Some were contained in unpublished notes for a course given at the University of Massachusetts [K] which have been recently cited [HP][P] [PP].

G. Pearlstein observed that they could be generalized to other Mumford-Tate groups and domains and used to describe the orbits of the real group at infinity [DFN][GGR][KeP1][KeP2]. I thank him for this observation and for his help and encouragement to have these results published.

2 Graded Polarized Mixed Hodge Structures

Let V be a complex vector space with a rational structure $V_{\mathbb{Q}}$, W an increasing filtration of V defined over $\mathbb{Q}$, and S a non-degenerate bilinear form on Gr^W, defined over $\mathbb{Q}$ and of parity $(-1)^l$ on Gr^W_l. Let

$$\mathcal{M} = \mathcal{M}(W,S).$$

be the set of decreasing filtrations F of V such that (W,F) is a mixed Hodge structure of given Hodge numbers and the graded pure Hodge structures are polarized by S. $\mathcal{M}$ is a complex manifold where the Lie group

$$G = \{g \in GL(V)^W : \ \ Gr^W(g) \in O(S, V_{\mathbb{R}})\}$$

This work was partially supported by CONICET, ANPCyT and Secyt (UNC) (Argentina)

acts transitively and effectively by holomorphic diffeomorphisms.

In general, the isotropy subgroups G^F, $F \in \mathcal{M}$, are not compact in the mixed case; therefore $\mathcal{M}$ cannot support $G-$invariant hermitian metrics. There are many metrics invariant under

$$G_{\mathbb{R}} = G \cap GL(V_{\mathbb{R}}),$$

but this group is not transitive on $\mathcal{M}$. Imposing invariance under certain complex unipotent groups singles out a special one.

The inclusions

$$G_{\mathbb{R}} \subset G \subset G_{\mathbb{C}}$$

are proper in general. Let $\mathfrak{g}_{\mathbb{R}}, \mathfrak{g}, \mathfrak{g}_{\mathbb{C}}$ be the corresponding Lie algebras (the notation should not generate confusion). Then for $j > 0$

$$W_{-j}(\mathfrak{g}) = W_{-j}(\mathrm{End}(V_{\mathbb{C}})).$$

Set $G_{-j} := \exp W_{-j}(\mathfrak{g})$. Then G_{-1} acts simply transitively among the gradings of W and

$$G = G_{\mathbb{R}} G_{-1}.$$

A bigrading of a mixed Hodge structure (W,F) is a decomposition $V = \bigoplus I^{p,q}$ such that

$$W_r = \bigoplus_{p+q \leq r} I^{p,q}, \quad F^p = \bigoplus_{a \geq p} I^{a,q}.$$

Deligne's choice

$$I^{p,q} = F^p \cap W_{p+q} \cap (\overline{F}^q \cap W_{p+q} + \overline{F}^{q-1} \cap W_{p+q-2} + \overline{F}^{q-2} \cap W_{p+q-3} + \cdots),$$

provides a functorial one such that the subspaces

$$\Lambda^{p,q} := \bigoplus_{a \leq p, b \leq q} I^{a,b}$$

satisfy

$$I^{p,q} \subset \overline{I}^{q,p} + \Lambda^{p-1,q-1}$$

or, equivalently,

$$\Lambda^{p,q} = \overline{\Lambda}^{q,p}.$$

These I^{pq} induce a bigrading

$$\mathfrak{g}_{\mathbb{C}} = \bigoplus I^{p,q}(\mathfrak{g}_{\mathbb{C}}).$$

Here

$$\Lambda^{-1,-1}(\mathfrak{g}_{\mathbb{C}}) \subset W_{-2}(\mathfrak{g}_{\mathbb{C}})$$

is a complex nilpotent subalgebra defined over $\mathbb{R}$. Let $L, L_{\mathbb{R}}$ be the connected subgroups of $G, G_{\mathbb{R}}$ such that

$$Lie(L) = \mathfrak{g} \cap \Lambda^{-1,-1}(\mathfrak{g}_{\mathbb{C}}), \quad Lie(L_{\mathbb{R}}) = \mathfrak{g}_{\mathbb{R}} \cap \Lambda^{-1,-1}(\mathfrak{g}_{\mathbb{C}}).$$

These are unipotent groups, L is complex, $L_{\mathbb{R}} \subset L$ is a real form of L, and for $g \in \exp(\Lambda^{-1,-1}(\mathfrak{g}))$,

$$I^{p,q}(gF,W) = gI^{p,q}(F,W).$$

By definition,

$$L(F) \subset G_{-2}.$$

The subspaces $E_r = \bigoplus_{p+q=r} I^{p,q}$ grade W and define an element $Y = Y(F) \in \mathfrak{gl}(V)$ by

$$Yv = rv \quad v \in E_r.$$

such that

$$\bar{Y} - Y \in W_{-2}(\mathfrak{gl}(V)).$$

Y itself will be called a grading of W.

Now fix W and write $\Lambda(F), L(F), Y(F)$, etc. to emphasize the dependence on the Hodge filtration.

3 Factorizations of G and $\mathcal{M}$

Theorem 3.1. *For any $F \in \mathcal{M}$ and $g \in G$ there exist $g_{\mathbb{R}} \in G_{\mathbb{R}}$, $\ell \in L(F)$ and $f \in G^F_{-1}$ such that*

$$g = g_{\mathbb{R}} \ell f.$$

Proof. Since $G = G_{\mathbb{R}} G_{-1}$, we may assume $g \in G_{-1}$. Recall that if $Y(F) \in \mathrm{End}(V_{\mathbb{R}})$, (W,F) is said to be split and that in that case $Y(F)$ is the unique real splitting of (W,F). Let $\mathcal{M}_{\mathbb{R}} \subset \mathcal{M}$ be the locus of split mixed Hodge structures. A "splitting operation" will be a map

$$e : \mathcal{M} \to G$$

satisfying

(i) $\widehat{F} := e(F)^{-1} F \in \mathcal{M}_{\mathbb{R}}$
(ii) $e(F) \in L(F)$
(iii) $e(F) = 1$ for $F \in \mathcal{M}_{\mathbb{R}}$
(iv) If $g \in G_{\mathbb{R}}$ then $e(gF) = ge(F)g^{-1}$ and $\widehat{gF} = g\widehat{F}$

An example is Deligne's $e(F) = \exp(\sqrt{-1}\delta)$ with $\delta \in \mathfrak{g}_{\mathbb{R}} \cap I^{-1,-1}(\mathfrak{g}_{\mathbb{C}})$, but sometimes other choices are called for (cf. [CKS][CDK]). Fix a specific e.

For $F \in \mathcal{M}$ let $\widehat{F} = e(F)^{-1} F$. Because both $Y(\widehat{F})$ and $Y(\widehat{gF})$ are real gradings of W, there must be

$$h \in \exp W_{-1}(\mathrm{End}(V_{\mathbb{R}})) = (G_{\mathbb{R}})_{-1}$$

such that $Ad(h^{-1}) Y(\widehat{F}) = Y(\widehat{gF})$. Then

$$Y(\widehat{F}) = Ad(h) Y(\widehat{gF}) = Y(h(\widehat{gF})) = Y(\widehat{hgF}).$$

The mixed Hodge structures $(W,\widehat{F})$ and $(W,\widehat{hgF})$ share real splittings and graded structures, so that the Hodge filtrations must agree:

$$\widehat{F} = \widehat{hgF}.$$

This implies

$$hgF = e(hgF)\widehat{hgF} = e(hgF)\widehat{F} = e(hgF)e(F)^{-1}F$$

and therefore $gF = h^{-1}e(hgF)e(F)^{-1}F$. It follows that

$$g = h^{-1}e(hgF)e(F)^{-1}f$$

for a unique $f \in G^F_{-1}$. Since $h \in G_{\mathbb{R}}$, $e(F)^{-1} \in L(F)$ and also

$$e(hgF) \in L(hgF) = L(\widehat{hgF}) = L(\hat{F}) = L(F),$$

the assertion follows. □

The decomposition is not unique, because $G_{\mathbb{R}} \cap L = L_{\mathbb{R}}$.

Corollary 3.2.

$$\mathcal{M} = G_{\mathbb{R}}L(F) \cdot F$$

In the terms of Theorem 3.1, the factor $g_{\mathbb{R}}$ satisfies

$$g_{\mathbb{R}}\hat{F} = \widehat{gF}$$

while $f \in G^F \cap G_{-1}$ arises from the fact that $g_{\mathbb{R}}$ must induce the same action as g on Gr^W in order to move $\hat{F}$ to $\widehat{gF}$.

The subspaces Λ satisfy

- $\Lambda^{p,q} = \bar{\Lambda}^{q,p}$
- $\Lambda^{p,q} \cap \Lambda^{p',q'} = \Lambda^{min(p,p'),min(q,q')}$
- $W_r = \sum_{p+q \le r} \Lambda^{p,q}$
- $Gr^W_r = \bigoplus_{p+q=r} Gr^W_r(\Lambda^{p,q})$ and this is a polarized Hodge structure of weight r.

Let $\mathcal{L}$ be the manifold of $\Lambda^{*,*}$ of given graded Hodge numbers satisfying these properties.

Theorem 3.3. *$F \mapsto (\Lambda(F), Y(F))$ is a diffeomorphism between $\mathcal{M}$ and the set of pairs (Λ, Y) s.t. $\Lambda \in \mathcal{L}$ and $Y \in \mathrm{End}(V_{\mathbb{C}})$ is a splitting of W preserving the $\Lambda^{p,q}$'s.*

Proof. From $I^{p,q}(F) = \Lambda^{p,q}(F) \cap E_{p+q}(Y(F))$ it follows that the map is injective. The properties of the Λ's imply it is surjective. □

One then gets a fibering

$$\lambda : \mathcal{M} \to \mathcal{L}$$

where the fiber $\mathcal{M}_{\Lambda_o}$ can be identified either with the F's s.t. $\Lambda(F) = \Lambda_o$, or with set of Y's stabilizing Λ_o.

Theorem 3.4. *$G_{\mathbb{R}}$ acts transitively on $\mathcal{M}_{\mathbb{R}}$.*

Proof. $G_{\mathbb{R}}$ acts transitively among the real gradings of W and among the graded Hodge structures, hence on $\mathcal{M}_{\mathbb{R}}$. □

Theorem 3.5.

$$\lambda(\mathcal{M}_{\mathbb{R}}) = \mathcal{L}$$

Proof. $e(F)F \in \mathcal{M}_{\mathbb{R}}$ and $e(F) \in L(F)$, hence $\lambda(e(F)F) = \lambda(F)$. □

Fix $F \in \mathcal{M}$, let $\widehat{F} = e(F)F$ and let $\widehat{Y}$ be the $\mathbb{R}$-split objects associated to F by any splitting map e. One has

$$\mathfrak{g}^{\widehat{Y}} = I^{0,0}(\mathfrak{g}, \widehat{F}) \oplus \bigoplus_{p \neq 0} I^{p,-p}(\mathfrak{g}, \widehat{F}).$$

Here $\mathfrak{k} = I^{0,0}(\mathfrak{g}, \widehat{F})$ is the isotropy subalgebra at $\widehat{F}$ and $\mathfrak{q} = \bigoplus_{p \neq 0} I^{p,-p}(\mathfrak{g}, \widehat{F})$ a linear complement defined over $\mathbb{R}$ and satisfying

$$[\mathfrak{k}, \mathfrak{q}] \subset \mathfrak{q}.$$

The following is a mixed analog of Cartan's decomposition $G = K \exp \mathfrak{p}$.

Theorem 3.6. *Any $g \in G$ can be decomposed as*

$$g = qwlf$$

where $q \in \exp(\mathfrak{q}_{\mathbb{R}})$, $w \in (G_{-1})_{\mathbb{R}}$, $l \in L(F)$, $f \in G^F_{-1}$.

Proof. Let K be the subgroup with Lie algebra $\mathfrak{k}$. Given $g \in G$ write $ge(F) = g_{\mathbb{R}} \ell f$ according to (3.1) applied to the split $\widehat{F}$. Decompose the first factor on the right hand side as $g_{\mathbb{R}} = g_{\mathbb{R}}^{\widehat{Y}} g_{-1}$ with $g_{\mathbb{R}}^{\widehat{Y}} \in G^{\widehat{Y}}$, $g_{-1} \in G_{-1}(\mathbb{R})$. In turn, according to the decomposition $\mathfrak{g}^{\widehat{Y}} = \mathfrak{k} \oplus \mathfrak{q}$, one can write $g_{\mathbb{R}}^{\widehat{Y}} = qk$ with $q \in \exp \mathfrak{q}$ and $k \in K$, so that $ge(F) = qkg_{-1}\ell f$. Therefore

$$g = qkg_{-1}\ell f e(F)^{-1} = q(kg_{-1}k^{-1})(k\ell k^{-1})e(F)^{-1}(e(F)(kf)e(F)^{-1})$$

The first factor is in $\exp \mathfrak{q}$. The second is in $(G_{-1})_{\mathbb{R}}$ because k is real. The third is in $L(F) = L(\widehat{F})$ because k fixes $\widehat{F}$. Finally, $e(F) \in L(F)$ and the last factor is in G^F because k and f are in $G^{\widehat{F}}$. □

4 A metric on $\mathcal{M}$

On $\mathcal{M}_{\mathbb{R}}$ one has the $G_{\mathbb{R}}$-invariant metric defined at each $F \in \mathcal{M}_{\mathbb{R}}$ by the identification $Gr^W \cong E(Y(F))$ for the unique real grading $Y(F)$. Explicitly, it is induced by the hermitian inner product on V

$$h_F(u,v) := \sigma_F(C_F u, \bar{v})$$

where σ^F is the bilinear form on V obtained by lifting the graded-polarizations via the grading $Y(F)$, and $C = C_F$ is the Weil operator $C_F(v) = i^{p-q}v$ for $v \in I^{p,q}(F)$. The metric on $\mathcal{M}_{\mathbb{R}}$ is regarded here as a function from $\mathcal{M}_{\mathbb{R}}$ to the cone of hermitian inner products on V, via the identification of $\mathcal{M}_{\mathbb{R}}$ as a subset of a product of Grassmannians of V. In this split case the metric is simply the sum of the Hodge metrics of the pure Hodge structures on the graded pieces. Invariance under $G_{\mathbb{R}}$ means that

$$h_{gF} = g \cdot h_F$$

$\forall g \in G_{\mathbb{R}}$.

The same formula with $I^{p,q}(F)$ denoting the Deligne bigrading of (W,F) extends that metric to a $G_{\mathbb{R}}$-invariant metric h on all of $\mathcal{M}$, but it is not G−invariant. Instead one has:

Theorem 4.1. *h is the unique extension such that*

$$h_{gF} = g \cdot h_F \quad \forall g \in L(F).$$

Proof. For $\Lambda \in \mathcal{L}$ let G_{-1}^{Λ} be the stabilizer of Λ in G_{-1} and $H_\Lambda \subset G_{-1}^{\Lambda}$ the subgroup with Lie algebra

$$\mathfrak{h}_\Lambda = \{X \in \mathfrak{g}_{-1}^{\Lambda} : X_{p,q} = \bar{X}_{q,p}\}$$

where $X_{p,q}$ is the component of X in $\mathfrak{gl}(\Lambda^{p,q}/\Lambda^{p-1,q-1})$. For emphasis, an $h \in H_\Lambda$ stabilizes Λ (and W), it is the identity in Gr^W, and it is defined over $\mathbb{R}$ on $\Lambda^{\cdot,\cdot}/\Lambda^{\cdot-1,\cdot-1}$. Also,

$$H_\Lambda = (H_\Lambda)_{\mathbb{R}} L(\Lambda), \quad (H_\Lambda)_{\mathbb{R}} \cap L(\Lambda) = L(\Lambda)_{\mathbb{R}}.$$

Lemma 4.2. *H_Λ acts simply transitively on each fiber $\mathcal{M}_\Lambda \hookrightarrow \mathcal{M} \to \mathcal{L}$.*

Proof of the Lemma: Any two real gradings Y_1, Y_2 of W are conjugate by a unique $g \in \exp(W_{-1}(\mathfrak{g}_{\mathbb{R}}))$. If both leave Λ invariant, the corresponding split F_1, F_2 are conjugate by g. But then $g\Lambda = g\Lambda_{F_1} = \Lambda_{gF_1} = \Lambda_{F_2} = \Lambda$, so that $g \in H_\Lambda$. It follows that $(H_\Lambda)_{\mathbb{R}}$ acts transitively on $\mathcal{M}_{\mathbb{R}}$. Let now $F_1, F_2 \in \mathcal{M}$. Fix a splitting map e and write $F_i = e_i\widehat{F}_i$ with $e_i \in L(F_i)$ and $\widehat{F}_i \in \mathcal{M}_{\mathbb{R}}$. If $h \in (H_\Lambda)_{\mathbb{R}}$ is such that $h\widehat{F}_1 = \widehat{F}_2$, then $F_2 = e_2 h e_1^{-1} F_1$. Together with $L(F)(H_\Lambda)_{\mathbb{R}} L(F) \subset H_\Lambda$, this implies that H_Λ acts transitively on $\mathcal{M}_\Lambda$. Viewing this as a space of gradings Y, $H_\Lambda \subset \exp(W_{-1}(\mathfrak{g}_\mathbb{R}))$ implies that only the identity can fix one such Y, which proves the Lemma.

Now we continue with the proof of Theorem 4.1. Choose a splitting map $e : \mathcal{M} \to G$ and write as before $\widehat{F} = e^{-1}(F)F$, $\widehat{Y} = Ad(e^{-1})(F)Y(F)$. Since e takes values in $L(F)$, for the required invariance to hold we must have

$$h_F = e(F) \cdot h_{\widehat{F}}.$$

Define h by this formula. It is hermitian, positive definite, and independent of the splitting map e. To see this let e_1, e_2 be any two and $\widehat{F}_i$, $\widehat{Y}_i$ the associated objects to e_i. Then $\widehat{F}_2 = e_2(F)^{-1}e_1(F)\widehat{F}_1$ and $\widehat{Y}_2 = Ad(e_2(F)^{-1}e_1(F))\widehat{Y}_1$. Since $e_j(F) \in L(F)$, $e_2(F)^{-1}e_1(F) \in L(F) \subset \exp(W_{-1}(\mathfrak{g}))$. But two gradings of W are conjugate by a unique element of the latter group, which must be real if the gradings are. Therefore $e_2(F)^{-1}e_1(F) \in G_\mathbb{R}$. Since the metric on $\mathcal{M}_\mathbb{R}$ is $G_\mathbb{R}$-invariant,

$$e_2 h_{F_2} = e_2 h_{e_2^{-1}e_1 F_1} = e_2 e_2^{-1} e_1 h_{F_1} = e_1 h_{F_1}.$$

It remains to prove the stated invariances. For $g \in G_\mathbb{R}$,

$$h_{gF} = e(gF) h_{\widehat{gF}} = (g e(F) g^{-1}) g h_{\widehat{F}} = g h_F.$$

For $g \in L(F)$, write $h_{gF} = e(gF) h_{\widehat{gF}}$. Since $\Lambda_{\widehat{gF}} = \Lambda_{gF} = \Lambda_F = \Lambda_{\widehat{F}}$, there exists a unique $a \in (H_\Lambda)_\mathbb{R}$ such that $\widehat{gF} = a\hat{F}$. We can now write

$$F = g^{-1} e(gF) a \widehat{F}.$$

Since $g^{-1}e(gF)a \in H_\Lambda$, the Lemma implies $g^{-1}e(gF)a = e(F)$. Therefore

$$h_{gF} = e(gF) h_{\widehat{gF}} = g e(F) a^{-1} h_{a\hat{F}} = g e(F) h_{\widehat{F}} = g h_F$$

□

A consequence of the proof independent of metrics, is that an orbit $\mathcal{O}$ of $L(\Lambda)$ in $\mathcal{M}_\Lambda$ is just an element of

$$H_\Lambda / L(\Lambda),$$

an abelian group. Moreover, $\mathcal{O} \cap \mathcal{M}_\mathbb{R}$ is a principal homogeneous space under $L(\Lambda)_\mathbb{R}$. For related results see [H][HP][HS][KNU][P][PP][U].

5 The metric at infinity

In a variation of pure structures the growth of the Hodge norm of a flat section at infinity is determined by the position of the section in the monodromy weight filtration $W(N)$, where N is the logarithm of the unipotent part of the monodromy. This was done in [S] by reducing the problem to period maps that come from nilpotent orbits, of the form $\exp(zN) \cdot F$ (in the one-variable case) with $\Im(z) >> 0$.

For nilpotent orbits of graded polarized mixed Hodge structure and the mixed Hodge metric defined above, one has the following. For $u, v \in V$, regarded as locally constant sections of a variation, write

$$h_z(u,v) = h_{\exp(zN)\cdot F}(u,v), \qquad ||v||_z = h_z(v,v)^{1/2}.$$

Let W' be the monodromy weight filtration of N relative to W [D] and m the highest weight of W. Then for $v \in V$, $|\Re(z)| < 1$ and $\Im(z) \to \infty$

$$v \in W_r \cap W'_s \quad \Leftrightarrow \quad ||v||_z \leq K|z|^{r-s+2m}.$$

The proof follows from Theorem 3.6 and can be found in my former's student thesis [G].

References

[CDK] E. Cattani, P. Deligne and A. Kaplan, *On the locus of Hodge classes*, Jour. A.M.S., **8**-2, 483-506 (1995).

[CKS] E. Cattani, A. Kaplan and W. Schmid, *Degeneration of Hodge structures*, Ann. of Math. **123**, 457-535 (1986).

[D] P. Deligne, *La conjecture de Weil, II*, IHES Publ. Math. No. 52 (1980)

[DFN] *Hodge Theory, Complex Geometry and Representation Theory*, R. Doran, G. Friedman and S. Nollet, Editors. Contemporary Math., AMS (2014)

[G] E. Gaze, *Degenerating variations of mixed Hodge structures*, PhD Thesis, http://scholarworks.umass.edu/open-access-dissertations/847/ (1996)

[GGK] M. Green, P. Griffiths and M. Kerr, *Mumford-Tate Groups and Domains*, arXiv: 1403.064v2 (2014)

[GGR] M. Green, P. Griffiths and C. Robles, *Extremal Degenerations of polarized Hodge structures*, AM-183, Princeton U. Press (2012)

[HP] T. Hayama and G. Pearlstein, *Asymptotics of degenerations of mixed Hodge structure*, http://arxiv.org/abs/1403.1971 (2013)

[H] C. Hertling, *Classifying spaces for polarized mixed Hodge structures and for Brieskorn lattices.*, Compositio Math. 116 (1999), no. 1, 1–37

[HS] C. Hertling, C. Sevenheck, *Curvature of classifying spaces for Brieskorn lattices.* J. Geom. Phys. 58 (2008), no. 11, 1591-1601.

[K] A. Kaplan, *Notes on the moduli of mixed Hodge structures* (1996)

[KNU] K. Kato, C. Nakayama, and S. Usui, *SL(2)-orbit theorem for degeneration of mixed Hodge structure.* J. Algebraic Geom. 17 (2008), no. 3, 401–479.

[KP] A. Kaplan and G. Pearlstein, *Singularities of Variations of Mixed Hodge Structure*, Asian J. of Math., 7, no. 3/2003, 307-336 (2004)

[KeP1] M. Kerr and G. Pearlstein, *Boundary Components of Mumford-Tate Domains*, arXiv:1210.5301 (2012)

[KeP2] M. Kerr and G. Pearlstein, *Naive boundary strata of nilpotent orbits Mumford-Tate Domains*, arXiv:1210.5301 (2012)

[P] G. Pearlstein, *On the asymptotic behavior of admissible variations of mixed Hodge structure*, Proc. Symp. on Algebraic Geometry in East Asia, Kyoto (2001)

[PP] G. Pearlstein and C. Peters, *Differential Geometry of the Mixed Hodge Metric*, arXiv:1407.4082 (2014)

[S] C. Schwarz, *Relative monodromy weight filtrations*, Math. Z., 236 (2001), 11-21.

[U] S. Usui, *Variation of mixed Hodge structure arising from family of logarithmic deformations, II: Classifying space.* Duke Math. J. 51 (1984), 851–875.

Supported by CONICET, ANPCyT and Secyt-UNC, Argentina.

CENTRO DE INVESTIGACIONES Y ESTUDIOS MATEMATICOS, UNIVERSIDAD NACIONAL DE CORDOBA, AND UNIVERSITY OF MASSACHUSETTS, AMHERST.

E-mail address kaplan@math.umass.edu, kaplan@famaf.unc.edu.ar

I.B Period maps and algebraic geometry

4

Deformation theory and limiting mixed Hodge structures

Mark Green and Phillip Griffiths

Outline

I Introduction

This paper was motivated by the following question: Recall that for a smooth projective variety X whose polarized Hodge structure on $H^n(X,\mathbb{Q})_{\mathrm{prim}}$ leads to a period point $\Phi(X) \in D$, the period domain for polarized Hodge structures of a fixed type, the differential

$$\Phi_* : T_X \operatorname{Def}(X) \to T_{\Phi(X)} D$$

from the tangent space to the Kuranishi space $\operatorname{Def}(X)$ to the tangent space $T_{\Phi(X)}D$ is expressed cohomologically by the inclusion $T_X \operatorname{Def}(X) \hookrightarrow \mathbb{E}\mathrm{xt}^1_{\mathcal{O}_X}\left(\Omega^1_X, \mathcal{O}_X\right) \cong H^1(T_X)$, and the resulting natural maps on the associated graded to $F^\bullet \mathbb{H}^m(\Omega^\bullet_X)$ induced from

$$\text{(I.1)} \qquad \mathbb{E}\mathrm{xt}^1_{\mathcal{O}_X}\left(\Omega^1_X, \mathcal{O}_X\right) \otimes \mathbb{H}^m(\Omega^\bullet_X) \to \mathbb{H}^m(\Omega^\bullet_X).$$[1]

Our question was how to extend this formalism to the case where X is singular, having singularities of the type that arise by semi-stable reduction in a family of projective varieties $\mathcal{X} \xrightarrow{f} S$ whose general member is smooth. This

[1] In more classical notation $\Theta_X = T_X$, this is the map

$$H^1(\Theta_X) \otimes H^{m-p}(\Omega^p_X) \to H^{m-p+1}\left(\Omega^{p-1}_X\right)$$

induced from the contraction $\Theta_X \otimes \Omega^p_X \to \Omega^{p-1}_X$.

question was studied by Friedman [Fr1] when $\dim S = 1$. We were interested in amplifying and extending his results, emphasizing the development of a formalism that lends itself to the computation of examples. In the course of trying to carry this out we have found that there is an interesting story surrounding the relationships among the various mixed Hodge structures associated to X and its 1st order neighborhood in $\mathcal{X}$, and one of the purposes of this paper has turned out to be to amplify and clarify these relationships in the context of deformation theory. Here for us the works [Fr2], [St1] and [St2] have been very important when $\dim S = 1$, as has the extension of [St1] to a general S by Fujisawa [Fu1], [Fu2]. In fact, this is a partly expository paper, drawn from, reinterpreting and building on the works [Fr1], [Fr2], [St1], [Zu], [PS], [St2], [Fu1], [Fu2], [CKS1], [KP2], [GGR] and others.

To address the question stated above one is led to focus on the singular variety X and its first order deformations. For the case when X is a normal crossing variety this is done in [Fr2]. Here motivated by the semi-stable reduction theorem in [AK] we shall assume more generally that

X is locally a product of normal crossing varieties.

This means locally in $\mathbb{C}^n$ with coordinates $x_1,\dots,x_n$ we have a sequence $1 \leqq i_1 < i_2 < \cdots < i_k \leqq n$ with index blocks $I_1 = \{1,\dots,i_1\}$, $I_2 = \{i_1+1,\dots,i_2\},\dots,$ and then X is locally given by

$$x_{I_1} = 0,\dots,x_{I_k} = 0 \tag{I.2}$$

where $x_{I_1} = x_1 \cdots x_{i_1}$, $x_{I_2} = x_{i_1+1} \cdots x_{i_2}, \dots$. The usual locally normal crossing variety is the case $k = 1$.[2] The deformation theory of such varieties is well understood [Pa], and for simplicity of exposition in this paper we shall abuse notation and set

$$T_X \operatorname{Def}(X) = \mathbb{E}\mathrm{xt}^1_{\mathcal{O}_X}(\Omega^1_X, \mathcal{O}_X).$$

The abuse of notation is because here the right-hand side is the space of deformations of X over $\Delta_\epsilon =: \operatorname{Spec}\mathbb{C}[\epsilon]$, $\epsilon^2 = 0$, so that it is only the Zariski tangent space to the Kuranishi space $\operatorname{Def}(X)$. In general there may be obstructions to lifting deformations defined over the Zariski tangent space, but

[2] There is an important distinction between the case when X is locally a normal crossing variety and when it is globally such. By a combination of blowings up and base changes the former may be reduced to the later, and for both theoretical and notational purposes this is generally done. For computational purposes the former is frequently more convenient; e.g., for irreducible nodal curves. In this paper we shall restrict to the global normal crossing case and its analogue when X is locally a product of normal crossing varieties. However, we expect that the discussion given below will remain valid in the more general case, and some of our examples are carried out in the local normal crossing case. The formalism in [De] and [St2] allows one to handle the general theory when X is locally a normal crossing variety.

this issue will play no role in what follows.[3] In fact, one of the main points is that the theory of limiting mixed Hodge structures for 1-parameter families depends only on the 1st order neighborhood of the singular variety, a point that is implicit in [Fr2] and explicit in a somewhat different form in [St2].

We shall make the crucial assumption that *there exists a* $\xi \in \mathbb{E}\mathrm{xt}^1_{\mathcal{O}_X}(\Omega^1_X, \mathcal{O}_X)$ *such that for every* $x \in X$ *the image* ξ_x *of* ξ *in the natural map*

$$\mathbb{E}\mathrm{xt}^1_{\mathcal{O}_X}(\Omega^1_X, \mathcal{O}_X) \to \mathrm{Ext}^1_{\mathcal{O}_X}(\Omega^1_X, \mathcal{O}_X)_x \tag{I.3}$$

smooths to 1st *order the singularity at* x. Equivalently, for every $x \in X$ the global deformations of X over Δ_ϵ map to smoothing deformations of the germ X_x of X at x. The smoothing deformations of (I.2) are given by $x_{I_j} = t_j$ and they have tangents $\sum_{i=1} \lambda_i \partial_{t_i}$ where all $\lambda_i \neq 0$. We denote by

$$T^0_X \mathrm{Def}(X) \subset T_X \mathrm{Def}(X)$$

the open set of all $\xi \in T_X \mathrm{Def}(X)$ whose localizations are smoothing deformations of X_x for every $x \in X$.

We define the pair (X, ξ) to be *projective* in case there is a very ample line bundle $L \to X$ such that L extends to $\mathcal{X}_\xi$. This can be expressed cohomologically in a standard way, and we shall assume it to always be the case.

A *limiting mixed Hodge structure* $(V, W_\bullet, F^\bullet)$ is given by a $\mathbb{Q}$-vector space V, a weight filtration $W_\bullet$ and Hodge filtration $F^\bullet$ that define a mixed Hodge structure, and where there exists a nilpotent $N \in \mathrm{End}(V)$ such that (i) $W_\bullet = W_\bullet(N)$ is the monodromy weight filtration, and (ii) for the integer m around which the monodromy weight filtration is centered, the $N^k : \mathrm{Gr}^{W_\bullet}_{m+k} V \xrightarrow{\sim} \mathrm{Gr}^{W_\bullet}_{m-k} V$ are isomorphisms for $0 \leq k \leq m$. The limiting mixed Hodge structure is *polarizable* if there exists a $Q : V \otimes V \to \mathbb{Q}$ with $Q(u,v) = (-1)^m Q(v,u)$, and an $N \in \mathrm{End}_Q(V)$ as above such that the *primitive spaces* $\ker N^{k+1}$ above are polarized Hodge structures via $Q_k(u,v) = \pm Q(N^k u, v)$ (cf. [CKS1]). Two limiting mixed Hodge structures $(V, W_\bullet, F^\bullet)$ and $(V, W'_\bullet, F'^\bullet)$ are *equivalent* if $W'_\bullet = W_\bullet$, and if $F'^\bullet = \exp(zN)F^\bullet$ for some $z \in \mathbb{C}$. We will denote by $[V, W_\bullet, F^\bullet]$ an equivalence class of limiting mixed Hodge structures.

We shall use the term *standard family* to mean that $\mathcal{X}_\Delta \to \Delta$ is a projective mapping where $\mathcal{X}_\Delta$ is smooth, the fibers $X_t = \pi^{-1}(t)$ are smooth for $t \neq 0$, and $X_0 = X$ is a reduced normal crossing variety.

THEOREM I: *Canonically associated to each* $\xi \in T^0_X \mathrm{Def}(X)$ *is a polarized limiting mixed Hodge structure* $(V_\xi, W_\bullet, F^\bullet_\xi)$. *In case X is a normal crossing variety and* ξ *is tangent to an arc* $\Delta \subset \mathrm{Def}(X)$ *giving a standard family* $\mathcal{X}_\Delta \xrightarrow{\pi}$

[3] One may make the blanket assumption that all 1st order deformations are unobstructed, and then at the end note that this assumption has never been used.

Δ with $\pi^{-1}(0) = X$, this polarized limiting mixed Hodge structure is the one associated to the family and $\xi \in T_0(\Delta)$.

This result is largely an amalgam and slight extension of those in [Fr2] and [St2]. A key point is to note that the data (X,ξ) gives a standard family $\mathcal{X}_\xi \to \Delta_\epsilon$, together with an extension

$$0 \to \mathcal{O}_X \to \Omega^1_{\mathcal{X}_\xi} \otimes \mathcal{O}_X \to \Omega^1_X \to 0 \tag{I.4}$$

of $\mathcal{O}_X$-modules.[4] A second key point is to show that, as explained in section III below, (I.4) gives an exact sequence

$$0 \to \mathcal{O}_X \to \Omega^1_{\mathcal{X}_\xi}(\log X) \otimes \mathcal{O}_X \to \Omega^1_{\mathcal{X}_\xi/\Delta_\epsilon}(\log X) \otimes \mathcal{O}_X \to 0, \tag{I.5}$$

where in the case of a standard family $\mathcal{X}_\Delta \to \Delta$ restricting to $\mathcal{X}_\xi \to \Delta_\epsilon$

$$\Omega^1_{\mathcal{X}_\xi/\Delta_\epsilon}(\log X) \otimes \mathcal{O}_X = \Omega^1_{\mathcal{X}_\Delta/\Delta}(\log X) \otimes \mathcal{O}_X.$$

The vector space in the limiting mixed Hodge structure is

$$V_\xi = \mathbb{H}^m\left(\Omega^\bullet_{\mathcal{X}_\xi/\Delta_\epsilon}(\log X) \otimes \mathcal{O}_X\right)$$

and $F^\bullet_\xi$ is induced from the "bête" filtration on $\Omega^\bullet_{\mathcal{X}_\xi/\Delta_\epsilon}(\log X) \otimes \mathcal{O}_X$. The monodromy logarithm is induced from the connecting homomorphisms arising from (I.5). The $\mathbb{Q}$-structure and properties of the monodromy logarithm and resulting monodromy weight filtration are more subtle to define and treat (cf. [St1], [Zu] and chapter 11 in [PS]).

We note that the usual ambiguity in either the Hodge filtration or the $\mathbb{Q}$-structure in the limiting mixed Hodge structure associated to $\mathcal{X} \to \Delta$, ambiguity that depends on a choice of parameter t, is removed by considering the data (X,ξ).

A subtle point, one that will be further explained below, is this: For X a smoothable normal crossing variety the singular locus D will have connected components D_a. Then we will see that $\mathrm{Ext}^1_{\mathcal{O}_X}(\Omega^1_X, \mathcal{O}_X) \cong \oplus \mathcal{O}_{D_a}$, and in the basic exact sequence (II.1) a global 1st order deformation $\xi \in \mathbb{E}\mathrm{xt}^1_{\mathcal{O}_X}(\Omega^1_X, \mathcal{O}_X)$ will induce $\xi_{D_a} \in H^0(\mathcal{O}_{D_a})$. The condition that ξ be to 1st order smoothing along D_a is that $\xi_{D_a} \neq 0$. *Then the equivalence class of the limiting mixed Hodge structure in Theorem I depends only on the ξ_{D_a} and not on the global ξ that maps to the ξ_{D_a}'s.* In fact, given a collection of non-zero ξ_{D_a}'s, we may construct a limiting mixed Hodge structure provided that there is a global smoothing ξ; the particular ξ does not matter.[5]

[4] We are here extending the notion of a standard family to include the smooth non-reduced scheme $\mathcal{X}_\xi$ with structure sheaf $\mathcal{O}_{\mathcal{X}_\xi}$ locally isomorphic to $\mathcal{O}_X[\epsilon]$. We will also say that fibers over Δ^*_e are smooth.

[5] As we hope to discuss further in a work in progress, this is related to the theorem of Cattani-Kaplan [CK] that the weight filtration $W_\bullet(N)$ is independent of N in the interior of a monodromy cone, and the result in [CKS1] that the equivalence class of the limiting mixed Hodge structure is independent of the direction of approach from the interior of the cone.

To handle several variable families we shall consider a vector

$$\boldsymbol{\xi} \in \mathbb{E}\mathrm{xt}^1_{\mathcal{O}_X}(\Omega^1_X, \mathcal{O}_X{}^\ell) = \overset{\ell}{\oplus} T_X \operatorname{Def}(X)$$

with the property that for $\lambda = (\lambda_1, \ldots, \lambda_\ell)$ and

$$\boldsymbol{\xi}_\lambda = \lambda_1 \xi_1 + \cdots + \lambda_\ell \xi_\ell, \qquad \lambda_i \neq 0$$

we have

$$\boldsymbol{\xi}_\lambda \in T^0_X \operatorname{Def}(X). \tag{I.6}$$

For $\Delta_\epsilon = \Delta_{\epsilon_1} \times \cdots \times \Delta_{\epsilon_\ell}$, it will be seen that we then have a family

$$\mathcal{X}_{\boldsymbol{\xi}_\lambda} \to \Delta_\epsilon$$

with smooth fibers over $\Delta^*_\epsilon = \Delta^*_{\epsilon_1} \times \cdots \times \Delta^*_{\epsilon_\ell}$. We think of this as a space of 1st order deformations that deform X to a "less singular" variety along the axes but which smooth X when we deform into the interior.[6] There is then a several variables analogue of Theorem I where the terms in the statement will be explained in the text.

THEOREM I′: *Associated to $\boldsymbol{\xi} \in \mathbb{E}\mathrm{xt}^1_{\mathcal{O}_X}(\Omega^1_X, \mathcal{O}_X{}^\ell)$ satisfying the condition that* (I.6) *holds, there is a several variable polarized limiting mixed Hodge structure $(V_{\boldsymbol{\xi}}, W_\bullet, F^\bullet_\epsilon)$ in the sense of [CKS1]. In case X is a normal crossing variety and $\boldsymbol{\xi}_\lambda$ is tangent to an arc $\Delta_\lambda \subset \operatorname{Def}(X)$, this polarized limiting mixed Hodge structure is the one associated to the standard family $\mathcal{X}_{\Delta_\lambda} \xrightarrow{\pi} \Delta_\lambda$ with $\pi^{-1}(0) = X$.*

As will be discussed below, for the last statement in the theorem the general case when X is locally of the form (I.3) seems to be open (cf. [Fu1], [Fu2]), and we will discuss a geometric reason for this.

Detailed proofs of Theorems I and I′, especially for the latter, will not be given below. The argument for Theorem I consists largely of proof analysis of those in the references [Fr2], [St1] and [St2] and will be addressed more fully in a work in progress. For Theorem I′, the construction of a mixed Hodge structure follows largely from our construction given below and [Fu1], [Fu2]. The construction of a *polarized limiting* mixed Hodge structure requires more work and will be taken up in the work in progress. We will however try to point out some of the key points in both of these arguments.

For the analogue of (I.1) we have

[6] In the paper [KN] the definition of a normal crossing variety with logarithmic structure is introduced. The presence of a logarithmic structure is equivalent to d-semi-stability in the sense of [Fr2] (cf. (II.6) below). A deformation theory for normal crossing varieties with logarithmic structure is then introduced. In the context of this paper this theory amounts to deformations of X that independently smooth the connected components of the singular locus D of X, modulo equisingular deformations. The log-geometry formalism nicely lends itself to computation of examples for Calabi-Yau varieties.

THEOREM II: *The class* $\xi \in \mathbb{E}\mathrm{xt}^1_{\mathcal{O}_X}(\Omega^1_X, \mathcal{O}_X)$ *in* (I.4) *defines a natural class* $\xi^{(1)} \in \mathbb{E}\mathrm{xt}^1_{\mathcal{O}_X}(\Omega^1_{\mathcal{X}_\xi/\Delta_\epsilon}(\log X) \otimes \mathcal{O}_X, \mathcal{O}_X)$ *corresponding to* (I.5), *and the* 1st *order variation of the limiting mixed Hodge structure is expressed as the natural mapping*

$$\mathbb{E}\mathrm{xt}^1_{\mathcal{O}_X}\left(\Omega^1_{\mathcal{X}_\xi/\Delta_\epsilon}(\log X) \otimes \mathcal{O}_X, \mathcal{O}_X\right) \to \mathrm{End}_{\mathrm{LMHS}}\, \mathbb{H}^m\left(\Omega^\bullet_{\mathcal{X}_\xi/\Delta_\epsilon}(\log X) \otimes \mathcal{O}_X\right).^7$$

Here, $\mathbb{H}^m\left(\Omega^\bullet_{\mathcal{X}_\xi/\Delta_\epsilon}(\log X) \otimes \mathcal{O}_X\right) = V_\xi$ is the vector space underlying the limiting mixed Hodge structure in Theorem I, and $\mathrm{End}_{\mathrm{LMHS}}$ means the endomorphisms of V_ξ that preserve the structure as a limiting mixed Hodge structure as explained below.

Again the terms in the statement will be explained in the text. An informal way to think about this result is this: Denoting by $\mathrm{Def}(X,\xi)$ the deformations of the pair (X,ξ), we have a natural *extended period mapping*

$$\mathrm{Def}(X,\xi) \to \check{D}$$

that assigns to $\xi \in T^0_X \mathrm{Def}(X)$ the well-defined point $F^\bullet_\xi \in \check{D}$, the dual space to the period domain D consisting of filtrations of V_ξ that satisfy only the 1st Hodge-Riemann bilinear relation. Then the map in Theorem II might be thought to be the differential

$$T_{(X,\xi)}\, \mathrm{Def}(X,\xi) \to T_{F^\bullet_\xi} \check{D} \tag{I.7}$$

of the extended period mapping. This is *not* the case, as will be made precise in Section IV below. The issue is more subtle in that ξ gives not only a well-defined limiting mixed Hodge structure, not just an equivalence class of such, but also defines a 1st order variation of that limiting mixed Hodge structure. This is the information in $\xi^{(1)}$. At first glance one might think that since it takes the tangent vector ξ to define $F^\bullet_\xi$, the information in $\xi^{(1)}$ which gives the variation of the entire limiting mixed Hodge structure would be of 2nd order. But this is not correct, and it was in trying to understand this point that we were led to many of the other topics in this paper.

We will however see by example that $\xi^{(1)}$ contains strictly more information than the differential at the origin of the Kato-Usui map [KU]

$$\Delta \to \Gamma_T \backslash D_N.$$

[7] The notation $\xi^{(1)}$ has been used because the construction of the sequence (I.5) from (I.4) resembles that of the construction of the first prolongation in the theory of exterior differential systems. The group $\mathbb{E}\mathrm{xt}^1_{\mathcal{O}_X}(\Omega^1_{\mathcal{X}_\xi/\Lambda_\epsilon}(\log X) \otimes \mathcal{O}_X, \mathcal{O}_X)$, which may be defined if there exists a logarithmic structure on X, appears naturally in the deformation theory of smooth logarithmic varieties (cf. [A†]).

Referring to footnote 17 below, in the setting of log-analytic geometry the important monograph [KU] contains a treatment of the differential of the period map at infinity for standard families $\mathcal{X}_\Delta \to \Delta$ (cf. Theorem 4.4.8). In case the ξ in Theorem II arises as the tangent vectors at the origin we believe that those results should be equivalent.

Here $D_N = D \cup B(N)$ is the period domain D with the boundary component $B(N)$ attached to D, where $B(N)$ consists of all equivalence classes of limiting mixed Hodge structures with monodromy logarithm N and where $\Gamma_T = \{T^{\mathbb{Z}}\}$ with $T = \exp N$ is the local monodromy group (cf. Section IV below for an explanation of the notations and terms used). It is in this sense that Theorem II provides an answer to our original question. The term "expressed" means that in examples $\mathbb{E}\mathrm{xt}^1_{\mathcal{O}_X}\big(\Omega^1_{\mathcal{X}_\xi/\Delta_\epsilon}(\log X) \otimes \mathcal{O}_X, \mathcal{O}_X\big)$ will have algebro-geometric meaning and the pairing is a cup-product. We will see by example that the additional information is non-trivial and somewhat subtle.[8]

As will be explained in Section V below, associated to a polarized limiting mixed Hodge structure is a *reduced limit period mapping* and distinguished point

$$F^\bullet_\infty \in \partial D$$

where $D = G_{\mathbb{R}}/H$ is a period domain ([KP1], [KP2], [GGK], [GG] and [GGR]). The boundary ∂D is stratified into finitely many $G_{\mathbb{R}}$-orbits and their geometry is a much studied and very interesting topic ([GGK], [FHW]).

On the other hand, the vector space $T_X\,\mathrm{Def}(X)$ is stratified by open sets $T^0_X\,\mathrm{Def}(X)_I$ contained in linear subspaces $T_X\,\mathrm{Def}(X)_I \subset T_X\,\mathrm{Def}(X)$. In the text we will explain this in case X is a normal crossing divisor, which is the only case for which thus far we have a result. Then the strata correspond to subsets of the set of connected components of the singular locus X_{sing} of X. The subspace $T^0_X\,\mathrm{Def}(X)$ is the open stratum of smoothing deformations; the other strata correspond to the components that are smoothed when X deforms in the directions of that strata. The opposite extreme to $T^0_X\,\mathrm{Def}(X)$ is the linear subspace $T_X\,\mathrm{Def}^{\mathrm{es}}(X) \cong H^1\left(\mathrm{Ext}^0_{\mathcal{O}_X}(\Omega^1_X, \mathcal{O}_X)\right)$ of equisingular deformations. It seems reasonable to expect, but we are not aware of a proof in the literature, that $\xi \in T^0_X\,\mathrm{Def}(X)_I$ corresponds to the limit in a variation of mixed Hodge structures over the punctured disc ([St-Zu]).

Leaving this important issue aside, we return to the deformation theory and limiting mixed Hodge structures in the several parameter case. In the study of limiting mixed Hodge structures over higher dimensional base spaces ([CKS1]) there are a number of cone structures that enter:

(i) the stratification of abelian subspaces $\mathfrak{A} \subset \mathfrak{g}^{\mathrm{nilp}}$ induced by the G-orbit structure on $\mathfrak{g}^{\mathrm{nilp}}$ ([Ro] and references cited therein);

[8] In very classical terms one may write the period matrix $\Omega(t)$ in block form where the blocks $\Omega_i(t)$ are polynomials in $\log t$ with holomorphic coefficients and where the remaining blocks $\Omega_\alpha(t)$ are holomorphic at $t = 0$. The differential of the map to $\Gamma_T \backslash D_N$ records the derivatives $\Omega'_\alpha(0)$ of the holomorphic terms, while (I.7) has the effect of regularizing the logarithmically divergent integrals that give the $\Omega_i(t)$ and then taking the linear part $\Omega'_i(0)$ at $t = 0$ of that regularization. The $\Omega'_\alpha(0)$ and $\Omega'_i(0)$ record the variation in the full extension data in the limiting mixed Hodge structure.

(ii) the stratification of nilpotent cones as in [CKS1] and [KU] (cf. [AMRT] for the classical weight one case);
(iii) the stratification of ∂D by $G_{\mathbb{R}}$-orbits and its relation to reduced limit period mappings [KP1], [KP2], [GG], [GGR] and [Ro] and work in progress by Kerr, Pearlstein and Robles; and
(iv) the stratification of $T_X \operatorname{Def}(X)$, as explained below for X a normal crossing divisor, and which we feel can reasonably be expected to extend to the case where X is locally a product of normal crossing divisors.

The basic known result, due to Robles [Ro], is that *the interiors of the strata in (ii) map to strata in (i)*, and as a consequence to strata in (iii).[9] Her argument makes full use of the deep properties of several variable nilpotent orbits [CKS1] and of the classification of $G_{\mathbb{R}}$-orbits in $\mathfrak{g}_{\mathbb{R}}^{\text{nilp}}$ (cf. the references in [Ro]). An algebro-geometric version of Robles' result might be that at the tangent space level strata in (iv) map to strata in (iii). The theorem to be described now is a partial result in this direction.

In the setting of the Cattani-Kaplan-Schmid theory there are defined nilpotent cones

$$\sigma = \operatorname{span}_{\mathbb{Q}>0}\{N_1,\dots,N_\ell\}$$

where the $N_i \in \mathfrak{g}^{\text{nilp}}$ are linearly independent commuting nilpotent transformations and several variable nilpotent orbits $(F^\bullet,\sigma)$. Here, $F^\bullet \in \check{D}$ and the conditions

- $\exp(z_1N_1 + \cdots + z_\ell N_\ell)\cdot F^\bullet \in D$ for all $\operatorname{Im} z_i \gg 0$;
- $[N_i, F^p] \subset F^{p-1}$

are satisfied. We denote by $\widetilde{B}(\sigma) \subset \check{D}$ the set of several variable nilpotent orbits, and by $B(\sigma)$ the equivalence classes of those orbits under reparametrization $z_i \to z_i + \lambda_i$. In [KP1], [GGK] and [GG] there are defined reduced limit period mappings for 1-dimensional cones, and the construction can be extended [KP2] to the general case to give the reduced limit period map

$$\Phi_\infty : B(\sigma) \to \partial D. \tag{I.8}$$

THEOREM III: *Let X be a normal crossing variety for which there exists a $\xi \in T_X \operatorname{Def}(X)$ that is nowhere vanishing along each component of X_{sing}. Then there exists a nilpotent cone*

$$\sigma_X \subset T_X \operatorname{Def}(X)/T_X^{\text{es}} \operatorname{Def}(X),$$

[9] For the interior of the full nilpotent cone, and suitably interpreted on its faces (where the Hodge filtration has been "pushed" by the complementary nilpotent elements), this result follows from [CKS1].

and a several variable polarized limiting mixed Hodge structure in the sense of [CKS1] with the property that under the reduced limit period mapping (I.8) $\widetilde{B}(\sigma_X)$ *maps to a* $G_{\mathbb{R}}$*-orbit in* ∂D.

This theorem follows from the construction of σ_X and the result of Robles mentioned above. As mentioned before, it is of interest to see if the construction of σ_X and the result can be extended to the faces of the cone σ_X.

For our next result we note that given a standard family $\mathfrak{X} \to \Delta$ there are the following four types of mixed Hodge structures that may be defined:

(i) the part of the mixed Hodge structure on $H^*(X)$ that comes from the limiting mixed Hodge structure;[10]
(ii) that part of the limiting mixed Hodge structure that may be defined in terms of X alone;
(iii) the limiting mixed Hodge structure, modulo reparametrizations $F^\bullet_{\lim} \sim \exp(zN) \cdot F^\bullet_{\lim}$ resulting from a change in parameter in the disc, associated to $\mathfrak{X} \to \Delta$;[11] and
(iv) the limiting mixed Hodge structure associated to (X,ξ), where $\xi \in T^0_X \operatorname{Def}(X)$ is the first order variation of X in $\mathfrak{X}$.

THEOREM IV: *There are strict implications*

$$\text{(iv)} \Longrightarrow \text{(iii)} \Longrightarrow \text{(ii)} \Longrightarrow \text{(i)}.$$

The term "strict implication" means that there is successively more information in (i), (ii), (iii), (iv); the precise meaning of this will be explained in the proof.

We will see that given an abstract X that is locally a normal crossing divisor, the condition that we may construct the data given in (ii) is that there exists a $\xi \in T^0_X \operatorname{Def}(X)$ that is smoothing to 1st order; the actual data will not depend on the particular ξ but rather will depend on the ξ_{D_a}'s as discussed above. A limiting mixed Hodge structure will decompose into N-strings under the action of the monodromy logarithm N. This decomposition may be pictured as

$$\begin{array}{ccccccc} H^0(-m) \longrightarrow & & \cdots & & \longrightarrow H^0(-1) \longrightarrow H^0 \\ & H^1(-m+1) \longrightarrow & \cdots & \longrightarrow & H^1(-1) \longrightarrow H^1 \\ & & \vdots & & \\ & & H^m & & \end{array} \tag{I.9}$$

where H^k is a pure Hodge structure of weight k.[12] We may think of (I.9) as giving the primitive decomposition in the associated graded to a polarized

[10] This is $\ker N$.
[11] This is by definition the same as an equivalence class of limiting mixed Hodge structures.
[12] If one thinks of N as being completed to an sl_2-triple, then the N-strings are composed from the irreducible pieces in the decomposition of the sl_2-module. The H^k's on the right end may themselves be Tate twists of lower weight Hodge structures.

limiting mixed Hodge structure, together with the iterated action of N on the primitive spaces. Then our result pertaining to (ii) is

THEOREM V: *The terms $H^{m-j}(-i)$, $0 \leqq i \leqq m-j$ in* (I.9), *together with the N-maps between them, may be constructed from X alone.*

We will also see for $[\xi] \in \mathbb{P}T^0_X\mathrm{Def}(X)$ with localizations ξ_{D_a} along the components of D_a of X_{sing}, we will have

$$\text{(iii)} \longleftrightarrow (X, [\xi_{D_a}]\text{'s}),$$

and where the brackets refer to the corresponding point in the designated projective space and the symbol "$\longleftrightarrow$" means that the data on each side are equivalent.

We hope that this result will clarify exactly what input is needed to be able to define the limiting mixed Hodge structures, or the parts thereof, that are associated to a degeneration $\mathcal{X} \to \Delta$ of a smooth projective variety.[13] All of (i)–(iv) require knowledge of at most the 1st order neighborhood of X in $\mathcal{X}$. It is worth noting that even though the central fiber X is in general not uniquely definable,[14] the ambiguity in the limiting mixed Hodge structures "washes out" in the constructions (ii), (iii), (iv).[15]

We note that the traditional approach in the study of the behavior of the polarized Hodge structures in a degenerating family of smooth projective varieties is to start with a family $\mathcal{X}^* \to \Delta^*$ with unipotent monodromy T. To this we may either associate a period mapping

$$\Phi : \Delta^* \to \Gamma_T \backslash D, \qquad \Gamma_T = T^{\mathbb{Z}},$$

and then by [Sc] to this period mapping associate an equivalence class of limiting mixed Hodge structures. Or more algebro-geometrically we may complete $\mathcal{X}^* \to \Delta^*$ to a standard family to which by [St1] we may associate the same equivalence class of limiting mixed Hodge structures.[16] In this paper we are *starting* with the central fiber X with only the assumptions that (a) X is projective and is locally a normal crossing divisor, or more generally that it is locally a product of normal crossing divisors, and (b) there exists

13 Its proof mainly consists of "proof analysis" of the construction of the limiting mixed Hodge structure in [St1], [Zu] and [St2]. Our main new point is to focus from the outset on the pair (X,ξ).

14 Exceptions include stable curves, principally polarized abelian varieties and marked $K3$ surfaces, all of which have "good" global moduli spaces.

15 We will not try to explain this precisely, but note that in the Clemens-Schmid exact sequence the effects of doing a modification to X cancel out and leave unchanged the terms with the limiting mixed Hodge structures. This phenomenon is of course familiar from Deligne's theory of mixed Hodge structures in which independence of the choice of smooth completions is established (cf. [PS] and the references cited therein).

16 More precisely, to each is associated an equivalence class of limiting mixed Hodge structures. In [St1] it is shown that the two equivalence classes of limiting mixed Hodge structures agree.

a $\xi \in T_X \operatorname{Def}(X)$ that is to 1st order smoothing and preserves the ample line bundle. We hope that this helps to explain the title of this work.

In what follows we shall use X to denote both a compact analytic variety and a germ of an analytic variety; we hope the context will make clear to which we are referring. When X is a compact analytic variety and $x \in X$ we shall denote by X_x the germ of analytic variety defined by localizing X at x.

The other notations we have used are either standard or will be noted where introduced. For our variety X we will have $\dim X = n$, and we shall generally consider cohomology and hypercohomology in degree m (e.g., $H^m(X^{[k]})$'s).

II Deformation theory

Our basic reference is [Pa], as summarized in [Fr2] for the normal crossing case and whose terminology and notations we shall generally follow.[17] For X either a compact analytic variety, or a germ of a reduced analytic variety, we shall denote by $\operatorname{Def}(X)$ the space parametrizing the corresponding family $\mathcal{X}_{\operatorname{Def}(X)} \xrightarrow{\pi} \operatorname{Def}(X)$ that is versal for germs of flat families $\mathcal{X} \xrightarrow{\pi} S$ with $\pi^{-1}(s_0) = X$. The Zariski tangent space to $\operatorname{Def}(X)$ is

$$T_X \operatorname{Def}(X) = \mathbb{E}\mathrm{xt}^1_{\mathcal{O}_X}(\Omega^1_X, \mathcal{O}_X).$$

As usual we think of $\xi \in T_X \operatorname{Def}(X)$ as giving a family

$$\mathcal{X}_\xi \to \Delta_\epsilon$$

where $\Delta_\epsilon = \operatorname{Spec} \mathbb{C}[\epsilon]$, $\epsilon^2 = 0$.

Of basic importance for us will be the exact sequence

$$\text{(II.1)} \quad \boxed{\begin{aligned} 0 &\to H^1\left(\mathcal{E}\mathrm{xt}^0_{\mathcal{O}_X}(\Omega^1_X, \mathcal{O}_X)\right) \to \mathbb{E}\mathrm{xt}^1_{\mathcal{O}_X}(\Omega^1_X, \mathcal{O}_X) \\ &\to H^0\left(\mathcal{E}\mathrm{xt}^1_{\mathcal{O}_X}(\Omega^1_X, \mathcal{O}_X)\right) \xrightarrow{\delta} H^2\left(\mathcal{E}\mathrm{xt}^0_{\mathcal{O}_X}(\Omega^1_X, \mathcal{O}_X)\right) \end{aligned}}$$

that results from the local to global spectral sequence for Ext. The image of the first map will be denoted by

$$T_X \operatorname{Def}^{\mathrm{es}}(X) \to T_X \operatorname{Def}(X);$$

[17] We note that the setting of log-analytic geometry is an alternate, and in many ways preferable, way to present this theory (cf. [A†], [KN], [KU] and the references cited in these works). For example, in this context the central concept of d-semi-stability (cf. (II.6) below) simply becomes the existence of a log structure. Moreover, a logarithmic deformation of a smoothable normal crossing variety remains smoothable; none of the "bad" components in $\operatorname{Def}(X)$ can arise. In the setting of logarithmic deformation theory, unobstructed deformations of X simply means independently smoothing the connected components of $D = X_{\mathrm{sing}}$. We have written this work in the traditional setting in part because this allows us more easily to connect with the other topics discussed.

it represents the Zariski tangent space to the equisingular, or locally trivial for the germs X_x in X, deformations. For $x \in X$ the image of the map

$$\mathbb{E}\mathrm{xt}^1_{\mathcal{O}_X}(\Omega^1_X, \mathcal{O}_X) \to \mathrm{Ext}^1_{\mathcal{O}_X}(\Omega^1_X, \mathcal{O}_X)_x$$

represents the 1st order deformation of the germ X_x of analytic variety induced by a global 1st order deformation of X.

In our situation where X is locally a product of normal crossing varieties given by (I.3) the local deformation theory is particularly harmonious. Taking first the case when X is a germ of a normal crossing variety given locally in $\mathbb{C}^{n+1}$ by

$$f(x) =: x_1 \cdots x_k = 0 \tag{II.2}$$

with versal deformation space $\mathcal{X} \subset \mathbb{C}^n \times \mathbb{C}$ given by $f(x) = t$, and with the notations

$$\begin{cases} X_i = \{x_i = 0\}, & \\ D_i = X_i \cap \left(\bigcup_{j \neq i} X_j\right) & \text{defined by the ideal } \{\partial_{x_i} f = x_1 \cdots \widehat{x_i} \cdots x_k\}, \\ D = \cup D_i = X_{\mathrm{sing}} & \\ X^{[\ell]} = \coprod_{|I| = \ell} X_I & \text{where } I = (i_1, \dots, i_\ell) \text{ with } 1 \leqq i_1 < \cdots < i_\ell \leqq n \\ & \text{and } X_I = X_{i_1 \cap \cdots \cap X_{i_\ell}} \end{cases}$$

we have as $\mathcal{O}_X$-modules

$$\mathrm{Ext}^i_{\mathcal{O}_X}(\Omega^1_X, \mathcal{O}_X) \cong \begin{cases} \ker\{T_{\mathcal{X}} \otimes \mathcal{O}_X \to \mathcal{O}_X(X)\} & i = 0, \\ (I_X/I_X^2)^* \otimes \mathcal{O}_D & i = 1, \\ 0 & i \geqq 2. \end{cases} \tag{II.3}$$

This follows from the Ext-sequence arising from the exact sequence

$$0 \to I_X/I_X^2 \to \Omega^1_{\mathcal{X}} \otimes \mathcal{O}_X \to \Omega^1_X \to 0 \tag{II.4}$$

which, setting $\varphi_i = \partial_{x_i} f\, dx_i = x_1 \cdots \widehat{x_i} \cdots x_k\, dx_i$, gives that $\Omega^1_{\mathcal{X}}$ is freely generated over $\mathcal{O}_X$, and that Ω^1_X is generated by $dx_1, \dots, dx_{n+1}$ subject to the defining relation

$$df = \sum_i \varphi_i = 0.$$

Assuming now that X is a complete algebraic variety that is locally a normal crossing variety given locally by (II.2), motivated by the middle equation in (II.3) and taking into account the scaling of f under $f \to uf$ where $u \in \mathcal{O}_X^*$ and following [Fr2], we may define the *infinitesimal normal bundle* by

$$\mathcal{O}_D(X) = \mathrm{Ext}^1_{\mathcal{O}_X}(\Omega^1_X, \mathcal{O}_X). \tag{II.5}$$

The point here is that, unless we are given a global embedding of X as a hypersurface in a smooth variety $\mathcal{X}$, we cannot define the normal bundle $\mathcal{O}_X(X)$, but we are able to intrinsically define what would be the restriction to D of the normal bundle of X in a smooth ambient space if such exists.

In more detail, we set

$$\mathcal{O}_{D_i}(-X) = \left(I_{D_i}/I_{D_i}^2\right) \otimes_{\mathcal{O}_{D_i}} \left(I_{X_i}/I_{X_i}I_{D_i}\right)$$
$$\mathcal{O}_D(-X) = \left(I_{X_1}/I_{X_1}J_D\right) \otimes_{\mathcal{O}_D} \cdots \otimes_{\mathcal{O}_D} \left(I_{X_k}/I_{X_k}J_D\right)$$

where I_{X_i} is the ideal sheaf of X_i in X and J_{D_i} is the ideal sheaf of D_i in X. The second equation then serves to define $\mathcal{O}_D(X)$ in agreement with (II.5) and we have

$$\operatorname{Ext}^1_{\mathcal{O}_X}\left(\Omega^1_X, \mathcal{O}_X\right) \cong \mathcal{O}_D(-X)^* \otimes \mathcal{O}_D$$

where $\mathcal{O}_D(-X) = I_X/I_X^2$ in case we have $X \subset \mathcal{X}$.

As in [Fr2], X is said to be *d-semi-stable* if

$$\mathcal{O}_D(X) \cong \mathcal{O}_D, \tag{II.6}$$

that is there exists a nowhere vanishing section of the line bundle $\mathcal{O}_D(X)$ over D.[18]

We shall assume throughout that X is d-semi-stable.

Returning to (II.1) and using (II.5) we have the map

$$\mathbb{E}\mathrm{xt}^1_{\mathcal{O}_X}\left(\Omega^1_X, \mathcal{O}_X\right) \to H^0\left(\mathcal{O}_D(X)\right),$$

which we denote by $\xi \to \xi_D$ where $\xi \in \mathbb{E}\mathrm{xt}^1_{\mathcal{O}_X}\left(\Omega^1_X, \mathcal{O}_X\right) = T_X \operatorname{Def}(X)$. We shall say that the 1st order deformation ξ of X is *smoothing* if ξ_D is nowhere vanishing. We shall also generally abuse terminology by dropping the "1st order," and we shall say that ξ_D is non-zero rather than nowhere vanishing. Given such a ξ we have a family $\mathcal{X}_\xi \to \Delta_\epsilon$, where $\Delta_\epsilon = \operatorname{Spec} \mathbb{C}[\epsilon]$ with $\epsilon^2 = 0$, where $\mathcal{X}_\xi$ is smooth and in which the fiber over 0 is X. As noted earlier, it may or may not be the case that $\mathcal{X}_\xi \to \Delta_\epsilon$ can be lifted to a family $\mathcal{X} \to \Delta$; this issue will play no role in what follows.

For later reference we note that the sheaf Ω^1_X of Kähler differentials is defined by (II.4), where the injectivity of the first map is a property of X as given by (II.2). It is not locally free as a sheaf of $\mathcal{O}_X$-modules, but rather has a torsion subsheaf

$$\tau^1_X \subset \Omega^1_X$$

[18] Here the point is that if we have $X \subset \mathcal{X}$ with $\mathcal{X}$ smooth and $\mathcal{X} \xrightarrow{\pi} \Delta$ with $\pi^{-1}(0) = X$, then the conormal bundle $\mathcal{O}(-X) = I_X/I_X^2$ is trivial. Thus if we just have $X \subset \mathcal{X}$ where $\mathcal{X}$ is smooth, a necessary condition for there to exist $\mathcal{X} \xrightarrow{\pi} \Delta$ as above is that $\mathcal{O}_X(X) \cong \mathcal{O}_X$. The d-semi-stability condition (I.6) is intrinsic to X and does not require the existence of an $\mathcal{X}$.

which is locally generated by the forms φ_i define above. Its support is $D = X_{\text{sing}}$, and as noted in [Fr2] since Ω^1_X/τ^1_X is locally free the above inclusion induces an isomorphism

$$\operatorname{Ext}^1_{\mathcal{O}_X}(\Omega^1_X, \mathcal{O}_X) \xrightarrow{\sim} \operatorname{Ext}^1_{\mathcal{O}_X}(\tau^1_X, \mathcal{O}_X),$$

which "explains" the identification (II.5).

Of importance for this work will be to consider the set A of connected components D_a, $a \in A$, of D. Recalling our blanket assumption that there exists a 1st order smoothing deformation of X, we will have for each $\alpha \in A$

$$\mathcal{O}_{D_\alpha}(X) \cong \mathcal{O}_{D_\alpha},$$

where the particular isomorphism depends on the choice of a non-zero section ξ of $\mathcal{O}_D(X)$. Thus if $\xi \in T_X \operatorname{Def}(X)$ with restriction ξ_{D_a} to $\mathcal{O}_{D_\alpha}(X)$, we see that *along the component D_α of $D = X_{\text{sing}}$ the deformation of X given by ξ is either everywhere smoothing or equisingular.* For each subset $B \subset A$ we set

$$T^B_X \operatorname{Def}(X) = \left\{\xi \in T_X \operatorname{Def}(X) : \xi_{D_\beta} = 0 \;\text{ for }\; \beta \in B\right\}.$$

Then $T^B_X \operatorname{Def}(X)$ corresponds to the deformations that are equisingular along the D_β for $\beta \in B$. The extremes are

- $B = \emptyset$ corresponds to the open set $T^0_X \operatorname{Def}(X)$ of smoothing deformations;
- $B = A$ corresponds to the space $T_X \operatorname{Def}^{\text{es}}(X)$ of equisingular deformations.

In a way that will be explained in detail later, this gives a stratification of $T_X \operatorname{Def}(X)$ and leads to the definition of the cone σ_X mentioned in the introduction.

Example: A simple example is when X is a nodal curve. The surjectivity of the map

$$\text{(II.7)} \qquad \mathbb{E}\text{xt}^1_{\mathcal{O}_X}(\Omega^1_X, \mathcal{O}_X) \to \bigoplus_{\alpha \in A} H^0(\mathcal{O}_{D_\alpha}(X)) \to 0$$

corresponds to individually smoothing the nodes.

Example: Suppose that X_0 is a singular variety with isolated singular points p_α given by $f_\alpha(x) = 0$. We may resolve the singularities to obtain X where D has connected components D_α. The versal deformation spaces given by $f_\alpha(x) = t_\alpha$ for the germ of X_α at p_α and for the inverse image X_α of p_α in X coincide (cf. [Pa]). The failure of surjectivity of the first map in

$$\mathbb{E}\text{xt}^1_{\mathcal{O}_X}(\Omega^1_X, \mathcal{O}_X) \to \bigoplus_{\alpha \in A} H^0(\mathcal{O}_{D_\alpha}(X)) \xrightarrow{\delta} H^2(\operatorname{Ext}^0_{\mathcal{O}_X}(\Omega^1_X, \mathcal{O}_X))$$

measures the obstruction to simultaneously smoothing the $p_\alpha \in X_0$.

When the p_a are ordinary double points the dual to the mapping δ in the sequence may be computed and leads to the conditions on the simultaneous smoothing of the nodes that may be lifted to a smoothing of X.[19]

In general when X may be smoothed but the connected components D_α may not be independently smoothed, the situation is more complicated and necessitates the blowing up of X. This is the situation where in the setting of logarithmic deformation theory there are obstructions and will be discussed at another time.

The example of K3 surfaces is discussed in [Fr1], [Fr2] and [KN].

We next turn to the local case where the germ of variety X is a product

$$X = X^1 \times \cdots \times X^k = \prod_{\mu \in U} X^\mu$$

of normal crossing varieties.[20] Letting $\pi_\mu : X \to X^\mu$ denote the projection, from the isomorphism of $\mathcal{O}_X$-modules

$$\Omega^1_X \cong \oplus \pi^*_\mu \Omega^1_{X^\mu}$$

leading to

$$\mathrm{Ext}^1_{\mathcal{O}_X}(\Omega^1_X, \mathcal{O}_X) \cong \bigoplus_\mu \mathrm{Ext}^1_{\mathcal{O}_{X^\mu}}(\Omega^1_{X^\mu}, \mathcal{O}_{X^\mu}) \otimes \mathcal{O}_X,$$

we may extend the local theory in the evident way. The sequence (II.4) now becomes

$$\begin{array}{ccccccccc} 0 & \longrightarrow & \mathcal{O}^\ell_X & \longrightarrow & \Omega^1_X \otimes \mathcal{O}_X & \longrightarrow & \Omega^1_X & \longrightarrow & 0 \\ & & \wr\| & & \wr\| & & \wr\| & & \\ & & \oplus \pi^*_\mu \mathcal{O}_{X^\mu} & & \oplus \pi^* \Omega_{X^\mu} \otimes \mathcal{O}_X & & \oplus \pi^*_\mu \Omega^1_{X_\mu}. & & \end{array}$$

For

$$\begin{cases} D^\mu = X^1 \times \cdots \times D_\mu \times \cdots \times X^k \\ D = \sum D^\mu = X_{\mathrm{sing}} \end{cases}$$

we have

$$\mathrm{Ext}^1_{\mathcal{O}_X}(\Omega^1_X, \mathcal{O}_X) \cong \bigoplus_\mu \pi^*_\mu \mathcal{O}_{D^\mu}.$$

The local versal deformation space is the product of the local versal deformation spaces for the factors.

[19] If the mapping (II.7) is surjective, then

$$\sigma_X \otimes \mathbb{R} \cong \mathrm{span}_{\mathbb{R}>0}\{i\xi_{D_1}, \ldots, i\xi_{(A)}\}.$$

The reason for the "i" in $i\xi_{D_a}$ is that if we think of ξ_{D_a} as giving a tangent vector to a one parameter family then $i\xi_{D_a}$ is supposed to suggest turning around the origin — i.e., monodromy — in the family.

[20] This case is treated in [Fu1] and [Fu2].

As in the normal crossing case one may intrinsically define an infinitesimal normal sheaf $\mathcal{N}$. In the stratification $X_{\text{sing},\ell}$ of X_{sing} by the number of singular factors in the local product of normal crossing varieties description given by (I.1), $\mathcal{N}$ is a coherent sheaf whose restriction to $X_{\text{sing},\ell}\backslash X_{\text{sing},\ell+1}$ is locally free of rank ℓ. The definition of d-semi-stability may then be extended. This will be done in the work in progress; the practical effect of assuming d-semi-stability is that the to be constructed locally defined sheaves of $\mathcal{O}_X$-modules $\Omega^1_{\mathcal{X}_\xi}\otimes\mathcal{O}_X$ and $\Omega^1_{\mathcal{X}_\xi/\Delta_\epsilon}(\log X)\otimes\mathcal{O}_X$ patch together to give global sheaves over all of X.

When we consider the global situation where X is locally a product as above, we retain our standing assumption that in the map

$$\mathbb{E}\mathrm{xt}^1_{\mathcal{O}_X}(\Omega^1_X,\mathcal{O}_X)\to H^0\left(\mathrm{Ext}^1_{\mathcal{O}_X}(\Omega^1_X,\mathcal{O}_X)\right)$$

there is $\xi\in\mathbb{E}\mathrm{xt}^1_{\mathcal{O}_X}(\Omega^1_X,\mathcal{O}_X)$ which is a smoothing deformation along each component of X_{sing}. This does *not* mean that for each germ $X_x\subset X$ the global deformations map *onto* the space of local smoothings of X_x. It does mean that there is a $\xi\in\mathbb{E}\mathrm{xt}^1_{\mathcal{O}_X}(\Omega^1_X,\mathcal{O}_X)$ whose local image in each

$$\mathrm{Ext}^1_{\mathcal{O}_{X^\mu}}\left(\Omega^1_{X^\mu},\mathcal{O}_{X^\mu}\right)\cong\mathcal{O}_{D^\mu}(X^\mu)$$

is non-vanishing. Then the above discussion regarding the connected components of D extends and will be taken up in a future work.

One significant difference in the local situation where the number l of local factors is strictly larger than one is this: For a 1-parameter smoothing family

$$\mathcal{X}_\Delta\to\Delta$$

with tangent ξ, the total space $\mathcal{X}_\Delta$ is *singular*. This can be seen already in the local situation

$$\begin{cases} xy=t_1\\ uv=t_2\end{cases}$$

where the disc is given by $t_1/t_2=\lambda\neq 0$. Then even though the total space $\mathcal{X}\to\Delta_1\times\Delta_2$ is smooth, the subvariety $\mathcal{X}_\Delta\subset\mathcal{X}$ is singular at the origin.

A final comment for this section: In the study of varieties that are locally products of normal crossing varieties, the necessary multi-index notations may obscure the essential points. Our experience has been that for normal crossings the two cases

$$\begin{cases} xy=0\\ uvw=0,\end{cases}$$

and for products of normal crossings the cases

$$xy=0,\quad uv=0$$

capture all the essential phenomena. The main subtlety seems to arise when we smooth the singularity to obtain $\mathcal{X}$, various exact sequences over $\mathcal{O}_{\mathcal{X}}$ fail to become exact when we restrict to X by tensoring with $\mathcal{O}_X$ and some care must be taken in the computations to keep track of this.

III Proofs of Theorems I and I′

This initial discussion is mainly local. We begin with a germ of normal crossing variety X given by (II.2). Given a non-zero

$$\xi \in \operatorname{Ext}^1_{\mathcal{O}_X}(\Omega^1_X, \mathcal{O}_X) \cong \mathcal{O}_D(X)$$

we denote by $\mathcal{X}_\xi \xrightarrow{\pi} \Delta_\epsilon$ the corresponding versal family

$$x_1 \cdots x_k = \epsilon, \qquad \epsilon^2 = 0$$

and write the extension as

$$0 \to \Omega^1_{\Delta_\epsilon} \otimes \mathcal{O}_X \to \Omega^1_{\mathcal{X}_\xi} \otimes \mathcal{O}_X \to \Omega^1_X \to 0.$$

Here, $\mathcal{O}_{\mathcal{X}_\xi}$ is locally isomorphic to $\mathcal{O}_X[\epsilon]$ and $\Omega^1_{\mathcal{X}_\xi}$ is the free $\mathcal{O}_{X_\xi}$-module generated by $dx_1, \ldots, dx_{n+1}, d\epsilon$ modulo the relation $d\epsilon = \sum_{i=1}^k \varphi_i$. Unless otherwise noted the tensor products are over $\mathcal{O}_{\mathcal{X}_\xi}$. We are setting $\Omega^1_{\Delta_\epsilon} = \pi^*\Omega^1_{\Delta_\epsilon}$ and are writing the sequence in this way to emphasize the scaling property with respect to ξ. Note that $\Omega^1_{X_\xi} \otimes \mathcal{O}_X$ is the $\mathcal{O}_X$-module with the same set of generators and defining relation, and where in computations we set $x_1 \cdots x_k = 0$ but do *not* set $d(x_1 \cdots x_k) = \sum_{i=1}^k \varphi_i = 0$.

We may as usual define the free $\mathcal{O}_{\mathcal{X}_\xi}$-module $\Omega^1_{\mathcal{X}_\xi}(\log X)$ with generators $dx_1/x_1, \ldots, dx_k/x_k, dx_{k+1}, \ldots, dx_{n+1}, d\epsilon/\epsilon$ modulo the relation $d\epsilon/\epsilon = \sum_{i=1}^k dx_i/x_i$. Then

$$\Omega^1_{\mathcal{X}_\xi}(\log X) \otimes \mathcal{O}_X$$

is freely generated over $\mathcal{O}_X$ with the same set of generators and defining relation.

We next define

$$\Omega^1_{\mathcal{X}_\xi/\Delta_\epsilon}(\log X) \otimes \mathcal{O}_X \tag{III.1}$$

to be the $\mathcal{O}_X$-module with the above generators and generating relation

$$\sum_i dx_i/x_i = 0.$$

We will describe this intrinsically in a moment. Here we note the crucial point that in the case of a global normal crossing variety X fixing a nowhere zero $\xi \in \mathbb{E}\mathrm{xt}^1_{\mathcal{O}_X}(\Omega^1_X, \mathcal{O}_X)$ *uniquely* locally determines a *normalized generator* $\epsilon - x_1 \cdots x_k$ of the ideal $I_{\mathcal{X}_\xi}$ of X_ξ: If we have $x_i' = u_i x_i$ where $u_i \in \mathcal{O}_X^*$, it follows from the equality

$$\epsilon - x_1 \cdots x_k = \epsilon - x_i' \cdots x_k'$$

of normalized generators that $u = u_1 \cdots u_k = 1$ so that $\sum dx'_i/x'_i = \sum dx_i/x_i$. This gives

(III.1) *may be defined by the pair* (X,ξ) *where* X *is a local normal crossing variety and* $\xi \in T_X \operatorname{Def}(X)$ *is non-vanishing.*[21]

One small point to notice is that the natural map

$$\text{(III.2)} \qquad \pi^*\Omega^1_\Delta \otimes \mathcal{O}_X \to \pi^*\Omega^1_\Delta(\log 0) \otimes \mathcal{O}_X$$

is *zero*; this is because

$$d\epsilon \otimes 1 \to \epsilon\left(\frac{d\epsilon}{\epsilon}\right) \otimes 1 = \frac{d\epsilon}{\epsilon} \otimes \epsilon = 0.$$

A related point is that there is a natural map of $\mathcal{O}_X$-modules

$$\Omega^1_X \to \Omega^1_{\mathcal{X}_\xi/\Delta_\epsilon}(\log X) \otimes \mathcal{O}_X$$

given on generators by $dx_i \to dx_i$, and then this map has kernel τ^1_X. Since $\Omega^1_{\mathcal{X}_\xi/\Delta_\epsilon}(\log X) \otimes \mathcal{O}_X$ is locally free we know that the above map must have kernel containing τ^1_X; computation shows that equality holds.

We follow the usual notations

$$X^{[1]} = \coprod_i X_i = \widetilde{X} \qquad \text{(normalization of } X\text{)}$$

$$X^{[2]} = \coprod_{i<j} X_i \cap X_j = \widetilde{X}_{\text{sing}} \qquad \text{(normalization of } X_{\text{sing}}\text{)}$$

$$X^{[3]} = \coprod_{i<j<k} X_i \cap X_j \cap X_k$$

$$\vdots$$

with maps

$$a_j : X^{[j]} \to X,$$

and where we set $a_1 = a : \widetilde{X} \to X$.

[21] In the setting of logarithmic geometry, to define a logarithmic structure on a normal crossing variety requires d-stability, and then the variety is log-smooth ([St2], [KN]).

(III.3) **Proposition (Basic Diagram):** *We have*

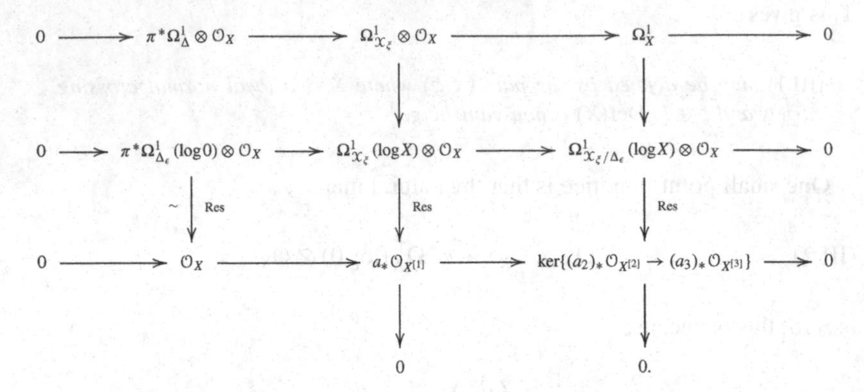

The right-hand map in the bottom row is the truncation of a resolution of $\mathcal{O}_X$ that is given in the comment immediately following the proof of this proposition.

Proof. We begin with the standard diagram of $\mathcal{O}_{\mathcal{X}_\xi}$-modules

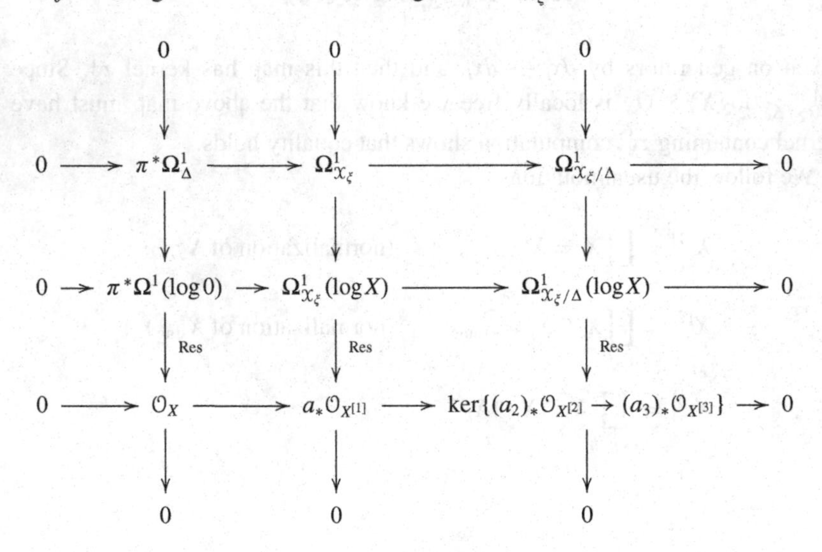

where the $\mathcal{O}_X$-modules on the bottom row are considered as $\mathcal{O}_{\mathcal{X}_\xi}$-modules by the restriction map $\mathcal{O}_{X_\xi} \to \mathcal{O}_X$. When we restrict to X by

$$\mathcal{F} \to \mathcal{F}\,|_X = \mathcal{F} \otimes_{\mathcal{O}_{\mathcal{X}}} \mathcal{O}_X$$

for an $\mathcal{O}_{\mathcal{X}}$-module $\mathcal{F}$, then as noted above we may lose exactness in certain places. Calculations in local coordinates gives the exactness in the basic diagram, where we note that

$$\Omega^1_{\mathcal{X}_\xi/\Delta_\epsilon} \otimes \mathcal{O}_X = \Omega^1_X$$

as the top row reduces to the defining relation $0 \to I_X/I_X^2 \xrightarrow{d} \Omega^1_{\mathfrak{X}_\xi} \otimes \mathcal{O}_X \to \Omega^1_X \to 0$ for Kähler differentials.

To give the flavor of the calculations we consider the simplest non-trivial case of $xy = \epsilon$. Then

- $\Omega^1_{\mathfrak{X}_\xi} \otimes \mathcal{O}_X$ is generated as an $\mathcal{O}_X$-module by $dx, dy, d\epsilon$ with the defining relation $xdy + ydx = d\epsilon$;
- $\Omega^1_{\mathfrak{X}_\xi}(\log X) \otimes \mathcal{O}_X$ is generated by dx/x, dy/y, $d\epsilon/\epsilon$ with the defining relation $dx/x + dy/y = d\epsilon/\epsilon$.

Any $\omega \in \Omega^1_{\mathfrak{X}_\xi} \otimes \mathcal{O}_X$ is of the form $f(x,y)dx + g(x,y)dy$, and using that $\otimes \mathcal{O}_X$ means setting "$xy = 0$" we see that ω may be normalized to be

$$\omega = \big(f_1(x) + f_2(y)\big)dx + \big(g_1(y) + g_2(y)\big)dy, \quad f_1(0) = g_2(0) = 0.$$

Similarly, $\varphi \in \Omega^1_{\mathfrak{X}_\xi}(\log X) \otimes \mathcal{O}_X$ may be normalized to be

$$\varphi = \big(a_1(x) + a_2(y)\big)\frac{dx}{x} + \big(b_1(x) + b_2(y)\big)\frac{dy}{y}, \qquad a_1(0) = b_2(0) = 0.$$

Then

$$\operatorname{Res}\varphi = a_2(y) \oplus b_1(x) \in (a_*)\mathcal{O}_{X^{[1]}}.$$

If $\operatorname{Res}\varphi = 0$, then writing $a_1(x) = x\widetilde{a_1}(x)$ and $b_2(y) = y\widetilde{b_2}(y)$ we have

$$\varphi = \widetilde{a_1}(x)dx + \widetilde{b_2}(y)dy \in \Omega^1_{\mathfrak{X}_\xi} \otimes \mathcal{O}_X.$$

A similar calculation gives the exactness of the right-hand column. For the case of a triple point $xyz = \epsilon$ the residue calculation is more complicated and is similar to (2.10) in [Fr2]. □

We want to make two comments on the basic diagram. The first is

The bottom row in the basic diagram is the truncation of the resolution

$$0 \to \mathcal{O}_X \to (a_1)_*\mathcal{O}_{X^{[1]}} \to (a_2)_*\mathcal{O}_{X^{[2]}} \to (a_3)_*\mathcal{O}_{X^{[3]}} \to \cdots$$

of $\mathcal{O}_X$.

The map $(a_k)_*\mathcal{O}_{X^{[k]}} \to (a_{k+1})_*\mathcal{O}_{X^{[k+1]}}$ is given by

$$u_{i_1\cdots i_k} \text{ on } X_{i_1} \cap \cdots \cap X_{i_k}$$

maps to

$$\sum_j (-1)^{j+1} u_{i_1\cdots\hat{i}_j\cdots i_{k+1}}\big|_{X_{i_1}\cap\cdots\cap X_{i_{k+1}}}.$$

This is standard (cf. [Fr2], [St1] and [Zu]). We note also the resolution

$$0 \to \mathbb{C}_X \to (a_1)_*\mathbb{C}_{X^{[1]}} \to (a_2)_*\mathbb{C}_{X^{[2]}} \to (a_3)_*\mathbb{C}_{X^{[3]}} \to \cdots$$

of the constant sheaf on X.

The second is that we list the main take-aways from the basic diagram:

(III.4)

(i) Given X and $\xi \in \mathbb{E}\mathrm{xt}^1_{\mathcal{O}_X}(\Omega^1_X, \mathcal{O}_X)$ with the property that ξ is non-zero along D, we may by definition construct an extension of $\mathcal{O}_X$-modules

$$0 \to \mathcal{O}_X \to \mathcal{F}_\xi \to \Omega^1_X \to 0.$$

(ii) From [Pa], we may actually construct a space $\mathcal{X}_\xi$ with structure sheaf $\mathcal{O}_{\mathcal{X}_\xi}$ locally isomorphic to $\mathcal{O}_X[\epsilon]$ giving a mapping $\mathcal{X}_\xi \to \Delta_\epsilon = \operatorname{Spec}\mathbb{C}[\epsilon]$ where
- $\mathcal{X}_\xi$ is smooth (this is the assumption that $\xi_D \neq 0$);
- $\mathcal{F}_\xi \cong \Omega^1_{\mathcal{X}_\xi} \otimes \mathcal{O}_X$ as $\mathcal{O}_X$-modules; this is the top row in the basic diagram.

(iii) We may then proceed, using $\Omega^1_{\mathcal{X}_\xi}$ as an $\mathcal{O}_{\mathcal{X}_\xi}$-module, to construct the remainder of the basic diagram; the inclusion map $\mathcal{O}_X \to \Omega^1_{\mathcal{X}_\xi}(\log X) \otimes \mathcal{O}_X$ is given by $1 \to d\epsilon/\epsilon$, and then the quotient defines the $\mathcal{O}_X$-module $\Omega^1_{\mathcal{X}_\xi/\Delta_\epsilon}(\log X) \otimes \mathcal{O}_X$.

(iv) From this we may, in the standard way, proceed to construct the complex $\left(\Omega^\bullet_{\mathcal{X}_\xi/\Delta_\epsilon}(\log X) \otimes \mathcal{O}_X, d\right)$; as will be noted below, the hypercohomology of this complex will give the complex vector space V_ξ and Hodge filtration $F^\bullet_\xi$ for the limiting mixed Hodge structure.

(v) The previous steps are either explicit or implicit in [Fr2]; the final steps to define the weight filtration and $\mathbb{Q}$-structure may then be carried out by the methods in [St2].

We will elaborate more on this at the end of the section.

Turning now to the case where X is locally a product of normal crossing varieties as given by (I.2), we may extend the discussion above with one significant change. Namely, in the local situation instead of a single smoothing deformation $\xi \in T^0_X \operatorname{Def}(X)$ we now need to be given a k-tuple

$$\boldsymbol{\xi} = (\xi_1, \dots, \xi_k)$$

where ξ_i smooths the factor X_i in X. Then for $\lambda = (\lambda_1, \dots, \lambda_k)$ with all $\lambda_i \neq 0$

$$\boldsymbol{\xi}_\lambda =: \sum \lambda_i \xi_i \in T_X \operatorname{Def}(X)$$

is a smoothing deformation of X. This is all local.

Globally we need to be given an ℓ-tuple $\boldsymbol{\xi} \in \overset{\ell}{\oplus} T_X \operatorname{Def}(X)$ such that locally around each $x \in X$ there is a k sub-tuple of $\boldsymbol{\xi}$ that satisfies the above condition.

We note again the difference when the number of local factors $k \geqq 2$; if $\boldsymbol{\xi}_\lambda$ is tangent to a family $\mathcal{X}_{\Delta_\lambda} \to \Delta_\lambda$, then

(III.5) *the total space $\mathcal{X}_{\Delta_\lambda}$ is singular.*

These singularities are of a standard form and may be resolved to give a standard family

$$\widetilde{\mathcal{X}}_{\Delta_\lambda} \to \widetilde{\Delta}_\lambda$$

where $\widetilde{\mathcal{X}}_{\Delta_\lambda}$ is smooth.

Setting $\Delta_{\epsilon_j} = \operatorname{Spec}\mathbb{C}[\epsilon_j]$ and $\Delta_\epsilon = \prod_j \Delta_{\epsilon_j}$, using the projection $X_1 \times \cdots \times X_k \xrightarrow{\pi_j} X_j$, we define

$$\text{(III.6)} \qquad \Omega^1_{\mathcal{X}_\xi/\Delta_\epsilon}(\log X) \otimes \mathcal{O}_X = \oplus \pi_j^* \Omega^1_{\mathcal{X}_j/\Delta_{\epsilon_j}}(\log X_j) \otimes \mathcal{O}_X.$$

In coordinates, for the case where X is given by

$$\begin{cases} xy = t_1 \\ uv = t_2 \end{cases}$$

so that I_X is generated by xy and uv, $\Omega^1_{\mathcal{X}_\xi/\Delta_\epsilon}(\log X) \otimes \mathcal{O}_X$ is generated as an $\mathcal{O}_X$-module by dx/x, dy/y, du/u, dv/v with the relations $dx/x + dy/y = 0$, $du/u + dv/v = 0$. This coordinate description extends in the evident way when X is given by (I.2).

Finally, we will relate this construction to that given in [Fu1], [Fu2]. We have

$$\text{(III.7)} \qquad \mathcal{X} \xrightarrow{\pi} S$$

where locally in $\mathbb{C}^{n+k}$ with coordinates $(x_1, \ldots x_n, t_1, \cdots, t_k)$ and using the notation (I.2), $\mathcal{X}$ is given by

$$\text{(III.8)} \qquad \begin{cases} x_{I_1} = t_1 \\ \quad\vdots \\ x_{I_k} = t_k \end{cases}$$

and π is the projection $(\mathbf{x}, \mathbf{t}) \to \mathbf{t}$. There are then normal crossing divisors $\mathcal{Y} \subset \mathcal{X}$ and $T \subset S$ such that (III.7) is a map

$$(\mathcal{X}, \mathcal{Y}) \to (S, T)$$

as defined in [Fu1], [Fu2]. If $\dim X = n$ and $\dim S = \ell$, then $\dim \mathcal{X} = n + \ell$. Locally S is embedded in $\mathbb{C}^k \times \mathbb{C}^{\ell-k}$ where the first k coordinates are the t_i above and the remaining $\ell - k$ coordinates are parameters. We note that π^{-1} (set of coordinate hyperplanes in $\mathbb{C}^k$) is a singular subvariety of $\mathcal{X}$. Globally, we will have divisors $D_1, \ldots, D_\ell$ on $\mathcal{X}$ such that locally $D_1, \ldots D_k$ are the inverse image under π of the coordinate hyperplanes $t_i = 0$ and $\mathcal{Y} = D_1 + \cdots + D_\ell$ is a reduced normal crossing divisor in $\mathcal{X}$ with $X = D_1 \cap \cdots \cap D_\ell$.

Discussion of the proofs of Theorems I and I′[22]

For the case when X is a local normal crossing variety, using (III.4) the essentials of the proof are in [Fr2] and [St2]. The sheaves

$$\wedge^\bullet_\xi = \wedge^\bullet \Omega^1_{\mathcal{X}_\xi/\Delta_\epsilon}(\log X) \otimes \mathcal{O}_X$$

form a filtered complex in the evident way, and

$$\begin{cases} V_\xi = \mathbb{H}^m(\wedge^\bullet_\xi) \\ F^\bullet V_\xi = F^\bullet \mathbb{H}^m(\wedge^\bullet_\xi) \end{cases}$$

defines the vector space and Hodge filtration for the limiting mixed Hodge structure. As usual, dating to [St1] (cf. also [Zu]), the construction of the monodromy weight filtration and $\mathbb{Q}$-structure are more subtle. These may be carried out by an adaptation of the methods in §5 in [St2].

More specifically, in [St1] and [Zu] associated to a standard family $\mathcal{X} \to \Delta$ several cohomological mixed Hodge complexes are constructed. One of these, denoted there by $A^\bullet$ (recalled in the proof of Theorem (VI) in Section IV below) leads to the limit mixed Hodge structure. Another of these, denoted by $L^\bullet$ in loc. cit., leads to the mixed Hodge structures on $H^*(X\backslash\mathcal{X})$ and on $H^*(\mathcal{X},\mathcal{X}\backslash X)$. In [St2], in the setting of log geometry which in his Section 5 corresponds to our (X,ξ), the analogue of $L^\bullet$, denoted there by $K^\bullet$, is constructed. Analysis of the construction leads to a cohomological mixed complex in our (X,ξ) setting that gives a limiting mixed Hodge structure on V_ξ.

For the general case where the central fiber X in a global map (III.7) is given locally by (III.8), in [Fu1], [Fu2] the methods of [St1] are extended to show that for $s \in S$ and $X_s = \pi^{-1}(s)$ the hypercohomology of the complexes $\Omega^\bullet_{\mathcal{X}/S}(\log \mathcal{Y}) \otimes \mathcal{O}_{X_s}$ give mixed Hodge structures. The adaptation of the calculations there extending the methods in [St2] to the several variable log-geometry setting that corresponds to our situation will then give the result. As we have no substantive content to add to what is implicit in [Fu1], [Fu2] and [St2] here we will not write out the details, but rather defer them to a later work.

An outstanding issue, as noted in [Fu1], [Fu2], for a family (III.7), is in what way the mixed Hodge structure constructed in [Fu1], [Fu2] using $\Omega^\bullet_{\mathcal{X}/S}(\log X) \otimes \mathcal{O}_X$ relates to the limiting mixed Hodge structure given along a disc Δ_λ in $S = \Delta_1 \times \cdots \times \Delta_\ell$ in [CKS1]. One main point may be (III.5). In case X is a local normal crossing variety, we have noted that $\mathcal{X}_{\Delta_\lambda}$ is smooth and $\mathcal{X}_{\Delta_\lambda} \to \Delta_\lambda$ is a standard family, so the result that the limiting mixed Hodge structures are the same is true in this case. Another outstanding matter is the construction of the monodromy logarithms N_i from the dt_i/t_i in the complexes

[22] In a work in progress we intend to provide details for this argument with emphasis on the local structure and how this relates to the results in [CK] and [CKS1].

constructed in [Fu1], [Fu2], and then to show that these give the structure as in [CKS1]. This also will be taken up in a later work.

Another issue, one that arises already when X is a normal crossing variety whose singular locus $D = \cup D_\alpha$ has connected components D_α, is this: In the exact sequence (II.1) when the mapping

$$\oplus H^0(\mathcal{O}_{D_\alpha}) \xrightarrow{\delta} H^2(\mathrm{Ext}^0_{\mathcal{O}_X}(\Omega^1_X, \mathcal{O}_X))$$

is non-zero, the Kuranishi space may be unobstructed but the D_α cannot be individually smoothed.[23] Suppose for example that there are three components so that projectively

$$\mathbb{P}(\oplus H^0(\mathcal{O}_{D_\alpha})) = \mathbb{P}^2$$

pictured as

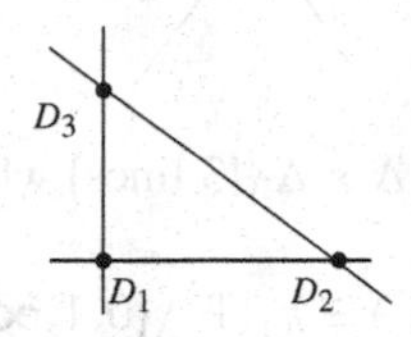

where the vertices correspond to the $H^0(\mathcal{O}_{D_\alpha})$. If $\dim(\ker\delta) = 2$, there are the following possibilities for the dotted line $L = \mathbb{P}(\ker\delta)$:

(i)

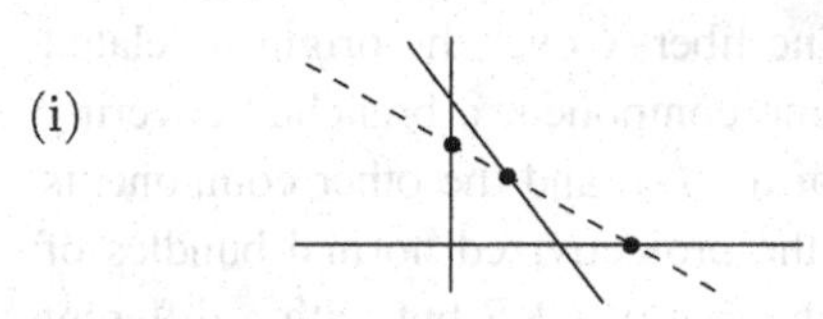

(ii)

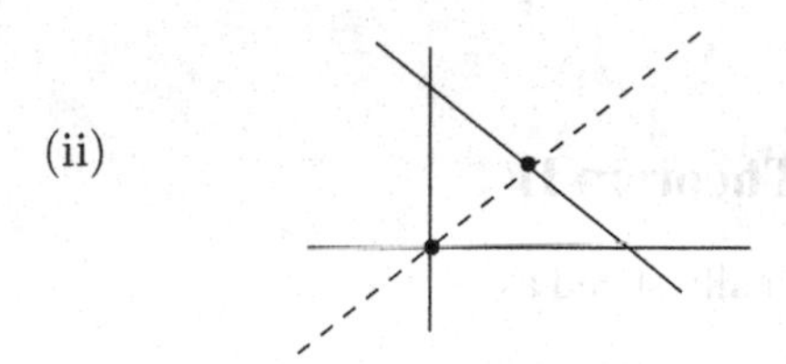

(iii)

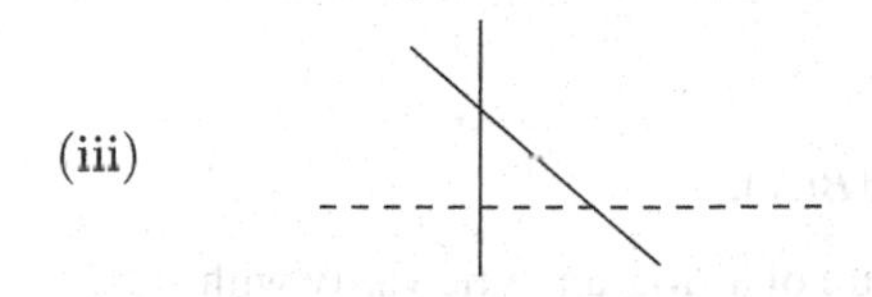

[23] In the logarithmic deformation theoretic context, there are non-zero obstructions in the logarithmic analogue of $T_X \mathrm{Def}(X)$.

Here, (iii) does not occur since we are assuming that X may be smoothed. For (ii), assuming that $T_X\,\mathrm{Def}(X)$ is unobstructed we have a 2-parameter family $\mathcal{X} \to \Delta \times \Delta$ where along one axis D_1 is smoothed while D_2 and D_3 deform equisingularly. Along the other axis a similar thing happens with the roles of D_1 and D_2, D_3 interchanged.

In case (i) we have a 2-parameter family with three axes along each of which one pair from D_1, D_2, D_3 deforms equisingularly while the remaining component of D is smoothed. Thus the picture of the tangent space to the 2-parameter family $\mathcal{X} \to S$ is

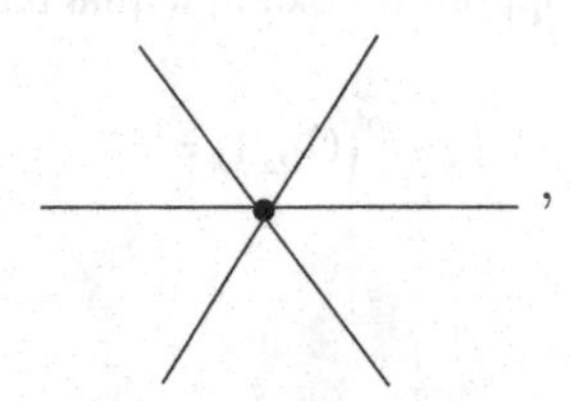

and for the family $\mathcal{X}^* \to S^* \cong \Delta \times \Delta \backslash \{3 \text{ lines}\}$ where the fibers are smooth we have

$$\pi_1(S^*) \cong \pi_1(\mathbb{P}^1 \backslash \{0, 1, \infty\}).$$

Thus to arrive at a CKS situation we have to blow up S at the origin and resolve singularities to arrive at a standard situation $\widetilde{X} \to \widetilde{S}$ where locally $\widetilde{S}$ is a $\Delta \times \Delta$ with singular fibers over the two axes. The fiber $\widetilde{X}$ over the origin is related to X in a standard way; it will have as one component a branched covering of the desingularization of X along two of the D_α, and the other components are easily described (if $X_3 = \emptyset$ they are the projectivized normal bundles of the D_α in $\mathcal{X}$). We are now back in the situation of CKS but with a different X. As in [KN] for interesting examples this complexity does not arise, and for theoretical purposes we can at least begin by assuming that $\delta = 0$ in (II.1).

IV Proof of Theorem II

We first will consider the question, informally stated as

What do we mean by T(LMHS)?

Here, "LMHS" is the set of limiting mixed Hodge structures with monodromy N.

Setting

$$D_N = D \cup B(N),$$

in [KU] there is defined on D_N the structure of a "log-analytic varity with slits." In particular, the tangent space $T_{[F^\bullet]}D_N$ to the underlying analytic variety at a point $[V, W_\bullet(N), F^\bullet]$ is defined, where the brackets denote the equivalence class of nilpotent orbits under the equivalence relation $F^\bullet \sim \exp(zN)\cdot F^\bullet$. This information may be refined if we do not pass to equivalence classes. Thus we define

$$\widetilde{D}_N = D \cup \widetilde{B}(N)$$

and seek to define $T_{F^\bullet}\widetilde{D}_N$. For the subspace $T^\bullet_F\widetilde{B}(N)$ of $T^\bullet_F\widetilde{D}_N$, recalling that $\widetilde{D}_N \subset \check{D}$ we set

$$\text{(IV.1)} \qquad T_{F^\bullet}\widetilde{D}_N = \{\tau \in T_{F^\bullet}\check{D} : \tau(F^p) \subset F^{p-1}\} = T_{F^\bullet}\check{D}.^{24}$$

We next consider the question

What is the algebro-geometric analogue of (IV.1)*?*

This means: What algebro-geometric object maps to (IV.1), extending what is given for a smooth X by (I.1)?

For this we recall that associated to a pair (X,ξ), where X is locally a product of normal crossing varieties and $\xi \in T^0_X\operatorname{Def}(X)$, is a limiting mixed Hodge structure whose underlying vector space is $\mathbb{H}^m(\Omega^\bullet_{\mathcal{X}_\xi/\Delta_\epsilon}(\log X)\otimes \mathcal{O}_X)$. With the identification $T_X\operatorname{Def}(X) = \mathbb{E}\mathrm{xt}^1_{\mathcal{O}_X}(\Omega^1_X,\mathcal{O}_X)$ in mind, we *define*

$$\text{(IV.2)} \qquad T_{(X,\xi)}\operatorname{Def}(X,\xi) = \mathbb{E}\mathrm{xt}^1_{\mathcal{O}_X}\left(\Omega^1_{\mathcal{X}_\xi/\Delta_\epsilon}(\log X)\otimes\mathcal{O}_X,\mathcal{O}_X\right)$$

where $\operatorname{Def}(X,\xi)$ is the set of deformations of the pair (X,ξ).[25] We shall not attempt here to give a precise definition of $\operatorname{Def}(X,\xi)$, but rather shall simply take (IV.2) as the definition of its tangent space. As partial justification, we observe that with this definition the obvious map

$$\operatorname{Def}(X,\xi) \to \operatorname{Def}(X),$$

together with the map $\Omega^1_X \to \Omega^1_{\mathcal{X}_\xi/\Delta_\epsilon}(\log X)$, give

$$\text{(IV.3)} \qquad \begin{array}{ccc} T_{(X,\xi)}\operatorname{Def}(X,\xi) & \longrightarrow & T_X\operatorname{Def}(X) \\ \| & & \| \\ \mathbb{E}\mathrm{xt}^1_{\mathcal{O}_X}\left(\Omega^1_{\mathcal{X}_\xi/\Delta_\epsilon}(\log X)\otimes\mathcal{O}_X,\mathcal{O}_X\right) & \longrightarrow & \mathbb{E}\mathrm{xt}^1_{\mathcal{O}_X}(\Omega^1_X,\mathcal{O}_X). \end{array}$$

Turning to the definition of the maps in Theorem II, from the middle row in the basic diagram (III.3) one may in the usual way infer the exact sequence of complexes

$$0 \to \Omega^{\bullet-1}_{\mathcal{X}_\xi/\Delta_\epsilon}(\log X)\otimes\mathcal{O}_X \to \Omega^\bullet_{\mathcal{X}_\xi}(\log X)\otimes\mathcal{O}_X \to \Omega^\bullet_{\mathcal{X}_\xi/\Delta_\epsilon}(\log X)\otimes\mathcal{O}_X \to 0.$$

[24] Essentially we are interpreting the additional infinitesimal information that is present if we consider D_N as a log-analytic variety.

[25] As was noted in the introduction, the right-hand side of (IV.2) appears naturally in logarithmic deformation theory.

The connecting homomorphism in the long exact hypercohomology sequence induces

$$\mathbb{H}^m\left(\Omega^{\bullet}_{\mathcal{X}_\xi/\Delta_\epsilon}(\log X)\otimes\mathcal{O}_X\right)\xrightarrow{\nabla_\xi}\mathbb{H}^{m+1}\left(\Omega^{\bullet-1}_{\mathcal{X}_\xi/\Delta_\epsilon}(\log X)\otimes\mathcal{O}_X\right)$$

which satisfies

$$\nabla_\xi F^p_\xi \subset F^{p-1}_\xi .$$

Then from (I.5) the element $\xi^{(1)}\in T_{(X,\xi)}\operatorname{Def}(X,\xi)$ gives an extension class in the above exact sequence of complexes, and using the identification

$$\mathbb{H}^{m+1}\left(\Omega^{\bullet-1}_{\mathcal{X}_\xi/\Delta_\epsilon}(\log X)\otimes\mathcal{O}_X\right)=\mathbb{H}^m\left(\Omega^{\bullet}_{\mathcal{X}_\xi/\Delta_\epsilon}(\log X)\otimes\mathcal{O}_X\right)$$

we obtain the map in the statement of Theorem II. The fact that we map to $\operatorname{End}_{\mathrm{LMHS}}$ is a consequence of the naturality of the construction of the limiting mixed Hodge structure.

The geometric picture to keep in mind is this: The kernel of the map (II.1)

$$\mathbb{E}\mathrm{xt}^1_{\mathcal{O}_X}(\Omega^1_X,\mathcal{O}_X)\to H^0\left(\mathcal{E}\mathrm{xt}^1_{\mathcal{O}_X}(\Omega^1_X,\mathcal{O}_X)\right)$$

represents the tangents to the equisingular deformations of X.[26] Modulo this kernel, the image of the above map reflects how the singularities are deforming. Given a smoothing deformation ξ, we may think of $\xi^{(1)}$ as giving us the infinitesimal change in this picture.[27] We will now illustrate this by example where it will be quite clear how the map in the statement of Theorem II gives information beyond that in the differential

$$T_{s_0}S\to T_{[F^\bullet]}D_N.$$

Here we are imagining a family $\mathcal{X}\to S$ where $S=\Delta^\ell$ and where the fibers are smooth over $S^*=\Delta^{*\ell}$ with commuting monodromy logarithm transformations $N_1,\dots,N_\ell$ around the axes. The corresponding nilpotent orbit is $\exp(z_1N_1+\cdots z_\ell N_\ell)\cdot F^\bullet$.

Example: This will be a simpler version of the example from the beginning of Section VI below, and we will use the notations from there. Then $F^\bullet$ is a single F given by the span of the columns in the matrices below:

$$F\longleftrightarrow\begin{pmatrix}1&&\\&1&\\&&1\\a_{11}&a_{12}&b_1\\a_{21}&a_{22}&b_2\\b_1&b_2&c\end{pmatrix},\qquad a_{12}=a_{21}$$

[26] Recall that we are assuming that under any non-smoothing deformation X' of X, including an equisingular one, the deformed X' remains smoothable; the condition for this is in [Fr2].
[27] Of course, there is more information than this in ξ.

$$[F] \longleftrightarrow \begin{pmatrix} 1 & & \\ & 1 & \\ & & 1 \\ 0 & a_{12} & b_1 \\ a_{21} & 0 & b_2 \\ b_1 & b_2 & c \end{pmatrix}$$

where the notation $\longleftrightarrow$ means "corresponds to." For $[F]$ we have normalized the point on the several variable nilpotent orbit by $a_{11} = a_{22} = 0$:

$$T_F\check{D} \longleftrightarrow \begin{pmatrix} da_{11} & da_{12} & db_1 \\ da_{21} & da_{22} & db_2 \\ db_1 & db_2 & dc \end{pmatrix}, \tag{IV.4}$$

$$T_F D_N \longleftrightarrow \begin{pmatrix} 0 & da_{12} & db_1 \\ da_{21} & 0 & db_2 \\ db_1 & db_2 & dc \end{pmatrix}.[28] \tag{IV.5}$$

Geometrically, the second contains the information in $T_X\mathcal{C} \subset T_X\partial\mathcal{M}_3$, while the first contains this information plus the information in the normal space to $\mathcal{C}$ in $\overline{\mathcal{M}}_3$; i.e., the refined direction of approach to X in the boundary of $\mathcal{M}_3$.

Still referring to the next section for the notations, the N-strings associated to the limiting mixed Hodge structure in this example may be written as

$$H^0(-1) \to H^0$$
$$H^1.$$

The extension data in $\mathrm{Ext}^1_{\mathrm{MHS}}(\mathrm{Gr}_1, G_0) = \mathrm{Ext}^1_{\mathrm{MHS}}\big(H^1(\tilde{X}), H^0(D)\big)$ corresponds to b_1, b_2, while that for $\mathrm{Ext}^1_{\mathrm{MHS}}\big(H^0(D)(-1), H^0(D)\big)$ corresponds to the 2×2 symmetric matrix (a_{ij}). In this case only the off-diagonal terms are invariant under $F^\bullet \to \exp(z_1N_1 + z_2N_2)\cdot F^\bullet$, while the diagonal terms require the choice of ξ.

From a cohomological perspective, the F in the limiting mixed Hodge structure is

$$H^0\left(\Omega^1_{\tilde{X}}(\log(p+q))\right)$$

and the matrix dF in (IV.4) is in

$$\mathrm{Hom}_s(F, \mathbb{C}^4/F) \cong \mathrm{Hom}_s\left(H^0\big(\Omega^1_{\tilde{X}}(\log(p+q)), H^1(\mathcal{O}_X)\big)\right)^{29}$$

where Hom_s are the symmetric maps. Under the inclusion

$$H^0(\Omega^1_{\tilde{X}}) \hookrightarrow H^0\left(\Omega^1_{\tilde{X}}(\log(p+q))\right)$$

[28] Here we recall that D_N is the set of equivalence classes of nilpotent orbit attached to D.

[29] Here we are identifying $H^1(\mathcal{O}_X)$ with V_ξ/F where $(V_\xi, W_\bullet(N), F)$ is the limiting mixed Hodge structure with $N = N_1 + N_2$.

the matrix (IV.5) contains a part in $\operatorname{Hom}\left(H^0(\Omega^1_{\widetilde{X}}), H^1(\mathcal{O}_X)\right)$. The term dc is in

$$\operatorname{Hom}\left(H^0\left(\Omega^1_{\widetilde{X}}\right), H^1(\mathcal{O}_{\widetilde{X}})\right)$$

and (db_1, db_2) belongs to

$$\operatorname{Hom}\left(H^0\left(\Omega^1_{\widetilde{X}}\right), H^1(\mathcal{O}_X)/H^1\left(\mathcal{O}_{\widetilde{X}}\right)\right).$$

An extreme example of the extra information is given by the genus 2 curve degenerations

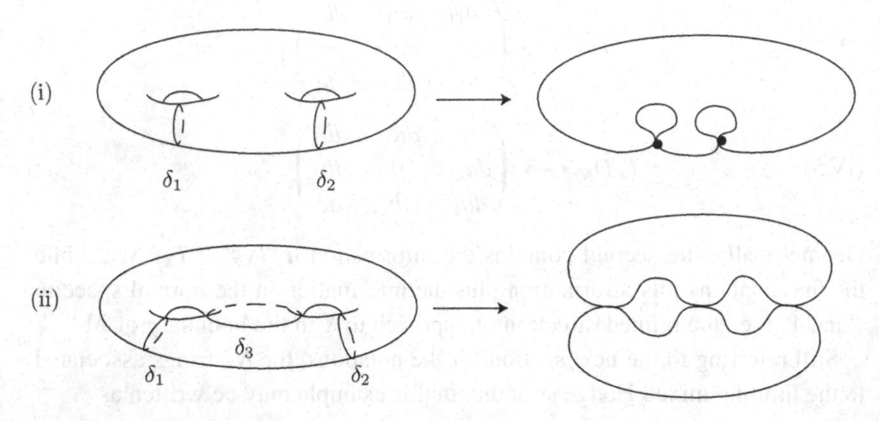

In each case the polarized limiting mixed Hodge structure is

$$\overset{2}{\oplus}\mathbb{Q}(-1) \to \overset{2}{\oplus}\mathbb{Q}$$

and the $\operatorname{Ext}^1_{\mathrm{PLMS}}(\bullet, \bullet)$ is given by a 2×2 symmetric case. For $T_{F^\bullet}\widetilde{D}_N$ and $T_{[F^\bullet]}D_N$ they are

	(i)	(ii)
$T_{F^\bullet}\check{D}_N$	$\begin{pmatrix} a & b \\ b & a \end{pmatrix}$	$\begin{pmatrix} a & b \\ b & a \end{pmatrix}$
$T_{[F^\bullet]}D_N$	$\begin{pmatrix} 0 & b \\ b & 0 \end{pmatrix}$	$\begin{pmatrix} 0 & 0 \\ 0 & 0 \end{pmatrix}$

This illustrates the additional information contained in considering the map $T_{(X,\xi)}\operatorname{Def}(X,\xi) \to T_{F^\bullet}\widetilde{D}_N$.

The above examples are of course special. However, the regularization of logarithmic integrals phenomenon they illustrate are fairly general. For instance, in [GGR] the generic degenerations of Hodge structures of odd weight $n = 2m+1$ are given, for $1 \leqq k \leqq n$, by a specialization

$$X_t \to X_0$$

where locally in $\mathbb{C}^{2m+2}$ X_0 is given by

$$x_1x_2 + \cdots + x_{2k-1}x_{2k} = 0$$

and thus has a double locus of codimension k. For $k = 1$ we have that $X_0 = X$ has a codimension one double $2m$-fold $X_{\text{sing}} = X_1$. For $k = m$, X_0 has an ordinary isolated quadratic singularity. For $k \geqq 2$ we have to blow up X_0 to achieve a standard family. The limiting mixed Hodge structures are

$$k = m \quad \begin{cases} H^0(-m-1) & \longrightarrow \quad H^0(-m) \qquad \dim H^0 = 1 \\ & \qquad \vdots \\ & \qquad H^n_m \end{cases}$$

$$k = 1 \quad \begin{cases} H^{2m-1}(-1) & \longrightarrow \quad H^{2m} \\ & \qquad \vdots \\ & \qquad H^n_1 \end{cases}$$

For H^n_m the Hodge numbers $h^{p,q}_k$ are the same as the $h^{p,q}$ for the original polarized Hodge structure on $H^n(X_t)$, except that

$$h^{m+k,m-k+1}_k = h^{m+k,m-k+1} - 1.$$

Geometrically we have a class ω_t in $H^0(\Omega^n_{X_t})$ that acquires a pole of order k along X_{sing}, and by a residue-type construction we end up with a class in $H^{n-m}\left(\Omega^{m-k+1}_{X[m]}\right)$. The above analysis of regularizing an integral

$$\lim_{t\to 0} \int_{\gamma_t} \omega_t$$

then will carry over. The details of this will be carried out in a future work.

Finally we would like to give a general cohomological description of the extra information in the map in Theorem II. We will do this in case X is a nodal curve; this description will extend to the general case when X has only an ordinary double locus $D = X_{\text{sing}}$. With this assumption the right-hand column in the basic diagram (III.3) gives a map

$$\Omega^1_{\mathcal{X}_\xi/\Delta_\epsilon}(\log X) \otimes \mathcal{O}_X \to \mathcal{O}_D,$$

which induces a map

$$\text{(IV.6)} \qquad \mathbb{E}\mathrm{xt}^1_{\mathcal{O}_X}(\mathcal{O}_D, \mathcal{O}_X) \to \mathbb{E}\mathrm{xt}^1_{\mathcal{O}_X}\left(\Omega^1_{\mathcal{X}_\xi/\Delta_\epsilon}(\log X) \otimes \mathcal{O}_X, \mathcal{O}_X\right).$$

The image of this map represents the "extra information" contribution to the map in Theorem II. The local result we need to describe this for X a germ given by $xy = 0$ is

(IV.7) $\mathrm{Ext}^1_{\mathcal{O}_X}(\mathcal{O}_D, \mathcal{O}_X) \cong \mathcal{O}_D$ *and is generated by the extension*

$$0 \to \mathcal{O}_X \to (a_1)_* \mathcal{O}_{X^{[1]}} \to \mathcal{O}_D \to 0.$$

In the case under consideration,

$$\mathbb{E}\mathrm{xt}^1_{\mathcal{O}_X}(\mathcal{O}_D, \mathcal{O}_X) \cong H^0\left(\mathrm{Ext}^1_{\mathcal{O}_X}(\mathcal{O}_D, \mathcal{O}_X)\right)$$

and there are the number of connected components of D additional parameters picked up in the additional information.

A cohomological formulation that identifies the N in a limiting mixed Hodge structure is this. Recall the bottom two rows in the basic diagram (III.3), where if we use the notation $\mathcal{S} = \ker\{(a_2)_* \mathcal{O}_{X^{[2]}} \to (a_3)_* \mathcal{O}_{X^{[3]}}\}$ and identify $\pi^* \Omega^1_{\Delta_\epsilon}(\log 0) \otimes \mathcal{O}_X$ with $\mathcal{O}_X$ we have

(IV.8)

$$\begin{array}{ccccccccc} 0 & \longrightarrow & \mathcal{O}_X & \longrightarrow & \Omega^1_{\mathcal{O}_X}(\log X) \otimes \mathcal{O}_X & \longrightarrow & \Omega^1_{\mathcal{X}/\Delta_\epsilon}(\log X) \otimes \mathcal{O}_X & \longrightarrow & 0 \\ & & \| & & \downarrow{\scriptstyle \mathrm{Res}} & & \downarrow{\scriptstyle \mathrm{Res}} & & \\ 0 & \longrightarrow & \mathcal{O}_X & \longrightarrow & (a)_* \mathcal{O}_{X^{[1]}} & \longrightarrow & \mathcal{S} & \longrightarrow & 0. \end{array}$$

This gives

$$\mathbb{E}\mathrm{xt}^1_{\mathcal{O}_X}(\mathcal{S}, \mathcal{O}_X) \xrightarrow{\mathrm{Res}^*} \mathbb{E}\mathrm{xt}^1_{\mathcal{O}_X}\left(\Omega^1_{\mathcal{X}/\Delta_\epsilon}(\log X) \otimes \mathcal{O}_X, \mathcal{O}_X\right) \to F^{-1}\mathrm{End}_{\mathrm{LMHS}}.$$

Then by interpreting the construction in [St1] we find that

> *the image of the extension class in the bottom row of* (IV.8) *is the monodromy logarithm* N.

Finally, we would like to point out the paper [Ca-Fe] in which the notion of an infinitesimal variation of Hodge structure at infinity is defined. Their definition pertains to equivalence classes of limiting mixed Hodge structures for several variable nilpotent orbits as in [CKS1]. In the above example the definition in [Ca-Fe] would record the data

$$\{N_1, N_2 : da_{12}, db_1, db_2, dc\}.$$

Roughly speaking, this data corresponds to $TB(N)$ and to the normal space to $B(N)$ in D_N, while that in Theorem II may be thought of as having the information in some sort of blow up of the normal space to $B(N)$ in D_N.

V Proof of Theorem III

The proof of Theorem III will be given following several preliminary discussions on the following topics:

- nilpotent orbits and the reduced limit period mapping;
- monodromy cone structure associated to a normal crossing variety;
- the differential of the reduced limit period mapping.

Nilpotent orbits and the reduced limit period mapping

We begin by recalling some definitions and results from [CKS1], [KP1], [KP2], [GGK], [GG] and [GGR], the last two of whose notations we shall generally follow. We let

- $\widetilde{B}(N) =$ set of nilpotent orbits $(F^\bullet, N)$.

Here, $D = G_{\mathbb{R}}/H$ is a Mumford-Tate domain embedded as an open $G_{\mathbb{R}}$-orbit in its compact dual $\check{D} = G_{\mathbb{C}}/P$. We assume that $G \subset \mathrm{Aut}(V, Q)$. The Mumford-Tate domain structure on D gives a realization of $\check{D}$ as a set of filtrations $F^\bullet = \{F^m \subset F^{m-1} \subset \cdots \subset F^0 = V_{\mathbb{C}}\}$ on the complexification of a $\mathbb{Q}$-vector space V. The monodromy logarithm $N \in \mathfrak{g}^{\mathrm{nilp}} \subset \mathrm{End}(V)$ is a nilpotent endomorphism of V that gives rise to the monodromy weight filtration, which we center at zero,

$$W_{-k}(N) \subset \cdots \subset W_0(N) \subset \cdots \subset W_k(N) = V, \qquad k \leqq m$$

where $N^{k+1} = 0$, $N^k \neq 0$ $(k \leqq m)$. The conditions to be a nilpotent orbit are

i $NF^p \subset F^{p-1}$;
ii $\exp(zN) \cdot F^\bullet \in D$ for $\mathrm{Im}\, z \gg 0$.

It is known and of central importance that ([CKS1])

(V.1) $(F^\bullet, N)$ *is a nilpotent orbit*

$\iff (V, W_\bullet(N), F^\bullet)$ *is a polarized limiting mixed Hodge structure* .

Here we recall that a limiting mixed Hodge structure $(V, W_\bullet(N), F^\bullet)$ is given by $F^\bullet$ and N where $W_\bullet(N)$ is the monodromy weight filtration and where $F^\bullet$ reduces on $\mathrm{Gr}_k^{W_\bullet(N)}$ a pure Hodge structure of weight k. All of our limiting mixed Hodge structures will be *polarized* by a $Q : V \otimes V \to \mathbb{Q}$ (cf. [Sc] and [CKS1]).

Two nilpotent orbits $(N, F^\bullet)$ and $(N, F'^\bullet)$ are said to be *equivalent* if

$$F'^\bullet = \exp(zN) \cdot F^\bullet$$

for some $z \in \mathbb{C}$; i.e., if they lie in the same $\exp(\mathbb{C}N)$ orbit in $\check{D}$. We let

- $B(N) = \exp(\mathbb{C}N)\backslash\widetilde{B}(N) =$ set of nilpotent orbits modulo equivalence.

Assuming that $N \neq 0$ there is a *reduced limit period mapping* (called a naïve limit in [KP1])

$$\Phi_\infty : B(N) \to \partial D, \tag{V.2}$$

whose image lies in a $G_{\mathbb{R}}$-orbit. The definition is

$$\Phi_\infty(F^\bullet,N) = \lim_{z\to\infty} \exp(zN)\cdot F^\bullet =: F^\bullet_\infty.$$

If we think of $\check{D}$ as embedded in a product of projective spaces via the Plücker embeddings of the individual subspaces $F^p \subset V_{\mathbb{C}}$, then since N is nilpotent the Plücker coordinates of $\exp(zN)\cdot F^p$ are polynomials in z and thus have a well-defined limit at $z=\infty$. In effect $\Phi_\infty(F^\bullet,N)$ picks out the highest powers of z in the Plücker coordinates of $\exp(zN)F^p$. An elementary general fact is that for any nilpotent N the vector field on $\check{D}$ induced by the action of the 1-parameter group $\exp(zN)$ vanishes to 2nd order at the limit point $F^\bullet_\infty$, so that the reduced limit period mapping is well defined on the quotient space $B(N)$ of $\widetilde{B}(N)$.

One of the important features of the reduced limit period mapping is

(V.3) *The mapping* (V.2) *factors*

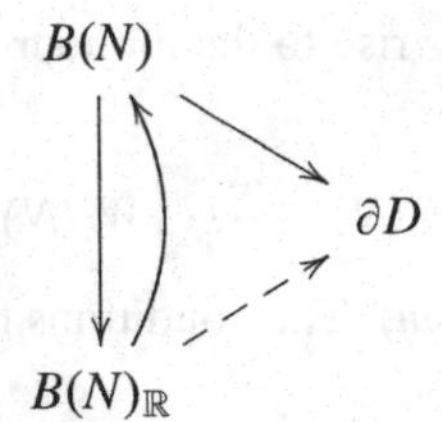

through the set $B(N)_{\mathbb{R}}$ of equivalence classes of $\mathbb{R}$-split limiting mixed Hodge structures.

- Associated to a mixed Hodge structure $(V,W_\bullet,F^\bullet)$ there is the canonical *Deligne bigrading*

(V.4) $$V_{\mathbb{C}} = \oplus I^{p,q}$$

where

$$\begin{cases} F^p = \bigoplus_{r\geqq p} I^{r,\bullet} \\ W_k = \bigoplus_{p+q\leqq k} I^{p,q} \\ I^{p,q} \equiv \overline{I}^{q,p} \ \text{modulo}\ W_{p+q-2}; \end{cases}$$

- The mixed Hodge structure is $\mathbb{R}$-*split* in case

$$I^{p,q} = \overline{I^{q,p}};$$

canonically associated to a mixed Hodge structure $(V,W_\bullet,F^\bullet)$ is an $\mathbb{R}$-split mixed Hodge structure $(V,W_\bullet,\widetilde{F}^\bullet)$;

- If $(V,W_\bullet(N),F^\bullet)$ is a limiting mixed Hodge structure, then so is the $\mathbb{R}$-split mixed Hodge structure $(V,W_\bullet(N),\widetilde{F}^\bullet)$, and conversely.

It follows from this last propery that we have the factorization (V.3). For the time being we will assume that

(V.5) $(V, W_\bullet(N), F^\bullet)$ *is* $\mathbb{R}$*-split.*

In this case the filtration $F^\bullet_\infty$ is related to $F^\bullet$ by

$$F^p_\infty = \bigoplus_{q \leqq m-p} I^{\bullet,q},$$

where m is the weight of the Hodge structure under consideration.

- If $(V, W_\bullet, F^\bullet)$ is a mixed Hodge structure, then the inclusion $\mathfrak{g} \subset \mathrm{End}_Q(V, V)$ induces on $\mathfrak{g}$ a mixed Hodge structure $(\mathfrak{g}, W_{\bullet,\mathfrak{g}}, F^\bullet_\mathfrak{g})$;
- Under the assumption (V.5) we have

$$\mathfrak{g}_\mathbb{C} = \oplus I^{p,q}_\mathfrak{g}$$

and

(V.6) $$F^p_{\mathfrak{g},\infty} = \bigoplus_{q \leqq p} I^{\bullet,q}_\mathfrak{g};$$

- the monodromy logarithm $N \in I^{-1,-1}_\mathfrak{g}$;
- with the identifications

$$\begin{cases} T_{F^\bullet_\mathfrak{g}} \check{D} = \bigoplus_{p \leqq -1} I^{p,\bullet}_\mathfrak{g} = F^{-1}_\mathfrak{g} \\ T_{F^\bullet_{\mathfrak{g},\infty}} \check{D} = \bigoplus_{q \geqq 1} I^{\bullet,q}_\mathfrak{g} = F^{-1}_{\mathfrak{g},\infty}; \end{cases}$$

(V.7) *the differential*

$$\Phi_{\infty,*} : T_{F^\bullet} B(N)_\mathbb{R} \to T_{F^\bullet_\infty} \check{D}$$

of the reduced limit period mapping is the identity on $I^{p,q}_\mathfrak{g}$ *for* $q \geq 1$ *and is zero on* $I^{p,q}_\mathfrak{g}$ *for* $q < 0$.

Pictorially, we picture $I^{p,q}_\mathfrak{g}$ in the (p,q) plane

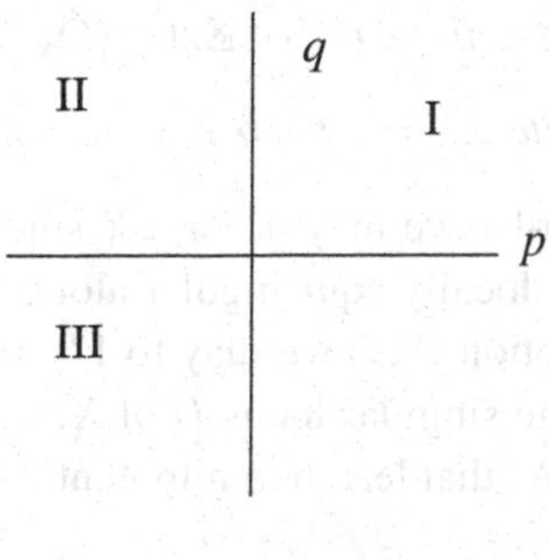

q

II I

p

III

Then

$$T_{F^\bullet}\widetilde{B}(N) \subseteq \mathrm{I} \cup \mathrm{II}$$

$$T_{F^\bullet_\infty}\check{D} \cong \mathrm{II} \cup \mathrm{III}$$

and $\Phi_{\infty,*}$ is the identity on the interior of II with

$$\begin{cases} \ker \Phi_{\infty,*} = \mathrm{I} \\ \operatorname{coker} \Phi_{\infty,*} = \mathrm{III}. \end{cases}$$

Monodromy cone structure associated to a normal crossing variety

More generally, associated to a nilpotent cone

$$\sigma = \operatorname{span}_{\mathbb{Q}\geqq 0}\{N_1,\ldots,N_\ell\}$$

with interior σ°, from [CKS1], [CKS2] there is an intricate and deep structure of nilpotent orbits, or equivalently limiting mixed Hodge structures in several variables. Among the properties of this structure are

- the monodromy weight filtration is independent of $N \in \sigma^0$ ([CK]);
- the limiting mixed Hodge structure associated to a nilpotent orbit $(F^\bullet, N)$ is independent of $N \in \sigma^\circ$;
- denoting by $\Delta^*(r)$ a punctured disc of radius r, in the manner described in [CKS1] on $\Delta^*(r_1) \times \cdots \times \Delta^*(r_\ell)$ there are several variable nilpotent orbits

  $$\exp(z_1 N_1 + \cdots + z_\ell N_\ell) \cdot F^\bullet, \qquad \operatorname{Im} z_i \gg 0$$

  which induce variations of mixed Hodge structure ([St-Zu]) on the axes in $\Delta(r_1) \times \cdots \times \Delta(r_\ell)$.

An important example of this cone structure is provided by a normal crossing variety X for which there exists a $\xi \in T_X \operatorname{Def}(X)$ such that

(V.8) $\quad \xi_{D_a} \neq 0$ *for each of the connected components* $D_a, a \in A,$ *of* D.

We will describe σ_X when the following special condition is satisfied:

(V.9) $\quad$ *for each* $a \in A$, *there is* $\xi_A \in \mathbb{E}\mathrm{xt}^1_{\mathcal{O}_X}(\Omega^1_X, \mathcal{O}_X)$ *such that*

$\xi_{A,a} \neq 0$ *while* $\xi_{A,b} = 0$ *for* $b \notin A$.

Geometrically, to first order we may deform X smoothing the component D_a of X_{sing} while remaining locally equisingular along the other components D_b, $b \neq a$. Under the assumption (V.9) we may to 1st order independently smooth the components D_a of the singular locus D of X. Then there are monodromy transformations N_a, $a \in N$, that lead to a nilpotent cone.

In general the map

$$\mathbb{E}\mathrm{xt}^1_{\mathcal{O}_X}(\Omega^1_X, \mathcal{O}_X) \to \underset{a\in A}{\oplus} H^0(\mathcal{O}_{D_a}) \tag{V.10}$$

will fail to be surjective and additional constructions are needed to obtain a set of monodromy cones described by the combinatorics of how the image of the mapping (V.10) meets the "coordinate axes" given by the right-hand term. The details of this will be given in the aforementioned work in progress. In that work we hope to also give the description of the cone in case X is locally a product of normal crossing varieties. In this situation the faces of the cone will correspond to where factors in the local product description become smoothed, as well as to where components in the stratification of X_{sing} become smoothed.

Proof of Theorem III. The proof now follows from a very particular case of Robles' result [Ro]. In this special case the argument is much simpler and goes as follows.

The first step is to identify the tangent spaces to the $G_{\mathbb{R}}$-orbit

$$\mathcal{O}_{F^\bullet_\infty} =: G_{\mathbb{R}} \cdot F^\bullet_\infty \subset \partial D.$$

This is done in [KP1] and later in [GGK] and [GG]; we shall follow the notations and indexing in the latter. From Section III.A we have for the real tangent space

$$\begin{aligned} T^{\mathbb{R}}_{F^\bullet_\infty} \mathcal{O}_{F^\bullet_\infty} &= \mathrm{Image}\{\mathfrak{g}_{\mathbb{R}} \to \mathfrak{g}_{\mathbb{C}}/F^0_\infty \mathfrak{g}_{\mathbb{C}}\} \\ &\cong \bigoplus_{\substack{q>0\\ p\leqq 0}} (\mathfrak{g}^{p,q} + q^{q,p}) \cap \mathfrak{g}_{\mathbb{R}} \oplus \bigoplus_{q\geqq p>0} (\mathfrak{g}^{p,q} \oplus q^{q,p}) \cap \mathfrak{g}_{\mathbb{R}} \\ &\cong \mathrm{Res}_{\mathbb{C}/\mathbb{R}} \Big\{ \bigoplus_{\substack{q>0\\ p\leqq 0}} \mathfrak{g}^{p,q} \Big\} \oplus \bigoplus_{q\geqq p>0} (\mathfrak{g}^{p,q} \oplus q^{q,p}) \cap \mathfrak{g}_{\mathbb{R}} \end{aligned}$$

where $\mathrm{Res}_{\mathbb{C}/\mathbb{R}}$ is the restriction of scalars from $\mathbb{C}$ to $\mathbb{R}$ that maps a complex vector space to the same space now considered as a vector space over $\mathbb{C} \subset \mathbb{R}$. From the discussion above we see that $T_{(X,\xi)}\,\mathrm{Def}(X,\xi)$ maps to the first factor, which is in the tangent space to the $G_{\mathbb{R}}$-orbit $\mathcal{O}_{F^\bullet_\infty}$. □

As noted above, what one would like is to show that the interiors of the faces of the cone also map under the reduced limit period mapping to $G_{\mathbb{R}}$-orbits that are in the closure of the image of σ_X.

VI The hierarchy of mixed Hodge structures

In this discussion we will restrict to a standard family $\mathcal{X} \to \Delta$. To this situation there are naturally associated four mixed Hodge structures:

(i) the mixed Hodge structure on $H^m(X)$;
(ii) that part of the limiting mixed Hodge structure that can be constructed from X alone;
(iii) the limiting mixed Hodge structure associated to $\mathcal{X} \to \Delta$, modulo the equivalence $F^\bullet \sim \exp(zN)\cdot F^\bullet$ arising from a change of parameter on Δ;[30]
(iv) the limiting mixed Hodge structure associated to the pair (X,ξ) where $\xi \in T_X \operatorname{Def}(X)$.

We recall from the introduction the

THEOREM IV: *In a manner to be explained in the proof, there are strict implications*

$$\text{(iv)} \Longrightarrow \text{(iii)} \Longrightarrow \text{(ii)} \Longrightarrow \text{(i)}.$$

Intuitively there is successively strictly less information in the data arising from the situations (iv), (iii), (ii), (i). Before turning to the proof we will illustrate the result in the following

Example:[31] Suppose that X_t is a smooth curve of genus $g=3$ whose limit X is an irreducible nodal curve whose normalization $\widetilde{X}$ has genus $\widetilde{g}=1$

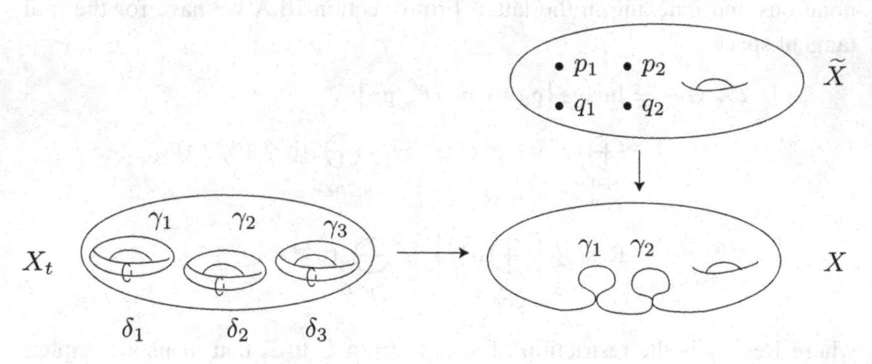

Setting $\ell(t) = (1/2\pi i)\log t$ and using the symplectic basis drawn above for $H_1(X_t,\mathbb{Z})$, the normalized period matrix is

$$\begin{pmatrix} 1 & & \\ & 1 & \\ & & 1 \\ \ell(t)+a_{11}(t) & a_{12}(t) & b_1(t) \\ a_{21}(t) & \ell(t)+a_{22}(t) & b_2(t) \\ b_1(t) & b_2(t) & c(t) \end{pmatrix}$$

[30] We have seen that this data will depend only on the 1st order neighborhood of X in $\mathcal{X}$.

[31] This is an extension to $g=3$ of the case $g=2$ in [Ca].

where the $a_{ij}(t) = a_{ji}(t)$, $b_i(t)$ and $c(t)$ are holomorphic in the disc and $\operatorname{Im} c(t) > 0$. With the choice t of parameter the nilpotent orbit is

$$\begin{pmatrix} 1 & & \\ & 1 & \\ & & 1 \\ \ell(t)+a_{11} & a_{12} & b_2 \\ a_{21} & \ell(t)+a_{22} & b_2 \\ b_1 & b_2 & c \end{pmatrix}$$

where $a_{ij} = a_{ij}(0)$, $b_i = b_i(0)$ and $c = c(0)$. Letting $\omega_i(t)$ be the holomorphic differentials on X with limits ω_i on X that pull up to $\widetilde{\omega}_i$ on $\widetilde{X}$, we have

- $\widetilde{\omega}_1, \widetilde{\omega}_2$ are differentials of the 3rd kind on $\widetilde{X}$ with divisor $p_i + q_i$ and $\operatorname{Res}_{p_i} \widetilde{\omega}_i = +1$, $\operatorname{Res}_{q_i} \widetilde{\omega}_i = -1$ for $i = 1, 2$;
- $\widetilde{\omega}_3$ is a holomorphic differential on $\widetilde{X}$.

Under a reparametrization $t' = e^{2\pi i\lambda} t$,

$$a_{ii}(t') = a_{ii}(t) + \lambda,$$

and all other entries in the period matrix evaluated at $t = 0$ are unchanged.

We note that implicit in the choice of symplectic basis is the monodromy weight filtration (here defined in terms of homology)

$$\begin{cases} W_{-1} = \operatorname{span}\{\delta_1, \delta_2\} \\ W_1/W_0 \cong \operatorname{span}\{\gamma_1, \gamma_2\} \\ W_0/W_{-1} \cong \operatorname{span}\{\delta_3, \gamma_3\}. \end{cases}$$

The entries in the above period matrix at $t = 0$ are (cf. [Ca])

(I_1) c is the period of the elliptic curve $\widetilde{X}$;

(I_2) b_i is the image of $\mathrm{AJ}_{\widetilde{X}}(p_i - q_i)$ in $J(\widetilde{X})$; this gives the extension data in

$$0 \to \mathrm{Gr}_0 \to \mathrm{Gr}_1 \to \mathrm{Gr}_1 / \mathrm{Gr}_0 \to 0$$

as described in [Ca];

(I_3) with suitable normalization of the $\widetilde{\omega}_i$,

$$a_{ij} = \int_{q_i}^{p_j} \widetilde{\omega}_i, \qquad i \neq j;$$

(I_4) finally, with a choice of parameter t we may uniquely define the improper integrals

$$\int_{q_i}^{p_i} \widetilde{\omega}_i.$$

This means that if $\gamma_{i,t}$ is the above curve on X_t for $t \neq 0$

$$\int_{\gamma_{i,t}} \omega_{i,t} = \ell(t) + a_{ii}(t),$$

and then on $\widetilde{X}$ we will have

$$\lim_{\begin{cases} p'_i \to p_i \\ q'_i \to q_i \end{cases}} \int_{q'_i}^{p'_i} \widetilde{\omega}_i = a_{ii}(0)$$

where the picture is

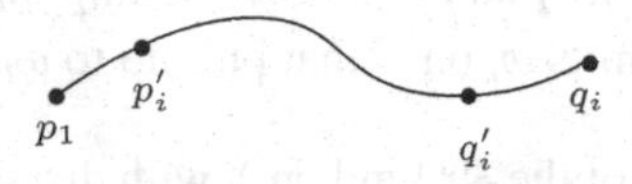

and where the logarithmic singularities at the endpoints cancel.

The a_{ij} for $i \neq j$ record the part of the "extension upon extension" data in Gr_2 / G_1 over $\mathrm{Gr}_1 / \mathrm{Gr}_0$ that is invariant under reparametrization, and the a_{ii} record the full extension data.

Algebro-geometrically the picture is the following. Denoting by $\mathcal{M}_g$ the moduli space of the stable curves of genus g, and by $\overline{\mathcal{M}}_g$ the Deligne-Mumford compactification, the curve X gives a point in

$$\partial \mathcal{M}_3 \subset \overline{\mathcal{M}}_3.$$

More specifically, X defines a point in a codimension-1 component $\mathcal{C}$ of the stratified variety $\partial \mathcal{M}_3$. Then

- $\dim \mathcal{C} = 4$ and c, b_1, b_2 and a_{12} are local coordinates in $\mathcal{C}$;
- a_{11} and a_{22} give normal parameters to $\mathcal{C}$ in $\overline{\mathcal{M}}_3$.

The difference between (ii) and (iii) in this case is that (iii) contains the information in the weight filtration, which is information that is not obtainable from that on X alone (see the subsequent discussion).

To explain (ii) we will picture a limiting mixed structure in terms of the N-strings as

$$\begin{array}{c} H^0(-m) \to \quad \cdot \quad \cdot \quad \cdot \quad \cdot \quad \to H^0(-1) \xrightarrow{N} H^0 \\ \vdots \\ H^{m-1}(-1) \to H^{m-1} \\ H^m \end{array}$$

where H^k is a pure Hodge structure of weight k. It is this presentation that is especially useful in the computation of examples [GG] and [GGR]. Equivalent data to the above are the following parts of a polarized limiting mixed Hodge structure $(V, W_\bullet(N), F^\bullet)$

- the Hodge structures $\mathrm{Gr}_k^{W_\bullet(N)}$;
- the iterated N operators on the N-strings.

It is known [GG], [Ro] that this data always arises from a non-unique limiting mixed Hodge structure.

Our main result, Theorem V in the introduction, is that, under the assumption that X is smoothable but with no $\xi \in T_X^0 \mathrm{Def}(X)$ singled out, we may compute the $H^{m-i}(-j)$'s above purely in terms of X alone. For this we will use the maps

$$\mathrm{Rest} : H^q\big(X^{[k]}\big) \to H^q\big(X^{[k+1]}\big)$$

obtained by the alternating sums of the restriction maps, and the suitably alternated Gysin maps

$$\mathrm{Gy} : H^q\big(X^{[k]}\big)(-1) \to H^{q+2}\big(X^{[k-1]}\big).$$

The result is

THEOREM VI: *Assuming that X is smoothable, there are complexes*

$$
\begin{array}{ccccc}
 & H^{q-4}\big(X^{[k+2]}\big)(-2) & & & \\
 & \oplus & & H^{q-2}\big(X^{[k+1]}\big)(-1) & \\
\cdots \longrightarrow & H^{q-2}\big(X^{[k]}\big)(-1) & \longrightarrow & \oplus & \longrightarrow H^q\big(X^{[k]}\big) \\
 & \oplus & & H^q\big(X^{[k-1]}\big) & \\
 & H^q\big(X^{[k-2]}\big) & & &
\end{array}
$$

such that for $0 \leqq j \leqq m-i$

$$H^{m-i}(-j) \cong H^*_{\mathrm{Rest}} H^*_{\mathrm{Gy}}\big(H^{m-i}\big(X^{[i+1]}\big)\big)(-j). \tag{VI.1}$$

The notation in (VI.1) means that the left-hand side is computed by the cohomology of complexes in the statement of the theorem at the spot designated by the right-hand side in (VI.1). A significant fact here is that

$$X \textit{ smoothable} \implies \mathrm{Rest} \circ \mathrm{Gy} = -\,\mathrm{Gy} \circ \mathrm{Rest}. \tag{VI.2}$$

The point is that this implication (VI.2) is generally not true unless X is smoothable, although the result does not depend on any particular smoothing.

A further significant point is that in taking the cohomology of the restriction sequences

$$H^*_{\mathrm{Gy}}\big(H^i\big(X^{[j-1]}\big)(-m)\big) \to H^*_{\mathrm{Gy}}\big(H^i\big(X^{[j]}\big)(-m)\big)$$
$$\to H^*_{\mathrm{Gy}}\big(H^i\big(X^{[j+1]}\big)(-m)\big)$$

we only put in $H^i(X^{[j]})(-k)$ if $0 \leqq k < j-1$.

Referring to (I.9) and Theorem V in the introduction, Theorem VI implies that result, and the N-maps in the N-strings are the twisted identity maps on the individual pieces as given by (VI.1).

For the proof of Theorem VI one uses the basic constriction introduced in [St1] and [Zu]. We will recall this for a standard family $\mathcal{X} \xrightarrow{\pi} \Delta$ and observe at the end that the vector space $\mathbb{H}^m\left(\Omega^\bullet_{\mathcal{X}/\Delta}(\log X) \otimes \mathcal{O}_X\right)$ has a filtration shifted down two steps by N, and the associated graded together with the mappings induced by N may be defined in terms of X alone. The N-strings that result are the ones expressed in the theorem. The upshot is that given an abstract normal crossing variety X we will be able to define the object that is defined in terms of X alone, provided only we assume that X is smoothable. This object is related to the limiting mixed Hodge structure associated to the standard family in the manner just described. It is interesting to note that the object so described will be *independent* of the smoothing of X, provided that one exists. This is a reflection of the result in [CKS1] that in a multi-parameter family the limiting mixed Hodge structure associated to the origin is independent of the direction of approach from the interior of the cone.

The construction in [St1] and [Zu] goes as follows: We define a bi-graded complex $A^{\bullet,\bullet}$ where

$$A^{p,q} = \Omega^{p+q+1}_{\mathcal{X}}(\log X)/\widetilde{W}_q\Omega^{p+q+1}_{\mathcal{X}}(\log X),$$

where $\widetilde{W}_q$ is the standard filtration given by

$$\widetilde{W}_q = \text{ differential forms with at most } q\ dx_i/x_i \text{ terms.}$$

The differentials are given by

$$\begin{cases} d' = \text{usual } d \\ d'' = \wedge dt/t. \end{cases}$$

The basic observation and definitions are

- The mapping $\Omega^p_{\mathcal{X}}(\log X) \xrightarrow{\wedge dt/t} A^{p,0}$ has co-kernel naturally isomorphic to $\Omega^p_{\mathcal{X}/\Delta}(\log X) \otimes \mathcal{O}_X$;
- $W_k A^{p,q} =: \widetilde{W}_{2q+k+1}\Omega^{p+q+1}_{\mathcal{X}}(\log X)/\widetilde{W}_q\Omega^{p+q+1}_{\mathcal{X}}(\log X)$;
- $F^r A^{p,q} = \begin{cases} A^{p,q} & \text{if } p \geqq r, \\ 0 & \text{if } p < r. \end{cases}$

Then with respect to $F^\bullet$ the map

$$\Omega^\bullet_{\mathcal{X}/\Delta}(\log X) \otimes \mathcal{O}_X \to A^\bullet$$

is a filtered quasi-isomorphism. By definition $W_\bullet A^\bullet$ induces the weight filtration on $\mathbb{H}^m(\Omega^\bullet_{\mathcal{X}/\Delta}(\log X)) \otimes \mathcal{O}_X$. We note that

$$\begin{aligned} \mathrm{Gr}^W_k A^{p,q} &= W_{k+2q+1}\Omega^{p+q+1}_{\mathcal{X}}(\log X) \big/ W_{k+2q}\Omega^{p+q+1}_{\mathcal{X}}(\log X) \\ &\cong \begin{cases} 0 & \text{if } k \leqq -(q+1), \\ \Omega^{p-q-k}_{X[2q+k+1]} & \text{if } k \geqq -q \end{cases} \end{aligned}$$

where the second isomorphism is by the iterated residue map.

An intermediate step to computing $\mathbb{H}^*(A)$ is to use the spectral sequence associated to $\widetilde{W}_\bullet$. For this spectral sequence where $A^i = \bigoplus_{p+q=i} A^{p,q}$

$$\text{(VI.3)} \qquad \begin{aligned} E_1^{a,b} &= H^{a+b}\left(\mathrm{Gr}^W_{-b} A^i\right) \\ &= H^{a+b}\left(\bigoplus_q \Omega^{i-2q+b}_{X[2q+1-b]}\right) \\ &= \bigoplus_{q=\min(0,b)} H^{i-2q+b,a+b}\left(X^{[2q+1-b]}\right). \end{aligned}$$

If $\dim X = n$, then $\dim X^{[2q+1-b]} = n+1-(2q+1-b) = n+b-2q$, from which we have

$$\begin{aligned} & H^{i-2q+b,a+b}\left(X^{[2q+1-b]}\right) \neq 0 \\ \implies & \begin{cases} i-2q+b \leqq n-2q+b = \dim X^{[2q+1-b]} \iff i \leqq n \\ a+b \leqq n-2q+b \iff a \leqq n-2q. \end{cases} \end{aligned}$$

These are the only inequalities other than $\max(a,b) \leqq q \leqq i$. Thus to have potentially non-zero $H^{r,s}\left(X^{[t]}\right)$ we need

$$\begin{matrix} i-2q+b = r & & i = t+r+1 & \\ a+b = s & \iff & b = 2q+1-t & (\implies b \equiv 1-t (\mathrm{mod}\, 2)) \\ t = 2q+1-b & & a = s+t-2q-1 & \end{matrix}$$

and $q \geqq \max(0,b)$ gives $b \geqq \max(1-t, 2b+1-t)$, which then gives

$$\begin{cases} t-1 \geqq b \geqq 1-t \\ \quad b \equiv 1-t (\mathrm{mod}\, 2) \end{cases}$$

(thus $t=1, b=0$; $t=2, b=-1,1$; $t=3, b=-2,0,2,\ldots$) for $r = i-t+1 = i+\dim X^{[t]} - n = \dim X^{[t]} - (n-i)$. Fixing t and noting $0 \leqq i \leqq n$ we then have

$$H^{r,*}\left(X^{[t]}\right) \text{ appears for one value of } i \text{ and } t \text{ values of } b.$$

The above gives the conclusion that for $H^n = \mathbb{H}^n(\text{LMHS})$, before cancellation in the spectral sequence

$$\begin{array}{lll} H^n(X^{[1]}) & \text{appears once} & (b=0) \\ H^{n-1}(X^{[2]}) & \text{appears once} & (b=-1,1) \\ H^{n-2}(X^{[3]}) & \text{appears three times} & (b=-2,0,2) \\ \vdots & & \end{array}$$

Sketch of the proof of Theorem IV. Referring now to (VI.3), by [St1] the spectral sequence degenerates at E_2. The E_1-terms are as indicated there. The d_1-map is, as noted in [Zu], "composed of various restriction and Gysin maps." After unwinding the indices, the d_1-complex turns out to be the one in the statement of the theorem.

We now come to the main point. Note that the individual terms and maps may be defined in terms of X alone. However in general we do *not* have

$$\text{(VI.4)} \qquad \text{Rest} \circ \text{Gy} = -\text{Gy} \circ \text{Rest}.$$

PROPOSITION: *The anti-commutativity commutativity* (VI.4) *holds if, and only if,* $\mathcal{O}_D(X)$ *is topologically trivial. This is the case if* X *is smoothable.*

Rather than give the formal argument we shall indicate by example in the simplest non-trivial cases why the result should be true.[32] We note that if X is smoothable the proposition is true. Our central point is that this sufficient condition is essentially also necessary. We say essentially, because (VI.4) is a purely topological fact which only requires that $\mathcal{O}_D(X)$ be topologically, but not necessarily analytically, trivial.

Let X be an irreducible surface having as singular locus a double curve C whose inverse image in $X^{[1]}$ is a disjoint union $X^{[1]} = C_1 \amalg C_2$ of two smooth curves. We will denote by $H^q(X^{[2]})_-$ the classes $\alpha \oplus -\alpha \in H^q(C_1 \amalg C_2)$. Then $H^q(X^{[2]})_- \cong H^q(C)$, but we put opposite signs on those in $H^q(C_1)$ and $H^q(C_2)$. Then we shall show

(i) $\mathcal{O}_D(X)$ is topologically trivial if, and only if, $C_1^2 = -C_2^2$;
(ii) the sequence

$$\text{(VI.5)} \qquad H^0(X^{[2]})_-(-1) \xrightarrow{\text{Gy}} H^2(X^{[1]}) \xrightarrow{\text{Rest}} H^2(X^{[2]})_-$$

is a complex if, and only if, $C_1^2 = -C_2^2$. The complex (VI.5) is the simplest non-trivial case of the sequences that appear in the statement of Theorem VI.

[32] The details of this argument will appear in the previously mentioned work in progress.

Denoting by $\eta_{C_i} \in H^2(X^{[1]})$ the fundamental class of C_i and by $[C_i]$ the fundamental class of C_i itself, the sequence (VI.5) is

$$1_{C_1} - 1_{C_2} \to \eta_{C_1} - \eta_{C_2} \longrightarrow \quad \{((C_1 - C_2)\cdot C_1)[C_1] + ((C_1 - C_2)\cdot C_2[C_2])\}$$

$$\| $$

$$\{C_1^2[C_1] - C_2^2[C_2]\}_-$$

$$\|$$

$$\left(\frac{C_1^2 + C_2^2}{2}\right)([C_1] - [C_2]),$$

which proves (ii).

As for (i) we have

$$N_D(X) \cong N_{C_1}(X^{[1]}) \otimes N_{C_2}(X^{[1]})^*.$$

In case X is still an irreducible surface a piece of the complex in Theorem VI is

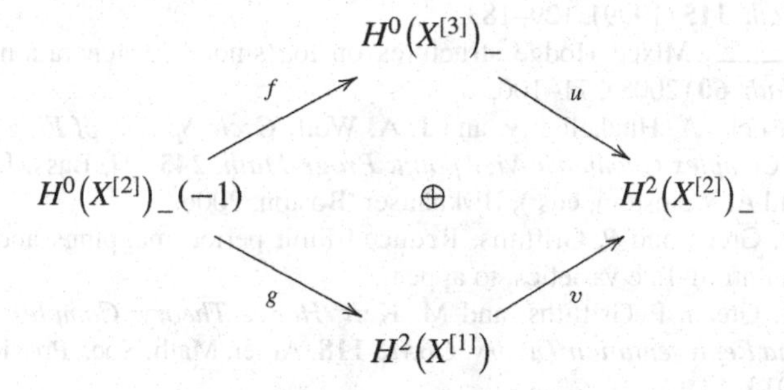

Here the minus sign on $(\ )_-$ refers to classes that transform by the sign of the induced action on cohomology given the labeling into even-odd of the components lying over a general point in the map $X^{[k]} \to X_k$. The conditions to have a complex are

$$u \cdot f + v \cdot g = 0,$$

which when worked out is a consequence of $\mathcal{O}_D(X) \cong \mathcal{O}_D$.

References

[A†] D. Abramovich, Q. Chen, D. Gillam, Y. Huang, M. Olsson, M. Satriano, and S. Sun, Logarithmic geometry and moduli, available at http://www.math.brown.edu/~abromovic/LOGGEOM/master.pdf.

[AK] D. Abramovich and K. Karu, Weak semistable reduction in characteristic 0, *Invent. Math* **139** (2000), 241–273.

[AMRT] A. Ash, D. Mumford, M. Rapoport, and Y. Tai, Smooth compactification of locally symmetric varieties, in *Lie Groups: History, Frontiers and Applications*, Vol. IV, Math. Sci. Press, Brookline, MA, 1975.

[Ca] J. A. Carlson, The geometry of the extension class of a mixed Hodge structure, *Proc. Sympos. Pure Math.* **46** (1987), 199–222.

[Ca-Fe] J. Fernandez and E. Cattani, Infinitesimal variations of Hodge structure at infinity, *Geom. Dedicata* **139** (2009), 299–312.

[CK] E. Cattani and A. Kaplan, Polarized mixed Hodge structures and the local monodromy of a variation of Hodge structure, *Invent. Math.* **67** (1982), 101–115.

[CKS1] E. Cattani, A. Kaplan, and W. Schmid, Degenerations of Hodge structures, *Ann. of Math.* **123** (1986), 457–535.

[CKS2] ———, L^2 and intersection cohomologies for a polarizable variation of Hodge structure, *Invent. Math.* **87** (1987), 217–252.

[De] P. Deligne, *Équations Differentielles à Points Singulars Réguliere, Springer Lect. Notes*, Springer-Verlag, New York, **163**, 1970.

[Fr1] R. Friedman, The period map at the boundary of muduli, in *Topics in Transcendental Algebraic Geometry*, 183–208, (P. Griffiths, ed.), Princeton Univ. Press, Princeton, NJ, 1984.

[Fr2] ———, Global smoothings of varieties with normal crossings, *Ann. of Math.* **118** (1983), 75–114.

[Fu1] T. Fujisawa, Limits of Hodge structures in several variables, *Compos. Math.* **115** (1999), 129–183.

[Fu2] ———, Mixed Hodge structures on log smooth degenerations, *Tohoku Math.* **60** (2008), 71–100.

[FHW] G. Fels, A. Huckelberry, and J. A. Wolf, *Cycle Spaces of Flag Domains, A Complex Geometric Viewpoint*, *Progr. Math.* **245**, (H. Bass, J. Oesterlé, and A. Weinstein, eds.), Birkhäuser, Boston, 2006.

[GG] M. Green and P. Griffiths, Reduced limit period mappings and orbits in Mumford-Tate varieties, to appear.

[GGK] M. Green, P. Griffiths, and M. Kerr, *Hodge Theory, Complex Geometry and Representation Theory*, *CBMS* **118**, Amer. Math. Soc., Providence, RI, 2013.

[GGR] M. Green, P. Griffiths, and C. Robles, Extremal degenerations of polarized Hodge structures, to appear.

[KN] Y. Kawamata and Y. Namikawa, Logarithmic deformations of normal crossing varieties smoothen of degenerate Calabi Yau varieties, *Invent. Math.* **118** (1994), 395–409.

[KP1] M. Kerr and G. Pearlstein, Boundary components of Mumford-Tate domains. arXiv:1210.5301.

[KP2] ———, Naïve boundary strata and nilpotent orbits, preprint, 2013.

[KU] K. Kato and S. Usui, *Classifying Spaces of Degenerating Polarized Hodge Structures*, *Ann. of Math. Stud.* **169**, Princeton Univ. Press, Princeton, NJ, 2009.

[Pa] V. P. Palamodov, Deformations of complex spaces, *Russian Math. Surveys* **31**, (1976), 128–197; translated from *Uspekhi Mat. Nauk.* **31** (1976), 129–194.

[PS] C. A. M. Peters and J. H. M. Steenbrink, *Mixed Hodge Structures*, *Ergeb. Math. Grenzgeb.* **52**, Springer-Verlag, New York, 2008.

[Ro] C. Robles, working notes at the IAS.

[Sc] W. Schmid, Variation of Hodge structure: the singularities of the period mapping, *Inv. Math.* **22** (1973), 211–319.

[St1] J. Steenbrink, Limits of Hodge structures, *Invent. Math.* **31** (1969), 229–257.

[St2] ______, Logarithmic embeddings of varieties with normal crossings and mixed Hodge structures, *Math. Ann.* **301** (1995), 105–118.

[St-Zu] J. Steenbrink and S. Zucker, Variation of mixed Hodge structure. I, *Invent. Math.* **80** (1985), 489–542.

[Zu] S. Zucker, Degeneration of Hodge bundles (after Steenbrink), in *Topics in Transcendental Algebraic Geometry*, 121–142, (P. Griffiths, ed.), Princeton Univ. Press, Princeton, NJ, 1984.

5

Studies of closed/open mirror symmetry for quintic threefolds through log mixed Hodge theory

Sampei Usui

Dedicated to Kazuhiro Konno and dedicated to James D. Lewis on his sixtieth birthday

ABSTRACT. We correct the definitions and descriptions of the integral structures in [30]. The previous flat basis in [ibid] is characterized by the Frobenius solutions and is integral in the first approximation by mean of the graded quotients of monodromy filtration, but it is not integral in the strict sense. In this article, we use $\hat{\Gamma}$-integral structure of Iritani in [7] for A-model. Using this precise version, we study open mirror symmetry for quintic threefolds through log mixed Hodge theory, especially the recent result on Néron models for admissible normal functions with non-torsion extensions in the joint work [14] with K. Kato and C. Nakayama. We understand asymptotic conditions as values in the fiber over a base point on the boundary of $S^{\log}$.

Contents

2010 *Mathematics Subject Classification*. Primary 14C30; Secondary 14D07, 32G20. Partially supported by JSPS. KAKENHI (B) No. 23340008.

0 Introduction

In a series of joint works with K. Kato and C. Nakayama, we are constructing a fundamental diagram which consists of various kind of partial compactifications of classifying space of mixed Hodge structures and their relations. We try to understand mirror symmetry in this framework of the fundamental diagram. In this paper, we first complete the insufficient results 3.5–3.6 in the previous paper of Usui [30] (see Remark in 2.6 below), and then study open mirror symmetry for quintic threefolds through log mixed Hodge theory, especially the fine moduli of log Hodge structures and Néron models over it.

0.1. Fundamental Diagram

For a classifying space D of Hodge structures of specified type, we have

$$\begin{array}{ccccccc} & & & & D_{\mathrm{SL}(2),\mathrm{val}} & \rightarrow & D_{\mathrm{BS},\mathrm{val}} \\ & & & & \downarrow & & \downarrow \\ \Gamma\backslash D_{\Sigma,\mathrm{val}} & \leftarrow & D^{\sharp}_{\Sigma,\mathrm{val}} & \stackrel{\psi}{\rightarrow} & D_{\mathrm{SL}(2)} & & D_{\mathrm{BS}} \\ \downarrow & & \downarrow & & & & \\ \Gamma\backslash D_{\Sigma} & \leftarrow & D^{\sharp}_{\Sigma} & & & & \end{array}$$

(Γ is a monodromy group) in pure case: [15], [16], [17]. For mixed case, we should extend to an amplified diagram: [9], [10], [12], [13], continuing.

0.2. Mirror symmetry for quintic threefolds

Let V be a quintic threefold in $\mathbf{P}^4$ and V°_ψ be its mirror family (cf. [1], Sect. 4.2). For simplicity, we denote the family V°_ψ simply by V° if there would be no confusions.

Mirror symmetry for the A-model of quintic threefold V and the B-model of its mirror V° was predicted by Candelas–de la Ossa–Green–Parks in the famous paper [2]. We recall two styles of the theorem 0.2.1 and 0.2.2

below. Every statement in the present paper is near the large radius point q_0 of the complexified Kähler moduli $\mathcal{KM}(V)$ and the maximally unipotent monodromy point p_0 of the complex moduli $\mathcal{M}(V^\circ)$.

Let $t := y_1/y_0$, $u := t/2\pi i$ be the canonical parameters and $q := e^t = e^{2\pi iu}$ be the canonical coordinate for B-model in 2.2 below and the respective ones for A-model in 2.3 below.

The following theorem is due to Givental [5] and Lian–Liu–Yau [20].

Theorem 0.2.1. (Potential). *The potentials of the two models coincide:* $\Phi^V_{\mathrm{GW}}(t) = \Phi^{V^\circ}_{\mathrm{GM}}(t)$.

Morrison [22] formulated the following style Theorem 0.2.2 and proved the theorem except integral structure. Iritani [7] defined $\hat{\Gamma}$-integral structure for A-model and proved the theorem completely for wider objects.

Theorem 0.2.2. (Variation of Hodge structure). *The isomorphism* $(q_0 \in \overline{\mathcal{KM}}(V)) \xleftarrow{\sim} (p_0 \in \overline{\mathcal{M}}(V^\circ))$ *of neighborhoods of the compactifications, by the canonical coordinates* $q = \exp(2\pi iu)$, *lifts to an isomorphism, over the punctured neighborhoods* $\mathcal{KM}(V) \xleftarrow{\sim} \mathcal{M}(V^\circ)$, *of polarized* $\mathbf{Z}$*-variations of Hodge structure with a specified section*

$$(\mathcal{H}^V, S, \nabla^{\mathrm{even}}, \mathcal{H}^V_{\mathbf{Z}}, F; T^3) \xleftarrow{\sim} (\mathcal{H}^{V^\circ}, Q, \nabla^{\mathrm{GM}}, \mathcal{H}^{V^\circ}_{\mathbf{Z}}, F; \tilde{\Omega}).$$

Our Theorem 0.2.3 below is equivalent to Theorem 0.2.1 and Theorem 0.2.2 by a log version [17], Theorem 2.5.14 of the nilpotent orbit theorem of Schmid [27] (this part of [30] is valid).

Theorem 0.2.3. (Log Hodge structure, Log pperiod map). *The isomorphism* $(q_0 \in \overline{\mathcal{KM}}(V)) \xleftarrow{\sim} (p_0 \in \overline{\mathcal{M}}(V^\circ))$ *of neighborhoods of the compactifications uniquely lifts to an isomorphism of B-model log variation of polarized Hodge structure with a specified section* $\tilde{\Omega}$ *for* V° *and A-model log variation of polarized Hodge structure with a specified section* T^3 *for* V, *whose restriction over the punctured* $\mathcal{KM}(V) \xleftarrow{\sim} \mathcal{M}(V^\circ)$ *coincides with the isomorphism of variations of polarized Hodge structure with specified sections in Theorem 0.2.2.*

This rephrases as follows. Let σ *be the common monodromy cone, transformed by a level structure into* End *of a reference fiber of the local system, for the A-model and for the B-model. Then, we have a commutative diagram of horizontal log period maps*

$$\begin{array}{ccc} (q_0 \in \overline{\mathcal{KM}}(V)) & \xleftarrow{\sim} & (p_0 \in \overline{\mathcal{M}}(V^\circ)) \\ & \searrow \quad \swarrow & \\ & ([\sigma, \exp(\sigma_{\mathbf{C}})F_0] \in \Gamma(\sigma)^{\mathrm{gp}} \backslash D_\sigma) & \end{array}$$

with extensions of specified sections in Theorem 0.2.2, where $[\sigma, \exp(\sigma_{\mathbf{C}})F_0]$ *is the class of the nilpotent orbit, regarded as a boundary point, and* $\Gamma(\sigma)^{\mathrm{gp}}\backslash D_\sigma$ *is the fine moduli of log Hodge structures of specified type which will be explained in Section 1 below.*

0.3. Open mirror symmetry for quintic threefolds

The following theorem is due to Walcher [31] and Morrison–Walcher [23].

Theorem 0.3.1. (Inhomogenous solutions). *The potentials of the two models coincide: Let* $\mathcal{L}$ *be the Picard–Fuchs differential operator for quintic mirror family (cf. 2.2 below). Let*

$$\mathcal{T}_A = \frac{u}{2} \pm \Big(\frac{1}{4} + \frac{1}{2\pi^2} \sum_{d\ odd} n_d q^{d/2}\Big)$$

be the A-model domainwall tension in [23], where the n_d *are open Gromov–Witten invariants, and*

$$\mathcal{T}_B = \int_{C_-}^{C_+} \Omega$$

be the B-model domainwall tension, where $C_\pm \subset V^\circ$ *are the disjoint smooth curves coming from the two conics in* $\{x_1 + x_2 = x_3 + x_4 = 0\} \cap V_\psi \subset \mathbf{P}^4(\mathbf{C})$ *[ibid].*

Then

$$\mathcal{L}(y_0(z)\mathcal{T}_A(z)) = \mathcal{L}(\mathcal{T}_B(z))\Big(= \frac{15}{16\pi^2}\sqrt{z}\Big) \quad \Big(z = \frac{1}{(5\psi)^5}\Big).$$

Concerning this, we have the following observations.

0.4. Log mixed Hodge structure, Log normal function

We describe for B-model. The same holds for A-model by 0.2–0.3 and the correspondence table in 2.5 below.

Put $\mathcal{H} := \mathcal{H}^{V^\circ}$ and $\mathcal{T} := \mathcal{T}_B$. We use $e^0 \in I^{0,0}$, $e^1 \in I^{1,1}$ which are a part of a basis of $\mathcal{H}_{\mathcal{O}}$ respecting the Deligne decomposition at p_0 (see 2.5 (3B)) and a part of flat basis $s^0 = e^0$, $s^1 = e^1 - (u-1)e^0$ of $\mathcal{H}_{\mathbf{Z}}$ (see 2.5 (7B)). To make the local monodromy of $\mathcal{T}$ unipotent, we take a double cover $z^{1/2} \mapsto z$. Let $L_{\mathbf{Q}}$ be the translated local system comparing to the trivial extension $\mathbf{Q}(-2) \oplus \mathcal{H}_{\mathbf{Q}}$ by $(0, -(\mathcal{T}/y_0)s^0)$ in $\mathcal{E}xt^1(\mathbf{Q}(-2), \mathcal{H}_{\mathbf{Q}})$. Let $J_{L_{\mathbf{Q}}}$ be the Néron model on a neighborhood S of p_0 in the $z^{1/2}$-plane which lies over $L_{\mathbf{Q}}$ in [14] (there is a difference of Tate twist). Then, $J_{L_{\mathbf{Q}}} = \mathcal{E}xt^1_{\mathrm{LMH}/S}(\mathbf{Z}(-2), \mathcal{H})$ (extension group of log mixed Hodge structures over S) in the present case (cf. [13], Corollary 6.1.6, and 1.8 below), and we have the following theorem.

Theorem 0.4.1. (Log normal function).

(1) *The normalized tension* $\mathcal{T}/y_0$ *is understood as a multi-valued truncated normal function by* $(\mathcal{T}/y_0)s^0$. *Then it lifts and extends uniquely to a single-valued log normal function* $S \to J_{L_{\mathbf{Q}}}$ *so that the corresponding exact sequence* $0 \to \mathcal{H} \to \tilde{\mathcal{H}} \to \mathbf{Z}(-2) \to 0$ *of log mixed Hodge structures over* S *is given by the liftings* $1_{\mathbf{Z}}$ *and* 1_F *in* $\tilde{\mathcal{H}}$ *of* $(2\pi i)^{-2} \cdot 1 \in \mathbf{Z}(-2)$ *respecting the lattice and the Hodge filtration, respectively, which are defined as follows:* $1_{\mathbf{Z}} := ((2\pi i)^{-2} \cdot 1, -(\mathcal{T}/y_0)s^0)$ *with* $(\mathcal{T}/y_0)s^0 \in \mathcal{H}_{\mathcal{O}^{\log}} = (\mathrm{gr}_3^W)_{\mathcal{O}^{\log}}$, *and* $1_F - 1_{\mathbf{Z}} := (\delta(\mathcal{T}/y_0))e^1 - (\mathcal{T}/y_0)e^0$, *where* $\delta := 2\pi i q d/dq$.
(2) *A splitting of the weight filtration* W *of the local system* $\tilde{\mathcal{H}}_{\mathbf{Z}}$, *i.e., a splitting compatible with the monodromy of the local system* $\tilde{\mathcal{H}}_{\mathbf{Z}}$, *is given by* $1_{\mathbf{Z}}^{spl} = 1_{\mathbf{Z}} + s^1/2$, *and the log normal function over it is given by* $1_F^{spl} - 1_{\mathbf{Z}}^{spl} = (\delta(\mathcal{T}/y_0))e^1 - (\mathcal{T}/y_0)e^0$, *where* δ *is as in (1).*
(3) *Theorem 0.3.1 says that the inverse of the normal function in (1) from its image is given by* $16\pi^2/15$ *times the Picard–Fuchs differential operator* $\mathcal{L}$.

Theorem 0.4.1 is proved in Section 3, and after these proofs some geometric backgrounds are discussed in Section 3.3.

0.5.

The organization of this paper is as follows. Section 1 is a summary of log mixed Hodge theory mainly from [17], [13] and [14], which is used to study mirror symmetry in later sections and also is expected to work as a brief guide of this theory. In Section 2, after preparations including $\hat{\Gamma}$-integral structure in [6] and [7], we give a correspondence table of closed mirror symmetry for quintic threefolds and their mirrors, which is the precision and the expansion of our previous paper [30], Sect. 3. In Section 3, we prove log mixed Hodge theoretic interpretation Theorem 0.4.1. We also give some discussions on the related geometries and local systems in Section 3.3.

Acknowledgments. The author thanks Kazuya Kato and Chikara Nakayama for series of joint works on log Hodge theory, from which he learns a lot and enjoys exciting studies. He thanks Hiroshi Iritani for pointing out insufficient parts in the previous paper [30]. He also thanks Yukiko Konishi and Satoshi Minabe, together with Iritani, for stimulating seminars on the present topic. The author thanks the referees for careful readings and useful comments.

Notation.
Fix $\Lambda := (H_0, W, (\langle\ ,\ \rangle_w)_w, (h^{p,q})_{p,q})$, where
H_0 is a free $\mathbf{Z}$-module of finite rank,

W is an increasing filtration on $H_{0,\mathbf{Q}} := \mathbf{Q} \otimes H_0$,
$\langle\, , \rangle_w$ is a non-degenerate $(-1)^w$-symmetric bilinear form on gr^W_w,
$(h^{p,q})_{p,q}$ is a set of Hodge numbers.
D : the classifying space of graded polarized mixed Hodge structures for the data Λ, consisting of all Hodge filtrations.
$\check{D}$: the "compact dual" of D.
$G_A := \mathrm{Aut}(H_{0,A}, W, (\langle\, , \rangle_w)_w)$, where $H_{0,A} := A \otimes H_0$ ($A = \mathbf{Z}, \mathbf{Q}, \mathbf{R}, \mathbf{C}$),
$\mathfrak{g}_A := \mathrm{Lie} G_A = \mathrm{End}(H_{0,A}, W, (\langle\, , \rangle_w)_w)$ ($A = \mathbf{Q}, \mathbf{R}, \mathbf{C}$).

1 Log mixed Hodge theory

This section is a summary of log mixed Hodge theory from [17], [13], and [14]. We write a general form of these results as a brief guide for future use. Section 1.8 is adapted for the use in Section 3. The corresponding results in [13] and [14] are written in more general settings.

1.1. Category $\mathcal{B}$(log)

Let S be a subset of an analytic space Z. The *strong topology of S in Z* is the strongest one among those topologies on S in which, for any analytic space A and any morphism $f : A \to Z$ with $f(A) \subset S$ as sets, $f : A \to S$ is continuous. S is regarded as a local ringed space by the pullback sheaf of $\mathcal{O}_Z$.

Let $\mathcal{B}$ be the category of local ringed spaces S over $\mathbf{C}$ which have an open covering $(U_\lambda)_\lambda$ satisfying the following condition: For each λ, there exist an analytic space Z_λ, and a subset S_λ of Z_λ such that, as local ringed space over $\mathbf{C}$, U_λ is isomorphic to an open subset of S_λ which is endowed with the strong topology in Z_λ and the inverse image of $\mathcal{O}_{Z_\lambda}$.

A *log structure* on a local ringed space S is a sheaf of monoids M on S together with a homomorphisim $\alpha : M \to \mathcal{O}_S$ such that $\alpha^{-1}\mathcal{O}_S^\times \xrightarrow{\sim} \mathcal{O}_S^\times$. *fs* log structure means, locally on the underlying space, the log structure has a chart which is finitely generated, integral and saturated.

Let $\mathcal{B}$(log) be the category of objects of $\mathcal{B}$ endowed with an fs log structure.

A log analytic space is called *log smooth* if, locally, it is isomorphic to an open set of a toric variety endowed with the canonical log structure. A *log manifold* is a log local ringed space over $\mathbf{C}$ which has an open covering $(U_\lambda)_\lambda$ satisfying the following condition: For each λ, there exist a log smooth fs log analytic space Z_λ, a finite subset I_λ of global log differential 1-forms $\Gamma(Z_\lambda, \omega^1_{Z_\lambda})$, and an isomorphism of log local ringed spaces over $\mathbf{C}$ between U_λ and an open subset of $S_\lambda := \{z \in Z_\lambda \mid \text{the image of } I_\lambda \text{ in the stalk } \omega^1_z \text{ is zero}\}$ in the strong topology in Z_λ.

1.2. Ringed space $(S^{\log}, \mathcal{O}_S^{\log})$

The ringed space $(S^{\log}, \mathcal{O}_S^{\log})$ was defined for fs log schemes by K. Kato and C. Nakayama in [8]. It was generalized for the category $\mathcal{B}(\log)$ in [15].

Let $S \in \mathcal{B}(\log)$. As a set define

$S^{\log} := \{(s,h) \mid s \in S, h : M_s^{\mathrm{gp}} \to \mathbf{S}^1$ homomorphism s.t. $h(u) = u/|u|$ if $u \in \mathcal{O}_{S,s}^{\times}\}$.

Endow $S^{\log}$ with the weakest topology such that the following two maps are continuous.

(1) $\tau : S^{\log} \to S, (s,h) \mapsto s$.

(2) For any open set $U \subset S$ and any $f \in \Gamma(U, M^{\mathrm{gp}})$, $\tau^{-1}(U) \to \mathbf{S}^1$, $(s,h) \mapsto h(f_s)$.

Then, τ is proper and surjective with fiber $\tau^{-1}(s) = (\mathbf{S}^1)^{r(s)}$, where $r(s)$ is the rank of $(M^{\mathrm{gp}}/\mathcal{O}_S^{\times})_s$ which varies with $s \in S$.

Define a sheaf $\mathcal{L}$ on $S^{\log}$ as the fiber product:

$$\begin{array}{ccccc} \mathcal{L} & \xrightarrow{\exp} & \tau^{-1}(M^{\mathrm{gp}}) & \ni & (f \text{ at } (s,h)) \\ \downarrow & & \downarrow & & \downarrow \\ \mathrm{Cont}(*, i\mathbf{R}) & \xrightarrow{\exp} & \mathrm{Cont}(*, \mathbf{S}^1) & \ni & h(f). \end{array}$$

Let $\iota : \tau^{-1}(\mathcal{O}_S) \to \mathcal{L}$ be a morphism induced from

$$\begin{array}{ccccc} f & \in & \tau^{-1}(\mathcal{O}_S) & \xrightarrow{\exp} & \tau^{-1}(\mathcal{O}_S^{\times}) \subset \tau^{-1}(M^{\mathrm{gp}}) \\ \downarrow & & \downarrow & & \downarrow \\ (f - \bar{f})/2 & \in & \mathrm{Cont}(*, i\mathbf{R}) & \xrightarrow{\exp} & \mathrm{Cont}(*, \mathbf{S}^1). \end{array}$$

Define

$$\mathcal{O}_S^{\log} := \frac{\tau^{-1}(\mathcal{O}_S) \otimes \mathrm{Sym}_{\mathbf{Z}}(\mathcal{L})}{(f \otimes 1 - 1 \otimes \iota(f) \mid f \in \tau^{-1}(\mathcal{O}_S))}.$$

Thus $\tau : (S^{\log}, \mathcal{O}_S^{\log}) \to (S, \mathcal{O}_S)$ is a morphism of ringed spaces over $\mathbf{C}$. For $s \in S$ and $t \in S^{\log}$ lying over s, let $t_j \in \mathcal{L}_t$ $(1 \le j \le r(s))$ be elements such that their images in $(M^{\mathrm{gp}}/\mathcal{O}_S^{\times})_s$ of $\exp(t_j)$ form a basis. Then, $\mathcal{O}_{S,t}^{\log} = \mathcal{O}_{S,s}[t_j\ (1 \le j \le r(s))]$ is a polynomial ring.

1.3. Toric variety

Toric varieties offer typical examples of $S^{\log}$ and also they are building blocks of fine moduli spaces of log mixed Hodge structures.

Let $\sigma \subset \mathfrak{g}_{\mathbf{R}}$ be a *nilpotent cone*, i.e., a sharp cone, $\sigma \cap (-\sigma) = \{0\}$, generated by a finite number of mutually commutative nilpotent elements. Assume that the cone generators of σ can be taken from $\mathfrak{g}_{\mathbf{Q}}$. Let Γ be a subgroup of $G_{\mathbf{Z}}$.

Define a monoid $\Gamma(\sigma) := \Gamma \cap \exp(\sigma)$ and the dual monoid $P(\sigma) := \Gamma(\sigma)^\vee = \mathrm{Hom}(\Gamma(\sigma), \mathbf{N})$. Define a toric variety and a torus by

$$\begin{aligned}\mathrm{toric}_\sigma :&= \mathrm{Spec}(\mathbf{C}[P(\sigma)])(\mathbf{C}) = \mathrm{Hom}\,(P(\sigma), \mathbf{C}^{\mathrm{mult}}) \supset \mathrm{torus}_\sigma : \\ &= \mathrm{Hom}\,(P(\sigma)^{\mathrm{gp}}, \mathbf{C}^\times),\end{aligned}$$

where $\mathbf{C}^{\mathrm{mult}}$ is $\mathbf{C}$ regarded as a monoid by multiplication and $P(\sigma)^{\mathrm{gp}}$ is the group generated by the monoid $P(\sigma)$. The exponential sequence $0 \to \mathbf{Z} \to \mathbf{C} \to \mathbf{C}^\times \to 1$ induces the universal covering of the torus

$$0 \to \mathrm{Hom}\,(P(\sigma)^{\mathrm{gp}}, \mathbf{Z}) \to \mathrm{Hom}\,(P(\sigma)^{\mathrm{gp}}, \mathbf{C}) \xrightarrow{\mathbf{e}} \mathrm{Hom}\,(P(\sigma)^{\mathrm{gp}}, \mathbf{C}^\times) \to 1,$$

where $\mathrm{Hom}\,(P(\sigma)^{\mathrm{gp}}, \mathbf{Z}) = \Gamma(\sigma)^{\mathrm{gp}}$ is considered as the fundamental group of torus_σ, and $\mathbf{e}(z \otimes \gamma) := e^{2\pi iz} \otimes \gamma$ $(z \in \mathbf{C},\ \gamma \in \Gamma(\sigma)^{\mathrm{gp}} = \mathrm{Hom}\,(P(\sigma)^{\mathrm{gp}}, \mathbf{Z}))$.

Fix the above cone σ. For a face ρ of σ, define $\tilde{P}(\rho) := \{l \in P(\sigma)^{\mathrm{gp}} \mid l(\rho) \geq 0\}$. Then we have an open covering

$$\mathrm{toric}_\sigma = \mathrm{Spec}(\mathbf{C}[P(\sigma)])(\mathbf{C}) = \bigcup_{\rho \prec \sigma} \mathrm{Spec}(\mathbf{C}[\tilde{P}(\rho)])(\mathbf{C}).$$

We now recall a stratification. Fix the above cone σ and let ρ be a face of the cone σ. Then, we have a homomorphism $P(\rho)\lambda P(\sigma)$ and hence a morphism $\mathrm{toric}_\rho \to \mathrm{toric}_\sigma$. The origin $0_\rho \in \mathrm{toric}_\rho$ is the monoid homomorphism $P(\sigma) \to \mathbf{C}^{\mathrm{mult}}$ sending 1 to 1 and all the other elements of $P(\rho)$ to 0, which is sent to a point of toric_σ by the above morphism. Then, as a set, we have a stratification into torus orbits

$$\mathrm{toric}_\sigma = \{\mathbf{e}(z)0_\rho \mid \rho \prec \sigma, z \in \sigma_{\mathbf{C}}/(\rho_{\mathbf{C}} + \log \Gamma(\sigma)^{\mathrm{gp}})\}.$$

Here $\mathbf{e}(c \log \gamma) := \mathbf{e}(c \otimes \gamma) = e^{2\pi ic} \otimes \gamma$ $(c \in \mathbf{C}, \gamma \in \Gamma(\sigma)^{\mathrm{gp}})$.

For $S := \mathrm{toric}_\sigma$, the polar coordinate $\mathbf{R}_{\geq 0} \times \mathbf{S}^1 \to \mathbf{R}_{\geq 0}\mathbf{S}^1 = \mathbf{C}$ induces $\tau : S^{\log} \to S$ as

$$\begin{aligned}\tau : S^{\log} &= \mathrm{Hom}\,(P(\sigma), \mathbf{R}^{\mathrm{mult}}_{\geq 0} \times \mathrm{Hom}\,(P(\sigma), \mathbf{S}^1) \\ &= \{(\mathbf{e}(iy)0_\rho, \mathbf{e}(x)) \mid \rho \prec \sigma, x \in \sigma_{\mathbf{R}}/(\rho_{\mathbf{R}} + \log \Gamma(\sigma)^{\mathrm{gp}}), y \in \sigma_{\mathbf{R}}/\rho_{\mathbf{R}}\} \\ &\to S = \mathrm{Hom}\,(P(\sigma), \mathbf{C}^{\mathrm{mult}}),\end{aligned}$$

$$\tau(\mathbf{e}(ib)0_\rho, \mathbf{e}(a)) = \mathbf{e}(a + ib)0_\rho.$$

Since $0 \to \rho_{\mathbf{R}}/\log \Gamma(\rho)^{\mathrm{gp}} \to \sigma_{\mathbf{R}}/\log \Gamma(\sigma)^{\mathrm{gp}} \to \sigma_{\mathbf{R}}/(\rho_{\mathbf{R}} + \log \Gamma(\sigma)^{\mathrm{gp}}) \to 0$ is exact, the fiber of τ, as a set, is described as

$$\tau^{-1}(\mathbf{e}(a+ib)0_\rho) = \{(\mathbf{e}(ib)0_\rho, \mathbf{e}(a+x)) \mid x \in \rho_{\mathbf{R}}/\log \Gamma(\rho)^{\mathrm{gp}}\} \simeq (\mathbf{S}^1)^r,$$

where $r = r(\rho) := \mathrm{rank}\,\rho$ varies with $\rho \prec \sigma$.

Let $H_\sigma = (H_{\sigma,\mathbf{Z}}, W, (\langle\ ,\ \rangle_w)_w)$ be the canonical local system endowed with the weight filtration and the polarizations on graded quotients on $S^{\log}$, which are given by the representation $\pi_1(S^{\log}) = \Gamma(\sigma)^{\mathrm{gp}} \subset G_{\mathbf{Z}} = \mathrm{Aut}(H_0, W, (\langle\ ,\ \rangle_w)_w)$.

1.4. Local systems on $(S^{\log}, \mathcal{O}_S^{\log})$

We recall three results about local systems on $(S^{\log}, \mathcal{O}_S^{\log}) \in \mathcal{B}(\log)$ from [17], Sect. 2.3.

Let L be a locally constant sheaf of abelian groups on $S^{\log}$. For $s \in S$ and $t \in S^{\log}$ lying over s, we call the action of $\pi_1(s^{\log}) = \pi_1(\tau^{-1}(s))$ on L_t the *local monodromy* of L at t. We say the local monodromy of L is *unipotent* if the local monodromy of L at t is unipotent for any $t \in S^{\log}$.

Let $s \in S$. Let $(q_j)_{1 \le j \le n}$ be a finite family of elements of $M_{S,s}^{\mathrm{gp}}$ whose image in $(M_S^{\mathrm{gp}}/\mathcal{O}_S^\times)_s$ is a free basis, and let $(\gamma_j)_{1 \le j \le n}$ be the dual basis of $\pi_1(s^{\log})$, that is $[\gamma_j, q_k] = (2\pi i)\delta_{jk}$ where $[\ ,\ ]$ is the pairing given by $\pi_1(s^{\log}) \simeq \mathrm{Hom}\,(M_s^{\mathrm{gp}}/\mathcal{O}_s^\times, \mathbf{Z})$.

Let L be a locally constant sheaf on $S^{\log}$ of free $\mathbf{Z}$-modules of finite rank. Let $s \in S$ and $t \in \tau^{-1}(s)$, and assume that the local monodromy of L at t is unipotent. For a fixed t, we denote L_0 the constant sheaf on $S^{\log}$ with fiber L_t. Let $L_{0,\mathbf{Q}} = \mathbf{Q} \otimes_A L_0$, and let $N_j : L_{0,\mathbf{Q}} \to L_{0,\mathbf{Q}}$ be the endomorphism of constant sheaf which is induced by the logarithm of the monodromy action of γ_j on the stalk L_t of the locally constant sheaf L. Lift q_j in $\Gamma(S, M_S^{\mathrm{gp}})$ (by replacing S by an open neighborhood of s), and let

$$\xi = \exp\Big(\sum_{j=1}^{n} (2\pi i)^{-1} \log(q_j) \otimes N_j\Big) : \mathcal{O}_S^{\log} \otimes_{\mathbf{Q}} L_{0,\mathbf{Q}} \xrightarrow{\sim} \mathcal{O}_S^{\log} \otimes_{\mathbf{Q}} L_{0,\mathbf{Q}}.$$

Note that the operator ξ depends on the choices of the branches of $\log(q_j)$ in $\mathcal{O}_S^{\log}$ locally on $S^{\log}$, but that the subsheaf $\xi^{-1}(1 \otimes L_0)$ of $\mathcal{O}_S^{\log} \otimes_{\mathbf{Z}} L_0$ is independent of the choices and hence is defined globally on $S^{\log}$.

The following proposition shows that the locally constant sheaf L is embedded in $\mathcal{O}_S^{\log} \otimes L_0$.

Proposition 1.4.1. ([17], Prop. 2.3.2). *Let the situation be as above. If we replace S by some open neighborhood of s, we have an isomorphism of $\mathcal{O}_S^{\log}$-modules*

$$\nu : \mathcal{O}_S^{\log} \otimes_A L \xrightarrow{\sim} \mathcal{O}_S^{\log} \otimes_A L_0$$

satisfying the following condition (1).

(1) *The restriction of ν to $L = 1 \otimes L$ induces an isomorphism of locally constant sheaves $\nu : L \xrightarrow{\sim} \xi^{-1}(1 \otimes L_0)$.*

If we take suitable branches $\log(q_j)_{t,0}$ *in* $\mathcal{O}^{\log}_{S,t}$ *of the germs* $\log(q_j)_t$ *at* t $(1 \le j \le n)$*, we can take an isomorphism* ν *which satisfies above* (1) *and also the following* (2).

(2) *The branch* $\xi_{t,0}$ *of the germ* ξ_t*, defined by the fixed branches* $\log(q_j)_{t,0}$ *of the germs* $\log(q_j)_t$*, satisfies* $\nu(1 \otimes v) = \xi^{-1}_{t,0}(1 \otimes v)$ *for any* $v \in L_t = L_0$.

The following proposition yields a log Hodge theoretic understanding [ibid, Sect. 2.3.6] of the canonical extension of Deligne in [3].

Proposition 1.4.2. ([17], Prop. 2.3.3). *Let* $S \in \mathcal{B}(\log)$ *and let* L *be a locally constant sheaf of finite dimensional* $\mathbf{C}$*-vector spaces on* $S^{\log}$.

(i) *If the local monodromy of* L *is unipotent, the* $\mathcal{O}_S$*-module* $\mathcal{M} := \tau_*(\mathcal{O}^{\log}_S \otimes_{\mathbf{C}} L)$ *is locally free of finite rank, and we have an isomorphism* $\mathcal{O}^{\log}_S \otimes_{\mathcal{O}_S} \mathcal{M} \xrightarrow{\sim} \mathcal{O}^{\log}_S \otimes_{\mathbf{C}} L$.

(ii) *Conversely, assume that there are a locally free* $\mathcal{O}_S$*-module* $\mathcal{M}$ *of finite rank on* S *and an isomorphism of* $\mathcal{O}^{\log}_S$*-modules* $\mathcal{O}^{\log}_S \otimes_{\mathcal{O}_S} \mathcal{M} \simeq \mathcal{O}^{\log}_S \otimes_{\mathbf{C}} L$*. Then the local monodromy of* L *is unipotent and* $\mathcal{M} \xrightarrow{\sim} \tau_*(\mathcal{O}^{\log}_S \otimes_{\mathbf{C}} L)$.

Proposition 1.4.3. ([17], Prop. 2.3.4). *Let* $S \in \mathcal{B}(\log)$*, let* L *be a locally constant sheaf on* $S^{\log}$ *of free* $\mathbf{Q}$*-modules of finite rank. Assume that the local monodromy of* L *is unipotent.*

(i) *There exists a unique* $\mathbf{Q}$*-homomorphism*

$$\mathcal{N} : L \to (M^{\mathrm{gp}}_S / \mathcal{O}^{\times}_S) \otimes L$$

satisfying the following condition (1).

(1) *For any* $s \in S$*, any* $t \in S^{\log}$ *lying over* s*, and any* $\gamma \in \pi_1(s^{\log})$*, if* $h_\gamma : (M^{\mathrm{gp}}_S / \mathcal{O}^{\times}_S)_s \to \mathbf{Z}$ *denotes the homomorphism corresponding to* γ *by* $\pi_1(s^{\log}) \simeq \mathrm{Hom}\,(M^{\mathrm{gp}}_s / \mathcal{O}^{\times}_s, \mathbf{Z})$*, the composition* $L_t \xrightarrow{\mathcal{N}} (M^{\mathrm{gp}}_S / \mathcal{O}^{\times}_S)_s \otimes L_t \xrightarrow{h_\gamma} L_t$ *coincides with the logarithm of the action of* γ *on* L_t.

(ii) *Assume that* S *is an fs log point* $\{s\}$*. Let*

$$\mathcal{N}' : L \to \omega^1_s \otimes L$$

be the composition of $\mathcal{N}$ *and the* $\mathbf{Q}$*-linear map* $M^{\mathrm{gp}}_s / \mathcal{O}^{\times}_s \otimes L \to \omega^1_s \otimes L, f \otimes v \mapsto (2\pi i)^{-1} d\log(f) \otimes v$*, and let* $1 \otimes \mathcal{N}' : \mathcal{O}^{\log}_s \otimes L \to \omega^{1,\log}_s \otimes L$ *be the* $\mathcal{O}^{\log}_s$*-linear homomorphism induced by* $\mathcal{N}'$*. Let* $\mathcal{M} := H^0(s^{\log}, \mathcal{O}^{\log}_s \otimes L) = \tau_*(\mathcal{O}^{\log}_s \otimes L)$*. Then the restriction* $\mathcal{M} \to \omega^1_s \otimes_{\mathbf{C}} \mathcal{M}$ *of* $d \otimes 1_L : \mathcal{O}^{\log}_s \otimes L \to \omega^{1,\log}_s \otimes L$ *coincides with the restriction of* $1 \otimes \mathcal{N}'$ *to* $\mathcal{M}$.

$\mathcal{N}$ in the above proposition is described as follows. Assume $L = \xi^{-1}(1 \otimes L_0)$ as in the first proposition. Then $\mathcal{N}(\xi^{-1}(1 \otimes v)) := \sum_{j=1}^{n} q_j \otimes \xi^{-1}(1 \otimes N_j v)$ for $v \in L_0$.

1.5. Graded polarized log mixed Hodge structure

Let $S \in \mathcal{B}(\log)$. A *pre-graded polarized log mixed Hodge structure on* S is a tuple $H = (H_{\mathbf{Z}}, W, (\langle\ ,\ \rangle_w)_w, H_{\mathcal{O}})$ consisting of a local system of $\mathbf{Z}$-free modules $H_{\mathbf{Z}}$ of finite rank on $S^{\log}$, an increasing filtration W of $H_{\mathbf{Q}} := \mathbf{Q} \otimes H_{\mathbf{Z}}$, a non-degenerate $(-1)^w$-symmetric $\mathbf{Q}$-bilinear form $\langle\ ,\ \rangle_w$ on gr^W_w, a locally free $\mathcal{O}_S$-module $H_{\mathcal{O}}$ on S, a specified isomorphism $\mathcal{O}_S^{\log} \otimes_{\mathbf{Z}} H_{\mathbf{Z}} \simeq \mathcal{O}_S^{\log} \otimes_{\mathcal{O}_S} H_{\mathcal{O}}$ (*log Riemann-Hilbert correspondence*), and a specified decreasing filtration $FH_{\mathcal{O}}$ of $H_{\mathcal{O}}$ such that $F^pH_{\mathcal{O}}$ and $H_{\mathcal{O}}/F^pH_{\mathcal{O}}$ are locally free. Put $F^p := \mathcal{O}_S^{\log} \otimes_{\mathcal{O}_S} F^pH_{\mathcal{O}}$. Then $\tau_* F^p = F^pH_{\mathcal{O}}$. For each integer w, the orthogonality condition $\langle F^p(\mathrm{gr}^W_w), F^q(\mathrm{gr}^W_w)\rangle_w = 0$ $(p+q > w)$ is imposed.

A pre-graded polarized log mixed Hodge structure on S is a *graded polarized log mixed Hodge structure on* S if its pullback to each $s \in S$ is a graded polarized log mixed Hodge structure on s in the following sense.

Let $(H_{\mathbf{Z}}, W, (\langle\ ,\ \rangle_w)_w, H_{\mathcal{O}})$ be a pre-graded polarized log mixed Hodge structure on a log point s. It is a *graded polarized log mixed Hodge structure* if it satisfies the following three conditions.

(1) (Admissibility). For each logarithm N of the local monodromy of the local system $(H_{\mathbf{R}}, W, (\langle\ ,\ \rangle_w)_w)$, there exists a W-relative N-filtration $M(N, W)$.

(2) (Griffiths transversality). For any integer p, $\nabla F^p \subset \omega_s^{1,\log} \otimes F^{p-1}$ is satisfied, where $\omega_s^{1,\log}$ is the sheaf of $\mathcal{O}_s^{\log}$-module of log differential 1-forms on $(s^{\log}, \mathcal{O}_s^{\log})$, and $\nabla = d \otimes 1_{H_{\mathbf{Z}}} : \mathcal{O}_s^{\log} \otimes H_{\mathbf{Z}} \to \omega_s^{1,\log} \otimes H_{\mathbf{Z}}$ is the log Gauss-Manin connection.

(3) (Positivity). For a point $t \in s^{\log}$ and a $\mathbf{C}$-algebra homomorphism $a : \mathcal{O}_{s,t}^{\log} \to \mathbf{C}$, define a filtration $F(a) := \mathbf{C} \otimes_{\mathcal{O}_{s,t}^{\log}} F_t$ on $H_{\mathbf{C},t}$. Then, $(H_{\mathbf{Z},t}(\mathrm{gr}^W_w), \langle\ ,\ \rangle_w, F(a))$ is a polarized Hodge structure of weight w in the usual sense if a is sufficiently twisted, i.e., for $(q_j)_{1 \le j \le n} \subset M_s$ inducing generators of $M_s/\mathcal{O}_s^{\times}$, $|\exp(a(\log q_j))| \ll 1$ for any j.

1.6. Nilpotent orbit

Let $\sigma \subset \mathfrak{g}_{\mathbf{R}}$ be a nilpotent cone (see 1.3). A subset $Z \subset \check{D}$ is σ*-nilpotent orbit* if the following (1)–(4) hold for $F \in Z$.

(1) $Z = \exp(\sigma_{\mathbf{C}})F$.
(2) For any $N \in \sigma$, there exists W-relative N-filtration $M(N, W)$.
(3) For any $N \in \sigma$ any p, $NF^p \subset F^{p-1}$.
(4) If $N_1, \ldots, N_n$ generate σ and $y_j \gg 0$ for any j, then $\exp(\sum_j iy_jN_j)F \in D$.

A *weak fan* Σ *in* $\mathfrak{g}_{\mathbf{Q}}$ is a set of nilpotent cones in $\mathfrak{g}_{\mathbf{R}}$, defined over $\mathbf{Q}$, which satisfies the following three conditions.

(5) Every $\sigma \in \Sigma$ is admissible relative to W.
(6) If $\sigma \in \Sigma$ and $\tau \prec \sigma$, then $\tau \in \Sigma$.

(7) If $\sigma, \sigma' \in \Sigma$ have a common interior point and if there exists $F \in \check{D}$ such that (σ, F) and (σ', F) generate nilpotent orbits, then $\sigma = \sigma'$.

Let Σ be a weak fan and Γ be a subgroup of $G_{\mathbf{Z}}$. Σ and Γ are *strongly compatible* if the following two conditions are satisfied.

(8) If $\sigma \in \Sigma$ and $\gamma \in \Gamma$, then $\mathrm{Ad}(\gamma)\sigma \in \Sigma$.
(9) For any $\sigma \in \Sigma$, σ is generated by $\log \Gamma(\sigma)$, where $\Gamma(\sigma) := \Gamma \cap \exp(\sigma)$.

1.7. Moduli of log mixed Hodge structures of type Φ

Let $\Phi := (\Lambda, \Sigma, \Gamma)$ be a data consisting of a Hodge data Λ (in Notation), a weak fan Σ and a subgroup Γ of $G_{\mathbf{Z}}$ such that Σ and Γ are strongly compatible (1.6).

Let $\sigma \in \Sigma$ and $S := \mathrm{toric}_\sigma$. Let $H_\sigma = (H_{\sigma,\mathbf{Z}}, W, (\langle\ ,\ \rangle_w)_w)$ be the canonical local system $H_{\sigma,\mathbf{Z}}$ endowed with the weight filtration W and the polarizations $\langle\ ,\ \rangle_w$ on the graded quotients gr^W_w ($w \in \mathbf{Z}$) over $S^{\log}$, which is determined by the representation $\Gamma \subset G_{\mathbf{Z}} = \mathrm{Aut}(H_0, W, (\langle\ ,\ \rangle_w)_w)$.

Let $\check{E}_\sigma := \mathrm{toric}_\sigma \times \check{D}$. The *universal pre-graded polarized log mixed Hodge structure H on $\check{E}_\sigma$* is given by H_σ together with the isomorphism $\mathcal{O}^{\log}_{\check{E}_\sigma} \otimes_{\mathbf{Z}} H_{\sigma,\mathbf{Z}} = \mathcal{O}^{\log}_{\check{E}_\sigma} \otimes_{\mathcal{O}_{\check{E}_\sigma}} H_{\mathcal{O}}$ (1.5), where $H_{\mathcal{O}} := \mathcal{O}_{\check{E}_\sigma} \otimes H_0$ is the free $\mathcal{O}_{\check{E}_\sigma}$-module coming from that on $\check{D}$ endowed with the universal Hodge filtration F.

Let $E_\sigma := \{x \in \check{E}_\sigma \mid H|_x$ is a graded polarized log mixed Hodge structure on $x\}$. Note that slits appear in E_σ because of log-pointwise Griffiths transversality 1.5 (2) and positivity 1.5 (3), or equivalently 1.6 (3) and 1.6 (4) respectively.

As a set, define $D_\Sigma := \{(\sigma, Z) \mid \text{nilpotent orbit}, \sigma \in \Sigma, Z \subset \check{D}\}$. Let $\sigma \in \Sigma$. Assume that Γ is neat. A structure as an object of $\mathcal{B}(\log)$ on $\Gamma \backslash D_\Sigma$ is introduced by a diagram:

$$\begin{array}{ccc} E_\sigma & \overset{\text{GPLMH}}{\subset} & \check{E} := \mathrm{toric}_\sigma \times \check{D} \\ \Big\downarrow {\scriptstyle \sigma_{\mathbf{C}}\text{-torsor}} & & \\ \Gamma(\sigma)^{\mathrm{gp}} \backslash D_\sigma & & \\ \Big\downarrow {\scriptstyle \text{loc. isom.}} & & \\ \Gamma \backslash D_\Sigma & & \end{array}$$

The action of $h \in \sigma_{\mathbf{C}}$ on $(\mathbf{e}(a)0_\rho, F) \in E_\sigma$ is $(\mathbf{e}(h+a)0_\rho, \exp(-h)F)$, and the projection is $(\mathbf{e}(a)0_\rho, F) \mapsto [\rho, \exp(\rho_{\mathbf{C}} + a)F]$.

Let $S \in \mathcal{B}(\log)$. A *log mixed Hodge structure of type Φ on S* is a pre-graded polarized log mixed Hodge structure $H = (H_{\mathbf{Z}}, W, (\langle\ ,\ \rangle_w)_w, H_{\mathcal{O}})$ endowed with Γ-level structure $\mu \in H^0(S^{\log}, \Gamma \backslash \mathcal{I}som((H_{\mathbf{Z}}, W, (\langle\ ,\ \rangle_w)_w), (H_0, W, (\langle\ ,\ \rangle_w)_w)))$ satisfying the following condition: For any point $s \in S$, any point $t \in \tau^{-1}(s) = s^{\log}$ and any representative $\tilde{\mu}_t : H_{\mathbf{Z},t} \xrightarrow{\sim} H_0$, there exists $\sigma \in \Sigma$ such that σ

contains $\tilde{\mu}_t \pi_1^+(s^{\log}) \tilde{\mu}_t^{-1}$ and $(\sigma, \tilde{\mu}_t(\mathbf{C} \otimes_{\mathcal{O}_{S,t}^{\log}} F_t))$ generates a nilpotent orbit. Here $\pi_1^+(s^{\log}) := \mathrm{Image}(\mathrm{Hom}((M_S/\mathcal{O}_S^\times)_s, \mathbf{N}) \hookrightarrow \pi_1(s^{\log}) \to \mathrm{Aut}(H_{\mathbf{Z},t}))$ is the local monodromy monoid of $H_{\mathbf{Z}}$ at s (cf. [17], Sect. 3.3.2). (Then, the smallest such σ exists.)

Theorem 1.7.1. *For a given data Φ, we have the following.*
(i) $\Gamma \backslash D_\Sigma \in \mathcal{B}(\log)$, *which is Hausdorff. If Γ is neat, $\Gamma \backslash D_\Sigma$ is a log manifold.*
(ii) *On $\mathcal{B}(\log)$, $\Gamma \backslash D_\Sigma$ represents a functor* LMH_Φ *of log mixed Hodge structures of type Φ.*

Log period map. *Given Φ. Let $S \in \mathcal{B}(\log)$. Then we have an isomorphism*

$$\mathrm{LMH}_\Phi(S) \tilde{\to} \mathrm{Map}(S, \Gamma \backslash D_\Sigma),$$
$$H \mapsto (S \ni s \mapsto [\sigma, \exp(\sigma_{\mathbf{C}}) \tilde{\mu}_t(\mathbf{C} \otimes_{\mathcal{O}_{S,t}^{\log}} F_t)])(t \in \tau^{-1}(s)),$$

which is functorial in S.

A log period map is a unified compactification of a period map and a normal function of Griffiths.

The above $\Gamma \backslash D_\Sigma$ is the fine moduli of log mixed Hodge structures of type Φ, whose underlying coarse moduli, in the sense of log points, is the set of equivalence classes of all nilpotent orbits of specified type.

1.8. Néron model for admissible normal function

We review some results from [14], Theorem 1.3, [13], Section 6.1, and [11], Section 8 adapted to the situation 0.4 in Introduction.

For a pure case $h^{p,q} = 1$ $(p+q = 3,\ p,q \geq 0)$ and $h^{p,q} = 0$ otherwise, a complete fan is constructed in [17], Section 12.3. For a mixed case $h^{p,q} = 1$ (the above (p,q), plus $(p,q) = (2,2)$) and $h^{p,q} = 0$ otherwise, over the above fan, a fan of Néron model for given admissible normal function is constructed in [14], Theorem 3.1, and we have a Néron model in the following sense.

Let $S \in \mathcal{B}(\log)$, $U := S_{\mathrm{triv}} \subset S$ (consisting of those points with trivial log structure), $H_{(-1)}$ be a polarized variation of Hodge structure of weight -1 (Tate-twisted by 2 for $\mathcal{H}$ in Introduction 0.4) on U and $L_{\mathbf{Q}}$ be a local system of $\mathbf{Q}$-vector spaces which is an extension of $\mathbf{Q}$ by $H_{(-1),\mathbf{Q}}$. An admissible normal function over U for $H_{(-1)}$ underlain by the local system $L_{\mathbf{Q}}$ can be regarded as an admissible variation of mixed Hodge structure which is an extension of $\mathbf{Z}$ by $H_{(-1)}$ and lies over local system $L_{\mathbf{Q}}$.

For any given unipotent admissible normal function over U as above, $H_{(-1)}$ and $L_{\mathbf{Q}}$ extend to a polarized log mixed Hodge structure on S and a local system on $S^{\log}$, respectively, denoted by the same symbols, and there is a relative log manifold $J_{L_{\mathbf{Q}}}$ over S which is strict over S (i.e., endowed with the pullback log structure from S) and which represents the following functor on $\mathcal{B}/S^\circ$ ($S^\circ \in \mathcal{B}$ is the underlying space of S):

$S' \mapsto$ {LMH H on S' satisfying $H(\mathrm{gr}^W_w) = H_{(w)}|_{S'}$ $(w = -1, 0)$ and $(*)$ below}/isom.

$(*)$ Locally on S', there is an isomorphism $H_{\mathbf{Q}} \simeq L_{\mathbf{Q}}$ on $(S')^{\log}$ preserving W.

Here $H_{(w)}|_{S'}$ is the pullback of $H_{(w)}$ by the structure morphism $S' \to S^\circ$, and S' is endowed with the pullback log structure from S.

Put $H' := H_{(-1)}$. In the present case, we have $J_{L_{\mathbf{Q}}} = \mathcal{E}xt^1_{\mathrm{LMH}/S}(\mathbf{Z}, H')$ by [13], Corollary 6.1.6. This is the subgroup of $\tau_*(H'_{\mathcal{O}^{\log}}/(F^0 + H'_{\mathbf{Z}}))$ restricted by admissibility condition and log-pointwise Griffiths transversality condition ([11], Section 8, cf. 1.5). Let $\tilde{J}_{L_{\mathbf{Q}}}$ be the pullback of $J_{L_{\mathbf{Q}}}$ by $\tau_*(H'_{\mathcal{O}^{\log}}/F^0) \to \tau_*(H'_{\mathcal{O}^{\log}}/(F^0 + H'_{\mathbf{Z}}))$, and $\bar{J}_{L_{\mathbf{Q}}}$ be the image of $\tilde{J}_{L_{\mathbf{Q}}}$ by $\tau_*(H'_{\mathcal{O}^{\log}}/F^0) \to \tau_*(H'_{\mathcal{O}^{\log}}/F^{-1})$. Then, by using the polarization, we have a commutative diagram:

$$\begin{array}{ccccc}
J_{L_{\mathbf{Q}}} = \mathcal{E}xt^1_{\mathrm{LMH}/S}(\mathbf{Z}, H') & \subset & \tau_*(H'_{\mathcal{O}^{\log}}/(F^0 + H'_{\mathbf{Z}})) & \overset{\mathrm{pol}}{\underset{\sim}{\to}} & \tau_*((F^0)^*/H'_{\mathbf{Z}}) \\
\uparrow & & \uparrow & & \uparrow \\
\tilde{J}_{L_{\mathbf{Q}}} & \subset & \tau_*(H'_{\mathcal{O}^{\log}}/F^0) & \overset{\mathrm{pol}}{\underset{\sim}{\to}} & \tau_*((F^0)^*) \\
\downarrow & & \downarrow & & \downarrow \\
\bar{J}_{L_{\mathbf{Q}}} & \subset & \tau_*(H'_{\mathcal{O}^{\log}}/F^{-1}) & \overset{\mathrm{pol}}{\underset{\sim}{\to}} & \tau_*((F^1)^*).
\end{array}$$

2 Quintic threefolds

Let V be a quintic threefold in $\mathbf{P}^4$ and let V°_ψ be its mirror family (cf. [1], Sect. 4.2).

In this section, we give a correspondence table of A-model for V and B-model for V°_ψ. This is a precision and an expansion of our previous [30], Sect. 3 by using $\hat{\Gamma}$-integral structure of Iritani [7]. We will use this table in Section 3 below.

2.1. Quintic mirror family

Following [21], [23], etc., we briefly recall the construction of the mirror family V°_ψ by quotient method. Let $V_\psi : f := \sum_{j=1}^5 x_j^5 - 5\psi \prod_{j=1}^5 x_j = 0$ $(\psi \in \mathbf{P}^1)$ be the Dwork pencil of quintics in $\mathbf{P}^4$. Let μ_5 be the group consisting of the fifth roots of the unity in $\mathbf{C}$. Then the group $G := \{(a_j) \in (\mu_5)^5 \mid a_1 \ldots a_5 = 1\}$ acts on V_ψ by $x_j \mapsto a_j x_j$. Let V°_ψ be a crepant resolution of quotient singularity of V_ψ/G (cf. [21], [23]). Divide further by the action $(x_1, \ldots, x_5) \mapsto (a^{-1}x_1, x_2, \ldots, x_5)$ and $\psi \mapsto a\psi$ $(a \in \mu_5)$.

2.2. Picard–Fuchs equation on the mirror V°

Let Ω be a 3-form on V°_ψ with a log pole over $\psi = \infty$ induced from

$$\left(\frac{5}{2\pi i}\right)^3 \mathrm{Res}_{V_\psi}\left(\frac{\psi}{f}\sum_{j=1}^{5}(-1)^{j-1}x_j dx_1 \wedge \cdots \wedge \widehat{dx_j} \wedge \cdots \wedge dx_5\right).$$

Let $z := 1/(5\psi)^5$ and $\theta := zd/dz$. Let

$$\mathcal{L} := \theta^4 - 5z(5\theta+1)(5\theta+2)(5\theta+3)(5\theta+4)$$

be the Picard–Fuchs differential operator for Ω, i.e., $\mathcal{L}\Omega = 0$ via the Gauss-Manin connection ∇. There are three special points of the complex moduli:

$z = 0$: maximally unipotent monodromy point,
$z = \infty$: Gepner point,
$z = 1/5^5$: conifold point.

At $z = 0$, the Picard–Fuchs differential equation $\mathcal{L}y = 0$ has the indicial equation $\rho^4 = 0$ (ρ is indeterminate), i.e., maximally unipotent. By the Frobenius method, we have a basis of solutions $y_j(z)$ $(0 \le j \le 3)$ as follows. Let

$$\tilde{y}(-z;\rho) := \sum_{n=0}^{\infty}\frac{\prod_{m=1}^{5n}(5\rho+m)}{\prod_{m=1}^{n}(\rho+m)^5}(-z)^{n+\rho}$$

be a solution of $\mathcal{L}(\tilde{y}(-z;\rho)) = \rho^4(-z)^\rho$, and let

$$\tilde{y}(-z;\rho) = y_0(z) + y_1(z)\rho + y_2(z)\rho^2 + y_3(z)\rho^3 + \cdots, \quad y_j(z) := \frac{1}{j!}\frac{\partial^j \tilde{y}(-z;\rho)}{\partial\rho^j}\Big|_{\rho=0}$$

be the Taylor expansion at $\rho = 0$. Then, y_j $(0 \le j \le 3)$ form a basis of homogeneous solutions for the linear differential equation $\mathcal{L}y = 0$. We have

$$y_0 = f_0 = \textstyle\sum_{n=0}^{\infty}\frac{(5n)!}{(n!)^5}z^n,$$

$$y_1 = f_0\log z + f_1 = y_0\log z + 5\textstyle\sum_{n=1}^{\infty}\frac{(5n)!}{(n!)^5}\left(\sum_{j=n+1}^{5n}\frac{1}{j}\right)z^n,$$

$$2!y_2 = f_0(\log z)^2 + 2f_1\log z + f_2,$$

$$3!y_3 = f_0(\log z)^3 + 3f_1(\log z)^2 + 3f_2\log z + f_3,$$

where all f_j are holomorphic functions in z with $f_0(0) = 1$ and $f_j(0) = 0$ for $j > 0$.

Define the canonical parameters by $t := y_1/y_0$, $u := t/2\pi i$, and the canonical coordinate by $q := e^t = e^{2\pi iu}$ which is a specific chart of the log structure given by the divisor $(z = 0)$ of a disc in $\mathbf{P}^1$ and gives a mirror map.

Write $z = z(q)$ which is holomorphic in q. Then we have

$$\log z = 2\pi\, iu - \frac{5}{y_0(z(q))} \sum_{n=1}^{\infty} \frac{(5n)!}{(n!)^5} \Big(\sum_{j=n+1}^{5n} \frac{1}{j} \Big) z(q)^n.$$

The Gauss-Manin potential of V_z° is

$$\Phi_{\mathrm{GM}}^{V^\circ} = \frac{5}{2} \Big(\frac{y_1}{y_0} \frac{y_2}{y_0} - \frac{y_3}{y_0} \Big).$$

Let $\tilde{\Omega} := \Omega / y_0$ and $\delta := 2\pi iqd/dq = du$. Then, the Yukawa coupling at $z = 0$ is

$$Y := - \int_{V^\circ} \tilde{\Omega} \wedge \nabla_\delta \nabla_\delta \nabla_\delta \tilde{\Omega} = \frac{5}{(1 + 5^5 z) y_0(z)^2} \Big(\frac{qdz}{zdq} \Big)^3.$$

2.3. *A-model of quintic V*

Let V be a general quintic hypersurface in $\mathbf{P}^4$. Let H be the cohomology class of a hyperplane section of V in $\mathbf{P}^4$, $K(V) = \mathbf{R}_{>0}H$ be the Kähler cone of V, and u be the coordinate of $\mathbf{C}H$. Put $t := 2\pi\, iu$. A complexified Kähler moduli is defined as

$$\mathcal{KM}(V) := (H^2(V, \mathbf{R}) + iK(V)) / H^2(V, \mathbf{Z}) \xrightarrow{\sim} \Delta^*, \quad uH \mapsto q := e^{2\pi\, iu}.$$

Let $C \in H_2(V, \mathbf{Z})$ be the homology class of a line on V.

For $\beta = dC \in H_2(V, \mathbf{Z})$, define $q^\beta := q^d$. The Gromov–Witten potential of V is defined as

$$\Phi_{\mathrm{GW}}^V := \frac{1}{6} \int_V (2\pi\, iuH)^3 + \sum_{0 \neq \beta \in H_2(V, \mathbf{Z})} N_d q^\beta = \frac{5}{6} (2\pi\, i)^3 u^3 + \sum_{d>0} N_d q^d.$$

Here the Gromov–Witten invariant N_d is

$$\overline{M}_{0,0}(\mathbf{P}^4, d) \xleftarrow{\pi_1} \overline{M}_{0,1}(\mathbf{P}^4, d) \xrightarrow{e_1} \mathbf{P}^4,$$

$$N_d := \int_{\overline{M}_{0,0}(\mathbf{P}^4, d)} c_{5d+1}(\pi_{1*} e_1^* \mathcal{O}_{\mathbf{P}^4}(5)).$$

Note that $N_d = 0$ if $d \le 0$. Let $N_d = \sum_{k|d} n_{d/k} k^{-3}$. Then $n_{d/k}$ is the instanton number. (n_l here is different from n_l in Theorem 0.3.1.)

The differentials of $\Phi = \Phi_{\mathrm{GW}}^V$ are computed easily:

$$\frac{d\Phi}{du} = \frac{5}{2} (2\pi\, i)^3 u^2 + (2\pi\, i) \sum_{d>0} N_d d q^d, \quad \frac{d^2\Phi}{du^2} = 5 (2\pi\, i)^3 u + (2\pi\, i)^2 \sum_{d>0} N_d d^2 q^d.$$

2.4. *Integral structure*

As we stated in Introduction, we consider everything near the large radius point q_0 and the maximally unipotent monodromy point p_0. Let S be a neighborhood disc of q_0 (resp. p_0) in $\overline{\mathcal{KM}}(V)$ (resp. $\overline{\mathcal{M}}(V^\circ)$) for A-model of V (resp. for B-model of V°), and let S^* be $S \smallsetminus \{q_0\}$ (resp. $S \smallsetminus \{p_0\}$) for A-model (resp. B-model) (see 2.2, 2.3). Endow S with the log structure associated to the divisor $S \smallsetminus S^*$.

The B-model variation of Hodge structure $\mathcal{H}^{V^\circ}$ is the usual variation of Hodge structure arising from the smooth projective family $f : X \to S^*$ of the quintic mirrors over the punctured neighborhood of p_0. Its integral structure is the usual one $\mathcal{H}^{V^\circ}_{\mathbf{Z}} = R^3 f_* \mathbf{Z}$. This is compatible with the monodromy weight filtration M around p_0. Define $M_{k,\mathbf{Z}} := M_k \cap \mathcal{H}^{V^\circ}_{\mathbf{Z}}$ for all k.

For the A-model $\mathcal{H}^V$ on S^*, the locally free sheaf on S^*, the Hodge filtration, and the monodromy weight filtration M around q_0 are given by $\mathcal{H}^V_{\mathcal{O}} := \mathcal{O}_{S^*} \otimes (\bigoplus_{0 \le p \le 3} H^{2p}(V))$, $F^p := \mathcal{O}_{S^*} \otimes H^{\le 2(3-p)}(V)$, and $M_{2p} := H^{\ge 2(3-p)}(V)$, respectively. Iritani defined $\hat{\Gamma}$-integral structure in more general setting in [7], Definition 3.6. In the present case, it is characterized as follows. Let H and C be a hyperplane section and a line on V, respectively. Then, in the present case, a basis of the $\hat{\Gamma}$-integral structure is given by $\{s(\mathcal{E}) \mid \mathcal{E} \text{ is } \mathcal{O}_V, \mathcal{O}_H, \mathcal{O}_C, \mathcal{O}_{\mathrm{pt}}\}$ [ibid, Example 6.18], where $s(\mathcal{E})$ is a unique ∇^{even}-flat section satisfying an asymptotic condition

$$s(\mathcal{E}) \sim (2\pi i)^{-3} e^{-2\pi i u H} \cdot \hat{\Gamma}(T_V) \cdot (2\pi i)^{\deg/2} \mathrm{ch}(\mathcal{E})$$

at the large radius point q_0 when $\mathrm{Im}(u) \to \infty$ for each fixed $\mathrm{Re}(u)$. Here, for the Chern roots $c(T_V) = \prod_{j=1}^3 (1+\delta_j)$, the Gamma class $\hat{\Gamma}(T_V)$ is defined by

$$\begin{aligned}\hat{\Gamma}(T_V) := \prod_{j=1}^{3} \Gamma(1+\delta_j) &= \exp(-\gamma\, c_1(V) + \sum_{k \ge 2} (-1)^k (k-1)! \zeta(k) \mathrm{ch}_k(T_V)) \\ &= \exp(\zeta(2)\mathrm{ch}_2(T_V) - 2\zeta(3)\mathrm{ch}_3(T_V))\end{aligned}$$

where γ is the Euler constant, and $\deg|_{H^{2p}(V)} := 2p$. The important point is that this class $\hat{\Gamma}(T_V)$ plays the role of a "square root" of the Todd class in Hirzebruch-Riemann-Roch ([6], 1; [7], 1, (13)). Denote this $\hat{\Gamma}$-integral structure by $\mathcal{H}^V_{\mathbf{Z}}$. This is compatible with the monodromy weight filtration M and we define $M_{k,\mathbf{Z}} := M_k \cap \mathcal{H}^V_{\mathbf{Z}}$ for all k. The above asymptotic relation is actually computed as

$$\begin{aligned}
s^0 :=& s(\mathcal{O}_{\mathrm{pt}}) = \frac{1}{5} H^3, \\
s^1 :=& s(\mathcal{O}_C) = \frac{1}{5}(2\pi i)^{-1} H^2 + \frac{1}{5}(-u+1) H^3, \\
s^2 :=& s(\mathcal{O}_H) \sim (2\pi i)^{-2} H + \frac{5}{2}(2\pi i)^{-1}\Big(-u - \frac{1}{2}\Big) H^2 + \Big(\frac{1}{2}u^2 + \frac{1}{2}u + \frac{7}{12}\Big) H^3,
\end{aligned}$$

$$s^3 := s(\mathcal{O}_V) \sim (2\pi i)^{-3} - (2\pi i)^{-2} uH + (2\pi i)^{-1}\Big(\frac{1}{2}u^2 + \frac{5}{12}\Big)H^2 + \Big(-\frac{1}{6}u^3 - \frac{5}{12}u + \frac{5i\zeta(3)}{\pi^3}\Big)H^3.$$

Fixing an isomorphism of VHS in (2) in Introduction, we also use s^p for the corresponding ∇-flat integral basis for the B-model $\mathcal{H}_{\mathbf{Z}}^{V^\circ}$ (vanishing cycles are used for B-model in [7], Theorems 6.9, 6.10, Example 6.18).

In both A-model case and B-model case, the integral structures $\mathcal{H}_{\mathbf{Z}}^{V}$ and $\mathcal{H}_{\mathbf{Z}}^{V^\circ}$ on S^* extend to the local systems of $\mathbf{Z}$-modules over $S^{\log}$ ([24]; [17], Proposition 2.3.5), still denoted $\mathcal{H}_{\mathbf{Z}}^{V}$ and $\mathcal{H}_{\mathbf{Z}}^{V^\circ}$, respectively.

Consider a diagram:

$$\begin{array}{ccc} \tilde{S}^{\log} := \mathbf{R} \times i(0,\infty] & \supset & \tilde{S}^* := \mathbf{R} \times i(0,\infty) \\ \downarrow & & \downarrow \\ S^{\log} & \supset & S^* \\ \tau \downarrow & & \\ S & & \end{array}$$

The coordinate u of $\tilde{S}^*$ extends over $\tilde{S}^{\log}$. Fix base points as $u_0 = 0 + i\infty \in \tilde{S}^{\log} \mapsto b := \bar{0} + i\infty \in S^{\log} \mapsto q = 0 \in S$, where $q = 0$ corresponds to q_0 for A-model and p_0 for B-model. Note that fixing a base point $u = u_0$ on $\tilde{S}^{\log}$ is equivalent to fixing a base point b on $S^{\log}$ and also a branch of $(2\pi i)^{-1}\log q$.

Let $B := \mathcal{H}_{\mathbf{Z}}^{V}(u_0) = \mathcal{H}_{\mathbf{Z}}^{V}(b)$ for A-model and $B := \mathcal{H}_{\mathbf{Z}}^{V^\circ}(u_0) = \mathcal{H}_{\mathbf{Z}}^{V^\circ}(b)$ for B-model.

2.5. *Correspondence table*

We use the mirror theorems in Introduction 0.2–0.3. Put $\Phi := \Phi_{\mathrm{GW}}^{V} = \Phi_{\mathrm{GM}}^{V^\circ}$ and fix an isomorphisim of VHS in 0.2.2 in Introduction (cf. 2.4).

(1A) *Polarization of A-model of V.*

$$S(\alpha,\beta) := (-1)^p (2\pi i)^3 \int_V \alpha \cup \beta \quad (\alpha \in H^{p,p}(V), \beta \in H^{3-p,3-p}(V)).$$

(1B) *Polarization of B-model of V°.*

$$Q(\alpha,\beta) := (-1)^{3(3-1)/2} \int_{V^\circ} \alpha \cup \beta = -\int_{V^\circ} \alpha \cup \beta \quad (\alpha,\beta \in H^3(V^\circ)).$$

(2A) ***Z**-basis compatible with monodromy weight filtration.*

Let $B := \mathcal{H}_{\mathbf{Z}}^{V}(u_0) = \mathcal{H}_{\mathbf{Z}}^{V}(b)$ be as in Section 2.4. Let $b^3 := s^3(u_0) = s(\mathcal{O}_V)(u_0)$, $b^2 := s^2(u_0) = s(\mathcal{O}_H)(u_0)$, $b^1 := s^1(u_0) = s(\mathcal{O}_C)(u_0)$ and $b^0 := s^0(u_0) = s(\mathcal{O}_{\mathrm{pt}})(u_0)$ be the basis of the fiber B at u_0 coming from ∇-flat integral basis in 2.4.

The endomorphism of $B_{\mathbf{Q}} := \mathbf{Q} \otimes B$ coming from the monodromy logarithm coincides with the cup product with $-2\pi iH$ where H is a hyperplane section of V ([7], Definition 3.6; cf. Proposition 1.4.3). Hence the above basis is compatible with the monodromy weight filtration M.

(2B) **Z**-*basis compatible with monodromy weight filtration.*

Let $B := \mathcal{H}_{\mathbf{Z}}^{V^\circ}(u_0) = \mathcal{H}_{\mathbf{Z}}^{V^\circ}(b)$, and b^0, b^1, b^2, b^3 be the basis of B corresponding to that in (2A) by the mirror symmetry 0.2.2 and 0.2.3 in Introduction.

The endomorphism of $B_{\mathbf{Q}}$ coming from the monodromy logarithm is denoted by N, and the above basis is compatible with the monodromy weight filtration M [ibid].

For both cases (2A) and (2B), we regard B as a constant sheaf on $S^{\log}$ and also on S, endowed with the associated filtrations M.

From the asymptotics of the basis s^p $(0 \le p \le 3)$ in 2.4, the matrix of the polarization pairings S in (1A) and Q in (1B) for the basis $b^p = s^p(u_0)$ is computed as

$$(S(b^p, b^q))_{p,q} = (Q(b^p, b^q))_{p,q} = \begin{pmatrix} 0 & 0 & 0 & -1 \\ 0 & 0 & 1 & -1 \\ 0 & -1 & 0 & -5 \\ 1 & 1 & 5 & 0 \end{pmatrix}.$$

(3A) *Sections compatible with Deligne decomposition and inducing* **Z**-*basis of* gr^M *for A-model of* V.

Let T^3, T^2, T^1, and T^0 be the basis of $\mathcal{H}_{\mathcal{O}}^V$ corresponding to the e^3, e^2, e^1, and e^0 in (3B) below by the mirror symmetry 0.2.2 and 0.2.3 in Introduction. Then $S(T^3, T^0) = 1$ and $S(T^2, T^1) = -1$. Hence T^3, T^2, $-T^0$, T^1 form a symplectic base for S in (1A).

Note that on gr^M they are

$$\mathrm{gr}_3^M(T^3) = 1 \in H^0(V, \mathbf{Z}), \quad \mathrm{gr}_2^M(T^2) = H \in H^2(V, \mathbf{Z}),$$

$$\mathrm{gr}_1^M(T^1) = C \in H^4(V, \mathbf{Z}), \quad \mathrm{gr}_0^M(T^0) = [\mathrm{pt}] \in H^6(V, \mathbf{Z}),$$

where H and C are the cohomology classes of a hyperplane section and a line on V, respectively. Abusing notation, we mean by C the Poincaré dual class of the homology class in 2.3.

(3B) *Sections compatible with Deligne decomposition and inducing* **Z**-*basis of* gr^M *for B-model of* V°.

We use Deligne decomposition [4]. We consider B in (2B) as a constant sheaf on $S^{\log}$. We have locally free $\mathcal{O}_S$-submodules $\mathcal{M}_{2p} := \tau_*(\mathcal{O}_S^{\log} \otimes_{\mathbf{Z}} M_{2p}B)$ and $\mathcal{F}^p$ in $\tau_*(\mathcal{O}_S^{\log} \otimes_{\mathbf{Z}} B) = \mathcal{O}_S \otimes_{\mathbf{Z}} B$ (canonical extension of Deligne in Proposition 1.4.2). The mixed Hodge structure of Hodge–Tate type $(\mathcal{M}, \mathcal{F})$

has decomposition:

$$\mathcal{O}_S \otimes_{\mathbf{Z}} B = \bigoplus_p I^{p,p}, \qquad I^{p,p} := \mathcal{M}_{2p} \cap \mathcal{F}^p \xrightarrow{\sim} \mathrm{gr}^{\mathcal{M}}_{2p}.$$

Transporting the basis b^p $(0 \le p \le 3)$ of B in (2B), regarded as sections of the constant sheaf B on $S^{\log}$, via isomorphism

$$I^{p,p} \xrightarrow{\sim} \mathcal{O}_S \otimes_{\mathbf{Z}} \mathrm{gr}^M_{2p} B$$

we define sections $e^p \in I^{p,p}$ $(0 \le p \le 3)$ over S. Then, e^3, e^2,$-e^0$, e^1 form a symplectic basis for Q in (1B), and $e^3 = \tilde{\Omega} = \Omega/y_0$ over S.

The asymptotic relation of the T^p in (3A) (resp. the e^p in (3B)) can be computed, via the s^p, from (7A) (resp. (7B)) below.

(4A) *A-model connection* $\nabla = \nabla^{even}$ *of* V.

Let $\delta = d/du = 2\pi iqd/dq$. The Dubrovin connection ∇ (cf. [1], Sect. 8.4) is characterized by

$$\nabla_\delta T^0 = 0, \quad \nabla_\delta T^1 = T^0, \quad \nabla_\delta T^2 = \frac{1}{(2\pi i)^3}\frac{d^3\Phi}{du^3}T^1 = \Big(5 + \frac{1}{(2\pi i)^3}\frac{d^3\Phi_{\mathrm{hol}}}{du^3}\Big)T^1,$$

$$\nabla_\delta T^3 = T^2.$$

∇ is flat, i.e., $\nabla^2 = 0$, and extends to a log connection over $S^{\log}$.

(4B) *B-model connection* $\nabla = \nabla^{\mathrm{GM}}$ *of* V°.

Let $\delta = d/du = 2\pi iqd/dq$. The Gauss-Manin connection ∇ is computed as

$$\nabla_\delta e^0 = 0, \quad \nabla_\delta e^1 = e^0, \quad \nabla_\delta e^2 = \frac{1}{(2\pi i)^3}\frac{d^3\Phi}{du^3}e^1 = \frac{5}{(1+5^5)y_0(z)^2}\Big(\frac{q}{z}\frac{dz}{dq}\Big)^3 e^1,$$

$$\nabla_\delta e^3 l = e^2.$$

∇ is flat, i.e., $\nabla^2 = 0$, and extends to a log connection over $S^{\log}$.

(5A) ∇*-flat basis of* $\mathcal{H}^V_{\mathbf{C}}$ *inducing* $\mathbf{Z}$*-basis of* gr^M.

$$\tilde{s}^0 := T^0, \quad \tilde{s}^1 := T^1 - uT^0, \quad \tilde{s}^2 := T^2 - \frac{1}{(2\pi i)^3}\frac{d^2\Phi}{du^2}T^1 + \frac{1}{(2\pi i)^3}\frac{d\Phi}{du}T^0,$$

$$\tilde{s}^3 := T^3 - uT^2 + \frac{1}{(2\pi i)^3}\Big(u\frac{d^2\Phi}{du^2} - \frac{d\Phi}{dlu}\Big)T^1 - \frac{1}{(2\pi i)^3}\Big(u\frac{d\Phi}{du} - 2\Phi\Big)T^0.$$

Then $\tilde{s}^3$, $\tilde{s}^2$,$-\tilde{s}^0$, $\tilde{s}^1$ form a symplectic basis for S in (1A).

(5B) ∇*-flat basis for* $\mathcal{H}^{V^\circ}_{\mathbf{C}}$ *inducing* $\mathbf{Z}$*-basis of* gr^M.

$$\tilde{s}^0 := e^0, \quad \tilde{s}^1 := e^1 - ue^0, \quad \tilde{s}^2 := e^2 - \frac{1}{(2\pi i)^3}\frac{d^2\Phi}{du^2}e^1 + \frac{1}{(2\pi i)^3}\frac{d\Phi}{du}e^0,$$

$$\tilde{s}^3 := e^3 - ue^2 + \frac{1}{(2\pi i)^3}\Big(u\frac{d^2\Phi}{du^2} - \frac{d\Phi}{du}\Big)e^1 - \frac{1}{(2\pi i)^3}\Big(u\frac{d\Phi}{du} - 2\Phi\Big)e^0.$$

Then $\tilde{s}^3$, $\tilde{s}^2$,$-\tilde{s}^0$, $\tilde{s}^1$ form a symplectic basis for Q in (1B).

For both cases (5A) and (5B), by using (4A) and (4B), the ∇-flat bases $\tilde{s}^p$ are determined inductively on $0 \le p \le 3$ from the T^p in (3A) and the e^p in (3B). These ∇-flat bases $\tilde{s}^p$ are characterized by the Frobenius solutions y_j $(0 \le j \le 3)$ in 2.2 such that $y_0 T^3$ in A-model and $y_0 e^3 = \Omega$ in B-model coincide with

$$y_0 \tilde{s}^3 + (2\pi i)^{-1} y_1 \tilde{s}^2 + 5(2\pi i)^{-2} y_2 \tilde{s}^1 + 5(2\pi i)^{-3} y_3 \tilde{s}^0.$$

(6A), (6B) *Relations of ∇-flat $\mathbf{Z}$-basis s^p and the ∇-flat basis $\tilde{s}^p$.*

$$s^0 = \tilde{s}^0, \quad s^1 = \tilde{s}^1 + \tilde{s}^0, \quad s^2 = \tilde{s}^2 - \frac{5}{2}\tilde{s}^1 + \frac{35}{12}\tilde{s}^0, \quad s^3 = \tilde{s}^3 + \frac{25}{12}\tilde{s}^1 + \frac{25 i \zeta(3)}{\pi^3}\tilde{s}^0.$$

(7A) *Expression of the T^p by the s^p over $S^{\log}$.*
It is computed that T^p are written by the ∇-flat $\mathbf{Z}$-basis s^p of $\mathcal{H}_{\mathbf{Z}}^V$ as follows.

$$T^0 = s^0, \quad T^1 = s^1 + (u-1)s^0,$$

$$T^2 = s^2 + \Big(\frac{1}{(2\pi i)^3}\frac{d^2\Phi}{du^2} + \frac{5}{2}\Big)s^1 + \Big(\frac{1}{(2\pi i)^3}\Big(u\frac{d^2\Phi}{du^2} - \frac{d\Phi}{du}\Big) - 5u - \frac{65}{12}\Big)s^0,$$

$$T^3 = s^3 + us^2 + \Big(\frac{1}{(2\pi i)^3}\frac{d\Phi}{du} + \frac{5}{2}u - \frac{25}{12}\Big)s^1 + \Big(\frac{1}{(2\pi i)^3}\Big(u\frac{d\Phi}{du} - 2\Phi\Big)$$
$$- \frac{65}{12}u + \frac{25}{12} - \frac{25i}{\pi^3}\zeta(3)\Big)s^0.$$

(7B) *Expression of the e^p by the s^p over $S^{\log}$.*
It is computed that e^p are written by the ∇-flat $\mathbf{Z}$-basis s^p of $\mathcal{H}_{\mathbf{Z}}^{V^\circ}$ as follows.

$$e^0 = s^0, \quad e^1 = s^1 + (u-1)s^0,$$

$$e^2 = s^2 + \Big(\frac{1}{(2\pi i)^3}\frac{d^2\Phi}{du^2} + \frac{5}{2}\Big)s^1 + \Big(\frac{1}{(2\pi i)^3}\Big(u\frac{d^2\Phi}{du^2} - \frac{d\Phi}{du}\Big) - 5u - \frac{65}{12}\Big)s^0,$$

$$e^3 = s^3 + us^2 + \Big(\frac{1}{(2\pi i)^3}\frac{d\Phi}{du} + \frac{5}{2}u - \frac{25}{12}\Big)s^1 + \Big(\frac{1}{(2\pi i)^3}\Big(u\frac{d\Phi}{du} - 2\Phi\Big)$$
$$- \frac{65}{12}u + \frac{25}{12} - \frac{25i}{\pi^3}\zeta(3)\Big)s^0.$$

(8A), (8B) *Relations of integral periods and Frobenius solutions.*
Let η_j $(0 \le j \le 3)$ be the integral periods defined by the condition that $y_0 T^3$ in A-model and $y_0 e^3 = \Omega$ in B-model coincide with $\eta_0 s^3 + \eta_1 s^2 + \eta_2 s^1 + \eta_3 s^0$. Then the relations in (6A), (6B) are interpreted as

$$\eta_0 = y_0, \quad \eta_1 = (2\pi i)^{-1} y_1, \quad \eta_2 = 5(2\pi i)^{-2} y_2 + \frac{5}{2}(2\pi i)^{-1} y_1 - \frac{25}{12} y_0,$$

$$\eta_3 = 5(2\pi i)^{-3} y_3 - 5(2\pi i)^{-2} y_2 + \frac{65}{12}(2\pi i)^{-1} y_1 + \Big(\frac{25}{12} - \frac{25 i \zeta(3)}{\pi^3}\Big) y_0.$$

Remark. The η_j coincide with the corresponding coefficients of the expression of $y_0 T^3$ in (7A) and of $y_0 e^3 = \Omega$ in (7B), and yield the same integral structure for periods given in [2] and [18]. Indeed, for the notation ϖ_j in [18], (8), they are related as $\varpi_0 = \eta_0$, $\varpi_1 = \eta_1$, $\varpi_2 = \eta_2 - 5\eta_1$, and $\varpi_3 = -\eta_3 - \eta_2 - 5\eta_1$.

2.6. Proofs of results in 2.5

Proofs of (4A) and (4B) in 2.5. We prove (4B). (4A) follows by mirror symmetry theorems in Introduction 0.2.

We improve the proof of [1], Prop. 5.6.1 carefully by a log Hodge theoretic understanding in 1.4 of the relation among a constant sheaf and the local system on $S^{\log}$, the canonical extension of Deligne on S, and the Deligne decomposition.

We investigate the Gauss-Manin connection ∇, corresponding to the local system $\mathcal{H}_{\mathbf{Z}}^{V^\circ}$, contracted with $\delta = 2\pi i q d/dq$. Since e^p maps to a $\nabla(\mathrm{gr}^{\mathcal{M}}_{2p})$-flat element of $\mathrm{gr}^{\mathcal{M}}_{2p}$, $\nabla_\delta(e^p)$ lies in $\mathcal{M}_{2p-1} = \mathcal{M}_{2p-2}$. But e^p is also an element of $\mathcal{F}^p$, so that $\nabla_\delta(e^p)$ lies in $\mathcal{F}^{p-1}$ by Griffiths transversality. This shows that $\nabla_\delta(e^p)$ is an element of $I^{p-1,p-1}$, and it follows that

$$\nabla_\delta(e^3) = Y_3 e^2, \quad \nabla_\delta(e^2) = Y_2 e^1, \quad \nabla_\delta(e^1) = Y_1 e^0, \quad \nabla_\delta(e^0) = 0$$

for some $Y_1, Y_2, Y_3 \in \mathcal{O}_S^{\log}$. However, since $Q(e^3, e^1) = 0$ by orthogonality of Hodge filtration, we have

$$\begin{aligned} 0 = \delta Q(e^3, e^1) &= Q(\nabla_\delta(e^3), e^1) + Q(e^3, \nabla_\delta(e^1)) = Y_3 Q(e^2, e^1) + Y_1 Q(e^3, e^0) \\ &= -Y_3 + Y_1, \end{aligned}$$

where the last equality follows from 2.5 (3B).

Since ∇ has a regular singular point and $\delta = 2\pi i q d/dq$, Y_1 is holomorphic over S. Considering over the log point p_0, we claim $Y_1(0) = \pm 1$. Since e^1 is taken to be the canonical extension over p_0, we have $\nabla_\delta(e^1) = N(e^1) = -e^0$ by [17], Prop. 2.3.4 (ii) (cf. Proposition 1.4.3). Replacing e^1, e^2 by $-e^1, -e^2$, we have $Y_1(0) = 1$.

Since we use the canonical coordinate q in 2.2, the arguments in [1], Sect. 5.6.4, Sect. 2.3, yield

$$q = \exp\Big(\int Y_1(q) \frac{dq}{q}\Big).$$

Taking logarithm of both sides and differentiating them by $d/d\log q$, we have $Y_1(q) = 1$, hence $\nabla_\delta e^1 = e^0$ and $\nabla_\delta e^3 = e^2$. Thus, relative to the basis

e^0, e^1, e^2, e^3 and using the canonical coordinate q, ∇_δ has the connection matrix

$$\begin{pmatrix} 0 & 1 & & \\ & 0 & Y & \\ & & 0 & 1 \\ & & & 0 \end{pmatrix}$$

where $Y = Y_2$.

The proofs in [ibid, Proof of Prop. 5.6.1] for the following assertions work well: e^3 is the normalized 3-form $\tilde{\Omega}$; the Picard–Fuchs equation for $\tilde{\Omega}$ is $\nabla_\delta^2(\nabla_\delta^2\tilde{\Omega}/Y) = 0$; Y is the Yukawa coupling. The notation e^0, e^1, e^2, e^3 (resp. T^0, T^1, T^2, T^3) in the present paper corresponds to e_3, e_2, e_1, e_0 in [ibid, p.105] (resp. T^0, T^1, T_1, T_0 in [ibid, Sect. 8.5.3]). □

Proofs of (5A), (5B), (6A), (6B), (7A), and (7B) in 2.5.

We use mirror symmetry theorems in Introduction 0.2.

From e^p in (3B) and ∇_δ in (4B), we produce $\tilde{s}^p$ inductively on $0 \le p \le 3$ as in (5B). These are transported as (5A) in A-model. For the last assertion in (5B) on the relation of ∇-flat basis $\tilde{s}^p$ and the basis of the Frobenius solutions y_j, since $y_0 T^3$, $y_0 e^3 = \Omega$, and the last expression in (5B) are killed by the operator $\mathcal{L}$, it is enough to show the equality on the fiber $B_\mathbf{C} := \mathbf{C} \otimes B$, i.e., the coincidence of the initial conditions. We work in A-model. By the asymptotics of the reverse relation of (5A) and of the expressions of y_j in 2.2, we have

$$\begin{aligned} y_0 T^3 &= y_0\tilde{s}^3 + y_0 u\tilde{s}^2 + y_0(2\pi i)^{-3}\frac{d\Phi}{du}\tilde{s}^1 + y_0(2\pi i)^{-3}(u\frac{d\Phi}{du} - 2\Phi)\tilde{s}^0 \\ &\sim y_0\tilde{s}^3 + y_0 u\tilde{s}^2 + \frac{5}{2}y_0 u^2\tilde{s}^1 + \frac{5}{6}y_0 u^3\tilde{s}^0 \\ &\sim \tilde{s}^3 + ((2\pi i)^{-1}\log z)\tilde{s}^2 + \frac{5}{2}((2\pi i)^{-1}\log z)^2\tilde{s}^1 + \frac{5}{6}((2\pi i)^{-1}\log z)^3\tilde{s}^0 \\ &\sim y_0\tilde{s}^3 + (2\pi i)^{-1}y_1\tilde{s}^2 + 5(2\pi i)^{-2}y_2\tilde{s}^1 + 5(2\pi i)^{-3}y_3\tilde{s}^0. \end{aligned}$$

To prove (6A), (6B), (7A), and (7B), we want to find $c^{10}, c^{21}, c^{20}, c^{32}, c^{31}, c^{30} \in \mathbf{C}$ such that, on the fiber B,

$$\tilde{s}^0(u_0) = b^0, \quad \tilde{s}^1(u_0) = b^1 + c^{10}\tilde{s}^0(u_0), \quad \tilde{s}^2(u_0) = b^2 + c^{21}\tilde{s}^1(u_0) + c^{20}\tilde{s}^0(u_0),$$

$$\tilde{s}^3(u_0) = b^3 + c^{32}\tilde{s}^2(u_0) + c^{31}\tilde{s}^1(u_0) + c^{30}\tilde{s}^0(u_0).$$

Then, since $\tilde{s}^p$ and s^p are ∇-flat, we have

$$s^0 = \tilde{s}^0, \; s^1 = \tilde{s}^1 - c^{10}\tilde{s}^0, \; s^2 = \tilde{s}^2 - c^{21}\tilde{s}^1 - c^{20}\tilde{s}^0, \; s^3 = \tilde{s}^3 - c^{32}\tilde{s}^2 - c^{31}\tilde{s}^1 - c^{30}\tilde{s}^0.$$

Express the e^p by the s^p by using the inverse expressions of the above and of (5B). Transporting these into A-model, we get expression of the T^p by the s^p

and the c^{jk}. Using Iritani asymptotics for the s^p in A-model in 2.4, we get

$$T^0 = \frac{1}{5}H^3, \quad T^1 = (2\pi i)^{-1}\frac{1}{5}H^2 + \frac{c^{10}}{5}H^3,$$

$$T^2 \sim (2\pi i)^{-2}H + (2\pi i)^{-1}\left(-\frac{1}{2} + \frac{c^{21}}{5}\right)H^2 + \left(\left(\frac{3}{2} - \frac{c^{21}}{5} + c^{10}\right)u + \left(\frac{7}{12} + \frac{c^{21}}{5} + \frac{c^{21}c^{10}}{5} + \frac{c^{20}}{5}\right)\right)H^3,$$

$$T^3 \sim (2\pi i)^{-3} + (2\pi i)^{-2}c^{32}H + (2\pi i)^{-1}\left(\left(-\frac{1}{2} + \frac{c^{21}}{5} - c^{32}\right)u + \left(\frac{5}{12} - \frac{c^{32}}{2} + \frac{c^{32}c^{21}}{5} + \frac{c^{31}}{5}\right)\right)H^2 + \left(\left(\frac{1}{6} - \frac{c^{31}}{5} + \frac{c^{20}}{5}\right)u + \left(\frac{5i\zeta(3)}{\pi^3} + \frac{c^{30}}{5}\right)\right)H^3.$$

Since $T^p \in \mathcal{F}^p$ by construction ((3A), (3B) in 2.5), we see that all coefficients of H^j in the above expressions of T^p are zero for $j+p>3$. Thus we get

$$c^{10} = -1,\ c^{21} = 5/2,\ c^{20} = -35/12,\ c^{32} = 0,\ c^{31} = -25/12,\ c^{30} = -25i\zeta(3)/\pi^3.$$

(6A), (6B), (7A), and (7B) follow from this and (5A), (5B).

We prove (8A), (8B). The same argument goes for both cases. We use the notation in B-model. By the definition of the η_j and (5B), $\Omega = \eta_0 s^3 + \eta_1 s^2 + \eta_2 s^1 + \eta_3 s^0 = y_0\tilde{s}^3 + (2\pi i)^{-1}y_1\tilde{s}^2 + 5(2\pi i)^{-2}y_2\tilde{s}^1 + 5(2\pi i)^{-3}y_3\tilde{s}^0$. Substituting (6B) and comparing the coefficients of $\tilde{s}^p$, we have expressions of the y_j by the η_j. Solving these for the η_j, we get (8B). □

Remark. It was pointed out by Hiroshi Iritani that the definitions and the descriptions of integral structures 3.5, 3.6 in [30] are insufficient. Actually, they were the first approximations of integral structures by means of gr^M, which are characterized by the Frobenius solutions as in the last statement of (5A) and (5B) in 2.5. The second proof in [ibid, 3.9] works well even in this approximation.

3 Proof of Theorem 0.4.1

In this section, we prove Theorem 0.4.1 in Introduction for open mirror symmetry of quintic threefolds. We prove it by constructing a normal function in log mixed Hodge theory for B-model in 3.1–3.2 below. This argument is applicable to the case of A-model by the theorems in Introduction 0.2 and 0.3 and the correspondence table in 2.5. We give some discussions on geometries and local systems in 3.3.

3.1. Proof of Theorem 0.4.1 over log disc S

We consider B-model. To make the monodromy of $\mathcal{T}_B$ unipotent, we take a double cover $z^{1/2} \mapsto z$. Let S be a neighborhood disc of p_0 in the $z^{1/2}$-plane endowed with log structure associated to the divisor p_0 in S. Denote by $\mathcal{H}$ and $\mathcal{T}$ the pullbacks of the log Hodge structure $\mathcal{H}^{V^\circ}$ and the tension $\mathcal{T}_B$ by the double covering, respectively.

We are looking for an extension $\tilde{\mathcal{H}}$:

$$0 \to \mathcal{H} \to \tilde{\mathcal{H}} \to \mathbf{Z}(-2) \to 0$$

of log mixed Hodge structures with $\mathrm{gr}_4^W \tilde{\mathcal{H}} = \mathbf{Z}(-2)$ and $\mathrm{gr}_3^W \tilde{\mathcal{H}} = \mathcal{H}$, which has liftings $1_{\mathbf{Z}}$ and 1_F of $(2\pi i)^{-2} \cdot 1 \in \mathbf{Z}(-2)$ in $\tilde{\mathcal{H}}$ respecting the lattice and the Hodge filtration, respectively, such that the tension $\mathcal{T}$ is described as

$$(1) \qquad Q(1_F - 1_{\mathbf{Z}}, \Omega) = \int_{C_-}^{C_+} \Omega = \mathcal{T},$$

where Q is the polarization of $\mathcal{H}$ coming from 2.5 (1B) and Ω is the 3-form from 2.2.

To find such a log mixed Hodge structure, we use the basis e^p $(0 \le p \le 3)$ respecting the Deligne decomposition of $(\mathcal{H}, \mathcal{M}, \mathcal{F})$ from 2.5 (3B), and the ∇-flat integral basis s^p $(0 \le p \le 3)$ from 2.5 (5B). We also use the integral periods η_j $(0 \le j \le 3)$ in 2.5 (8B). Note that these players are already extended and live together over $S^{\log}$.

Let the local system $L_{\mathbf{Q}}$ and the Néron model $J_{L_{\mathbf{Q}}}$ be as in 0.4 in Introduction (see also 1.8). Then $J_{L_{\mathbf{Q}}} = \mathcal{E}xt^1_{\mathrm{LMH}/S}(\mathbf{Z}(-2), \mathcal{H})$, and let $1_{\mathbf{Z}} := ((2\pi i)^{-2} \cdot 1, -(\mathcal{T}/\eta_0)s^0) \in \tilde{\mathcal{H}}_{\mathbf{Z}}$ be a lifting of $(2\pi i)^{-2} \cdot 1 \in \mathbf{Z}(-2) = (\mathrm{gr}_4^W)_{\mathbf{Z}}$, where $(\mathcal{T}/\eta_0)s^0 \in \mathcal{H}_{\mathcal{O}^{\log}} = (\mathrm{gr}_3^W)_{\mathcal{O}^{\log}}$. In particular, the connection $\nabla = \nabla^{\mathrm{GM}}$ on $\mathcal{H}$ is extended over $\tilde{\mathcal{H}}$ by $\nabla(1_{\mathbf{Z}}) = 0$.

To find 1_F, we write $1_F - 1_{\mathbf{Z}} = ae^3 + be^2 + ce^1 - (\mathcal{T}/\eta_0)e^0$ with $a, b, c \in \mathcal{O}_S^{\log}$ by using (1). The Griffiths transversality condition on $1_F - 1_{\mathbf{Z}}$ is understood as vanishing of the coefficient of e^0 in $\nabla_\delta(1_F - 1_{\mathbf{Z}})$. Using 2.5 (4B), we have

$$\nabla_\delta(1_F - 1_{\mathbf{Z}}) = (\delta a)e^3 + (a + \delta b)e^2 + \left(b \frac{1}{(2\pi i)^3} \frac{d^3\Phi}{du^3} + \delta c\right)e^1 + (c - \delta(\mathcal{T}/\eta_0))e^0.$$

Hence, the above condition is equivalent to $c = \delta(\mathcal{T}/\eta_0)$ and a, b arbitrary. Using the relation "modulo F^2," we can take $a = b = 0$. Thus

$$1_F = 1_{\mathbf{Z}} + (\delta(\mathcal{T}/\eta_0))e^1 - (\mathcal{T}/\eta_0)e^0.$$

The pair $1_{\mathbf{Z}}$ and 1_F yields the desired element of $\mathcal{E}xt^1_{\mathrm{LMH}/S}(\mathbf{Z}(-2), \mathcal{H})$, hence $1_F - 1_{\mathbf{Z}}$ yields the desired log normal function. Theorem 0.4.1 (1) is proved.

Next, we will find a splitting of the weight filtration W of the local system $L_{\mathbf{Q}}$, i.e., a splitting of W which is compatible with the local monodromy of the local system $L_{\mathbf{Q}}$. We use the monodromy table (3.14) in [31]. This is computed for A-model but applicable also for B-model by the theorems in Introduction 0.2–0.3 and in Section 2.5. Let T_∞^2 be the monodromy around $p_0 : z^{1/2} = 0$ and put $N := \log(T_\infty^2)$. By [ibid], $N(\mathcal{T}/\eta_0) = -1$ hence $N(1_{\mathbf{Z}}) = s^0$. On the other hand, we have $N(s^1) = -2s^0$. (Here we use the rotation of the monodromy as $\log z \mapsto \log z + 2\pi i$.) Define

$$1_{\mathbf{Z}}^{\mathrm{spl}} := 1_{\mathbf{Z}} + \frac{1}{2}s^1 = \left((2\pi i)^{-2}\cdot 1, \frac{1}{2}s^1 - \frac{\mathcal{T}}{\eta_0}s^0\right) \in \tilde{\mathcal{H}}_{\mathbf{Q}}.$$

Then $N(1_{\mathbf{Z}}^{\mathrm{spl}}) = 0$, and this gives the desired splitting of W of the local system $L_{\mathbf{Q}}$.

A lifting 1_F^{spl} for $1_{\mathbf{Z}}^{\mathrm{spl}}$, respecting the Hodge filtration, is computed as before and we get

$$(2) \qquad 1_F^{\mathrm{spl}} = 1_{\mathbf{Z}}^{\mathrm{spl}} + \left(\delta\left(\frac{\mathcal{T}}{\eta_0}\right)\right)e^1 - \frac{\mathcal{T}}{\eta_0}e^0.$$

The pair $1_{\mathbf{Z}}^{\mathrm{spl}}$ and 1_F^{spl} yields the desired element of $\mathcal{E}xt^1_{\mathrm{LMH}/S}(\mathbf{Z}(-2), \mathcal{H})$ which splits the weight filtration W of the local system $L_{\mathbf{Q}}$. Note that $1_F^{\mathrm{spl}} - 1_{\mathbf{Z}}^{\mathrm{spl}} = 1_F - 1_{\mathbf{Z}} = (\delta(\mathcal{T}/\eta_0))e^1 - (\mathcal{T}/\eta_0)e^0$. Theorem 0.4.1 (2) is proved.

Theorem 0.4.1 (3) follows immediately from the above results.

We add a remark that the W-relative N-filtration $M = M(N,W)$ on $H_{\mathbf{R}}$ in the admissibility condition 1.5 (1) is given by

$$M_{-1} = 0 \subset M_0 = M_1 = \mathbf{R}s^0 \subset M_2 = M_3 = M_1 + \mathbf{R}s^1$$

$$\subset M_4 = M_5 = M_3 + \mathbf{R}s^2 + \mathbf{R}1_{\mathbf{Z}} \subset M_6 = \tilde{\mathcal{H}}_{\mathbf{R}} = M_5 + \mathbf{R}s^3.$$

3.2. Proofs of (1) and (2) in Theorem 0.4.1 over log point p_0

We still consider B-model. We show here that (1) and (2) in Theorem 0.4.1 have meanings just over the log point p_0 and that the computations in their proofs become simpler.

Recall that

$$(1) \qquad \mathcal{T} = -\frac{\eta_1}{2} - \frac{\eta_0}{4} + a_0\tau \qquad \left(a_0 := \frac{15}{\pi^2},\ \tau : \text{tau function}\right)$$

from [31]. We substitute $z^{1/2} = 0$ to $\mathcal{T}$ carefully as follows. Recall $\eta_1 = \eta_0 u$ from 2.2 and $u = x + iy$ from 2.4. Write $v := x + i\infty$ and define

$$\mathcal{T}(0) := -\frac{v}{2} - \frac{1}{4} + a_0 \qquad \text{in } \mathcal{O}_{p_0}^{\log} = \mathbf{C}[v].$$

We abuse the notation e^p and s^p also for their restrictions over the log point p_0, and so they live together over $p_0^{\log} = (\mathbf{S}^1, \mathbf{C}[v])$.

Similarly as in 0.4 in Introduction, but using now $\mathcal{T}(0)s^0$ instead of $(\mathcal{T}/\eta^0)s^0$ because $\eta^0(0) = 1$, we define a local system $L_{\mathbf{Q}}$ and a Néron model $J_{L_{\mathbf{Q}}}$ lying over $L_{\mathbf{Q}}$. Let $\tilde{\mathcal{H}}$ be an extension of log mixed Hodge structures over the log point p_0, we are looking for, like in 3.1, and let $1_{\mathbf{Z}} := ((2\pi i)^{-2} \cdot 1, -\mathcal{T}(0)s^0)$ be a lifting of $(2\pi i)^{-2} \cdot 1 \in \mathbf{Z}(-2) = (\mathrm{gr}_4^W)_{\mathbf{Z}}$ in $\tilde{\mathcal{H}}_{\mathbf{Z}}$. Hence the connection ∇ on $\mathcal{H}$ is extended over $\tilde{\mathcal{H}}$ by $\nabla(1_{\mathbf{Z}}) = 0$. Note that both $2\pi iq\frac{d}{dq}$ and $2\pi iz\frac{d}{dz}$ coincide with $\frac{d}{dv}$ now, which is denoted by δ. To find 1_F, write $1_F - 1_{\mathbf{Z}} = ae^3 + be^2 + ce^1 - \mathcal{T}(0)e^0$ $(a,b,c \in \mathbf{C}[v],\ \eta_0(0) = 1)$ and compute $\nabla_\delta(1_F - 1_{\mathbf{Z}})$ as in 3.1. Then, by the Griffiths transversality, we have $c = -1/2$, a and b arbitrary. By the relation "modulo F^2," a and b can be reduced to 0. Thus, we have

$$1_F = 1_{\mathbf{Z}} + (\delta\mathcal{T}(0))e^1 - \mathcal{T}(0)e^0 = 1_{\mathbf{Z}} - \frac{1}{2}e^1 + \Big(\frac{v}{2} + \frac{1}{4} - a_0\Big)e^0.$$

The pair $1_{\mathbf{Z}}$ and 1_F yields the desired element of $\mathcal{E}xt^1_{\mathrm{LMH}/S}(\mathbf{Z}(-2), \mathcal{H})$. Theorem 0.4.1 (1) is proved.

The splitting of the weight filtration W of the local system $L_{\mathbf{Q}}$ is computed as in 3.1 but more simply, and we define

$$1_{\mathbf{Z}}^{\mathrm{spl}} := 1_{\mathbf{Z}} + \frac{1}{2}s^1 = \Big((2\pi i)^{-2} \cdot 1, \frac{1}{2}s^1 - \mathcal{T}(0)s^0\Big) \in \tilde{\mathcal{H}}_{\mathbf{Q}}.$$

Similarly, a lifting 1_F^{spl} for $1_{\mathbf{Z}}^{\mathrm{spl}}$ is computed simply, and we get

$$1_F^{\mathrm{spl}} = 1_{\mathbf{Z}}^{\mathrm{log}} + (\delta\mathcal{T}(0))e^1 - \mathcal{T}(0)e^0.$$

Theorem 0.4.1 (2) is proved. □

Remark. Note that 0.4.1 (3) does not have meaning in the present context. This is because tau function disappears except its constant term when $z^{1/2} = 0$ is substituted. That is, in this step, we lose the transcendental data of the tension $\mathcal{T}$, contained as the extension of its underlying local system, from which we can recover the position of the quintic mirror in its complex moduli space.

3.3. *Discussions on geometries and local systems*

We discuss here the relation with geometries and local systems considered in [31] and [23]. Forgetting Hodge structures, we consider only local systems corresponding to the monodromy of integral periods and tensions.

Let V_ψ and V_ψ° be a quintic threefold and its mirror from 2.1. Let S be a small neighborhood in the z-plane (z in 2.2) of the maximal unipotent monodromy point p_0 endowed with the log structure associated to the divisor p_0.

We first consider B-model. Let the setting be as in [23], Sect. 4. For $z \neq 0$ near 0, i.e., near p_0, let V_z° be the mirror quintic and $C_{+,z} \cup C_{-,z}$ be the disjoint union of smooth rational curves on V_z° coming from the two conics contained in $V_\psi \cap \{x_1+x_2=x_3+x_4=0\} \subset \mathbf{P}^4(\mathbf{C})$. From the relative homology sequence for $(V_z^\circ, (C_{+,z} \cup C_{-,z}))$, we have

$$(1)\quad 0 \to H_3(V_z^\circ;\mathbf{Z}) \to H_3(V_z^\circ,(C_{+,z} \cup C_{-,z});\mathbf{Z}) \xrightarrow{\partial} \mathbf{Z}([C_{+,z}]-[C_{-,z}]) \to 0,$$

where $\mathbf{Z}([C_{+,z}] - [C_{-,z}])$ is $\mathrm{Ker}\,(H_2(C_{+,z} \cup C_{-,z};\mathbf{Z}) \to H_2(V_z^\circ;\mathbf{Z}))$. The monodromy T_∞ around p_0 interchanges $C_{+,z}$ and $C_{-,z}$.

Respecting the sequence (1), we take a family of cycles Poincaré duality isomorphic to the flat integral basis s^p $(0 \le p \le 3)$ in 2.4 and a family of chains joining from $C_{-,z}$ to $C_{+,z}$ (a choice up to integral cycles and up to half twists), and over them integrate the family of 3-forms $\Omega(z)$ with log pole over $z=0$ (z in the punctured disc in the z-plane) in 2.2, then we have a family of vectors $(\eta_0,\eta_1,\eta_2,\eta_3,\mathcal{T})$ consisting of periods (2.5 (8B)) and a tension. This corresponds to the data in [31], [23] (cf. Remark in 2.5). Since $T_\infty(\mathcal{T}) = -(\mathcal{T}+\eta_1+\eta_0)$ by [31], (3.14), we find $\mathcal{T}+\frac{1}{2}\eta_1+\frac{1}{4}\eta_0 = \frac{15}{\pi^2}\tau$ (see 3.2 (1)) is an eigenvector of the monodromy T_∞ with eigenvalue -1.

The family of sequences (1) $(z \neq 0)$ forms an exact sequence of local systems of $\mathbf{Z}$-modules. Pulling this back to S^* in 3.1 by the double cover $z^{1/2} \mapsto z$, we have a sequence with unipotent local monodromy and its extension over $S^{\log}$. Applying Tate twist (-3) and Poincaré duality isomorphism to the left and the right ends of this exact sequence, we have a local system L' over $S^{\log}$ which is an extension of $\mathbf{Z}(-2)$ by $\mathcal{H}_\mathbf{Z}$:

$$(2)\qquad 0 \to \mathcal{H}_\mathbf{Z} \to L' \to \mathbf{Z}(-2) \to 0.$$

Take a lifting $1_\mathbf{Z} := ((2\pi i)^{-2}\cdot 1, (\mathcal{T}/\eta_0)s^0)$ in L' of $(2\pi i)^{-2}\cdot 1 \in \mathbf{Z}(-2)$, and extend ∇ on $\mathcal{H}_\mathbf{Z}$ over L' by $\nabla(1_\mathbf{Z}) = 0$. We look for a ∇-flat T_∞^2-invariant element associated to $1_\mathbf{Z}$. This is done as in 3.1, and we get $1_\mathbf{Z}^{\mathrm{spl}} := 1_\mathbf{Z} - (s^1/2)$. Thus we know that L' coincides with $\tilde{\mathcal{H}}_\mathbf{Z}$ in 3.1.

For the relative monodromy weight filtration $M = M(N,W)$ of L', we see that $1_\mathbf{Z} \in M_4$ and $s^1 \in M_2$ are the smallest filters containing each element in question. Taking the graded quotients by M of the sequence (2), we have

$$(3)\qquad \begin{gathered} \mathrm{gr}_6^M \mathcal{H}_\mathbf{Z} \xrightarrow{\sim} \mathrm{gr}_6^M L', \\ 0 \to \mathrm{gr}_4^M \mathcal{H}_\mathbf{Z} \to \mathrm{gr}_4^M L' \to \mathbf{Z}(-2) \to 0, \\ 0 \to \mathrm{gr}_2^M \mathcal{H}_\mathbf{Z} \to \mathrm{gr}_2^M L' \to \text{(2-torsion)} \to 0, \\ \mathrm{gr}_0^M \mathcal{H}_\mathbf{Z} \xrightarrow{\sim} \mathrm{gr}_0^M L'. \end{gathered}$$

Here we abuse the notation M also for the monodromy filtration on $\mathcal{H}_\mathbf{Z}$, because it coincides with the restriction of $M = M(N,W)$ to $\mathcal{H}_\mathbf{Z}$. The 2-torsion in the third sequence of (3) corresponds to a half twist of chains from C_- to C_+. Standing on a half integral point and looking at the integral points nearby, we have two orientations. These correspond to the two orientations of a half twist of the chains, and also correspond to $\mathcal{T}_\pm := \pm(\frac{15}{\pi^2}\tau - \frac{\eta_0}{4}) - \frac{\eta_1}{2}$ in [31]. $\mathcal{T}_-$ is different from $-\mathcal{T}_+$ by the complementary half twist, i.e., $\mathcal{T}_+ + \mathcal{T}_- = -\eta_1$.

By using mirror symmetries in 0.2–0.4, or more precisely, by the results in Section 2.5 and Section 3.1, $\mathcal{H}_{\mathbf{Z}} = \mathcal{H}_{\mathbf{Z}}^{V^\circ}$, $\mathcal{T} = \mathcal{T}_B$, $1_{\mathbf{Z}}$, $\nabla = \nabla^{\mathrm{GM}}$, $1_{\mathbf{Z}}^{\mathrm{spl}}$, and $M = M(N,W)$ of B-model are transformed to the corresponding $\mathcal{H}_{\mathbf{Z}} = \mathcal{H}_{\mathbf{Z}}^{V}$, $\mathcal{T} = y_0\mathcal{T}_A$, $1_{\mathbf{Z}}$, $\nabla = \nabla^{\mathrm{even}}$, $1_{\mathbf{Z}}^{\mathrm{spl}}$, and $M = M(N,W)$ of A-model, and the exact sequences (2) and (3) of B-model are transformed to the corresponding exact sequences of A-model.

It is interesting to study the relations of these exact sequences with the geometries of Fermat quintic $V = V_\psi$ with $\psi = 0$ and its Lagrangian submanifold $Lg := V \cap \mathbf{P}^4(\mathbf{R})$ in [31], 2.1; [23], 3.

Remark.

(1) The argument in 3.3 can be performed even over the log point p_0.

(2) [26] and [19] are related with the topics in this subsection.

References

[1] D. A. Cox, S. Katz, *Mirror symmetry and algebraic geometry* MathSurveys and Monographs **68**, AMS, 1999, pp. 469. MR 2000d:14048.

[2] P. Candelas, C. de la Ossa, P. S. Green, L. Parks, *A pair of Calabi-Yau manifolds as an exactly soluble superconformal theory,* Nuclear Physics **B 358** (1991), 21-74.

[3] P. Deligne, *Équations differentielles à points singuliers réguliers* Lect. Notes in Math. No. 163, Springer-Verlag, 1970, MR 54♯5232.

[4] _____, *Local behavior of Hodge structures at infinity,* in Mirror Symmetry II (B. Greene and S.-T. Yau, eds.), AMS/IP Stud. Adv. Math. **1**, 1997, 683–699. MR 98a:14015.

[5] A. B. Givental, *Equivariant Gromov–Witten invariants*, Internat. Math. Res. Notes **13** (1996), 613–663.

[6] H. Iritani, *An integral structure in quantum cohomology*, Adv. Math. **222 (3)** (2009), 1016–1079.

[7] _____, *Quantum cohomology and periods*, Ann. Inst. Fourier (Grenoble) **61 no.7** (2011), 2909 – 2958.

[8] K. Kato, C. Nakayama, *Log Betti cohomology, log étale cohomology, and log de Rham cohomology of log schemes over* **C**, Kodai Math. J. **22** (1999), 161–186. MR 2000i:14023.

[9] K. Kato, C. Nakayama, S. Usui, SL(2)*-orbit theorem for degeneration of mixed Hodge structure*, J. Algebraic Geometry **17** (2008), 401–479. MR 2009b:14020.

[10] _____, *Classifying spaces of degenerating mixed Hodge structures, I: Borel–Serre spaces*, Advanced Studies in Pure Math. **54**: Algebraic Analysis and Around, 2009, 187–222. MR 2010g:14010.

[11] _____, *Log intermediate Jacobians*, Proc. Japan Acad. **86, Ser. A** (2010), 73–78.

[12] _____, *Classifying spaces of degenerating mixed Hodge structures, II: Spaces of* SL(2)*-orbits*, Kyoto J. Math. **51-1**: Nagata Memorial Issue (2011), 149–261. MR 2012f:14012.

[13] _____, *Classifying spaces of degenerating mixed Hodge structures, III: Spaces of nilpotent orbits*, J. Algebraic Geometry, **22** (2013), 671–772.

[14] _____, *Néron models for admissible normal functions*, Proc. Japan Acad. **90, A** (2014), 6–10.

[15] K. KATO, S. USUI, *Logarithmic Hodge structures and classifying spaces* (summary), in CRM Proc. and Lect. Notes: The Arithmetic and Geometry of Algebraic Cycles, (NATO Advanced Study Institute / CRM Summer School 1998: Banff, Canada), **24** (1999), 115–130. MR 2001e:14009

[16] _____, *Borel-Serre spaces and spaces of* SL(2)*-orbits*, Advanced Studies in Pure Math. **36**: Algebraic Geometry 2000, Azumino, 2002, 321–382. MR 2004f:14021.

[17] _____, *Classifying spaces of degenerating polarized Hodge structures*, Ann. Math. Studies, Princeton Univ. Press **169**, 2009, pp. 288. MR 2009m:14012.

[18] G. LAPORTE, J. WALCHER, *Monodromy of an inhomogenoues Picard–Fuchs equation*, SIGMA **8** (2012), 056, 10 pages.

[19] S. LI, B. LIAN, S.-T. YAU, *Picard–Fuchs equations for relative periods and Abel-Jacobi map for Calabi-Yau hypersurfaces*, Amer. J. Math. **134-5** (2012), 1345-1384. MR 2975239.

[20] B. LIAN, K. LIU, S.-T. YAU, *Mirror principle I*, Asian J. Math. **1** (1997), 729–763. MR 99e:14062.

[21] D. MORRISON, *Mirror symmetry and rational curves on quintic threefolds: A guide for mathematicians*, J. of AMS **6-1** (1993), 223–247. MR 93j:14047.

[22] _____, *Mathematical aspects of mirror symmetry*, in Complex algebraic geometry (Park City, UT, 1993), IAS/Park City Math. Ser. **3** (1997), 265–327. MR 98g:14044.

[23] D. MORRISON, J. WALCHER, *D-branes and normal functions*, Adv. Theor. Math. Phys. **13-2** (2009), 553–598. MR 2010b:14081.

[24] A. OGUS, *On the logarithmic Riemann-Hilbert correspondences*, Documenta Math. Extra volume: Kazuya Kato's Fiftieth birthday, 2003, 655–724.

[25] R. PANDHARIPANDE, *Rational curves on hypersurfaces* [after A. Givental], Séminaire Bourbaki 848, Astérisque **252** (1998), Exp. No. 848, 5, 307–340. MR 2000e:14094.

[26] R. PANDHARIPANDE, J. SOLOMON, J. WALCHER, *Disk enumeration on the quintic 3-fold*, J. Amer. Math. Soc. **21-4** (2008), 1169–1209. MR 2009j:14075.

[27] W. SCHMID, *Variation of Hodge structure: The singularities of the period mapping*, Invent. Math. **22** (1973), 211–319. MR 52♯3157.

[28] S. USUI, *Variation of mixed Hodge structure arising from family of logarithmic deformations II: Classifying space*, Duke Math. J. **51-4** (1984), 851–875. MR 86h:14005.

[29] _____, *Generic Torelli theorem for quintic-mirror family*, Proc. Japan Acad. **84, A** (2008), 143–146. MR 2010b:14012.

[30] _____, *A study of mirror symmetry through log mixed Hodge theory*, Hodge Theory, Complex Geometry, and Representation Theory, Contemporary Math., AMS **608** (2014), 285–311.

[31] J. WALCHER, *Opening mirror symmetry on the quintic*, Commun. Math. Phys. **276** (2007), 671–689. MR 2008m:14111.

Sampei USUI
Graduate School of Science
Osaka University
Toyonaka, Osaka, 560-0043, Japan
usui@math.sci.osaka-u.ac.jp

6

The 14th case VHS via K3 fibrations

Adrian Clingher, Charles F. Doran, Jacob Lewis,
Andrey Y. Novoseltsev, and Alan Thompson

ABSTRACT. We present a study of certain singular one-parameter subfamilies of Calabi-Yau threefolds realized as anticanonical hypersurfaces or complete intersections in toric varieties. Our attention to these families is motivated by the Doran-Morgan classification of variations of Hodge structure which can underlie families of Calabi-Yau threefolds with $h^{2,1} = 1$ over the thrice-punctured sphere. We explore their torically induced fibrations by M-polarized K3 surfaces and use these fibrations to construct an explicit geometric transition between an anticanonical hypersurface and a nef complete intersection through a singular subfamily of hypersurfaces. Moreover, we show that another singular subfamily provides a geometric realization of the missing "14th case" variation of Hodge structure from the Doran-Morgan list.

Contents

A. Clingher was supported by Simons Foundation grant no. 208258 and by a Bitdefender Invited Professor scholarship from IMAR.

C. F. Doran and A. Y. Novoseltsev were supported by the Natural Sciences and Engineering Resource Council of Canada (NSERC), the Pacific Institute for the Mathematical Sciences, and the McCalla Professorship at the University of Alberta.

J. Lewis was supported in part by NSF grant OISE-0965183.

A. Thompson was supported in part by NSERC and in part by a Fields Institute Ontario Postdoctoral Fellowship with funding provided by NSERC and the Ontario Ministry of Training, Colleges and Universities.

1 Introduction

In their paper [DM06], Doran and Morgan give a classification of the possible variations of Hodge structure that can underlie families of Calabi-Yau threefolds with $h^{2,1} = 1$ over the thrice-punctured sphere. They find fourteen possibilities. At the time of publication of [DM06], explicit families of Calabi-Yau threefolds realising thirteen of these cases were known and are given in [DM06, Table 1]. The aim of this paper is to give a geometric example which realizes the fourteenth and final case (henceforth known as the **14th case**) from their classification, and to study its properties.

By analogy with other examples (see [DM06, Section 4.2]), one might expect that the 14th case variation of Hodge structure should be realized by the mirror of a complete intersection of bidegree $(2,12)$ in the weighted projective space $\mathbb{WP}(1,1,1,1,4,6)$. However, this ambient space is not Fano, so the Batyrev-Borisov mirror construction cannot be applied to obtain such a mirror family.

Instead, Kreuzer and Sheidegger [KKRS05] suggest working with a slightly different ambient space, given by a non-crepant blow up of $\mathbb{WP}(1,1,1,1,4,6)$. However, the complete intersection Calabi-Yau threefold of bidegree $(2,12)$

in this ambient space has $h^{1,1} = 3$, so its mirror will have $h^{2,1} = 3$, making it unsuitable as a candidate for the 14th case on the Doran-Morgan list.

We take as the starting point for our discussion this mirror family with $h^{2,1} = 3$, which we denote by Y. We begin by constructing it explicitly and exploring its properties, noting that it has a singular one-parameter subfamily Y_1 with GKZ series matching that predicted for the 14th case. Moreover, we find that Y has a torically induced fibration by K3 surfaces given as anticanonical hypersurfaces in the three-dimensional space polar to $\mathbb{WP}(1,1,4,6)$.

The existence of this fibration suggests that a model for the 14th case variation of Hodge structure could also be constructed by considering Calabi-Yau threefolds admitting fibrations by anticanonical hypersurfaces in the space polar to $\mathbb{WP}(1,1,4,6)$. A computer search for such fibrations yields a second candidate family, Z, which is mirror to a hypersurface of degree 24 in the weighted projective space $\mathbb{WP}(1,1,2,8,12)$. The threefolds in this family also have $h^{2,1} = 3$.

This poses the natural question of whether the two families Y and Z are related. A careful study of the geometry of each shows that the generic fibre of their K3 fibrations is polarized by the lattice $M := H \oplus E_8 \oplus E_8$. Lattice polarized K3 surfaces of this type were studied by [CD07] and [CDLW09]; using their results we are able to show that the K3 fibrations on Y and Z are very closely related. With this relationship as a guide, we proceed to find an explicit geometric transition between the two, by degenerating Z to a singular subfamily Z_2 and blowing up to obtain Y.

Having shown that the families Y and Z are in fact two halves of the same picture, we turn our focus back to the family Y and, in particular, its singular subfamily Y_1. The main results of this paper are Lemma 6.1 and Theorem 6.4, which describe the mixed Hodge structure on the cohomology groups of Y_1. Together they imply:

Theorem 1.1. *The mixed Hodge structure on the cohomology groups $H^i(Y_1)$ is pure unless $i = 3$ and the weight filtration $W_\bullet$ on $H^3(Y_1)$ has the following description:*

$$\mathrm{Gr}_k^{W_\bullet} H^3(Y_1) = 0 \quad \text{for } k \neq 2,3$$
$$\dim_{\mathbb{C}} W_2(H^3(Y_1)) = 2$$

Furthermore, as Y_1 varies in its one-parameter subfamily, the third graded piece $\mathrm{Gr}_3^{W_\bullet} H^3(Y_1)$ admits a pure variation of Hodge structure of weight 3 and type $(1,1,1,1)$, which realizes the 14th case variation of Hodge structure.

A crucial step in the proof of this theorem involves quotienting the threefold Y_1 by a certain involution and resolving to obtain a new family of Calabi-Yau threefolds W. This new family is closely related to Y_1 and is interesting in its own right. The K3 fibration on Y_1 induces a fibration of W by Kummer surfaces, via the mechanism discussed in [CD07]. We digress to show how W may be constructed directly as a family of Kummer surfaces using the techniques of [DHNT13]; this provides a great deal of insight into the geometry of this new threefold.

Finally, we conclude by discussing the mirrors of the families Y and Z considered here, and compute a mirror for the geometric transition between them. In particular, we find that the singular subfamily Y_1 may naturally be seen as the mirror of a complete intersection of bidegree $(2,12)$ in $\mathbb{WP}(1,1,1,1,4,6)$, as originally expected. We note, however, that Y_1 does not admit a Calabi-Yau resolution, despite the fact that its mirror admits a Calabi-Yau smoothing. This provides a counterexample to a conjecture of Morrison [Mor99].

The structure of this paper is as follows. In Section 2, we give an overview of relevant results from toric geometry that will be used throughout the rest of the paper; this also serves the function of setting up the notation that will be used in what follows. Then in Section 3 we construct the families Y and Z and study some of their basic properties, these are summarized by Propositions 3.1 and 3.2. In particular, Proposition 3.1 shows that Y has a subfamily Y_1 with GKZ series matching that predicted for the 14th case variation of Hodge structure.

We begin Section 4 with a detailed study of the K3 fibrations on Y and Z, using results of [CD07] and [CDLW09]. From this, we show that the parameters describing the K3 fibration on Y match with those describing the K3 fibration on a certain subfamily Z_2 of Z, and those describing the K3 fibration on the subfamily Y_1 match with those describing the K3 fibration on a further subfamily Z_1 of Z_2. These subfamilies Y_1, Z_1 and Z_2 of Y and Z are studied in Section 5; their properties are summarized by Propositions 5.1 and 5.2. Using these results, we show in Proposition 5.3 that the families Y and Z are in fact related by a geometric transition, through the subfamily Z_2, and that this transition also relates the subfamilies Y_1 and Z_1.

Section 6 begins with a result, Lemma 6.1, which describes the mixed Hodge structure on the cohomology groups of the subfamily Y_1. However, there is an ambiguity in the description of the weight filtration on $H^3(Y_1)$ given by this lemma, which prevents us from drawing any conclusions. To resolve this, we study the action of certain involutions on Y, which restrict to the subfamily Y_1. The resolved quotient of Y_1 by one such involution gives a new Calabi-Yau threefold, W, with geometry closely related to that of Y_1. Proposition 6.2 and Corollary 6.3 describe this geometry; from them we are able to deduce Theorem 6.4 and Corollary 6.5, which complete the description of the subfamily Y_1. In particular, we find that Y_1 realizes the 14th case variation of Hodge structure that we have been seeking.

In Section 7 we digress to discuss the geometry of the Calabi-Yau threefolds W. The K3 fibration on Y_1 induces a K3 fibration on W, the general fibre of which is a Kummer surface. We use this K3 fibration, along with the methods of [DHNT13], to construct a model for W directly. The geometric nature of this construction provides a great deal of insight into the geometry of W and demonstrates the utility of the methods of [DHNT13] in a practical example.

Finally, in Section 8 we compute mirrors for Y and Z, their subfamilies Y_1, Z_1 and Z_2, and exhibit a mirror for the geometric transition between them. In particular, we find that Y_1 may be seen as a mirror to a complete intersection of degree $(2,12)$ in $\mathbb{WP}(1,1,1,1,4,6)$, as expected. Finally, we show that Y_1 and its mirror provide a counterexample to a conjecture of Morrison [Mor99].

Most of the computer-aided computations performed for this work were done in the free open source mathematics software Sage [S$^+$15], specifically using the toric geometry framework developed by Braun and Novoseltsev [BN12]. At the time of this writing, it is necessary to use a small extra patch[1], but we plan to integrate it into the official release of Sage. An interested reader may easily reproduce all computations used in this article following the presented code, as well as adjust them to suit his or her own needs. Typesetting of code snippets and some of the complicated expressions was greatly facilitated by SageTEX [D$^+$12].

1.1. Acknowledgements

The authors would like to thank Andrew Harder and Ursula Whitcher for their assistance with some of the calculations in Section 7. A. Clingher would like to thank the Institute of Mathematics “Simion Stoilow” of the Romanian Academy in Bucharest, the Institut des Hautes Études Scientifique in Bures-sur-Yvette and the Max Planck Institute for Mathematics in Bonn for their hospitality and support during the academic year 2012-2013. A portion of this work was completed while A. Thompson was in residence at the Fields Institute Thematic Program on Calabi-Yau Varieties: Arithmetic, Geometry and Physics; he would like to thank the Fields Institute for their support and hospitality.

2 Toric Geometry Review and Notation

In this section we recall the basic facts of toric geometry in order to set up notation that will be used in the rest of the paper; we refer to the wonderful book [CLS11] for details. We also briefly describe Batyrev's

[1] Available as a git branch at https://github.com/novoselt/sage/commits/Cayley

construction of mirror families of Calabi-Yau anticanonical hypersurfaces in toric varieties [Bat94] and its generalization to nef complete intersections by Batyrev and Borisov [BB96b]. We conclude with a discussion of fibrations of Calabi-Yau varieties that can be obtained from natural fibrations of their ambient spaces.

2.1. Toric Varieties

Let M and N be dual lattices of rank n, i.e. $M \simeq N \simeq \mathbb{Z}^n$ as free Abelian groups, where M is identified with $\operatorname{Hom}_{\mathbb{Z}}(N,\mathbb{Z})$ and N with $\operatorname{Hom}_{\mathbb{Z}}(M,\mathbb{Z})$ using the natural pairing $\langle\cdot,\cdot\rangle : M \times N \to \mathbb{Z}$. The real vector space associated to a lattice will be denoted by the subscript $\mathbb{R}$, e.g. the real vector space associated to the lattice N is $N_{\mathbb{R}} = N \otimes_{\mathbb{Z}} \mathbb{R}$. The torus of a lattice N is defined to be $T_N = N \otimes_{\mathbb{Z}} \mathbb{C}^* \simeq (\mathbb{C}^*)^n$, it is an affine variety with group structure corresponding to componentwise multiplication in $(\mathbb{C}^*)^n$ and character lattice M.

A cone $\sigma \subset N_{\mathbb{R}}$ will always be convex, polyhedral, and rational. Its dual cone will be denoted by $\sigma^\vee \subset M_{\mathbb{R}}$. The affine toric variety corresponding to a cone σ will be denoted by $U_\sigma = \operatorname{Spec}\left(\mathbb{C}[\sigma^\vee \cap M]\right)$.

For a fan Σ in $N_{\mathbb{R}}$, we will denote by X_Σ the corresponding toric variety and by $\Sigma(k)$ the set of its k-dimensional cones. In particular, $\Sigma(1)$ is the set of rays and for each ray $\varrho \in \Sigma(1)$ we have a primitive integral generator $v_\varrho \in N$, a homogeneous coordinate z_ϱ, and a prime torus-invariant Weil divisor D_ϱ given by $\{z_\varrho = 0\}$.

Any toric divisor $D = \sum_\varrho a_\varrho D_\varrho$ has an associated polyhedron given by

$$P_D = \left\{u \in M_{\mathbb{R}} : \langle u, v_\varrho \rangle \geqslant -a_\varrho \text{ for all } \varrho \in \Sigma(1)\right\}.$$

For any $m \in P_D \cap M$, the D-homogenization of the character χ^m is the monomial

$$z^{\langle m,D\rangle} = \prod_\varrho z_\varrho^{\langle m,v_\varrho\rangle + a_\varrho},$$

and the global sections of D are

$$\Gamma(X_\Sigma, \mathcal{O}_{X_\Sigma}(D)) = \bigoplus_{m \in P_D \cap M} \mathbb{C}\chi^m.$$

Of particular interest to us are the toric canonical divisor $K_{X_\Sigma} = -\sum_\varrho D_\varrho$ and the global sections of the anticanonical divisor, in the following special case.

2.2. Gorenstein Fano Toric Varieties

Gorenstein Fano toric varieties are in bijection (after taking into account equivalences) with reflexive polytopes. A reflexive polytope is a full-dimensional lattice polytope $\Delta \subset M_{\mathbb{R}}$ that contains the origin in its interior and has polar

$$\Delta^\circ = \{v \in N_{\mathbb{R}} : \langle u,v\rangle \geqslant -1 \text{ for all } u \in \Delta\} \subset N_{\mathbb{R}}$$

that is also a lattice polytope. Since $(\Delta^\circ)^\circ = \Delta$, there is a polar duality between reflexive polytopes in $M_{\mathbb{R}}$ and reflexive polytopes in $N_{\mathbb{R}}$, and even an inclusion-reversing bijective correspondence between the faces of Δ and Δ°.

Given a reflexive polytope $\Delta \subset M_{\mathbb{R}}$, its normal fan in $N_{\mathbb{R}}$ will be denoted by Σ_Δ. To avoid towers of subscripts, we will simply write X_Δ instead of X_{Σ_Δ} for the toric variety associated to this fan. The vertices of Δ° coincide with the primitive integral generators of rays of Σ_Δ, so one can also think of Σ_Δ as the face fan of Δ° (with cones over faces of Δ°).

Definition 2.1. Let $\Delta \subset M_{\mathbb{R}}$ be a reflexive polytope and let Σ be a subdivision of Σ_Δ. If all rays of Σ are generated by (some of) the boundary lattice points of Δ°, it is a **crepant subdivision**. If Σ is also simplicial and the corresponding toric variety X_Σ is projective, it is a **projective crepant subdivision**. If furthermore *all* boundary lattice points of Δ° generate rays of Σ, it is a **maximal projective crepant subdivision**.

Such subdivisions correspond to *maximal projective crepant partial desingularizations (MPCP-desingularizations)*, introduced by Batyrev in [Bat94]. They lead to a very concrete construction of Calabi-Yau orbifolds.

Theorem 2.2. *Let Δ be a reflexive polytope of dimension n. A generic anticanonical hypersurface in $X = X_\Delta$, i.e. a generic section $f \in \Gamma(X, \mathcal{O}_X(-K_X))$, is a Calabi-Yau variety of dimension $n-1$. A generic anticanonical hypersurface in $X = X_\Sigma$, where Σ is a projective crepant subdivision of Σ_Δ, is a Calabi-Yau orbifold of dimension $n-1$.*

Proof. See [CK99, Proposition 4.1.3]. □

For a reflexive polytope $\Delta \subset M_{\mathbb{R}}$, projective crepant subdivisions of Σ_Δ correspond to special triangulations of the boundary of Δ°. For any such subdivision the polytope of the anticanonical divisor is easily seen to be Δ itself. This means that in the equations of the anticanonical Calabi-Yau hypersurfaces, the variables correspond to lattice points of Δ° while the (coefficients of) monomials correspond to lattice points of Δ:

$$\sum_{m \in \Delta \cap M} a_m \prod_{v_\varrho \in \partial\Delta^\circ \cap N} z_\varrho^{\langle m, v_\varrho\rangle + 1} = 0.$$

Polar duality of reflexive polytopes means that the roles of Δ and Δ° in the above discussion can be reversed, leading to another family of hypersurfaces in another toric variety. Batyrev showed [Bat94] that if $\dim \Delta = 4$, then generic anticanonical hypersurfaces in MPCP-desingularizations of X_Δ are *smooth* Calabi-Yau threefolds and the exchange $\Delta \leftrightsquigarrow \Delta^\circ$ corresponds to the exchange $h^{1,1} \leftrightsquigarrow h^{2,1}$ of the Hodge numbers of the anticanonical hypersurfaces of the two families, making them candidates for mirror pairs.

2.3. Nef Complete Intersections

Batyrev and Borisov generalized the construction of Calabi-Yau varieties as anticanonical hypersurfaces in toric varieties to the case of complete intersections associated to nef-partitions of reflexive polytopes [Bor93, BB96b].

Definition 2.3. Let $\Delta \subset M_\mathbb{R}$ be a reflexive polytope. A **nef-partition** is a decomposition of the vertex set V of $\Delta^\circ \subset N_\mathbb{R}$ into a disjoint union

$$V = V_0 \sqcup V_1 \sqcup \cdots \sqcup V_{r-1}$$

such that all divisors $E_i = \sum_{v_\varrho \in V_i} D_\varrho$ are Cartier. Equivalently, let $\nabla_i \subset N_\mathbb{R}$ be the convex hull of the vertices from V_i and the origin. These polytopes form a nef-partition if their Minkowski sum $\nabla \subset N_\mathbb{R}$ is a reflexive polytope.

The **dual nef-partition** is formed by the polytopes $\Delta_i \subset M_\mathbb{R}$ of the E_i, which give a decomposition of the vertex set of $\nabla^\circ \subset M_\mathbb{R}$; their Minkowski sum is Δ.

"Nef-partition" may refer to any of the following decompositions:

1. $V(\Delta^\circ)$ into a disjoint union of V_i,
2. Δ into a Minkowski sum of Δ_i,
3. the anticanonical divisor of X_Δ into a sum of E_i.

Each of these decompositions can be easily translated into the others. Some care should be taken only to avoid mixing a nef-partition and its dual.

It follows from the definition that polar duality of reflexive polytopes switches convex hull and Minkowski sum for dual nef-partitions:

$$\begin{aligned}
\Delta^\circ &= \operatorname{Conv}(\nabla_0, \nabla_1, \ldots, \nabla_{r-1}), \\
\nabla &= \nabla_0 + \nabla_1 + \cdots + \nabla_{r-1}, \\
\Delta &= \Delta_0 + \Delta_1 + \cdots + \Delta_{r-1}, \\
\nabla^\circ &= \operatorname{Conv}(\Delta_0, \Delta_1, \ldots, \Delta_{r-1}).
\end{aligned}$$

Given a nef-partition of an n-dimensional reflexive polytope consisting of r-parts, generic sections of the divisors E_i determine an $(n-r)$-dimensional complete intersection Calabi-Yau variety. In [BB96a] Batyrev and Borisov show that such varieties corresponding to dual nef-partitions have mirror-symmetric *stringy* Hodge numbers, which we will denote by $h_{st}^{p,q}(Y)$.

2.4. Torically Induced Fibrations

Let N and N' be lattices, Σ be a fan in $N_{\mathbb{R}}$, and Σ' be a fan in $N'_{\mathbb{R}}$. There is a bijection between toric morphisms $\varphi\colon X_\Sigma \to X_{\Sigma'}$ (i.e. morphisms of toric varieties preserving the group structure of their tori) and fan morphisms $\widetilde{\varphi}\colon \Sigma \to \Sigma'$ (i.e. lattice homomorphisms $\widetilde{\varphi}\colon N \to N'$ compatible with the fan structure: the linear extension $\widetilde{\varphi}_{\mathbb{R}}\colon N_{\mathbb{R}} \to N'_{\mathbb{R}}$ of $\widetilde{\varphi}$ maps each cone $\sigma \in \Sigma$ into a single cone $\sigma' \in \Sigma'$).

Consider the special case when the lattice homomorphism is surjective, i.e. we have an exact sequence of lattices

$$0 \to N_0 \to N \xrightarrow{\widetilde{\varphi}} N' \to 0,$$

where $N_0 = \ker \widetilde{\varphi}$. Let $\Sigma_0 = \{\sigma \in \Sigma : \sigma \subset (N_0)_{\mathbb{R}}\}$. We can consider Σ_0 either as a fan in $N_{\mathbb{R}}$ or as a fan in $(N_0)_{\mathbb{R}}$, giving two corresponding toric varieties $X_{\Sigma_0,N}$ (a dense subset of X_Σ) and X_{Σ_0,N_0}.

As discussed in [CLS11, § 3.3], there is a clear relation between these two varieties,

$$X_{\Sigma_0,N} \simeq X_{\Sigma_0,N_0} \times T_{N'}.$$

In fact $X_{\Sigma_0,N} = \varphi^{-1}(T_{N'})$, so part of X_Σ is a fibre bundle over $T_{N'}$, with fibres X_{Σ_0,N_0}. Moreover, if Σ is *split by* Σ' *and* Σ_0, then the whole of X_Σ is a fibre bundle over $X_{\Sigma'}$ (see [CLS11, Definition 3.3.18 and Theorem 3.3.19]). However, splitting is a very strong condition on fans, so instead of imposing it we will work with more general fibrations than fibre bundles.

Definition 2.4. Let $\varphi : X \to Y$ be a morphism between two varieties. Then φ is a **fibration** if it is surjective and all of its fibres have the same dimension $\dim X - \dim Y$.

In [HLY02, Proposition 2.1.4] the authors provide a detailed description of the fibres of *arbitrary* toric morphisms[2]. However, we are primarily interested in fibrations $\varphi\colon X_\Sigma \to X_{\Sigma'}$, as they may induce fibrations $\varphi|_Y\colon Y \to X_\Sigma$ of Calabi-Yau subvarieties Y realized as anticanonical hypersurfaces or nef complete intersections in X_Σ. This fibration condition may prevent Σ from being "too refined", leading to singularities of Y, but it may still be possible to compose φ with a crepant resolution of singularities in such a way that Y becomes smooth and the restriction of the composition to Y is still a fibration.

We now describe a strategy for searching for such toric fibrations. Let $\Delta \subset M_{\mathbb{R}}$ be a reflexive polytope and let Σ be a crepant subdivision of Σ_Δ. Suppose that $\widetilde{\varphi}\colon \Sigma \to \Sigma'$ is a fibration (meaning that $\varphi\colon X_\Sigma \to X_{\Sigma'}$ is a fibration in the above sense). As before, its fibre is determined by the subfan Σ_0

[2] Note that while the notation in [HLY02] is very similar to ours, the authors sometimes *implicitly* assume that toric varieties in question are *complete*.

of Σ in the sublattice $N_0 = \ker \widetilde{\varphi}$ of N. Since we would like the fibres of Calabi-Yau subvarieties to be lower-dimensional Calabi-Yau varieties, it is natural to require that Σ_0 *is also associated to a reflexive polytope*, i.e. that it is a crepant subdivision of Σ_∇ for some $\nabla \subset (M_0)_\mathbb{R}$, where M_0 is the dual lattice of N_0. In this case ∇° is a "slice" of Δ° by a linear subspace, so one can search for such slices of Δ° and take $\widetilde{\varphi}$ to be the projection along the linear subspace of ∇°.

We can also reformulate this problem in dual terms: $M_0 = M/(N_0)^\perp$ and the condition that ∇° is inside Δ° implies that the image of Δ in M_0 is inside ∇. So alternatively one can look for "projections" of Δ that are reflexive. Here we recall that the origin is the only interior lattice point of any reflexive polytope, so all lattice points of Δ must be projected onto either the origin or the boundary of the projection. Due to this restriction a "large" Δ with many lattice points is less likely to have any fibrations than a "small" one.

3 Models

3.1. Complete Intersections

In [DM06] Doran and Morgan have shown that there are 14 possible classes of variations of Hodge structure which can be associated to families of Calabi-Yau threefolds Y with $h^{2,1} = 1$. They have provided explicit examples for all but one of these classes and given some suggestions on how one could construct an example for the last class (the "14th case" referred to in the title of this paper).

By analogy with other examples, one could hope to start with a complete intersection with $h^{1,1} = 1$ in the weighted projective space $\mathbb{WP}(1,1,1,1,4,6)$. Unfortunately this ambient space is not Fano, so the Batyrev-Borisov mirror construction based on nef-partitions, described in Section 2.3, cannot be applied to obtain a family with $h^{2,1} = 1$. Instead, Kreuzer and Scheidegger have suggested working with a slightly different ambient space, a non-crepant blow-up of $\mathbb{WP}(1,1,1,1,4,6)$, which is Fano and has a family of complete intersections corresponding to a suitable nef-partition, so the mirror transition is possible (See [KKRS05, Section 8 and Appendix E.2] for some discussion of this example). We will construct and explore this mirror family below using Sage.

Let $\Delta \subset M_\mathbb{R}$ be a 5-dimensional reflexive polytope with polar given by

```
sage: Delta5_polar = LatticePolytope([(1,-1,0,0,0),
    (-1,1,0,0,0), (-1,-1,0,0,0), (-1,-1,2,0,0),
    (12,0,-1,-1,-1), (0,12,-1,-1,-1), (0,0,-1,-1,-1),
    (0,0,11,-1,-1), (0,0,-1,2,-1), (0,0,-1,-1,1)],
    lattice=ToricLattice(5))
```

i.e. the vertices of Δ° are given by the columns of the following matrix

$$(3.1) \qquad \left(\begin{array}{r|r|r|r|r|r|r|r|r|r} 1 & -1 & -1 & -1 & 12 & 0 & 0 & 0 & 0 & 0 \\ -1 & 1 & -1 & -1 & 0 & 12 & 0 & 0 & 0 & 0 \\ 0 & 0 & 0 & 2 & -1 & -1 & -1 & 11 & -1 & -1 \\ 0 & 0 & 0 & 0 & -1 & -1 & -1 & -1 & 2 & -1 \\ 0 & 0 & 0 & 0 & -1 & -1 & -1 & -1 & -1 & 1 \end{array}\right)_N .$$

The nef-partition we are interested in is "in agreement" with the block structure of this matrix: one part is formed by the first four vertices and the other by the last six.

```
sage: np = NefPartition([0]*4+[1]*6, Delta5_polar)
sage: np
Nef-partition {0, 1, 2, 3} U {4, 5, 6, 7, 8, 9}
```

Let Σ be a crepant subdivision of Σ_Δ and $X = X_\Sigma$ be the corresponding crepant partial resolution of X_Δ. The choice of this resolution will depend on our needs, but for the moment we only ensure that Σ is simplicial. We will use y_i to denote homogeneous coordinates on X, with i being the index of the corresponding point of Δ°. For future use we also introduce parameters c, d, and e into the base field of X. Finally, we let $Y \subset X$ be a generic member of the family of complete intersections corresponding to the nef-partition above.

```
sage: X5 = CPRFanoToricVariety(np.Delta(),
    make_simplicial=True, coordinate_names="y+",
    base_field=QQ["c,d,e"].fraction_field())
sage: Y = X5.nef_complete_intersection(np)
```

The defining polynomials of Y are

$$(3.2) \qquad g_0 = a_0 y_0^2 y_4^{12} + a_1 y_1^2 y_5^{12} + a_2 y_0 y_1 y_2 y_3,$$

$$g_1 = b_4 y_4^6 y_5^6 y_6^6 y_7^6 + b_5 y_4^4 y_5^4 y_6^4 y_7^4 y_8 + b_3 y_2^2 y_6^{12} + b_2 y_3^2 y_7^{12}$$
$$(3.3) \qquad + b_7 y_4^3 y_5^3 y_6^3 y_7^3 y_9 + b_6 y_4^2 y_5^2 y_6^2 y_7^2 y_8^2 + b_8 y_4 y_5 y_6 y_7 y_8 y_9 + b_0 y_8^3 + b_1 y_9^2.$$

There are 12 parameters in these equations, but their number can be significantly reduced. First of all, we can use a change of variables to set $b_5 = b_6 = b_7 = 0$. To see this more easily, we switch to an affine chart.

```
sage: Yap = Y.affine_patch(21)
sage: g1 = Yap.defining_polynomials()[1]
```

In this chart g_1 takes the form

$$g_1 = b_0 y_8^3 + b_3 y_2^2 + b_2 y_3^2 + b_6 y_8^2 + b_8 y_8 y_9 + b_1 y_9^2 + b_5 y_8 + b_7 y_9 + b_4$$

and making a substitution $y_8 = y_8 + c$, $y_9 = y_9 + d + ey_8$ does not lead to any new monomials.

```
sage: X5 = Y.ambient_space()
sage: X5.inject_coefficients();
sage: X5.inject_variables();
sage: g1s = g1.subs(y8=y8+c, y9=y9+d+e*y8)
sage: g1s.monomials()
[y8^3, y2^2, y3^2, y8^2, y8*y9, y9^2, y8, y9, 1]
sage: g1s.monomial_coefficient(y8^2)
e^2*b1 + 3*c*b0 + e*b8 + b6
sage: g1s.monomial_coefficient(y9)
2*d*b1 + c*b8 + b7
sage: g1s.monomial_coefficient(y8)
3*c^2*b0 + 2*d*e*b1 + c*e*b8 + 2*c*b6 + e*b7 + d*b8 + b5
```

From this we see that one can pick c, d, and e to make the coefficients of the 3 monomials y_8, y_8^2 and y_9 vanish, leaving only 9 parameters. Since we can also scale both polynomials and 5 of the variables, we can further reduce the number of parameters to 2.

On the other hand, computing stringy Hodge numbers of Y (one can use the generating function from [BB96a] implemented in PALP [KS04] or the closed form expressions from [DN10]) we obtain

$$h_{st}^{1,1}(Y) = 243, \qquad h_{st}^{2,1}(Y) = 3.$$

If we take X to be a MPCP-desingularization of X_Δ, then Y is generically smooth and its stringy Hodge numbers coincide with its regular ones, so the dimension of the space of complex deformations of Y is 3. The fact that we have reduced the number of parameters in the defining polynomials of Y to 2 suggests that the dimension of the space of polynomial deformations of Y is $h_{poly}^{2,1}(Y) = 2$. To confirm this observation we compute the toric part of $h_{st}^{1,1}(Y^\circ)$ for the Batyrev-Borisov mirror Y° of Y inside a MPCP-desingularization X° of X_∇ (it is easier to work with a maximal resolution in this case, since ∇° is much smaller than Δ°, i.e. it has only a few lattice points.)

```
sage: X5m = CPRFanoToricVariety(np.nabla(),
    make_simplicial=True, coordinate_points="all")
sage: X5m.is_smooth()
True
sage: Ym = X5m.nef_complete_intersection(np.dual(),
    monomial_points="vertices")
sage: H = Ym.ambient_space().cohomology_ring()
sage: H.gens()
```

```
([z5], [z5], [4*z5 + 2*z6 + z7 + z9], [6*z5 + 3*z6 + 2*z7 + z8
    + z9], [z5], [z5], [z5], [z6], [z7], [z8], [z9])
```

Here we have restricted the monomials used in the defining equations of Y° to save some time, since the nef divisors corresponding to it have hundreds of monomial sections and we only need its cohomology class. Note that X° is smooth, so Y° is (generically) smooth as well, giving $h^{1,1}_{st}(Y^\circ) = h^{1,1}(Y^\circ)$. To compute $h^{1,1}_{tor}(Y^\circ)$ (the contribution of $H^{1,1}(X^\circ)$ to $H^{1,1}(Y^\circ)$), we just need to consider the intersections of the cohomology class of Y° with the generators of the cohomology ring of X°. As can be seen from the above output, this ring can be generated by the last five of "all generators", each of which corresponds to a torus-invariant subvariety of codimension 1, e.g. `[z5]` corresponds to $\{z_5 = 0\}$, whose cohomology class is equivalent to $\{z_0 = 0\}$, $\{z_1 = 0\}$, and $\{z_4 = 0\}$.

```
sage: Ym_c = Ym.cohomology_class()
sage: Ym_c
[24*z5^2 - 6*z6^2 + 8/3*z7*z8 - 2/3*z8^2 - 13*z6*z9 - 3/2*z9^2]
sage: [Ym_c * g for g in H.gens()[-5:]]
[[24*z5^3 + 3*z6^3 + 14/9*z7*z8^2 + 1/9*z8^3 + 19/2*z6^2*z9 +
    3/8*z9^3], [-z6^2*z9], [0], [0], [0]]
```

From this we conclude that $h^{1,1}_{tor}(Y^\circ) = h^{2,1}_{poly}(Y) = 2$.

As pointed out in [DM06], a certain subfamily of the complete intersections Y has the desired hypergeometric series corresponding to the 14th case. To describe that subfamily precisely in our setting we will compute the GKZ series of Y, following the algorithm outlined in [CK99, Section 5.5] and [KKRS05, Appendix A].

It is easy to check that the monomials that can be eliminated in the equations for Y correspond to all points of Δ_i that are neither vertices of Δ_i nor the origin (whilst this is true for our particular case, in general such information cannot be easily determined and one has to perform computations in the cohomology ring as above). Let's use this information to reconstruct our varieties:

```
sage: X5 = CPRFanoToricVariety(np.Delta(),
    make_simplicial=True, coordinate_names="y+")
sage: Y = X5.nef_complete_intersection(np,
    monomial_points="vertices+origin")
sage: X5m = CPRFanoToricVariety(np.nabla())
sage: X5m.Mori_cone().rays()
[ 0  1]
[ 0  1]
[ 2  0]
[ 3  0]
[ 0  1]
```

```
[ 0  1]
[ 1 -2]
[-6 -2]
in Ambient free module of rank 8 over the principal ideal
    domain Integer Ring
```

We now use the generators of the Mori cone of X° to construct moduli parameters of Y. These generators are given as elements of the row span of the Gale transform of the fan of X°, the i-th element of each generator corresponds to the i-th ray of this fan, except for the last one which corresponds to the origin and is equal to the negative sum of other entries. Note that for complete intersections we need to take such sums for each part of the nef-partition separately; the following code adds them in a way that is compatible with the order of coefficients of Y in Sage, with a_2 and b_8 corresponding to the origin (the coefficient-monomial correspondence for the newly constructed Y is the same as in (3.2) and (3.3), since the coefficient indices come from the internal enumeration of polytope lattice points).

```
sage: coefs = Y.ambient_space().base_ring().gens()
sage: coefs
(a0, a1, a2, b0, b1, b2, b3, b4, b8)
sage: degrees = [[[ray[i] for i in p] for p in
    np.dual().parts()] for ray in X5m.Mori_cone()]
sage: degrees = [flatten([p + [-sum(p)] for p in degs]) for
    degs in degrees]
sage: matrix(degrees)
[ 0  0  0  2  3  0  0  1 -6]
[ 1  1 -2  0  0  1  1 -2  0]
sage: B = [prod(c^d for c, d in zip(coefs, ds)) for ds in
    degrees]
```

We find that the modular parameters of Y are

$$B_0 = \frac{b_0^2 b_1^3 b_4}{b_8^6}, \qquad B_1 = \frac{a_0 a_1 b_2 b_3}{a_2^2 b_4^2},$$

and its GKZ series is

$$\sum_{m,n} \frac{(2m)!(6n)!}{(m!)^4(2n)!(3n)!(n-2m)!} B_1^m B_0^n,$$

where the summation is over all integers m and n such that the arguments of all factorials are non-negative. Making a substitution $(n-2m) \to n$, we can sum over all non-negative integers:

$$\sum_{m,n\in\mathbb{Z}_{\geqslant 0}} \frac{(2m)!(12m+6n)!}{(m!)^4(4m+2n)!(6m+3n)!n!} B_1^m B_0^{2m+n}.$$

To simplify the description of Y further, we now scale its defining polynomials and coordinates to set all coefficients to 1 except for $b_3 = \xi_0$ and $b_4 = \xi_1$.

```
sage: Y = X5.nef_complete_intersection(np, monomial_points=
    "vertices+origin",coefficients=[[1,1,1],[1,1,1,"xi0",
    "xi1",1]])
```

The defining polynomials of Y become

$$g_0 = y_0^2 y_4^{12} + y_1^2 y_5^{12} + y_0 y_1 y_2 y_3, \tag{3.4}$$

$$g_1 = \xi_1 y_4^6 y_5^6 y_6^6 y_7^6 + \xi_0 y_2^2 y_6^{12} + y_3^2 y_7^{12} + y_4 y_5 y_6 y_7 y_8 y_9 + y_8^3 + y_9^2, \tag{3.5}$$

and the GKZ series takes the form

$$\sum_{m,n\in\mathbb{Z}_{\geqslant 0}} \frac{(2m)!(12m+6n)!}{(m!)^4(4m+2n)!(6m+3n)!n!} \xi_0^m \xi_1^n.$$

Comparing this series with the one given in the end of [DM06], we see that the subfamily of interest is $\xi_1 = 0$, leading to the series

$$\sum_{m\in\mathbb{Z}_{\geqslant 0}} \frac{(2m)!(12m)!}{(m!)^4(4m)!(6m)!} \xi_0^m.$$

This subfamily, henceforth denoted by Y_1, will be of further interest in our pursuit of a geometric model for the 14th case; we will come back to it in Section 5.2.

We complete this subsection by showing that the complete intersections Y admit a natural fibration structure, which will allow us to access many of their properties. Going back to the definition of Δ° in (3.1), we observe that, in addition to its column decomposition into a nef-partition, it has a "natural" row decomposition. Indeed, the projection onto the first two coordinates corresponds to a toric fibration $\widetilde{\alpha} : X \to B$ over a 2-dimensional toric variety B, as long as we pick a compatible resolution of X_Δ. To get such a resolution, we start with the face fan of Δ°, take its minimal subdivision compatible with the projection, and then subdivide it to get a simplicial fan, so that X is an orbifold.

```
sage: m = matrix([(1,0), (0,1)] + [(0,0)]*3)
sage: Delta2_polar = LatticePolytope([(1,-1), (-1,1), (-1,-1),
    (1,0), (0,1)])
sage: B2 = CPRFanoToricVariety(Delta_polar=Delta2_polar,
    coordinate_names="u+")
```

```
sage: Sigma5 = FaceFan(Delta5_polar)
sage: Sigma5.nrays(), Sigma5.ngenerating_cones()
(10, 14)
sage: Sigma5 = FanMorphism(m, Sigma5, B2.fan(),
    subdivide=True).domain_fan()
sage: Sigma5.nrays(), Sigma5.ngenerating_cones()
(10, 22)
sage: X5 = CPRFanoToricVariety(np.Delta(),
    charts=[C.ambient_ray_indices() for C in Sigma5],
    coordinate_names="y+", make_simplicial=True, check=False)
sage: alpha = FanMorphism(m, X5.fan(), B2.fan())
sage: alpha.is_fibration()
True
```

In homogeneous coordinates

$$\widetilde{\alpha}\colon [y_0 : \cdots : y_9] \mapsto [u_0 : \cdots : u_4] = \left[y_0 : y_1 : y_2 y_3 : y_4^{12} : y_5^{12}\right].$$

Note that the hypersurface defined by the polynomial g_0 in (3.4) depends only on the variables involved in the projection map. This means that we can interpret $g_0 = 0$ as a defining equation of a *curve* $C = \widetilde{\alpha}(\{g_0 = 0\}) \subset B$ and $g_1 = 0$ as a defining equation of a *surface* in each fibre of $\widetilde{\alpha}$, in other words, $\widetilde{\alpha}$ induces a fibration of the complete intersection Y over C.

Generic fibres of $\widetilde{\alpha}\colon X \to B$ correspond to the fan whose rays are generated by the last four vertices of Δ°. The polytope spanned by these vertices is the last (the 4318-th) 3-dimensional reflexive polytope in the Kreuzer-Skarke list (included in Sage), with its normal fan corresponding to $\mathbb{WP}(1,1,4,6)$:

```
sage: vertices = [r[2:] for r in alpha.kernel_fan().rays()]
sage: p = LatticePolytope(vertices)
sage: p.is_reflexive()
True
sage: p.index()
4318
sage: NormalFan(p).rays()
[ 1  0  0 -1]
[ 0  1  0 -4]
[ 0  0  1 -6]
in 3-d lattice N
```

Summarizing the results of this section, we find:

Proposition 3.1. *The complete intersection model Y is generically a smooth Calabi-Yau threefold having $h^{2,1}(Y) = 3$ and $h^{2,1}_{poly}(Y) = 2$. There is a torically induced fibration $\tilde{\alpha}\colon Y \to C$ onto a curve C whose generic fibre* *is an anticanonical hypersurface in the three-dimensional space polar to* $\mathbb{WP}(1,1,4,6)$.

Furthermore, the GKZ series of the subfamily Y_1 *obtained by setting the modular parameter* $\xi_1 = 0$ *coincides with the GKZ series predicted for the 14th case VHS.*

3.2. Anticanonical Hypersurfaces

In the previous section we were able to represent the complete intersection Y as a fibration over a curve with fibres living in the space polar to $\mathbb{WP}(1,1,4,6)$. This suggests that the 14th case could also be realized by a family of anticanonical hypersurfaces in a four-dimensional space which can be fibred by the same toric varieties.

To try to find such a family, we searched[3] for Fano varieties fibred by the space polar to $\mathbb{WP}(1,1,4,6)$ among those whose anticanonical hypersurfaces have small $h^{2,1}$, with the extra condition that the torically induced fibration is "balanced": this means that the same 3-dimensional reflexive polytopes can play the roles of both slices and projections, as described in Section 2.4. The space polar to $\mathbb{WP}(1,1,2,8,12)$ satisfies these requirements; below we study the family of anticanonical hypersurfaces in it.

Let $\Delta \subset M_{\mathbb{R}}$ be a 4-dimensional reflexive polytope given by

```
sage: Delta4 = LatticePolytope([(1,0,0,0), (0,1,0,0),
    (0,0,1,0), (0,0,0,1), (-1,-2,-8,-12)])
```

i.e. the vertices of Δ and of $\Delta^\circ \subset N_{\mathbb{R}}$ are given by columns of the following matrices

$$\begin{pmatrix} 1 & 0 & 0 & 0 & -1 \\ 0 & 1 & 0 & 0 & -2 \\ 0 & 0 & 1 & 0 & -8 \\ 0 & 0 & 0 & 1 & -12 \end{pmatrix}_M, \qquad \begin{pmatrix} 23 & -1 & -1 & -1 & -1 \\ -1 & -1 & 11 & -1 & -1 \\ -1 & 2 & -1 & -1 & -1 \\ -1 & -1 & -1 & -1 & 1 \end{pmatrix}_N.$$

Let Σ be a crepant subdivision of Σ_Δ and $X = X_\Sigma$ be the corresponding crepant partial resolution of X_Δ. As before the choice of this resolution will depend on our needs; for now we add only one extra ray in addition to the vertices of Δ°, specifically the ray corresponding to the midpoint $(11,-1,-1,-1)$

[3] We performed this search using Sage, but have chosen not to include the corresponding code in the text due to its length. The algorithm used simply runs over all hyperplanes that intersect a given polytope in a sublattice polytope, with a few technical tricks to make this search happen in a reasonable timeframe and to store the results of big searches compactly. The interested reader may find a Sage worksheet implementing this search attached to the arXiv submission of this paper (note, however, that in order to run this worksheet it is necessary to install the optional package `polytopes_db_4d`; beware of the 8.7GB download size!)

between the 0-th and the 3-rd vertices, which is the 16-th point in the internal enumeration in Sage:

```
sage: Delta4.polar().point(16)
N(11, -1, -1, -1)
sage: X4 = CPRFanoToricVariety(Delta=Delta4,
    coordinate_points=range(5)+[16])
sage: B1 = toric_varieties.P1("s,t")
sage: beta = FanMorphism(matrix(4,1,[1,1,4,6]), X4.fan(),
    B1.fan())
sage: beta.is_fibration()
True
```

This extra ray is necessary to make Σ compatible with the projection onto the line in the direction $(1,1,4,6)$. If $B = \mathbb{P}^1$ with coordinates $[s:t]$ corresponding to the (unique) complete fan on this line and $\widetilde{\beta}: X \to B$ is the toric morphism associated to this projection, then

$$\widetilde{\beta}: [z_0 : z_1 : z_2 : z_3 : z_4 : z_{16}] \mapsto [s:t] = \left[z_0^{12} : z_3^{12}\right]. \tag{3.6}$$

Let $Z \subset X$ be a generic anticanonical hypersurface. Its defining polynomial is

$$h = a_0 z_0^{24} z_{16}^{12} + a_5 z_0^{12} z_3^{12} z_{16}^{12} + a_4 z_3^{24} z_{16}^{12} + a_6 z_0^6 z_2^6 z_3^6 z_{16}^6$$
$$+ a_1 z_2^{12} + a_{10} z_0 z_1 z_2 z_3 z_4 z_{16} + a_2 z_1^3 + a_3 z_4^2.$$

Scaling the whole polynomial and four independent coordinates we can eliminate 5 out of 8 parameters. Using Batyrev's formulas for the Hodge numbers of anticanonical hypersurfaces we check that $h^{2,1}(Z) = h^{2,1}_{poly}(Z) = 3$, so we should indeed have 3 independent parameters. Summarizing, we find:

Proposition 3.2. *The hypersurface model Z is generically a smooth Calabi-Yau threefold with* $h^{2,1}(Z) = h^{2,1}_{poly}(Z) = 3$. *There is a torically induced fibration* $\widetilde{\beta}: Z \to B \cong \mathbb{P}^1$ *whose general fibre is an anticanonical hypersurface in the three-dimensional space polar to* $\mathbb{WP}(1,1,4,6)$.

Anticanonical hypersurfaces inside the space polar to $\mathbb{WP}(1,1,2,8,12)$ were extensively studied by Billó et al. [BDF⁺98]; in order to conveniently use their results we will match our toric description with theirs (they considered hypersurfaces in $\mathbb{WP}(1,1,2,8,12)$ with extra symmetries, which allow taking the quotient under a certain group action).

First, we rewrite the polynomial of Z in its "fibred" form, thinking of it as a polynomial in z_1, z_2, z_4, z_{16} only and working in a chart with $t = z_3 = 1$:

$$\left[a_0 s^2 + a_5 s + a_4\right] z_{16}^{12} + \left[a_6 z_0^6\right] z_2^6 z_{16}^6 + a_1 z_2^{12} + \left[a_{10} z_0\right] z_1 z_2 z_4 z_{16} + a_2 z_1^3 + a_3 z_4^2.$$

Now we can compare our representation with equation (4.19) in [BDF+98]:

$$W^{(2)}(x;B',\psi_0,\psi_1) = \frac{1}{12}(B'x_0^{12}+x_3^{12})+\frac{1}{3}x_4^3+\frac{1}{2}x_5^2-\psi_0x_0x_3x_4x_5-\frac{1}{6}\psi_1x_0^6x_3^6.$$

We see that the matching of coordinates and coefficients is

$$z_1 = x_4, \qquad z_2 = x_3, \qquad z_4 = x_5, \qquad z_0z_{16} = x_0,$$

$$a_1 = \frac{1}{12}, \qquad a_2 = \frac{1}{3}, \qquad a_3 = \frac{1}{2}, \qquad a_6 = -\frac{1}{6}\psi_1, \qquad a_{10} = -\psi_0,$$

and

$$a_0s + \frac{a_4}{s} + a_5 = \frac{1}{12}B'.$$

To match these remaining parameters we use the definition of B' given by equation (3.18) in [BDF+98]:

$$B' = \frac{1}{2}\left(B\zeta + \frac{B}{\zeta} - 2\psi_s\right),$$

where ζ is an affine coordinate on the base of the fibration, so

$$s = \zeta, \qquad a_0 = \frac{B}{24}, \qquad a_4 = \frac{B}{24}, \qquad a_5 = -\frac{\psi_s}{12}.$$

We can use these parameters in Sage as follows:

```
sage: var("B,psi0,psi1,psi_s");
sage: Z = X4.anticanonical_hypersurface(
    coefficients=[B/24,1/12,1/3,1/2,B/24,-psi_s/12,-psi1/6,-psi0])
```

Now the defining polynomial of Z has the form

$$\frac{B}{24}z_0^{24}z_{16}^{12} - \frac{\psi_s}{12}z_0^{12}z_3^{12}z_{16}^{12} + \frac{B}{24}z_3^{24}z_{16}^{12} - \frac{\psi_1}{6}z_0^6z_2^6z_3^6z_{16}^6$$
$$+\frac{1}{12}z_2^{12} - \psi_0z_0z_1z_2z_3z_4z_{16} + \frac{1}{3}z_1^3 + \frac{1}{2}z_4^2.$$

One of the four parameters in this representation is redundant, e.g. we can set $\psi_0 = 1$.

3.3. Geometric Transitions

Both the complete intersections Y and anticanonical hypersurfaces Z presented above have $h^{2,1} = 3$, while the original goal in their construction was to obtain families with $h^{2,1} = 1$ that could provide geometric examples of the 14th case. In fact, in the hypersurface case it was known in advance that we would "fail", since it is known that there are only 5 reflexive polytopes yielding Calabi-Yau threefolds with $h^{2,1} = 1$ and [DM06] showed that they already provide examples for other classes of Hodge structure variations.

However, we can still try to obtain our desired families by using subfamilies of the constructed ones. Of course, simply fixing some of the parameters does not change the Hodge numbers of the Calabi-Yau threefolds in question, but if these threefolds were to become singular we could try to resolve the singularities and hope that $h^{2,1} = 1$ holds for the resolved family. So we are now looking for geometric transitions from the already constructed families to some new ones with, hopefully, "correct" Hodge numbers.

The existence of such geometric transitions was first suggested by Clemens [Cle83a, Cle83b] and later expanded upon by Friedman [Fri86]. In this subsection we give a precise definition and a basic classification of geometric transitions, following [Ros11].

Definition 3.3. Let Y and $\widetilde{Y}$ be smooth Calabi-Yau threefolds. They are connected by a **geometric transition** if there exist a normal variety $\overline{Y}$, a birational contraction $\varphi\colon Y \to \overline{Y}$, and a complex deformation (smoothing) of $\overline{Y}$ to $\widetilde{Y}$. It is a **primitive geometric transition** if φ cannot be factored into birational morphisms of normal varieties. It is a **conifold transition** if $\overline{Y}$ has only conifold singularities (ordinary double points). It is a **trivial geometric transition** if $\widetilde{Y}$ is a deformation of Y.

Theorem 3.4. *Let $\varphi\colon Y \to \overline{Y}$ be a primitive contraction of a smooth Calabi-Yau threefold Y to a normal variety $\overline{Y}$. Let E be the exceptional locus of φ. Then φ is of one of the following three types:*

Type I *φ is small, E may be reducible and is composed of finitely many rational curves;*

Type II *φ contracts a divisor to a point, E is irreducible and is a generalized del Pezzo surface;*

Type III *φ contracts a divisor to a smooth curve C, E is irreducible and is a conic bundle over C.*

Proof. See [Wil92, Wil93], the given formulation is [Ros11, Theorem 1.9]. □

Definition 3.5. A primitive geometric transition is of type I, II, or III, if the corresponding birational contraction is of type I, II, or III, respectively.

The existence of geometric transitions relating Calabi-Yau threefolds with different Hodge numbers has interesting consequences for their moduli spaces. In particular, it has been suggested [Rei87, Fri91] that geometric transitions could be used to connect together the moduli spaces of all compact complex threefolds with trivial canonical bundle. Moreover, it is also natural to ask how geometric transitions interact with the string-theoretic phenomenon of mirror symmetry [Dix88, CLS90, GP90] and this has been the object of a great deal of study; a good summary of this theory is given by Morrison [Mor99]. We will return to the question of mirror symmetry in our context in Section 8.

However, before we use geometric transitions to construct new families, we first find that our two models Y and Z are in fact connected via a type III geometric transition. To see this explicitly, we begin by comparing their fibration structures.

4 Matching the Models

4.1. K3 Fibrations

The generic fibres of the fibrations induced by $\widetilde{\alpha}$ and $\widetilde{\beta}$ constructed earlier are anticanonical hypersurfaces inside the three-dimensional space polar to $\mathbb{WP}(1,1,4,6)$, i.e. they are generically two-dimensional Calabi-Yau varieties: K3 surfaces. The choice of toric ambient space for them induces a lattice polarization, in this case by the lattice $M = H \oplus E_8 \oplus E_8$, where H is the hyperbolic lattice of rank 2 and E_8 is the unique even negative-definite unimodular lattice of rank 8. Such M-polarized K3 surfaces (the name "M-polarized" is a bit unfortunate in the toric context, but it should not cause too much confusion) were originally studied by Shioda and Inose [SI77, Ino78] and have recently been revisited in [Shi06], [CD07] and [CDLW09], we start this section with a summary of their properties.

Definition 4.1. An **M-polarization** on a K3 surface X is a primitive lattice embedding $i\colon M \hookrightarrow \mathrm{NS}(X)$, such that the image $i(M)$ in the Néron-Severi lattice $\mathrm{NS}(X)$ contains a pseudo-ample class (corresponding to an effective nef divisor with positive self-intersection).

Definition 4.2. Let X be a K3 surface. An involution i on X is a **Nikulin involution** if $i^*\omega = \omega$ for any holomorphic 2-form ω on X.

Theorem 4.3. *Let X be an M-polarized K3 surface. Then*

1. *X is isomorphic to the minimal resolution of a quartic surface in $\mathbb{P}^3$ given by*

$$y^2zw - 4x^3z + 3axzw^2 + bzw^3 - \frac{1}{2}(dz^2w^2 + w^4) = 0;$$

2. *the parameters a, b, and d in the above equation specify a unique point $(a,b,d) \in \mathbb{WP}(2,3,6)$ with $d \neq 0$;*
3. *X canonically corresponds to a pair of elliptic curves $\{E_1, E_2\}$;*

4. *the modular parameters of X and* $\{E_1, E_2\}$ *are related by*
$$\pi = j(E_1)j(E_2) = \frac{a^3}{d} \quad \text{and} \quad \sigma = j(E_1) + j(E_2) = \frac{a^3 - b^2 + d}{d};$$
5. *generically there are exactly two isomorphism classes of elliptic fibrations with section on X: the "standard" fibration, which has an* $H \oplus E_8 \oplus E_8$ *polarization realized via two type II* exceptional fibres, and the "alternate" fibration which has an* $H \oplus D_{16}$ *polarization realized via a type* I_{12}^* *exceptional fibre;*
6. *there exists a Nikulin involution on X.*

Proof. See [Shi06], [CD07, Theorem 1.1, Corollary 1.3, Section 3] and [CDLW09, Theorems 3.1, 3.2]. □

Proposition 4.4. *An anticanonical hypersurface in the space polar to the weighted projective* $\mathbb{WP}(1,1,4,6)$ *is an M-polarized K3 surface defined by*
$$\lambda_0 x_0^{12} + \lambda_4 x_0^6 x_1^6 + \lambda_1 x_1^{12} + \lambda_5 x_0 x_3 x_1 x_2 + \lambda_3 x_3^3 + \lambda_2 x_2^2 = 0.$$

It is related to the normal form given in Theorem 4.3 by
$$a^3 = \frac{1}{12^6 \Lambda_0^2 \Lambda_1},\ b^2 = \frac{(6 \cdot 12^2 \Lambda_0 - 1)^2}{12^6 \Lambda_0^2 \Lambda_1},\ d = 1,\ \text{with } \Lambda_0 = \frac{\lambda_2^3 \lambda_3^2 \lambda_4}{\lambda_5^6},\ \Lambda_1 = \frac{\lambda_0 \lambda_1}{\lambda_4^2}.$$

Proof. See [CDLW09, Section 3.4][4]. □

Both the standard and alternate elliptic fibrations have toric realizations on the anticanonical hypersurfaces in the above proposition. Here we will point out a few key facts about this toric picture that will be used later; for a detailed treatment of the combinatorics of torically induced elliptic fibrations of K3 surfaces we refer the reader to [PS97] or [Roh04].

In our case the toric correction term is zero, meaning that the intersection pattern of divisors on a generic anticanonical hypersurface, which have three linear relations between them, is given by the 1-skeleton of the reflexive polytope in the lattice N. Two directions in N determine the two fibrations over $\mathbb{P}^1$, and the ADE diagrams of the fibres over zero and infinity are given by the parts of the 1-skeleton that are projected into the interiors of rays of the $\mathbb{P}^1$ fan. Here is an explicit construction of these diagrams:

```
sage: Delta3 = LatticePolytope([(1,0,0), (0,1,0), (0,0,1),
    (-1,-4,-6)])
sage: P = Delta3.polar()
```

[4] There was a typo in the preprint version of [CDLW09] posted on arXiv: the numerator of the expression for b^2 was not squared.

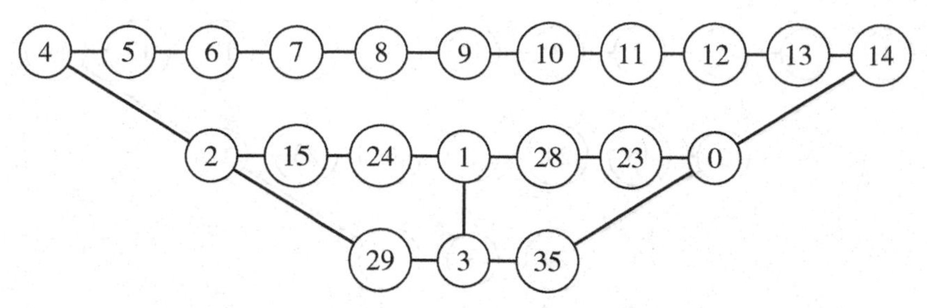

Figure 1. Intersection of toric divisors on an M-polarized K3 surface.

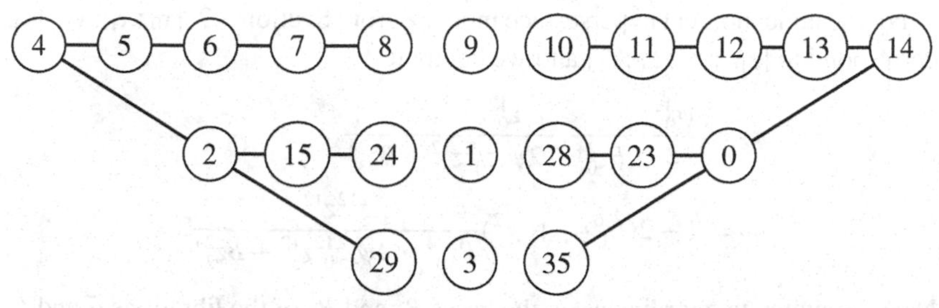

Figure 2. ADE diagram of the standard fibration on an M-polarized K3 surface.

```
sage: sum(e.ninterior_points() * ed.ninterior_points() for e,
    ed in zip(Delta3.edges(), P.edges()))
0
sage: G = P.skeleton()
sage: d_std = vector([1,2,3])
sage: E8E8 = G.subgraph(edge_property=lambda e:
    (P.point(e[0])*d_std) * (P.point(e[1])*d_std) > 0)
sage: d_alt = vector([0,1,1])
sage: D16 = G.subgraph(edge_property=lambda e:
    (P.point(e[0])*d_alt) * (P.point(e[1])*d_alt) > 0)
```

The graphs constructed above are shown in Figure 1 to Figure 3 on pages 187–188. In these graphs, the labels correspond to the indices of lattice points of the polytope P. For the standard fibration the toric divisor corresponding to point 9 gives a uniquely defined section, and for the alternate fibration there are two toric sections corresponding to points 24 and 28.

Next we will see how this theory can be applied to our models Y and Z. For the complete intersection model Y described in Section 3.1 the role of $[x_0 : x_1 : x_2 : x_3]$ in Proposition 4.4 is played by $[y_6 : y_7 : y_9 : y_8]$. Using this we obtain the following expressions for the modular parameters of a fibre:

$$\pi_Y = \frac{y_4^{12} y_5^{12}}{12^6 \xi_0 y_2^2 y_3^2}, \qquad \sigma_Y = 1 + \left(\xi_1 - 3 \cdot 12^2 \xi_1^2\right) \frac{y_4^{12} y_5^{12}}{12^3 \xi_0 y_2^2 y_3^2}.$$

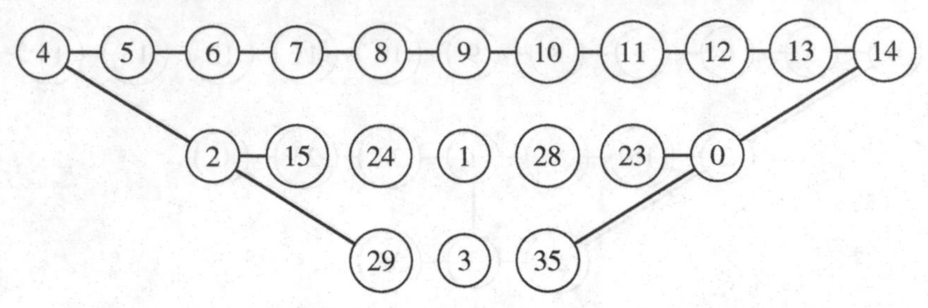

Figure 3. ADE diagram of the alternate fibration on an M-polarized K3 surface.

For the anticanonical hypersurface model Z from Section 3.2, $[x_0 : x_1 : x_2 : x_3]$ correspond to $[z_{16} : z_2 : z_4 : z_1]$ and we compute

$$\pi_Z = \frac{\psi_0^{12}}{2} \cdot \frac{z_0^{12} z_3^{12}}{B z_0^{24} - 2\psi_s z_0^{12} z_3^{12} + B z_3^{24}},$$

$$\sigma_Z = 1 - 2(\psi_0^6 \psi_1 + \psi_1^2) \frac{z_0^{12} z_3^{12}}{B z_0^{24} - 2\psi_s z_0^{12} z_3^{12} + B z_3^{24}}.$$

Next we switch to coordinates on the bases B^2 and B^1 of the fibrations $\widetilde{\alpha}$ and $\widetilde{\beta}$ respectively:[5]

$$\pi_Y = \frac{u_3 u_4}{12^6 \xi_0 u_2^2},$$

$$\sigma_Y = 1 + \left(\xi_1 - 3 \cdot 12^2 \xi_1^2\right) \frac{u_3 u_4}{12^3 \xi_0 u_2^2},$$

$$\pi_Z = \frac{\psi_0^{12}}{2} \cdot \frac{st}{B s^2 - 2\psi_s st + B t^2}, \tag{4.1}$$

$$\sigma_Z = 1 - 2(\psi_0^6 \psi_1 + \psi_1^2) \frac{st}{B s^2 - 2\psi_s st + B t^2}. \tag{4.2}$$

Finally, recall that Y is fibred not over B^2, but over a curve $C \subset B^2$ corresponding to g_0. It is easy to see from (3.4), that g_0 is the pullback of

$$u_0^2 u_3 + u_1^2 u_4 + u_0 u_1 u_2. \tag{4.3}$$

Using the fan of B^2, shown in Figure 4 on the facing page, we can see that C does not intersect the divisors corresponding to u_0 and u_1, e.g. if $u_0 = 0$ then $u_1^2 u_4 \neq 0$, since there are no cones containing rays corresponding to u_0 and u_1 or u_4.

This means that C is isomorphic to the curve in $\mathbb{P}^2$ with coordinates $[u : v : w] = [u_3 : u_4 : u_2]$, given by $u + v + w = 0$, i.e. $C \simeq \mathbb{P}^1$ with coordinates $[u : v]$.

[5] We have used the same notation Δ, Σ, X, and B in Section 3.1 and Section 3.2 to refer to different objects. When we need to consider them together and it is necessary to distinguish them, we will use their dimensions as superscripts. Names in Sage examples always include these dimensions to allow reusing of objects in later sections.

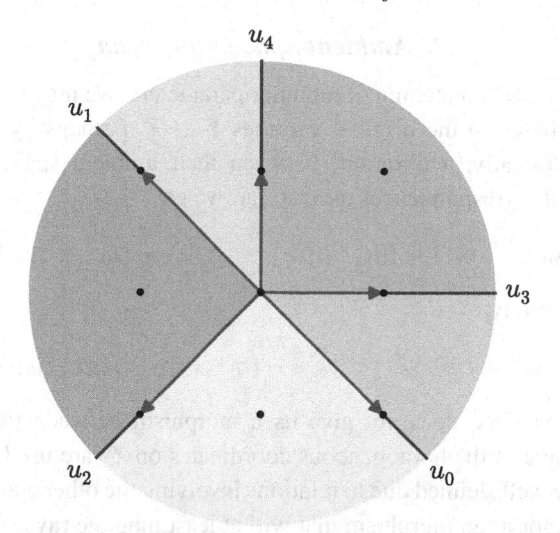

Figure 4. Base B^2 of the fibration $\widetilde{\alpha}$.

In terms of these new coordinates the modular parameters of a fibre are

$$\pi_Y = \frac{1}{12^6 \xi_0} \cdot \frac{uv}{(u+v)^2}, \tag{4.4}$$

$$\sigma_Y = 1 + \frac{\xi_1 - 3 \cdot 12^2 \xi_1^2}{12^3 \xi_0} \cdot \frac{uv}{(u+v)^2}. \tag{4.5}$$

Comparing expressions for π and σ for the complete intersection model Y, (4.4) and (4.5), with analogous expressions for the anticanonical hypersurface model Z, (4.1) and (4.2), we see that they are quite similar. This similarity is even more striking if we let $\psi_s = -B$ in (4.1) and (4.2), to obtain:

$$\pi_Z = \frac{\psi_0^{12}}{2B} \cdot \frac{st}{(s+t)^2}, \tag{4.6}$$

$$\sigma_Z = 1 - \frac{2(\psi_0^6 \psi_1 + \psi_1^2)}{B} \cdot \frac{st}{(s+t)^2}. \tag{4.7}$$

In fact, the expressions (4.4) and (4.5) and the expressions (4.6) and (4.7) would be exactly the same if we let $[u:v] = [s:t]$ and

$$\xi_0 = \frac{2B}{\left(12\psi_0^2\right)^6}, \qquad \xi_1 = -\frac{4\psi_1}{\left(12\psi_0^2\right)^3}. \tag{4.8}$$

Finally, we note also that with respect to the fibre modular parameters π_Y and σ_Y, the subfamily of complete intersections with $\xi_1 = 0$ is very special: in this case $\sigma_Y \equiv 1$, so all K3 fibres correspond to "complementary" elliptic curves with $j(E_1) + j(E_2) = 1$.

4.2. Ambient Space Morphism

Given such a perfect matching of modular parameters, we may try to construct a morphism between the original varieties $Y \to Z$, perhaps by constructing a morphism (ideally, a fibration) between their ambient spaces $X^5 \to X^4$. Combining all correspondences used so far, we get

$$[y_0 : \cdots : y_9] \leftrightsquigarrow ([u : v], [x_0 : \cdots : x_3]) \leftrightsquigarrow [z_0 : \cdots : z_{16}],$$

or, more concretely,

$$[y_4 : y_8 : y_7 : y_5 : y_9 : y_6] \leftrightsquigarrow [z_0 : z_1 : z_2 : z_3 : z_4 : z_{16}].$$

This correspondence does not give us a morphism between toric varieties, since only some of the homogeneous coordinates on X^5 are used and they are unlikely to be well-defined due to relations involving the other coordinates. But we may look for a fan morphism that will at least map the rays corresponding to these selected y_i to the rays corresponding to the indicated z_j.

```
sage: m_y = Delta5_polar.vertices_pc()(4,8,7,5,9,6).matrix()
sage: m_z = Delta4.polar().points_pc()(0,1,2,3,4,16).matrix()
sage: m = m_y.solve_right(m_z)
```

This computation shows that if such a fan morphism exists, it must be given by the matrix `m`

$$\begin{pmatrix} 1 & 0 & 0 & 0 \\ -1 & 0 & 0 & 0 \\ -1 & 1 & 0 & 0 \\ -4 & 0 & 1 & 0 \\ -6 & 0 & 0 & 1 \end{pmatrix},$$

which is unique since `m_y` has maximal rank. If this matrix defines a fibration, then rays of Σ^5 must be mapped onto rays of Σ^4 or the origin.

```
sage: rays = [r*m for r in Sigma5.rays() if not (r*m).is_zero()]
sage: from sage.geometry.cone import normalize_rays
sage: rays = set(normalize_rays(rays, None))
sage: points = sorted(Delta4.polar().points_pc().index(r) for r
    in rays)
sage: points
[0, 1, 2, 3, 4, 16, 331, 333, 334]
```

Since all necessary image rays can be generated by lattice points of $\left(\Delta^4\right)^\circ$, we can pick Σ^4 to be a crepant subdivision of Σ_{Δ^4}. One can then subdivide the current Σ^5 (the subdivision of Σ_{Δ^5} compatible with the fibration α) to obtain a fibration over X^4, but, unfortunately, this subdivision of Σ_{Δ^5} will not be crepant.

To fix this problem, we go back to the defining polynomial (4.3) of the curve $C \subset B^2$ and the fan of B^2, shown in Figure 4 on page 189. We have already established that C does not intersect the divisors corresponding to u_0 and u_1. Now we note that it also does not contain the point with $u_3 = u_4 = 0$, since this would imply that one of the other coordinates is zero, which is not possible. This means that C is completely contained within the part of B^2 corresponding to the fan generated by the rays of u_2, u_3, and u_4. But then $Y \subset X^5$ is completely contained within the toric variety corresponding to Σ^5 *without* all cones that contain rays corresponding to either y_0, or y_1, or y_4 and y_5 together. Starting with such a subfan Σ^5_{part} of Σ^5, we can find a subdivision that leads to a fibration $\widetilde{\Phi}\colon X^5_{\text{part}} \to X^4$ with affine lines as generic fibres.

```
sage: selected = []
sage: for sigma in flatten(Sigma5.cones()):
...       indices = sigma.ambient_ray_indices()
...       if (0 in indices or
...           1 in indices or
...           4 in indices and 5 in indices):
...           continue
...       selected.append(sigma)
sage: Sigma5_part = Fan(cones=selected, rays=Sigma5.rays(),
    discard_faces=True)
sage: rays = [ray*m for ray in Sigma5_part.rays() if not
    (ray*m).is_zero()]
sage: rays = set(normalize_rays(rays, None))
sage: points4 = sorted(Delta4.polar().points_pc().index(r) for
    r in rays)
sage: points4
[0, 1, 2, 3, 4, 16, 334]
sage: Delta4.polar().point(334)
N(-1, 1, 0, 0)
sage: X4 = CPRFanoToricVariety(Delta4,
    coordinate_points=points4)
sage: Phi = FanMorphism(m, Sigma5_part, X4.fan(),
    subdivide=True)
sage: Phi.is_fibration()
True
sage: Phi.kernel_fan().rays()
[-1]
[-1]
[ 0]
[ 0]
[ 0]
in Sublattice <N(1, 1, 0, 0, 0)>
```

```
sage: all(ray in Delta5_polar.points_pc() for ray in
    Phi.domain_fan().rays())
True
sage: points5 = [Delta5_polar.points_pc().index(ray) for ray in
    Phi.domain_fan().rays()]
sage: points5
[2, 3, 4, 5, 6, 7, 8, 9, 752]
sage: Delta5_polar.point(752)
N(0, 0, 1, 0, 0)
sage: Phi.domain_fan().is_simplicial()
True
sage: X5_part = ToricVariety(Phi.domain_fan(),
     coordinate_names="y+", coordinate_indices=points5)
```

Now not only can all necessary rays in the codomain be generated by lattice points of $\left(\Delta^4\right)^\circ$, but also all necessary rays in the domain can be generated by lattice points of $\left(\Delta^5\right)^\circ$. We can therefore think of X^5_{part} as an open subset in some crepant partial desingularization of X_{Δ^5}, whilst $Y \subset X^5_{\text{part}}$ is still a Calabi-Yau variety.

From the computational point of view, we can no longer use the framework of CPR-Fano toric varieties in Sage to represent X^5_{part}; instead we have to use generic toric varieties. We will however still use the same convention for naming coordinates: the missing y_0 and y_1 now reflect the fact that we are interested only in charts where they are non-zero, in which case both can be set to 1 using relations between coordinates.

In homogeneous coordinates we get

$$\begin{aligned}\widetilde{\Phi}\colon [y_2 : \cdots : y_9 : y_{752}] \mapsto {} & [z_0 : z_1 : z_2 : z_3 : z_4 : z_{16} : z_{334}] \\ & = [y_4 : y_8 : y_7 : y_5 : y_9 : y_6 : y_3^2 y_{752}],\end{aligned}$$

which also does not involve y_2, since it corresponds to the only ray of the kernel fan. This map can be used to pull back Z and compare it with Y. We delay this comparison, however, since for analysis of singularities it is convenient to perform a few more subdivisions of the underlying fans first.

5 Singular Subfamilies

5.1. Hypersurfaces

Generically, complete intersections Y and anticanonical hypersurfaces Z in MPCP-desingularizations of their ambient spaces are smooth, since the singular locus in this case has codimension at least four. However, in this paper we are also interested in the following subfamilies:

1. complete intersections Y with $\xi_1 = 0$, corresponding to the subfamily with the desired GKZ series, we will denote a generic member of this subfamily as Y_1 to emphasize dependence on a single parameter only;

2. hypersurfaces Z with $\psi_s = -B$, corresponding to the subfamily whose K3 fibration can be "perfectly matched" with the K3 fibration of complete intersections Y, we will denote a generic member of this subfamily as Z_2;
3. hypersurfaces Z with $\psi_s = -B$ and $\psi_1 = 0$, corresponding to the subfamily of the above subfamily with an analogue of the $\xi_1 = 0$ restriction, we will denote a generic member of this subfamily as Z_1.

Generic members of these subfamilies may be singular, we analyze their singularities using results of [BDF+98] and computer software (Sage interfacing with Magma [BCP97] for computing the singular loci of affine varieties). We begin in this subsection with an analysis of the singularities of the hypersurfaces Z_2 and Z_1.

The singular locus in the moduli space of hypersurfaces Z is summarized in relations (4.39) of [BDF+98]:

$$\begin{aligned}
S^{\pm}_{a1}: &\quad (\psi_0^6 + \psi_1)^2 + \psi_s = \pm B,\\
S^{\pm}_{a2}: &\quad \psi_1^2 + \psi_s = \pm B,\\
S^{\pm}_{b}: &\quad \psi_s = \pm B,\\
S_0: &\quad 0 = B.
\end{aligned}$$

Here relations $S^{\pm}_{a1}$ and $S^{\pm}_{a2}$ are actually the same, in the sense that they are switched by an appropriate change of coordinates (there is a finite group action on the simplified polynomial moduli space of hypersurfaces). We see that the hypersurfaces Z_2 are singular (and, therefore, cannot be isomorphic to the full two-parameter family of complete intersections Y), since the condition S^{-}_{b} is satisfied. The condition $\psi_1 = 0$ makes relations $S^{\pm}_{a2}$ and $S^{\pm}_{b}$ the same, but does not impose singularities on its own.

An explicit chart-by-chart check for singularities of Z_2 using our current fan Σ^4 and ignoring the orbifold structure of X^4 (i.e. ignoring singularities of Z_2 which are inevitable due to the ambient space structure) reveals a singular locus of dimension 1 in all charts involving z_{334}. To study this singular locus, it is convenient to put it into smooth charts of X^4; in order to do this we need to subdivide Σ^4 further.

In fact, when doing this one finds that it is better to start afresh with Σ_{Δ^4}. Recall that adding the ray corresponding to the 16-th point $v_{16} = (11,-1,-1,-1)_N$ of $(\Delta^4)^{\circ}$, which is the midpoint between the vertices $v_0 = (23,-1,-1,-1)_N$ and $v_3 = (-1,-1,-1,-1)_N$, was necessary to ensure compatibility with the fibration $\widetilde{\beta}\colon X^4 \to B^1 \simeq \mathbb{P}^1$. We also had to add $v_{334} = (-1,1,0,0)_N$ to allow for a fibration $\widetilde{\Phi}\colon X^5_{\mathrm{part}} \to X^4$. This is the only interior point of the triangular face on $v_1 = (-1,-1,2,-1)_N$, $v_2 = (-1,11,-1,-1)_N$, and $v_4 = (-1,-1,-1,1)_N$. The face fan of this triangle (in the spanned affine sublattice with v_{334} being the origin) is the fan of $\mathbb{WP}(1,2,3)$, and the three interior points of its edges correspond to $v_{251} = (-1,7,0,-1)_N$,

$v_{276} = (-1,3,1,-1)_N$, and $v_{325} = (-1,5,-1,0)_N$. Adding all these rays is sufficient to resolve all singularities of the 3-dimensional cone on v_1, v_2, and v_4, but the resulting subcones are still faces of singular 4-dimensional cones. This is also reflected in the homogeneous coordinate representation of $\widetilde{\beta}$ (3.6): the coordinates on the base B^1 correspond to the 12th powers of coordinates on X^4 (yet the defining equation of Z involves first powers of all variables). The problem is that v_0 and v_3 are "too far away" from the slice hyperplane defining the projection to B^1. We can remedy the situation by adding two more points right "above" and "below" the face on v_1, v_2, and v_4 (which is completely contained in the slice hyperplane), namely $v_{168} = (-1,10,-1,-1)_N$ and $v_{170} = (1,10,-1,-1)_N$.

Using consecutive star-like subdivisions (which are used in Sage for automatic insertion of rays), it turns out that the best sequence is the following:

1. add v_{16} to allow the fibration $\widetilde{\beta}\colon X^4 \to B^1$;
2. add v_{168} and v_{170} to "improve" this fibration;
3. add v_{334} to allow the fibration $\widetilde{\Phi}\colon X^5_{\text{part}} \to X^4$;
4. add v_{251}, v_{276}, and v_{325} to cover the divisor of z_{334} by smooth charts.

```
sage: X4 = CPRFanoToricVariety(Delta4,
    coordinate_points=[0,1,2,3,4,16,168,170,334,251,276,325])
sage: beta = FanMorphism(matrix(4,1,[1,1,4,6]), X4.fan(),
    B1.fan())
sage: beta.is_fibration()
True
sage: all(sigma.is_smooth() for sigma in X4.fan() if (-1,1,0,0)
    in sigma)
True
sage: Z = X4.anticanonical_hypersurface(
    coefficients=[B/24,1/12,1/3,1/2,B/24,-psi_s/12,-psi1/6,-psi0])
sage: X4 = Z.ambient_space()
sage: Z2 = X4.anticanonical_hypersurface(
    coefficients=[B/24,1/12,1/3,1/2,B/24,B/12,-psi1/6,-psi0])
sage: Z1 = X4.anticanonical_hypersurface(
    coefficients=[B/24,1/12,1/3,1/2,B/24,B/12,0,-psi0])
```

The ray matrix of the fan of X^4 is now

$$\left(\begin{array}{r|r|r|r|r|r|r|r|r|r|r|r} 23 & -1 & -1 & -1 & -1 & 11 & -1 & 1 & -1 & -1 & -1 & -1 \\ -1 & -1 & 11 & -1 & -1 & -1 & 10 & 10 & 1 & 7 & 3 & 5 \\ -1 & 2 & -1 & -1 & -1 & -1 & -1 & -1 & 0 & 0 & 1 & -1 \\ -1 & -1 & -1 & -1 & 1 & -1 & -1 & -1 & 0 & -1 & -1 & 0 \end{array}\right)_N$$

and the fibration $\widetilde{\beta}\colon X^4 \to B^1$ takes the form

$$z \mapsto [s:t] = \left[z_0^{12} z_{170} : z_3^{12} z_{168}\right],$$

so working in affine charts we may treat, say, z_{170} as the K3 fibre parameter.

Since the singularities of Z_2 are located in charts involving z_{334}, it is natural to represent its defining polynomial as

$$h_2 = q_2 z_{334} + r_2, \tag{5.1}$$

where

$$q_2 = \frac{1}{12} z_2^{12} z_{168}^{11} z_{170}^{11} z_{251}^{8} z_{276}^{4} z_{325}^{6} z_{334} - \frac{\psi_1}{6} z_0^6 z_2^6 z_3^6 z_{16}^6 z_{168}^6 z_{170}^6 z_{251}^4 z_{276}^2 z_{325}^3$$
$$- \psi_0 z_0 z_1 z_2 z_3 z_4 z_{16} z_{168} z_{170} z_{251} z_{276} z_{325} + \frac{1}{3} z_1^3 z_{251} z_{276}^2 + \frac{1}{2} z_4^2 z_{325}, \tag{5.2}$$

$$r_2 = \frac{B}{24} z_{16}^{12} \left(z_3^{12} z_{168} + z_0^{12} z_{170}\right)^2. \tag{5.3}$$

In this form it is easy to see that the fibre of $\widetilde{\beta}\colon Z_2 \to B^1$ (which is generically a K3 surface) over $[s:t] = [-1:1]$ splits into two components, corresponding to $q_2 = 0$ and $z_{334} = 0$. The intersection of these components is a curve C_2, which is the singular locus of Z_2. In the affine chart $(z_1, z_4, z_{170}, z_{334})$ the defining equations of C_2 take the form

$$\frac{z_1^3}{3} + \psi_0 z_1 z_4 + \frac{z_4^2}{2} - \frac{\psi_1}{6} = 0, \qquad z_{170} = -1, \qquad z_{334} = 0,$$

which is (generically) a smooth elliptic curve. To see what type of singularities occur in this locus, we translate variables from $(z_1, z_4, z_{170}, z_{334})$ to $(R+u_4, S+u_1, u_2-1, u_3)$, where $(z_1, z_4) = (R, S)$ satisfies the first defining equation of C_2 above, to get a hypersurface that is singular at the origin. A further power series substitution brings the singularities into the standard form $u_1^2 + u_2^2 + u_3^2 + (\ldots)u_4$, which is a cA_1 compound Du Val singularity. The hypersurfaces Z_2 thus have cA_1 singularities along C_2.

If we now pass to the subfamily Z_1, the structure of the singularities remains mostly the same, except that the curve of singularities of Z_1, let's call it C_1 for this subfamily, develops a singularity of its own and becomes a nodal elliptic curve. In the same affine chart as before the position of the node is $(0, 0, -1, 0)$. Summarizing, we find:

Proposition 5.1. *The subfamily of hypersurface models Z_2 with $\psi_s = -B$ are generically singular Calabi-Yau threefolds with a smooth elliptic curve C_2 of cA_1 singularities.*

The subfamily of hypersurface models Z_1 with $\psi_s = -B$ and $\psi_1 = 0$ are generically singular Calabi-Yau threefolds with a nodal elliptic curve C_1 of cA_1 singularities.

5.2. Desingularization

In this subsection we complete our analysis of the singularities of subfamilies by determining the singularities of the subfamily Y_1. We then keep an old promise and use the fibration $\widetilde{\Phi}\colon X^5_{\text{part}} \to X^4$ to pull back families of hypersurfaces Z to X^5_{part} then compare them with complete intersections Y.

Since we have changed the resolution used for X^4, we need to reconstruct both X^5_{part} and $\widetilde{\Phi}$. We do this using Σ^5_{part} as a starting point, constructed in Section 4.2.

```
sage: Phi = FanMorphism(m, Sigma5_part, X4.fan(),
    subdivide=True)
sage: [Delta5_polar.points_pc().index(ray) for ray in
    Phi.domain_fan().rays()]
[2, 3, 4, 5, 6, 7, 8, 9, 32, 630, 469, 109, 752, 667]
```

We see that X^5_{part} can still be realized as an open subset of a crepant partial desingularization of X_{Δ^5}. In homogeneous coordinates

$$\begin{aligned}\widetilde{\Phi}\colon y \mapsto & \left[z_0 \colon z_1 \colon z_2 \colon z_3 \colon z_4 \colon z_{16} \colon z_{168} \colon z_{170} \colon z_{334} \colon z_{251} \colon z_{276} \colon z_{325}\right] \\ & = \left[y_4 \colon y_8 \colon y_7 \colon y_5 \colon y_9 \colon y_6 \colon y_{109} \colon y_{32} \colon y_3^2 y_{752} \colon y_{469} \colon y_{630} \colon y_{667}\right].\end{aligned}$$

Considering the subfamily of complete intersections $Y_1 \subset X^5_{\text{part}}$, it is possible to determine that there is a singular point $(0,0,0,0;-1)$ in the chart $(y_2, y_3, y_8, y_9; y_{32})$ (recall that the generating cones of Σ^5_{part} are 4-dimensional, so the corresponding affine charts have a torus factor without a canonical choice of coordinates; in such cases we will use the coordinate corresponding to some suitable ray of the total fan and separate it from the canonically chosen coordinates by ";"). Note that this point is mapped by $\widetilde{\Phi}$ to the singular point of the curve of singularities $C_1 \subset Z_1$. It is also a singular point of X^5_{part} itself, since the cone on rays corresponding to y_2, y_3, y_8, and y_9 is not smooth. To fix this we will perform one last subdivision, by inserting the ray corresponding to the midpoint $(-1,-1,1,0,0)$ between ray generators of y_2 and y_3.

```
sage: Sigma = Phi.domain_fan().subdivide([(-1,-1,1,0,0)])
sage: all(sigma.is_smooth() for sigma in Sigma if (-1,-1,1,0,0)
    in sigma)
True
sage: Phi = FanMorphism(m, Sigma, X4.fan())
sage: Phi.is_fibration()
True
sage: points5 = [Delta5_polar.points_pc().index(ray) for ray in
    Phi.domain_fan().rays()]
sage: points5
[2, 3, 4, 5, 6, 7, 8, 9, 32, 630, 469, 109, 752, 667, 745]
```

The ray matrix of the new fan is

$$\left(\begin{array}{r|r|r|r|r|r|r|r|r|r|r|r|r|r|r} -1 & -1 & 12 & 0 & 0 & 0 & 0 & 0 & 1 & 0 & 0 & 0 & 0 & 0 & -1 \\ -1 & -1 & 0 & 12 & 0 & 0 & 0 & 0 & 0 & 0 & 0 & 1 & 0 & 0 & -1 \\ 0 & 2 & -1 & -1 & -1 & 11 & -1 & -1 & 10 & 3 & 7 & 10 & 1 & 5 & 1 \\ 0 & 0 & -1 & -1 & -1 & -1 & 2 & -1 & -1 & 1 & 0 & -1 & 0 & -1 & 0 \\ 0 & 0 & -1 & -1 & -1 & -1 & -1 & 1 & -1 & -1 & -1 & -1 & 0 & 0 & 0 \end{array}\right)_N$$

and $\widetilde{\Phi}$ is still a fibration with coordinate representation

$$\begin{aligned} \widetilde{\Phi}\colon y \mapsto & \left[z_0 : z_1 : z_2 : z_3 : z_4 : z_{16} : z_{168} : z_{170} : \quad z_{334} \quad : z_{251} : z_{276} : z_{325}\right] \\ = & \left[y_4 : y_8 : y_7 : y_5 : y_9 : y_6 : y_{109} : y_{32} : y_3^2 y_{745} y_{752} : y_{469} : y_{630} : y_{667}\right]. \end{aligned} \tag{5.4}$$

To construct Y as a subvariety of X^5_{part} in Sage, we first construct Y in the full space X^5 using the same coordinates as for X^5_{part} (plus y_0 and y_1), then we use its equations to obtain a subvariety of X^5_{part}. (In the code we refer to it as `Y_part`, but in the text we continue using Y only since mathematically these are the same varieties.)

```
sage: X5 = CPRFanoToricVariety(np.Delta(),
    coordinate_points=[0,1]+points5, coordinate_names="y+")
sage: Y = X5.nef_complete_intersection(np, monomial_points=
    "vertices+origin",coefficients=[[1,1,1],[1,1,1,"xi0",
    "xi1",1]])
sage: X5_part = ToricVariety(Phi.domain_fan(),
    coordinate_names= "y+", coordinate_indices=points5,
    base_field=Y.base_ring())
sage: X5_part.inject_coefficients();
sage: X5_part.inject_variables();
sage: S = X5_part.coordinate_ring()
sage: Y_part =
    X5_part.subscheme([sum(S(g.monomial_coefficient(m)) *
    S(m.subs(y0=1, y1=1)) for m in g.monomials()) for g in
    Y.defining_polynomials()])
```

The defining polynomials of Y in the new X^5_{part} are

$$g_0 = y_5^{12} y_{109} + y_4^{12} y_{32} + y_2 y_3 y_{745}, \tag{5.5}$$

$$\begin{aligned} g_1 = & \, y_3^2 y_7^{12} y_{109}^{11} y_{469}^8 y_{630}^4 y_{32}^{11} y_{667}^6 y_{752}^2 y_{745} \\ & + \xi_1 y_4^6 y_5^6 y_6^6 y_7^6 y_{109}^6 y_{469}^4 y_{630}^2 y_{32}^6 y_{667}^3 y_{752} + \xi_0 y_2^2 y_6^{12} y_{745} \\ & + y_4 y_5 y_6 y_7 y_8 y_9 y_{109} y_{469} y_{630} y_{32} y_{667} y_{752} + y_8^3 y_{469} y_{630}^2 y_{752} + y_9^2 y_{667} y_{752}. \end{aligned} \tag{5.6}$$

There are *two* singular points on the complete intersections Y_1, occurring at the points $\left(\pm 1/\sqrt{\xi_0}, 0, 0, 0; -1\right)$ in the chart $(y_2, y_8, y_9, y_{745}; y_{32})$, which is smooth and does not induce any singularities on its subvarieties. To find their types, consider the defining polynomials of Y_1 in this chart:

$$g_0 = y_2 y_{745} + y_{32} + 1,$$
$$g_1 = y_{32}^{11} y_{745} + \xi_0 y_2^2 y_{745} + y_{32} y_8 y_9 + y_8^3 + y_9^2,$$

then use g_0 (which is smooth) to eliminate y_{32} from g_1 and shift variables to one of the indicated singular points. The leading terms of the remaining polynomial are quadratic and can be converted into the form $u_1^2 + u_2^2 + u_3^2 + u_4^2$. Both points are therefore conifold singularities on Y_1. Thus, in summary, we have:

Proposition 5.2. *The subfamily of complete intersection models Y_1 with $\xi_1 = 0$ are generically singular Calabi-Yau threefolds having two isolated nodes.*

We finally turn our attention to the explicit comparison of Y and the pullback of Z under the fibration $\widetilde{\Phi}$, combined with the first defining equation of Y (recall that the fibration map $\alpha : X^5 \to B^2$ induces the K3 fibration of Y over the curve in B^2 corresponding to $g_0 = 0$). Since in Section 4.1 we matched modular parameters of the K3 fibres of Y and Z_2, we start with the subfamily Z_2.

It is clear that if we use $\widetilde{\Phi}$ (5.4) to pullback h_2 (5.1) we will not get g_1 (5.6), due to a mismatch in the number of monomials and different coefficients, even taking into account the parameter correspondence (4.8). However, the pullback of r_2 (5.3) is

$$\frac{B}{24}\left(y_5^{12} y_{109} + y_4^{12} y_{32}\right)^2 y_6^{12} = \frac{B}{24}\left(y_2 y_3 y_{745}\right)^2 y_6^{12},$$

where equality follows from vanishing of g_0 (5.5). Now the number of monomials is no longer an issue, but we can see that $y_3^2 y_{745}$ is a factor of $\widetilde{\Phi}^*(h_2)$. Further comparison of $\widetilde{\Phi}^*(h_2)$ and g_1 reveals that, subject to $g_0 = 0$,

$$\Psi^*(h_2) = \frac{y_3^2 y_{745}}{48} g_1,$$

where Ψ is $\widetilde{\Phi}$ precomposed with the following coordinate scaling (which cannot be realized as a toric morphism):

$$y_6 = \frac{y_6}{\psi_0 \sqrt{12}}, \qquad y_8 = \frac{y_8}{2}, \qquad y_9 = -\frac{y_9}{\sqrt{12}}, \qquad y_{752} = \frac{y_{752}}{2}.$$

Using Ψ to pullback the defining polynomial h_3 of a generic member Z of the full 3-parameter family of hypersurfaces, we obtain an isomorphism of Z with the complete intersection defined by $\Psi^*(h_3)$ and g_0. Indeed,

$$\Psi^*(h_3) = \Psi^*\left(h_2 - \frac{\psi_s + B}{12}\left(z_0 z_3 z_{16}\right)^{12} z_{168} z_{170}\right)$$
$$= \frac{y_3^2 y_{745}}{48} g_1 - \frac{\psi_s + B}{12^7 \psi_0^{12}}\left(y_4 y_5 y_6\right)^{12} y_{32} y_{109},$$

and $\Psi^*(h_3) = 0$ implies that $y_3 y_{745} \neq 0$, since otherwise it would be necessary to have $y_4 y_5 y_6 y_{32} y_{109} = 0$, which is impossible: the following computation shows that this set of variables does not appear together in any of the covering charts of X^5_{part}.

```
sage: charts = [set(S.gen(j) for j in
      sigma.ambient_ray_indices()) for sigma in X5_part.fan()]
sage: any(chart.intersection([y3, y745]) and
      chart.intersection([y4, y5, y6, y32, y109]) for chart in
      charts)
False
```

Thus $g_0 = 0$ can be solved for y_2, which is the only fibre variable. This 3-parameter family of complete intersections does not correspond to a nef-partition of Δ^5, the easiest way to see this is to note that it shares one of its equations with Y and, in the case of two-part nef-partitions (and corresponding nef complete intersections), each part completely determines the other.

Coming back to the subfamily Z_2, we see that these singular hypersurfaces are pulled back to the union of Y (which is generically smooth) and two toric divisors intersected with $\{g_0 = 0\}$. The composition $\widetilde{\beta} \circ \Psi$ maps both of these intersections to $[s:t] = [-1:1]$.

Next we look at the preimage under Ψ of the curve of singularities C_2 of Z_2 inside of Y. This preimage satisfies the following conditions:

$$g_0 = g_1 = \Psi^*(z_{334}) = \Psi^*(r_2) = \Psi^*(q_2) = 0.$$

Conditions $g_0 = \Psi^*(r) = 0$ imply $y_2 y_3 y_{745} = 0$, while $\Psi^*(z_{334}) = 0$ means that $y_3 y_{745} y_{752} = 0$. But y_2 and y_{752} cannot vanish simultaneously (we check this below), so we must have $y_3 y_{745} = 0$. We also have $g_1 - 24 y_{752} \Psi^*(q_2) = \zeta_0 y_2^2 y_6^{12} y_{745}$ and y_3 cannot vanish simultaneously with y_2 or y_6, so $y_{745} = 0$. Finally, y_{745} and y_{752} cannot vanish simultaneously, so the above conditions are equivalent to

$$g_0 = g_1 = y_{745} = 0$$

and we see that the preimage of C_2 in Y is a (generically smooth) surface S_2. Therefore, we have obtained a geometric transition from Y to Z through the singular subfamily Z_2! In fact, $S_2 \simeq C_2 \times \mathbb{P}^1$ and we have obtained a primitive geometric transition of type III in terms of Definition 3.5. It just remains to verify the assertions made about the simultaneous vanishing of coordinates:

```
sage: any(chart.issuperset([y2, y752]) for chart in charts)
False
sage: any(chart.issuperset([y2, y3]) for chart in charts)
False
```

```
sage: any(chart.issuperset([y3, y6]) for chart in charts)
False
sage: any(chart.issuperset([y745, y752]) for chart in charts)
False
```

The last thing that we should do in this section is to use the map Ψ to compare the two subfamilies Z_1 and Y_1. We have already determined that the hypersurfaces Z_1 have a curve of singularities C_1, which itself has a singular point $(0,0,-1,0)$ in the chart $(z_1,z_4,z_{170},z_{334})$. Let $S_1 \subset Y_1$ be the preimage of C_1 under Ψ (defined by the same equations as $S_2 \subset Y_2$), it is a singular surface. If we consider the preimage of the singular point in Y_1, then in addition to the defining equations of S_1 we need to impose the condition $y_8 = y_9 = 0$. This leads to

$$y_8 = y_9 = y_{745} = y_5^{12} y_{109} + y_4^{12} y_{32} = 0,$$

which defines a projective line containing both nodes of Y_1. The singular locus of the surface S_1 is precisely this line, which is given by $(*,0,0,0;-1)$ in charts $(y_{2\text{ or }3}, y_8, y_9, y_{745}; y_{32})$. Thus, we find:

Proposition 5.3. *The models Y and Z are related by a geometric transition of type III through the singular subfamily Z_2. Furthermore, this induces a geometric transition between the subfamilies Y_1 and Z_1. We have a diagram*

$$\begin{array}{ccccc} & & Y & \rightsquigarrow & Y_1 \\ & & \downarrow & & \downarrow \\ Z & \rightsquigarrow & Z_2 & \rightsquigarrow & Z_1 \end{array} \tag{5.7}$$

where wiggly arrows denote degeneration and straight arrows denote blow-ups.

6 Involutions

Now we have constructed two families of threefolds, Y and Z, along with their singular subfamilies Y_1 and Z_1, and showed that they are in fact two halves of the same picture: one can move from one to the other by means of a geometric transition. Furthermore, from Proposition 3.1 we know that the GKZ series of the singular threefolds with two nodes in the subfamily Y_1 agrees with the expected GKZ series for examples of the 14th case. However, we find:

Lemma 6.1. *The mixed Hodge structure on the cohomology groups $H^i(Y_1)$ of Y_1 is pure unless $i = 3$ and the weight filtration $W_\bullet$ on $H^3(Y_1)$ has the following* *description:*

$$\mathrm{Gr}_k^{W_\bullet} H^3(Y_1) = 0 \quad \text{for } k \neq 2,3$$
$$\dim_{\mathbb{C}} W_2(H^3(Y_1)) \in \{1,2\}$$

Furthermore, Y_1 is factorial if and only if $\dim_{\mathbb{C}} W_2(H^3(Y_1)) = 2$.

Proof. With the exception of the statement about $\dim_{\mathbb{C}} W_2(H^3(Y_1))$, the statements about the mixed Hodge structure on $H^i(Y_1)$ are an easy consequence of Proposition 3.10 and Corollary 3.13 from [NS95].

To obtain the statement about $\dim_{\mathbb{C}} W_2(H^3(Y_1))$ we use [NS95, Proposition 3.10]. Recall from Proposition 5.2 that Y_1 is singular at precisely two nodes. Let U be a contractible Stein open neighbourhood of either node. Let $\mathrm{Weil}(U)$ (resp. $\mathrm{Cart}(U)$) denote the group of Weil (resp. Cartier) divisors on U. Then [Mil68] shows that $\mathrm{Weil}(U)/\mathrm{Cart}(U) \cong \mathbb{Z}$. [NS95, Proposition 3.10] thus gives

$$\dim_{\mathbb{C}} W_2(H^3(Y_1)) = 2 - \sigma(Y_1),$$

where $\sigma(Y_1)$ is the rank of the group of Weil modulo Cartier divisors on Y_1. To prove that this is either 1 or 2, we simply need to show that $\sigma(Y_1) \neq 2$.

Note that Y_1 is smoothable, as it smooths to Y. Let $\tilde{Y}_1$ be a small (not necessarily projective) resolution of Y_1 with exceptional curves C_1 and C_2 Then, by [Nam02, Theorem 2.5], there is a non-trivial relation $\alpha_1[C_1] + \alpha_2[C_2] = 0$ in $H_2(\tilde{Y}_1, \mathbb{C})$. This implies that $\sigma(Y_1) \neq 2$, as required.

We conclude by showing that Y_1 is factorial if and only if $\dim_{\mathbb{C}} W_2(H^3(Y_1)) = 2$. Note that $\dim_{\mathbb{C}} W_2(H^3(Y_1)) = 2$ if and only if $\sigma(Y_1) = 0$, which occurs if and only if Y_1 is $\mathbb{Q}$-factorial. But, since $\mathrm{Weil}(U)/\mathrm{Cart}(U)$ is torsion-free in a neighbourhood of each node, Y_1 is $\mathbb{Q}$-factorial if and only if it is factorial. □

In order to show which of the two cases from Lemma 6.1 hold, we will prove that Y_1 is factorial. In order to do this, we note that it is easier to show factoriality for the single node family of threefolds obtained as the quotient of Y_1 by a certain involution. We will then use factoriality of this family to deduce factoriality of Y_1. To this end, we begin this section by studying several involutions on the complete intersections Y.

One of these involutions, i_1, is easily visible from the vertices of $(\Delta^5)^\circ$ or rays of Σ^5: it is the exchange of the first two affine coordinates. In homogeneous coordinates it is realized by

$$y_4 \leftrightarrow y_5, \qquad y_{32} \leftrightarrow y_{109},$$

with the other y_i unchanged, and in terms of the K3 fibration of Y over $\mathbb{P}^1$ induced by $\widetilde{\alpha}$, as discussed in Section 4.1, it corresponds to $[u:v] \leftrightarrow [v:u]$.

Another involution, i_2, is given in homogeneous coordinates *on the unresolved toric variety* X_{Δ^5} by

$$y_2 \leftrightarrow y_3, \qquad y_6 \leftrightarrow y_7,$$

with the other y_i unchanged, followed by the rescaling map

$$y_2 \mapsto \frac{y_2}{\sqrt{\xi_0}}, \qquad y_3 \mapsto y_3\sqrt{\xi_0}.$$

The resolution that we are currently using for X^5 is not suitable for a clean realization of the first part of i_2 as a toric morphism, however for a suitable refinement of the fan the underlying lattice map is given by the matrix

$$m_N = \begin{pmatrix} 1 & 0 & -1 & 0 & 0 \\ 0 & 1 & -1 & 0 & 0 \\ 0 & 0 & -1 & 0 & 0 \\ 0 & 0 & -4 & 1 & 0 \\ 0 & 0 & -6 & 0 & 1 \end{pmatrix}. \tag{6.1}$$

For "unsuitable" refinements we can view i_2 as a birational map, which is certainly defined on the torus and may extend to some of the lower dimensional orbits as well. This involution corresponds to the same involution on the $\mathbb{P}^1$ base of one of the elliptic fibrations on the M-polarized K3 fibres as i_1 did on the $\mathbb{P}^1$ base of the K3 fibration of Y.

Yet another involution, i_3, is given in homogeneous coordinates by

$$y_9 \mapsto -y_9 - y_4y_5y_6y_7y_8y_{109}y_{469}y_{630}y_{32},$$

with the other y_i unchanged. It comes from either of the two elliptic fibrations on the M-polarized K3 surfaces, as the $y \mapsto -y$ involution in the coordinates of the Weierstrass normal form. This is the same involution as the one described in [BDF+98, Section 4.2.1].[6]

It follows from the representations of these three involutions in homogeneous coordinates that they commute with each other, thus it is possible to take their compositions to obtain new involutions, forming a group isomorphic to $\mathbb{Z}_2^3$.

[6] There it is realized as $x_5 \mapsto -x_5$ in the "alternative gauge". One can convert it to the gauge used in Section 3.2 via formulas (4.13) and (4.15) in [BDF+98] (note that those formulas contain a typo in the expression for $\widetilde{\psi}_4$: the term $-4b_4\lambda_4'^2\lambda_4\psi_0$ should not contain ψ_0, while the term $4\lambda_5'\lambda_4'\lambda_{03}$ should).

6.1. Action on the Singular Locus

Of particular interest is the action of the involutions on the singular locus of Y_1 and its related subvarieties. To analyse this action we can work in the chart $(y_2, y_8, y_9, y_{745}; y_{32})$, used before.

It is very easy to restrict the last involution, i_3, to the chosen chart; we obtain $y_9 \mapsto -y_9 - y_8 y_{32}$. On the exceptional K3 fibre $y_{32} = -1$ containing the nodes, this becomes $y_9 \mapsto y_8 - y_9$. The line passing through the nodes with $y_8 = y_9 = 0$ is a part of the fixed point locus.

In order to get the chart representation of i_1 we need to know the relationship between y_{109} and the variables of the chosen chart.

```
sage:
    Delta5_polar.points_pc()(2,8,9,745,32,109).matrix().kernel()
Free module of degree 6 and rank 1 over Integer Ring
Echelon basis matrix:
[ 11   4   6 -10   1   1]
```

This shows that there is a scaling action by $\mathbb{C}^*$ on the homogeneous coordinates y given by

$$\left[\lambda^{11} y_2 : \lambda^4 y_8 : \lambda^6 y_9 : \lambda^{-10} y_{745} : \lambda y_{32} : \lambda y_{109}\right].$$

Using this to eliminate y_{109}, we get

$$\left[y_{109}^{-11} y_2 : y_{109}^{-4} y_8 : y_{109}^{-6} y_9 : y_{109}^{10} y_{745} : y_{109}^{-1} y_{32} : 1\right],$$

so the chart representation of i_1 is

$$(y_2, y_8, y_9, y_{745}; y_{32}) \mapsto \left(y_{32}^{-11} y_2, y_{32}^{-4} y_8, y_{32}^{-6} y_9, y_{32}^{10} y_{745}; y_{32}^{-1}\right),$$

which reduces to

$$(y_2, y_8, y_9, y_{745}; -1) \mapsto (-y_2, y_8, y_9, y_{745}; -1), \text{ i.e. } y_2 \mapsto -y_2,$$

on the exceptional K3 fibre containing the singular locus. Note that this action exchanges the nodes of Y_1.

We now perform a similar computation for i_2 (the first step is to determine the indices of points of $\left(\Delta^5\right)^\circ$ which are images under the action by the matrix (6.1) of the points corresponding to our current homogeneous coordinates y):

```
sage: points5
[2, 3, 4, 5, 6, 7, 8, 9, 32, 630, 469, 109, 752, 667, 745]
sage: [Delta5_polar.points_pc().index(Delta5_polar.point(p)*mN)
    for p in points5]
[3, 2, 4, 5, 7, 6, 8, 9, 21, 626, 461, 98, 750, 661, 745]
sage: Delta5_polar.points_pc()(2,8,9,745,32,3).matrix().kernel()
```

```
Free module of degree 6 and rank 1 over Integer Ring
Echelon basis matrix:
[ 1  0  0 -2  0  1]
sage:
    Delta5_polar.points_pc()(2,8,9,745,32,21).matrix().kernel()
Free module of degree 6 and rank 1 over Integer Ring
Echelon basis matrix:
[ 11   0   0 -11   1  -1]
```

leading to the chart representation

$$i_2\colon (y_2, y_8, y_9, y_{745}; y_{32}) \mapsto \left(\xi_0^{-1} y_2^{-1} y_{32}^{11}, y_8, y_9, \xi_0 y_2^2 y_{32}^{-11} y_{745}; y_{32}\right).$$

On the exceptional K3 fibre this reduces to

$$i_2\colon (y_2, y_8, y_9, y_{745}; -1) \mapsto \left(-\xi_0^{-1} y_2^{-1}, y_8, y_9, -\xi_0 y_2^2 y_{745}; -1\right)$$

which also exchanges the nodes of Y_1.

6.2. Fixed Point Loci

We will determine the fixed point loci of the involutions in the same affine chart as before. The defining polynomials of Y in this chart are

$$g_0 = y_2 y_{745} + y_{32} + 1,$$
$$g_1 = y_{32}^{11} y_{745} + \xi_1 y_{32}^6 + \xi_0 y_2^2 y_{745} + y_{32} y_8 y_9 + y_8^3 + y_9^2.$$

For i_1 it is easy to see that the fixed point locus has two components, one is determined by $y_{32} = 1$ and the other by $y_{32} = -1, y_2 = 0$; both correspond to surfaces on Y.

For i_2 the fixed point locus is determined by

$$y_2 = \xi_0^{-1} y_2^{-1} y_{32}^{11} \quad \text{and} \quad y_{745} = \xi_0 y_2^2 y_{32}^{-11} y_{745},$$

which is equivalent to the single condition

$$\xi_0 y_2^2 = y_{32}^{11},$$

which cuts out a surface in Y.

Finally, for i_3 the fixed point locus is defined by $2y_9 + y_8 y_{32} = 0$, which again determines a surface in Y.

Each of the i_j changes the sign of the holomorphic $(3,0)$-form on Y (this can be seen from [BDF+98, Section 6]), so an interesting subgroup of the involutions is formed by their pairwise compositions, which preserve the sign of the holomorphic $(3,0)$-form.

The composition $i_2 \circ i_3$ gives, in fact, a Nikulin involution on the smooth K3 fibres of Y with exactly 8 fixed points in each of them. More explicitly,

$$i_2 \circ i_3 \colon (y_2, y_8, y_9, y_{745}; y_{32}) \mapsto \left(\xi_0^{-1} y_2^{-1} y_{32}^{11}, y_8, -y_9 - y_8 y_{32}, \xi_0 y_2^2 y_{32}^{-11} y_{745}; y_{32}\right),$$

so the fixed point locus is determined by intersection of the fixed point loci of i_2 and i_3 considered separately:

$$\xi_0 y_2^2 = y_{32}^{11} \qquad \text{and} \qquad 2y_9 + y_8 y_{32} = 0.$$

Specifying the K3 fibration base coordinate y_{32} generically gives a two-fold ambiguity in y_2, which allows us to uniquely determine y_{745} from g_0 and get a cubic equation for y_8 from g_1. This gives us 6 fixed points: two more should be located outside this chart. It does not seem possible, however, to locate them in the other charts of our resolution of X_{Δ^5}, due to the fact that i_2 is only a birational map, not a morphism.

Let us work out the 6 visible points in more detail treating $y_{32} \in \mathbb{C}^*$ as a parameter; this will be useful for the analysis we will perform later in Section 7.1. In this case

$$\begin{aligned}
y_2 &= \pm \frac{y_{32}^6}{\sqrt{\xi_0 y_{32}}}, \\
y_{745} &= -\frac{y_{32}+1}{y_2}, \\
0 &= y_8^3 - \frac{y_{32}^2}{4} y_8^2 + 2y_{32}^6 \left(\frac{\xi_1}{2} \mp \sqrt{\xi_0 y_{32}} \mp \frac{\xi_0}{\sqrt{\xi_0 y_{32}}}\right), \\
y_9 &= -\frac{y_8 y_{32}}{2},
\end{aligned}$$

with a consistent choice of signs.

Next we consider the composition

$$i_1 \circ i_2 \colon (y_2, y_8, y_9, y_{745}; y_{32}) \mapsto \left(\xi_0^{-1} y_2^{-1}, y_{32}^{-4} y_8, y_{32}^{-6} y_9, \xi_0 y_2^2 y_{32}^{-1} y_{745}; y_{32}^{-1}\right).$$

Its fixed point locus is concentrated in fibres corresponding to $y_{32} = \pm 1$, here the involution takes the form

$$(y_2, y_8, y_9, y_{745}; \pm 1) \mapsto \left(\xi_0^{-1} y_2^{-1}, y_8, y_9, \pm \xi_0 y_2^2 y_{745}; \pm 1\right),$$

where all alternating signs must be chosen to be the same. Thus the fixed points satisfy the conditions

$$\xi_0 y_2^2 = 1 \qquad \text{and} \qquad \xi_0 y_2^2 y_{745} = \pm y_{745}.$$

In the case $y_{32} = -1$ the second condition yields $y_{745} = 0$, which also follows from $g_0 = 0$, so on Y the fixed point locus is determined by $y_{32} = \pm 1$ and $\xi_0 y_2^2 = 1$. This corresponds to four elliptic curves, two of which become nodal with nodes coinciding with the nodes of Y_1 in the case $\xi_1 = 0$.

The last pairwise composition is

$$i_1 \circ i_3 \colon (y_2, y_8, y_9, y_{745}; y_{32}) \mapsto \left(y_{32}^{-11} y_2, y_{32}^{-4} y_8, -y_{32}^{-6} y_9 - y_{32}^{-5} y_8, y_{32}^{10} y_{745}; y_{32}^{-1}\right).$$

Its fixed point locus is also located in K3 fibres with $y_{32} = \pm 1$. For $y_{32} = 1$ the only other condition is $2y_9 + y_8 = 0$, while for $y_{32} = -1$ we need both $2y_9 - y_8 = 0$ and $y_2 = 0$. Taking into account the defining equations of Y, we see that we for $y_{32} = 1$ we get an elliptic curve, while for $y_{32} = -1$ the component of the fixed point locus is a 3-section of the elliptic fibration on the surface defined by $y_2 = 0$.

We also observe that the intersection of the fixed point loci of these compositions on Y is given by

$$y_{32} = 1, \quad \xi_0 y_2^2 = 1, \quad 2y_9 + y_8 = 0,$$

which coincides with the fixed point locus of the Nikulin involution in the K3 fibre corresponding to $y_{32} = 1$.

6.3. Quotient by the Nikulin involution

By the results of Section 6.2, a complete intersection Y admits a fibrewise Nikulin involution $i_2 \circ i_3$. If we quotient Y_1 by this involution and perform a crepant partial desingularization of the resulting quotient singularities, we obtain a new Calabi-Yau threefold W. By [CD07, Theorem 3.13], the K3 fibration on Y_1 induces a fibration on W whose general fibres are Kummer surfaces and, furthermore, if $\{E_1, E_2\}$ are the pair of elliptic curves canonically associated to a fibre of Y_1 by Theorem 4.3, the corresponding fibre of W is $\mathrm{Kum}(E_1 \times E_2)$ (i.e. the Kummer surface associated to $E_1 \times E_2$). This threefold W will be constructed in a more direct way in Section 7.

Proposition 6.2. *W is a Calabi-Yau threefold that moves in a 1-parameter family, the general member of which is factorial and smooth away from a single node. The mixed Hodge structure on the cohomology groups $H^i(W)$ of W is pure unless $i = 3$ and the weight filtration $W_\bullet$ on $H^3(W)$ has the following description:*

$$\mathrm{Gr}_k^{W_\bullet} H^3(W) = 0 \quad \text{for } k \neq 2, 3$$
$$\dim_{\mathbb{C}} W_2(H^3(W)) = 1$$

Proof. Note that quotient of Y_1 by the fibrewise Nikulin involution introduces a curve of A_1 singularities, which are resolved by a single crepant blow-up. The two nodes of Y_1 are identified by the quotient to give a single node on W.

With this in place, the remainder of the proof proceeds in the same way as the proof of Lemma 6.1. Note here that the involution $i_2 \circ i_3$ is defined on the family of smooth threefolds Y, not just on Y_1, and the resolved quotient of Y by this involution gives a smoothing of W. □

It is immediate from factoriality that W does not admit a small projective resolution. In fact, more is true:

Corollary 6.3. *W does not admit a small symplectic resolution. In particular, the singularity of W cannot be resolved to give a smooth Calabi-Yau threefold.*

Proof. Since W does not admit a small projective resolution, [CS05, Corollary 8] shows that the Lagrangian 3-sphere obtained by smoothing the single node cannot be nullhomologous. But, by [STY02, Theorem 2.9], this implies that W does not admit a small symplectic resolution.

The statement about non-existence of Calabi-Yau resolutions is immediate from the fact that W has terminal singularities and no small resolution. Thus any resolution of W will contribute positively to the canonical divisor, violating the Calabi-Yau condition. □

From Proposition 6.2 we can deduce factoriality of Y_1. This will in turn allow us to show that Y_1 provides the geometric realisation of the 14th case variation of Hodge structure that we have been searching for.

Theorem 6.4. *The singular Calabi-Yau threefold Y_1 is factorial, so the weight filtration $W_\bullet$ on $H^3(Y_1)$ has $\dim_{\mathbb{C}} W_2(H^3(Y_1)) = 2$. Furthermore, as Y_1 varies in its one-parameter subfamily, the graded piece $\mathrm{Gr}_3^{W_\bullet} H^3(Y_1)$ admits a pure variation of Hodge structure of weight 3 and type $(1,1,1,1)$, which realizes the 14th case variation of Hodge structure.*

Proof. Clingher and Doran [CD07][CD11] have shown that the fibrewise Nikulin involution $i_2 \circ i_3$ is a *geometric* 2*-isogeny* on the fibres of Y_1, i.e. there exists a second fibrewise Nikulin involution on W such that the resolved quotient of W by this second involution is isomorphic to Y_1. As W is factorial and $\mathbb{Q}$-factoriality is preserved under finite surjective morphisms [KM98, Lemma 5.16] and resolutions, we thus have that Y_1 must be $\mathbb{Q}$-factorial. Then, from the proof of Lemma 6.1, we see that Y_1 is $\mathbb{Q}$-factorial if and only if it is factorial.

Given this, the fact that $\dim_{\mathbb{C}} W_2(H^3(Y_1)) = 2$ follows from Lemma 6.1. The statement about the Hodge decomposition of $\mathrm{Gr}_3^{W_\bullet} H^3(Y_1)$ then follows from calculation of $b_3(Y_1) = 6$, using [NS95, Theorem 3.2].

Finally, recall from Proposition 3.1 that the GKZ series of Y_1 matches that predicted for the 14th case VHS. Furthermore, it follows from the classification of Doran and Morgan [DM06] that the 14th case VHS is the only one with $h^{2,1} = 1$ having this GKZ series. The variation of Hodge structure on $\mathrm{Gr}_3^{W_\bullet} H^3(Y_1)$ must therefore be the 14th case VHS. □

Again, factoriality implies that Y does not admit a small projective resolution. Indeed, as for W, a stronger statement can be made:

Corollary 6.5. *Y_1 does not admit a small symplectic resolution. In particular, the singularities of Y_1 cannot be resolved to give a smooth Calabi-Yau threefold.*

Proof. Given that $b_3(Y) = 8$ and $b_3(Y_1) = 6$, [Ros06, Theorem 3.2] shows that the two vanishing cycles in Y are homologically independent. But, by [STY02, Theorem 2.9], this implies that Y_1 does not admit a small symplectic resolution. The statement about non-existence of Calabi-Yau resolutions is proved in exactly the same way as the corresponding statement in Corollary 6.3. □

Remark 6.6. We note that the statement of [Ros06, Theorem 3.2] technically requires that Y_1 have a small projective resolution. However, by examining the proof it is easy to see that the assertion about the number of homologically independent vanishing cycles is independent of this assumption.

7 The Forward Construction

7.1. Undoing the Kummer construction.

The one-parameter family of Calabi-Yau threefolds W is very closely related to Y_1; for instance, the quotient-resolution procedure does not affect the GKZ series, so W provides a second example of a one-parameter family with GKZ series equal to that predicted for the 14th case. It could thus prove fruitful to study the geometry of W explicitly. In order to do this, we apply the methods of [DHNT13] to produce an explicit model for W and, in doing so, illustrate the usage of these methods in a concrete example.

Recall that W admits a fibration whose general fibres are Kummer surfaces and, furthermore, if $\{E_1, E_2\}$ are the pair of elliptic curves canonically associated to a fibre of Y_1 by Theorem 4.3, the corresponding fibre of W is $\mathrm{Kum}(E_1 \times E_2)$ (i.e. the Kummer surface associated to $E_1 \times E_2$). Our aim is to find a method by which W can be constructed directly by performing a fibrewise Kummer construction on a threefold fibred by products of elliptic curves.

In order to do this, we begin by using the methods of [DHNT13, Section 4] to undo the Kummer construction on W, i.e. find a family of Abelian surfaces $\mathcal{A}$, such that the application of the Kummer construction fibrewise to $\mathcal{A}$ yields a birational model of W. Unfortunately, as we shall see, this is not possible directly: instead, we will need to proceed to a finite cover of W first.

We begin with some setup. Let $U \subset \mathbb{P}^1$ be the open set over which the fibres of the K3 fibration $Y_1 \to \mathbb{P}^1$ induced by $\tilde{\alpha}$ are nonsingular and let $p \in U$ be a general point. Applying the method of [BDF+98, Section 4.3] to the affine description of Y_1 given in Section 6.2 (and recalling that Y_1 is obtained by setting $\xi_1 = 0$ in the equations for Y), we see that the singular fibres of Y_1 occur over $y_{32} \in \{0, -1, \alpha, \beta, \infty\}$ where, as before, y_{32} is an affine coordinate on the base $\mathbb{P}^1$ of the K3 fibration, and α and β are given by

$$\alpha, \beta = \frac{2 - 12^6\xi_0 \pm 2\sqrt{1 - 12^6\xi_0}}{12^6\xi_0}.$$

For now we will assume $\xi_0 \notin \{0, \frac{1}{12^6}, \infty\}$, as degenerate behaviour occurs at these points, although the case $\xi_0 = \frac{1}{12^6}$ will be discussed later in Section 7.3. U is thus a copy of $\mathbb{P}^1$ with these five points removed.

Let $Y_{1,U} \to U$ denote the restriction of $Y_1 \to \mathbb{P}^1$ to U. Then we have:

Lemma 7.1. *$Y_{1,U}$ is an M-polarized family of K3 surfaces over U, in the sense of [DHNT13, Definition 2.1].*

Proof. From the discussion in Section 4.1, we already know that the fibres of $Y_{1,U}$ are M-polarized. In order to complete the proof of the lemma, we just need to show that the polarization is invariant under monodromy around the punctures in U. To do this, we first note that the lattice M is generated by the classes of the divisors given in Figure 1. As these divisors are toric, they are invariant under monodromy in U. Thus the lattice M is also. □

Proceeding now to W, we see that the restricted family of K3 surfaces $W_U \to U$ has fibres isomorphic to $\mathrm{Kum}(E_1 \times E_2)$. Let $\mathrm{NS}(W_p)$ denote the Néron-Severi lattice of the fibre of $W_U \to U$ over p. Then [DHNT13, Theorem 3.3] shows that W_U is an $(\mathrm{NS}(W_p), G)$-polarized family of K3 surfaces, in the sense of [DHNT13, Definition 2.4], where G is a finite group controlling the action of monodromy in U on $\mathrm{NS}(W_p)$.

By [DHNT13, Proposition 4.1], we may undo the Kummer construction on W if G is trivial. Unfortunately, this will prove not to be the case, so we will need to proceed to a finite cover of W to kill the action of G. In order to do this, we use the methods of [DHNT13, Section 4.3] to find a finite cover $V \to U$ so that the action of monodromy in V on $\mathrm{NS}(W_p)$ is trivial.

We begin by noting that, by the discussion in [DHNT13, Section 4.3], the action of monodromy on $\mathrm{NS}(W_p)$ may be computed from its action on the eight exceptional curves $\{F_1, \ldots, F_8\}$ in W_p arising from the blow-up of the eight fixed points of the Nikulin involution.

The arrangement of these eight curves may be described in terms of a certain elliptic fibration on W_p. By [CD07, Proposition 3.10], the fibrewise Nikulin involution $i_2 \circ i_3$ on Y_1 acts on the alternate fibration as translation by a section, so the alternate fibration induces an elliptic fibration Ψ on W_p. As the alternate fibration is invariant under monodromy in $Y_{1,U}$, the induced fibration Ψ is invariant under mondromy in W_U. Furthermore, the fact that $\sigma = 1$ for Y_1 implies that [DHNT13, Assumption 4.6] holds, so Ψ has four sections and seven singular fibres: one of Kodaira type I_6^* and six of Kodaira type I_2.

Given this description, we can locate the exceptional curves F_i in relation to the fibration Ψ. The I_6^* fibre contains two such curves, which we label F_1 and F_2, and each I_2 fibre contains precisely one such curve, giving $\{F_3,\dots,F_8\}$. Furthermore, of the four sections

- one intersects none of the F_i,
- one intersects all of $\{F_3,\dots,F_8\}$,
- one intersects F_1 and precisely three of $\{F_3,\dots,F_8\}$ (say F_3, F_4, F_5), and
- one intersects F_2 and the other three $\{F_3,\dots,F_8\}$ (say F_6, F_7, F_8).

By [DHNT13, Proposition 4.8], the action of monodromy around a loop in U must either

1. fix both F_1 and F_2, in which case the sets $\{F_3,F_4,F_5\}$ and $\{F_6,F_7,F_8\}$ are preserved, or
2. swap F_1 and F_2, in which case the sets $\{F_3,F_4,F_5\}$ and $\{F_6,F_7,F_8\}$ are interchanged.

In order to see which of these cases occurs, we will compute the action of monodromy on the two sets $\{F_3,F_4,F_5\}$ and $\{F_6,F_7,F_8\}$. For clarity of notation, we will use coordinates $[u:v]$ on the $\mathbb{P}^1$ base of the fibrations $W \to \mathbb{P}^1$ and $Y_1 \to \mathbb{P}^1$, where the affine coordinate $y_{32} = \frac{u}{v}$; note that this agrees with the notation of equations (4.4) and (4.5). We have:

Proposition 7.2. *The action of monodromy in U on* $\mathrm{NS}(W_p)$ *around:*

- $[u:v] \in \{[-1:1],[\alpha:1],[\beta:1]\}$ *fixes both F_1 and F_2.*
- $[u:v] \in \{[0:1],[1:0]\}$ *interchanges F_1 and F_2.*

Proof. As noted above, it is enough to show that the two sets $\{F_3,F_4,F_5\}$ and $\{F_6,F_7,F_8\}$ are preserved by monodromy around $\{[-1:1],[\alpha:1],[\beta:1]\}$ and interchanged by monodromy around $\{[0:1],[1:0]\}$. Furthermore, recall that the F_i arise as the blow up of the eight fixed points of the Nikulin involution, so in order to understand their behaviour under monodromy on W_U it is sufficient to study the action of monodromy on $Y_{1,U}$ on these eight fixed points.

In Section 6.2 an affine description for Y_1 was computed and the locations of the six fixed points corresponding to the curves $F_3,\dots,F_8$ within it were calculated explicitly. In particular, we found that the y_2-coordinates of these fixed points satisfy

$$y_2 = \pm\frac{y_{32}^6}{\sqrt{\xi_0 y_{32}}},$$

with each choice of sign corresponding to three fixed points. These two sets of three fixed points are precisely those corresponding to the two sets of curves

$\{F_3, F_4, F_5\}$ and $\{F_6, F_7, F_8\}$. Therefore, these two sets are interchanged by monodromy if and only if monodromy switches the sign of y_2 in the expression above. But this happens precisely for monodromy around $y_{32} \in \{0, \infty\}$, i.e. around $[u : v] \in \{[0{:}1],[1{:}0]\}$. □

We therefore see that $\mathrm{NS}(W_p)$ is *not* fixed under monodromy in U. This presents an obstruction to undoing the Kummer construction on W. To resolve this, we pass to a cover of W.

Define W' to be the threefold fibred over $\mathbb{P}^1$ (with base coordinate $[s : t]$) obtained as the pull-back of $W \to \mathbb{P}^1$ by the map $f\colon \mathbb{P}^1 \to \mathbb{P}^1$ given by $[s : t] \mapsto [s^2 : t^2] = [u : v]$. The fibration $W' \to \mathbb{P}^1$ has eight singular fibres, over

$$[s : t] \in \{[0 : 1], [i : 1], [-i : 1], [\alpha' : 1], [-\alpha' : 1], [\beta' : 1], [-\beta' : 1], [1 : 0]\},$$

where $(\alpha')^2 = \alpha$ and $(\beta')^2 = \beta$. Let

$$U' = \mathbb{P}^1 \setminus \{[0 : 1], [i : 1], [-i : 1], [\alpha' : 1], [-\alpha' : 1], [\beta' : 1], [-\beta' : 1], [1 : 0]\};$$

note that $U' = f^{-1}(U)$.

Let W'_p denote a fibre of W' over a point $p \in U'$. Then W'_p is smooth and isomorphic to $W_{f(p)}$, so we may identify the eight curves $\{F_1, \ldots, F_8\}$ in W'_p and, as before, the action of monodromy in U' on $\mathrm{NS}(W'_p)$ may be computed from its action these eight curves. Furthermore, by construction the action of monodromy in U' on $\mathrm{NS}(W'_p)$ fixes F_1 and F_2 and so, by [DHNT13, Proposition 4.8], the sets $\{F_3, F_4, F_5\}$ and $\{F_6, F_7, F_8\}$ are preserved.

It remains to compute the action of monodromy on these two sets. To do this, we use the method from the proof of Proposition 7.2 and track the locations of the fixed points of the Nikulin involution. Taking the double cover $W' \to W$ corresponds to making a change of coordinates $y_{32} = \frac{u}{v} = \frac{s^2}{t^2} =: x^2$ in the equations for these fixed points in the affine description of Section 6.2, giving

$$y_2 = \pm \frac{x^{11}}{\sqrt{\xi_0}},$$

$$y_{745} = \mp \sqrt{\xi_0} \frac{-x^2 + 1}{x^{11}},$$

$$0 = y_8^3 - \frac{1}{4} x^4 y_8^2 \mp 2\sqrt{\xi_0} x^{11} \mp 2\sqrt{\xi_0} x^{13},$$

$$y_9 = -\frac{y_8 x^2}{2},$$

where we make a consistent choice of signs throughout.

For a given choice of signs, there are three solutions to these equations. These three solutions correspond to the three divisors $\{F_3, F_4, F_5\}$ for one choice of sign and the divisors $\{F_6, F_7, F_8\}$ for the other. Once a choice of signs has been made, y_2 and y_{745} are uniquely determined by the first two

Table 1. *Action of monodromy on* $\{F_3, F_4, F_5\}$ *and* $\{F_6, F_7, F_8\}$.

Monodromy around $[s:t]$	Action on $\{F_3, F_4, F_5\}$	Action on $\{F_6, F_7, F_8\}$
$[0:1]$	(123)	(456)
$[i:1]$	(12)	(45)
$[-i:1]$	(13)	(45)
$[\alpha':1]$	(23)	id
$[-\alpha':1]$	id	(46)
$[\beta':1]$	(23)	id
$[-\beta':1]$	id	(56)
$[1:0]$	(123)	(456)

equations, the third equation gives three possibilities for y_8, and the fourth equation uniquely determines y_9. Thus we see that the action of monodromy on these equations is completely determined by its action on the third equation. We have therefore shown:

Lemma 7.3. *The action of monodromy in* U' *on the set of divisors* $\{F_3, F_4, F_5\}$ *is identical to its action on the roots of the cubic equation*

$$0 = t^{13} y_8^3 - \frac{1}{4} s^4 t^9 y_8^2 - 2\sqrt{\xi_0} s^{11} t^2 - 2\sqrt{\xi_0} s^{13}$$

and its action on the set of divisors $\{F_6, F_7, F_8\}$ *is identical to its action on the roots of the cubic equation*

$$0 = t^{13} y_8^3 - \frac{1}{4} s^4 t^9 y_8^2 + 2\sqrt{\xi_0} s^{11} t^2 + 2\sqrt{\xi_0} s^{13}.$$

This monodromy can be calculated using the `monodromy` command in *Maple*'s `algcurves` package (this calculation was performed with the generic value $\xi_0 = 1$). Using the base point $[s:t] = [-1:10]$, Table 1 shows the action of monodromy around anticlockwise loops about each of the eight singular fibres in W' on each of the sets $\{F_3, F_4, F_5\}$ and $\{F_6, F_7, F_8\}$, expressed as permutations in S_3 (in cycle notation). It is easy to see that the permutations listed in Table 1 generate the full group $S_3 \times S_3$, so monodromy acts as $S_3 \times S_3$ on the sets $\{F_3, F_4, F_5\}$ and $\{F_6, F_7, F_8\}$ in $W'_{[-1:10]}$.

Therefore, in order to undo the Kummer construction, we need to pull-back W' to a further 36-fold cover $g \colon C \to \mathbb{P}^1$. This cover is constructed as follows: the 36 preimages of the point $[-1:10] \in \mathbb{P}^1$ are labelled by elements of $S_3 \times S_3$ and monodromy around each of the points $[s:t]$ listed in Table 1 acts on these labels as composition with the corresponding permutation. This action extends to an action of $S_3 \times S_3$ on the whole of C. The map g has ramification index 2 at all points over $[s:t] \in \{[i:1], [-i:1], [\alpha':1], [-\alpha':1], [\beta:1], [-\beta':1]\}$ and ramification index 3 at all points over $[s:t] \in \{[1:0], [0:1]\}$.

Let $W'' \to C$ denote the pull-back of $W' \to \mathbb{P}^1$ under this 36-fold cover, let $V = g^{-1}(U')$ and let W''_V denote the restriction of W'' to V. Then the argument above shows that the Néron-Severi group of a general fibre of $W''_V \to V$ is fixed under the action of monodromy in V. Therefore, [DHNT13, Proposition 4.1] shows that we may undo the Kummer construction on W''_V to obtain a threefold $\mathcal{A}_V \to V$ fibred by products of elliptic curves $E_1 \times E_2$.

7.2. Elliptic Surfaces

The aim of the remainder of this section is to directly construct a birational model for $\mathcal{A}_V \to V$ as a product of elliptic surfaces and to explain how this can be used to construct a birational model for $W \to \mathbb{P}^1$.

Begin by letting $\mathcal{A}_p$ be a fibre of $\mathcal{A}_V \to V$ over a point $p \in V$, then $\mathcal{A}_p$ is isomorphic to a product of smooth elliptic curves $E_1 \times E_2$. Furthermore, by Theorem 4.3 and equations (4.4) and (4.5), the j-invariants of these elliptic curves are given by the roots of the quadratic equation

$$j^2 - j + \frac{uv}{(u+v)^2 12^6 \xi_0} = 0, \tag{7.1}$$

where $[u : v] = f \circ g(p) \in \mathbb{P}^1$.

We claim that $\mathcal{A}_V$ is isomorphic over V to a fibre product of elliptic surfaces. This will be proved in several steps, the first of which is to show that the fibration on $\mathcal{A}_V$ admits a section.

Lemma 7.4. *The fibration $\mathcal{A}_V \to V$ admits a section $\sigma : V \to \mathcal{A}_V$.*

Proof. Note that the fibres of W''_V over V are all smooth Kummer surfaces. Furthermore, by the results of Section 7.1, we see that the Néron-Severi group of a fibre W''_p of W''_V is invariant under monodromy in V. The sixteen exceptional (-2)-curves arising from the Kummer construction on W''_p are thus fixed under monodromy in V, so sweep out sixteen divisors on W''_V. On progression to $\mathcal{A}_V$, these divisors are contracted to give sections of $\mathcal{A}_V \to V$. □

Using this, we can now prove the following proposition.

Proposition 7.5. *$\mathcal{A}_V \to V$ is isomorphic over V to a fibre product $\mathcal{E}_1 \times_C \mathcal{E}_2$ of minimal elliptic surfaces $\mathcal{E}_{1,2} \to C$ with section. Furthermore, the j-invariants of the elliptic curves E_1 and E_2 forming the fibres of $\mathcal{E}_1$ and $\mathcal{E}_2$ over a point $p \in C$ are related by $j(E_1) + j(E_2) = 1$*

Proof. Let $\mathcal{A}_p = E_1 \times E_2$ be a fibre of $\mathcal{A}_V$ over a point $p \in V$. The section $\sigma : V \to \mathcal{A}_V$ defined in Lemma 7.4 intersects $\mathcal{A}_p$ at a single point; we identify E_1 and E_2 with the two elliptic curves in $\mathcal{A}_p$ passing through this point. Then we may deform E_1 and E_2 along paths in $\sigma(V)$ to obtain elliptic curves in the other fibres of $\mathcal{A}_V$. Over any simply connected subset of V, this gives a decomposition of $\mathcal{A}_V$ as a fibre product of two elliptic surfaces with section. To show that this decomposition extends to all of V, we have to show that the action of monodromy in V cannot switch E_1 and E_2.

Let γ be a path in V that begins at $p \in \mathcal{A}_p$, loops around one of the punctures in V, then returns to p. If we deform E_1 along $\sigma(\gamma)$, we obtain an elliptic curve $\gamma(E_1)$ in $\mathcal{A}_p$ passing through $\sigma(p)$. Thus, $\gamma(E_1)$ must be either E_1 or E_2.

We can determine which it is by tracking its j-invariant using equation (7.1). The discriminant of this quadratic vanishes singly at the points $[u:v] \in \{[\alpha:1],[\beta:1]\}$ and doubly at $[u:v]=[0:1]$, which are outside V, so we cannot have $j(E_1)=j(E_2)$ at p. We can therefore determine $\gamma(E_1)$ by comparing its j-invariant to $j(E_1)$ and $j(E_2)$.

If γ is a loop around one of the punctures in V above $\{[0:1],[-1:1]\} \subset \mathbb{P}^1$, then the discriminant of (7.1) does not vanish inside $f \circ g(\gamma)$ and we must have $j(\gamma(E_1)) = j(E_1)$, i.e. $\gamma(E_1) = E_1$. If γ is a loop around one of the punctures in V above $[1:0] \in \mathbb{P}^1$, then $f \circ g(\gamma)$ loops three times around a point where the discriminant of (7.1) vanishes doubly, so again we must have $j(\gamma(E_1)) = j(E_1)$ and $\gamma(E_1) = E_1$. Finally, if γ is a loop around one of the punctures in V above $\{[\alpha:1],[\beta:1]\} \subset \mathbb{P}^1$, then $f \circ g(\gamma)$ loops twice around a point where the discriminant of (7.1) vanishes singly, so we must have $j(\gamma(E_1)) = j(E_1)$ and $\gamma(E_1) = E_1$.

Thus, $\gamma(E_1) = E_1$ for any loop γ in V. So E_1 sweeps out an elliptic surface $p_1 \colon \mathcal{E}_{1,V} \to V$ over V, with section given by σ. Similarly, E_2 also sweeps out an elliptic surface $p_2 \colon \mathcal{E}_{2,V} \to V$ with the same section, so we see that $\mathcal{A}_V$ is isomorphic to the fibre product $\mathcal{E}_{1,V} \times_V \mathcal{E}_{2,V}$.

Now, by [Nak88, Theorem 2.5], there are unique extensions of $p_i \colon \mathcal{E}_{i,V} \to V$ to minimal elliptic surfaces $p_i \colon \mathcal{E}_i \to C$ over C, for $i=1,2$. By construction, we have an isomorphism between $\mathcal{A}_V$ and $\mathcal{E}_1 \times_C \mathcal{E}_2$ over V. Finally, the statement about the j-invariants is an easy consequence of equation (7.1). □

Thus to construct a birational model for $\mathcal{A}_V$, it is enough to construct the elliptic surfaces $\mathcal{E}_1$ and $\mathcal{E}_2$. In order to do this, we begin by studying their properties. Recall that, by the discussion in Section 7.1, there is an action of $S_3 \times S_3$ on C. Let $i \colon C \to C$ be the involution defined by the transposition (23). Then we have:

Lemma 7.6. *The involution i induces an isomorphism $\mathcal{E}_1 \to \mathcal{E}_2$.*

Proof. As $\mathcal{E}_1$ and $\mathcal{E}_2$ are minimal, any birational map $\mathcal{E}_1 \to \mathcal{E}_2$ is an isomorphism by [Mir89, Proposition II.1.2]. So suffices to show that i induces an isomorphism over the open set V. To do this, we show that the fibre of $\mathcal{E}_1$ over a point $p \in V$ is isomorphic to the fibre of $\mathcal{E}_2$ over $i(p)$, for then we can define an isomorphism $\mathcal{E}_{1,V} \to \mathcal{E}_{2,V}$ by $(p,e) \mapsto (i(p),e)$, where e is any point in the fibre of $\mathcal{E}_1$ over p.

So let E_k denote the fibre of $\mathcal{E}_k$ over $p \in V$ and let E'_k denote the fibre of $\mathcal{E}_k$ over $i(p)$. By equation (7.1), we see that the j-invariants of $\{E_1,E_2\}$ are equal to those of $\{E'_1,E'_2\}$. So to show that $E_1 \cong E'_2$, we just have to show that $j(E_1) = j(E'_2)$.

To do this, we perform a calculation similar to the one used to prove Proposition 7.5. Let γ be a loop in $f \circ g(V)$ that begins at $f \circ g(p)$, loops once around $[\alpha : 1] \in \mathbb{P}^1$, then returns to $f \circ g(p)$. Then one of the preimages γ' of γ under $(f \circ g)$ is a path from p to $i(p)$. As in the proof of Proposition 7.5, if we deform $E_1, E_2 \subset \mathcal{A}_p$ along $\sigma(\gamma')$ we obtain the fibres of $\mathcal{E}_1, \mathcal{E}_2$ over $i(p)$. Since γ loops once around a point where the discriminant of (7.1) vanishes singly, such a deformation swaps the j-invariants of the fibres of $\mathcal{E}_1$ and $\mathcal{E}_2$. So $j(E_1) = j(E'_2)$ and $j(E_2) = j(E'_1)$, as required. □

Using this lemma, we see that in order to construct a birational model for $\mathcal{A}_V$ it is enough to construct a birational model for the elliptic surface $\mathcal{E}_1$. As we can compute the fibrewise j-invariant of $\mathcal{E}_1$ from equation (7.1), it just remains to study the forms of the singular fibres. These fibres will be calculated in the next section, but before that we conclude this section with a useful result about $\mathcal{E}_1$.

Lemma 7.7. *Let G denote the group of order* 18 *obtained by intersecting $S_3 \times S_3 \subset S_6$ with the alternating group A_6. Then the action of G on C induces automorphisms of $\mathcal{E}_1$.*

Proof. The proof of this lemma proceeds using the same arguments used to prove Lemma 7.6. □

7.3. Singular Fibres

The aim of this subsection is to compute the types of the singular fibres appearing in $\mathcal{E}_1$. We begin by noting that, by equation (7.1), we have $j(\mathcal{E}_1) = \infty$ with multiplicity two at the points lying over $[u:v] = [1:-1]$ and $j(\mathcal{E}_1) \in \{0,1\}$ with multiplicity six at the points lying over $[u:v] \in \{[0:1],[1:0]\}$. Without loss of generality, we may assume that $j(\mathcal{E}_1) = 1$ over $[u:v] = [0:1]$ and $j(\mathcal{E}_1) = 0$ over $[u:v] = [1:0]$ (making the opposite choice simply corresponds to interchanging the labels of $\mathcal{E}_1$ and $\mathcal{E}_2$). This data determines the singular fibres of $\mathcal{E}_1$ up to quadratic twists.

To determine these fibres exactly, we study the action of monodromy on the cohomology of the fibres of $\mathcal{E}_1$. More precisely, let p be a point in V and let E_i denote the fibre of $\mathcal{E}_i$ over p. Let γ be a loop in V that starts at p, travels once around a puncture in V, then returns to p. Then monodromy around γ

acts on $H^1(E_1,\mathbb{Z})$ and the precise form of this action determines the type of the singular fibre enclosed by γ.

Note that the fibre of $\mathcal{A}_V \to V$ over p is isomorphic to $E_1 \times E_2$. By the discussion at the end of [CD07, Section 3.5], we see that there is a canonical isomorphism

$$H^1(E_1,\mathbb{Z}) \otimes H^1(E_2,\mathbb{Z}) \cong i(M)^{\perp} \subset H^2(Y_{1,p},\mathbb{Z}),$$

where $Y_{1,p}$ is the fibre of $Y_1 \to \mathbb{P}^1$ over $f \circ g(p)$ (which is a smooth M-polarized K3 surface) and $i\colon M \to H^2(Y_{1,p},\mathbb{Z})$ is the lattice embedding defining the M-polarisation on $Y_{1,p}$ (see Definition 4.1).

Thus we can obtain information about the action of monodromy around γ on $H^1(E_1,\mathbb{Z})$ by studying the action of monodromy around $f \circ g(\gamma)$ on $i(M)^{\perp}$. To do this, we will study the Picard-Fuchs equation of Y_1. This equation has been computed in the degenerate case $\xi_0 = \frac{1}{12^6}$ by Chen, Doran, Kerr and Lewis in [CDKL11, Section 5.2]. In order to use their result in our computations, we perform a brief study of this case.

So suppose $\xi_0 = \frac{1}{12^6}$. The construction detailed in Section 7.1 proceeds much as before, except in this case we find that $\alpha = \beta = 1$. Due to this, the ramification points of the 36-fold cover $g\colon C \to \mathbb{P}^1$ lying over $[u : v] \in \{[\alpha : 1], [\beta : 1]\}$ collide and g splits into two disjoint 18-fold covers (which are preserved by even permutations in $S_3 \times S_3 \subset S_6$ and exchanged by odd ones). However, this splitting does not affect the ramification behaviour over $[u : v] \in \{[0 : 1], [-1 : 1], [1 : 0]\}$, so the singular fibres of $\mathcal{E}_1$ and $\mathcal{E}_2$ over these points are unaffected.

In this setting, Chen, Doran, Kerr and Lewis show that the Picard-Fuchs equation on Y_1 splits as a product of the second-order ODE's

$$f_1''(r) + \frac{3r+1}{2r(r+1)} f_1'(r) + \frac{5}{144r(r+1)} f_1(r) = 0,$$

$$f_2''(r) + \frac{3r+1}{2r(r+1)} f_2'(r) + \frac{5}{144r^2(r+1)} f_2(r) = 0,$$

where $r = \frac{u}{v}$ is an affine parameter on the base $\mathbb{P}^1$. Furthermore, this splitting corresponds exactly with the splitting of $i(M)^{\perp}$ as $H^1(E_1,\mathbb{Z}) \otimes H^1(E_2,\mathbb{Z})$ so we can use it to study the singular fibres of $\mathcal{E}_1$.

We recognize the first of these equations as the differential equation for the hypergeometric function ${}_2F_1(\frac{1}{12}, \frac{5}{12}; 1 \mid r+1)$. Matrix generators for the monodromy group with respect to this system are well-known (see, for instance, [IKSY91, Section 2.4]), so we can immediately deduce the action of monodromy around a point $[u : v] \in \mathbb{P}^1$ on $H^1(E_1,\mathbb{Z})$. The results are shown in Table 2.

Note, however, that this table does not give the action of monodromy in C on $H^1(E_1,\mathbb{Z})$. To obtain that, we have to take account of the cover $(f \circ g)\colon C \to \mathbb{P}^1$.

Table 2. *Action of monodromy around a point* $[u:v] \in \mathbb{P}^1$ *on* $H^1(E_1, \mathbb{Z})$.

Point $[u:v] \in \mathbb{P}^1$	Monodromy
$[0:1]$	$i\begin{bmatrix} 0 & 1 \\ -1 & 0 \end{bmatrix}$
$[-1:1]$	$\begin{bmatrix} 1 & 1 \\ 0 & 1 \end{bmatrix}$
$[1:0]$	$i\begin{bmatrix} 0 & 1 \\ -1 & -1 \end{bmatrix}$

Table 3. *Singular fibres of* $\mathcal{E}_1$ *over points* $p \in C$.

$f \circ g(p)$	j-invariant	Multiplicity of j	Monodromy	Fibre
$[0:1]$	1	6	$\begin{bmatrix} 1 & 0 \\ 0 & 1 \end{bmatrix}$	I_0
$[-1:1]$	∞	2	$\begin{bmatrix} 1 & 2 \\ 0 & 1 \end{bmatrix}$	I_2
$[1:0]$	0	6	$\begin{bmatrix} -1 & 0 \\ 0 & -1 \end{bmatrix}$	I_0^*

Let γ be a loop in C around one of the preimages $(f \circ g)^{-1}[0:1]$. Then $f \circ g(\gamma)$ is a loop in $\mathbb{P}^1$ that encircles $[0:1]$ six times, so the action of monodromy around γ on $H^1(E_1, \mathbb{Z})$ is as the action of a loop encircling $[0:1] \in \mathbb{P}^1$ raised to the sixth power. Similarly, the action of monodromy around a loop in C about one of the preimages $((f \circ g)^{-1}[-1:1]$ (resp. $((f \circ g)^{-1}[1:0]$) is as the action of a loop encircling $[-1:1] \in \mathbb{P}^1$ (resp. $[1:0] \in \mathbb{P}^1$) raised to the power of two (resp. six). The resultant monodromy matrices and the corresponding singular fibres of $\mathcal{E}_1$ are given in Table 3.

7.4. The forward construction

The last thing we will discuss is how to use this information to construct a birational model for $W \to \mathbb{P}^1$ and what this process can tell us about W.

Our beginning data is the two covers $C \xrightarrow{g} \mathbb{P}^1 \xrightarrow{f} \mathbb{P}^1$ (which can be determined from ξ_0) and the elliptic surface with section $\mathcal{E}_1$ on C. Note that $S_3 \times S_3$ acts on C and that the group G of order 18 given by the intersection of $S_3 \times S_3$ with A_6 acts on $\mathcal{E}_1$.

We first construct $\mathcal{E}_2$ as $i(\mathcal{E}_1)$, where $i\colon C \to C$ is the involution defined by the transposition (23) in $S_3 \times S_3$. Then we perform the fibrewise Kummer construction on the fibre product $\mathcal{E}_1 \times_C \mathcal{E}_2$ to obtain a birational model $\mathcal{F}$ for W''.

The next step is to perform a quotient by $S_3 \times S_3$ to obtain a birational model for W'. However, the action of $S_3 \times S_3$ is not the obvious one induced by the action of $S_3 \times S_3$ on $\mathcal{E}_1 \times_C \mathcal{E}_2$ (if it were, we would be able to undo the Kummer construction on W', which we have previously shown to be impossible). Instead, we compose this action with the fibrewise automorphism induced by the action of $S_3 \times S_3$ on the sets of curves $\{F_3, F_4, F_5\}$ and $\{F_6, F_7, F_8\}$.

Remark 7.8. As the fibrewise Kummer construction defines a natural double Kummer pencil on smooth fibres of $\mathcal{F}$, we can use the results of Kuwata and Shioda [KS08, Section 5.2] to define the elliptic fibration Ψ on $\mathcal{F}$. The curves $\{F_3, \ldots, F_8\}$ are then the components of the I_2 fibres that do not meet a chosen section.

Once $\{F_3, \ldots, F_8\}$ are known, the automorphisms permuting them may be calculated explicitly as compositions of the symplectic automorphisms $f_{r'}$ from Keum and Kōndo [KK01, Section 4.1]. We will give an example of such calculations in Example 7.9.

Quotienting by this $S_3 \times S_3$ action, we obtain a birational model for $W' \to \mathbb{P}^1$. Finally, there is a natural involution identifying the fibres over $[s:t]$ and $[-s:t]$ in this model. Quotienting by this involution, we obtain the model for $W \to \mathbb{P}^1$ we desire.

We conclude this section with an example showing how the action of $S_3 \times S_3$ can be computed around a singular fibre; this will allow us to gather information about some of the singular fibres in W. For simplicity we consider the fibres over $[\alpha : 1]$ and $[\beta : 1]$; the remaining fibres have too many components to allow such simple explicit calculations to be performed.

Example 7.9. We will calculate the form of the singular fibre in our birational model over $[\alpha : 1]$ (the calculation for $[\beta : 1]$ is identical). Let Δ denote a small disc in $\mathbb{P}^1$ centred at α, and let Δ' denote one of the connected components of $(f \circ g)^{-1}(\Delta)$. Then the map $(f \circ g)\colon \Delta' \to \Delta$ is a double cover, ramified over $0 \in \Delta$.

Over Δ' the elliptic surface $\mathcal{E}_1$ has no singular fibres. The involution i acts on Δ' to exchange the sheets of the double cover $(f \circ g)$. Define $\mathcal{E}_2 := i(\mathcal{E}_1)$. Note that the fibres of E_1 and E_2 over $(f \circ g)^{-1}(0) \in \Delta'$ are isomorphic.

Perform the fibrewise Kummer construction on $\mathcal{E}_1 \times_{\Delta'} \mathcal{E}_2$ to obtain $\mathcal{F} \to \Delta'$. Each fibre of $\mathcal{F}$ comes equipped with 24 distinguished (-2)-curves forming a **double Kummer pencil** $\{G_i, H_j, E_{ij} \mid 0 \leq i,j \leq 3\}$ (as defined in [CD07,

Definition 3.18]). As the involution i exchanges $\mathcal{E}_1$ and $\mathcal{E}_2$, it acts on this double Kummer pencil to exchange $G_i \leftrightarrow H_i$ and $E_{ij} \leftrightarrow E_{ji}$.

Now use [KS08, Section 5.2] to define the elliptic fibration Ψ on $\mathcal{F}$. In this fibration, the curves $\{F_3,\dots,F_8\}$ are the components of the I_2 fibres that do not meet a chosen section (which we take to be G_2). This set divides naturally into two triples, $\{F_3,F_4,F_5\}$ and $\{F_6,F_7,F_8\}$. By [KS08, Section 5.2], the curves $\{F_3,F_4,F_5\}$ (up to labelling) are given in the double Kummer pencil by

$$F_3 \sim E_{33}$$

$$F_4 \sim \frac{1}{2}\left(\sum_{i=0}^{3}(G_i+H_i)+\sum_{i,j}E_{ij}-2E_{00}-2E_{12}-2E_{21}\right)$$

$$F_5 \sim \sum_{i=0}^{3}(G_i+H_i)+\sum_{i,j}E_{ij}-2E_{00}-E_{13}-E_{31}-E_{12}-E_{21}-E_{22}$$

For simplicity, we will assume that the fixed curves swapped by monodromy around $0 \in \Delta$ are F_3 and F_4 (the other choices give the same result, but the calculations are substantially longer). By the results of [KK01, Section 4.1], this switching is realized by a symplectic involution φ of the Kummer surface that takes

$$\varphi(G_i) = H_i$$

$$\varphi(H_i) = G_i$$

$$\varphi(E_{ij}) = \begin{cases} E_{ji} & \text{if } (i,j) \notin \{(0,0),(1,2),(2,1),(3,3)\} \\ E_{ji}+D & \text{if } (i,j) \in \{(0,0),(1,2),(2,1),(3,3)\} \end{cases}$$

where D denotes the divisor

$$D := \frac{1}{2}\left(\sum_{i=0}^{3}(G_i+H_i)+\sum_{i,j}E_{ij}-2E_{00}-2E_{12}-2E_{21}-2E_{33}\right).$$

The involution defining the map from $\mathcal{F}$ to the birational model for W is realized as the composition of i with the fibrewise involution given by φ. The fixed points of this involution all lie in the fibre of $\mathcal{F}$ over $(f\circ g)^{-1}(0)$. This fibre is isomorphic to $\mathrm{Kum}(E\times E)$ for some smooth elliptic curve E.

In the fibre of $\mathcal{F}$ over $(f\circ g)^{-1}(0)$, the I_2 fibres in Ψ that are exchanged by the fibrewise involution φ collide to give an I_4 fibre. Let D_1 and D_2 denote the two disjoint (-2)-curves in this I_4 fibre that do not meet the sections $\{G_2,G_3,H_2,H_3\}$. Then $D = D_1 + D_2$, but D_1 and D_2 cannot be expressed as a combination of divisors from the double Kummer pencil $\{G_i,H_j,E_{ij}\}$ (the Néron-Severi group of a generic fibre of $\mathcal{F}$ has rank 18 and is spanned by the divisors in the double Kummer pencil, but this rank jumps to 19 on the special fibre over $(f \circ g)^{-1}(0)$ and the double Kummer pencil no longer spans). The involution φ, which is defined in terms of the double Kummer pencil, is not well-defined on D_1 and D_2, so they must be contracted before we can quotient.

After performing this contraction, the threefold $\mathcal{F}$ has two nodes and its fibre over $(f \circ g)^{-1}(0)$ has two A_1 singularities. The involution φ acts on the remaining divisors in $\mathrm{Kum}(E \times E)$ to exchange $G_i \leftrightarrow H_i$ and $E_{ij} \leftrightarrow E_{ji}$. But this is precisely the same as the action of the involution i. So the composition $(\varphi \circ i)$ acts trivially on the fibre of $\mathcal{F}$ over $(f \circ g)^{-1}(0)$.

After performing the quotient by $(\varphi \circ i)$, we find that the threefold total space is smooth over the disc Δ. Its fibre over $0 \in \Delta$ is isomorphic to the singular K3 surface obtained from $\mathrm{Kum}(E \times E)$ by contracting the two (-2)-curves D_1 and D_2. This fibre therefore has two A_1 singularities.

8 Moduli Spaces and the Mirror Map

The aim of this final section is to discuss the mirrors to the families Y and Z, along with their subfamilies Y_1, Z_2 and Z_1. As has already been discussed in Section 3.1 and Section 3.2, the mirror Y° of Y is known to be a family of $(2,12)$-complete intersections inside a blow-up of $\mathbb{WP}(1,1,1,1,4,6)$ and the mirror Z° of Z is known to be a family of degree 24 hypersurfaces in a blow-up of $\mathbb{WP}(1,1,2,8,12)$.

We will exhibit a candidate mirror for the degeneration of Z to Z_2 and Z_1 in terms of the Kähler moduli of Z°, and discuss the mirror of the geometric transition between Y and Z. This will enable us to match the components of the discriminant locus in the complex moduli space of Z to the boundary components in the Kähler cone of Z°. These considerations finally lead us to conclude that Y_1 and its mirror provide a counterexample to a conjecture of Morrison [Mor99].

8.1. Complex and Kähler Moduli

We begin by discussing the complex moduli of Z and the Kähler moduli of Z°. As noted in Section 3.2, the complex moduli are controlled by four parameters, B, ψ_0, ψ_1 and ψ_s. The discriminant locus was originally calculated by [BDF+98] and is given in Section 5.1. It splits into three parts:

- The locus $B = 0$, which we call S_0. From Equations (4.1) and (4.2), we see that this corresponds to the locus where the M-polarized K3 fibration on Z becomes isotrivial.
- The locus $\psi_s = \pm B$, which we call S_b. This corresponds to the degenerate family Z_2.

- The loci defined by the relations $(\psi_0^6+\psi_1)^2+\psi_s=\pm B$ and $\psi_1^2+\psi_s=\pm B$. These two relations are switched by an appropriate change of coordinates, so only give a single locus in the complex moduli space, which we call S_a.

Note that the intersection $S_a\cap S_b$ corresponds to the degenerate family Z_1.

Next we discuss the Kähler moduli of Z°. This has been studied by both Scheidegger [Sch01, Appendix C.3] and Hosono, Klemm, Theisen and Yau [HKTY97]; we will use Scheidegger's notation in the following discussion.

A generic degree 24 hypersurface in $\mathbb{WP}(1,1,2,8,12)$ has an elliptic curve C of A_1 singularities along with an exceptional $\mathbb{Z}_4$ point lying on this curve. The mirror Z° of Z is given by the crepant resolution of these singularities and contains two exceptional components: a ruled surface E over the curve C and a Hirzebruch surface $F\cong\mathbb{F}_2$ coming from the blow up of the exceptional point. Along with the hyperplane section H, these classes span the Kähler cone of Z°.

By [HKTY97, Equation (A.40)], the Kähler cone of Z° is given in terms of these classes by

$$\{t_HH+t_EE+t_FF \mid t_H+t_F>0,\ t_E<0,\ t_E-2t_F>0\}.$$

Scheidegger shows that a basis of the Mori cone of Z° is given by the classes h, d and l, where h is the class of a section of E (and is dual to H), d is the class of a fibre of F, and l is the class of a fibre of E (which is also the class of the (-2)-section of F). Proceeding to a boundary component of the Kähler cone corresponds to contracting one of these classes, as follows:

- The boundary component $t_E=0$ is spanned by H and F. Scheidegger shows that $H.l=F.l=0$, so l is contracted on this component. This corresponds to the birational morphism contracting E along its ruling onto the curve C. The (-2)-section in F is also contracted, giving a cone $\mathbb{F}_2^0$.
- The boundary component $t_E-2t_F=0$ is spanned by H and $F+2E$. Scheidegger shows that $H.d=(F+2E).d=0$, so d is contracted on this component. This corresponds to the birational morphism contracting F along its ruling.
- The boundary component $t_H+t_F=0$ is spanned by E and $H-F$. Scheidegger shows that $E.h=(H-F).h=0$, so h is contracted on this component. This does not give rise to a birational contraction.

8.2. The Mirror Map

Let Z_2° denote the threefold obtained by contracting the divisor E in Z° along its ruling; it may also be seen as the threefold obtained by blowing-up the $\mathbb{Z}_4$ point in a generic degree 24 hypersurface in $\mathbb{WP}(1,1,2,8,12)$. Along with the hyperplane class H, it contains one exceptional component F, which is isomorphic to a cone $\mathbb{F}_2^0$. As explained above, the Kähler modulus of Z_2° lies on the boundary component $t_E = 0$, which is spanned by the classes of H and F.

Now let Z_1° denote the threefold obtained from Z_2° by contracting F. Then Z_1° is isomorphic to a generic degree 24 hypersurface in $\mathbb{WP}(1,1,2,8,12)$. Its Kähler modulus lies on the ray given by the intersection of the boundary components $t_E = 0$ and $t_E - 2t_F = 0$, which is spanned by the class of H, the hyperplane section.

We claim that Z_2° and Z_1° are mirror to Z_2 and Z_1 respectively. To justify this, we exhibit a mirror for the geometric transition between Y and Z.

Recall from Section 5.2 that we may obtain a geometric transition from Z to Y by first degenerating to Z_2, then blowing-up the curve of singularities in Z_2. Furthermore, this transition extends to Z_1: we may degenerate Z_2 further to Z_1, then blow-up the curve of singularities in Z_1 to obtain the (singular) threefold Y_1. This relationship was summarized by Diagram (5.7).

Next, we discuss the mirror picture. Note first that we may embed Z_1° into $\mathbb{WP}(1,1,1,1,4,6)$ as a (non-generic) complete intersection of type $(2,12)$, via the degree two Veronese embedding $\mathbb{WP}(1,1,2,8,12) \hookrightarrow \mathbb{WP}(2,2,2,2,8,12)$. This admits a partial smoothing to a generic complete intersection of type $(2,12)$ in $\mathbb{WP}(1,1,1,1,4,6)$, which should be thought of as mirror to Y_1 and will be denoted by Y_1°; note that this exhibits Y_1 as mirror to a complete intersection of type $(2,12)$ in $\mathbb{WP}(1,1,1,1,4,6)$, precisely as expected. The threefold Y_1° has an isolated Gorenstein canonical singularity, which may be blown-up once to obtain the smooth Calabi-Yau threefold Y°, which is the mirror of Y. Kobayashi [Kob98, Theorem 1] has shown that the exceptional locus of this resolution is isomorphic to $\mathbb{P}^1 \times \mathbb{P}^1$.

By considering the blow-up of a family of threefolds in $\mathbb{WP}(1,1,1,1,4,6)$, we may naturally see Z_2° as a degeneration of Y°, where the generic $\mathbb{P}^1 \times \mathbb{P}^1$ exceptional locus degenerates to $\mathbb{F}_2^0$ in the limit. We thus have a diagram, where wiggly arrows denote degeneration and straight arrows denote blow-ups:

$$
\begin{array}{ccccc}
 & & Y^\circ & \longrightarrow & Y_1^\circ \\
 & & \wr\downarrow & & \wr\downarrow \\
Z^\circ & \longrightarrow & Z_2^\circ & \longrightarrow & Z_1^\circ
\end{array}
\tag{8.1}
$$

Diagrams (5.7) and (8.1) should be thought of as mirror to one another. In particular, we see that there is a mirror geometric transition from Z° to Y°, given by first contracting to Z_2°, then smoothing to Y°. In fact, this mirror correspondence between the geometric transitions from Y to Z and from Z° to Y° provides an explicit example of the kind of mirror correspondence discussed by Mavlyutov [Mav11, Section 5].

From this, we can deduce the mirror correspondence between the components of the discriminant locus in the complex moduli space of Z and the components of the boundary of the Kähler cone of Z°. We find:

- The locus S_b in the complex moduli space of Z, which corresponds to the degenerate family Z_2, must be mirror to the boundary component $t_E = 0$ in the Kähler moduli space of Z°, which corresponds to Z_2°.
- The locus S_a in the complex moduli space of Z must be mirror to the boundary component $t_E - 2t_F = 0$ in the Kähler moduli space of Z°, as the intersection $S_a \cap S_b$ corresponds to the degenerate family Z_1, and the mirror family Z_1° corresponds to the boundary ray $t_E = t_E - 2t_F = 0$.
- By elimination, the remaining locus S_0 in the complex moduli space of Z must be mirror to the boundary component $t_H + t_F = 0$ in the Kähler moduli space of Z°.

We conclude with a brief note on a conjecture of Morrison [Mor99], which states that a singular Calabi-Yau space has a Calabi-Yau resolution if and only if its mirror has a Calabi-Yau smoothing. However, in our case we find that, by a theorem of Gross [Gro97, Theorem 5.8], the fact that the exceptional locus in the mirror Y° is isomorphic to $\mathbb{P}^1 \times \mathbb{P}^1$ implies that Y_1° has a Calabi-Yau smoothing. But, by Corollary 6.5, Y_1 does not admit a Calabi-Yau resolution. This provides a counterexample to Morrison's conjecture.

References

[Bat94] Victor V. Batyrev. Dual polyhedra and mirror symmetry for Calabi-Yau hypersurfaces in toric varieties. *J. Algebraic Geom.*, 3(3):493–535, 1994. arXiv:alg-geom/9310003v1.

[BB96a] Victor V. Batyrev and Lev A. Borisov. Mirror duality and string-theoretic Hodge numbers. *Invent. Math.*, 126(1):183–203, 1996.

[BB96b] Victor V. Batyrev and Lev A. Borisov. On Calabi-Yau complete intersections in toric varieties. In *Higher-dimensional complex varieties (Trento, 1994)*, pages 39–65. de Gruyter, Berlin, 1996. arXiv:alg-geom/9412017v1.

[BCP97] Wieb Bosma, John Cannon, and Catherine Playoust. The Magma algebra system. I. The user language. *J. Symbolic Comput.*, 24(3-4):235–265, 1997. Computational algebra and number theory (London, 1993).

[BDF+98] Marco Billó, Frederik Denef, Pietro Frè, Igor Pesando, Walter Troost, Antoine Van Proeyen, and Daniela Zanon. The rigid limit in special Kähler geometry — from K3-fibrations to special Riemann surfaces: a detailed case study. *Classical Quantum Gravity*, 15(8):2083–2152, 1998.

[BN12] Volker Braun and Andrey Y. Novoseltsev. *Toric varieties framework for Sage*. The Sage Development Team, 2012. http://www.sagemath.org/doc/reference/sage/schemes/toric/variety.html.

[Bor93] Lev A. Borisov. Towards the mirror symmetry for Calabi-Yau complete intersections in Gorenstein toric Fano varieties. `arXiv:alg-geom/9310001v1`, 1993.

[CD07] Adrian Clingher and Charles F. Doran. Modular invariants for lattice polarized $K3$ surfaces. *Michigan Math. J.*, 55(2):355–393, 2007.

[CD11] Adrian Clingher and Charles F. Doran. Note on a geometric isogeny of K3 surfaces. *Int. Math. Res. Not. IMRN*, (16):3657–3687, 2011.

[CDKL11] X. Chen, C. Doran, M. Kerr, and J. Lewis. Normal functions, Picard-Fuchs equations, and elliptic fibrations on K3 surfaces. `arXiv:1108.2223v3 [math.AG]`, 2011.

[CDLW09] Adrian Clingher, Charles F. Doran, Jacob Lewis, and Ursula Whitcher. Normal forms, $K3$ surface moduli, and modular parametrizations. In *Groups and symmetries*, volume 47 of *CRM Proc. Lecture Notes*, pages 81–98. Amer. Math. Soc., Providence, RI, 2009. `arXiv:0712.1880v1 [math.AG]`.

[CK99] David A. Cox and Sheldon Katz. *Mirror symmetry and algebraic geometry*, volume 68 of *Mathematical Surveys and Monographs*. American Mathematical Society, Providence, RI, 1999.

[Cle83a] Herbert Clemens. Double solids. *Adv. in Math.*, 47(2):107–230, 1983.

[Cle83b] Herbert Clemens. Homological equivalence, modulo algebraic equivalence, is not finitely generated. *Inst. Hautes Études Sci. Publ. Math.*, (58):19–38 (1984), 1983.

[CLS90] P. Candelas, M. Lynker, and R. Schimmrigk. Calabi-Yau manifolds in weighted $\mathbf{P}_4$. *Nuclear Phys. B*, 341(2):383–402, 1990.

[CLS11] David A. Cox, John Little, and Hal Schenck. *Toric Varieties*, volume 124 of *Graduate Studies in Mathematics*. American Mathematical Society, Providence, RI, 2011.

[CS05] A. Corti and I. Smith. Conifold transitions and Mori theory. *Math. Res. Lett.*, 12(5-6):767–778, 2005.

[D+12] Dan Drake et al. *SageTeX package for LaTeX(Version 2.3.3)*, 2012. https://bitbucket.org/ddrake/sagetex.

[DHNT13] C. F. Doran, A. Harder, A. Y. Novoseltsev, and A. Thompson. Families of lattice polarized K3 surfaces with monodromy. Published online by Int. Math. Res. Notices (2015). http://dx.doi.org/10.1093/imrn/rnv071

[Dix88] Lance J. Dixon. Some world-sheet properties of superstring compactifications, on orbifolds and otherwise. In *Superstrings, unified theories and cosmology 1987 (Trieste, 1987)*, volume 4 of *ICTP Ser. Theoret. Phys.*, pages 67–126. World Sci. Publ., Teaneck, NJ, 1988.

[DM06] Charles F. Doran and John W. Morgan. Mirror symmetry and integral variations of Hodge structure underlying one-parameter families of Calabi-Yau threefolds. In *Mirror symmetry. V*, volume 38 of *AMS/IP Stud. Adv. Math.*, pages 517–537. Amer. Math. Soc., Providence, RI, 2006. `arXiv:math/0505272v1 [math.AG]`.

[DN10] Charles F. Doran and Andrey Y. Novoseltsev. Closed form expressions for Hodge numbers of complete intersection Calabi-Yau threefolds in toric varieties. In *Mirror Symmetry and Tropical Geometry*, volume 527 of *Contemporary Mathematics*, pages 1–14. Amer. Math. Soc., Providence, RI, 2010. `arXiv:0907.2701v2 [math.CO]`.

[Fri86] Robert Friedman. Simultaneous resolution of threefold double points. *Math. Ann.*, 274(4):671–689, 1986.

[Fri91] Robert Friedman. On threefolds with trivial canonical bundle. In *Complex geometry and Lie theory (Sundance, UT, 1989)*, volume 53 of *Proc. Sympos. Pure Math.*, pages 103–134. Amer. Math. Soc., Providence, RI, 1991.

[GP90] B. R. Greene and M. R. Plesser. Duality in Calabi-Yau moduli space. *Nuclear Phys. B*, 338(1):15–37, 1990.

[Gro97] M. Gross. Deforming Calabi-Yau threefolds. *Math. Ann.*, 308(2):187–220, 1997.

[HKTY97] S. Hosono, A. Klemm, S. Theisen, and S.-T. Yau. Mirror symmetry, mirror map and applications to complete intersection Calabi-Yau spaces. In *Mirror symmetry, II*, volume 1 of *AMS/IP Stud. Adv. Math.*, pages 545–606. Amer. Math. Soc., Providence, RI, 1997.

[HLY02] Yi Hu, Chien-Hao Liu, and Shing-Tung Yau. Toric morphisms and fibrations of toric Calabi-Yau hypersurfaces. *Adv. Theor. Math. Phys.*, 6(3):457–506, 2002. `arXiv:math/0010082v2 [math.AG]`.

[IKSY91] Katsunori Iwasaki, Hironobu Kimura, Shun Shimomura, and Masaaki Yoshida. *From Gauss to Painlevé*. Aspects of Mathematics, E16. Friedr. Vieweg & Sohn, Braunschweig, 1991. A modern theory of special functions.

[Ino78] Hiroshi Inose. Defining equations of singular $K3$ surfaces and a notion of isogeny. In *Proceedings of the International Symposium on Algebraic Geometry (Kyoto Univ., Kyoto, 1977)*, pages 495–502. Kinokuniya Book Store, Tokyo, 1978.

[KK01] J. Keum and S. Kōndo. The automorphism groups of Kummer surfaces associated with the product of two elliptic curves. *Trans. Amer. Math. Soc.*, 353(4):1469–1487, 2001.

[KKRS05] Albrecht Klemm, Maximilian Kreuzer, Erwin Riegler, and Emanuel Scheidegger. Topological string amplitudes, complete intersection Calabi-Yau spaces and threshold corrections. *J. High Energy Phys.*, (5):023, 116 pp. (electronic), 2005. `arXiv:hep-th/0410018v2`.

[KM98] J. Kollár and S. Mori. *Birational geometry of algebraic varieties*, volume 134 of *Cambridge Tracts in Mathematics*. Cambridge University Press, Cambridge, 1998.

[Kob98] M. Kobayashi. A special Lagrangian 3-torus as a real slice. In *Integrable systems and algebraic geometry (Kobe/Kyoto, 1997)*, pages 315–319. World Sci. Publ., River Edge, NJ, 1998.

[KS04] Maximilian Kreuzer and Harald Skarke. PALP: a package for analysing lattice polytopes with applications to toric geometry. *Comput. Phys. Comm.*, 157(1):87–106, 2004. `arXiv:math/0204356v1 [math.NA]`.

[KS08] M. Kuwata and T. Shioda. Elliptic parameters and defining equations for elliptic fibrations on a Kummer surface. In *Algebraic geometry in East Asia—Hanoi 2005*, volume 50 of *Adv. Stud. Pure Math.*, pages 177–215. Math. Soc. Japan, Tokyo, 2008.

[Mav11] A. R. Mavlyutov. Degenerations and mirror contractions of Calabi-Yau complete intersections via Batyrev-Borisov mirror symmetry. `arXiv:0910.0793 v2 [math.AG]`, February 2011.

[Mil68] J. Milnor. *Singular points of complex hypersurfaces*. Annals of Mathematics Studies, No. 61. Princeton University Press, Princeton, N.J., 1968.

[Mir89] R. Miranda. *The basic theory of elliptic surfaces*. Dottorato di Ricerca in Matematica. [Doctorate in Mathematical Research]. ETS Editrice, Pisa, 1989.

[Mor99] D. R. Morrison. Through the looking glass. In *Mirror symmetry, III (Montreal, PQ, 1995)*, volume 10 of *AMS/IP Stud. Adv. Math.*, pages 263–277. Amer. Math. Soc., Providence, RI, 1999.

[Nak88] N. Nakayama. On Weierstrass models. In *Algebraic geometry and commutative algebra, Vol. II*, pages 405–431. Kinokuniya, Tokyo, 1988.

[Nam02] Y. Namikawa. Stratified local moduli of Calabi-Yau threefolds. *Topology*, 41(6):1219–1237, 2002.

[NS95] Y. Namikawa and JHM Steenbrink. Global smoothing of Calabi-Yau threefolds. *Inventiones Mathematicae*, 122(1):403–419, 1995.

[PS97] Eugene Perevalov and Harald Skarke. Enhanced gauge symmetry in type II and F-theory compactifications: Dynkin diagrams from polyhedra. *Nuclear Phys. B*, 505(3):679–700, 1997. arXiv:hep-th/9704129v2.

[Rei87] Miles Reid. The moduli space of 3-folds with $K = 0$ may nevertheless be irreducible. *Math. Ann.*, 278(1-4):329–334, 1987.

[Roh04] Falk Rohsiepe. Lattice polarized toric K3 surfaces. arXiv:hep-th/0409290v1, 2004.

[Ros06] Michele Rossi. Geometric transitions. *J. Geom. Phys.*, 56(9):1940–1983, 2006.

[Ros11] Michele Rossi. Homological type of geometric transitions. *Geom. Dedicata*, 151:323–359, 2011.

[S+15] William A. Stein et al. *Sage Mathematics Software (Version 6.5)*. The Sage Development Team, 2015. http://www.sagemath.org.

[Sch01] E. Scheidegger. *D-branes on Calabi-Yau spaces*. PhD thesis, Ludwig-Maximilian University of Munich, July 2001.

[Shi06] Tetsuji Shioda. Kummer sandwich theorem of certain elliptic $K3$ surfaces. *Proc. Japan Acad. Ser. A Math. Sci.*, 82(8):137–140, 2006.

[SI77] T. Shioda and H. Inose. On singular $K3$ surfaces. In *Complex analysis and algebraic geometry*, pages 119–136. Iwanami Shoten, Tokyo, 1977.

[STY02] I. Smith, R. P. Thomas, and S.-T. Yau. Symplectic conifold transitions. *J. Differential Geom.*, 62(2):209–242, 2002.

[Wil92] P. M. H. Wilson. The Kähler cone on Calabi-Yau threefolds. *Invent. Math.*, 107(3):561–583, 1992.

[Wil93] P. M. H. Wilson. Erratum: "The Kähler cone on Calabi-Yau threefolds" [Invent. Math. **107** (1992), no. 3, 561–583; MR1150602 (93a:14037)]. *Invent. Math.*, 114(1):231–233, 1993.

Department of Mathematics and Computer Science, University of Missouri – St. Louis, St. Louis, MO, 63121, USA
E-mail address: clinghera@umsl.edu

Department of Mathematical and Statistical Sciences, 632 CAB, University of Alberta, Edmonton, Alberta, T6G 2G1, Canada
E-mail address: charles.doran@ualberta.ca

Fakultät für Mathematik, Universität Wien, Garnisongasse 3/14, 1090 Wien, Austria
E-mail address: Jacobml@u.washington.edu

Department of Mathematical and Statistical Sciences, 632 CAB, University of Alberta, Edmonton, Alberta, T6G 2G1, Canada
E-mail address: novoselt@ualberta.ca

Fields Institute, 222 College Street, Toronto, Ontario, M5T 3J1, Canada
E-mail address: amthomps@ualberta.ca

PART II

Algebraic cycles and normal functions

7

A simple construction of regulator indecomposable higher Chow cycles in elliptic surfaces

Masanori Asakura*

Dedicated to the 60th birthday of Professor James D. Lewis

1 Introduction

Let $\mathrm{CH}^j(X,i)$ be Bloch's higher Chow groups of a projective smooth variety X over $\mathbb{C}$. A higher Chow cycle $z \in \mathrm{CH}^j(X,i)$ is called *indecomposable* if it does not belong to the image of the map of the product

$$\mathrm{CH}^1(X,1) \otimes \mathrm{CH}^{j-1}(X,i-1) \longrightarrow \mathrm{CH}^j(X,i).$$

Of particular interest to us is $\mathrm{CH}^2(X,1)$. For $A = \mathbb{Q}$ or $\mathbb{R}$, we say $z \in \mathrm{CH}^2(X,1)$ *A-regulator indecomposable* if the regulator class $\mathrm{reg}(z) \in H^3_{\mathscr{D}}(X,A(2))$ in the Deligne-Beilinson cohomology group with coefficients in A does not belong to the image of $H^1_{\mathscr{D}}(X,\mathbb{Z}(1)) \otimes H^2_{\mathscr{D}}(X,A(1)) \cong \mathbb{C}^\times \otimes \mathrm{CH}^1(X) \otimes A$. In other words, z is A-regulator indecomposable if and only if

$$\mathrm{reg}(z) \neq 0 \in \mathrm{Ext}^1_{\mathrm{MHS}}(A, H^2_{\mathrm{ind}}(X,A(2))), \quad H^2_{\mathrm{tr}}(X,A) := H^2(X,A)/\mathrm{NS}(X) \otimes A.$$

Obviously $\mathbb{R}$-reg. indecomp. $\Longrightarrow$ $\mathbb{Q}$-reg. indecomp. $\Longrightarrow$ indecomposable.

Quite a lot of examples of $\mathbb{Q}$ or $\mathbb{R}$-regulator indecomposable cycles are obtained by many people ([1], [3], [4], [5], [6] and more).

In this note we construct $\mathbb{R}$-regulator indecomposable cycles for X an elliptic surface which satisfies certain conditions. The main theorem is the following.

*Supported by the Grant-in-aid No. 24540001 (C) and partially by the Grant-in-aid No. 23224001 (S) for Scientific Research, JSPS.

Theorem 1.1 *Let S be a smooth irreducible curve over $\mathbb{C}$. Let*

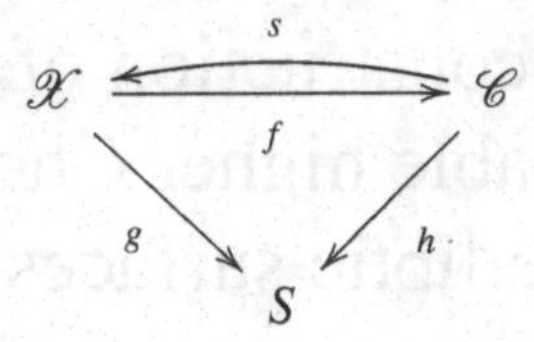

be an elliptic fibration over S with a section s. This means that g and h are projective smooth morphisms of relative dimension 2 and 1 respectively, and the general fiber of f is an elliptic curve. For a point $t \in S$ we denote $X_t = g^{-1}(t)$ or $C_t = h^{-1}(t)$ the fibers over t. Assume that the following conditions hold.

1. *Let η be the generic point of S. Then there is a split multiplicative fiber $D_\eta = f^{-1}(P) \subset X_\eta$ of Kodaira type I_n, $n \geq 1$ (see [8] VII, §5 for the terminology of "split multiplicative fiber").*
2. *Let $\mathscr{D} \subset \mathscr{X}$ be the closure of D_η. Then there is a closed point $0 \in S(\mathbb{C})$ such that the specialization $D_0 := \mathscr{D} \times_{\mathscr{X}} X_0$ is multiplicative of type I_m with $m > n$.*

Then the composition

$$\mathrm{CH}^1(D_t, 1) \longrightarrow \mathrm{CH}^2(X_t, 1) \xrightarrow{\mathrm{reg}} \mathrm{Ext}^1_{\mathrm{MHS}}(\mathbb{R}, H^2(X_t, \mathbb{R}(2))/\mathrm{NF}(X_t) \otimes \mathbb{R})$$

is non-zero for a general $t \in S(\mathbb{C})$. Here $\mathrm{NF}(X_t) \subset \mathrm{NS}(X_t)$ denotes the subgroup generated by components of singular fibers and the section $s(C_t)$. In particular, if $\mathrm{NF}(X_t) \otimes \mathbb{Q} = \mathrm{NS}(X_t) \otimes \mathbb{Q}$, then there is a $\mathbb{R}$-regulator indecomposable higher Chow cycle.

The key assumption is a "degeneration of D_t."

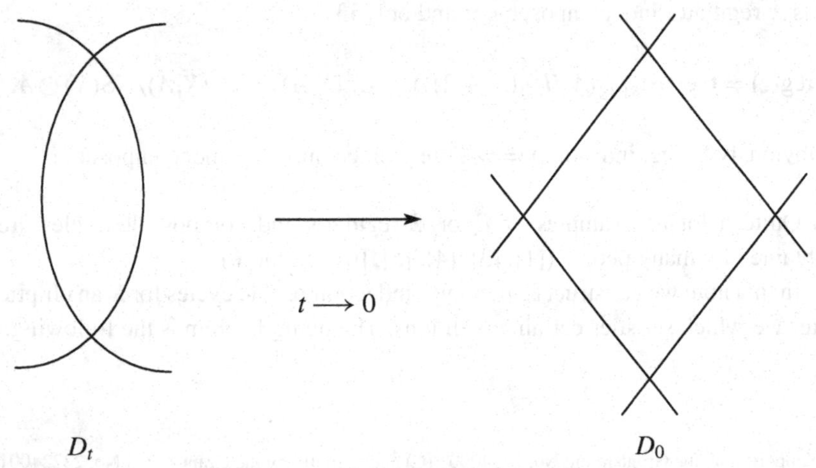

Figure : Degeneration of I_2 to I_4.

In §3, we will apply Theorem 1.1 to construct a $\mathbb{R}$-reg. indecomp. cycle in a self-product of elliptic curves. However, to do it in more general situation, the computation of the Picard number might be an obstacle. Indeed it is easy to compute the rank of $\mathrm{NF}(X_t)$, whereas there is no general method to do it for $\mathrm{NS}(X_t)$, and usually it is done by case-by-case analysis.

Acknowledgment. The author is grateful to Professors James D. Lewis and Matt Kerr for reading the first draft carefully and giving many comments.

2 Proof of Theorem 1.1

We keep the notation and assumption in Theorem 1.1.

2.1. Step 1: Construction of a higher Chow cycle

Lemma 2.1 *Let $f : X \to C$ be an elliptic fibration over a field K of characteristic$\neq 2,3$, with a (fixed) section s. Let $D = f^{-1}(P)$ be a split multiplicative fiber, and let $i : D \hookrightarrow X$. Then there is an exact sequence*

$$\mathrm{CH}^1(\widetilde{D},1)_{\mathbb{Q}} \longrightarrow \mathrm{CH}^1(D,1)_{\mathbb{Q}} \xrightarrow{v} \mathbb{Q} \longrightarrow 0 \tag{2.1}$$

where $\widetilde{D} \to D$ is the normalization. There exists a higher Chow cycle $Z \in \mathrm{CH}^1(D,1)$ such that $v(Z) \neq 0$ and $i_*(Z) \in \mathrm{CH}^2(X,1)$ is vertical to generators of $\mathrm{NF}(X)$. Here "vertical to a curve $E \subset X$" means that it lies in the kernel of the composition $\mathrm{CH}^2(X,1) \to \mathrm{CH}^2(\widetilde{E} \times_K \overline{K},1) \to \mathrm{CH}^1(\mathrm{Spec}\overline{K},1) \otimes \mathbb{Q} = \overline{K}^\times \otimes \mathbb{Q}$ where the first map is the pull-back and the second map is the transfer with respect to $\widetilde{E} \times_K \overline{K} \to \mathrm{Spec}\overline{K}$.

Proof. We omit to show the exact sequence (2.1) (easy exercise). We show the existence of Z. In this proof, we use K-groups rather than higher Chow groups. Let $D = \sum_{i=1}^n D_i$ be the irreducible decomposition. It is enough to construct $Z \in K_1'(D)^{(1)}$ such that $v(Z) \neq 0$ and $i_*(Z)$ is vertical to each D_i and $s(C)$ because it is obviously vertical to the other fibral divisors. Since D is split multiplicative, each D_i is geometrically irreducible and the singularities of D are K-rational. Therefore we may assume $K = \overline{K}$ by the standard norm argument.

It is enough to show that the image of the composition

$$K_1'(D)^{(1)} \xrightarrow{i_*} K_1(X)^{(2)} \xrightarrow{i^*} K_1(\widetilde{D})^{(2)} \cong (K^\times \otimes \mathbb{Q})^{\oplus n}$$

coincides with that of

$$K_1'(\widetilde{D})^{(1)} \xrightarrow{i_*} K_1(X)^{(2)} \xrightarrow{i^*} K_1(\widetilde{D})^{(2)} \cong (K^\times \otimes \mathbb{Q})^{\oplus n}.$$

Indeed, the above implies that there is a cycle $Z_0 \in K_1'(D)^{(1)}$ such that $v(Z_0) \neq 0$ and $i_*(Z_0)$ is vertical to each D_i. Let $f^* : K_1(K) \to K_1'(D)^{(1)}$. Then $Z := Z_0 + f^*(\lambda)$ for suitable $\lambda \in K^\times$ can be vertical to the section $s(C)$ and D_i. Moreover $v(Z) = v(Z_0) \neq 0$.

To do the above we may replace X with $\hat{X} = f^{-1}(\mathrm{Spec}K[[s]])$ the formal neighborhood around D. Then it is enough to show that there is $Z_0 \in K_1'(D)^{(1)}$ such that $v(Z_0) \neq 0$ and $i_*(Z_0) = 0$ in $K_1(\hat{X})^{(2)}$. To do this, we may further replace $\hat{X}$ with $\hat{X}_n$ the minimal desingularization of $\hat{X} \times \mathrm{Spec}K[[s^{1/n}]]$ for some $n \geq 1$ due to a commutative diagram

$$\begin{array}{ccccc} \mathbb{Q} & \xleftarrow{v} & K_1'(D_n)^{(1)} & \xrightarrow{i_*} & K_1(\hat{X}_n)^{(2)} \\ \downarrow{\scriptstyle n} & & \downarrow{\scriptstyle \phi_*} & & \downarrow{\scriptstyle \phi_*} \\ \mathbb{Q} & \xleftarrow{v} & K_1'(D)^{(1)} & \xrightarrow{i_*} & K_1(\hat{X})^{(2)} \end{array}$$

where $\phi : \hat{X}_n \longrightarrow \hat{X}$. Thus we can assume $\hat{X}$ is defined by a Weierstrass equation

$$y^2 = x^3 + x^2 + c(s), \quad c(s) \in sK[[s]].$$

Then letting $\partial : K_2(\hat{X} \setminus D)^{(2)} \to K_1'(D)^{(1)}$ be the boundary map, we put

$$Z_0 := \partial \left\{ \frac{y-x}{y+x}, \frac{-c(s)}{x^3} \right\}.$$

This satisfies $v(Z_0) \neq 0$ and $i_*(Z_0) = 0$ in $K_1(\hat{X})^{(2)}$. □

By Lemma 2.1, there is a higher Chow cycle $Z_\eta \in \mathrm{CH}^1(D_\eta, 1)$ such that $v(Z_\eta) \neq 0$ and it is vertical to $\mathrm{NF}(X_\eta)$. We use the same symbol "Z_η" for $i_*(Z_\eta) \in \mathrm{CH}^2(X_\eta, 1)$ with $i : D_\eta \hookrightarrow X_\eta$ since it will be clear from the context what is meant.

Let $\mathscr{Z}^* \subset \mathscr{X}^*$ be the closure of Z_η in $\mathscr{X}^* := g^{-1}(S^*)$ for some nonempty Zariski open $S^* \subset S$. Then the goal is to show the nonvanishing of

$$\mathrm{reg}(Z_t) \neq 0 \in \mathrm{Ext}^1_{\mathrm{MHS}}(\mathbb{R}, H^2(X_t, \mathbb{R}(2))), \quad Z_t := \mathscr{Z}^*|_{X_t} \tag{2.2}$$

for a general $t \in S^*(\mathbb{C})$. Indeed, since Z_t is vertical to $\mathrm{NF}(X_t)$, the above implies the desired nonvanishing

$$\mathrm{reg}(Z_t) \neq 0 \in \mathrm{Ext}^1_{\mathrm{MHS}}(\mathbb{R}, H^2(X_t, \mathbb{R}(2))/\mathrm{NF}(X_t) \otimes \mathbb{R}).$$

2.2. Step 2 : Boundary of Z_η

Lemma 2.2 *Let $\partial : \mathrm{CH}^2(X_\eta,1) \to \mathrm{CH}^1(X_0)$ be the boundary map arising from the localization exact sequence, and let $cl : \mathrm{CH}^1(X_0) \to H^2(X_0, \mathbb{Q}(1))$ the cycle map. Then $cl \circ \partial(Z_\eta) \neq 0$.*

Proof. We show that $\partial(Z_\eta)$ is not numerically equivalent to zero. We may assume $S = \mathrm{Spec} R$ where R is a DVR, with a closed point 0 and generic point η. Let $D'_\eta \subset D_\eta$ and $D'_0 \subset D_0$ be the unique chain of rational curves which forms Neron polygons. Then there are exactly n reduced components (resp. m reduced components) in D'_η (resp. D'_0), as it is of Kodaira type I_n (resp. I_m). Since $m > n$, there is a reduced irreducible component $E'_\eta \subset D'_\eta$ such that its specialization E'_0 has at least two reduced components. Let $\mathscr{E}' \subset \mathscr{D}$ be the closure of E'_η and $j : \widetilde{\mathscr{E}} \to \mathscr{E}'$ the normalization. Let $\widetilde{E}_0 := \widetilde{\mathscr{E}} \times_S \{0\} = \sum_{j=1}^q r_j C_j$ be the special fiber.

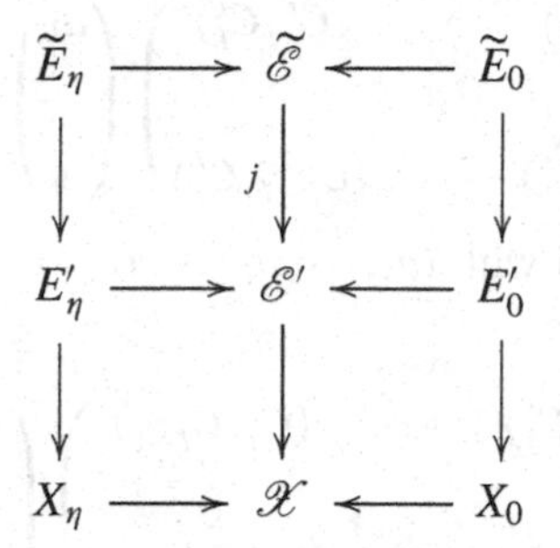

The generic fiber $\widetilde{E}_\eta := \widetilde{\mathscr{E}} \times_S \{\eta\}$ is a smooth irreducible rational curve. Let $T_1, T_2 \subset \widetilde{\mathscr{E}}$ be the inverse image of the intersection locus of the Neron polygon D'_η. Let T_1 hits a component of $\widetilde{E}_0$, say C_1. Then T_2 hits another component (say C_2), since the image $j(\widetilde{E}_0) \subset D'_0$ has at least two components. One has $r_1 = r_2 = 1$, namely C_1 and C_2 are reduced components, since T_1 and T_2 are sections of $\widetilde{\mathscr{E}} \to S$. Moreover the intersection points $(T_1 \cap C_1)$ and $(T_2 \cap C_2)$ are nonsingular points in $\widetilde{\mathscr{E}}$.

Let f be a rational function on E_η such that the divisor $\mathrm{div}_{E_\eta}(f) = T_1 - T_2$ and hence

$$\mathrm{div}_{\widetilde{\mathscr{E}}}(f) = T_1 - T_2 + \sum_{j=1}^{q} a_j C_j. \tag{2.3}$$

Then

$$\partial(Z_\eta) = \sum_{j=1}^{q} a_j C_j + N \cdot j(\widetilde{E}_0) + (\text{other components of } D_0) \in \mathrm{NS}(X_0)$$

for some $N \in \mathbb{Z}$ by the construction of Z_η. Therefore it is enough to show $a_1 \neq a_2$.

We take a desingularization $\rho : \mathscr{E} \to \widetilde{\mathscr{E}}$ (if necessary). Then

$$\mathrm{div}_{\mathscr{E}}(f) = T_1 - T_2 + \sum_{j=1}^{q} a_j C'_j + \sum_{j=q+1}^{\ell} a_j C'_j. \tag{2.4}$$

where C'_j ($j \le q$) are the strict transform of C_j and C'_j ($j > q$) are the exceptional curves. We may assume T_i intersects with C'_i in $\mathscr{E}$ for $i = 1, 2$, since $T_i \cap C_i$ are nonsingular points of $\widetilde{\mathscr{E}}$. Renumbering C'_j, we may assume T_1 intersects with C'_1 and T_2 intersects with C'_ℓ. We then want to show $a_1 \neq a_\ell$. By the intersection theory, (2.4) yields

$$\begin{cases} 1 + \sum_{j=1}^{\ell} a_j (C'_1, C'_j) = 0 \\ \sum_{j=1}^{\ell} a_j (C'_i, C'_j) = 0 \qquad 2 \le \forall i \le \ell - 1 \end{cases}$$

namely

$$\begin{pmatrix} (C'_1, C'_1) & \cdots & (C'_1, C'_\ell) \\ \vdots & & \vdots \\ (C'_{\ell-1}, C'_1) & \cdots & (C'_{\ell-1}, C'_\ell) \end{pmatrix} \begin{pmatrix} a_1 \\ \vdots \\ a_\ell \end{pmatrix} = \begin{pmatrix} -1 \\ \vdots \\ 0 \end{pmatrix} \tag{2.5}$$

By replacing $(a_1, \cdots, a_\ell)$ with $(a_1, \cdots, a_\ell) + c(r_1, \cdots, r_\ell)$ for some $c \in \mathbb{Q}$, we may assume $a_\ell = 0$. Then

$$A \begin{pmatrix} a_1 \\ \vdots \\ a_{\ell-1} \end{pmatrix} = \begin{pmatrix} (C'_1, C'_1) & \cdots & (C'_1, C'_{\ell-1}) \\ \vdots & & \vdots \\ (C'_{\ell-1}, C'_1) & \cdots & (C'_{\ell-1}, C'_{\ell-1}) \end{pmatrix} \begin{pmatrix} a_1 \\ \vdots \\ a_{\ell-1} \end{pmatrix} = \begin{pmatrix} -1 \\ \vdots \\ 0 \end{pmatrix} \tag{2.6}$$

By Zariski's lemma ([2] III (8.2)), one has $\det A < 0$ and $\det A_{11} < 0$ where A_{11} is the cofactor matrix. Therefore one has $a_1 = -\det A_{11} / \det A \neq 0$, the desired assertion. This completes the proof. □

2.3. *Step 3 : Extension of admissible variations of MHS's*

Let

$$\mathrm{reg}(\mathscr{Z}^*) \in H^3_{\mathscr{D}}(\mathscr{X}^*, \mathbb{Q}(2))$$

be the Deligne-Beilinson cohomology class where $\mathscr{Z}^*$ and $\mathscr{X}^* = g^{-1}(S^*)$ are as in the end of §2.1. Lemma 2.2 together with the commutative diagram

$$\begin{array}{ccccc} \mathrm{CH}^2(\mathscr{X}^*, 1) & \xrightarrow{\partial} & \mathrm{CH}^1(X_0) & & \\ \downarrow \mathrm{reg} & & \downarrow cl & & \\ H^3_{\mathscr{D}}(\mathscr{X}^*, \mathbb{Q}(2)) & \xrightarrow{\mathrm{Res}} & H^2_{\mathscr{D}}(X_0, \mathbb{Q}(1)) & \xrightarrow{\cong} & H^2(X_0, \mathbb{Q}(1)) \cap H^{1,1} \end{array} \tag{2.7}$$

yields nonvanishing $\mathrm{Res}\circ\mathrm{reg}(\mathscr{Z}^*) = cl\circ\partial(\mathscr{Z}^*) \neq 0$ where Res denotes the residue map on Deligne-Beilinson cohomology. Let $R^i g_*\mathbb{Q}(2)$ denotes an admissible variations of mixed Hodge structures on S^* (cf. [7] §14.4.1). Since $\mathrm{Ext}^i_{\mathrm{VMHS}(S^*)}(\mathbb{Q}, R^{3-i}g_*\mathbb{Q}(2)) = 0$ for $i \neq 1,2$, there is a surjective map

$$H^3_{\mathscr{D}}(\mathscr{X}^*, \mathbb{Q}(2)) \longrightarrow \mathrm{Ext}^1_{\mathrm{VMHS}(S^*)}(\mathbb{Q}, H_{\mathbb{Q}}), \quad H_{\mathbb{Q}} := R^2 g_*\mathbb{Q}(2)$$

to the extension group of admissible VMHS's on S^*. Hence $\mathrm{reg}(\mathscr{Z}^*)$ induces an exact sequence

$$0 \longrightarrow H_{\mathbb{Q}} \longrightarrow V_{\mathbb{Q}} \longrightarrow \mathbb{Q} \longrightarrow 0. \tag{2.8}$$

The residue map in (2.7) factors as follows

$$\begin{aligned} H^3_{\mathscr{D}}(\mathscr{X}^*, \mathbb{Q}(2)) &\to \mathrm{Ext}^1_{\mathrm{VMHS}(S^*)}(\mathbb{Q}, H_{\mathbb{Q}}) \to \mathrm{Ext}^1_{S^*}(\mathbb{Q}, H_{\mathbb{Q}}) \\ &\cong H^1(S^*, H_{\mathbb{Q}}) \to H^2(X_0, \mathbb{Q}(1)) \end{aligned}$$

where $\mathrm{Ext}^\bullet_{S^*}$ denotes the extension group of abelian sheaves on S^* with analytic topology, and the last arrow is the residue map on Betti cohomology. Therefore nonvanishing $\mathrm{Res}\circ\mathrm{reg}(\mathscr{Z}^*) \neq 0$ implies that (2.8) is nontrivial as extension of abelian sheaves on S^*, and hence so is the exact sequence

$$0 \longrightarrow H_{\mathbb{R}} \longrightarrow V_{\mathbb{R}} \longrightarrow \mathbb{R} \longrightarrow 0 \tag{2.9}$$

obtained by tensoring with $\mathbb{R}$.

The following lemma finishes the proof of Theorem 3.1.

Lemma 2.3 *Let $\Delta \subset S$ be a small neighborhood of 0, and put $\Delta^* = \Delta \setminus \{0\}$. For $t \in \Delta^*$ such that $0 < |t| \ll 1$, the extension*

$$0 \longrightarrow H_{\mathbb{R},t} \longrightarrow V_{\mathbb{R},t} \longrightarrow \mathbb{R} \longrightarrow 0 \tag{2.10}$$

is non-trivial where $H_{\mathbb{R},t}$ etc. denotes the fiber at t. Hence the nonvanishing (2.2) follows.

Before proving Lemma 2.3, we note that "$\mathbb{Q}$-regulator indecomposability of Z_t" is immediate from the fact that (2.8) is non-trivial. Let $H_{\mathscr{O}} := \mathscr{O}_{S^*} \otimes H_{\mathbb{Q}}$ and $F^\bullet$ be the Hodge bundles. Put $J_{\mathbb{Q}} := H_{\mathscr{O}}/(F^2 + H_{\mathbb{Q}})$ and

$$J^h_{\mathbb{Q}} := \mathrm{Ker}[J_{\mathbb{Q}} \xrightarrow{\nabla} \Omega^1_{S^*} \otimes H_{\mathscr{O}}/F^1]$$

the sheaf of horizontal sections where ∇ denotes the Gauss-Manin connection. As is well-known, there is the injective map

$$\imath : \mathrm{Ext}^1_{\mathrm{VMHS}(S^*)}(\mathbb{Q}, H_{\mathbb{Q}}) \hookrightarrow \Gamma(S^*, J^h_{\mathbb{Q}}), \tag{2.11}$$

and $\nu_{\mathscr{Z}^*} := \imath(\mathrm{reg}(\mathscr{Z}^*)) \in \Gamma(S^*, J^h_{\mathbb{Q}})$ is called the *normal function* associated to $\mathscr{Z}^*$. Since the zero locus of the normal function is at most a countable set, we have $\nu_{\mathscr{Z}^*}(t) \neq 0$ for a general t, and hence that Z_t is $\mathbb{Q}$-regulator indecomposable. However to obtain the "$\mathbb{R}$-regulator indecomposability" in the same way, we need to show that the injectivity of (2.11) remains true if we replace $J^h_{\mathbb{Q}}$ with $J^h_{\mathbb{R}}$, and we don't know how to prove it in general. We prove Lemma 2.3 in a different way.

Proof of Lemma 2.3.

Let T be the local monodromy around Δ^*. The action on $H_t = H^2(X_t, \mathbb{Q}(2))$ is trivial, whereas that on $V_{\mathbb{Q},t}$ is non-trivial. Indeed, let $\{e_{1,t}, \ldots, e_{m,t}\}$ be a basis of $H_{\mathbb{Q},t}$ and $\{e_{0,t}, e_{1,t}, \ldots, e_{m,t}\}$ be a basis of $V_{\mathbb{Q},t}$ such that $e_{0,t} \notin H_{\mathbb{Q},t}$. Let $N = T - 1$ be the log monodromy. Then one has $(2\pi i)^{-1} N e_{0,t} = \mathrm{Res}(\mathrm{reg}(\mathscr{Z}^*)) = cl[\partial \mathscr{Z}^*] \neq 0$ under the natural isomorphism $H_{\mathbb{Q},t} \otimes \mathbb{Q}(-1) \cong H_{\mathbb{Q},0} \otimes \mathbb{Q}(-1) = H^2(X_0, \mathbb{Q}(1))$.

Let us fix a frame $z_t \in H_{\mathbb{Q},t}$ which satisfies that $z|_{t=0} = cl[\partial \mathscr{Z}^*] \in H^2(X_0, \mathbb{C})$. Then the *admissibility* of V yields that there are holomorphic functions $u_i(t)$ on Δ^* with at most meromorphic singularities at $t = 0$ such that

$$f_0 = e_{0,t} + \frac{\log(t)}{2\pi i} z_t + \sum_i u_i(t) e_{i,t}$$

belongs to the Hodge bundle $\Gamma(\Delta^*, F^0 V_{\mathscr{O}})$. Moreover, by adding some $\theta \in \Gamma(\Delta^*, F^2 H_{\mathscr{O}})$, one has

$$\tilde{f}_0 := f_0 + \theta = e_{0,t} + \frac{\log(t)}{2\pi i} z_t + \sum_i \tilde{u}_i(t) e_{i,t}$$

with $\tilde{u}_i(t)$ holomorphic at $t = 0$. Therefore the extension data of (2.8) is given as follows

$$(\text{extension class of } V|_{\Delta^*}) = \frac{\log(t)}{2\pi i} z_t + \sum_i \tilde{u}_i(t) e_{i,t} \in H_{\mathscr{O}}/(F^2 + H_{\mathbb{Q}}). \quad (2.12)$$

Let $\mathscr{A}_{\Delta^*}$ be the sheaf of C^∞-functions on Δ^*, and let $c : H_{\mathbb{R}} \otimes_{\mathbb{R}} \mathscr{A}_{\Delta^*} \to H_{\mathbb{R}} \otimes_{\mathbb{R}} \mathscr{A}_{\Delta^*}$ the complex conjugation given by $c(x \otimes f) = x \otimes \bar{f}$. Let

$$H_{\mathscr{O}}/(F^2 + H_{\mathbb{R}}) \xrightarrow{1-c} H_{\mathbb{R}} \otimes \mathscr{A}_{\Delta^*}/(1-c)F^2 \cong (H_{\mathbb{R}} \cap H^{1,1}) \otimes \mathscr{A}_{\Delta^*}. \quad (2.13)$$

Then the extension class (2.12) goes to

$$\frac{\log|t|}{\pi} z_t + \sum_i (\text{bounded function}) e_{i,t} \in (H_{\mathbb{R}} \cap H^{1,1}) \otimes \mathscr{A}_{\Delta^*} \quad (2.14)$$

via (2.13). This does not vanish for $0 < |t| \ll 1$ as $\log|t| \to -\infty$. This completes the proof of Lemma 2.3. □

The proof of Theorem 1.1 is done.

3 Example : Self-Product of elliptic curves

Recall the following theorem due to Lewis and Gordon[1].

Theorem 3.1 (Lewis - Gordon, [4] Thm.1) *Let $X = E \times E$ be a product of elliptic curves over $\mathbb{C}$. If E is sufficiently general, then there exists a $\mathbb{R}$-regulator indecomposable cycle $z \in \mathrm{CH}^2(X,1)$* [2].

We here give an alternative proof as an application of Theorem 1.1. Let

$$E_a : y^2 = F(x) = x(x-1)(x-a), \quad a \in \mathbb{C} \setminus \{0,1\}$$

be an elliptic curve over $\mathbb{C}$. Let $X_a := \mathrm{Km}(E_a \times E_a)$ be the *Kummer surface* associated to the product $E_a \times E_a$. The birational model of X_a is given by the double sextic

$$X_a : w^2 = F(x_1)F(x_2) = x_1x_2(x_1-1)(x_2-1)(x_1-a)(x_2-a).$$

Changing the variables $x := x_1$, $t := x_1x_2$ and $y := wx_1$, we get

$$X_a : y^2 = t(x-1)(x-t)(x-a)(ax-t). \tag{3.1}$$

This gives an elliptic fibration

$$f : X_a \longrightarrow \mathbb{P}^1, \quad (x,y,t) \longmapsto t \tag{3.2}$$

where t is the affine parameter of $\mathbb{P}^1$. This is naturally extended to a family over $S = \mathbb{A}_{\mathbb{C}} \setminus \{0,1\}$

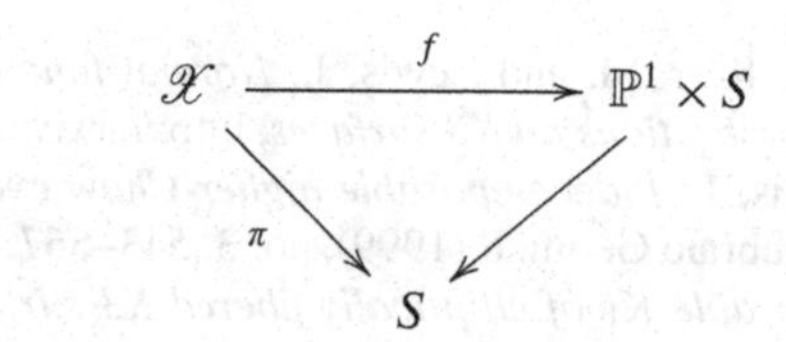

in which $X_a = \pi^{-1}(a)$. We fix a section s of $(x,y) = (1,0)$.

Let us look at the singular fibers in (3.2). The singular fibers are located at $t = 1, a, a^2, 0, \infty$. Two additive fibers appear at $t = 0, \infty$ and both are of Kodaira type I_2^*. If $a \neq -1$, the singular fibers at $t = 1, a, a^2$ are multiplicative of type I_2, I_4, I_2 respectively, and if $a = -1$, two fibers at $t = \pm 1$ are multiplicative of type I_4. Let $\mathrm{NF}(X_a) \subset \mathrm{NS}(X_a)$ be the subgroup generated by irreducible components of singular fibers and the section s. The rank of $\mathrm{NF}(X_a)$ is 19 if

[1] Though there was an error in their proof, Türkmen recently corrected it [9].

[2] [4] Thm.1 deals with a general product $E_1 \times E_2$, though their techniques allow to handle the case of a self-product $E \times E$ as well.

$a \neq -1$, and 20 if $a = -1$. As is well-known,

$$\operatorname{rankNS}(X_a) = \begin{cases} 19 & E_a \text{ has no CM} \\ 20 & E_a \text{ has a CM.} \end{cases} \tag{3.3}$$

In particular

$$\mathrm{NF}(X_a) \otimes \mathbb{Q} = \mathrm{NS}(X_a) \otimes \mathbb{Q} \text{ if and only if } E_a \text{ has no CM.} \tag{3.4}$$

If $a \neq -1$, there is the isomorphism

$$H^2(X_a, \mathbb{Q})/\mathrm{NF}(X_a)_{\mathbb{Q}} \cong \mathrm{Sym}^2 H^1(E_a, \mathbb{Q}). \tag{3.5}$$

We now apply Theorem 1.1 for $\mathscr{D} = f^{-1}(1)$ and $X_{-1} = \pi^{-1}(-1)$. As we see in the above, D_η is a split multiplicative fiber of type I_2 and $D_{-1} := \mathscr{D} \times_{\mathscr{X}} X_{-1}$ is of type I_4, so the conditions (1) and (2) are satisfied. We obtain a higher Chow cycle $Z_a \in \mathrm{CH}^2(X_a, 1)$ arising from $f^{-1}(1)$, and this is $\mathbb{R}$-regulator indecomposable for a general a by Thm. 1.1 and (3.4). By (3.5), this gives a $\mathbb{R}$-regulator indecomposable cycle in a self product $E_a \times E_a$.

Remark 3.2 *The real regulator* $\mathrm{reg}(Z_a)$ *(as a function of a) is studied in detail in [3] §6.4.*

References

[1] Asakura, M.: *A formula for Beilinson's regulator map on K_1 of a fibration of curves having a totally degenerate semistable fiber*, http://arxiv.org/abs/1310.2810.

[2] Barth, W., Peters, C., Van de Ven, A. *Compact complex surfaces*. Springer-Verlag, Berlin.

[3] Chen, X., Doran, C., Kerr, M. and Lewis, J.: *Normal functions, Picard-Fuchs equations and elliptic fibrations on K3 surfaces*, http://arxiv.org/abs/1108.2223.

[4] Gordon, B. and Lewis, J.: *Indecomposable higher Chow cycles on products of elliptic curves*. J. Algebraic Geom. **8** (1999), no. 3, 543–567.

[5] Kerr, M.: *Indecomposable K_1 of elliptically fibered K3 surfaces: A tale of two cycles*, to appear in Fields Inst. Comm.

[6] Müller-Stach, S.: *Constructing indecomposable motivic cohomology classes on algebraic surfaces*, J. Algebraic Geom. **6** (1997), 513–543.

[7] Peters, C. and Steenbrink, J. *Mixed Hodge structures*. A Series of Modern Surveys in Mathematics, **52**. Springer-Verlag, Berlin, 2008.

[8] Silverman, J.: *The arithmetic of elliptic curves*. Grad. Texts in Math. **106**, New York, Springer 1986.

[9] Türkmen, Inan Utku.: *Regulator indecomposable cycles on a product of elliptic curves*. Canad. Math. Bull. **56** (2013), no. 3, 640–646.

Department of Mathematics, Hokkaido University, Sapporo 060-0810, JAPAN
asakura@math.sci.hokudai.ac.jp

8

A relative version of the Beilinson-Hodge conjecture

Rob de Jeu, James D. Lewis, and Deepam Patel

ABSTRACT. Let $k \subseteq \mathbb{C}$ be an algebraically closed subfield, and $\mathcal{X}$ a variety defined over k. One version of the Beilinson-Hodge conjecture that seems to survive scrutiny is the statement that the Betti cycle class map $\mathrm{cl}_{r,m} : H^{2r-m}_{\mathcal{M}}(k(\mathcal{X}),\mathbb{Q}(r)) \to \hom_{\mathrm{MHS}}\big(\mathbb{Q}(0), H^{2r-m}(k(\mathcal{X})(\mathbb{C}),\mathbb{Q}(r))\big)$ is surjective, that being equivalent to the Hodge conjecture in the case $m = 0$. Now consider a smooth and proper map $\rho : \mathcal{X} \to \mathcal{S}$ of smooth quasi-projective varieties over k, and where η is the generic point of $\mathcal{S}$. We anticipate that the corresponding cycle class map is surjective, and provide some evidence in support of this in the case where $\mathcal{X} = \mathcal{S} \times X$ is a product and $m = 1$.

1 Introduction

The results of this paper are aimed at providing some evidence in support of an affirmative answer to a question first formulated in [SJK-L, Question 1.1], now upgraded to the following:

Conjecture 1.1. *Let $\rho : \mathcal{X} \to \mathcal{S}$ be a smooth proper map of smooth quasi-projective varieties over a subfield $k = \overline{k} \subseteq \mathbb{C}$, with $\eta = \eta_{\mathcal{S}}$ the generic point of $\mathcal{S}/k$. Further, let $r, m \geq 0$ be integers. Then*

$$\mathrm{cl}_{r,m} : \mathrm{CH}^r(\mathcal{X}_\eta, m; \mathbb{Q}) = H^{2r-m}_{\mathcal{M}}(\mathcal{X}_\eta, \mathbb{Q}(r)) \to \hom_{\mathrm{MHS}}\big(\mathbb{Q}(0), H^{2r-m}(\mathcal{X}_\eta(\mathbb{C}),\mathbb{Q}(r))\big),$$

is surjective.

Here

$$H^{2r-m}(\mathcal{X}_\eta(\mathbb{C}),\mathbb{Q}(r)) := \lim_{\substack{\to \\ U \subset \mathcal{S}/k}} H^{2r-m}(\rho^{-1}(U)(\mathbb{C}),\mathbb{Q}(r)),$$

2000 *Mathematics Subject Classification.* Primary: 14C25, 19E15; Secondary: 14C30.

Key words and phrases. Chow group, algebraic cycle, regulator, Beilinson-Hodge conjecture, Abel-Jacobi map, Bloch-Beilinson filtration.

The authors are grateful to the Netherlands Organisation for Scientific Research (NWO) The second author also acknowledges partial support by a grant from the Natural Sciences and Engineering Research Council of Canada.

is a limit of mixed Hodge structures (MHS), for which one should not expect finite dimensionality, and for any smooth quasi-projective variety W/k, we identify motivic cohomology $H^{2r-m}_{\mathcal{M}}(W,\mathbb{Q}(r))$ with Bloch's higher Chow group $\mathrm{CH}^r(W,m;\mathbb{Q}) := \mathrm{CH}^r(W,m)\otimes\mathbb{Q}$ (see [Bl1]). Note that if $\mathcal{S} = \mathrm{Spec}(k)$, and $m = 0$, then $\mathcal{X} = X_k$ is smooth, projective over k. Thus in this case Conjecture 1.1 reduces to the (classical) Hodge conjecture. The motivation for this conjecture stems from the following:

Firstly, it is a generalization of a similar conjecture in [dJ-L, (§1, statement (S3)], where $\mathcal{X} = \mathcal{S}$, based on a generalization of the Hodge conjecture (classical form) to the higher K-groups, and inspired in part by Beilinson's work in this direction.

In passing, we hope to instill in the reader that any attempt to deduce Conjecture 1.1 from [dJ-L, §1, statement (S3)] seems to be hopelessly naive, and would require some new technology. To move ahead with this, we eventually work in the special situation where $\mathcal{X} = S\times X$ is a product, with $S = \mathcal{S}$ and X smooth projective, $m = 1$, and employ some motivic input, based on reasonable pre-existing conjectures.

Secondly, as a formal application of M. Saito's theory of mixed Hodge modules (see [A], [K-L], [SJK-L] and the references cited there), one could conceive of the following short exact sequence:

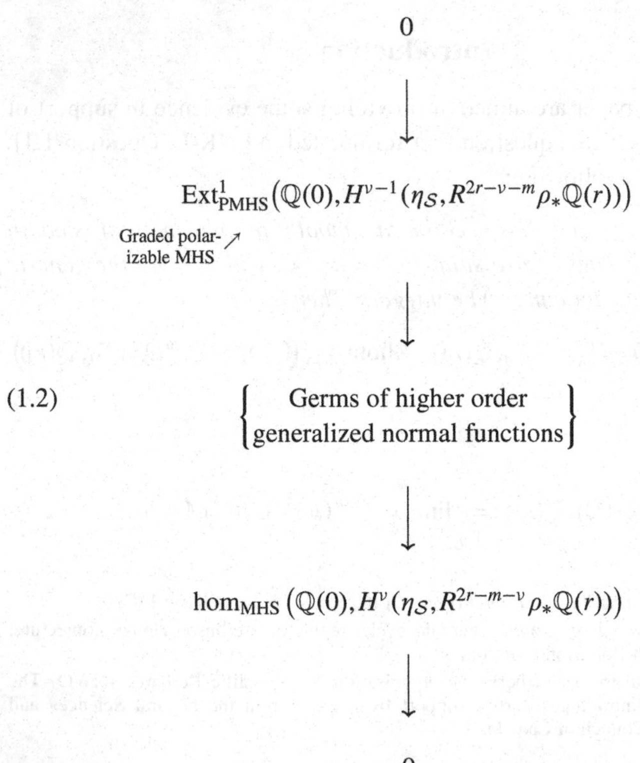

0
$\mathrm{Ext}^1_{\mathrm{PMHS}}\big(\mathbb{Q}(0), H^{\nu-1}(\eta_{\mathcal{S}}, R^{2r-\nu-m}\rho_*\mathbb{Q}(r))\big)$
Graded polar-izable MHS
(1.2)
Germs of higher order generalized normal functions
$\hom_{\mathrm{MHS}}\big(\mathbb{Q}(0), H^{\nu}(\eta_{\mathcal{S}}, R^{2r-m-\nu}\rho_*\mathbb{Q}(r))\big)$
0

(Warning: As mentioned earlier, passing to the generic point $\eta_{\mathcal{S}}$ of $\mathcal{S}$ is a limit process, which implies that the spaces above need not be finite dimensional over $\mathbb{Q}$. This particularly applies to the case $m \geq 1$, where there are residues.) The key point is, is there lurking a generalized Poincaré existence theorem for higher normal functions? Namely, modulo the "fixed part" $\mathrm{Ext}^1_{\mathrm{PMHS}}\big(\mathbb{Q}(0), H^{\nu-1}(\eta_{\mathcal{S}}, R^{2r-\nu-m}\rho_*\mathbb{Q}(r))\big)$, are these normal functions cycle-induced? In another direction, this diagram is related to a geometric description of the notion of a Bloch-Beilinson (BB) filtration. As a service to the reader, and to make sense of this all, we elaborate on all of this.

1. For the moment, let us replace ($\eta_{\mathcal{S}}$ by $\mathcal{S}$, (ν,m) by $(1,0)$ in diagram (1.2), and where $\mathcal{S}$ is chosen to be a curve). Then this diagram represents the schema of the original Griffiths program aimed at generalizing Lefschetz's famous $(1,1)$ theorem, via normal functions.[1] This program was aimed at solving the Hodge conjecture inductively. Unfortunately, the lack of a Jacobi inversion theorem for the Jacobian of a general smooth projective variety involving a Hodge structure of weight > 1 led to limited applications towards the Hodge conjecture. However the qualitative aspects of his program led to the non-triviality of the now regarded Griffiths group. In that regard, the aforementioned diagram represents a generalization of this idea to the higher K-groups of $\mathcal{X}$ and the general fibers of $\rho : \mathcal{X} \to \mathcal{S}$.

2. The notion of a BB filtration, first suggested by Bloch and later fortified by Beilinson, tells us that for any X/k smooth projective and $r, m \geq 0$, there should be a descending filtration

$$\{F^{\nu}\mathrm{CH}^r(X,m;\mathbb{Q}) \mid \nu = 0,\ldots,r\},$$

whose graded pieces can be described in terms of extension datum, viz.,

$$Gr_F^{\nu}\mathrm{CH}^r(X,m;\mathbb{Q}) \simeq \mathrm{Ext}^{\nu}_{\mathcal{MM}}(\mathrm{Spec}(k), h^{2r-\nu-m}(X)(r)),$$

where $\mathcal{MM}$ is a conjectural category of mixed motives and $h^{\bullet}(X)(\bullet)$ is motivic cohomology.[2] Although there were many excellent candidate BB filtrations proposed by others over the years, a few are derived from the point of view of "spreads", in the case $k = \overline{\mathbb{Q}}$ (see [A], [Lew1], [GG]) as well as a conjectural description in terms of normal functions (see [K-L], [Lew3]). Namely, if $X/\mathbb{C}$ is smooth and projective, then there is a field K of finite transcendence degree over $\overline{\mathbb{Q}}$ and a smooth and proper spread $\mathcal{X} \xrightarrow{\rho} \mathcal{S}$ of smooth quasi-projective varieties over $\overline{\mathbb{Q}}$, such that if η is the generic point of $\mathcal{S}$, then K can be identified with $\overline{\mathbb{Q}}(\eta)$ via a suitable embedding $\overline{\mathbb{Q}}(\eta) \hookrightarrow \mathbb{C}$;

[1] Technically speaking, Griffiths worked with normal functions that extended to the boundary $\overline{\mathcal{S}}\backslash\mathcal{S}$, but let's not go there.

[2] The original formulation involved only the case $m = 0$; this is just a natural extension of those ideas to the higher K-groups of X.

moreover with respect to that embedding, $X/\mathbb{C} = \mathcal{X}_\eta \times_{\overline{\mathbb{Q}(\eta)}} \mathbb{C}$. Diagram (1.2) then provides yet another schema of describing a candidate BB filtration in terms of normal functions.

As indicated earlier, we focus our attention mainly on the case $m = 1$ (K_1 case), and provide some partial results in the case where $\mathcal{X} = S \times X$ is a product, with $S = \mathcal{S}$, and X smooth projective. Our main results are Theorems 4.4, 6.7 and 6.11.

We wish the express our gratitude to the referee for the careful reading of this paper, and the various suggestions for improvements.

2 Notation

- $\bullet_0$ Unless specified to the contrary, all varieties are defined over $\mathbb{C}$.
- $\bullet_1$ $\mathbb{Q}(m)$ is the Tate twist with Hodge type $(-m,-m)$.
- $\bullet_2$ For a mixed Hodge structure (MHS) H over $\mathbb{Q}$, we put $\Gamma(H) = \hom_{\mathrm{MHS}}(\mathbb{Q}(0),H)$ and $J(H) = \mathrm{Ext}^1_{\mathrm{MHS}}(\mathbb{Q}(0),H)$.
- $\bullet_3$ The higher Chow groups $\mathrm{CH}^r(W,m)$ for a quasi-projective variety W over a field k are defined in [Bl1]. Let us assume W/k is regular. An abridged definition of $\mathrm{CH}^r(W,1)$, viz., in the case $m = 1$ is given by:

$$\mathrm{CH}^r(W,1) = \frac{\ker\left(\sum_{\mathrm{cd}_W Z=r-1}^{Z \text{ irred}} (k(Z)^\times, Z) \xrightarrow{\mathrm{div}} z^r(W)\right)}{\text{Image of tame symbol}},$$

where $z^r(W)$ is the free abelian group generated by (irreducible) subvarieties of codimension r in W; moreover, the denominator admits this description. If $V \subset W$ is an irreducible subvariety of codimension $r-2$, and $f,g \in k(V)^\times$, then the tame symbol is given as:

$$T(\{f,g\}_V) = \sum_{\mathrm{cd}_V D=1} (-1)^{\nu_D(f)\nu_D(g)} \left(\frac{f^{\nu_D(g)}}{g^{\nu_D(f)}}\right)_D,$$

as D ranges through all irreducible codimension one subvarieties of V, $\nu_D(\cdot)$ is the order of a zero or pole, and $(\cdots)_D$ means the restriction to the generic point of D. The "Image of the tame symbol" is the subgroup generated by $T(\{f,g\}_V)$, as V ranges in W and f,g range through $k(V)^\times$.

- $\bullet_4$ Assume W in $\bullet_3$ is also smooth of dimension d_W, and let $Z \subset W$ be irreducible of codimension $r-1$, $f \in k(W)^\times$), where $k \subset \mathbb{C}$ is a subfield. Then the Betti class map

$$\mathrm{cl}_{r,1} : \mathrm{CH}^r(W,1;\mathbb{Q}) \to \Gamma\left(H^{2r-1}(W,\mathbb{Q}(r))\right) \subset H^{2r-1}(W,\mathbb{Q}(r))$$
$$\simeq \left[H_c^{2d_W-2r+1}(W,\mathbb{Q}(d_W-r))\right]^\vee \subset \left[H_c^{2d_W-2r+1}(W,\mathbb{C})\right]^\vee$$

is induced by the current

$$(Z,f) \mapsto \frac{1}{(2\pi\mathrm{i})^{\mathrm{d_w - r + 1}}} \int_Z d\log(f) \wedge \omega, \quad \{\omega\} \in H_c^{2\mathrm{d_w}-2r+1}(W,\mathbb{C}).$$

3 What is known

In this section, we summarize some of the results in [SJK-L], where $r \geq m = 1$. The setting is the following diagram

$$\begin{array}{ccc} \mathcal{X} & \hookrightarrow & \overline{\mathcal{X}} \\ \downarrow \rho & & \downarrow \overline{\rho} \\ \mathcal{S} & \hookrightarrow & \overline{\mathcal{S}} \end{array}$$

where $\overline{\mathcal{X}}$ and $\overline{\mathcal{S}}$ are nonsingular complex projective varieties, $\overline{\rho}$ is a dominating flat morphism, $D \subset \overline{\mathcal{S}}$ a divisor, $\mathcal{Y} := \overline{\rho}^{-1}(D)$, $\mathcal{S} := \overline{\mathcal{S}} \backslash D$, $\mathcal{X} := \overline{\mathcal{X}} \backslash \mathcal{Y}$ and $\rho := \overline{\rho}|_{\mathcal{X}}$.[3] There is a short exact sequence

$$(3.1) \qquad 0 \to \frac{H^{2r-1}(\overline{\mathcal{X}},\mathbb{Q}(r))}{H_{\mathcal{Y}}^{2r-1}(\overline{\mathcal{X}},\mathbb{Q}(r))} \to H^{2r-1}(\mathcal{X},\mathbb{Q}(r)) \to H_{\mathcal{Y}}^{2r}(\overline{\mathcal{X}},\mathbb{Q}(r))^{\circ} \to 0,$$

where with regard to the former term in (3.1), $H_{\mathcal{Y}}^{2r-1}(\overline{\mathcal{X}},\mathbb{Q}(r))$ is identified with its image in $H^{2r-1}(\overline{\mathcal{X}},\mathbb{Q}(r))$, and $H_{\mathcal{Y}}^{2r}(\overline{\mathcal{X}},\mathbb{Q}(r))^{\circ} := \ker\big(H_{\mathcal{Y}}^{2r}(\overline{\mathcal{X}},\mathbb{Q}(r)) \to H^{2r}(\overline{\mathcal{X}},\mathbb{Q}(r))\big)$. One has a corresponding diagram

(3.2)

$$\begin{array}{ccccc} \mathrm{CH}^r(\mathcal{X},1;\mathbb{Q}) & \longrightarrow & \mathrm{CH}^r_{\mathcal{Y}}(\overline{\mathcal{X}};\mathbb{Q})^{\circ} & \xrightarrow{\alpha_{\mathcal{Y}}} & \mathrm{CH}^r_{\mathrm{hom}}(\overline{\mathcal{X}};\mathbb{Q}) \\ \downarrow \mathrm{cl}^{\mathcal{X}}_{r,1} & & \downarrow \beta_{\mathcal{Y}} & & \downarrow \underline{AJ}^{\overline{\mathcal{X}}} \\ \Gamma\big(H^{2r-1}(\mathcal{X},\mathbb{Q}(r))\big) & \hookrightarrow & \Gamma\big(H^{2r}_{\mathcal{Y}}(\overline{\mathcal{X}},\mathbb{Q}(r))^{\circ}\big) & \longrightarrow & J\left(\frac{H^{2r-1}(\overline{\mathcal{X}},\mathbb{Q}(r))}{H^{2r-1}_{\mathcal{Y}}(\overline{\mathcal{X}},\mathbb{Q}(r))}\right) \end{array}$$

well-known to commute by an extension class argument, and where $\underline{AJ}^{\overline{\mathcal{X}}}$ is the corresponding "reduced" Abel-Jacobi map. Further, the definition of $\mathrm{CH}^r_{\mathcal{Y}}(\overline{\mathcal{X}};\mathbb{Q})^{\circ}$ is the obvious one, being the cycles in $\mathrm{CH}^{r-1}(\mathcal{Y};\mathbb{Q})$ that are homologous to zero on $\overline{\mathcal{X}}$.

[3] While we assume that the base field is $\mathbb{C}$, the results here are valid for varieties over an algebraically closed field $k \subset \mathbb{C}$.

Remark 3.3. Poincaré duality gives an isomorphism of MHS:

$$H^{2r}_{\mathcal{Y}}(\overline{\mathcal{X}},\mathbb{Q}(r)) \simeq H_{2\dim\overline{\mathcal{X}}-2r}(\mathcal{Y},\mathbb{Q}) \otimes \mathbb{Q}(r-\dim\overline{\mathcal{X}}).$$

Thus $\ker\beta_{\mathcal{Y}} = \mathrm{CH}^{r-1}_{\mathrm{hom}}(\mathcal{Y};\mathbb{Q})$, viz., the subspace of cycles in $\mathrm{CH}^{r-1}(\mathcal{Y};\mathbb{Q})$ that are homologous to zero on $\mathcal{Y}$.

Let us assume that $\beta_{\mathcal{Y}}$ is surjective, as is the case if the (classical) Hodge conjecture holds. If we apply the snake lemma, we arrive at

$$\mathrm{coker}(\mathrm{cl}^{\mathcal{X}}_{r,1}) \simeq \frac{\ker\left[\underline{AJ^{\overline{\mathcal{X}}}}|_{\mathrm{Image}(\alpha_{\mathcal{Y}})} : \mathrm{Image}(\alpha_{\mathcal{Y}}) \to J\left(\frac{H^{2r-1}(\overline{\mathcal{X}},\mathbb{Q}(r))}{H^{2r-1}_{\mathcal{Y}}(\overline{\mathcal{X}},\mathbb{Q}(r))}\right)\right]}{\alpha_{\mathcal{Y}}\big(\ker(\beta_{\mathcal{Y}})\big)}.$$

Now take the limit over all $D \subset \overline{S}$ to arrive at an induced cycle map:

$$\mathrm{cl}^{\eta}_{r,1} : \mathrm{CH}^r(\mathcal{X}_\eta, 1;\mathbb{Q}) \to \Gamma\big(H^{2r-1}(\mathcal{X}_\eta,\mathbb{Q}(r))\big). \tag{3.4}$$

where η is the generic point of $\overline{S}$. We arrive at:

$$\frac{\Gamma\big(H^{2r-1}(\mathcal{X}_\eta,\mathbb{Q}(r))\big)}{\mathrm{cl}^{\eta}_{r,1}\big(\mathrm{CH}^r(\mathcal{X}_\eta,1;\mathbb{Q})\big)} \simeq \frac{\ker\left[\mathcal{K} \overset{AJ}{\to} J\left(\frac{H^{2r-1}(\overline{\mathcal{X}},\mathbb{Q}(r))}{N^1_{\overline{S}}H^{2r-1}(\overline{\mathcal{X}},\mathbb{Q}(r))}\right)\right]}{N^1_{\overline{S}}\mathrm{CH}^r(\overline{\mathcal{X}};\mathbb{Q})}, \tag{3.5}$$

where $\mathcal{K} := \ker[\mathrm{CH}^r_{\mathrm{hom}}(\overline{\mathcal{X}};\mathbb{Q}) \to \mathrm{CH}^r(\mathcal{X}_\eta;\mathbb{Q})]$, $N^q_{\overline{S}}\mathrm{CH}^r(\overline{\mathcal{X}}) \subseteq \mathrm{CH}^r_{\mathrm{hom}}(\overline{\mathcal{X}};\mathbb{Q})$ is the subspace consisting of classes $\{\xi\}$ such that there is a (pure) codimension q subscheme $D \subset \overline{S}$ together with a representative ξ of $\{\xi\}$, for which $|\xi| \subset \overline{\rho}^{-1}(D)$ and (ignoring twists) the fundamental class of ξ is zero in $H_{2\dim\overline{\mathcal{X}}-2r}(\overline{\rho}^{-1}(D),\mathbb{Q})$. (Compare this with the second sentence of Remark 3.3 when $q = 1$.) Finally, $N^q_{\overline{S}}H^{2r-1}(\overline{\mathcal{X}},\mathbb{Q}(r)) \subseteq H^{2r-1}(\overline{\mathcal{X}},\mathbb{Q}(r))$ is the union of the images of $H^{2r-1}_{\overline{\rho}^{-1}(D)}(\overline{\mathcal{X}},\mathbb{Q}(r)) \to H^{2r-1}(\overline{\mathcal{X}},\mathbb{Q}(r))$, where D runs through the (pure) codimension q subschemes of $\overline{S}$.

A relatively simple argument, found in [SJK-L], yields the following.

Proposition 3.6. *Under the assumption of the Hodge conjecture,* (3.5) *becomes:*

$$\frac{\Gamma\big(H^{2r-1}(\mathcal{X}_\eta,\mathbb{Q}(r))\big)}{\mathrm{cl}^{\eta}_{r,1}\big(\mathrm{CH}^r(\mathcal{X}_\eta,1;\mathbb{Q})\big)} \simeq \frac{N^1_{\overline{S}}\mathrm{CH}^r(\overline{\mathcal{X}};\mathbb{Q}) + \ker\big[\mathcal{K} \overset{AJ}{\to} J\big(H^{2r-1}(\overline{\mathcal{X}},\mathbb{Q}(r))\big)\big]}{N^1_{\overline{S}}\mathrm{CH}^r(\overline{\mathcal{X}};\mathbb{Q})}.$$

Example 3.7. *Suppose that* $\overline{\mathcal{X}} = \overline{S}$ *with* $\overline{\rho}$ *the identity. In this case Proposition 3.6 becomes:*

$$\frac{\Gamma\big(H^{2r-1}(\mathbb{C}(\overline{\mathcal{X}}),\mathbb{Q}(r))\big)}{\mathrm{cl}_{r,1}\big(\mathrm{CH}^r(\mathrm{Spec}(\mathbb{C}(\overline{\mathcal{X}})),1;\mathbb{Q})\big)} \simeq \frac{N^1\mathrm{CH}^r(\overline{\mathcal{X}};\mathbb{Q}) + \mathrm{CH}^r_{AJ}(\overline{\mathcal{X}};\mathbb{Q})}{N^1\mathrm{CH}^r(\overline{\mathcal{X}};\mathbb{Q})}, \tag{3.8}$$

where $N^1\mathrm{CH}^r(\overline{\mathcal{X}};\mathbb{Q})$ *is the subgroup of cycles, that are homologous to zero on codimension* 1 *subschemes of* $\overline{\mathcal{X}}$, *and* $\mathrm{CH}^r_{AJ}(\overline{\mathcal{X}};\mathbb{Q})$ *are cycles in the kernel of the Abel-Jacobi map* $AJ : \mathrm{CH}^r(\overline{\mathcal{X}};\mathbb{Q}) \to J\big(H^{2r-1}(\overline{\mathcal{X}},\mathbb{Q}(r))\big)$. *According to Jannsen [Ja1, p. 227], there is a discussion that strongly hints that the right hand side of* (3.8) *should be zero. In light of [Lew2], we conjecturally believe this to be true. In particular, since* $\mathrm{Spec}(\mathbb{C}(\overline{\mathcal{X}}))$ *is a point, this implies that* $\Gamma\big(H^{2r-1}(\mathbb{C}(\overline{\mathcal{X}}),\mathbb{Q}(r))\big) = 0$ *for* $r > 1$. *The reader can easily check that*

$$\mathrm{cl}_{r,1}\big(\mathrm{CH}^r(\mathrm{Spec}(\mathbb{C}(\overline{\mathcal{X}})),1;\mathbb{Q})\big) = \Gamma\big(H^{2r-1}(\mathbb{C}(\overline{\mathcal{X}}),\mathbb{Q}(r))\big),$$

holds unconditionally in the case $r = \dim\overline{\mathcal{X}}$, *that being well known in the case* $r = \dim\mathcal{X} = 1$, *and for* $r = \dim\mathcal{X} > 1$, *from the weak Lefschetz theorem for affine varieties.*

Example 3.9. [SJK-L] *Here we give some evidence that the RHS (hence LHS) of Proposition 3.6 is zero. Suppose* $\overline{\mathcal{X}} = X \times \overline{S}$, $(\overline{S} := \overline{\mathcal{S}})$, *and let us assume the condition*

$$\mathrm{CH}^r(\overline{\mathcal{X}};\mathbb{Q}) = \bigoplus_{\ell=0}^{r} \mathrm{CH}^\ell(\overline{S};\mathbb{Q}) \otimes \mathrm{CH}^{r-\ell}(X;\mathbb{Q}). \tag{3.10}$$

An example situation is when $\overline{S}$ *is a flag variety, such as a projective space; however conjecturally speaking, this condition is expected to hold for a much broader class of examples. If we further assume the Hodge conjecture, then as a consequence of Proposition 3.6, we arrive at:*

Corollary 3.11. *Under the assumptions of the Künneth condition in* (3.10) *above, and the Hodge conjecture, with* $\overline{\mathcal{X}} = \overline{S} \times X$, *if*

$$\mathrm{cl}_{\ell,1}\big(\mathrm{CH}^\ell(\mathrm{Spec}(\mathbb{C}(\overline{S})),1;\mathbb{Q})\big) = \Gamma\big(H^{2\ell-1}(\mathbb{C}(\overline{S}),\mathbb{Q}(\ell))\big),$$

holds for all $\ell \le r$, *then the map* $\mathrm{cl}^\eta_{r,1}$ *in* (3.4) *is surjective.*

4 The split case and rigidity

4.1. Base a curve

In this section, we observe that the Beilinson-Hodge conjecture (Conjecture 1.1), in the special case of a split projection with base given by a curve, holds under the assumption of the Hodge conjecture on the fibre. Let X be a smooth projective variety and C a smooth curve. Let $\pi : C \times X \to C$ denote the projection morphism.

Proposition 4.1. *Let* X *and* C *be as above.*

1. *If* $m > 1$, *then* $\mathrm{CH}^r(C \times X, m) \to \Gamma(H^{2r-m}(C \times X, \mathbb{Q}(r)))$ *is surjective.*
2. *If* $m = 1$, *then* $\mathrm{CH}^r(C \times X, m) \to \Gamma(H^{2r-m}(C \times X, \mathbb{Q}(r)))$ *is surjective if the Hodge conjecture holds for* X *in codimension* $r - 1$.

In particular, the Beilinson Hodge conjecture (Conjecture 1.1) for $\pi : C \times X \to C$ holds unconditionally if $m > 1$ and, if $m = 1$, then it holds under the assumption of the Hodge conjecture for X.

We begin with some preliminary reductions. By the Künneth decomposition we can identify $H^{2r-m}(C \times X, \mathbb{Q})$ with

$$\bigoplus_{i=0}^{2} H^i(C,\mathbb{Q}) \otimes H^{2r-m-i}(X,\mathbb{Q}).$$

For $i = 0$ and 2, $H^i(C,\mathbb{Q}) \otimes H^{2r-m-i}(X,\mathbb{Q})(r)$ is pure of weight $-m$ by the purity of $H^{2r-m-i}(X,\mathbb{Q})$ and that of $H^0(C,\mathbb{Q})$ as well as $H^2(C,\mathbb{Q})$. Therefore $\Gamma(H^i(C,\mathbb{Q}) \otimes H^{2r-m-i}(X,\mathbb{Q})(r)) = 0$ for $m > 0$ and $i \neq 1$. The same also holds for $i = 1$ if C is projective. But in general we have the following.

Lemma 4.2. *With notation and assumptions as above,*

$$\Gamma\big(H^1(C,\mathbb{Q}) \otimes H^{2r-2}(X,\mathbb{Q})(r)\big) = \Gamma(H^1(C,\mathbb{Q}(1))) \otimes \Gamma(H^{2r-2}(X,\mathbb{Q}(r-1))).$$

Proof. Let H denote $H^{2r-2}(X,\mathbb{Q}(r-1))$. Setting $W_j = W_j H^1(C,\mathbb{Q}(1))$ gives the following commutative diagram of mixed Hodge structures with exact rows:

$$\begin{array}{ccccccccc}
0 & \longrightarrow & W_{-1} \otimes H & \longrightarrow & W_0 \otimes H & \longrightarrow & Gr_0^W \otimes H & \longrightarrow & 0 \\
 & & \uparrow & & \uparrow & & \uparrow & & \\
0 & \longrightarrow & W_{-1} \otimes \Gamma(H) & \longrightarrow & W_0 \otimes \Gamma(H) & \longrightarrow & Gr_0^W \otimes \Gamma(H) & \longrightarrow & 0
\end{array}$$

Since $W_{-1} \otimes H$ and $W_{-1} \otimes \Gamma(H)$ have negative weights, their Γ's are trivial. It follows that we have a commutative diagram with exact rows:

$$\begin{array}{ccccccc}
0 & \longrightarrow & \Gamma(W_0 \otimes H) & \longrightarrow & \Gamma(Gr_0^W \otimes H) & \longrightarrow & J(W_{-1} \otimes H) \\
 & & \uparrow & & \uparrow & & \uparrow \\
0 & \longrightarrow & \Gamma(W_0 \otimes \Gamma(H)) & \longrightarrow & \Gamma(Gr_0^W \otimes \Gamma(H)) & \longrightarrow & J(W_{-1} \otimes \Gamma(H))
\end{array}$$

Since Gr_0^W is pure Tate of weight 0, the middle vertical is an isomorphism. We also have an injection $H^1(\overline{C},\mathbb{Q}) \to H^1(C,\mathbb{Q})$, identifying W_{-1} with $H^1(\overline{C},\mathbb{Q}(1))$. Using semi-simplicity of polarized Hodge structures we see that the natural injection

$$W_{-1} \otimes \Gamma(H) \to W_{-1} \otimes H$$

is split, so that we obtain a split injection

$$J(W_{-1} \otimes \Gamma(H)) \to J(W_{-1} \otimes H).$$

Finally, noting that $\Gamma(W_0 \otimes \Gamma(H)) = \Gamma(W_0) \otimes \Gamma(H)$ and using the snake lemma gives the desired result. □

Proof of Proposition 4.1. If $m > 1$, then by a weight argument

$$\Gamma(H^1(C,\mathbb{Q}) \otimes H^{2r-m-1}(X,\mathbb{Q})(r)) = 0$$

and the surjectivity is trivial. For r arbitrary, and $m = 1$, first note that $\mathrm{CH}^1(C,1)$ maps surjectively to $\Gamma(H^1(C,\mathbb{Q}(1)))$. Under the assumption of the Hodge conjecture for X in codimension $r-1$, one also has that

$$\mathrm{CH}^{r-1}(X,0;\mathbb{Q}) \to \Gamma\left(H^{2r-2}(X,\mathbb{Q}(r-1))\right)$$

is surjective. Since the natural morphism

$$\mathrm{CH}^1(C,1) \otimes \mathrm{CH}^{r-1}(X,0) \to \mathrm{CH}^r(C \times X,1)$$

induced by pullback to $C \times X$ followed by the cup product is compatible with the tensor product in the Künneth decomposition, the previous remarks and Lemma 4.2 show that $\mathrm{cl}_{r,1}$ surjects onto $\Gamma(H^1(C,\mathbb{Q}) \otimes H^{2r-2}(X,\mathbb{Q})(r))$. The last claim in the proposition follows from the first two by taking limits over open $U \subset C$.

□

Remark 4.3. Note that in the situation above, the surjectivity in Conjecture 1.1 holds for every open $U \subset C$, and, in particular, one does not need to pass to the generic point. However, in general, in the examples of [dJ-L, Section 5] one has to.

4.2. Base a product of two curves

In this section, we prove the strong form (i.e., without passing to the generic point) of the Beilinson-Hodge conjecture (Conjecture 1.1) for $r = 2$ and $m = 1$ in the special case of a split projection with base given by a product of two curves, under a certain rigidity assumption (see below). More precisely, let X be smooth projective, and $\overline{C}_1$, $\overline{C}_2$ smooth projective curves, with non-empty open $C_j \subsetneq \overline{C}_j$. Let $\overline{S} = \overline{C}_1 \times \overline{C}_2$, $S = C_1 \times C_2$, $\Sigma_j = \overline{C}_j \backslash C_j$, and $E = \overline{S} \backslash S = \Sigma_1 \times \overline{C}_2 \cup \overline{C}_1 \times \Sigma_2$. Finally, let $\mathcal{X} = S \times X$ and let $\pi : \mathcal{X} \to S$ denote the canonical projection map.

Theorem 4.4. *Let X and S be as above. If $H^2(\overline{S},\mathbb{Q})$ does not have a non-zero $\mathbb{Q}$ subHodge structure contained in $H^{2,0}(\overline{S}) \oplus H^{0,2}(\overline{S})$, then*

$$\mathrm{cl}_{2,1} : \mathrm{CH}^2(S \times X,1;\mathbb{Q}) \to \Gamma\left(H^3(S \times X,\mathbb{Q}(2))\right)$$

is surjective.

Note that if we were to replace S with $C_1 \times \overline{C}_2$ or $S = \overline{C}_1 \times C_2$, then the result is already part of Proposition 4.1. Again this holds even if the C_i are complete by reduction to the case of one curve in the base.

We begin with some preliminary reductions. First observe that, as $\dim S = 2$, by the weak Lefschetz theorem for affine varieties, the Künneth decomposition of $H^3(S \times X, \mathbb{Q})$ is

$$H^0(S,\mathbb{Q}) \otimes H^3(X,\mathbb{Q}) \oplus H^1(S,\mathbb{Q}) \otimes H^2(X,\mathbb{Q}) \oplus H^2(S,\mathbb{Q}) \otimes H^1(X,\mathbb{Q}),$$

and we shall deal with the three summands separately (after twisting with $\mathbb{Q}(2)$).

For the first term, note that $H^0(S,\mathbb{Q}) \otimes H^3(X,\mathbb{Q})(2)$ is pure of weight -1 by the purity of $H^3(X,\mathbb{Q})$, so that $\Gamma(H^0(S,\mathbb{Q}) \otimes H^3(X,\mathbb{Q})(2)) = 0$.

Lemma 4.5. *With notation as above,*

$$\Gamma\big(H^1(S,\mathbb{Q}) \otimes H^2(X,\mathbb{Q})(2)\big) = \Gamma(H^1(S,\mathbb{Q}(1))) \otimes \Gamma\big(H^2(X,\mathbb{Q}(1))\big)$$

and the image of

$$\mathrm{cl}_{2,1} : \mathrm{CH}^2(S \times X, 1; \mathbb{Q}) \to \Gamma\big(H^3(S \times X, \mathbb{Q}(2))\big)$$

contains the direct summand $\Gamma\big(H^1(S,\mathbb{Q}) \otimes H^2(X,\mathbb{Q})(2)\big)$.

Proof. The equality is obtained as in the proof of Lemma 4.2. Furthermore, the proof of the surjectivity of $\mathrm{cl}_{2,1}$ is similar to the proof of Proposition 4.1, noting that the remarks about the compatibility of the cup product of pullbacks along the projections $S \times X \to S$ and $S \times X \to X$ is compatible with the tensor product in the Künneth decomposition yield that the image of the elements we obtain are in $\Gamma\big(H^1(S,\mathbb{Q}) \otimes H^2(X,\mathbb{Q})(2)\big)$. We leave the details to the reader. □

Before proving Theorem 4.4, we note that by Lemma 4.5 and the two paragraphs preceding it, it now suffices to show that the image of $\mathrm{cl}_{2,1}$ followed by the projection $H^3(S \times X, \mathbb{Q}) \to H^2(S,\mathbb{Q}) \otimes H^1(X,\mathbb{Q})$ contains $\Gamma\big(H^2(S,\mathbb{Q}) \otimes H^1(X,\mathbb{Q})(2)\big)$. The following lemma describes this last space in terms of the Abel-Jacobi map. First put

$$H^3_E(\overline{S},\mathbb{Q})^\circ = \ker\big[H^3_E(\overline{S},\mathbb{Q}) \to H^3(\overline{S},\mathbb{Q})\big].$$

Lemma 4.6. *One has these identifications:*
(i) $\Gamma\big(H^2(S,\mathbb{Q}) \otimes H^1(X,\mathbb{Q})(2)\big) \simeq$

$$\ker\big[\Gamma\,(H^3_E(\overline{S},\mathbb{Q}(2))^\circ \otimes H^1(X,\mathbb{Q})) \to J(H^1(\overline{C}_1,\mathbb{Q}(1)) \otimes H^1(\overline{C}_2,\mathbb{Q}(1)) \otimes H^1(X,\mathbb{Q}))\big].$$

(ii) $\Gamma\big(H^3_E(\overline{S},\mathbb{Q}(2))^\circ \otimes H^1(X,\mathbb{Q})\big) \simeq$

$$\Gamma\Big(\big[H^1(\overline{C}_1,\mathbb{Q})\otimes H^0_{\deg 0}(\Sigma_2,\mathbb{Q})\bigoplus H^0_{\deg 0}(\Sigma_1,\mathbb{Q})\otimes H^1(\overline{C}_2,\mathbb{Q})\big]\otimes H^1(X,\mathbb{Q}(1))\Big).$$

Proof. Part (i): Observe that

$$\frac{H^2(\overline{S},\mathbb{Q})}{H^2_E(\overline{S},\mathbb{Q})} = H^1(\overline{C}_1,\mathbb{Q})\otimes H^1(\overline{C}_2,\mathbb{Q}), \quad H^2(S,\mathbb{Q}) = H^1(C_1,\mathbb{Q})\otimes H^1(C_2,\mathbb{Q}),$$

as $H^2(C_j) = 0$. There is a short exact sequence:

$$0 \to \frac{H^2(\overline{S},\mathbb{Q})}{H^2_E(\overline{S},\mathbb{Q})} \to H^2(S,\mathbb{Q}) \to H^3_E(\overline{S},\mathbb{Q})^\circ \to 0.$$

This in turn gives rise to a short exact sequence:

$$0 \to H^1(\overline{C}_1,\mathbb{Q}(1))\otimes H^1(\overline{C}_2,\mathbb{Q}(1))\otimes H^1(X,\mathbb{Q}) \to$$

$$H^1(C_1,\mathbb{Q}(1))\otimes H^1(C_2,\mathbb{Q}(1))\otimes H^1(X,\mathbb{Q}) \to H^3_E(\overline{S},\mathbb{Q}(2))^\circ \otimes H^1(X,\mathbb{Q}) \to 0.$$

So, using purity,

$$\Gamma\big(H^1(C_1,\mathbb{Q}(1))\otimes H^1(C_2,\mathbb{Q}(1))\otimes H^1(X,\mathbb{Q})\big)$$

can be identified with

$$\ker\big[\Gamma\big(H^3_E(\overline{S},\mathbb{Q}(2))^\circ\otimes H^1(X,\mathbb{Q})\big)\to J\big(H^1(\overline{C}_1,\mathbb{Q}(1))\otimes H^1(\overline{C}_2,\mathbb{Q}(1))\otimes H^1(X,\mathbb{Q})\big)\big].$$

Part (ii): Poincaré duality gives an isomorphism of MHS

(4.7)

$$H^3_E(\overline{S},\mathbb{Q}(2)) \simeq H_1(E,\mathbb{Q}), \text{ hence } H^3_E(\overline{S},\mathbb{Q}(2))^\circ \simeq \ker\big(H_1(E,\mathbb{Q}) \to H_1(\overline{S},\mathbb{Q})\big).$$

Moreover the Mayer-Vietoris sequence gives us the exact sequence

$$0 \to \big[H_1(\overline{C}_1\times\Sigma_2,\mathbb{Q})\oplus H_1(\Sigma_1\times\overline{C}_2,\mathbb{Q})\big]\otimes H^1(X,\mathbb{Q})$$
$$\to H_1(E,\mathbb{Q})\otimes H^1(X,\mathbb{Q}) \to H_0(\Sigma_1\times\Sigma_2,\mathbb{Q})\otimes H^1(X,\mathbb{Q}).$$

But $\Gamma\big(H_0(\Sigma_1\times\Sigma_2,\mathbb{Q})\otimes H^1(X,\mathbb{Q})\big) = 0$; moreover one has a commutative diagram

$$\begin{array}{ccc}
H_1(\overline{C}_1\times\Sigma_2,\mathbb{Q})\oplus H_1(\Sigma_1\times\overline{C}_2,\mathbb{Q}) & \hookrightarrow & H_1(E,\mathbb{Q}) \\
\| & & \\
H_1(\overline{C}_1,\mathbb{Q})\otimes H_0(\Sigma_2,\mathbb{Q})\oplus H_0(\Sigma_1,\mathbb{Q})\otimes H_1(\overline{C}_2,\mathbb{Q}) & & \downarrow \\
\downarrow & & \\
H_1(\overline{C}_1,\mathbb{Q})\otimes H_0(\overline{C}_2,\mathbb{Q})\oplus H_0(\overline{C}_1,\mathbb{Q})\otimes H_1(\overline{C}_2,\mathbb{Q}) & \simeq & H_1(\overline{S},\mathbb{Q}).
\end{array}$$

Hence from this and (4.7), $\Gamma\big(H^3_E(\overline{S},\mathbb{Q}(2))^\circ \otimes H^1(X,\mathbb{Q})\big)$ can be identified with

$$\Gamma\Big(\big[H^1(\overline{C}_1,\mathbb{Q})\otimes H^0_{\deg 0}(\Sigma_2,\mathbb{Q})\bigoplus H^0_{\deg 0}(\Sigma_1,\mathbb{Q})\otimes H^1(\overline{C}_2,\mathbb{Q})\big]\otimes H^1(X,\mathbb{Q}(1))\Big).$$

□

Proof of Theorem 4.4. Note that by the Lefschetz (1,1) theorem, $\Gamma\big(H^1(\overline{C}_j,\mathbb{Q})\otimes H^1(X,\mathbb{Q})(1)\big)$ is algebraic. Let us assume for the moment that there exists B in $\Gamma\big(H^0_{\deg 0}(\Sigma_1,\mathbb{Q}) \otimes H^1(\overline{C}_2,\mathbb{Q}) \otimes H^1(X,\mathbb{Q}(1))\big)$ of the form $B = \xi \times D$, where $D \subset \overline{C}_2 \times X$ is an irreducible curve, $\xi \in H^0_{\deg 0}(\Sigma_1,\mathbb{Q})$, and that B is in the kernel of the Abel-Jacobi map in Lemma 4.6(i). Notice that the inclusion $H^1(\overline{C}_1,\mathbb{Q}(1))\otimes\mathbb{Q}(1)\cdot[D] \hookrightarrow H^1(\overline{C}_1,\mathbb{Q}(1))\otimes H^1(\overline{C}_2,\mathbb{Q}(1))\otimes H^1(X,\mathbb{Q})$ defines a splitting, and hence an inclusion

$$J\big(H^1(\overline{C}_1,\mathbb{Q}(1))\big) \hookrightarrow J\big(H^1(\overline{C}_1,\mathbb{Q}(1))\otimes H^1(\overline{C}_2,\mathbb{Q}(1))\otimes H^1(X,\mathbb{Q})\big).$$

By applying Abel's theorem to $\overline{C}_1$, it follows that there exists $f \in \mathbb{C}(\overline{C}_1 \times D)^\times$ for which $(f) = \xi \times D = B$, thus supplying the necessary element in $\mathrm{CH}^2(S \times X,1;\mathbb{Q})$. The same story holds if we replace D by any divisor with non-trivial image in the Neron-Severi group. Using a basis for the Neron-Severi group of $\overline{C}_2 \times X$, one sees that the kernel of the Abel-Jacobi map restricted to $\Gamma\big(H^0_{\deg 0}(\Sigma_1,\mathbb{Q})\otimes H^1(\overline{C}_2,\mathbb{Q})\otimes H^1(X,\mathbb{Q}(1))\big)$ is in the image of $\mathrm{cl}_{2,1}$. A similar story holds separately for A in $\Gamma\big(H^1(\overline{C}_1,\mathbb{Q})\otimes H^0_{\deg 0}(\Sigma_2,\mathbb{Q})\otimes H^1(X,\mathbb{Q}(1))\big)$ in the kernel of the Abel-Jacobi map.

The more complicated issue is the case where $A+B$ in

$$\Gamma\Big(\big[H^1(\overline{C}_1,\mathbb{Q})\otimes H^0_{\deg 0}(\Sigma_2,\mathbb{Q})\bigoplus H^0_{\deg 0}(\Sigma_1,\mathbb{Q})\otimes H^1(\overline{C}_2,\mathbb{Q})\big]\otimes H^1(X,\mathbb{Q}(1))\Big)$$

is in the kernel of the Abel-Jacobi map. The problem boils down to the following. There are two subHodge structures V_1, V_2 of $H^1(\overline{C}_1,\mathbb{Q}(1)) \otimes H^1(\overline{C}_2,\mathbb{Q}(1)) \otimes H^1(X,\mathbb{Q})$, where $V_1 = H^1(\overline{C}_1,\mathbb{Q}(1)) \otimes \Gamma\big(H^1(\overline{C}_2,\mathbb{Q}(1)) \otimes H^1(X,\mathbb{Q})\big)$, and $V_2 \simeq H^1(\overline{C}_2,\mathbb{Q}(1))\otimes\Gamma\big(H^1(\overline{C}_1,\mathbb{Q}(1))\otimes H^1(X,\mathbb{Q})\big)$ is defined similarly. If their intersection V is trivial, then

$$J(V_1)\oplus J(V_2) \hookrightarrow J\big(H^1(\overline{C}_1,\mathbb{Q}(1))\otimes H^1(\overline{C}_2,\mathbb{Q}(1))\otimes H^1(X,\mathbb{Q})\big),$$

so $A+B$ in the kernel of the Abel-Jacobi map implies that A, B are in the kernel, and from our earlier discussion it follows that then

$$\mathrm{cl}_{2,1} : \mathrm{CH}^2(S\times X,1;\mathbb{Q}) \to \Gamma H^3(S\times X,\mathbb{Q}(2))$$

is surjective. If V is non-trivial, then from types we see that $V(-2)$ is contained in

$$\{H^{1,0}(\overline{C}_1)\otimes H^{1,0}(\overline{C}_2)\otimes H^{0,1}(X)\}\bigoplus\{H^{0,1}(\overline{C}_1)\otimes H^{0,1}(\overline{C}_2)\otimes H^{1,0}(X)\}$$

inside $H^2(\overline{S},\mathbb{Q}) \otimes H^1(X,\mathbb{Q})$. Tensoring this with $H^{2d-1}(X,\mathbb{Q}(d))$, where $d = \dim X$, and applying the cup product,

$$H^1(X,\mathbb{Q}) \times H^{2d-1}(X,\mathbb{Q}(d)) \xrightarrow{\cup} H^{2d}(X,\mathbb{Q}(d)),$$

followed by the identification $H^{2d}(X,\mathbb{Q}(d)) \simeq \mathbb{Q}(0)$, we find that $V(-2)$ results in a non-trivial $\mathbb{Q}$-subHodge structure of $H^2(\overline{S},\mathbb{Q})$ contained in $H^{2,0}(\overline{S}) \bigoplus H^{0,2}(\overline{S})$. □

We conclude this section with some discussion of the rigidity condition appearing in Theorem 4.4.

Example 4.8. *If $\overline{C}_1$, $\overline{C}_2$ are elliptic curves, then the existence of a $\mathbb{Q}$-dimension 2 Hodge structure $V \subset H^{2,0}(\overline{C}_1 \times \overline{C}_2) \oplus H^{0,2}(\overline{C}_1 \times \overline{C}_2)$ is equivalent to that the Néron-Severi group of $\overline{C}_1 \times \overline{C}_2$, has the maximal rank 4, which is well known to be equivalent to $\overline{C}_1$ and $\overline{C}_2$ being isogenous and of CM type by the (same) imaginary quadratic field ([H](IV, Thm. 4.19)).*

This example leads to:

Question 4.9. *Let W be a smooth projective surface. If $H^{2,0}(W) \oplus H^{0,2}(W)$ is the complexification of a non-zero $\mathbb{Q}$-subHodge structure of $H^2(W,\mathbb{Q})$, can W be obtained by base extension from a surface defined over $\overline{\mathbb{Q}}$?*

Proposition 4.10. *Suppose X is a K3 surface or an abelian surface. Then the answer to the previous question is positive.*

Proof. In both cases, $H^{0,2}$ is one dimensional, so the assumption implies that $H^{2,0}(W) \oplus H^{0,2}(W)$ arises from a $\mathbb{Q}$-subHodge structure of $H^2(W,\mathbb{Q})$. Since the former is a Hodge structure of type $(1,0,1)$, it follows that its Mumford-Tate group is abelian. In the case of an abelian surface, this implies that the Mumford-Tate group of the abelian variety is abelian, and therefore the abelian surface has complex multiplication. On the other hand, every CM abelian variety over $\mathbb{C}$ is defined over number field. If W is a $K3$ surface, then it has maximal Picard rank, hence is well-known to be rigid, *a fortiori,* defined over $\overline{\mathbb{Q}}$. □

Remark 4.11. In spite of the mild "rigidity" assumption in Theorem 4.4, any attempt to extend the theorem to the generic point of S, without introducing some conjectural assumptions, seems incredibly difficult.

5 Generalities

Before proceeding further, and to be able to move further ahead, we explain some necessary assumptions. *For the remainder of this paper (with the occasional reminder), we assume the following:*

Assumptions 5.1. (i) The Hodge conjecture.
(ii) The Bloch-Beilinson conjecture on the injectivity of the Abel-Jacobi map for Chow groups of smooth projective varieties defined over $\overline{\mathbb{Q}}$ (see [Lew2, Conj 3.1 and §4]).

To spare the reader a time consuming search of multiple sources by many others, we refer mostly to [Lew2], for all the necessary statements and details. Let $Z/\mathbb{C}$ be any smooth projective variety of dimension d_0, $r \geq 0$, and put

$$\mathrm{CH}^r_{AJ}(Z;\mathbb{Q}) = \ker\big(AJ : \mathrm{CH}^r_{\mathrm{hom}}(Z;\mathbb{Q}) \to J\big(H^{2r-1}(Z,\mathbb{Q}(r))\big).$$

As mentioned in §1, we recall the notion of a descending Bloch-Beilinson (BB) filtration $\{F^\nu \mathrm{CH}^r(Z;\mathbb{Q})\}_{\nu\geq 0}$, with $F^0\mathrm{CH}^r(Z;\mathbb{Q}) = \mathrm{CH}^r(Z;\mathbb{Q})$, $F^1\mathrm{CH}^r(Z;\mathbb{Q}) = \mathrm{CH}^r_{\mathrm{hom}}(Z;\mathbb{Q})$, $F^{r+1}\mathrm{CH}^r(Z;\mathbb{Q}) = 0$, and satisfying a number of properties codified for example in [Ja2, §11], [Lew2, §4]. There is also the explicit construction of a candidate BB filtration by Murre, based on a conjectured Chow-Künneth decomposition, and subsequent conjectures in [M], which is equivalent to the existence of a BB filtration as formulated in [Ja2, §11]). Further, Jannsen also proved that the BB filtration is unique if it exists. The construction of the filtration in [Lew1] (and used in [Lew2]) relies on Assumptions 5.1, which if true, provides the existence of a BB filtration, and hence is the same filtration as Murre's, by the aforementioned uniqueness.

All candidate filtrations seem to show that $F^2\mathrm{CH}^r(Z;\mathbb{Q}) \subseteq \mathrm{CH}^r_{AJ}(Z;\mathbb{Q})$. The following is considered highly non-trivial:

Conjecture 5.2. $\mathrm{CH}^r_{AJ}(Z;\mathbb{Q}) = F^2\mathrm{CH}^r(Z;\mathbb{Q})$.

In light of Assumptions 5.1, this is equivalent to the surjectivity of

$$\mathrm{cl}_{r,1} : \mathrm{CH}^r(\mathrm{Spec}(\mathbb{C}(Z)),1;\mathbb{Q}) \twoheadrightarrow \Gamma\big(H^{2r-1}(\mathbb{C}(Z),\mathbb{Q}(r))\big)$$

as in [dJ-L, (S3)], provided both statements apply to all smooth projective $Z/\mathbb{C}$. For a proof, see [Lew2, Thm 1.1].

One of the key properties of the BB filtration is the factorization of graded pieces of that filtration through the Grothendieck motive. Let Δ_Z in $\mathrm{CH}^{d_0}(Z\times Z)$ be the diagonal class, with cohomology class $[\Delta_Z]$ in $H^{2d_0}(Z\times Z,\mathbb{Z}(d_0))$. Write $\Delta_Z = \sum_{p+q=2d_0} \Delta_Z(p,q)$ in $\mathrm{CH}^{d_0}(Z\times Z;\mathbb{Q})$ such that

$$\bigoplus_{p+q=2d_0} [\Delta_Z(p,q)] \in \bigoplus_{p+q=2d_0} H^p(Z,\mathbb{Q})\otimes H^q(Z,\mathbb{Q})(d_0) = H^{2d_0}(Z\times Z,\mathbb{Q}(d_0)).$$

is the Künneth decomposition of $[\Delta_Z]$. Then

$$\Delta_Z(p,q)_*\big|_{Gr^\nu_F \mathrm{CH}^r(Z;\mathbb{Q})}$$

is independent of the choice of $\Delta_Z(p,q)$. (This is essentially due to the fact that $F^1\mathrm{CH}^r(Z;\mathbb{Q}) = \mathrm{CH}^r_{\mathrm{hom}}(Z;\mathbb{Q})$, and functoriality properties of the BB filtration.) Furthermore,

$$(5.3) \qquad \Delta_Z(2d_0 - 2r + \ell, 2r - \ell)_*\big|_{Gr_F^\nu \mathrm{CH}^r(Z;\mathbb{Q})} = \delta_{\ell,\nu}\mathrm{id}_{Gr_F^\nu \mathrm{CH}^r(Z;\mathbb{Q})}$$

with $\delta_{i,j}$ the Kronecker delta. Consequently,

$$(5.4) \qquad \Delta_Z(2d_0 - 2r + \nu, 2r - \nu)_*\mathrm{CH}^r(Z;\mathbb{Q}) \simeq Gr_F^\nu \mathrm{CH}^r(Z;\mathbb{Q}),$$

and accordingly there is a non-canonical decomposition

$$\mathrm{CH}^r(Z;\mathbb{Q}) = \bigoplus_{\nu=0}^{r} \Delta_Z(2d_0 - 2r + \nu, 2r - \nu)_*\mathrm{CH}^r(Z;\mathbb{Q}).$$

In summary, we will view the kernel of the Abel-Jacobi map in terms of $Gr_F^\nu \mathrm{CH}^r(Z;\mathbb{Q})$ for $\nu \geq 2$, i.e., under Conjecture 5.2,

$$\mathrm{CH}^r_{AJ}(Z;\mathbb{Q}) = F^2\mathrm{CH}^r(Z;\mathbb{Q}) \simeq \bigoplus_{\nu=2}^{r} Gr_F^\nu \mathrm{CH}^r(Z;\mathbb{Q}),$$

(to re-iterate, non-canonically).

Remark 5.5. Murre's Chow-Künneth decomposition [M] is by definition a decomposition of the diagonal class

$$\Delta_Z = \sum_{p+q=2d_0} \Delta_Z(p,q) \text{ in } \mathrm{CH}^{d_0}(Z \times Z;\mathbb{Q}),$$

where the $\Delta_Z(p,q)$'s are commuting, pairwise orthogonal idempotents. By Beilinson and Jannsen, such a lift from a corresponding decomposition on cohomology is possible if $\mathrm{CH}^{d_0}_{\mathrm{hom}}(Z \times Z;\mathbb{Q})$ is a nilpotent ideal under composition, which is a consequence of Assumptions 5.1. It should be pointed out that such $\Delta_Z(p,q)$'s are still not unique.

The following will play an important role in Section 6.

Proposition 5.6. *Under Assumptions 5.1, the map*

$$\Xi_{Z,*} := \bigoplus_{\nu=2}^{r} \Delta_Z(2d_0 - 2r + \nu, 2r - \nu)_* : F^2\mathrm{CH}^r(Z;\mathbb{Q}) \to F^2\mathrm{CH}^r(Z;\mathbb{Q}),$$

is an isomorphism. Moreover, if the $\Delta_Z(p,q)$'s are chosen as in Murre's Chow-Künneth decomposition (viz., in Remark 5.5), then it is the identity.

Proof. If $r = 2$, this is obvious, as $F^r\mathrm{CH}^r(Z;\mathbb{Q}) = Gr_F^r\mathrm{CH}^r(Z;\mathbb{Q})$. So assume $r > 2$. The 5-lemma, together with the diagram

$$\begin{array}{ccccccccc} & & F^r\mathrm{CH}^r(Z;\mathbb{Q}) & & & & & & \\ & & \| & & & & & & \\ 0 & \to & Gr_F^r\mathrm{CH}^r(Z;\mathbb{Q}) & \to & F^{r-1}\mathrm{CH}^r(Z;\mathbb{Q}) & \to & Gr_F^{r-1}\mathrm{CH}^r(Z;\mathbb{Q}) & \to & 0 \\ & & \Xi_{Z,*}\downarrow \| & & \Xi_{Z,*}\downarrow & & \Xi_{Z,*}\downarrow \| & & \\ 0 & \to & Gr_F^r\mathrm{CH}^r(Z;\mathbb{Q}) & \to & F^{r-1}\mathrm{CH}^r(Z;\mathbb{Q}) & \to & Gr_F^{r-1}\mathrm{CH}^r(Z;\mathbb{Q}) & \to & 0 \end{array}$$

tells us that the middle vertical arrow is an isomorphism. By an inductive-recursive argument, we arrive at another 5-lemma argument:

$$\begin{array}{ccccccccc} 0 & \to & F^3\mathrm{CH}^r(Z;\mathbb{Q}) & \to & F^2\mathrm{CH}^r(Z;\mathbb{Q}) & \to & Gr_F^2\mathrm{CH}^r(Z;\mathbb{Q}) & \to & 0 \\ & & \Xi_{Z,*}\downarrow \wr & & \Xi_{Z,*}\downarrow & & \Xi_{Z,*}\downarrow \| & & \\ 0 & \to & F^3\mathrm{CH}^r(Z;\mathbb{Q}) & \to & F^2\mathrm{CH}^r(Z;\mathbb{Q}) & \to & Gr_F^2\mathrm{CH}^r(Z;\mathbb{Q}) & \to & 0 \end{array}$$

which implies the isomorphism in the proposition.

For the second statement, clearly $\Delta_Z = \oplus_{\ell=2r-2d_0}^{2r} \Delta_Z(2d_0 - 2r + \ell, 2r - \ell)$ induces the identity. But from (5.4) we see that the terms with $\ell \geq r+1$ do not contribute because $Gr_F^\ell\mathrm{CH}^r(Z;\mathbb{Q}) = 0$. Also, from (5.3) we find that if $\ell \neq \nu$, then $\Delta_Z(2d_0 - 2r + \ell, 2r - \ell)$ maps $F^\nu\mathrm{CH}^r(Z;\mathbb{Q})$ into $F^{\nu+1}\mathrm{CH}^r(Z;\mathbb{Q})$. But if $\ell < \nu$ and this correspondence is idempotent then we can iterate this, finding it kills $F^\nu\mathrm{CH}^r(Z;\mathbb{Q})$ as $F^{r+1}\mathrm{CH}^r(Z;\mathbb{Q}) = 0$. Taking $\nu = 2$, we are done. □

So to understand more about $\mathrm{CH}^r_{AJ}(Z;\mathbb{Q})$, it makes sense to study

$$\Delta_Z(2d_0 - 2r + \nu, 2r - \nu)_*\mathrm{CH}^r_{AJ}(Z;\mathbb{Q}),$$

for $2 \leq \nu \leq r$.

6 Main results

In this section, we will be assuming Assumptions 5.1 as well as Conjecture 5.2. *We shall also assume that we are working with Murre's Chow-Künneth decomposition.* Furthermore, $\overline{S}$ and X are assumed smooth and projective, with $\dim \overline{S} = N$, $\dim X = d$. Then we have the results of Section 5 for $Z = \overline{S} \times X$, with dimension $d_0 = N + d$.

Remark 6.1. This remark is critical to understanding our approach to the main results in the remainder of this paper. We want to take some earlier results, in particular Proposition 5.6, one step further. Let us assume the notation and setting in Proposition 3.6, for $\overline{\mathcal{X}} = \overline{S} \times X$, with $\mathcal{K}$ as in (3.5). Our goal is to show that the RHS of Proposition 3.6 is zero; in particular, that

$$\Xi_0 := \ker\big(AJ : \mathcal{K} \to J\big(H^{2r-1}(\overline{S} \times X, \mathbb{Q}(r))\big)\big)$$

is contained in $N^1_{\overline{S}}\mathrm{CH}^r(\overline{S} \times X; \mathbb{Q}(r))$. Proposition 5.6 shows that the induced map

$$(6.2) \qquad \bigoplus_{\nu=2}^{r} \Delta_{\overline{S}\times X}(2(N+d) - 2r + \nu, 2r - \nu)_* : \Xi_0 \to \Xi_0,$$

is the identity.

Below we only consider $\ell \leq N$ since ultimately we will be passing to the generic point $\eta_{\overline{S}}$ of $\overline{S}$, hence through an affine $S \subset \overline{S}$, where we apply the affine weak Lefschetz theorem, with $D = \overline{S} \backslash S$. Although not needed, we could also take $\ell \geq 2$ because the cases $\ell = 0, 1$ can be proved as in the proof of Proposition 4.1(2).

Note that

$$(6.3) \;\; \Gamma\big(H^{2r-1}(S \times X, \mathbb{Q}(r))\big) = \bigoplus_{\ell=1}^{N} \Gamma\left(H^{\ell}(S, \mathbb{Q}(1)) \otimes H^{2r-\ell-1}(X, \mathbb{Q}(r-1))\right),$$

hence we break down our arguments involving each of the N terms on the RHS of (6.3). This is similar to how we handled things in §4. Note that if we apply the Künneth projector $[\Delta_{\overline{S}} \otimes \Delta_X(2d - 2r + \ell + 1, 2r - \ell - 1)]_*$ to the short exact sequence in (3.1) of §3 (with $\mathcal{Y} = D \times X$), where the action of the aforementioned Künneth projector on $H^{2r-1}(S \times X, \mathbb{Q}(r))$ is given by $\mathrm{Pr}_{13,*}\big(\mathrm{Pr}^*_{23}[\Delta_X(2d - 2r + \ell + 1, 2r - \ell - 1)] \cup \mathrm{Pr}^*_{12}(-)\big)$, observing that both Pr_{13}, $\mathrm{Pr}_{12} : S \times X \times X \to S \times X$ are proper and flat, we end up with the short exact sequence:

$$\begin{aligned} 0 \to &\left\{\frac{H^{\ell}(\overline{S}, \mathbb{Q}(1))}{H^{\ell}_D(\overline{S}, \mathbb{Q}(1))}\right\} \otimes H^{2r-\ell-1}(X, \mathbb{Q}(r-1)) \\ &\to H^{\ell}(S, \mathbb{Q}(1)) \otimes H^{2r-\ell-1}(X, \mathbb{Q}(r-1)) \\ &\to H^{\ell+1}_D(\overline{S}, \mathbb{Q}(1))^{\circ} \otimes H^{2r-\ell-1}(X, \mathbb{Q}(r-1)) \to 0, \end{aligned}$$

where

$$H^{\ell+1}_D(\overline{S}, \mathbb{Q})^{\circ} = \ker\big(H^{\ell+1}_D(\overline{S}, \mathbb{Q}) \to H^{\ell+1}(\overline{S}, \mathbb{Q})\big).$$

This accordingly modifies the bottom row of (3.2) of §3 in the obvious way. As a reminder, the Künneth components cycle representatives $\Delta_X(2d - \bullet, \bullet)$

of the diagonal class Δ_X (and of $\Delta_{\overline{S}}$, hence the product $\Delta_{\overline{S}\times X} = \Delta_{\overline{S}} \otimes \Delta_X$) are now assumed chosen in the sense of Murre (see Remark 5.5). So we can likewise apply the projector $\Delta_{\overline{S}} \otimes \Delta_X(2d-2r+\ell+1, 2r-\ell-1)$ to the top row of (3.2) of §3, and arrive at a modified commutative diagram (3.2), based on $\ell = 1,...,N$. We can be more explicit here. For the sake of brevity, let us denote $\Delta_X(2d-2r+\ell+1, 2r-\ell-1)$ in $\mathrm{CH}^d(X\times X;\mathbb{Q})$ here with P, and let $\mathcal{Y} = D\times X$ for some codimension one subscheme $D \subset \overline{S}$. Then $\Delta_{\overline{S}}\otimes P$ acts on $\mathrm{CH}^r(\overline{S}\times X, 1;\mathbb{Q})$ in a natural way, and its action on $\mathrm{CH}^r_{D\times X}(\overline{S}\times X;\mathbb{Q})$ is given as follows. First of all, $\mathrm{CH}^r_{D\times X}(\overline{S}\times X;\mathbb{Q}) = \mathrm{CH}^{r-1}(D\times X;\mathbb{Q})$. There are proper flat maps $D\times X\times X$ given by projections $Pr_{12}: D\times X\times X \to D\times X$, $Pr_{23}: D\times X\times X \to X\times X$ and $Pr_{13}: D\times X\times X\to D\times X$. For γ in $\mathrm{CH}^{r-1}(D\times X;\mathbb{Q})$, $Pr_{12}^*(\gamma)$ in $\mathrm{CH}^{r-1}(D\times X\times X;\mathbb{Q})$ is defined (flat pullback). For $P\in \mathrm{CH}^d(X\times X;\mathbb{Q})$, the intersection $Pr_{12}^*(\gamma)\bullet Pr_{23}^*(P)\in \mathrm{CH}^{r-1+d}(D\times X\times X;\mathbb{Q})$ is likewise well defined [F, §2]. The action then is given by $Pr_{13,*}\big(Pr_{12}^*(\gamma)\bullet Pr_{23}^*(P)\big)$ in $\mathrm{CH}^{r-1}(D\times X;\mathbb{Q}) = \mathrm{CH}^r_{D\times X}(\overline{S}\times X;\mathbb{Q})$. The action of $\Delta_{\overline{S}}\otimes P$ on $\mathrm{CH}^r_{\mathrm{hom}}(\overline{S}\times X;\mathbb{Q})$ is clear. Finally, by an elementary Hodge theory argument, one arrives at a modified version of Proposition 3.6. Specifically, $\Delta_{\overline{S}}\otimes P$ acts naturally on all terms on the RHS of the display in Proposition 3.6, making use of functoriality of the Abel-Jacobi map, and operates naturally on the LHS, as clearly evident in the above discussion following (6.3). As ℓ ranges from $1,...,N$, both sides of the aforementioned display decompose accordingly into a direct sum.

We need to determine what $\Delta_{\overline{S}}\otimes \Delta_X(2d-2r+\ell+1, 2r-\ell-1)$ does to γ, which is an algebraic cycle of codimension r (dimension $N+d-r$) on $\overline{S}\times X$, but supported on $D\times X$. Decomposing $\Delta_{\overline{S}}$, we can write

$$\Delta_{\overline{S}\times X}(2(N+d)-2r+\nu, 2r-\nu),$$

as a sum of

$$\Delta_{\overline{S}}(2N+\nu-\ell-1, \ell+1-\nu)\otimes\Delta_X(2d-2r+\ell+1, 2r-\ell-1). \tag{6.4}$$

We recall that we have $2\le \nu\le r$, and, as indicated earlier, only consider ℓ with $2\le \ell\le N$. Before stating our next result, it is helpful to introduce the following, which includes those $\overline{S}$ which are complete intersections in projective space or more generally a Grassmannian.

Lemma 6.5. *Suppose that $\overline{S}$ is a variety of dimension N such that for $i\neq N$:*

$H^i(\overline{S},\mathbb{Q})$ is zero for i odd and generated by algebraic cycles for i even.

Then $\overline{S}$ admits a Chow-Künneth decomposition in the sense of [M], Remark 5.5 with the supports of the Künneth projectors compatible with the supports of the cohomology classes in $H^\bullet(\overline{S},\mathbb{Q})$. Specifically, for $j\neq N$, $\Delta_{\overline{S}}(2N-2j, 2j)$ is contained in the image of $\mathrm{CH}^{N-j}(\overline{S};\mathbb{Q})\otimes\mathrm{CH}^j(\overline{S};\mathbb{Q})$ under pullback to $\mathrm{CH}^N(\overline{S}\times\overline{S};\mathbb{Q})$ and taking the product.

Proof. For $j = 0,\dots,N$, let W_{2N-2j} in $\mathrm{CH}^{N-j}(\overline{S},\mathbb{Q})$ and V_{2j} in $\mathrm{CH}^{j}(\overline{S},\mathbb{Q})$ be algebraic cycles such that $\lambda_j := \deg\big(\langle W_{2N-2j}, V_{2j}\rangle_{\overline{S}}\big) \neq 0$. Then

$$\big\{W_{2N-2k} \times V_{2k}\big\} \circ \big\{W_{2N-2j} \times V_{2j}\big\} = \begin{cases} 0 & \text{if } k \neq j \\ \lambda_j \cdot \{W_{2N-2j} \times V_{2j}\} & \text{if } k = j \end{cases}$$

To see this, compute $\big\{W_{2N-2k} \times V_{2k}\big\} \circ \big\{W_{2N-2j} \times V_{2j}\big\}$ as

$$Pr_{13,*}\big(\langle Pr_{12}^*(W_{2N-2k} \times V_{2k}), Pr_{23}^*(W_{2N-2j} \times V_{2j})\rangle_{\overline{S}\times\overline{S}\times\overline{S}}\big) \tag{6.6}$$

in $\mathrm{CH}^N(\overline{S} \times \overline{S};\mathbb{Q})$. If $j = k$ the statement is clear. If $k > j$ then $V_{2k} \cap W_{2N-2j} = 0$ by codimension, and if $k < j$ then this intersection has dimension at least 1, so (6.6) has codimension bigger than N. In either case it is trivial. This principle allows us to define, for $i \neq N$, mutually orthogonal idempotents π_i in $\mathrm{CH}^N(\overline{S} \times \overline{S};\mathbb{Q})$ with $[\pi_i]$ in $H^{2N-i}(\overline{S},\mathbb{Q}) \otimes H^i(\overline{S},\mathbb{Q})(N)$. We can take $\pi_i = 0$ if $i \neq N$ odd, so that $[\pi_i] = [\Delta(2N-i,i)] = 0$ by our assumption on the odd cohomology groups. For even $i = 2j \neq N$, we can arrange that $[\pi_{2j}] = [\Delta(2N-2j,2j)]$ by using the assumption about the even cohomology groups being generated by algebraic cocyles, as well as the non-degeneracy of the intersection product. Put $\pi_N = \Delta_{\overline{S}} - \sum_{i\neq N}\pi_i$. Because the π_i for $i \neq N$ are mutually orthogonal idempotents by construction, the same holds if we use all $\pi_0,\dots,\pi_{2N}$. Because $[\pi_i] = [\Delta_{\overline{S}}(2N-i,i)]$ for $i \neq N$, it follows from the definition of π_N that $[\pi_N] = [\Delta_{\overline{S}}(N,N)]$ as well. □

Theorem 6.7. *Suppose that $\overline{S}$ is a variety of dimension N such that for $i \neq N$:*

$H^i(\overline{S},\mathbb{Q})$ is zero for i odd and generated by algebraic cycles for i even.

Given Assumptions 5.1 and Conjecture 5.2, then

$$\mathrm{CH}^r(X_{\eta_{\overline{S}}}, 1;\mathbb{Q}) \to \Gamma\big(H^{2r-1}(X_{\eta_{\overline{S}}}, \mathbb{Q}(r))\big),$$

is surjective.

Proof. We shall use the Chow-Künneth decomposition of $\Delta_{\overline{S}}$ as in Lemma 6.5. Observe that with regard to (6.4),

$$\Delta_{\overline{S}}(2N+\nu-\ell-1, \ell+1-\nu) = \Delta_{\overline{S}}(N,N) \Leftrightarrow \ell+1-\nu = N.$$

But $\ell \leq N$ and $\nu \geq 2$, so this never happens. Also, the situation where $\ell + 1 - \nu = 0$ does not contribute. Namely, remember that $\gamma \in \mathrm{CH}^r_{D\times X}(\overline{S} \times X;\mathbb{Q})^\circ$ maps to a class in $\mathrm{CH}^r_{AJ}(\overline{S} \times X;\mathbb{Q})$. Since $|\gamma| \subset D \times X$, $\Delta_{\overline{S}}(2N,0) = \{p\} \times \overline{S}$ for some $p \in \overline{S}$, so that $D \times \overline{S}$ doesn't meet $\Delta_{\overline{S}}(2N,0)$ for a suitable choice of p, it follows in this case that

$$\big(\Delta_{\overline{S}}(2N,0) \otimes \Delta_X(2d-2r+\nu, 2r-\nu)\big)_*(\gamma) = 0.$$

For $1 \le \ell+1-\nu \le N-1$, and γ in $\mathcal{K}$, we have that

$$\big(\Delta_{\overline{S}}(2N+\nu-\ell-1,\ell+1-\nu)\otimes\Delta_X(2d-2r+\ell+1,2r-\ell-1)\big)_*(\gamma)$$

is in $N^1_{\overline{S}}\mathrm{CH}^r(\overline{S}\times X;\mathbb{Q})$. This is immediate from the fact that γ is null-homologous on $\overline{S}\times X$ and the support of the Chow-Künneth components here, as described in Lemma 6.5. Hence by Proposition 3.6 (more precisely, the incarnation of Proposition 3.6 in the discussion following (6.3)), and (6.2) of Remark 6.1,

$$\mathrm{CH}^r(X_{\eta_{\overline{S}}},1;\mathbb{Q}) \to \Gamma\big(H^{2r-1}(X_{\eta_{\overline{S}}},\mathbb{Q}(r))\big),$$

is surjective. □

6.1. Grand finale

For our final main result, we again consider $\overline{S}=\overline{C}_1\times\cdots\times\overline{C}_N$, a product of smooth complete curves (cf. Section 4.2). As before, we restrict ourselves to $2\le\ell\le N$. Let us write

$$\Delta_{\overline{C}_j}=e_j\times\overline{C}_j+\Delta_{\overline{C}_j}(1,1)+\overline{C}_j\times e_j,$$

where $e_j\in\overline{C}_j$ and $\Delta_{\overline{C}_j}(1,1)$ is defined by the equality. Consider the decomposition

$$\Delta_{\overline{S}}=\Delta_{\overline{C}_1}\otimes\cdots\otimes\Delta_{\overline{C}_N}=\bigotimes_{j=1}^{N}\big\{e_j\times\overline{C}_j+\Delta_{\overline{C}_j}(1,1)+\overline{C}_j\times e_j\big\}. \tag{6.8}$$

Note that $\Delta_{\overline{S}}(D)=D$. It needs to be determined what the RHS of (6.8) does to D, and more precisely, what $\Delta_{\overline{S}}\otimes\Delta_X(2d-2r+\ell+1,2r-\ell-1)$ does to γ, which is an algebraic cycle of dimension $N+d-r$ supported on $D\times X$. Now up to permutation, the RHS of (6.8) is made up of terms of the form

$$(\Delta_{\overline{C}_j}(1,1))^{\otimes_{j=1}^{k_1}}\otimes(\{e_j\times\overline{C}_j\})^{\otimes_{j=k_1+1}^{k_1+k_2}}\otimes(\{\overline{C}_j\times e_j\})^{\otimes_{j=k_1+k_2+1}^{N}}, \tag{6.9}$$

which is in the $(k_1+2k_2,2N-k_1-2k_2)$-component of $\Delta_{\overline{S}}$. Because in (6.4) we want $\Delta_{\overline{S}}(2N+\nu-\ell-1,\ell+1-\nu)$, we have

$$2N+\nu-\ell-1=k_1+2k_2. \tag{6.10}$$

Clearly, we have $0\le k_1+k_2\le N$, $2\le\nu\le r$, and we are restricting ourselves to $2\le\ell\le N$. Notice that if $k_1+k_2<N$, we arrive at the situation where a correspondence in (6.8), which when tensored with Δ_X, takes γ to an element of $N^1_{\overline{S}}\mathrm{CH}^r(\overline{S}\times X;\mathbb{Q})$. If $k_1+k_2=N$, then from (6.10), $N+\nu-\ell-1=k_2\le N$, and hence $\nu\le\ell+1$. As in the proof of Theorem 6.7, we can ignore the case $\nu=\ell+1$.

Theorem 6.11. *Under Assumption 5.1, and Conjecture 5.2, if* $\overline{S} = \overline{C}_1 \times \cdots \times \overline{C}_N$ *is a product of smooth complete curves and X a smooth projective variety, then for any* $r \geq 1$,

$$\mathrm{CH}^r(X_{\eta_{\overline{S}}}, 1; \mathbb{Q}) \to \Gamma\big(H^{2r-1}(X_{\eta_{\overline{S}}}, \mathbb{Q}(r))\big),$$

is surjective.

Proof. We will prove this by induction on $N \geq 1$, the case $N = 1$ being part of Proposition 4.1(2). It will be crucial that no part of $\Delta_{\overline{S}}(N,N)$ occurs because $\ell \leq N$ and $\nu \geq 2$ imply $\ell + 1 - \nu < N$. We shall argue on the summands of $\Delta_{\overline{S}}$ that, up to a permutation, are as in (6.9). The reductions preceding the theorem allow us to assume $k_1 + k_2 = N$, and that $1 \leq k_2 \leq N - 1$. We see (6.9) is

$$\Xi := \Delta_{\overline{C}_1}(1,1) \otimes \cdots \otimes \Delta_{\overline{C}_{k_1}}(1,1) \otimes \{e_{k_1+1} \times \overline{C}_{k_1+1}\} \otimes \cdots \otimes \{e_N \times \overline{C}_N\}$$

which is in the $(N + k_2, N - k_2)$-component of $\Delta_{\overline{S}}$. Now let $D \subset \overline{S}$ have codimension 1. By choosing $\{e_{k_1+1}, ..., e_N\}$ appropriately, we can assume that

$$D' := \big|\Xi[D]\big| \subseteq E \times \overline{C}_{k_1+1} \times \cdots \times \overline{C}_N$$

where $E \subset \overline{C}_1 \times \cdots \times \overline{C}_{k_1}$ has codimension 1. Let $\gamma \in \mathrm{CH}^r_{AJ}(\overline{S} \times X; \mathbb{Q}) = F^2\mathrm{CH}^r(\overline{S} \times X; \mathbb{Q})$, supported on $D \times X$, represent a class

$$[\gamma] \in \Gamma\big(H^{\ell+1}_D(\overline{S}, \mathbb{Q}(1))^\circ \otimes H^{2r-\ell-1}(X, \mathbb{Q}(r-1))\big),$$

where

$$H^{\ell+1}_D(\overline{S}, \mathbb{Q}(1))^\circ = \ker\big(H^{\ell+1}_D(\overline{S}, \mathbb{Q}(1)) \to H^{\ell+1}(\overline{S}, \mathbb{Q}(1))\big).$$

Then taking note of (6.4), together with functoriality of the Abel-Jacobi map,

$$\gamma' := \Xi_*(\gamma) \in \mathrm{CH}^r_{AJ}(\overline{S} \times X; \mathbb{Q}),$$

is supported on $D' \times X$. Indeed, we have

$$[\gamma'] \in \Gamma\big(H^{\ell+1}_{D'}(\overline{S}, \mathbb{Q}(1))^\circ \otimes H^{2r-\ell-1}(X, \mathbb{Q}(r-1))\big).$$

By the properties of the BB filtration, and in light of Remark 6.1, we can reduce to the case where $\gamma = \gamma'$ and $D = D'$. Notice that we have a Künneth decomposition (indexed by j).

$$H^{\ell+1}_{D'}(\overline{S}, \mathbb{Q})^\circ = \bigoplus_{j=1}^{2k_1-1} H^{j+1}_E(\overline{C}_1 \times \cdots \times \overline{C}_{k_1}, \mathbb{Q})^\circ \otimes H^{\ell-j}(\overline{C}_{k_1+1} \times \cdots \times \overline{C}_N, \mathbb{Q}),$$

and $[\gamma']$ decomposes accordingly. By fixing a j, and taking the jth component of $[\gamma']$, we may assume $[\gamma']$ lies in

$$\Gamma\big(H^{j+1}_E(\overline{C}_1 \times \cdots \times \overline{C}_{k_1}, \mathbb{Q}(1))^\circ \otimes H^{\ell-j}(\overline{C}_{k_1+1} \times \cdots \times \overline{C}_N, \mathbb{Q}) \otimes H^{2r-\ell-1}(X, \mathbb{Q}(r-1))\big).$$

Now let's put $\overline{S}_0 = \overline{C}_1 \times \cdots \times \overline{C}_{k_1}$ and $X_0 = \overline{C}_{k_1+1} \times \cdots \times \overline{C}_N \times X$. Then $\overline{S}_0 \times X_0 = \overline{S} \times X$, and $\gamma' \in \mathrm{CH}^r_{AJ}(\overline{S}_0 \times X_0; \mathbb{Q})$ is supported on $E \times X_0$. Further, $\dim \overline{S}_0 < \dim \overline{S}$. Then by Proposition 3.6, and induction on N,

$$\gamma' \in N^1_{\overline{S}_0}\mathrm{CH}^r(\overline{S}_0 \times X_0; \mathbb{Q}) \subset N^1_{\overline{S}}\mathrm{CH}^r(\overline{S} \times X; \mathbb{Q}).$$

As mentioned above, the same applies to γ , and we are done. □

References

[A] M. Asakura, *Motives and algebraic de Rham cohomology,* in: The Arithmetic and Geometry of Algebraic Cycles, Proceedings of the CRM Summer School, June 7-19, 1998, Banff, Alberta, Canada (Editors: B. Gordon, J. Lewis, S. Müller-Stach, S. Saito and N. Yui), NATO Science Series **548** (2000), Kluwer Academic Publishers.

[Bl1] S. Bloch, *Algebraic cycles and higher K-theory,* Advances in Math. **61** (1986), 267-304.

———, *The moving lemma for higher Chow groups,* Jour. of Alg. Geometry **3** (1994), 537-568.

[dJ-L] R. de Jeu, J. D. Lewis, (with an appendix by M. Asakura), *Beilinson's Hodge conjecture for smooth varieties.* J. K-Theory **11** (2013), no. 2, 243-282.

[F] W. Fulton, *Rational equivalence of singular varieties,* Inst. Hautes Études Sci. Publ. Math. No. **45** (1975), 147-167.

[GG] Mark Green and Phillip Griffiths, *Hodge-Theoretic Invariants for Algebraic Cycles,* Int. Math. Res. Not. 2003, no. **9**, 477-510.

[H] Robin Hartshorne, *Al;gebraic Geometry,* Graduate Texts in Mathematics **52**, (1977), Springer-Verlag, Springer-Verlag, New York, Heildelberg, Berlin.

[Ja1] U. Jannsen, *Equivalence relations on algebraic cycles,* in: The Arithmetic and Geometry of Algebraic Cycles, Proceedings of the CRM Summer School, June 7-19, 1998, Banff, Alberta, Canada (Editors: B. Gordon, J. Lewis, S. Müller-Stach, S. Saito and N. Yui), NATO Science Series **548** (2000), 225-260, Kluwer Academic Publishers.

[Ja2] ———, *Mixed Motives and Algebraic K-Theory.* Springer-Verlag.

[K-L] M. Kerr, J. D. Lewis, *The Abel-Jacobi map for higher Chow groups, II,* Invent. Math. **170**, (2007), no. 2, 355-420.

[SJK-L] S. J. Kang, J. D. Lewis, *Beilinson's Hodge conjecture for* K_1 *revisited.* In: Cycles, Motives and Shimura varieties, 197-215, Tata Inst. Fund. Res. Stud. Math., Tata Inst. Fund. Res., Mumbai, 2010.

[Lew0] J. D. Lewis, A Survey of the Hodge Conjecture. Second edition. Appendix B by B. Brent Gordon. CRM Monograph Series, **10**. American Mathematical Society, Providence, RI, 1999. xvi+368 pp.

[Lew1] ———, *A filtration on the Chow groups of a complex projective variety,* Compositio Math. **128** (2001), no. 3, 299-322.

[Lew2] ———, *Abel-Jacobi equivalence and a variant of the Beilinson-Hodge conjecture,* J. Math. Sci. Univ. Tokyo **17** (2010), 179-199.

[Lew3] ______, *Arithmetic normal functions and filtrations on Chow groups,* Proc. Amer. Math. Soc. **140** (2012), no. 8, 2663-2670.

[M] J. P. Murre, *On a conjectural filtration on the Chow groups of an algebraic variety.* Mathematical Institute, Univ. of Leiden. Report W 92-06, April 1992.

Faculteit Exacte Wetenschappen, Afdeling Wiskunde, VU University Amsterdam, The Netherlands
E-mail address: r.m.h.de.jeu@vu.nl

632 Central Academic Building, University of Alberta, Edmonton, Alberta T6G 2G1, Canada
E-mail address: lewisjd@ualberta.ca

Faculteit Exacte Wetenschappen, Afdeling Wiskunde VU University Amsterdam, The Netherlands
E-mail address: deeppatel1981@gmail.com

9

Normal functions and spread of zero locus

Morihiko Saito

ABSTRACT. If there is a topologically locally constant family of smooth algebraic varieties together with an admissible normal function on the total space, then the latter is constant on any fiber if this holds on some fiber. Combined with spreading out, it implies for instance that an irreducible component of the zero locus of an admissible normal function is defined over k if it has a k-rational point where k is an algebraically closed subfield of the complex number field with finite transcendence degree. This generalizes a result of F. Charles that was shown in case the normal function is associated with an algebraic cycle defined over k.

Introduction

Let k be an algebraically closed subfield of $\mathbf{C}$ with finite transcendence degree. Let X be a smooth complex variety defined over k. Let $\mathbf{H}$ be an admissible variation of mixed Hodge structure of strictly negative weights on X (see [Ka], [SZ]). Let ν be an admissible normal function of $\mathbf{H}$, which is a holomorphic section of the family of Jacobians $J(\mathbf{H})$ satisfying some good properties (see [GGK], [Sa3]). Let Z be an irreducible component of the zero locus $\nu^{-1}(0)$ of ν. This is algebraic as a corollary of [BP], [KNU], [Sch2] (see [BPS]).

Assume Z is not defined over k. Let K be the (minimal) field of definition of Z. This is the smallest subfield $K \subset \mathbf{C}$ containing k and such that Z is defined over K, see [Gro2, Cor. 4.8.11] (and also [We]). Let R be a finitely generated k-subalgebra of K whose field of fractions is K. Set

$$S := \operatorname{Spec} R \otimes_k \mathbf{C}.$$

The dimension of S coincides with the relative transcendence degree of K over k, and is called the *transcendence degree* of Z over k. We may assume that S is sufficiently small by replacing R without changing K. Then S is smooth, and there is a closed subvariety

$$Y \subset X \times S,$$

defined over k and having the morphisms

$$f : Y \to S, \quad \pi : Y \to X,$$

induced by the projections and such that π induces an isomorphism

$$Y_{s_0} := f^{-1}(s_0) \xrightarrow{\sim} Z \subset X,$$

where s_0 is the k-generic point of S corresponding to the inclusion $R \hookrightarrow K \hookrightarrow \mathbf{C}$. (In this paper a point of a complex algebraic variety means always a closed point.)

Let $\overline{Z}^k$ denote the k-Zariski closure of Z in X, i.e. the smallest Zariski closed subset of X defined over k and containing Z. This coincides with the Zariski closure of $\pi(Y) \subset X$, see Remark (2.2)(iii) below. Set $Y_s := Y \cap \big(X \times \{s\}\big) \subset X$ for $s \in S$. We have the following.

Theorem 1. *Assume S is sufficiently small. Then the induced morphism $\pi : Y \to \overline{Z}^k$ is dominant and quasi-finite so that $\dim \overline{Z}^k = \dim Z + \dim S$. Moreover there is a Zariski closed subset $\Sigma \subset S$ such that $\{s_0\}$ is a connected component of Σ and we have $Y_s \subset \nu^{-1}(0)$ if $s \in \Sigma$, and $Y_s \subset X \setminus \nu^{-1}(0)$ if $s \notin \Sigma$.*

Here no condition is assumed about the relation between the normal function ν and the field k, although there is a rather strong restriction coming from the assumption that the non-constant normal function ν is defined on a smooth complex algebraic variety X defined over k (see Remark (3.8)(i) below). The quasi-finiteness of π implies that the restriction of the family over any curve C on S has no fixed point (locally on C). It might be possible that Y_s for some $s \in \Sigma \setminus \{s_0\}$ is contained in another irreducible component of $\nu^{-1}(0)$ intersecting Z (although no such examples are explicitly known, see also Remark (3.8)(ii) below). Note that Σ is smooth by Theorem 2 below (by shrinking S if necessary). However, it is unclear whether every irreducible component of the zero locus has the same dimension.

For the proof of Theorem 1, we show the following.

Proposition 1. *In the notation and the assumption of Theorem 1, there is a resolution of singularities $Y' \to Y$ such that the induced morphism $f' : Y' \to S$ is smooth and topologically locally trivial over S, and the pull-back of the normal function ν to Y' by π' coincides with the pull-back of some normal function of $f'_*\pi'^*\mathbf{H}$ by f', where f', π' denote the compositions of $Y' \to Y$ with f, π respectively.*

Here $f'_*\pi'^*\mathbf{H}$ is an admissible variation of mixed Hodge structure (see [Sa1]). This assertion easily follows from the theory of admissible normal function explained in Section 1.

Theorem 1 implies the following corollary (which was shown by F. Charles [Ch2] using a completely different method [Ch1] if ν is associated with an algebraic cycle defined over k).

Corollary 1. If Z contains a k-rational point z, then Z is defined over k.

In the case Z is a point (which is important for the proof of the Hodge conjecture [Sch1]) Corollary 1 and Theorem 1 say nothing. In fact, S is a k-Zariski open subset of $\overline{Z}^k$, and π is the natural inclusion in this case.

In the situation of Theorem 1, we say that Z has *maximal transcendence degree* over k if $\dim S = \operatorname{codim}_X Z$. Note that the last condition is equivalent to $X = \overline{Z}^k$ if X is irreducible.

Theorem 2. *In the notation of Theorem* 1, *assume Z has maximal transcendence degree over k. Then Z is smooth, and does not intersect the other irreducible components of* $\nu^{-1}(0)$.

I would like to thank F. Charles, B. Kahn, and M. Levine for useful discussions about the subject of this paper. In fact, the original idea was inspired by the discussions with them. This work is partially supported by Kakenhi 24540039.

In Section 1 we review some basics of admissible normal functions. In Section 2 we recall the notion of spread out. In Section 3 we prove the main theorems after recalling some facts from algebraic or analytic geometry.

1 Admissible normal functions

In this section we review some basics of admissible normal functions.

1.1. Absolute case. Let X be a smooth complex algebraic variety. Let $\mathbf{H}$ be an admissible variation of mixed Hodge structure of strictly negative weights on X, see [Ka], [SZ]. (Here the underlying $\mathbf{Z}$-local system is assumed torsion-free.) We have the family of Jacobians $J(\mathbf{H})$ over X. Its fiber at $x \in X$ is set-theoretically identified with

$$J(\mathbf{H}_x)\big(:= \mathbf{H}_{x,\mathbf{C}}/(F^0\mathbf{H}_{x,\mathbf{C}} + \mathbf{H}_{x,\mathbf{Z}})\big) = \operatorname{Ext}^1_{\mathrm{MHS}}(\mathbf{Z}, \mathbf{H}_x),$$

where $\mathbf{H}_x$ is the fiber of $\mathbf{H}$ at x, and MHS is the abelian category of graded-polarizable mixed $\mathbf{Z}$-Hodge structures [De] (see [Ca] for the last isomorphism).

Let $\mathrm{NF}^{\mathrm{ad}}(X, \mathbf{H})$ be the group of admissible normal functions of $\mathbf{H}$ (see [Sa3] and also [GGK]). We have a canonical isomorphism

$$\mathrm{NF}^{\mathrm{ad}}(X, \mathbf{H}) = \operatorname{Ext}^1_{\mathrm{VMHS}(X)}(\mathbf{Z}_X, \mathbf{H}),$$

where VMHS(X) is the abelian category of admissible variations of mixed Hodge structure on X (assumed always graded-polarizable). We have the following exact sequence (see e.g. [Sa2, Theorem 3.6]):

$$0 \to J\big(H^0(X,\mathbf{H})\big) \to \mathrm{NF}^{\mathrm{ad}}(X,\mathbf{H}) \xrightarrow{cl} H^1(X,\mathbf{H}), \tag{1.1.1}$$

where $H^i(X,\mathbf{H}) \in \mathrm{MHS}$ by [Sa1]. It denotes also its underlying $\mathbf{Z}$-module as in the case of the last term of (1.1.1). The image of ν by cl in (1.1.1) is called the cohomology class of ν, and is defined by considering the extension class of underlying local systems.

If $cl(\nu) = 0$, then ν is the pull-back of an element of the first term of (1.1.1) by $X \to pt$, and is called *constant* on X.

1.2. Relative case. Let $f : Y \to S$ be a smooth morphism of smooth connected complex algebraic varieties. Set $Y_s := f^{-1}(s)$ for $s \in S$. Let $\mathbf{H}$ be an admissible variation of mixed Hodge structure of strictly negative weights on Y. Assume f is topologically locally trivial over S. Then $\mathcal{H}^0 f_*\mathbf{H}$ is an admissible variation of mixed Hodge structure of strictly negative weights on S with

$$f^*\mathcal{H}^0 f_*\mathbf{H} \subset \mathbf{H},$$

and $\mathcal{H}^1 f_*\mathbf{H}$ is an admissible variation of mixed Hodge structure on S (by using the stability of mixed Hodge modules by cohomological direct images, see [Sa1]). Moreover, there is an exact sequence

$$0 \to \mathrm{NF}^{\mathrm{ad}}(S,\mathcal{H}^0 f_*\mathbf{H}) \xrightarrow{f^*} \mathrm{NF}^{\mathrm{ad}}(Y,\mathbf{H}) \to \Gamma(S,\mathcal{H}^1 f_*\mathbf{H}), \tag{1.2.1}$$

where the last term means the global sections of the underlying local system of $\mathcal{H}^1 f_*\mathbf{H}$. The image of an admissible normal function ν of $\mathbf{H}$ to the last term is the cohomology class of ν, and is denoted by $cl(\nu)$. The exactness of (1.2.1) is proved by taking the restriction to each fiber of f and reducing to (1.1.1). If $cl(\nu)$ vanishes, then ν is an admissible normal function of $f^*\mathcal{H}^0 f_*\mathbf{H} \subset \mathbf{H}$, and coincides with the pull-back of some admissible normal function of $\mathcal{H}^0 f_*\mathbf{H}$.

Proposition 1.3. *Let ν be an admissible normal function of $\mathbf{H}$. If the restriction of ν to the fiber Y_{s_0} of f at some $s_0 \in S$ is constant in the sense of* (1.1)*, then the restriction of ν to any fiber Y_s is constant, and ν is the pull-back of some admissible normal function of $\mathcal{H}^0 f_*\mathbf{H}$.*

Proof. The assumption implies that the section of $\mathcal{H}^1 f_*\mathbf{H}$ corresponding to $cl(\nu)$ vanishes at s_0, and hence $cl(\nu) = 0$. So the assertion follows from the exact sequence (1.2.1).

2 Spread of closed subvarieties

In this section we recall the notion of spread out.

2.1. Spread out. Let X be a complex algebraic variety. Let k be an algebraically closed subfield of $\mathbf{C}$ with finite transcendence degree. We say that X is defined over k, if there is a k-variety X_k endowed with an isomorphism

$$X = X_k \otimes_k \mathbf{C},$$

where $\otimes_k\mathbf{C}$ means the base change by $\operatorname{Spec}\mathbf{C} \to \operatorname{Spec} k$. We say that a point x of X is *k-generic* if there is no proper closed subvariety V of an irreducible component of X such that $x \in V$ and V is defined over k, see also [We]. We say that U is a *k-Zariski open* subset if its complement is a closed subvariety defined over k.

Let X be a complex algebraic variety defined over k, and Z be a closed complex algebraic subvariety of X. There is a complex smooth affine variety S together with a closed subvariety Y of $X \times S$ which are all defined over k and such that Z coincides with the fiber

$$Y_{s_0} := Y \cap \big(X \times \{s_0\}\big) \subset X,$$

for some k-generic point s_0 of S. We say that Y is a *spread* of Z over k. We will denote respectively by π and f the morphisms of Y induced by the first and second projections from $X \times S$.

More precisely, $Z \subset X$ is defined over some subfield K of $\mathbf{C}$ which is finitely generated over k. Let R be a finitely generated k-subalgebra of K whose filed of fractions is K. Set $S_k := \operatorname{Spec} R$, $S := \operatorname{Spec} R \otimes_k \mathbf{C}$. If R is sufficiently large, then there is a closed k-subvariety

$$Y_k \hookrightarrow X_{S_k} := X_k \times_k S_k,$$

such that its base change by $\overline{\eta} : \operatorname{Spec}\mathbf{C} \to S_k$ is identified with the natural inclusion $Z \hookrightarrow X$. Here $\overline{\eta}$ is induced by the inclusions $R \hookrightarrow K \hookrightarrow \mathbf{C}$, and induces a morphism $R \otimes_k \mathbf{C} \to \mathbf{C}$ defining $s_0 \in S$. We may assume S smooth by shrinking S_k if necessary.

Remarks 2.2

(i) There is the smallest subfield $K_Z \subset \mathbf{C}$ such that Z is defined over it, see [Gro2, Cor. 4.8.11]. This field K_Z is called the *minimal field of definition* of $Z \subset X$ over k.

(ii) The minimum of the dimension of S for all the spreads of Z is called the *transcendence degree* of Z over k, and is denoted by $\operatorname{tr.deg}_k Z$. This coincides with $\operatorname{tr.deg}_k K_Z$, that is the relative transcendence degree over k of the minimal field of definition K_Z.

(iii) Assume $\dim S$ is minimal (i.e. $\dim S = \operatorname{tr.deg}_k Z$). Then the field of fractions of the affine ring R of S_k is a finite extension of the minimal field of definition K_Z. Moreover we have the equality

$$\overline{\pi(Y)} = \overline{Z}^k,$$

where the left-hand side is the closure of $\pi(Y)$ in X, and the right-hand side is the k-Zariski closure of Z in X, that is the minimal closed k-subvariety of X containing Z. Indeed, take the closure of Y in $X \times \overline{S}$ with $\overline{S}$ a compactification of S defined over k. Its image by the projection $X \times \overline{S} \to X$ is a k-Zariski closed subvariety containing $\pi(Y)$ as a dense subset (even in the classical topology). So the Zariski closure of $\pi(Y)$ is defined over k, and hence contains the k-Zariski closure of Z. On the other hand, we can replace X with the k-Zariski closure of Z for the construction of spread of Z. This implies the opposite inclusion. So the desired equality follows.

Examples 2.3

(i) Assume $k = \overline{\mathbf{Q}}$, $X = \mathbf{C}^2$ and $Z = \{y = cx\} \subset \mathbf{C}^2$ where c is a transcendental number. In this case Z contains the origin, which is a k-rational point, and the fibers Y_s of the spread Y of Z are given by $\{y = sx\} \subset \mathbf{C}^2$ where s is identified with the coordinate of $S = \mathbf{C}$.

(ii) Assume $k = \overline{\mathbf{Q}}$, $X = \mathbf{C}^3$, and Z is the line $C(P_c)$ passing through the origin and the point

$$P_c := (1 - c^2, 2c, 1 + c^2) \in \mathbf{C}^3.$$

Here c is a transcendental number and $C(P_c)$ means the cone of $[P_c] \in \mathbf{P}^2$. In this case we have

$$\overline{Z}^k = \{x^2 + y^2 = z^2\} \subset \mathbf{C}^3,$$

and the fibers Y_s of the spread Y of Z are given by the lines $C(P_s)$ where s is identified with the coordinate of $S = \mathbf{C}$ and P_s is defined similarly to P_c. This is closely related to the parametrization of the projective curve $\{x^2 + y^2 = z^2\} \subset \mathbf{P}^2$ by $\mathbf{P}^1$ using the intersection with the lines $\{y = s(x+z)\}$ for $s \in \mathbf{C}$.

3. Proof of the main theorems

In this section we prove the main theorems after recalling some facts from algebraic or analytic geometry.

3.1. Some consequence of Weierstrass preparation theorem. Let $\pi : Y \to X$ be a morphism of complex analytic spaces. Assume $y \in Y$ is an isolated point of $\pi^{-1}\pi(y)$. Then there is an open neighborhood V_y of y in Y which is finite over an open neighborhood U_x of $x := \pi(y)$ in X.

Indeed, we may assume that X is a closed analytic subset of a polydisk Δ^m, and Y is a closed analytic subset of $\Delta^n \times X$ defined by holomorphic functions $h_1, \dots, h_r$, where $\pi : Y \to X$ is induced by the projection $\Delta^{n+m} \to \Delta^m$ (by using the graph embedding), and the origins of the polydisks Δ^{n+m} and Δ^n respectively correspond to y and x. Since y is isolated in $\pi^{-1}(x)$, we may assume that the restriction of h_1 to $\Delta^n \times \{0\}$ is not identically zero, and then $h_1(x_1, 0, \dots, 0)$ is not identically zero. Let W be the closed subset of Δ^{n+m} defined by h_1. By the Weierstrass preparation theorem, the projection $W \to \Delta^{n+m-1}$ is a finite morphism by shrinking the polydisks if necessary. Moreover, the image of an analytic space by a finite (or more generally, projective) morphism is a closed analytic subspace (by using the direct image of the structure sheaf). So the assertion follows by induction on n. (If $\pi : Y \to X$ is algebraically defined, this assertion also follows from Remark (ii) below.)

3.2. Grothendieck's version of Zariski's main theorem. Let $\pi : Y \to X$ be a quasi-projective morphism of algebraic varieties defined over k. By [Gro1, Thm. 4.4.3] the quasi-finite locus Y^{qf} of π is a k-Zariski open subset of Y. Here the quasi-finite locus means the subset consisting of $y \in Y$ which is isolated in $\pi^{-1}\pi(y)$. (Note that this assertion is local on X.) Moreover the morphism $\pi^{\mathrm{qf}} : Y^{\mathrm{qf}} \to X$ induced by π can be extended to a finite morphism $\widetilde{\pi} : \widetilde{Y^{\mathrm{qf}}} \to X$ defined over k in such a way that Y^{qf} is a dense k-Zariski open subvariety of $\widetilde{Y^{\mathrm{qf}}}$. This is Grothendieck's version of Zariski's main theorem.

3.3. Proof of Proposition 1. Let Y, S be as in (2.1) where we assume that $\dim S$ is minimal. Let $\overline{X}$ be a smooth compactification of X defined over k. Let $\overline{Y}$ be the closure of

$$Y \subset X \times S \quad \text{in} \quad \overline{X} \times S,$$

with $\overline{\pi}, \bar{f}$ the morphisms of $\overline{Y}$ induced by the first and second projections from $\overline{X} \times S$. These are also defined over k. Note that $\overline{\pi}^{-1}(X) = Y$ since Y is closed in $X \times S$.

Let $\rho : \overline{Y}' \to \overline{Y}$ be a resolution of singularities defined over k and inducing an isomorphism over the smooth part Y^{sm} of Y. We may assume that the inverse image of $D := \overline{Y} \setminus Y^{\mathrm{sm}}$ by ρ is a divisor with simple normal crossings on $\overline{Y}'$ such that any intersections of irreducible components of $\rho^{-1}(D)$ are smooth over S by assuming S sufficiently small. Here we may also assume that $\rho^{-1}(\overline{Y} \setminus Y)$ is a divisor by using a blow-up. Then the above assertion holds also for $\rho^{-1}(\overline{Y} \setminus Y)$, since the latter is a divisor contained in $\rho^{-1}(\overline{Y} \setminus Y^{\mathrm{sm}})$.

Set $\mathbf{H}' := \pi'^{*}\mathbf{H}$, and $\nu' := \pi'^{*}\nu$. Then $\mathcal{H}^1 f'_* \mathbf{H}'$ is an admissible variation of mixed Hodge structure on S, and $cl(\nu')$ vanishes so that ν' comes from an admissible normal function ν'' of $f'_* \mathbf{H}'$, see Remark (1.3). So the assertion follows. This finishes the proof of Proposition 1.

3.4. Proof of Theorem 1. Let Σ be the zero locus of ν'' on S. Note that $s_0 \in \Sigma$ and the zero locus of ν' coincides with $f^{-1}(\Sigma)$. So the fiber $Y_s \subset X$ of f over any $s \in \Sigma$ is contained in $\nu^{-1}(0)$, and Y_s does not intersect $\nu^{-1}(0)$ if $s \notin \Sigma$. Note that this implies that Y_{s_0} is contained in the quasi-finite locus Y^{qf} of $\pi : Y \to X$ if Σ is 0-dimensional at s_0. We can show the last property as follows.

If Σ has positive dimension at s_0, then $Y_s = Z$ for any s in the connected component Σ_0 of Σ containing s_0 (since Y_s must contain at least a non-empty smooth open subset of Z in the classical topology by using Proposition 1). Take a general hyperplane section H defined over k and intersecting Σ_0 (by using the fact that any non-empty Zariski open subset of complex projective space has a k-rational point). Then we may replace S with the smallest closed subvariety of H defined over k and containing a point of $H \cap \Sigma_0$. But this contradicts the minimality of $\dim S$. So Σ must be 0-dimensional at s_0.

It now remains to show the quasi-finiteness of $\pi : Y \to X$ by assuming S sufficiently small. We have $Y_{s_0} \subset Y^{\text{qf}}$ by the above argument. So the proof of Theorem 1 is reduced to Lemma (3.5) below with W the non-quasi-finite locus $Y \setminus Y^{\text{qf}}$, since the latter is a closed subvariety of Y defined over k by (3.2). (Note that finitely generated fields are countable and algebraic closures of countable fields are also countable.)

Lemma 3.5. *Let $f : Y \to S$ be a morphism of complex algebraic varieties defined over a countable field k, where S is irreducible. Let W be a closed subvariety of Y defined over k, and s_0 be a k-generic point of S. Assume $W \cap f^{-1}(s_0) = \emptyset$. Then $f(W)$ is contained in a proper closed subvariety of S defined over k.*

Proof. We may assume W irreducible. The hypothesis implies that

$$W \cap f^{-1}(s) = \emptyset,$$

for any k-generic point $s \in S$ by using the action of $\mathrm{Aut}(\mathbf{C}/k)$. Let S'' denote the union of all the proper closed subvarieties of S defined over k. This is a countable union, i.e.,

$$S'' = \bigcup_{i \in \mathbf{N}} S''_i,$$

with S''_i proper closed subvarieties of S defined over k ($i \in \mathbf{N}$), since k is a countable field. Moreover the set of k-generic points of S coincides with $S \setminus S''$ (see also [SaSch, Prop. 3.2]). So we get

$$W \subset f^{-1}(S'').$$

Hence $W = W \cap f^{-1}(S'')$ is a countable union of closed subvarieties

$$W_i := W \cap f^{-1}(S''_i) \subset W \quad (i \in \mathbf{N}).$$

This implies that $W = W_i$ for some i, and the assertion follows. This finishes the proofs of Lemma (3.5) and Theorem 1.

3.6. Proof of Corollary 1. Assume Z is not defined over k. Apply Theorem 1 to Z, where $\dim S > 0$. We then see that s_0 in Theorem 1 is the image of a k-rational point of Y by the morphism $f : Y \to S$. For this, consider

$$(z, s_0) \in X \times S,$$

which belongs to Y, and is moreover k-rational. Indeed, it is a connected component of the intersection defined over k

$$Y \cap \big(\{z\} \times S\big) \subset X \times S,$$

by using the quasi-finiteness of $\pi : Y \to X$ (and k is algebraically closed). This is a contradiction since s_0 is a k-generic point of S. So the assertion follows.

3.7. Proof of Theorem 2. We have $\dim X = \dim Y$ by assumption. Let $D' \subset X$ denote the discriminant locus of $\pi' : Y' \to X$ in Proposition 1, i.e. D' is the smallest Zariski closed subset such that π' is étale over the complement of D'. This is defined over k. Note also that D' contains the image of the singular locus of Y. Since the k-Zariski closure of Z is X, we get $Z \not\subset D'$, i.e.,

$$Z \setminus D' \neq \emptyset. \tag{3.7.1}$$

Assume Z is singular. Let z be any singular point of Z, and z' be the corresponding point of Y_{s_0} by π. In the classical topology, there is an open neighborhood $V_{z'}$ of z' in Y which is finite over an open neighborhood U_z of z in X by the morphism $\pi_z : V_{z'} \to U_z$, induced by π, see (3.1). The morphism π_z is locally biholomorphic over $U_z \setminus D'$, and induces an isomorphism over $Z \cap U_z$ by using Theorem 1. Here we assume V_z sufficiently small so that

$$f(V_{z'}) \cap \Sigma = \{s_0\}.$$

Then the degree of π_z must be 1 since $(U_z \cap Z) \setminus D' \neq \emptyset$ by (3.7.1). So $V_{z'}$ must be smooth. Applying this argument to any singular point of Z, we see that Y is smooth on a neighborhood of Y_{s_0}. However, this contradicts the assumption that the geometric generic fiber Y_{s_0} of $f : Y \to S$ (which is isomorphic to Z) is singular. Thus Z is smooth, and the first assertion is proved.

It is then easy to see that Z cannot intersect the other components of $\nu^{-1}(0)$, since $\pi(Y)$ contains an open neighborhood of Z (at least in the classical topology). This finishes the proof of Theorem 2.

Remarks 3.8. (i) The quasi-finiteness of π in Theorem 1 does not hold for the spread of an arbitrary closed irreducible subvariety Z of a smooth complex algebraic variety X defined over k, see Example (2.3)(i). In this example, any admissible variation of mixed Hodge structure on $X = \mathbf{C}^2$ is constant, and so is any admissible normal function on X. Here we cannot replace X with the self-product of an elliptic curve E, since $\mathrm{End}(E)$ is discrete (more precisely, it is isomorphic to $\mathbf{Z}$ or $\mathbf{Z}^2$). If we replace X with the "cone" of a k-Zariski open subset of $\mathbf{P}^1$ having a non-constant normal function on it, then X cannot contain the origin (as long as it is a variety), and we get no contradictions. Note also that the cone of an abelian variety is quite singular although it is complete and has many non-constant normal functions on it.

(ii) There is an example of a closed subvariety W of a smooth complex algebraic variety X defined over k such that $W \subset \overline{W}_1^k$ with W_1 an irreducible component of W and some fiber Y_s of the spread Y of W_1 coincides with another irreducible component of W intersecting W_1. Consider, for instance, the case $k = \overline{\mathbf{Q}}$, $X = \mathbf{C}^3$ and $W = \bigcup_{0 \leqslant i < r} W_i$ with $W_i = C(P_{c_i})$ in the notation of Example (2.3)(ii), where the c_i are transcendental numbers. In this case we have

$$\overline{W}_i^k = \{x^2 + y^2 = z^2\} \subset \mathbf{C}^3.$$

Note that there is no non-constant admissible normal function on X.

References

[BP] Brosnan, P. and Pearlstein, G., On the algebraicity of the zero locus of an admissible normal function, Compos. Math. 149 (2013), 1913–1962.

[BPS] Brosnan, P., Pearlstein, G. and Schnell, C., The locus of Hodge classes in an admissible variation of mixed Hodge structure, C. R. Math. Acad. Sci. Paris 348 (2010), 657–660.

[Ca] Carlson, J., Extensions of mixed Hodge structures, in Journées de Géométrie Algébrique d'Angers 1979, Sijthoff-Noordhoff Alphen a/d Rijn, 1980, pp. 107–128.

[Ch1] Charles, F., On the zero locus of normal functions and the étale Abel-Jacobi map, Int. Math. Res. Notices 12 (2010), 2283–2304.

[Ch2] Charles, F., On the zero locus of l-adic normal functions (talk at Regulators III Conference in Barcelona, 2010).

[De] Deligne, P., Théorie de Hodge II, Publ. Math. IHES 40 (1971), 5–58.

[GGK] Green, M., Griffiths, P. and Kerr, M., Néron models and limits of Abel-Jacobi mappings, Compos. Math. 146 (2010), 288–366.

[Gro1] Grothendieck, A., Eléments de géométrie algébrique, III-1, Publ. Math. IHES 11, 1961.

[Gro2] Grothendieck, A., Eléments de géométrie algébrique, IV-2, Publ. Math. IHES 24, 1965.

[Ka] Kashiwara, M., A study of variation of mixed Hodge structure, Publ. Res. Inst. Math. Sci. 22 (1986), 991–1024.

[KNU] Kato, K., Nakayama, C. and Usui, S., Analyticity of the closures of some Hodge theoretic subspaces, Proc. Japan. Acad. Ser. A, Math. Sci. 87 (2011), 167–172.

[Sa1] Saito, M., Mixed Hodge modules, Publ. RIMS, Kyoto Univ. 26 (1990), 221–333.

[Sa2] Saito, M., Extension of mixed Hodge modules, Compos. Math. 74 (1990), 209–234.

[Sa3] Saito, M., Admissible normal functions, J. Alg. Geom. 5 (1996), 235–276.

[SaSch] Saito, M. and Schnell, C., Fields of definition of Hodge loci (in these proceedings).

[Sch1] Schnell, C., Two observations about normal functions, Clay Math. Proc. 9 (2010), 75–79.

[Sch2] Schnell, C., Complex analytic Néron models for arbitrary families of intermediate Jacobians, Inv. Math., 188 (2012), 1–81.

[SZ] Steenbrink, J.H.M. and Zucker, S., Variation of mixed Hodge structure, I, Inv. Math. 80 (1985), 489–542.

[We] Weil, A., Foundation of algebraic geometry, AMS 1946.

RIMS Kyoto University, Kyoto 606-8502 Japan

10

Fields of definition of Hodge loci

Morihiko Saito and Christian Schnell

ABSTRACT. We show that an irreducible component of the Hodge locus of a polarizable variation of Hodge structure of weight 0 on a smooth complex variety X is defined over an algebraically closed subfield k of finite transcendence degree if X is defined over k and the component contains a k-rational point. We also prove a similar assertion for the Hodge locus inside the Hodge bundle if the Hodge bundle together with the connection is defined over k. This is closely related with the theory of absolute Hodge classes. The proof uses the spread of the Hodge locus, and is quite similar to the case of the zero locus of an admissible normal function.

Introduction

Let k be an algebraically closed subfield of $\mathbf{C}$ with finite transcendence degree. Let X be a smooth complex algebraic variety defined over k. Let $\mathbf{H}$ be a polarizable variation of $\mathbf{Z}$-Hodge structure of weight 0 on X with underlying $\mathbf{Z}$-local system $\mathbf{H}_{\mathbf{Z}}$ torsion-free. Let $\mathrm{Hdg}_X(\mathbf{H}) \subset X$ denote the union of the Hodge loci of all local sections of $\mathbf{H}_{\mathbf{Z}}$. It is a countable union of closed algebraic subvarieties of X by Cattani, Deligne, and Kaplan [CaDeKa]. In this paper we prove the following.

Theorem 1. *Let Z be an irreducible component of* $\mathrm{Hdg}_X(\mathbf{H})$*. If Z contains a k-rational point z_0 of X, then Z is defined over k as a closed subvariety of X.*

The argument is essentially the same as in the case of the zero locus of an admissible normal function [Sa3], where the spread of closed subvarieties is used in an essential way.

Let V be the algebraic Hodge bundle over X associated to $\mathbf{H}$ such that the associated algebraic connection $\nabla : \mathcal{O}_X(V) \to \Omega^1_X \otimes_{\mathcal{O}_X} \mathcal{O}_X(V)$ has regular singularities at infinity, where $\mathcal{O}_X(V)$ is the sheaf of local sections of V (see [De1]). The last condition is satisfied in the geometric case as is well-known, where the Hodge bundle is given by the relative de Rham cohomology sheaves and the connection by the Gauss-Manin connection. (Note that the algebraic structure on V cannot be obtained correctly without using this regular singularity at infinity.) In this paper we also consider the Hodge locus $\mathrm{Hdg}_V(\mathbf{H})$ insider V, which is used in the theory of absolute Hodge (de Rham cohomology) classes, see [De4], [ChSch], [Vo2]. It is a countable union of closed algebraic subvarieties of V by [CaDeKa].

Theorem 2. *Assume the Hodge bundle V and the connection ∇ are defined over k. Then an irreducible component Y of $\mathrm{Hdg}_V(\mathbf{H})$ is defined over k as a closed subvariety of V if it contains a k-rational point y_0 of V.*

The assumption on the Hodge bundle and the connection is satisfied if these are associated with a smooth projective morphism of complex algebraic varieties $f : \mathcal{Y} \to X$ defined over k, where $\mathbf{H}_\mathbf{Z} := R^{2m} f_* \mathbf{Z}_\mathcal{Y}(m)/\mathrm{torsion}$, see [De2] for the Tate twist (m). (Without this Tate twist, the Hodge locus in V cannot be defined over k unless $2\pi i \in k$. This is related with the notion of weakly absolute Hodge class in [Vo2].) It seems rather difficult to extend Theorems 1 and 2 to the case where "an irreducible component" is replaced with "a connected component" in the statements.

Let $k_0 \subset k$ be a finitely generated subfield such that X is defined over k_0. For Z in Theorem 1, let $k_Z \supset k_0$ be the minimal field of definition of $Z \subset X$ over k_0, see [Gro3, Cor. 4.8.11]. Let $\mathcal{S}^{ac}(Z)$ denote the set of algebraically closed subfields $K \subset \mathbf{C}$ such that $K \supset k_0$ and Z has a K-rational point. It has a partial ordering by inclusion. We define similarly k_Y and $\mathcal{S}^{ac}(Y)$ for Y in Theorem 2. Combining Theorems 1 and 2 with [Gro3, Cor. 4.8.11], we get the following.

Theorem 3. *In the above notation, there is the minimal element of $\mathcal{S}^{ac}(Z)$, and it coincides with the algebraic closure $\overline{k}_Z$ of k_Z. A similar assertion holds for $\mathcal{S}^{ac}(Y)$ and k_Y if the hypothesis of Theorem 2 is satisfied.*

If we define $\mathcal{S}^{fg}(Z)$, $\mathcal{S}^{fg}(Y)$ by using finitely generated subfields instead of algebraically closed subfields in the definition of $\mathcal{S}^{ac}(Z)$, $\mathcal{S}^{ac}(Y)$, then the assertion of Theorem 3 holds with $\overline{k}_Z, \overline{k}_Y, \mathcal{S}^{ac}(Z), \mathcal{S}^{ac}(Y)$ replaced respectively by $k_Z, k_Y, \mathcal{S}^{fg}(Z), \mathcal{S}^{fg}(Y)$, provided that we consider everything modulo finite extensions (that is, K and K' are identified if K' is a finite extension of K).

Theorem 2 is closely related with the theory of absolute Hodge classes (see [De4], [Ja], [ChSch], [Vo2]). In this paper we consider only *de Rham* cohomology classes as in [ChSch], [Vo2] (and not so-called *adelic* cohomology classes as in [De4]). Assume $f : \mathcal{Y} \to X$ is defined over $k_0 = \mathbf{Q}$ so that V is also defined over $\mathbf{Q}$. (Here X is not necessarily connected, since a $\mathbf{Q}$-variety $X_\mathbf{Q}$ satisfying $X = X_\mathbf{Q} \times_{\mathrm{Spec}\,\mathbf{Q}} \mathrm{Spec}\,\mathbf{C}$ is not necessarily absolutely irreducible.) Let $V(\overline{\mathbf{Q}})$ denote the set of $\overline{\mathbf{Q}}$-rational points of V. Set $\mathcal{Y}_{x_0} := f^{-1}(x_0)$ for a point $x_0 \in X$. Then $H^{2m}(\mathcal{Y}_{x_0}, \mathbf{C})$ is canonically identified with V_{x_0}. Let α be a Hodge class in $H^{2m}(\mathcal{Y}_{x_0}, \mathbf{C}) = V_{x_0}$, and $Z(\alpha, \overline{\mathbf{Q}}) \subset V$ be the closure of α in the $\overline{\mathbf{Q}}$-Zariski topology. Let $Y(\alpha)$ be an irreducible component of $\mathrm{Hdg}_V(\mathbf{H})$ passing through α. From Theorem 2 we can deduce the following.

Corollary 1. *In the above notation the following three conditions are equivalent*:

(a) $Y(\alpha)$ *is defined over* $\overline{\mathbf{Q}}$ *as a closed subvariety of* V.

(a)′ $Y(\alpha) \cap V(\overline{\mathbf{Q}}) \neq \emptyset$.

(a)″ $Z(\alpha, \overline{\mathbf{Q}}) \subset Y(\alpha)$.

It is conjectured that these equivalent three conditions always hold in the situation of Corollary 1. (In fact, if the Hodge conjecture is true, then the conditions in Corollary 1 and Corollary 2 below would always hold, see [Vo2] and also Remarks (2.3) and (3.6)(iv) below.) We have furthermore the following corollaries which are closely related with results in [Vo2].

Corollary 2. *If the equivalent three conditions in Corollary* 1 *are satisfied, then the following three conditions are equivalent*:

(b) $Y(\alpha)^\sigma \subset \mathrm{Hdg}_V(\mathbf{H})$ $(\forall \sigma \in \mathrm{Gal}(\overline{\mathbf{Q}}/\mathbf{Q}))$.

(b)′ *For some* $\beta \in Y(\alpha) \cap V(\overline{\mathbf{Q}})$, *we have* $\beta^\sigma \in \mathrm{Hdg}_V(\mathbf{H})$ $(\forall \sigma \in \mathrm{Gal}(\overline{\mathbf{Q}}/\mathbf{Q}))$.

(b)″ α *is absolutely Hodge.*

Corollary 3. *Assume* x_0 *is a* $\mathbf{Q}$*-generic point of* X. *Then we have the equivalences*

$$(\mathrm{a}) + (\mathrm{b}) \Longleftrightarrow (\mathrm{b})' \Longleftrightarrow (\mathrm{b})''.$$

Moreover, if these equivalent conditions are satisfied, then

(c) $Y(\alpha)$ *is finite étale over* X.

Note that condition (b)′ logically contains condition (a)′, and $Y(\alpha)^\sigma$ for $\sigma \in \mathrm{Gal}(\overline{\mathbf{Q}}/\mathbf{Q})$ in condition (b) is meaningful only under condition (a). Here $Y(\alpha)^\sigma$ denotes the image of $Y(\alpha)$ by the contravariant (or right) action of $\sigma \in \mathrm{Gal}(\overline{\mathbf{Q}}/\mathbf{Q})$ (which is the inverse of the usual action so that $(Y(\alpha)^\sigma)^{\sigma'} = Y(\alpha)^{\sigma\sigma'}$). We can take σ from $\mathrm{Aut}(\mathbf{C})$ since condition (a) implies that $Y(\alpha)^\sigma$ depends only on the image of σ in $\mathrm{Gal}(\overline{\mathbf{Q}}/\mathbf{Q})$. Moreover we can take σ from $G_\alpha \backslash G$ with $G_\alpha \subset G := \mathrm{Gal}(\overline{\mathbf{Q}}/\mathbf{Q})$ the stabilizer of $Y(\alpha)$, and $G_\alpha \backslash G$ is finite. Note that the subgroup $G_\alpha \subset G$ corresponds to the smallest number field $k_{Y(\alpha)} \subset \overline{\mathbf{Q}}$ such that $Y(\alpha)$ is defined over $k_{Y(\alpha)}$ (by using the functoriality of the Galois descent). In condition (b)′, we can similarly take σ from $H_\beta \backslash G$ with $H_\beta \subset G$ the stabilizer of β, and $H_\beta \backslash G$ is finite.

The implication (a) + (b) $\Longrightarrow$ (b)″ is closely related to Principle B in [De4] (see also [ChSch] for the de Rham cohomology class version) in case condition (c) is satisfied. In fact, it is also possible to reduce the implication (a) + (b) $\Longrightarrow$ (b)″ to Principle B by using the base change under a certain étale morphism defined over $\mathbf{Q}$ in this case (where Theorem 2 is used), see Remark (3.6)(i) below.

We thank A. Tamagawa for answering our question on the reference for the invariance of geometric fundamental groups under base changes by extensions of algebraically closed fields. We also thank the referee for useful comments. The first-named author is partially supported by Kakenhi 24540039. The second-named author is partially supported by NSF grant DMS-1331641, and is very grateful to Daniel Huybrechts for the opportunity to spend the academic year 2013–2014 at the University of Bonn.

In Section 1 we review some basic facts from the theories of variations of Hodge structure and algebraic fundamental groups. In Section 2 we prove the main theorems. In Section 3 we explain the relation with the theory of absolute Hodge classes.

Conventions. In this paper, a variety means a separated scheme of finite type over a field of characteristic 0 although we consider only its closed points. So it is a variety in a classical sense, although it can be reducible or non-reduced in general. However, we need only reduced varieties in this paper (since no non-reduced varieties appear), and the reader may assume that the varieties in this paper are reduced.

For a complex algebraic variety X, we use the classical topology for the constant sheaf $\mathbf{Z}_X$, and the Zariski topology for the structure sheaf $\mathcal{O}_X$.

We say that a complex algebraic variety X is defined over a subfield k of $\mathbf{C}$, if there is a variety X_k over k together with an isomorphism $X = X_k \times_{\operatorname{Spec} k} \operatorname{Spec} \mathbf{C}$. If k is a subfield of finite transcendence degree, then $\mathbf{C}$ is sometimes viewed as a universal domain in the sense of [We].

1 Preliminaries

In this section we review some basic facts from the theories of variations of Hodge structure and algebraic fundamental groups.

1.1. Variations of Hodge structure on singular varieties. Let Y be a complex algebraic variety. Let L be a torsion-free $\mathbf{Z}$-local system on Y^{an}, and F be a finite filtration of $\mathcal{O}_{Y^{\mathrm{an}}} \otimes_{\mathbf{Z}} L$ by vector subbundles. We will say that $\big(L, (\mathcal{O}_{Y^{\mathrm{an}}} \otimes_{\mathbf{Z}} L, F)\big)$ is a polarizable variation of Hodge structure on Y if so is its pull-back to a desingularization of Y.

Let η be a global section of L. Assuming Y connected, we have the following

$$\text{(1.1.1)} \qquad \eta \text{ has type } (0,0) \text{ everywhere if it has type } (0,0) \text{ at one point.}$$

This is reduced to the Y nonsingular case, where the assertion is well-known. More precisely, it is due to P. Griffiths [Gri, Theorem 7.1] in the Y complete case, and P. Deligne [De2, Corollary 4.1.2] in the $\mathbf{H}$ geometric case. In general, the assertion can be shown by using the existence of a canonical mixed Hodge structure on $H^0(Y,\mathbf{H})$ together with the property that the restriction morphism $H^0(Y,\mathbf{H}) \to \mathbf{H}_y$ is a morphism of mixed Hodge structure for any $y \in Y$, see

[Zu] for the Y curve case (where the last property does not seem to be stated explicitly) and [Sa1] in general. For the proof of (1.1.1), the result in [Zu] may be enough since the assertion can be reduced to the curve case (provided that the reader can verify the above property by himself where he must at least understand the definition of L^2 complexes in the Poincaré metric).

1.2. Invariance of geometric fundamental groups by base changes. Let X_k be an algebraic variety over an algebraically closed subfield k of $\mathbf{C}$, and X be the base change of X_k by $k \hookrightarrow \mathbf{C}$. There is a canonical isomorphism between the algebraic fundamental groups of X_k and X (see for instance [NaTaMo]). This assertion follows from the theory of Grothendieck on algebraic fundamental groups by using [Gro4, Exp. XIII, Prop. 4.6], according to A. Tamagawa.

By the above invariance together with a well-known comparison theorem for topological and complex algebraic coverings [Gro4, Exp. XII, Cor. 5.2], we get the following:

Let $\pi : Y \to X$ be a topological finite covering. If X has a structure of complex algebraic variety defined over $k \subset \mathbf{C}$, then Y has a unique structure of complex algebraic variety defined over k which is compatible with the one on X via π.

Remark 1.3. In the case X is smooth affine, it is rather easy to prove the last assertion except for the uniqueness of the structure over k. In fact, a complex algebraic covering $\pi : Y \to X$ can be obtained from an analytic one by applying GAGA to a normal variety which is finite over a smooth compactification of X. To get a structure of k-variety, consider a finite étale morphism $\pi' : \mathcal{Y} \to X \times S$ of complex algebraic varieties defined over k such that its restriction over $X \times \{s_0\}$ is isomorphic to $\pi : Y \to X$, where S is a complex affine variety defined over k and s_0 is a k-generic point of S. We may assume that there is a relative smooth compactification of $\mathcal{Y}$ over S such that its boundary is a family of divisors with normal crossings over S (shrinking S if necessary). These are all defined over k. Then it is enough to restrict π' over $X \times \{s\}$ for some k-rational point $s \in S$.

2 Proofs of Theorems 1 and 2

In this section we prove the main theorems.

2.1. Proof of Theorem 1. The assertion is reduced to the X affine case by taking a k-affine open subset of X containing z_0 and by using the closure. Let Z^o denote the smooth part of Z. There is a finite étale morphism $\pi_{Z^o} : \widetilde{Z}^o \to Z^o$ such that

$$\mathbf{H}_{\widetilde{Z}^o,\mathbf{Z}} := \pi_{Z^o}^* \mathbf{H}_{\mathbf{Z}}|_{Z^o}$$

has a global section $\eta_{\widetilde{Z}^o}$ which has everywhere type (0,0). Here π_{Z^o} is finite by using the polarization restricted to the type (0,0) part (on which the Weil operator is trivial), see [CaDeKa].

The morphism π_{Z^o} can be extended to a finite morphism

$$\pi_Z : \widetilde{Z} \to Z \text{ with } \widetilde{Z} \text{ normal.}$$

This may be obtained, for instance, by considering a smooth variety which is proper over X and contains $\widetilde{Z}^o$ as a dense open subvariety, and then using the Stein factorization (i.e., Spec of the direct image of its structure sheaf, see [Ha]).

Then $\eta_{\widetilde{Z}^o}$ is uniquely extended to a global section $\eta_{\widetilde{Z}}$ of

$$\mathbf{H}_{\widetilde{Z},Z} := \pi_Z^* \mathbf{H}_Z|_Z.$$

Indeed, $\eta_{\widetilde{Z}^o}$ is a global section defined on the dense Zariski-open subset $\widetilde{Z}^o$ of $\widetilde{Z}$, and $\widetilde{Z}$ is analytic-locally irreducible by using the condition that $\widetilde{Z}$ is normal.

Since Z is affine and π_Z is finite, we see that $\widetilde{Z}$ is also affine. So there is a closed embedding $\widetilde{Z} \hookrightarrow \mathbf{C}^n$, which implies a closed embedding

$$\iota : \widetilde{Z} \hookrightarrow \widetilde{X} := X \times \mathbf{C}^n,$$

such that its composition with the projection $pr_1 : \widetilde{X} \to X$ coincides with the composition of π_Z with $Z \hookrightarrow X$. Note that $\widetilde{X}$ and the projection $pr_1 : \widetilde{X} \to X$ are defined over k. Moreover, we may assume that there is a k-rational point $\widetilde{z}_0 \in \widetilde{Z}$ over z_0 by changing the affine coordinates of $\mathbf{C}^n$ if necessary. (The above argument can be simplified slightly in case the vector bundle V of the variation of Hodge structure $\mathbf{H}$ is defined over k. In fact, $\widetilde{X}$ can be replaced with V which is trivialized by shrinking X, and it is unnecessary to assume $\widetilde{Z}$ to be normal as long as the section can be extended over $\widetilde{Z}$.)

Consider a spread of $\widetilde{Z} \subset \widetilde{X}$ defined over k

$$\widetilde{Z}_S \subset \widetilde{X} \times S,$$

satisfying

$$\widetilde{Z} = \widetilde{Z}_{S,s_0} \quad \text{in } \widetilde{X}, \tag{2.1.1}$$

where S is an integral (that is, irreducible and reduced) complex affine variety defined over k with $s_0 \in S$ a k-generic point, and

$$\widetilde{Z}_{S,s} := \widetilde{Z}_S \cap \big(\widetilde{X} \times \{s\}\big) \subset \widetilde{X} \quad \text{for } s \in S,$$

(see for instance [Sa3]). We have

$$z_0 \in \widetilde{Z}_{S,s} \quad \text{for any } s \in S, \tag{2.1.2}$$

since

$$\widetilde{Z}_S \cap \big(\{z_0\} \times S\big) = \{z_0\} \times S.$$

Indeed, the left-hand side is defined over k and contains s_0.

There is a sufficiently small open neighborhood U of s_0 in classical topology such that $\eta_{\widetilde{z}}$ can be extended uniquely to a global section $\eta_{\widetilde{Z}_U}$ of

$$\mathbf{H}_{\widetilde{Z}_U,\mathbf{Z}} := \rho^*_{\widetilde{Z}_U}\mathbf{H}_{\mathbf{Z}},$$

where

$$\rho_{\widetilde{Z}_U} : \widetilde{Z}_U := \widetilde{Z}_S \cap (\widetilde{X} \times U) \to X$$

is the composition of the canonical morphisms $\widetilde{Z}_U \to \widetilde{X} \to X$. This can be shown, for instance, by taking a desingularization of a partial compactification of $\widetilde{Z}_S$ which is proper over S and is defined over k, where we assume that the total transform of the boundary is a divisor with simple normal crossings over S, see also [Sa3]. (Here it is also possible to apply the generic base change theorem in [De3] together with (1.1.1). One can also use a Whitney stratification defined over k, which may have been known to many people for a long time, see [Ma], [Ve], [Te], [Li], [Ar], etc., although it seems more difficult to verify all the necessary arguments by oneself. Note that it is much easier to construct controlled vector fields in the normal crossing case than in the general singularity case.)

The restriction of $\eta_{\widetilde{Z}_U}$ to $\widetilde{Z}_s$ has everywhere type (0,0) for any $s \in U$ by (1.1.1) and (2.1.2) (In fact, it is a global section of a polarizable variation of Hodge structure, and has type (0,0) at $\widetilde{z}_0$.) This implies

$$Z = pr_1(\widetilde{Z}_U), \tag{2.1.3}$$

since Z is a union of irreducible components of $\mathrm{Hdg}_X(\mathbf{H})$, where the inclusion $\subset$ is clear by (2.1.1). We then get

$$Z = pr_1(\widetilde{Z}_S), \tag{2.1.4}$$

by considering the pull-backs of local defining functions of Z in X to $\widetilde{Z}_S$, which vanish on $\widetilde{Z}_U \subset \widetilde{Z}_S$. Since $pr_1(\widetilde{Z}_S)$ is defined over k, Theorem 1 is proved.

2.2. Proof of Theorem 2. Let Z and z_0 respectively denote the image of Y and y_0 in X. By [CaDeKa], Y is finite over X, and Z is a closed subvariety of X. We first consider the case

(2.2.1) $\quad Z$ is an irreducible component of $\mathrm{Hdg}_X(\mathbf{H}) \subset X$.

(In general we may only have an inclusion of Z in an irreducible component of $\mathrm{Hdg}_X(\mathbf{H})$, where it is unclear if Z is defined over k.) By (2.2.1) and Theorem 1, Z is defined over k. Take a desingularization $\rho : Z' \to Z$ defined over k. Set

$V_Z := V \times_X Z$, and similarly for $V_{Z'}$. We have the commutative diagram

$$\begin{array}{ccc} Y & \overset{\rho''}{\longleftarrow} & Y' \\ \cap & & \cap \\ V_Z & \overset{\rho'}{\longleftarrow} & V_{Z'} \\ {\scriptstyle p}\downarrow & & \downarrow{\scriptstyle p'} \\ Z & \overset{\rho}{\longleftarrow} & Z' \end{array}$$

where Y' is the proper transform of Y so that Y is the image of Y' by ρ'. We have to show that there is a k-rational point $y_0' \in Y'$ with $\rho''(y_0') = y_0$. (This is non-trivial since only the lower half of the diagram is defined over k.)

Take a general curve $C \subset Z$ defined over k and passing through $z_0 := p(y_0)$. More precisely, this is given by an intersection of general hyperplane sections of Z defined over k and passing through z_0. Then $p^{-1}(C) \cap Y$ is a curve passing through y_0, since the induced morphism $Y \to Z$ is an isomorphism over its image analytic-locally on Y. Let C' be the proper transform of C in Z'. We see that the proper transform Y' of Y contains a k-rational point given by $(y_0, z_0') \in V \times_X C'$ if we choose an appropriate k-rational point $z_0' \in C'$ over $z_0 \in C$.

By replacing X, V, and Y respectively with Z', $V \times_X Z'$, and Y', the assertion is then reduced to the case

$$Z = X. \tag{2.2.2}$$

Here Y is identified with a *multivalued* section of $\mathbf{H_Z}$ which has everywhere type $(0,0)$. In particular, it is nonsingular, and is finite étale over X. Hence it is defined over k by the theory of Grothendieck on fundamental groups [Gro4], see (1.2) above.

We have to show moreover

$$\text{the inclusion } Y \hookrightarrow V \text{ is defined over } k. \tag{2.2.3}$$

For this we may assume that Y is a *univalued* section of $\mathbf{H_Z}$ which has everywhere type $(0,0)$ on X, by using a commutative diagram as above. (More precisely, using the finite étale morphism $\pi : Y \to X$ defined over k, we replace X with Y, V with its base change by π, and Y with an irreducible component of its base change by π.)

Consider a spread of $Y \subset V$ defined over k

$$Y_{S'} \subset V \times S',$$

satisfying

$$Y_{S',s_0'} = Y \quad \text{in } V, \tag{2.2.4}$$

where S' is a complex affine variety defined over k with $s'_0 \in S'$ a k-generic point, and

$$Y_{S',s'} := Y_{S'} \cap \big(V \times \{s'\}\big) \subset V \text{ for } s' \in S',$$

(see for instance [Sa3]). By the same argument as in the proof of (2.1.2), we have

$$y_o \in Y_{S',s'} \text{ for any } s' \in S'. \tag{2.2.5}$$

We show that there is a k-Zariski open subset

$$U' \subset X \times S'$$

containing $X \times \{s'_0\}$ and such that the restriction of the projection $Y_{S'} \to X \times S'$ over U' is an isomorphism. Calculating the differential of $Y_{S'} \to X \times S'$, we first get a non-empty k-open subvariety $U'' \subset Y_{S'}$ which is étale over $X \times S'$. By [Gro2, Theorem 4.4.3], we have an open immersion $U'' \hookrightarrow \overline{U}''$ and a finite morphism $\overline{U}'' \to X \times S'$, which are defined over k and such that their composition coincides with the canonical morphism $U'' \to X \times S'$. Set $D = \overline{U}'' \setminus U''$. Its image D' in $X \times S'$ is defined over k and hence cannot contain $X \times \{s'_0\}$. (Indeed, if D' contains $X \times \{s'_0\}$, then $D' = X \times S'$ by restricting D' to $\{x\} \times S'$ for any k-rational point x of X. But this is a contradiction.) So there is $(x, s'_0) \in X \times \{s'_0\}$ which is not contained in D'. This implies that the finite morphism $\overline{U}'' \to X \times S'$ has degree 1 and

$$D' \cap (X \times \{s'_0\}) = \emptyset,$$

(by replacing U'' if necessary). So it induces an isomorphism

$$U'' \xrightarrow{\sim} U' := (X \times S') \setminus D'.$$

We thus get a section η of $\mathcal{O}_{U'}(V_{U'})$ defined over k, where $V_{U'} \to U'$ is the base change of $V \to X$ by the projection $U' \to X$.

We have the induced relative connection

$$\nabla' : \mathcal{O}_{U'}(V_{U'}) \to \Omega^1_{U'/S'} \otimes_{\mathcal{O}_{U'}} \mathcal{O}_{U'}(V_{U'}).$$

Consider the zero locus of

$$\nabla' \eta \in \Gamma\big(U', \Omega^1_{U'/S'} \otimes_{\mathcal{O}_{U'}} \mathcal{O}_{U'}(V_{U'})\big).$$

It is a Zariski-closed subset of U', which is defined over k and contains $X \times \{s'_0\}$. Hence it coincides with U'. (This is shown by restricting to $\{x\} \times S'$ for any k-rational point x of X.) This implies that $Y_{S',s'}$ in (2.2.5) is contained in the image of a horizontal section of $V \to X$ for any $s' \in S'$ (by shrinking S' if necessary). Then (2.2.5) implies that $Y_{S',s'}$ is contained in Y. So Theorem 2 is proved in case (2.2.1) is satisfied.

In case the image of Y in X is a proper subvariety of an irreducible component Z' of $\mathrm{Hdg}_X(\mathbf{H})$, we can replace X with a desingularization of Z', and $\mathbf{H}$ with its pull-back (by using the same argument as above where C is a general curve contained in the image of Y in X). Moreover, we can divide $\mathbf{H}$ by the maximal subvariation of $\mathbf{Z}$-Hodge structure of type $(0,0)$ over X contained in $\mathbf{H}$. Then we can proceed by induction on the rank of $\mathbf{H}$. This finishes the proof of Theorem 2.

Remark 2.3. Condition (a) in Corollary 1 is satisfied if there is a subset $\Sigma \subset Y(\alpha)$ which is not contained in any countable union of proper closed subvarieties of $Y(\alpha)$ and such that the Hodge conjecture holds for any $\beta \in \Sigma$. In fact, for an irreducible component S of the Hilbert scheme of $\mathcal{Y} \to X$ which has a flat family over it whose fibers are closed subschemes of codimension m in fibers of $\mathcal{Y} \to X$, we have the cycle map $S \to V$ over X, and the above condition implies that $Y(\alpha)$ coincides with the image of $S \times S'$ for certain two components S, S', where the morphism is given by $cl(\xi_s) - cl(\xi'_{s'})$ for closed points s, s' of S, S' if we denote by ξ_s the associated algebraic cycle over s, and similarly for $\xi'_{s'}$. This assertion follows from the countability of the irreducible components of the Hilbert scheme (using the Hilbert polynomials as is well-known, see [Gro1]). Then condition (a) holds since the Hilbert scheme of $\mathcal{Y} \to X$ is isomorphic to the base change of that for $\mathcal{Y}_k \to X_k$. (The last property follows from the definition of Hilbert schemes using representable functors.) Here we also need the fact that the cycle map can be defined by using relative de Rham cohomology sheaves (or algebraic D-modules) so that it is compatible with the base change by $k \hookrightarrow \mathbf{C}$, see [Ja], [Sa2].

It is well-known that condition (b)$''$ in Corollary 2 is satisfied if the Hodge conjecture holds for α, see [De4]. (This can be verified by using a construction as in Remark (3.6)(ii) below).

3 Relation with absolute Hodge classes

In this section we explain the relation with the theory of absolute Hodge classes.

3.1. Corollaries 1, 2 and 3 can be generalized naturally to the case where $\mathbf{Q}$ is replaced by a finitely generated subfield k of $\mathbf{C}$ (so k is not algebraically closed in this section). Here "absolutely Hodge" means that a Hodge class is "absolutely Hodge over k"; that is, the image of the Hodge class by the action of any element of $\mathrm{Aut}(\mathbf{C}/k)$ (instead of $\mathrm{Aut}(\mathbf{C}) = \mathrm{Aut}(\mathbf{C}/\mathbf{Q})$) is still a Hodge class. (Here k is countable. In fact, if k is countable, then $k[x]$, $k(x)$ and an algebraic closure of k are countable, where x is an algebraically independent variable.)

Assume X, $\mathcal{Y}$, V in Corollary 1 are base changes of X_k, $\mathcal{Y}_k$, V_k by $k \hookrightarrow \mathbf{C}$, and moreover

$$X_k := \operatorname{Spec} A_k, \quad V_k := \operatorname{Spec} B_k,$$

by restricting $\mathcal{Y}$, V over an affine open subvariety of X defined over k. Here A_k is a k-algebra of finite type, and $B_k = \operatorname{Sym}^{\bullet}_{A_k} M_k^{\vee}$ (the symmetric algebra of the dual $M_k^{\vee}$ of M_k over A_k) with

$$M_k := \Gamma(X_k, R^{2m} f_* \Omega^{\bullet}_{\mathcal{Y}_k/X_k}).$$

We assume that X_k is smooth over k. However, it is not necessarily absolutely irreducible, and X is not necessarily connected. Here X_k is absolutely irreducible over $k' := k(X_k) \cap \overline{k}$ which is a finite extension of k. (In fact, the k-variety X_k is actually a variety over $k' \supset k$; that is, the structure morphism $X_k \to \operatorname{Spec} k$ naturally factors through $\operatorname{Spec} k'$, since $\Gamma(X_k, \mathcal{O}_{X_k})$ contains k' by using the normality of X_k. Moreover k' is algebraically closed in the function field of X_k. The absolute irreducibility of X_k over k' then follows from the theory of regular extensions by Weil, see, for instance, [La]. Note that X_k is isomorphic to an irreducible component of $X_{k'} := X_k \times_{\operatorname{Spec} k} \operatorname{Spec} k'$, although the number of irreducible components of $X_{k'}$ does not necessarily coincide with that of X, which is equal to the extension degree $[k' : k]$, unless k'/k is Galois.) So one can assume X connected by replacing X with a connected component if k can be replaced by k'. (This non-connectivity of X is related to the existence of a variety Z over $\mathbf{C}$ which is not homeomorphic to Z^{σ} for some $\sigma \in \operatorname{Aut}(\mathbf{C})$.)

Shrinking further X_k if necessary, we have free generators $v_1, \ldots, v_r$ of M_k over A_k. They induce an isomorphism (k-Zariski-locally on X)

$$V = X \times \mathbf{C}^r.$$

Assume furthermore

$$A_k = k[x_1, \ldots, x_n]/\mathfrak{p},$$

with $\mathfrak{p}$ a prime ideal. Then X is identified with the zero locus of $\mathfrak{p}$ in $\mathbf{C}^n$ so that each morphism of k-algebras $\psi : A_k \to \mathbf{C}$ corresponds to

$$\xi = (\xi_1, \ldots, \xi_n) \in X \subset \mathbf{C}^n \text{ with } \xi_j = \psi(x_j) \ (\forall j \in [1, n]).$$

The action of $\sigma \in \operatorname{Aut}(\mathbf{C}/k)$ on V is expressed by

$$V \ni (\xi_1, \ldots, \xi_n; \zeta_1, \ldots, \zeta_r) \mapsto (\xi_1^{\sigma}, \ldots, \xi_n^{\sigma}; \zeta_1^{\sigma}, \ldots, \zeta_r^{\sigma}) \in V,$$

where ξ^{σ} for $\xi \in \mathbf{C}$ denotes the contravariant (or right) action of σ defined by the inverse of the natural action of σ on $\mathbf{C}$ so that $(\xi^{\sigma})^{\sigma'} = \xi^{\sigma\sigma'}$ holds. (If $Z = \operatorname{Spec} \mathbf{C}[x]/(f_1, \ldots, f_r)$ with $f_i = \sum_{\nu} a_{i,\nu} x^{\nu}$ and $a_{i,\nu} \in \mathbf{C}$, then we have $Z^{\sigma} = \operatorname{Spec} \mathbf{C}[x]/(f_1^{\sigma}, \ldots, f_r^{\sigma})$ with $f_i^{\sigma} = \sum_{\nu} a_{i,\nu}^{\sigma} x^{\nu}$. Here Z^{σ} is a scheme which is isomorphic to Z as a scheme, but whose structure over $\operatorname{Spec} \mathbf{C}$ is given by the composition with $\sigma^* : \operatorname{Spec}\mathbf{C} \to \operatorname{Spec}\mathbf{C}$. Hence Z^σ in this paper means $Z^{\sigma^{-1}}$ in [ChSch].)

Let $\eta \in M_{\mathbf{C}} := M_k \otimes_k \mathbf{C}$. This can be identified with

$$(g_1,\dots,g_r) \in A^n_{\mathbf{C}} \quad \text{such that} \quad \eta = \textstyle\sum_i g_i v_i,$$

where $A_{\mathbf{C}} := A_k \otimes_k \mathbf{C}$. The value of the section η at $\xi = (\xi_1,\dots,\xi_n) \in X$ is identified with

$$(\xi; g_1(\xi),\dots,g_r(\xi)) \in X \times \mathbf{C}^r.$$

Its image by the action of $\sigma \in \operatorname{Aut}(\mathbf{C}/k)$ is given by

$$(\xi^\sigma; g_1(\xi)^\sigma,\dots,g_r(\xi)^\sigma) \in X \times \mathbf{C}^r.$$

If $\eta \in M_k$, or equivalently, if $g_i \in A_k$ $(\forall i)$, then the image coincides with

$$(\xi^\sigma; g_1(\xi^\sigma),\dots,g_r(\xi^\sigma)) \in X \times \mathbf{C}^r,$$

and is still contained in the section η (as it should be since η is defined over k). This argument together with a construction in Remark (3.6)(ii) below can be used to show that an algebraic cycle class is absolutely Hodge.

Let $\alpha \in V$. This defines a prime ideal $\mathfrak{p}_\alpha \subset B_k$ as in [We]. However, $\operatorname{Spec} B_k/\mathfrak{p}_\alpha$ is not necessarily absolutely irreducible. Let $Z(\alpha,k)$ denote the zero locus of $\mathfrak{p}_\alpha$ in V. In other words, $Z(\alpha,k)$ is the closure of α in the k-Zariski topology of V. It is a finite union of irreducible closed subvarieties defined over the algebraic closure $\overline{k}$ of k in $\mathbf{C}$, and we have

(3.1.1) $\operatorname{Gal}(\overline{k}/k)$ acts transitively on the set of irreducible components of $Z(\alpha,k)$.

Here the number of irreducible components of $Z(\alpha,k)$ is equal to the extension degree $[k_\alpha : k]$ with k_α the algebraic closure of k in the field of fractions of $B_k/\mathfrak{p}_\alpha$ (by using the theory of regular extensions as is explained in the beginning of this section).

Let $Z(\alpha,\overline{k}) \subset V$ denote the zero locus of the ideal of $B_k \otimes_k \overline{k}$ associated with α. This is the smallest irreducible closed subvariety of V containing α and defined over $\overline{k}$. In other words, $Z(\alpha,\overline{k})$ is the closure of α in the $\overline{k}$-Zariski topology. We have moreover

(3.1.2) $Z(\alpha,\overline{k})$ coincides with the irreducible component of $Z(\alpha,k)$ containing α.

Here the uniqueness of the component follows from a well-known assertion that the singular locus of $Z(\alpha,k)$ is defined over k. Note that, if $G'_\alpha \subset \operatorname{Gal}(\overline{k}/k)$ denotes the stabilizer of $Z(\alpha,\overline{k})$, then the subgroup G'_α corresponds to the finite extension $k_\alpha \supset k$, where k_α is as in a remark after (3.1.1). Moreover this corresponding field k_α coincides with the minimal field of definition of $Z(\alpha,\overline{k})$

in V by using the functoriality of the Galois descent (which we apply to a finite Galois extension of k_α over which $Z(\alpha,\overline{k})$ is defined).

Let x_0 be the image of α in X. If this is a k-generic point of X, then it defines the zero ideal of A_k, and $\operatorname{Spec} B_k/\mathfrak{p}_\alpha$ is dominant over $\operatorname{Spec} A_k$. So we get

(3.1.3) $\quad \dim Z(\alpha,k) \geqslant \dim X$ if x_0 is a k-generic point.

We have the following well-known assertion, which shows that it is actually enough to consider all the embeddings $K \hookrightarrow \mathbf{C}$ (instead of all the automorphisms of $\mathbf{C}$) over k in order to see whether α is absolutely Hodge over k, where $K \subset \mathbf{C}$ is a fixed subfield containing k and over which α is defined, see also [Vo2], [ChSch].

Proposition 3.2. *The set of points of V which are conjugate to α by automorphisms of $\mathbf{C}$ over k is the complement in $Z(\alpha,k)$ of the countable union of all the proper closed subvarieties defined over k.*

Proof. Each point β of $Z(\alpha,k)$ corresponds to a morphism of k-algebras

$$\psi_\beta : B_k/\mathfrak{p}_\alpha \to \mathbf{C}.$$

This is not necessarily injective. It holds if and only if β is not contained in any proper closed subvariety of $Z(\alpha,k)$ defined over k.

By the definition of $\mathfrak{p}_\alpha$, we have the injectivity of

$$\psi_\alpha : B_k/\mathfrak{p}_\alpha \hookrightarrow \mathbf{C},$$

and $\alpha^\sigma \in V$ for $\sigma \in \operatorname{Aut}(\mathbf{C}/k)$ corresponds to the composition

$$B_k/\mathfrak{p}_\alpha \overset{\psi_\alpha}{\hookrightarrow} \mathbf{C} \overset{\sigma^{-1}}{\longrightarrow} \mathbf{C},$$

where σ^{-1} appears in order to get the contravariant action. Note that $\mathfrak{p}_\alpha = \mathfrak{p}_{\alpha^\sigma}$ since B_k is a quotient of a polynomial ring over k.

Let K be the field of fractions of $B_k/\mathfrak{p}_\alpha$. Proposition (3.2) is then reduced to the following well-known assertion:

(3.2.1) For any two inclusions $\psi_a : K \hookrightarrow \mathbf{C}$ over k ($a = 1,2$), there is $\sigma \in \operatorname{Aut}(\mathbf{C}/k)$ with $\sigma \circ \psi_1 = \psi_2$.

(This can be shown by using the theory of transcendence bases.) This finishes the proof of Proposition (3.2).

We prove Corollaries 1, 2 and 3 with $\mathbf{Q}$ replaced by k, where "absolutely Hodge" means "absolutely Hodge over k." The reader may assume $k = \mathbf{Q}$.

3.3. Proof of Corollary 1. The implications (a) $\Longrightarrow$ (a)$''$ $\Longrightarrow$ (a)$'$ are clear, and we have (a)$'$ $\Longrightarrow$ (a) by Theorem 2. So Corollary 1 is proved.

3.4. Proof of Corollary 2. Conditions (b) clearly implies (b)$''$ by using condition (a)$''$. Assume condition (b)$''$ holds, i.e., α is absolutely Hodge over k.

By Proposition (3.2) we have

$$Z(\alpha,k) \subset \mathrm{Hdg}_V(\mathbf{H}), \tag{3.4.1}$$

by using the fact that $\mathrm{Hdg}_V(\mathbf{H})$ is locally a finite union of closed analytic subsets (since it is contained in the $\mathbf{Z}$-lattice). Note that (3.4.1) follows from (b)$''$ without assuming (a). Combining (3.4.1) with condition (a)$''$, we get condition (b)$'$.

Assume condition (b)$'$. We see that $Y(\alpha)^\sigma$ in condition (a)$''$ is contained in the vector subbundle corresponding to the Hodge filtration $F^0\mathcal{O}_X(V)$ since the latter is defined over k. We have to show that $Y(\alpha)^\sigma$ is also contained in the analytic subset of V corresponding to $\mathbf{H}_\mathbf{Z}$. Here $Y(\alpha)^\sigma$ can be identified with a flat section of V locally on X in k-étale topology by using the fact that the Gauss-Manin connection is defined over k. In fact, let $\widetilde{Y}(\alpha)$ be the normalization of the k-closure of $Y(\alpha)$ in V, which is identified with the disjoint union

$$\bigsqcup_{i=1}^d Y(\alpha)^{\sigma_i},$$

where the σ_i are representatives of $G_\alpha \backslash G$ with $G_\alpha \subset G := \mathrm{Gal}(\overline{k}/k)$ the stabilizer of $Y(\alpha)$, and $d := |G_\alpha \backslash G|$. There is a canonical étale morphism defined over k

$$\widetilde{\pi} : \widetilde{Y}(\alpha) \to X,$$

which factors through

$$\bigcup_i Y(\alpha)^{\sigma_i} \subset V.$$

This defines a section η of the vector bundle

$$V \times_X \widetilde{Y}(\alpha) \to \widetilde{Y}(\alpha),$$

which is the base change of $V \to X$ by $\widetilde{\pi}$. By definition the image of the section η in V coincides with $\bigcup_i Y(\alpha)^{\sigma_i}$. This section η is flat, since it is defined over k and is flat over the connected component

$$Y(\alpha) = Y(\alpha)^{\sigma_1} \subset \widetilde{Y}(\alpha),$$

where $\sigma_1 = id$. So we get the desired flatness of $Y(\alpha)^\sigma$ since $Y(\alpha)^\sigma = Y(\alpha)^{\sigma_i}$ for some i. This implies that $Y(\alpha)^\sigma$ is contained in the analytic subset of V corresponding to $\mathbf{H}_\mathbf{Z}$, since $Y(\alpha)^\sigma$ is contained in the subset at $\beta^\sigma \in Y(\alpha)^\sigma$ by condition (b)$'$. So condition (b) holds. Thus Corollary 2 is proved.

3.5. Proof of Corollary 3. By Corollaries 1 and 2, it remains to show (b)$'' \Longrightarrow$ (a)$''$ + (c). Assume condition (b)$''$. Then (3.4.1) holds. Together with (3.1.3) this implies

$$\dim Z(\alpha,k) = \dim X = \dim Y(\alpha).$$

Hence $Y(\alpha)$ is finite étale over X, and condition (c) holds. This implies that $Y(\alpha)$ is the unique irreducible component of $\mathrm{Hdg}_V(\mathbf{H})$ containing α, and we get

$$Z(\alpha,\overline{k}) = Y(\alpha). \tag{3.5.1}$$

So condition (a)$''$ also holds. Thus Corollary 3 is proved.

Remarks 3.6. (i) We can also prove the implication (b)$' \Longrightarrow$ (b)$''$ by using Principle B (see Remark (iii) below) where the base change by the étale morphism $\widetilde{\pi} : \widetilde{Y}(\alpha) \to X$ as in (3.3) is also used. Here we need Theorem 2 to show that $Y(\alpha)$ is defined over $\overline{k}$.

(ii) If $\alpha \in V$ is associated with an algebraic cycle and if the cycle is extended to a family of algebraic cycles over X defined over k, then it corresponds to an element of M_k which corresponds to a section of $V \to X$ defined over k. The second condition is satisfied if we take a finitely generated k-subalgebra A'_k of $\mathbf{C}$ over which $\mathcal{Y}_{x_0}$ and the cycles are defined, and if we replace $f : \mathcal{Y} \to X$ by the family over $\mathrm{Spec}\, A'_k \otimes_k \mathbf{C}$, see [Vo1], [Sa2]. (Here $\mathrm{Spec}\, A'_k$ is not necessarily absolutely irreducible, and $\mathrm{Spec}\, A'_k \otimes_k \mathbf{C}$ is not necessarily connected.) The latter construction can be used to show that an algebraic cycle class is absolutely Hodge.

(iii) Let $f : \mathcal{Y} \to X$ and V be as in Corollary 1. Let η be a flat section of V defined on a connected component of X. Assume X_k is irreducible. Then Principle B in [De4], [ChSch] asserts that, if the value of η at one point is absolutely Hodge over k, then the value at any point is. This follows from the assertion (1.1.1) which is a consequence of the global invariant cycle theorem [De2] in this case. Indeed, (1.1.1) together with the assumption of Principle B implies that the image η^σ of the section η by the action of $\sigma \in \mathrm{Aut}(\mathbf{C}/k)$ has type $(0,0)$ at every point.

(iv) If there is a subset $\Sigma \subset Y(\alpha)$ which is not contained in any countable union of proper closed subvarieties of $Y(\alpha)$ and such that any $\beta \in \Sigma$ is absolutely Hodge, then conditions (a) and (b) would hold in the notation of Corollaries 1 and 2, where $\mathbf{Q}$ is replaced with k as in (3.1). In fact, consider all the closed subvarieties of V defined over $\overline{k}$ and not entirely containing $Y(\alpha)$. These are countable, and there is $\beta \in \Sigma$ which is not contained in any such varieties, and is not a singular point of $\mathrm{Hdg}_V(\mathbf{H})$. Then the $\overline{k}$-Zariski closure of β in V is contained in $Y(\alpha)$ and its images by the action of $\mathrm{Gal}(\overline{k}/k)$ are contained in $\mathrm{Hdg}_V(\mathbf{H})$, since β is absolutely Hodge over k. (Here the condition that β is a smooth point of $\mathrm{Hdg}_V(\mathbf{H})$ implies that $Y(\alpha)$ is the only irreducible component containing it.) If this $\overline{k}$-closure does not coincide with $Y(\alpha)$, then this contradicts the condition that β is not contained in any closed subvariety of V which is defined over $\overline{k}$ and does not contain entirely $Y(\alpha)$. So conditions (a) and (b) follow. (See also [Vo2].)

References

[Ar] Arapura, D., An abelian category of motivic sheaves, Adv. Math. 233 (2013), 135–195.

[CaDeKa] Cattani, E., Deligne, P. and Kaplan, A., On the locus of Hodge classes, J. Amer. Math. Soc. 8 (1995), 483–506.

[ChSch] Charles, F. and Schnell, Ch., "Absolute Hodge classes". In *Hodge theory*, Mathematical Notes, 49, Princeton Univ. Press, Princeton, NJ, 2014. pp. 469–530.

[De1] Deligne, P., Equations différentielles à points singuliers réguliers, Lect. Notes in Math. 163, Springer, Berlin, 1970.

[De2] Deligne, P., Théorie de Hodge II, Publ. Math. IHES 40 (1971), 5–58.

[De3] Deligne, P., Théorème de finitude en cohomologie ℓ-adique, Lect. Notes in Math. 569, Springer, Berlin, 1977, pp. 233–261.

[De4] Deligne, P., Hodge cycles on abelian varieties (notes by J. S. Milne), Lect. Notes in Math. 900, Springer, Berlin, 1982, pp. 9–100.

[Gri] Griffiths, P. A., Periods of integrals on algebraic manifolds III. Some global differential-geometric properties of the period mapping, Publ. Math. IHES 38 (1970), 125–180.

[Gro1] Grothendieck, A., Techniques de construction et théorèmes d'existence en géométrie algébrique IV: les schémas de Hilbert, Séminaire Bourbaki, 1960–1961, Exp. 221, pp. 249–276.

[Gro2] Grothendieck, A., Eléments de géométrie algébrique, III-1, Publ. Math. IHES 11, 1961.

[Gro3] Grothendieck, A., Eléments de géométrie algébrique, IV-2, Publ. Math. IHES 24, 1965.

[Gro4] Grothendieck, A., Revêtements étales et groupe fondamental, Séminaire de Géométrie Algébrique 1, Lect. notes in Math. 224, Springer, Berlin, 1971.

[Ha] Hartshorne, R., Algebraic Geometry, Springer, Berlin, 1977.

[Ja] Jannsen, U., Mixed motives and algebraic K-theory, Lect. Notes in Math. 1400, Springer, Berlin, 1990.

[La] Lang, S., Algebra, Addison-Wesley, 1993.

[Li] Lipman, J., Equisingularity and simultaneous resolution of singularities, in Resolution of singularities (Obergurgl, 1997), Progr. Math. 181, Birkhäuser, Basel, 2000, pp. 485–505.

[Ma] Mather, J., Notes on topological stability (Harvard, 1970), Bull. Amer. Math. Soc. (N.S.) 49 (2012), 475–506.

[NaTaMo] Nakamura, H., Tamagawa, A. and Mochizuki, S., The Grothendieck conjecture on the fundamental groups of algebraic curves, Sugaku Expositions 14 (2001), 31–53.

[Sa1] Saito, M., Mixed Hodge modules, Publ. RIMS, Kyoto Univ. 26 (1990), 221–333.

[Sa2] Saito, M., Arithmetic mixed sheaves, Inv. Math. 144 (2001), 533–569.

[Sa3] Saito, M., Normal functions and spread of zero locus (in these proceedings).

[Te] Teissier, B., Variétés polaires, II, Multiplicités polaires, sections planes, et conditions de Whitney, Lect. Notes in Math. 961, Springer, Berlin, 1982, pp. 314–491.

[Ve] Verdier, J.-L., Stratifications de Whitney et théorème de Bertini-Sard, Inv. Math. 36 (1976), 295–312.

[Vo1] Voisin, C., Transcendental methods in the study of algebraic cycles, Lect. Notes in Math. 1594, Sringer, Berlin, 1994, pp. 153–222.

[Vo2] Voisin, C., Hodge loci and absolute Hodge classes, Compos. Math. 143 (2007), 945–958.

[We] Weil, A., Foundation of algebraic geometry, AMS 1946.

[Zu] Zucker, S., Hodge theory with degenerating coefficients. L_2 cohomology in the Poincaré metric, Ann. Math. (2) 109 (1979), 415–476.

RIMS Kyoto University, Kyoto 606-8502 Japan
E-mail address: `msaito@kurims.kyoto-u.ac.jp`

Department of Mathematics, Stony Brook University, Stony Brook, NY 11794, USA
E-mail address: `cschnell@math.sunysb.edu`

11

Tate twists of Hodge structures arising from abelian varieties

Salman Abdulali

ABSTRACT. We consider the category of Hodge substructures of the cohomology of abelian varieties, and ask when a Tate twist of such a Hodge structure belongs to the same category.

1 Introduction

A (rational) Hodge structure $V_{\mathbb{C}} = \bigoplus_{p+q=n} V^{p,q}$ is said to be *effective* if $V^{p,q} = 0$ unless p, $q \geq 0$, and, it is said to be *geometric* if it is isomorphic to a Hodge substructure of $H^n(X, \mathbb{Q})$ for some smooth, projective variety X over $\mathbb{C}$. For $m \in \mathbb{Z}$, the Tate twist $V(m)$ is the Hodge structure of weight $n-2m$ defined by $V(m)^{p,q} = V^{p+m,q+m}$.

A geometric Hodge structure must be effective and polarizable, but not conversely (Grothendieck [17, p. 300, 2nd footnote]). Any effective and polarizable Hodge structure of weight 1 is the first cohomology of an abelian variety, and hence geometric. In [7] we have shown that any Hodge structure of CM-type is geometric. These are the only known criteria for an abstract Hodge structure to be geometric [15, p. 305].

The general Hodge conjecture as formulated by Grothendieck [17] implies that any effective Tate twist of a geometric Hodge structure is again geometric. In a series of papers [1–9] we have shown that, for certain abelian varieties A, every effective Tate twist of a Hodge structure in the cohomology of A is isomorphic to a Hodge structure occurring in the cohomology of some abelian variety. Moreover, we have used this to prove the general Hodge conjecture for certain abelian varieties. We have also shown the existence of a Hodge structure which occurs in the cohomology of an abelian variety, but which has an effective Tate twist that does not occur in the cohomology of *any* abelian variety [5, Theorem 5.5, p. 926].

2010 *Mathematics Subject Classification.* Primary 14C30, 14K20.

We now outline the contents and organization of this paper. We begin in §2 with a summary of the definitions and basic properties of the Hodge and Lefschetz groups of an abelian variety, and the Kuga fiber varieties associated with them. In §3 we formulate our problem in representation theoretic form. In the following four sections we review our results for various classes of abelian varieties. In §8, we discuss the abelian varieties for which the general Hodge conjecture is known; the full list of such abelian varieties is given in Appendix A—in all these cases the general Hodge conjecture follows from known cases of the usual Hodge conjecture and the results included in this article. §9 is a remark on the open cases.

The results mentioned above have appeared in various papers with inconsistent definitions, notations and terminology. Our goal in this article is to give a unified presentation, with simplifications and generalizations where possible. Certain errors crept into the earlier articles; Appendix B alerts the reader to them.

Notations and conventions. All representations are finite-dimensional and algebraic. The derived group of a group G is denoted by G'. All abelian varieties are over $\mathbb{C}$. For an abelian variety A, we let $D(A) = \mathrm{End}(A) \otimes \mathbb{Q}$ be its endomorphism algebra, $L(A)$ its Lefschetz group, $G(A)$ its Hodge group, $L'(A)$ the derived group of $L(A)$, and, $G'(A)$ the derived group of $G(A)$. For a finite field extension E of a field F, we let $\mathrm{Res}_{E/F}$ be the restriction of scalars functor, from varieties over E to varieties over F. For an algebraic or topological group G, we denote by G^0 the connected component of the identity.

2 Algebraic groups and abelian varieties

2.1. Abelian varieties

Let A be an abelian variety over $\mathbb{C}$, and let $V = H_1(A, \mathbb{Q})$. As a complex manifold, A is the torus $V_{\mathbb{R}}/V_{\mathbb{Z}}$ with a complex structure J on $V_{\mathbb{R}}$, where $V_{\mathbb{Z}}$ is the lattice $H_1(A, \mathbb{Z})$. We view V as a left vector space over the endomorphism algebra $D = D(A) = \mathrm{End}(A) \otimes \mathbb{Q}$.

Recall that a polarization of A induces an involution $a \mapsto \bar{a}$ on D, and a Riemann form β, i.e., an alternating $\mathbb{Q}$-bilinear form on V such that $\beta(x, Jy)$ is symmetric and positive definite, $\beta(V_{\mathbb{Z}}, V_{\mathbb{Z}}) \subset \mathbb{Z}$, and such that $\beta(ax, y) = \beta(x, \bar{a}y)$ for $a \in D$. We use the nondegenerate form β to identify V with its dual $H^1(A, \mathbb{Q})$, and thus $\bigwedge^n V$ with $H^n(A, \mathbb{Q})$.

2.2. Hodge group

The Hodge group (or Special Mumford-Tate group) of A is defined to be the smallest $\mathbb{Q}$-algebraic subgroup $G = G(A)$ of $GL(V)$ such that $G(\mathbb{R})$ contains $\exp(\theta J)$ for all $\theta \in \mathbb{R}$ (Mumford [28]). It is a connected, reductive group characterized by the property that its invariants in $H^\star(A^k,\mathbb{Q})$ are precisely the Hodge classes for any positive integer k.

2.3. Lefschetz group

The Lefschetz group $L(A)$ is defined to be the centralizer of $\mathrm{End}(A)$ in $Sp(V,\beta)$; it is characterized by the property that for any positive integer k, the subring of $H^\star(A^k,\mathbb{Q})$ generated by the classes of divisors equals $H^\star(A^k,\mathbb{C})^{L(A)_\mathbb{C}} \cap H^\star(A^k,\mathbb{Q})$ (Milne [27, Theorem 3.2, p. 656] and Murty [31, §3.6.2, p. 93]). Since any divisor class is a Hodge class, it follows that $G(A) \subset L(A) \subset Sp(V,\beta)$.

2.4. Classification

Assume, now, that A is simple. Then by Albert's classification, D is one of the following [37]:

type I: a totally real number field F;
type II: a totally indefinite quaternion algebra over a totally real number field F;
type III: a totally definite quaternion algebra over a totally real number field F;
type IV: a division algebra over a CM-field E. In this case let F be the maximal totally real subfield of E.

In each case there exists a unique F-bilinear form $T\colon V \times V \to D$ such that $\beta(x,y) = \mathrm{Tr}_{D/\mathbb{Q}} T(x,y)$, $T(ax,by) = aT(x,y)\bar{b}$, and, $T(y,x) = -\overline{T(x,y)}$ for all x, $y \in V$, a, $b \in D$ [38, Lemma 1.2, p. 162]. The Lefschetz group is then the restriction of scalars, from F to $\mathbb{Q}$, of the unitary group of T:

$$L(A) = \mathrm{Res}_{F/\mathbb{Q}}\, U(T) = \mathrm{Res}_{F/\mathbb{Q}}\, \mathrm{Aut}_D(V,T). \tag{2.1}$$

Let S be the set of embeddings of F into $\mathbb{R}$. We can write

$$L(A)_\mathbb{R} = \prod_{\alpha\in S} L_\alpha \qquad \text{and} \qquad V_\mathbb{R} = \bigoplus_{\alpha\in S} V_\alpha, \tag{2.2}$$

where L_α acts trivially on $V_{\alpha'}$ unless $\alpha = \alpha'$. Then L_α and its action on V_α are given as follows [30]:

type I: $L_\alpha = Sp(V_\alpha,\beta_\alpha)$ is a symplectic group acting via its standard representation on V_α;
type II: L_α is a symplectic group acting on V_α as two copies of the standard representation;
type III: $L_{\alpha,\mathbb{C}}$ is an orthogonal group acting on $V_{\alpha,\mathbb{C}}$ as two copies of the standard representation;

type IV: $L_\alpha = U(p_\alpha, q_\alpha)$, and $L_{\alpha,\mathbb{C}} \cong GL_m(\mathbb{C})$ acts on $V_{\alpha,\mathbb{C}}$ as the direct sum of the standard representation and its contragredient.

2.5. Kuga fiber varieties

We review below the construction of Kuga fiber varieties. These are families of abelian varieties which include the solutions to fine moduli problems for abelian varieties with additional structures such as Hodge cycles (Mumford [28]) or PEL-structures (polarizations, endomorphisms, level structures) (Shimura [39]).

Let β be a nondegenerate alternating form on a finite-dimensional vector space V over $\mathbb{Q}$, and L a lattice in V. The symplectic group $Sp(V,\beta)$ is a reductive group of hermitian type, which acts by conjugation on the Siegel space $\mathfrak{S}$ consisting of all complex structures J on $V_\mathbb{R}$ such that $\beta(x,Jy)$ is symmetric and positive definite. If Γ' is a torsion-free arithmetic subgroup of $Sp(V,\beta)$, then $\mathcal{V}' = \Gamma\backslash\mathfrak{S}$ is a quasiprojective algebraic variety, and there exists a universal family of abelian varieties $\mathcal{A}' \to \mathcal{V}'$ such that the fiber over $P \in \mathcal{V}'$ is the torus $V_\mathbb{R}/L$ with the complex structure J, where J is a point of $\mathfrak{S}$ lying over P.

Now let G be a semisimple $\mathbb{Q}$-algebraic group of hermitian type, Γ a torsion-free arithmetic subgroup of $G(\mathbb{Q})$, K a maximal compact subgroup of $G(\mathbb{R})^0$, and, $X = G(\mathbb{R})^0/K$ the corresponding bounded symmetric domain. Suppose given a representation $\rho\colon G \to Sp(V,\beta)$ such that $\rho(\Gamma) \subset \Gamma'$, and an equivariant holomorphic map $\tau : X \to \mathfrak{S}$, between the symmetric domains of the two groups. The family of abelian varieties $\mathcal{A} \to \mathcal{V} = \Gamma\backslash X$ obtained as the pullback of the universal family $\mathcal{A}' \to \mathcal{V}'$ is called a *Kuga fiber variety*, We refer the reader to [1] or [33] for details.

For A as in §2.1, the semisimple parts of the groups $G(A)$ and $L(A)^0$ are of hermitian type, and their inclusions into $Sp(V,\beta)$ define Kuga fiber varieties which are the Hodge families of Mumford [28] and (generalizations of) the PEL-families of Shimura [39], respectively.

2.6. Abelian varieties of PEL-type

We say that an abelian variety A is of PEL-*type* if the semisimple parts of $G(A)$ and $L(A)^0$ are equal. Thus a simple abelian variety is of PEL-type if and only if it is a general member of a PEL-family of abelian varieties (see [1, §1 and §4.6]). This means in essence that for each $k \geq 1$, A^k has no Hodge classes other than those it is required to have by virtue of its endomorphisms. We note that all abelian varieties of CM-type, all abelian varieties of dimension less than 4, and, all abelian varieties for which the general Hodge conjecture is currently known, are of PEL-type.

3 Dominating varieties

We say that a Hodge structure V of weight w is *fully twisted* if V is effective, and $V^{w,0} \neq 0$. Thus V is fully twisted if and only if V is effective, but $V(1)$ is not effective. More generally, for a subfield F of $\mathbb{C}$, we say that an F-subspace U of V_F is *fully twisted* if $U_{\mathbb{C}} \cap V^{w,0} \neq 0$.

Definition 3.1. A smooth, projective, complex algebraic variety A is said to be *dominated* by a class $\mathcal{X}$ of smooth, projective, complex algebraic varieties if, given any irreducible Hodge structure V in the cohomology of A, there exists a fully twisted Hodge structure V' in the cohomology of some $X \in \mathcal{X}$ such that V' is isomorphic to a Tate twist of V.

This definition is motivated by the following observation of Grothendieck (for a proof, see the proof of [1, Proposition 2.1, p. 243]).

Proposition 3.2 (Grothendieck [17, p. 301]). *Let A be a smooth projective variety over $\mathbb{C}$ which is dominated by $\mathcal{X}$. If the usual Hodge conjecture holds for $A \times X$ for each $X \in \mathcal{X}$, then the general Hodge conjecture holds for A.*

Hodge structures in the cohomology of an abelian variety A correspond to $G(A)$-submodules of $H^\star(A,\mathbb{Q})$. Thus A is dominated by $\mathcal{X}$ if and only if given any irreducible $G(A)$-submodule, V, of $H^\star(A,\mathbb{Q})$, there exist $B \in \mathcal{X}$, and a fully twisted $G(B)$-submodule V' of $H^n(B,\mathbb{Q})$ for some n, such that V and V' are isomorphic as $G(A \times B)$-modules. (Note that $G(A \times B)$ is a subgroup of $G(A) \times G(B)$, so it makes sense to consider V and V' as $G(A \times B)$-modules.)

Since it is usually easier to work over $\mathbb{C}$ than over $\mathbb{Q}$, it is natural to modify the above by extending scalars. Given a subfield F of $\mathbb{C}$, which will usually be either $\mathbb{Q}$ or $\mathbb{C}$, we say that A is *F-dominated* by a class $\mathcal{X}$ of abelian varieties if, given any irreducible $G(A)_F$-submodule, V, of $H^\star(A,F)$, there exist $B \in \mathcal{X}$, and a fully twisted $G(B)_F$-submodule V' of $H^n(B,F)$ for some n, such that V and V' are isomorphic as $G(A \times B)_F$-modules. In particular, A is dominated by $\mathcal{X}$ if and only if A is $\mathbb{Q}$-dominated by $\mathcal{X}$. We then have the following results:

Lemma 3.3 (Abdulali [8, Lemma 2, p. 1165]). *If an abelian variety A is F-dominated by $\mathcal{X}$ for some subfield F of $\mathbb{C}$, then A is dominated by $\mathcal{X}$.*

Theorem 3.4 (Abdulali [8, Propositions 3 and 4, p. 1166]). *Let A and B be abelian varieties such that $G(A \times B) = G(A) \times G(B)$.*

1. *If A is $\mathbb{C}$-dominated by $\mathcal{X}$ and B is $\mathbb{C}$-dominated by $\mathcal{Y}$, then $A \times B$ is $\mathbb{C}$-dominated by $\mathcal{X} \cdot \mathcal{Y}$.*
2. *If A is $\mathbb{C}$-dominated by $\mathcal{X}$ and B is dominated by $\mathcal{Y}$, then $A \times B$ is dominated by $\mathcal{X} \cdot \mathcal{Y}$.*

Here, $\mathcal{X} \cdot \mathcal{Y} = \{X \times Y \mid X \in \mathcal{X},\ Y \in \mathcal{Y}\}$.

4 Semisimple Hodge groups

We begin with abelian varieties with semisimple Hodge groups, without factors of type III. Abelian varieties of type III will be dealt with in §7.

Theorem 4.1 (Abdulali [1, Theorem 5.1, p. 348]). *Let A be an abelian variety of* PEL*-type. Suppose that the Hodge group of A is semisimple and A has no factors of type* III. *Then A is* $\mathbb{C}$*-dominated by the set of powers of itself. The usual Hodge conjecture for A implies the general Hodge conjecture for all powers of A.*

Idea of proof. We illustrate the idea of proof of this theorem in the simplest case: a generic abelian variety A of dimension n. Let β be a Riemann form for A. Then the Hodge group of A is the symplectic group $G = Sp(V,\beta)$, acting on $V = H^1(A,\mathbb{Q})$ via the standard representation. By the representation theory of the symplectic group, for $2 \leq k \leq \dim A$, we have

$$\bigwedge^k V = \pi_k \oplus \bigwedge^{k-2} V,$$

where π_k is an irreducible representation of G. Thus

$$H^k(A,\mathbb{Q}) = V_k \oplus H^{k-2}(A,\mathbb{Q})(-1),$$

where V_k is equivalent to π_k as a G-module. Since $H^{k-2}(A,\mathbb{C})(-1)$ cannot contain any $(k,0)$-forms, it follows that all $(k,0)$-forms are in $V_{k,\mathbb{C}}$. For each irreducible representation π_k of G in the cohomology of A we have found a fully twisted Hodge structure V_k in $H^\star(A,\mathbb{Q})$ with π_k equivalent to V_k, proving that A is dominated by itself. Since the usual Hodge conjecture is known for $A \times A$, this proves the general Hodge conjecture for A (due originally to Mattuck [26]). □

The abelian varieties for which the general Hodge conjecture follows from this theorem are listed in Appendix A (see item (1)).

5 The CM-case

A polarizable Hodge structure is said to be of CM-type if its Hodge group is commutative. As shown by Mumford [29], an abelian variety A is of CM-type if and only if the Hodge structure $H^1(A,\mathbb{Q})$ is of CM-type.

Theorem 5.1 (Abdulali [7, Theorem 3, p. 159]). *Every effective Hodge structure of* CM*-type occurs in the cohomology of an abelian variety of* CM*-type.*

Theorem 5.2 (Abdulali [7, Theorem 4, p. 159]). *Any abelian variety of* CM*-type is dominated by the class of all* CM *abelian varieties. If A is a simple abelian variety of* CM*-type with* CM *by E, then any power of A is dominated by the set of products of abelian varieties with* CM *by E.*

Theorem 5.3 (Abdulali [7, Proposition 5, p. 160]). *Let $E_1,\dots,E_m$ be* CM*-fields whose Galois closures are linearly disjoint over $\mathbb{Q}$. For each $i=1,\dots,m$, let A_i be an abelian variety with complex multiplication by E_i, and let $\mathcal{X}_i$ be a class of abelian varieties which dominates A_i. Then $A=\prod_{i=1}^m A_i$ is dominated by $\mathcal{X}=\{\prod_{i=1}^m X_i \mid X_i\in\mathcal{X}_i\}$.*

Using Theorem 5.2, we can prove the general Hodge conjecture for certain abelian varieties (see Appendix A, item (3)). We note that Hazama [19, 20] had independently proved that the usual Hodge conjecture for all CM abelian varieties implies the general Hodge conjecture for all CM abelian varieties.

For further discussion of Theorem 5.1 see [16].

6 Abelian varieties of type IV

We have dealt above with the extreme cases where the Hodge group is either semisimple or commutative. The general situation turns out to be quite subtle. In light of Theorem 5.1, it is tempting to rephrase the definition of $\mathbb{C}$-domination of abelian varieties in terms of just the semisimple part $G'(A)$ of the Hodge group. One might, for example, say that A is "semisimply dominated" by $\mathcal{X}$ if, given any irreducible $G'(A)_{\mathbb{C}}$-submodule V of $H^\star(A,\mathbb{C})$, there exists a fully twisted $G'(B)_{\mathbb{C}}$-submodule V' of $H^\star(B,\mathbb{C})$ for some $B\in\mathcal{X}$, which is equivalent to V as a $G'(A\times B)_{\mathbb{C}}$-module. It is tempting to hope that in this situation A will be dominated by abelian varieties of the form $B\times C$, where $B\in\mathcal{X}$ and C is of CM-type. This is unfortunately not true. It turns out that we require not just V', but all of its Galois conjugates, to be fully twisted. The formal definition follows.

Definition 6.1. An abelian variety A is *semidominated* by a set $\mathcal{X}$ of abelian varieties if, given any nontrivial irreducible representation ρ of $G'(A)_{\mathbb{C}}$ such that ρ occurs in $H^n(A,\mathbb{C})$ for some n, there exist $A_\rho\in\mathcal{X}$, a positive integer c_ρ, and, $V_\rho\subset H^{c_\rho}(A_\rho,\mathbb{C})$, such that

1. V_ρ is a $G(A_\rho)_{\mathbb{C}}$-submodule of $H^{c_\rho}(A_\rho,\mathbb{C})$,
2. the action of $G'(A\times A_\rho)_{\mathbb{C}}$ on V_ρ is equivalent to $\rho\circ p_1$, where

$$G'(A)\times G'(A_\rho)\supset G'(A\times A_\rho)\xrightarrow{p_1}G'(A)$$

   is the projection to the first factor, and,
3. for each $\sigma\in\mathrm{Aut}(\mathbb{C})$, the conjugate $(V_\rho)^\sigma$ contains a nonzero $(c_\rho,0)$-form.

Theorem 6.2 (Abdulali [9, Theorem 7]). *Let A be an abelian variety semidominated by $\mathcal{X}$. Then A is dominated by the set of abelian varieties of the form $B \times C$, where $B \in \mathcal{X}$, and C is of* CM*-type.*

Remark 6.3. The converse of Theorem 6.2 is false. For example, the abelian varieties in Theorem 7.1 are dominated by the set of powers of themselves, but are not semidominated by any class of abelian varieties [9, Remark 5].

Lemma 6.4 (Abdulali [9, Lemma 6]). *If A is semidominated by $\mathcal{X}$ and B is semidominated by $\mathcal{Y}$, and if $G'(A \times B) = G'(A) \times G'(B)$, then $A \times B$ is semidominated by*

$$\mathcal{X} \cdot \mathcal{Y} = \{X \times Y \mid X \in \mathcal{X},\ Y \in \mathcal{Y}\}.$$

In the next two theorems we use these ideas to show that certain abelian varieties of type IV are dominated by abelian varieties. In a few cases we can also deduce the general Hodge conjecture (see Appendix A, item (4)).

Theorem 6.5 (Abdulali [8, Theorem 10, p. 1167]). *Let A be an abelian variety of* PEL*-type such that each simple factor of A is of type* IV*. Then we can write $G'(A)(\mathbb{R}) \cong \prod_{\alpha \in S} SU(p_\alpha, q_\alpha)$. Assume that for each $\alpha \in S$ we have $|p_\alpha - q_\alpha| = 1$. Then A is semidominated by the set of powers of A.*

Theorem 6.6 (Abdulali [9, Theorem 10]). *Let A be an abelian variety such that each simple factor of $G'(A)(\mathbb{R})$ is isomorphic to $SU(p,1)$ for some $p > 1$. Then A is dominated by abelian varieties.*

Idea of proof. Using Lemma 6.4, we reduce to the case of a power of a simple abelian variety A_0. Then $A_0 = \mathcal{A}_P$ is the fiber at a general point P of a Hodge family $\mathcal{A} \to \mathcal{V}$. For $k = 1, \ldots, p$, Satake [32] constructs Kuga fiber varieties $\mathcal{A}_k \to \mathcal{V}$, from symplectic representations $\rho_k \colon G'(A) \to Sp(V_k, \beta_k)$ such that a simple factor of $G'(A)(\mathbb{R})$ acts as the direct sum of $\bigwedge^k$ and its contragredient. Let A_k be the fiber over P of $\mathcal{A}_k$. It follows from Satake's classification [32] that $\mathcal{A}$ is isogenous to $\mathcal{A}_k$ and hence A_0 is isogenous to A_k for some k. Let

$$\mathcal{X} = \left\{A_1^{n_1} \times \cdots \times A_p^{n_p} \mid n_i \geq 0\right\}.$$

We then show that every member of $\mathcal{X}$ is semidominated by $\mathcal{X}$, and hence dominated by abelian varieties. □

Remark 6.7. The families of abelian varieties $\mathcal{A}_k \to \mathcal{V}$ in the proof of Theorem 6.6 are not, in general, PEL-families, so the abelian varieties in $\mathcal{X}$ are not, in general, of PEL-type.

7 Abelian varieties of type III

We now consider a simple abelian variety A of PEL-type whose endomorphism algebra is a totally definite quaternion algebra $D = D(A)$ over a totally real number field F. We use the notations of §2, so that $T\colon V \times V \to D$ is now a quaternionic skew-hermitian form. Let $m = \dim_D V$. The Hodge group $G = G(A)$ is the connected component of the identity of $\mathrm{Res}_{F/\mathbb{Q}} \mathrm{Aut}(V,T)$, and $G(\mathbb{R}) = \prod_{\alpha \in S} G_\alpha$, where each G_α is isomorphic to the group of type D_m called $SO^\star(2m)$ by Helgason [21, p. 445], and $SU^-(m,\mathbb{H})$ by Satake [33, Exercise 2, p. 278].

The discriminant, $\mathrm{discr}\, T$, plays a crucial role here. Theorem 7.1 deals with the cases where $\mathrm{discr}\, T$ is not a square. When $\mathrm{discr}\, T$ is a square, we deal with the cases $m = 4$ and $m > 4$ in Theorems 7.2 and 7.3, respectively.

Theorem 7.1 (Abdulali [5, Theorem 4.1, p. 922]). *Let A be a simple abelian variety of PEL-type such that $D(A)$ is a totally definite quaternion algebra over $\mathbb{Q}$. Let T be the skew-hermitian form determined by a polarization of A. If* $\mathrm{discr}\, T$ *is not a square, then any power of A is dominated by the set of powers of A, and the usual Hodge conjecture for A implies the general Hodge conjecture for all powers of A.*

Outline of proof. We have $V_{\mathbb{C}} = W \oplus \overline{W}$, with $G_{\mathbb{C}}$ acting as the standard representation on both W and $\overline{W}$, and, $\dim W = 2m$. As explained in [5, p. 923], $W = W^{1,0} \oplus W^{0,1}$, where $W^{1,0}$ is an m-dimensional space of $(1,0)$-forms, and $W^{0,1}$ is an m-dimensional space of $(0,1)$-forms. Let $\{u_1,\dots,u_m\}$ be a basis of $W^{1,0}$.

Let $\mathfrak{g}$ be the Lie algebra of $G_{\mathbb{C}}$. We denote by $\mu_1,\dots,\mu_m$, the fundamental weights of $\mathfrak{g}$ with respect to a Cartan algebra $\mathfrak{h}$ (see [5, §3.3, p. 921] and [44, p. 329]). For $1 \le j \le m-2$, $\bigwedge^j W$ is an irreducible $\mathfrak{g}$-module with highest weight μ_j which contains the $(j,0)$-form $u_1 \wedge \dots \wedge u_j$. The irreducible $\mathfrak{g}$-module $\bigwedge^{m-1} W$ has highest weight $\mu_{m-1} + \mu_m$, and contains the $(m-1,0)$-form $u_1 \wedge \dots \wedge u_{m-1}$. However, $\bigwedge^m W$ is the direct sum of two irreducible $\mathfrak{g}$-modules, say U_1 and U_2, having highest weights $2\mu_{m-1}$ and $2\mu_m$ respectively. The $(m,0)$-form $u_1 \wedge \dots \wedge u_m$ is in U_2, while U_1 contains no $(m,0)$-forms. The assumption that $\mathrm{discr}\, T$ is not a square implies that the weights $2\mu_{m-1}$ and $2\mu_m$ belong to the same orbit of $\mathrm{Gal}(\overline{\mathbb{Q}}/\mathbb{Q})$. Hence if a Hodge structure U is such that $U_{\mathbb{C}}$ contains U_1, then it must also contain U_2, and thus be fully twisted.

Now let M be an irreducible Hodge structure in the cohomology of a power of A, and M_0 an irreducible $\mathfrak{g}$-submodule of $M_{\mathbb{C}}$. Let μ be the highest weight of M_0, and write $\mu = a_1\mu_1 + \dots + a_m\mu_m$, for some nonnegative integers $a_1,\dots,a_m$. Since the representation occurs in the tensor algebra generated by $V_{\mathbb{C}}$, we must have $a_{m-1} \equiv a_m \pmod 2$.

If $a_m \geq a_{m-1}$, we have

$$\mu = a_1\mu_1 + \cdots + a_{m-2}\mu_{m-2} + a_{m-1}(\mu_{m-1} + \mu_m) + b_m(2\mu_m)$$

with $b_m = a_m - a_{m-1} \geq 0$. We have seen above that for each of the weights $\mu_1, \ldots, \mu_{m-2}, \mu_{m-1} + \mu_m, 2\mu_m$, there exists an irreducible and fully twisted $\mathfrak{g}$-module in the cohomology of A with that highest weight. Hence, taking tensor products, we obtain a fully twisted $\mathfrak{g}$-module M_1 of highest weight μ in the cohomology of some power of A. An irreducible Hodge structure containing M_1 is fully twisted and isomorphic to a Tate twist of M.

If $a_m < a_{m-1}$, we have

$$\mu = a_1\mu_1 + \cdots + a_{m-2}\mu_{m-2} + a_{m-1}(\mu_{m-1} + \mu_m) + c_m(2\mu_{m-1})$$

with $c_m = a_{m-1} - a_m \geq 0$. This is a Galois conjugate of

$$\mu' = a_1\mu_1 + \cdots + a_{m-2}\mu_{m-2} + a_{m-1}(\mu_{m-1} + \mu_m) + c_m(2\mu_m)$$

so we can apply the arguments of the previous paragraph to a conjugate of M_0 contained in $M_{\mathbb{C}}$. □

Theorem 7.1 implies the general Hodge conjecture for some abelian varieties of type III (see Appendix A, item (2)).

Theorem 7.2 (Abdulali [5, Theorem 5.3, p. 925]). *Let A be a simple abelian variety of* PEL*-type of type* III. *Let T be the skew-hermitian form determined by a polarization of A. If* $\operatorname{discr} T$ *is a square and* $\dim_{D(A)} H_1(V, \mathbb{Q}) = 4$, *then there exists an abelian variety B such that each power of A is $\mathbb{C}$-dominated by the set of abelian varieties of the form $A^i \times B^j$.*

Idea of proof. Assume, for simplicity, that the center of $D(A)$ is $\mathbb{Q}$. The spin group $\mathrm{Spin}(6,2)$ is a 2-fold covering of $SO^\star(8)$, so their symmetric domains are isomorphic (cf. [33, Exercise 1, p. 289]). We thus get two Kuga fiber varieties $\mathcal{A} \to \mathcal{V}$ and $\mathcal{B} \to \mathcal{V}$ over the same arithmetic variety $\mathcal{V}$, where $\mathcal{A} \to \mathcal{V}$ is the PEL-family having A as the fiber over a point P, and, $\mathrm{Spin}(6,2)$ acts on the first cohomology of $B = \mathcal{B}_P$ via the spin representation with highest weight μ_3. Thus $H^2(B \times B, \mathbb{C})$ contains a fully twisted $\mathrm{Spin}(6,2)$-module which is equivalent to the representation U_1 constructed in the proof of Theorem 7.1. □

Theorem 7.3 (Abdulali [5, Theorem 5.5, p. 926]). *Let A be a simple abelian variety of* PEL*-type of type* III. *Let T be the skew-hermitian form determined by a polarization of A. If* $\operatorname{discr} T$ *is a square, and* $\dim_{D(A)} H_1(V, \mathbb{Q}) > 4$, *then A is not dominated by abelian varieties.*

Sketch of proof. Assume, for simplicity, that the center of $D(A)$ is $\mathbb{Q}$. Let A be the fiber at P of the PEL-family $\mathcal{A} \to \mathcal{V}$. Let M be an irreducible Hodge structure in $H^m(A,\mathbb{Q})$ such that $M_{\mathbb{C}}$ contains the representation U_1 constructed in the proof of Theorem 7.1. Then $M(1)$ is effective, but representation theory shows that a representation equivalent to U_1 cannot occur in $H^{m-2}(A^k,\mathbb{C})$ for any positive integer k [5, Lemma 3.3.1, p. 921]. This shows that A is not dominated by the set of powers of A. Next, we prove that if $M(1)$ were to occur in the cohomology of an abelian variety B, then we could take B to be the fiber over P of a Kuga fiber variety $\mathcal{B} \to \mathcal{V}$. Finally, we invoke Satake's classification of Kuga fiber varieties [32, 33] to rule out the existence of such an abelian variety. □

8 The general Hodge conjecture

8.1. Some special cases

In Appendix A we give a list of abelian varieties for which the general Hodge conjecture can be unconditionally proved using the methods of this paper. These include the following interesting special cases:

1. Combining cases (1a), (3c), and, (5) of Appendix A, we obtain the general Hodge conjecture for any product of three (or fewer) elliptic curves.
2. An abelian surface is either a product of two elliptic curves, or, of CM-type (Case (3d)), or, its Hodge group is simple and it belongs to Case (1a). Thus the general Hodge conjecture is true for any power of an abelian surface.
3. Bardelli [10] proves the general Hodge conjecture for various abelian 3-folds. All of these are included in our list.

8.2. An open problem

Let E be a quadratic imaginary number field, and A an abelian m-fold of PEL-type with $D(A) = E$. Then the Lefschetz group of A is a $\mathbb{Q}$-form of $U(p,q)$, where (p,q) is the signature of the hermitian form determined by a polarization of A. The Weil Hodge structure $W \subset H^m(A,\mathbb{Q})$ is a 2-dimensional Hodge structure of type $\{(p,q),(q,p)\}$. Schoen [35, §6] has proved the general Hodge conjecture for W in the following cases:

1. $m = 4$, $(p,q) = (3,1), E = \mathbb{Q}(\sqrt{-1})$, and,
2. $m = 6$, $(p,q) = (5,1), E = \mathbb{Q}(\sqrt{-3})$.

W is of CM-type, and isomorphic to a Hodge substructure of C^w where C is an elliptic curve with CM by E, and, $w = |p-q|$ [6]. The full general Hodge conjecture for (powers of) these abelian varieties is open, and perhaps approachable using known methods.

9 Concluding remark

I have given above large classes of abelian varieties which are dominated by abelian varieties, and one example of an abelian variety not dominated by abelian varieties. I believe that the exception is the rule, and "most" abelian varieties not considered above are *not* dominated by abelian varieties—and expect to provide evidence for this in a future paper.

Appendix A Known Cases

The following is a list of abelian varieties for which the general Hodge conjecture is currently known. All these cases can be proved using the methods of this paper.

1. An abelian variety A of PEL-type such that $G(A)$ is semisimple, and, each simple factor of A is one of the following (see Theorem 4.1):
   (a) Any simple abelian variety A_0 of PEL-type of type I or type II. The usual Hodge conjecture for all powers of A_0 is due to Kuga [25, pp. 79–80]. The general Hodge conjecture was first proved by Hazama [18] and Tankeev [41]. Special cases have been proved by Mattuck [26], Knight [22,23], Gordon [13,14], and, Tankeev [42,43].
   (b) A simple $2m$-dimensional abelian variety A_0 of PEL-type such that $D(A_0) = \mathbb{Q}(\sqrt{-d})$, and polarization given by a hermitian form H of signature (m,m), in the following cases:
      i. $m = 2$ and $d \in \{1,3\}$;
      ii. $m = 3$, $d \in \{1,3\}$, and, $\operatorname{discr} H = -1$.
   The usual Hodge conjecture for these abelian varieties is due to Schoen [34,36], van Geemen [12], and Koike [24].
2. Any power of one of the following abelian varieties of type III:
   (a) A 4-dimensional abelian variety A of type III such that $D(A)$ contains a square root of -3 or -1 [5, Corollary 4.3].
   (b) A 6-dimensional abelian variety A of PEL-type of type III such that $D(A)$ contains a square root of -3 or -1, and the polarization is given by a skew-hermitian form of discriminant -1 [2, Example 5.1].
3. The following abelian varieties of CM-type [7, §4, pp. 160-161]:
   (a) Any power of an abelian variety A with CM by a CM-field E such that $[\overline{E} : \overline{F}] = 2^d$, where F is the maximal totally real subfield of E, $d = [F : \mathbb{Q}]$, and, bars denote Galois closure (Tankeev [41, Theorem 2, p. 180]). Dodson [11, Proposition 2.2.2, p. 82] provides examples of such CM-fields.
   (b) Any power of an elliptic curve of CM-type [40].
   (c) Any abelian variety of the form $A_1^j \times A_2^k \times A_3^\ell$, where A_1, A_2, A_3 are elliptic curves of CM-type.

   (d) Any power of an abelian surface of CM-type.
4. Any abelian variety of PEL-type which is isogenous to a product of abelian varieties of the following types (Abdulali [8, Theorem 14, p. 1169]):
   (a) a simple 3-dimensional abelian variety with endomorphism algebra either $\mathbb{Q}(\sqrt{-1})$ or $\mathbb{Q}(\sqrt{-3})$, and with a polarization given by a hermitian form of signature $(2,1)$;
   (b) a simple 5-dimensional abelian variety with endomorphism algebra either $\mathbb{Q}(\sqrt{-1})$ or $\mathbb{Q}(\sqrt{-3})$, and with a polarization given by a hermitian form of signature $(3,2)$;
   (c) an elliptic curve with CM by either $\mathbb{Q}(\sqrt{-1})$ or $\mathbb{Q}(\sqrt{-3})$.
5. An abelian variety isogenous to a product $A \times B$, where A is as in Case (1), and B is in one of the other cases above (Theorem 3.4).

Appendix B Errata to earlier papers

I thank Chad Schoen for kindly pointing out to me that Proposition 4.4.1 of [1] is false. This requires several modifications to subsequent papers.

B.1. Abelian varieties and the general Hodge conjecture [1]

1. page 341, line 7 from bottom: replace $V(-r)$ by $V(r)$.
2. page 345, line 4: replace $(-r)$ by (r).
3. Proposition 4.4.1 is false. It should be replaced by [8, Propositions 3 and 4, p. 1166] and [7, Proposition 5, p. 160] (see Theorems 3.4 and 5.3 of this paper.) Consequently,
   (a) Theorem 5.1 requires the additional hypothesis that there is at most one simple factor of type III.
   (b) Theorem 6.1 requires the additional hypothesis that the CM-fields corresponding to the factors of CM-type are linearly disjoint.

B.2. Filtrations on the cohomology of abelian varieties [3]

1. page 7, line 10: replace $V(-r)$ by $V(r)$.
2. §7 of [3] is corrected, updated, and expanded by Appendix A of the current paper. In particular, [3, 7.1 (2), 7.1 (4), and, 7.2.1] are incorrect.
3. page 10, line 20: replace $U(-1)$ by $U(1)$.

B.3. Hodge structures on abelian varieties of CM-type [4]

Theorem 3.1 as stated is false. Consequently, some of the examples in Section 4 are incorrect. Correct results and examples may be found in [7], and are summarized in §5 of this paper.

B.4. Hodge structures on abelian varieties of type III [5]

Add to the hypotheses of Theorem 4.1 that the center of $D(A)$ is $\mathbb{Q}$ (see Theorem 7.1 of this paper).

B.5. Hodge structures on abelian varieties of type IV [6]

Theorems 3.1 and 3.2 require the additional hypothesis that the signature (p,q) of the hermitian form satisfies $|p-q|=1$. Hence Remark 3.4 is invalid, but, Corollary 3.3 is correct as stated.

References

[1] Salman Abdulali, *Abelian varieties and the general Hodge conjecture*, Compositio Math. **109** (1997), 341–355. MR98m:14008

[2] ———, *Abelian varieties of type III and the Hodge conjecture*, Internat. J. Math. **10** (1999), 667–675. MR2000g:14013

[3] ———, *Filtrations on the cohomology of abelian varieties*, The Arithmetic and Geometry of Algebraic Cycles (Banff, AB, 1998) (B. B. Gordon, J. D. Lewis, S. Müller-Stach, S. Saito, and N. Yui, eds.), CRM Proc. Lecture Notes, vol. 24, Amer. Math. Soc., Providence, RI, 2000, pp. 3–12. MR2001d:14011

[4] ———, *Hodge structures on abelian varieties of CM-type*, J. Reine Angew. Math. **534** (2001), 33–39. MR2002c:14019

[5] ———, *Hodge structures on abelian varieties of type III*, Ann. of Math. (2) **155** (2002), 915–928. MR2003g:14008

[6] ———, *Hodge structures on abelian varieties of type IV*, Math. Z. **246** (2004), 203–212. MR2004k:14013

[7] ———, *Hodge structures of CM-type*, J. Ramanujan Math. Soc. **20** (2005), 155–162. MR2006g:14016

[8] ———, *Tate twists of Hodge structures arising from abelian varieties of type IV*, J. Pure Appl. Algebra **216** (2012), 1164–1170. MR2875335

[9] ———, *Hodge structures associated to SU(p,1)*, Comm. Algebra **43** (2015), 4209–4220.

[10] Fabio Bardelli, *A footnote to a paper by A. Grothendieck (the Grothendieck generalized Hodge conjecture for some geometric families of abelian threefolds)*, Proceedings of the Geometry Conference (Milan and Gargnano, 1987), Rend. Sem. Mat. Fis. Milano **57** (1987), 109–124. MR90k:14049

[11] Bruce Dodson, *Solvable and nonsolvable CM-fields*, Amer. J. Math. **108** (1986), 75–93. MR87h:11053

[12] Bert van Geemen, *Theta functions and cycles on some abelian fourfolds*, Math. Z. **221** (1996), 617–631. MR97e:14012

[13] B. Brent Gordon, *Topological and algebraic cycles in Kuga-Shimura varieties*, Math. Ann. **279** (1988), 395–402. MR89j:11059

[14] ———, *Algebraic cycles and the Hodge structure of a Kuga fiber variety*, Trans. Amer. Math. Soc. **336** (1993), 933–947. MR93f:14003

[15] Mark Green, Phillip Griffiths, and Matt Kerr, *Mumford-Tate domains*, Boll. Unione Mat. Ital. (9) **3** (2010), 281–307. MR2666359

[16] ———, *Mumford-Tate Groups and Domains: Their Geometry and Arithmetic*, Annals of Mathematics Studies, vol. 183, Princeton University Press, Princeton, NJ, 2012. MR2918237

[17] Alexander Grothendieck, *Hodge's general conjecture is false for trivial reasons*, Topology **8** (1969), 299–303. MR40#5624

[18] Fumio Hazama, *The generalized Hodge conjecture for stably nondegenerate abelian varieties*, Compositio Math. **93** (1994), 129–137. MR95d:14011

[19] ———, *General Hodge conjecture for abelian varieties of CM-type*, Proc. Japan Acad. Ser. A Math. Sci. **78** (2002), 72–75. MR2003e:14005

[20] ———, *On the general Hodge conjecture for abelian varieties of CM-type*, Publ. Res. Inst. Math. Sci. **39** (2003), 625–655. MR2004k:14014

[21] Sigurdur Helgason, *Differential Geometry, Lie Groups, and Symmetric Spaces*, Graduate Studies in Mathematics, vol. 34, American Mathematical Society, Providence, RI, 2001. Corrected reprint of the 1978 original. MR1834454 (2002b:53081)

[22] A. J. Knight, *A classification of the cycles on an abelian variety*, Proc. London Math. Soc. (3) **13** (1963), 184–192. MR26#6165

[23] ———, *On general Riemann matrices and their attached abelian varieties*, Proc. London Math. Soc. (3) **20** (1970), 573–596. MR41#8430

[24] Kenji Koike, *Algebraicity of some Weil Hodge classes*, Canad. Math. Bull. **47** (2004), 566–572. MR2099754

[25] Michio Kuga, *Fibred variety over symmetric space whose fibres are abelian varieties*, Proc. U.S.-Japan Seminar in Differential Geometry (Kyoto, 1965), Nippon Hyoronsha, Tokyo, 1966, pp. 72–81. MR36#315

[26] Arthur Mattuck, *Cycles on abelian varieties*, Proc. Amer. Math. Soc. **9** (1958), 88–98. MR20#5207

[27] James S. Milne, *Lefschetz classes on abelian varieties*, Duke Math. J. **96** (1999), 639–675. MR99m:14017

[28] David Mumford, *Families of abelian varieties*, Algebraic Groups and Discontinuous Subgroups (Boulder, Colo., 1965) (A. Borel and G. D. Mostow, eds.), Proc. Sympos. Pure Math., vol. 9, Amer. Math. Soc., Providence, RI, 1966, pp. 347–351. MR34#5828

[29] ———, *A note of Shimura's paper "Discontinuous groups and abelian varieties,"* Math. Ann. **181** (1969), 345–351. MR40#1400

[30] V. Kumar Murty, *Exceptional Hodge classes on certain abelian varieties*, Math. Ann. **268** (1984), 197–206. MR85m:14063

[31] ———, *Hodge and Weil classes on abelian varieties*, The Arithmetic and Geometry of Algebraic Cycles (Banff, AB, 1998) (B. B. Gordon, J. D. Lewis, S. Müller-Stach, S. Saito, and N. Yui, eds.), NATO Sci. Ser. C Math. Phys. Sci., vol. 548, Kluwer Acad. Publ., Dordrecht, 2000, pp. 83–115. MR2001d:14013

[32] Ichiro Satake, *Symplectic representations of algebraic groups satisfying a certain analyticity condition*, Acta Math. **117** (1967), 215–279. MR35#6694

[33] ———, *Algebraic Structures of Symmetric Domains*, Publ. Math. Soc. Japan, vol. 14 (Kanô Mem. Lect. 4), Iwanami Shoten, Tokyo, and Princeton Univ. Press, Princeton, NJ, 1980. MR82i:32003

[34] Chad Schoen, *Hodge classes on self-products of a variety with an automorphism*, Compositio Math. **65** (1988), 3–32. MR89c:14013

[35] ———, *Cyclic covers of* $\mathbf{P}^v$ *branched along* $v+2$ *hyperplanes and the generalized Hodge conjecture for certain abelian varieties*, Arithmetic of complex manifolds (Erlangen, 1988) (W.-P. Barth and H. Lange, eds.), Lecture Notes in Math., vol. 1399, Springer, Berlin, 1989, pp. 137–154. MR90m:14009

[36] ———, *Addendum to: Hodge classes on self-products of a variety with an automorphism*, Compositio Math. **114** (1998), 329–336. MR99m:14021

[37] Goro Shimura, *On analytic families of polarized abelian varieties and automorphic functions*, Ann. of Math. (2) **78** (1963), 149–192. MR27#5934

[38] ———, *On the field of definition for a field of automorphic functions*, Ann. of Math. (2) **80** (1964), 160–189. MR29#4739

[39] ———, *Moduli of abelian varieties and number theory*, Algebraic Groups and Discontinuous Subgroups (Boulder, Colo., 1965) (Armand Borel and G. D. Mostow, eds.), Proc. Sympos. Pure Math., vol. 9, Amer. Math. Soc., Providence, RI, 1966, pp. 312–332. MR38#5789

[40] Tetsuji Shioda, *What is known about the Hodge conjecture?*, Algebraic Varieties and Analytic Varieties (Tokyo, 1981) (S. Iitaka, ed.), Adv. Stud. Pure Math., vol. 1, Kinokuniya, Tokyo, 1983, pp. 55–68. MR84k:14007

[41] S. G. Tankeev, *Abelian varieties and the general Hodge conjecture*, Izv. Ross. Akad. Nauk Ser. Mat. **57** (1993), 192–206; English transl., Russian Acad. Sci. Izv. Math. **43** (1994), 179–191. MR95j:14011

[42] ———, *Algebraic cycles on an abelian variety without complex multiplication*, Izv. Ross. Akad. Nauk Ser. Mat. **58** (1994), 103–126; English transl., Russian Acad. Sci. Izv. Math. **44** (1995), 531–553. MR95j:14062

[43] ———, *Cycles on abelian varieties and exceptional numbers*, Izv. Ross. Akad. Nauk Ser. Mat. **60** (1996), 159–194; English transl., Izv. Math. **60** (1996), 391–424. MR97f:14045

[44] V. S. Varadarajan, *Lie Groups, Lie Algebras, and their Representations*, Prentice-Hall, Englewood Cliffs, NJ, 1974. MR51#13113

Department of Mathematics, East Carolina University, Greenville, NC 27858, USA
E-mail address: abdulalis@ecu.edu

12

Some surfaces of general type for which Bloch's conjecture holds

C. Pedrini and C. Weibel

ABSTRACT. We give many examples of surfaces of general type with $p_g = 0$ for which Bloch's conjecture holds, for all values of $K_2 \neq 9$. Our surfaces are equipped with an involution.

Let S be a smooth complex projective surface with $p_g(S) = 0$. Bloch's conjecture states that the Albanese map $A_0(S)_0 \to \mathrm{Alb}(S)$ is an isomorphism, where $A_0(S)_0$ is the Chow group of 0-cycles of degree 0 on S. It is known for all surfaces except those of general type (see [BKL]). For a surface S of general type with $p_g(S) = 0$ we also have $q(S) = 0$, i.e., $\mathrm{Alb}(S) = 0$ and the canonical divisor satisfies $1 \leq K^2 \leq 9$.

In the decades since this conjecture was formulated, surfaces of general type have become somewhat better understood. Two key developments have been (i) the results of S. Kimura on finite dimensional motives in [Ki] and (ii) the notion of the *transcendental motive* $t_2(S)$ which was introduced in [KMP]. This includes the theorem that if S is a surface with $p_g(S) = q(S) = 0$ then Bloch's conjecture holds for S iff $t_2(S) = 0$; see Lemma 1.5.

In this paper we give motivic proofs of Bloch's conjecture for several examples of surfaces of general type for each value of K^2 between 1 and 8. This includes some numerical Godeaux surfaces, classical Campedelli surfaces, Keum-Naie surfaces, Burniat surfaces and Inoue's surfaces. All these surfaces carry an involution, and many were previously known to satisfy Bloch's conjecture. We can say nothing about the remaining case $K^2 = 9$, because a surface of general type with $p_g = 0$ and $K^2 = 9$ has no involution ([DMP, 2.3]).

Bloch's conjecture is satisfied by all surfaces whose minimal models arise as quotients $C_1 \times C_2/G$ of the product of two curves of genera ≥ 2 by the action of a finite group G. A partial classification of these surfaces has been given in [BCGP] and [BCG, 0.1]; the special case where G acts freely only occurs when $K_S^2 = 8$. We also show in Corollary 7.8 that Bloch's conjecture holds for surfaces with an involution σ for which $K^2 = 8$ and S/σ is rational. The only known examples with $K^2 = 7$ are Inoue's surfaces, which satisfy Bloch's conjecture (either by our Theorem 8.1 or Bauer [Bau]), and a new family, recently constructed by Y. Chen in [Ch], who also shows that these surfaces satisfy Bloch's conjecture. Burniat surfaces with $K^2 = 6$ satisfy Bloch's conjecture (either by our Theorem 9.1 or Bauer-Catanese [BC1, BC2]); this provides also examples with $K^2 = 3,4,5$. Propositions 9.3 and 9.5 give examples of surfaces with $K^2 = 6$ which are not Burniat, and which satisfy Bloch's conjecture. Other examples with $K^2 = 3,4$ are given in Section 6. Some examples with $K^2 = 1,2$ are treated in Sections 4–5.

Notation. We will work in the (covariant) category $\mathcal{M}_{\text{rat}}(k)$ of Chow motives with coefficients in $\mathbb{Q}$, where the morphisms from $h(X)$ to $h(Y)$ are just the elements of $A^d(X \times Y)$, $d = \dim(Y)$. There is a covariant functor h from smooth projective varieties to $\mathcal{M}_{\text{rat}}(k)$, sending a morphism $f : X \to Y$ to the map $f^t : h(X) \to h(Y)$ determined by the graph Γ_f. Here, and in the rest of this paper, $A^i(X \times Y)$ denotes the Chow group $CH^i(X \times Y) \otimes \mathbb{Q}$ of codimension i cycles modulo rational equivalence, with $\mathbb{Q}$ coefficients.

1 Preliminaries

Let $f : X \to Y$ be a finite morphism between smooth projective varieties of dimension d. Then Γ_f^t determines a map $h(Y) \to h(X)$ and the composition $\Gamma_f \circ {}^t\Gamma_f$ is $\deg(f)$ times the identity of $h(Y)$.

Lemma 1.1. *$f : X \to Y$ be a finite morphism between smooth projective varieties of dimension d, and set $p = ({}^t\Gamma_f \circ \Gamma_f)/\deg(f)$. Then $p : h(X) \to h(X)$ is idempotent, and expresses $h(Y)$ as a direct summand of $h(X)$ in $\mathcal{M}_{\text{rat}}(k)$.*

Proof. We compute: $p \circ p = {}^t\Gamma_f^t \circ (\Gamma_f \circ {}^t\Gamma_f) \circ \Gamma_f / \deg(f)^2 = p$. □

Example 1.2. Suppose that σ is an involution on X and $Y = X/\sigma$ is smooth. Then $p = (1+\sigma)/2$ is represented by the correspondence $(\Delta_X + \Gamma_\sigma)/2$, because ${}^t\Gamma_f \circ \Gamma_f = \Delta_X + \Gamma_\sigma$. In particular, $h(Y) = h(X)^\sigma$.

More generally, suppose that a finite group G acts on X and that $Y = X/G$ is smooth. Then $p = (\sum g)/|G|$ is idempotent, and expresses $h(Y)$ as the direct summand $h(X)^G$ of $h(X)$ in $\mathcal{M}_{\text{rat}}(k)$.

Let $\rho(S)$ denote the rank of $NS(S)$. The Riemann-Roch Theorem gives a well known formula for ρ:

Lemma 1.3. *If $p_g(S) = q(S) = 0$, then $\rho(S) = 10 - K_S^2$.*

Proof. The topological Euler characteristic of S is $\deg(c_2) = 2 + \rho(S)$. Since the Euler characteristic of $\mathcal{O}_S$ is 1, the Riemann-Roch Theorem ([Fu, 15.2.2]) yields $12 = K_S^2 + \deg(c_2)$. Thus $\rho(S) = 10 - K_S^2$. □

Remark 1.3.1. If S is a minimal surface of general type, then $K_S^2 > 0$ [BHPV, VII.2.2]. Since $\rho(S) \geq 1$, we derive the inequality $1 \leq K_S^2 \leq 9$. We also have $\deg c_2 > 0$, see [BHPV, VII.1.1]. By Noether's formula $1 - q(S) + p_g(S)$ is positive, hence if $p_g(S) = 0$ then also $q(S) = 0$.

The algebraic motive $h_2^{\text{alg}}(S)$. If S is a surface, the Neron-Severi group $NS(S)$ determines a summand of the motive of S (with $\mathbb{Q}$ coefficients). To construct it, choose an orthogonal basis $\{E_1, \cdots, E_\rho\}$ for the $\mathbb{Q}$-vector space $NS(S)_\mathbb{Q}$, where the self-intersections E_i^2 are nonzero. Then the correspondences $\epsilon_i = \frac{[E_i \times E_i]}{(E_i)^2}$ are orthogonal and idempotent, so

$$\pi_2^{\text{alg}}(S) = \sum_{1 \leq i \leq \rho} \frac{[E_i \times E_i]}{(E_i)^2}$$

is also an idempotent correspondence. Since $\{E_i/(E_i)^2\}$ is a dual basis to the $\{E_i\}$, it follows from [KMP, 7.2.2] that $\pi_2^{\text{alg}}(S)$ is independent of the choice of basis. We set $M_i = (S, \epsilon_i, 0)$ and $h_2^{\text{alg}}(S) = (S, \pi_2^{\text{alg}}(S)) = \oplus M_i$. In fact, $M_i \simeq \mathbb{L}$ for all i by [KMP, 7.2.3], so we have isomorphisms $h_2^{\text{alg}}(S) \cong \mathbb{L}^{\oplus \rho}$ and $H^2(h_2^{\text{alg}}(S)) \cong NS(S)_\mathbb{Q}$.

The transcendental motive t_2. We also need a description of the transcendental motive t_2 of a surface S. It is well known that the motive $h(S)$ has a Chow-Künneth decomposition as $\sum_0^4 h_i(S)$, where $H^i(S) = H^i(h_i(S))$. The middle factor $h_2(S)$ further decomposes as $h_2(S) = h_2^{\text{alg}}(S) \oplus t_2(S)$; see [KMP].

The factor $t_2(S) = (S, \pi_2^{\text{tr}}, 0)$ is called the *transcendental part* of the motive (cf. [KMP, 7.2.3]). This terminology is justified by the following result, implicit in [CG, p. 289], which identifies $H^2(t_2(S)) = \pi_2^{tr} H^2(S)$ with the Hodge-theoretic group $H_{\text{tr}}^2(S)$.

Lemma 1.4. *Under the intersection pairing on $H^2(S, \mathbb{C})$ the orthogonal complement $H_{tr}^2(S)$ of $NS(S) \otimes \mathbb{C}$ is $\pi_2^{tr} H^2(S)$.*

The following result was established in [KMP, 7.4.9 & 7.6.11].

Lemma 1.5. *If S is a smooth projective complex surface, then the following are equivalent:*

(a) $t_2(S) = 0$ holds;
(b) $p_g(S) = 0$ and the motive $h(S)$ is finite dimensional;
(c) $p_g(S) = 0$ and S satisfies Bloch's conjecture.

We will often use without comment the fact that t_2 is a birationally invariant functor on the category of smooth 2-dimensional varieties; this is proven in [KMP, 7.8.11].

Lemma 1.6. *If a finite group G acts on a surface S, and Y is a desingularization of S/σ, then $t_2(Y) = t_2(S)^G$.*

Proof. Because t_2 is a birational invariant, we may blow up S to assume that $Y = S/\sigma$. (See (2.1) below). Since $h(S) \to h(Y)$ sends $t_2(S)$ to $t_2(Y)$, and $h(Y) = h(S)^G$ by Example 1.2, the result follows. □

Here is the "enough automorphisms" criterion of Inose and Mizukami. Our argument depends upon the motivic Lemma 1.5.

Lemma 1.7 (Inose–Mizukami [IM, 1.2]). *Suppose that a finite group G acts on S, and that Y is a desingularization of S/G. If $\sum g = 0$ in* $\mathrm{End}\, h(S)$ *then $p_g(Y) = 0$ and Bloch's conjecture holds for Y.*

Proof. Because $t_2(S)$ is a birational invariant, we may blow up S to assume that $Y = S/G$. By 1.2 and 1.6, $h(Y) = h(S)^G$ and $t_2(Y) = t_2(S)^G$, so if $p = (\sum g)/|G| = 0$ in $\mathrm{End}\, h(S)$ then $p = 0$ in $\mathrm{End}\, t_2(S)$, i.e., $t_2(Y) = 0$. By Lemma 1.5, this implies that $p_g(Y) = 0$ and Bloch's conjecture holds for Y. □

2 Involutions on surfaces

Let S be a smooth projective surface with an involution σ. The fixed locus consists of a 1-dimensional part D (a union of smooth curves, possibly empty) and $k \geq 0$ isolated fixed points $\{P_1, \ldots, P_k\}$. The images Q_i of the P_i are nodes on the quotient surface S/σ, and S/σ is smooth elsewhere. To resolve these singularities, let X denote the blow-up of S at the set of isolated fixed points; σ lifts to an involution on X (which we will still call σ), and the quotient $Y = X/\sigma$ is a desingularization of S/σ. The images $C_1, \ldots, C_k$ in Y of the exceptional divisors of X are disjoint nodal curves, i.e., smooth rational curves with self-intersection -2. In summary, we have a commutative diagram

$$\begin{array}{ccc} X & \xrightarrow{h} & S \\ \downarrow \pi & & \downarrow f \\ Y & \xrightarrow{g} & S/\sigma. \end{array} \tag{2.1}$$

The image f_*D is a smooth curve on S/σ, disjoint from the singular points Q_i, and its proper transform $B' = g^*(f_*D)$ in Y is smooth and disjoint from the curves C_i. It follows that $\pi : X \to Y$ is a double cover with smooth branch locus $B = B' + \sum C_i$. As such, π is determined by a line bundle L on Y such that $2L \equiv B$.

Lemma 2.2. $2L^2 = D^2 - k$.

Proof. Because the curves C_i on X have $C_i^2 = -2$ and are disjoint from $B' = g^*(f_*D)$ we have $(B')^2 = 2D^2$. Since $2L \equiv B' + \sum C_i$ we have $4L^2 = (B')^2 + \sum C_i^2 = 2D^2 - 2k$. □

Remark 2.2.1. If S is a minimal surface of general type and $p_g(S) = 0$, it is proven in [CCM, 3.3] that $k \geq 4$ and the linear system $2K_Y + L$ has dimension $l(2K_Y + L) = (K_S^2 + 4 - k)/2$.

Since $t_2(-)$ is a birational invariant for smooth projective surfaces, the maps $h: X \to S$ and $\pi: X \to Y$ induce a morphism

$$\theta: t_2(S) \cong t_2(X) \to t_2(Y).$$

By Lemma 1.1 and Example 1.2, θ is the projection onto the direct summand $t_2(S)^\sigma$ of $t_2(S)$, and $A_0(Y)$ is the direct summand of $A_0(X)$ fixed by σ.

Proposition 2.3. *Let S be a smooth surface $q(S)=0$. If σ is an involution on S then:*

(i) $t_2(S) \cong t_2(Y) \Longleftrightarrow \bar{\sigma} = +1$ *in* $\mathrm{End}_{\mathcal{M}_{\mathrm{rat}}}(t_2(S))$
(ii) $t_2(Y) = 0 \Longleftrightarrow \bar{\sigma} = -1$ *in* $\mathrm{End}_{\mathcal{M}_{\mathrm{rat}}}(t_2(S))$.

Here $\bar{\sigma}$ is the endomorphism of $t_2(S)$ induced by σ

Proof. Since we have $t_2(Y) = t_2(S)^\sigma$ by Lemma 1.6, the projection onto $t_2(Y)$ is given by the idempotent endomorphism $e = (\bar{\sigma} + 1)/2$ of $t_2(S)$. Since $t_2(S) \cong t_2(Y)$ is equivalent to $e = 1$, and $t_2(Y) = 0$ is equivalent to $e = 0$, the result follows. □

Remark 2.3.1. If S is a smooth minimal surface of general type with $p_g(S) = 0$ and an involution σ, such that the minimal model W of Y is either an Enriques surface, a rational surface or a surface of Kodaira dimension equal to 1, then by [GP2] we have $t_2(Y) = 0$.

Definition 2.4. A *bidouble cover* $V \xrightarrow{f} X$ between smooth projective surfaces is a finite flat Galois morphism with Galois group $\mathbb{Z}/2 \times \mathbb{Z}/2$. By [Pa91], in order to define f it is enough to give smooth divisors D_1, D_2, D_3 in X with pairwise transverse intersections and no common intersections, and line bundles L_1, L_2, L_3 such that $2L_i \equiv D_j + D_k$ for each permutation (i,j,k) of $(1,2,3)$. We will frequently use the fact that every nontrivial element σ of G is an involution on V.

Theorem 2.5. *Let S be surface of general type with $p_g(S) = 0$ which is the smooth minimal model of a bidouble cover of a surface. Let Y_i denote the desingularization of S/σ_i, where $\sigma_1, \sigma_2, \sigma_3$ are the non-trivial involutions of S associated to the bidouble cover. If $t_2(Y_i) = 0$ for $i = 1,2,3$ then $t_2(S) = 0$.*

Proof. By Lemma 1.5, $t_2(S)^\sigma = t_2(Y_i) = 0$. Thus each σ_i acts as multiplication by -1 on $t_2(S)$. But since $\sigma_1\sigma_2 = \sigma_3$, σ_3 must act as multiplication by $(-1)^2 = +1$. Since $+1 = -1$ on $t_2(S)$, $t_2(S) = 0$. □

Remark 2.5.1. Theorem 2.5 is the special case where σ_1, σ_2 and σ_3 act as -1 of Lemma 1.7 (the "enough automorphisms" criterion of Inose and Mizukami). (Compare Proposition 1.3 in [Bau].)

3 Composed involutions

If S is a smooth minimal surface of general type, then the linear system $|2K_S|$ determines the bicanonical map $\Phi_2 : S \to \mathbb{P}^N$, where $N = \dim(2K_S)$. We say that Φ_2 is *composed* with an involution σ if Φ_2 factors through the map $S \to S/\sigma$.

Recall that k denotes the number of isolated fixed points, and that Y is a resolution of S/σ, as in (2.1).

Theorem 3.1. *Let S be a smooth minimal surface of general type with $p_g(S) = 0$, supporting an involution σ. Then*

(1) Φ_2 is composed with σ iff $k = K_S^2 + 4$.
If the bicanonical map Φ_2 is composed with σ then
(2) S/σ is either rational or it is birational to an Enriques surface.
(3) $-4 \le K_Y^2 \le 0$, and $K_Y^2 = 0$ iff K_Y is numerically zero.

Proof. This is proven in [CCM, 3.6(iv) and 3.7(iv,v)]. □

Corollary 3.2. *Let S be a minimal surface of general type with $p_g(S) = 0$ and let X be the image of the bicanonical map Φ_2. If $\Phi_2 : S \to X$ is a bidouble cover, then S satisfies Bloch's conjecture.*

Proof. The assumption that the image X of the bicanonical map of S is a surface and that Φ_2 is a morphism imply that $K_S^2 \ge 2$ (see [MP02, 2.1]) and that the degree of Φ_2 is ≤ 8. If Φ_2 is a bidouble cover then all three nontrivial involutions σ_i of S (see 2.4) are composed with the bicanonical map. By Theorem 3.1, $t_2(S/\sigma_i) = 0$ so Theorem 2.5 implies that $t_2(S) = 0$. That is, S satisfies Bloch's conjecture. □

Recall that $\rho(S)$ denotes the rank of the Néron-Severi group $NS(S)$.

Corollary 3.3. *Let S be a minimal complex surface of general type with $p_g(S) = 0$ and an involution σ. Let D be the 1-dimensional part of the fixed locus S^σ. Then*

(1) σ acts as the identity on $H^2(S, \mathbb{Q})$ iff $D^2 = K_S^2 - 8$.
(2) If the bicanonical map is composed with σ and X, Y are as in diagram (2.1), then $D^2 = K_S^2 + 2K_Y^2$ and

$$K_S^2 - 8 \le D^2 \le K_S^2.$$

When $D^2 = K_S^2$, S/σ is birational to an Enriques surface.

Proof. Since $p_g(S) = 0$ we have $H^2(S,\mathbb{Q}) \cong NS(S)_{\mathbb{Q}}$. Let t denote the trace of σ on $H^2(S,\mathbb{Q})$; it is at most $\rho(S) = \dim H^2(S,\mathbb{Q})$. By [DMP, 4.2], $t = 2 - D^2$. Since $\rho(S) = 10 - K_S^2$ by 1.3, we deduce that $D^2 \geq K_S^2 - 8$, with equality iff $t = \rho(S)$, i.e., iff σ acts as the identity on $H^2(S,\mathbb{Q})$.

Assume now that σ is composed with the bicanonical map. By Theorem 3.1, σ has $k = K_S^2 + 4$ fixed points. By [CCM, 3.3(ii) and 3.6(iii)], the line bundle L introduced after (2.1) satisfies $L^2 + 2 = -K_Y \cdot L = +K_Y^2$. We also have $2L^2 = D^2 - k$, by Lemma 2.2. Therefore

$$D^2 = K_S^2 + 2K_Y^2.$$

Since $-4 \leq K_Y^2 \leq 0$ by Theorem 3.1, we get $K_S^2 - 8 \leq D^2 \leq K_S^2$. Theorem 3.1 also implies that if $K_Y^2 = 0$ then (since Y is not rational) the minimal model of Y is an Enriques surface, i.e., $2K_Y = 0$. □

Remark 3.3.1. From Lemma 1.3, $\rho(X) = \rho(S) + k$ equals $10 - K_S^2 + k$. If the bicanonical map is composed with σ then $k = K_S^2 + 4$ by 3.1, so we get $\rho(X) = 14$. Since $Y = X/\sigma$, we have $p_g(Y) = q(Y) = 0$, $NS(Y)_{\mathbb{Q}} = NS(X)_{\mathbb{Q}}^{\sigma}$ and hence $\rho(Y) \leq \rho(X)$. Since $K_Y^2 \leq 0$ and $\rho(Y) = 10 - K_Y^2$ we get the bounds

$$10 \leq \rho(Y) \leq 14.$$

The equality $\rho(Y) = 14 = \rho(X)$ corresponds to the case $K_Y^2 = -4$, in which case $D^2 = K_S^2 - 8$.

Example 3.4 ($K_S^2 = 8$)**.** Let S be a minimal surface of general type with $p_g(S) = 0$ and $K_S^2 = 8$, having an involution σ. Then $\rho(S) = 10 - K_S^2 = 2$ (Lemma 1.3), and the Hodge Index Theorem implies that the involution σ acts as the identity on $H^2(S,\mathbb{Q})$. By Corollary 3.3, $D^2 = K_S^2 - 8 = 0$.

The following facts are established in [DMP, 4.4]. The number k of the fixed points is one of 4,6,8,10,12. The bicanonical map is composed with σ iff $k = 12$ and in that case $Y = X/\sigma$ is rational (see 3.1). If $k = 10$ then Y is a rational surface and $\rho(Y) = \rho(X) = 12$. If $k = 8$ the Kodaira dimension of Y is 1. Finally, if $k = 4, 6$ the quotient surface Y is of general type.

Carlos Rito has analyzed the situation where Φ_2 is *not* composed with σ. We extract the following result from [Ri2, Thm. 2]. Given a surface S with involution, we construct a resolution $Y \to S/\sigma$ and $X \to Y$ with smooth branch locus B as in (2.1); by blowing down curves on Y, we obtain a minimal resolution W of S/σ. Let $\bar{B}$ denote the image of B under the proper map $Y \to W$.

Theorem 3.5. *[Rito] Let S be a smooth minimal surface of general type with $p_g(S) = 0$, and σ an involution on S with k isolated fixed points. Suppose that the bicanonical map is not composed with σ. Let W be a minimal model of the* *resolution Y of S/σ and let $\bar{B}$ be as above. If W has Kodaira dimension 2, then one of the following holds:*

1. $K_S^2 = 2K_W^2$, *and $\bar{B}$ is a disjoint union of 4 nodal curves;*
2. $4 \le K_S^2 \le 8$, $K_W^2 = 1$, $\bar{B}^2 = -12$, $K_W \cdot \bar{B} = 2$, $k = K_S^2$ *and $\bar{B}$ has at most one double point;*
3. $6 \le K_S^2 \le 8$, $K_W^2 = 2$, $\bar{B}^2 = -12$, $K_W \cdot \bar{B} = 2$, $k = K_S^2 - 2$ *and $\bar{B}$ is smooth.*

Corollary 3.6. *Let S be a smooth minimal surface of general type with $p_g(S) = 0$, having a nontrivial involution σ. If $K_S^2 = 3$ then the resolution Y of S/σ is not of general type. In particular, S/σ satisfies Bloch's conjecture.*

Proof. If the bicanonical map of S is composed with σ, then Theorem 3.1(2) states that Y is either rational or its minimal model is an Enriques surface. If the bicanonical map of S is *not* composed with σ, and Y is of general type (has Kodaira dimension 2), then Theorem 3.5 shows that K_S^2 cannot be 3. □

Example 3.7. Here are some examples in which the minimal model W of S/σ is of general type, and the divisorial part D of S^σ satisfies $D = 0$. Let k be the number of isolated fixed points of σ.

1) The Barlow surface W in [Bar] is of general type with $K_W^2 = 1$ and $p_g(W) = 0$. It is the minimal model of S/σ, where S is a Catanese surface with $K_S^2 = 2$, by factoring with an involution σ which has only 4 isolated fixed points. Hence $D = 0$ so that $\bar{B} = \sum_{1 \le i \le k} C_i$ is the disjoint union of 4 nodal curves. Therefore W satisfies condition (1) of Theorem 3.5. In [Bar] it is proved that $t_2(W) = 0$.
2) Another example comes from [MP01a, Ex. 4.3]; cf. [DMP, 4.4(i)]. Using Pardini's method for the group $G = (\mathbb{Z}/2)^4$ (as in Lemma 7.6 below), one first constructs two smooth G-covers C_1, C_2 of $\mathbb{P}^1$ of genus 5 as in [MP01a, Ex. 4.3]. The authors choose a subgroup Γ of $G \times G$ with $|\Gamma| = 16$ acting freely on $C_1 \times C_2$; the quotient $S = (C_1 \times C_2)/\Gamma$ is the desired minimal surface of general type with $K_S^2 = 8$ and $p_g(S) = q(S) = 0$.

For this S, it is shown in *loc. cit.* that the bicanonical map is birational, hence not composed with any involution. The construction of S starts with a basis of G and constructs the G-covers $C_i \to \mathbb{P}^1$ so that any of the 11 other nontrivial elements of G acts freely on both curves, and acts on S (via the diagonal action) with $D = 0$ and $k = K_S \cdot D + 4 = 4$ fixed points.

Since $D = 0$ and $k = 4$, a minimal model W of S/σ has $K_W^2 = 4$ by [DMP, 4.4(i)]. Thus W satisfies condition (1) of Theorem 3.5. In fact, the branch locus of $X \to Y$ is the disjoint union of the 4 lines over the fixed points of σ, so $\bar{B}$ is the disjoint union of the 4 nodal curves C_i.

Since the motives $h(C_i)$ are finite dimensional, so are $h(C_1 \times C_2)$ and $h(S) = h(C_1 \times_G C_2)$. Hence $t_2(S)$ is also finite dimensional. As $p_g(S) = 0$, Lemma 1.5 implies that S satisfies Bloch's conjecture.

4 Numerical Godeaux surfaces ($K^2 = 1$)

In the next few sections, we will give examples of complex surfaces S of general type, having an involution σ, with $p_g(S) = 0$ and $3 \le K_S^2 \le 7$, for which Bloch's conjecture holds. In this section and the next we deal with the cases $K_S^2 = 1, 2$.

Complex surfaces of general type with $p_g = q = 0$ and $K_S^2 = 1$ are called *numerical Godeaux surfaces*. Examples of Godeaux surfaces are obtained as quotients of a Catanese surface under the action of a finite group. (A Catanese surface is a minimal surface V of general type with $p_g(V) = 0$ and $K_V^2 = 2$).

Barlow gave two such examples of numerical Godeaux surfaces S in the 1980's, both with $\pi_1(S) = \mathbb{Z}/5$ and $\pi_1(S) = \{1\}$, and showed in [Bar] that Bloch's conjecture holds for these S.

By a *typical Godeaux surface* we mean the quotient surface of the Fermat quintic in $\mathbb{P}^3$ by a cyclic group of order 5. It has $\rho(S) = 9$ and $K_S^2 = 1$. These surfaces were first shown to satisfy Bloch's conjecture by Inose-Mizukami [IM, Thm. 1]. From a motivic viewpoint, S satisfies Bloch's conjecture because it has a finite dimensional motive; see [GP1].

We now turn to numerical Godeaux surfaces with an involution σ. These were classified in [CCM]; if S/σ is rational they are either of *Campedelli type* or *Du Val type*; if S/σ is an Enriques surface they are of *Enriques type*. This exhausts all cases, since (by [CCM, 4.5]) the bicanonical map is composed with σ and σ has exactly $k = 5$ fixed points; as noted in Theorem 3.1, this implies that S/σ is either rational or birational to an Enriques surface.

If S is of Du Val type, there is an étale double cover $\tilde{S}$ of S with $p_g(\tilde{S}) = 1$ and $K^2 = 2$; see [CCM, 8.1]. Since $\pi_1(\tilde{S})$ is the kernel of $\pi_1(S) \to \mathbb{Z}/2$, we also have $q(\tilde{S}) = 0$. In addition, $\mathrm{Pic}(\tilde{S}) = NS(\tilde{S})$ is either torsion free or has a torsion subgroup of order 2.

Theorem 4.1. *Let S be a numerical Godeaux surface of Du Val type such that the torsion subgroup of* Pic(S) *has order 2. Let $\tilde{S} \to S$ be the étale double cover of S associated to the non trivial 2-torsion element in* Pic(S). *If $\tilde{S}$ has bicanonical map of degree 4, then S satisfies Bloch's conjecture.*

Proof. By [CD, §3], $\mathrm{Pic}(\tilde{S})$ is torsion free and the image of the bicanonical map $\Phi : \tilde{S} \to \mathbb{P}^3$ is a smooth quadric. By Lemma 4.2 below, $\tilde{S}$ has a finite dimensional motive; this implies that $h(S)$ is also finite dimensional. As $p_g(S) = 0$, Lemma 1.5 implies that S satisfies Bloch's conjecture. □

Lemma 4.2 is based upon the Catanese–Debarre paper [CD].

Lemma 4.2. *Let S be a surface of general type with $p_g(S) = 1$, $q(S) = 0$ and $K_S^2 = 2$. Suppose that the bicanonical map $\Phi : S \to \mathbb{P}^3$ has degree 4 and is a morphism onto a smooth quadric Σ. Then the motive of S is finite dimensional.*

Proof. As shown in [CD, 3.2], the bicanonical map $S \to \Sigma \subset \mathbb{P}^3$ is a bidouble cover with Galois group $G = \{1, \sigma_1, \sigma_2, \sigma_3\}$. Set $S_i = S/\sigma_i$, with $i = 1,2,3$. By [CD, 3.1], both S_1 and S_2 are rational surfaces, while the minimal smooth surface M over S_3 is a K3 surface. Moreover S_3 has 10 nodal points, and their inverse images in M are 10 disjoint smooth rational curves $F_1, \cdots, F_{10}$.

Since $t_2(S)^{\sigma_i} = t_2(S_i)$ and $t_2(S_1) = t_2(S_2) = 0$, it follows that σ_1 and σ_2 act as -1 to $t_2(S)$. Hence $\sigma_3 = \sigma_1 \circ \sigma_2$ acts as $+1$. It follows that $t_2(S) = t_2(S_3) = t_2(M)$.

On S_3, the involutions $\bar{\sigma}_1$ and $\bar{\sigma}_2$ agree and induce an involution σ on M. Since $S_3/\bar{\sigma}_1 = \Sigma$, M^σ is the union of the ten curves F_i. Because Σ is a rational surface, $H^{2,0}(M)^\sigma = H^{2,0}(\Sigma) = 0$, so σ acts as -1 on $H^{2,0}(M) \cong \mathbb{C}$. It follows from [Zh, 3.1] that the Neron-Severi group of M has rank $\rho(M) = 20$. By Theorems 2 and 3 of [Ped], this implies that the motive $h(M)$ is finite dimensional in $\mathcal{M}_{rat}$. In particular $t_2(M) = t_2(S)$ is finite dimensional. Since $h_0(S), h_4(S)$ and $h_2^{alg}(S)$ are finite dimensional, it follows that $h(S)$ is finite dimensional. □

5 Numerical Campedelli surfaces ($K^2 = 2$)

Complex surfaces of general type with $p_g = q = 0$ and $K_S^2 = 2$ are called *numerical Campedelli surfaces.*

Now let S be a numerical Campedelli surface with an involution σ; these have been classified in [CMP]. By [CCM, 3.32(i,iv)], σ has either $k = 4$ or $k = 6$ isolated fixed points. By Theorem 3.1, the bicanonical map $\Phi : S \to \mathbb{P}^2$ is composed with σ iff $k = 6$; in this case Corollary 3.3 yields $-6 \le D^2 \le 2$. (This result appeared in [CMP, 3.1(ii)]).

Campedelli surfaces with fundamental group $(\mathbb{Z}/2)^3$ satisfy Bloch's conjecture, by the results in [IM]. Catanese surfaces have $p_g = q = 0$, $K2_S = 2$ and fundamental group $\mathbb{Z}/5\mathbb{Z}$. In [Vois, 2.4], Voisin proves that Bloch's conjecture holds for a family of Catanese and on surfaces constructed from 5×5 symmetric matrices $M(a)$, with $a \in \mathbb{P}^{11}$, of linear forms on $\mathbb{P}^3$ satisfying certain conditions.

Numerical Campedelli surfaces which arise as a $(\mathbb{Z}/2)^3$-cover of $\mathbb{P}^2$ branched along 7 lines are called *Classical Campedelli surfaces.* They were constructed by Campedelli, and later by Kulikov in [Ku]. All 7 of the nontrivial involutions of classical Campedelli surfaces are composed with bicanonical map. The following result was first established in [IM, Thm. 3].

Theorem 5.1. *Classical Campedelli surfaces satisfy Bloch's conjecture.*

Proof. The automorphism group of S coincides with $(\mathbb{Z}/2)^3$; see [Ku, Thm. 4.2]. By [CMP, 5.1], the bicanonical map is composed with every nontrivial involution σ of S and each quotient S/σ is either rational or else it is birationally equivalent to an Enriques surface, so $t_2(S/\sigma) = 0$. Fix a subgroup

$H = \langle \sigma_1, \sigma_2 \rangle$ of automorphisms, so that S is a bidouble cover of S/H. By Theorem 2.5 (applied to $S \to S/H$) we get $t_2(S) = 0$, and hence S satisfies Bloch's conjecture. □

Here is one such family of classical Campedelli surfaces. We fix 7 distinct lines L_i in $\mathbb{P}^2$ such that at most 3 pass through the same point, and enumerate the nontrivial elements of $G = (\mathbb{Z}/2)^3$ as $g_1, \dots, g_7$, so that g_1, g_2, g_3 generate G and if $g_i + g_j + g_k = 0$ then L_i, L_j and L_k do not pass through the same point. Fix characters χ_1, χ_2, χ_3 generating G^*. By [Pa91], the equations $2\mathcal{L}_i = \sum_{j=1}^{7} \epsilon_{ij} L_j$ $(i = 1,2,3)$ determine a normal G-cover V of $\mathbb{P}^2$ as long as ϵ_{ij} is 1 if $\chi_i(g_j) = -1$ and zero otherwise. (See Example 1 of [CMP, §5], or [Ku, 4.1].) The surface S is obtained by resolving the singular points of V, which only lie over the triple intersection points of the lines L_i in the plane.

6 The case $K_S^2 = 3, 4$

Let S be a minimal surface of general type with $p_g(S) = 0$ and $K_S^2 = 3$. For each nontrivial involution σ of S, Corollary 3.6 and Lemma 1.5 imply that $t_2(S/\sigma) = 0$. Therefore if S is birational to a bidouble cover, then S satisfies Bloch's conjecture by Theorem 2.5.

Example 6.1 ($\mathbf{K_S^2 = 3}$). Rito gives an example in [Ri1, 5.2] in which S is the minimal model of a bidouble cover of $\mathbb{P}^2$ and $K_S^2 = 3$. Write $\sigma_1, \sigma_2, \sigma_3$ for the 3 involutions of S corresponding to the bidouble cover. Rito shows that the bicanonical map is not composed with σ_1 or σ_2 but is composed with σ_3. If W_i is a minimal model of the desingularization of S/σ_i, he shows that W_1 is an Enriques surface, W_2 has Kodaira dimension 1 and W_3 is a rational surface. Therefore $t_2(W_i) = 0$ for $i = 1,2,3$. As remarked above, this implies that S satisfies Bloch's conjecture.

Similar examples with $4 \le K_S^2 \le 7$ have been constructed in [Ri2].

Y. Neum and D. Naie [Na] constructed a family of surfaces of general type with $K^2 = 4$ as double covers of an Enriques surface Y with 8 nodes. I. Bauer and F. Catanese proved in [BC3] that the connected component of the moduli space corresponding to this family is irreducible, normal, unirational of dimension 6. In Remark 3.3 of *op. cit.* they noticed that Keum's proof, given in the unpublished manuscript [Ke], of Bloch's conjecture for a subfamily of dimension 4 of the connected component of the moduli space of all Keum-Naie surfaces, can be extended to the whole family. We will now give a motivic proof of Bloch's conjecture for these surfaces.

Here is the relevant part of the construction of Keum-Naie surfaces given in [BC3]. Fix elliptic curves E_1, E_2 and points $a_i \in E_i$ of order 2. The group

$G = (\mathbb{Z}/2)^2$ acts freely on $E_1 \times E_2$ using the involutions

$$\gamma_1(z_1,z_2) = (z_1 + a_1, -z_2) \; ; \; \gamma_2(z_1,z_2) = (-z_1, z_2 + a_2).$$

and the Enriques surface is $Y = (E_1 \times E_2)/G$. The automorphism $\gamma_3(z_1,z_2) = (-z_1 + a_1, z_2)$ commutes with G and the quotient of Y by γ_3 is a del Pezzo surface Σ in $\mathbb{P}^4$.

In [BC3, 1.4], a G-invariant double cover $\tilde{X} \xrightarrow{\pi} E_1 \times E_2$ is constructed. If $\tilde{S}$ is a minimal resolution of singularities of $\tilde{X}$ then the quotient $S = \tilde{S}/G$ is a Keum-Naie surface with $K_S^2 = 4$ and $p_g(S) = q(S) = 0$. By [BC3, 4.1], the bicanonical map $\Phi_2 : S \to \mathbb{P}^4$ of S has degree 4 and its image is Σ.

Theorem 6.2. *Let S be a Keum-Naie surface with $K_S^2 = 4$ and $p_g(S) = 0$. Then S satisfies Bloch's conjecture.*

Proof. Because the étale degree 2 map $\tilde{S} \to E_1 \times E_2$ is G-equivariant, it induces an étale degree 2 map $p : S \to Y$. Let $H \cong (\mathbb{Z}/2)^2$ be the group of automorphisms of S generated by γ_3 and the involution associated to p. Since the composition of p with $Y \to \Sigma$ is the bicanonical map Φ_2, all three nontrivial elements h_i of H are composed with Φ_2. Hence $t_2(S/\sigma_i) = 0$ for $i = 1,2,3$ by Theorem 3.1. From Theorem 2.5 we get $t_2(S) = 0$. □

7 Fibrations on surfaces

A surface S is said to be a *product-quotient surface* if it is birational to a quotient $(C_1 \times C_2)/G$ of the product $C_1 \times C_2$ of two curves of genera ≥ 2 by the action of a finite group G. Since the motives $h(C_i)$ are finite dimensional, so are $h(C_1 \times C_2)$ and $h(S) = h(C_1 \times C_2)^G$. Since t_2 is a birational invariant, $t_2(S)$ is also finite dimensional. If $p_g(S) = 0$, this implies that S satisfies Bloch's conjecture (by Lemma 1.5).

A complete classification is given in [BCGP] of surfaces S with $p_g(S) = q(S) = 0$, whose canonical models arise as product-quotient surfaces. If G acts freely then the quotient surface is minimal of general type and it is said to be *isogenous to a product*. If S is isogenous to a product then $K_S^2 = 8$, see [BCG, 0.1] or [BCP, Thm. 4.3].

In the case when G acts freely on both C_1 and C_2, then the projection $C_1 \times_G C_2 \to C_2/G$ has fibers C_1. More generally, a fibration $S \to B$ from a a smooth projective surface onto a smooth curve is said to be *isotrivial* if the smooth fibers are mutually isomorphic.

Theorem 7.1. *Let S be a complex surface of general type with $p_g = 0$ which has a fibration $S \xrightarrow{\pi} B$ with B a smooth curve of genus b, and general fibre a* *curve F of genus $g \geq 1$. If π is isotrivial, then $t_2(S) = 0$, and S satisfies Bloch's conjecture.*

Proof. By [Se, 2.0.1] there is a finite group G acting on the fiber F and a Galois cover $C \to B$ so that $B = C/G$ and S is birational to $S' = F \times_G C$. □

One source of isotrivial fibrations comes from the following observation of Beauville. Let S be a smooth projective complex surface and $S \to B$ a fibration with general fibre a smooth curve F of genus $g \geq 1$. Let b denote the genus of the curve B. Beauville proved in [Be] that

$$K_S^2 \geq 8(b-1)(g-1), \tag{7.2}$$

and if equality holds then the fibration is isotrivial.

Let S be a smooth complex surface with a fibration $f : S \to \mathbb{P}^1$ with general fiber F a smooth curve of genus $g(F)$. In many cases, we can contruct a finite map $h : C \to \mathbb{P}^1$ with C smooth such that the normalization X of $C \times_{\mathbb{P}^1} S$ is nonsingular and the map $\tilde{h} : X \to S$ is étale. Because $\tilde{h}$ is étale, X is smooth and $K_X = \tilde{h}^* K_S$, we have $K_X^2 = \deg(\tilde{h}) K_S^2$. This information is summarized in the commutative diagram:

$$\begin{array}{ccc} X & \xrightarrow{\tilde{h}} & S \\ \big\downarrow{\scriptstyle f_C} & & \big\downarrow{\scriptstyle f} \\ C & \xrightarrow{h} & \mathbb{P}^1. \end{array} \tag{7.3}$$

Lemma 7.4. *Suppose that the singular fibers of a fibration $S \to \mathbb{P}^1$ consist only in double fibers, over points $P_1, \dots, P_r$ of $\mathbb{P}^1$, and that there is a smooth cover $h : C \to \mathbb{P}^1$ branched along the P_k, with $h^{-1}(P_k)$ consisting only in double points. Then, in the diagram (7.3), X is nonsingular and fibered over C, and $\tilde{h}$ is an étale map.*

Proof. The base change, $C \times_{\mathbb{P}^1} S \to S$ is étale except over the double fibers, where it is a simple normal crossings divisor. Therefore the normalization X of $C \times_{\mathbb{P}^1} S$ is étale over S. □

Example 7.5. Suppose that $f : S \to \mathbb{P}^1$ has an even number of singular fibers, all double fibers. Let C be the double cover of $\mathbb{P}^1$ branched at the points of $\mathbb{P}^1$ supporting the double fibers of f. Then the hypotheses of Lemma 7.4 are satisfied, so we have a diagram (7.3) with $\tilde{h}$ étale.

Pardini's method. We recall Pardini's method (from [Pa91, (2.21)]) for producing a smooth cover $C \to \mathbb{P}^1$, branched along a given set of r points P_k, together with a faithful action of the group $G = (\mathbb{Z}/2)^s$ on C so that $C/G = \mathbb{P}^1$. Fix linearly independent 1-dimensional characters $\chi_1, \dots, \chi_s$ of $G = (\mathbb{Z}/2)^s$ and enumerate the $2^s - 1$ cyclic subgroups H_j of G in some order. Define integers ϵ_{ij} to be 1 if the character χ_i is nontrivial on H_j and 0 otherwise.

Suppose that we can partition the P_k into $2^s - 1$ subsets D_j of cardinality n_j, and find integers $L_1, \dots, L_s$ such that $2L_i = \sum \epsilon_{ij} n_j$ for all i. Regarding each D_j as an effective divisor of degree n_j and the L_i as the degrees of line bundles, this yields a family of "reduced building data" in the sense of [Pa91, Prop. 2.1]. Pardini constructs a G-cover of $\mathbb{P}^1$ from this data in *loc. cit.* Since the fiber over each P_k has 2^{s-1} double points, the ramification divisor on C has degree $2^{s-1}r$ and C has genus $2^{s-2}r + 1 - 2^s$, by Hurwitz' theorem. The case $s = 1$ recovers the classical result that any even number of points in $\mathbb{P}^1$ forms the branch locus of a double cover. Here is the case $s = 2$:

Lemma 7.6. *Let $P_1, \cdots P_r$ be distinct points on $\mathbb{P}^1$, $r \geq 3$. Then there exists a smooth curve C and a $(\mathbb{Z}/2)^2$-cover $h : C \to \mathbb{P}^1$ such that the ramification divisor on C has degree $2r$ and C has genus $r - 3$.*

Proof. We use Pardini's method with $s = 2$. We need to partition the points into three subsets of cardinalities n_i, corresponding to the three subgroups H_1, H_2, H_3 of order 2 in $(\mathbb{Z}/2)^2$. As above, we need to solve the equations $n_1 + n_2 + n_3 = r$, $n_2 + n_3 = 2L_1$ and $n_1 + n_3 = 2L_2$ for positive integers n_i and L_j. If r is even and at least 4, we can take $n_1 = n_2 = 2$ and $n_3 = r - 4$; if r is odd and at least 3, we can take $n_1 = n_3 = 1$ and $n_2 = r - 2$. □

Theorem 7.7. *Let S be a minimal surface of general type with $p_g(S) = 0$. Suppose that there exists a fibration $f : S \to \mathbb{P}^1$ with general fiber F and r double fibers as singular fibers. Then*

$$K_S^2 \geq 2(r-4)(g(F) - 1).$$

If equality holds then S satisfies Bloch's conjecture.

Proof. By Lemma 7.6 we can find a $(\mathbb{Z}/2)^2$-cover $h : C \to \mathbb{P}^1$ branched over the r points of $\mathbb{P}^1$ corresponding to the double fibres of f. We have seen that the degree of the ramification divisor of h on C is $2r$, and that C has genus $r - 3$. By Lemma 7.4 we can find a square (7.3) with $\tilde{h} : X \to S$ étale. By Beauville's bound (7.2),

$$K_X^2 \geq 8(r-4)(g(F) - 1),$$

and if equality holds then the fibration $X \to C$ is isotrivial. Since $K_X^2 = 4K_S^2$, we have

$$K_S^2 = 2^{-2} K_X^2 \geq 2(r-4)(g(F) - 1),$$

and if equality holds then (by Theorem 7.1) $t_2(S) = 0$ and S satisfies Bloch's conjecture. □

Corollary 7.8. *Let S be a minimal complex surface of general type with $K_S^2 = 8$. If S has an involution σ such that S/σ is rational, then S satisfies Bloch's conjecture.*

Proof. By [Pa03, Thm. 2.2], S has a fibration $S \to \mathbb{P}^1$ with general fiber F and r double fibers, such that either $r = 6$ and $g(F) = 3$ or $r = 5$ and $g(F) = 5$. In both cases there is an equality $K_S^2 = 2(r-4)(g(F)-1)$. By Theorem 7.7, S satisfies Bloch's conjecture. □

8 Inoue's surface with $K^2 = 7$

Until recently, the only known family of examples of surfaces S of general type with $K_S^2 = 7$ and $p_g(S) = 0$ was constructed by M. Inoue in [I]. It is a quotient of a complete intersection in the product of four elliptic curves, by a free action of $\mathbb{Z}/5$. (Another family was found recently by Y. Chen in [Ch]; see Remark 8.4). In this section, we use Theorem 8.1 to show that Bloch's conjecture holds holds for Inoue's surfaces; Bauer's recent preprint [Bau] also gives a proof of Theorem 8.1, and her proof is similar to ours.

An alternative description of Inoue's surface as a bidouble cover of a rational surface was given in [MP01a]. Let Π be the blow up of $\mathbb{P}^2$ in 6 points $P_1,\dots,P_6$ as in [MP01a, Ex. 4.1] and let $\pi: X \to \Pi$ be the bidouble cover with branch locus $D = D_1 + D_2 + D_3$, where

$$D_1 = \Delta_1 + F_2 + L_1 + L_2\ ;\ D_2 = \Delta_2 + F_3\ ;$$

$$D_3 = \Delta_3 + F_1 + F_1' + L_3 + L_4.$$

Here Δ_i ($i = 1,2,3$) is the strict transform in Π of the diagonal lines P_1P_3, P_2P_4 and P_5P_6, respectively; L_i is the strict transform of the line between P_i and P_{i+1} of the quadrilateral P_1,P_2,P_3,P_4; F_1 is the strict transform of a general conic through P_2,P_4,P_5,P_6 and $F_1' \in |F_1|$; F_2 is the strict transform of a general conic through P_1,P_3,P_5,P_6; and F_3 is that of a general conic through P_1,P_2,P_3,P_4.

The image of the morphism $f: \Pi \to \mathbb{P}^3$ given by $|-K_\Pi|$ is a cubic surface $V \subset \mathbb{P}^3$; f contracts each L_i to a point A_i, and is an isomorphism on $\Pi \setminus \cup_i L_i$. The (set-theoretic) inverse image in X of the 4 lines L_i is the disjoint union of two (-1)-curves $E_{i,1}$ and $E_{i,2}$. The surface S is obtained by contracting these eight exceptional curves on X; the results in [MP01a] show that S is a surface of general type with $p_g(S) = 0$, $K_S^2 = 7$, that the bicanonical map $\Phi_2: S \to \mathbb{P}^7$ has degree 2, and the bicanonical map is one of the maps $S \to S/\sigma$ associated to the bidouble cover.

Theorem 8.1. *Inoue's surface S with $K_S^2 = 7$ satisfies Bloch's conjecture.*

Proof. Let $\sigma_1, \sigma_2, \sigma_3$ be the nontrivial involutions of X over Π, or equivalently, of S over V. We will determine the number k_i of isolated fixed points of σ_i on S in Lemma 8.2 below. Let Y_i be the desingularization of S/σ_i given by (2.1). For σ_1 we have $k_1 = 11 = K_S^2 + 4$ so, by Theorem 3.1, Φ_2 is composed with the involution σ_1 and Y_1 is either rational or birational to an Enriques surface. In particular, $t_2(Y_1) = 0$.

Since k_2 and k_3 are less than $K_S^2 + 4$, Φ_2 is *not* composed with either σ_2 or σ_3, by Theorem 3.1. By Theorem 3.5, the minimal models of Y_2 and Y_3 cannot have Kodaira dimension 2, because $K_S^2 = 7$ is odd and $k_2 = k_3 = 9 = K_S^2 + 2$. It follows from [BKL] that Y_2 and Y_3 satisfy Bloch's conjecture, and so $t_2(Y_2) = t_2(Y_3) = 0$. By Theorem 2.5, this shows that $t_2(S) = 0$ and finishes the proof. □

Lemma 8.2. *The involutions σ_i on S have $k_1 = 11$, $k_2 = 9$ and $k_3 = 9$ fixed points, respectively.*

Proof. There is a smooth bidouble cover $p : S \to V$, where V is obtained from Π by contracting the curves L_i to 4 singular points $A_1, \cdots A_4$. Hence we get a commutative diagram

$$\begin{array}{ccc} X & \xrightarrow{\pi} & \Pi \\ {\scriptstyle g}\downarrow & & \downarrow{\scriptstyle f} \\ S & \xrightarrow{p} & V \end{array}$$

where $g(E_{i,j}) = Q_{i,j}$ and $p(Q_{i,j}) = A_i$ for $i = 1, \cdots, 4$ and $j = 1, 2$. The divisors on Π satisfy the following relations, see [MP01a, Ex.4.1]:

$$-K_\Pi \equiv \Delta_1 + \Delta_2 + \Delta_3; \quad F_i \equiv \Delta_{i+1} + \Delta_{i+2}, i \in \mathbb{Z}/3;$$

$$\Delta_i \cdot L_j = 0; \quad \Delta_i \cdot F_j = 2\delta_{ij}.$$

From these relations it is easy to calculate that the number k_i' of isolated fixed points of σ_i on X is: $k_1' = (D_2 \cdot D_3) = 7$; $k_2' = (D_1 \cdot D_3) = 9$; and $k_3' = (D_1 \cdot D_2) = 5$. The surface S may have additional isolated fixed points $Q_{i,j}$, because some of the curves $E_{i,j}$ may be fixed on X.

By construction the divisorial part of the fixed locus of σ_i on X lies over D_i. Since $L_1 + L_2$ is contained in D_1, the fixed locus of σ_1 on X contains the four (-1)-curves $E_{i,1}$ $(i = 1, 2)$ lying over L_1 and L_2. Their images $Q_{i,1}$ and $Q_{i,2}$ are fixed points of σ_1 on S, so σ_1 has $k_1 = k_1' + 4 = 11$ isolated fixed points on S. Similarly, since L_3 and L_4 are contained in D_3 the fixed locus of σ_3 on S contains the four (-1)-curves $E_{3,j}$ and $E_{4,j}$, so σ_3 has $k_3 = k_3' + 4 = 9$ isolated fixed points on S. Finally, we have $k_2 = k_2' = 9$ because none of the L_i are contained in the divisorial part D_2 of σ_2. □

Remark 8.3. In the recent preprint [LS12], Lee and Shin consider the case of the Inoue's surface S with $K_S^2 = 7$ and $p_g(S) = 0$ and compute $K_{W_i}^2$, where

W_i is a minimal model of S/σ_i ($i = 1,2,3$). In particular they show that W_1 and W_3 (hence S/σ_1 and S/σ_3) are rational, while W_2 (and hence S/σ_2) is birational to an Enriques surface. Theorem 8.1 also follows from this via Theorem 2.5.

Remark 8.4. Recently Y. Chen in [Ch] has produced a family of surfaces of general type with $K^2 = 7$ and $p_g = 0$, whose bicanonical map is not composed with any involution on S. Chen has verified that Bloch's conjecture holds for these surfaces.

9 Burniat surfaces and surfaces with $K^2 = 6$, $K^2 = 7$

In this section we show that Burniat surfaces satisfy Bloch's conjecture and consider other surfaces with $K^2 = 6$.

Burniat surfaces

Burniat surfaces are certain surfaces of general type with $2 \leq K_S^2 \leq 6$. It is proven in [BC1, Thm. 2.3] that Burniat surfaces are exactly the surfaces constructed by Inoue in [I] as the $(\mathbb{Z}/2)^3$-quotient of an invariant hypersurface in the product of 3 elliptic curves.

To construct them, one first forms a del Pezzo surface Π as the blowup of the plane at 3 non-collinear points P_1, P_2, P_3, see [Pet]. Then one forms a bidouble cover $\bar{S}$ of Π whose branch locus is the union of the exceptional curves and the proper transform of 9 other lines. Then S is the minimal resolution of $\bar{S}$, and $\bar{S}$ has $k = 6 - K_S^2$ singular points.

If $K_S^2 = 6$ (so $S = \bar{S}$), these are called *primary* Burniat surfaces. If $K_S^2 = 4,5$ (so $k = 1,2$) S is called *secondary*; if $K_S^2 = 2,3$ (so $k = 3,4$) S is called *tertiary*. see [BC1, 2.2].

An important feature of the Burniat surfaces S is that their bicanonical map is the composition of a bidouble cover $S \to \Pi'$ onto a normal del Pezzo surface Π' (a blowup of Π) followed by the anticanonical embedding of Π' into $\mathbb{P}^6$. (See [MP01, 3.1] and [BC1, 4.1]). Thus Φ_2 is composed with each of the three nontrivial involutions σ_i on S.

The following result was first proven by Inoue-Mizukami in [IM, Thm. 2]; cf. [BC2, 4.4].

Theorem 9.1. *Every Burniat surface S satisfies Bloch's conjecture.*

Proof. As noted above, the bicanonical map Φ_2 is the composition of a bidouble cover $S \to \Pi'$ with the anticanonical embedding of Π'. Thus Φ_2 is composed with each of the three nontrivial involutions σ_i on S associated to this cover. By Theorem 3.1 each S/σ_i is either rational or birational to an Enriques surface, so in particular a desingularization W_i of S/σ_i has $t_2(W_i) = 0$. From Theorem 2.5 we have $t_2(S) = 0$, viz., S satisfies Bloch's conjecture. □

Remark 9.1.1. As noted in [BC1, §5], there is only one Burniat surface with $K_S^2 = 2$, and it is a surface in the 6-dimensional family of classical Campedelli surfaces with $\pi_1(S) = (\mathbb{Z}/2)^3$. Thus this surface fits into the discussion in Section 5, where we also noted that such surfaces satisfy Bloch's conjecture.

Surfaces with $K^2 = 6$

We now consider a minimal surface S of general type with $p_g(S) = 0$ and $K_S^2 = 6$. Either the bicanonical map $\Phi_2 : S \to \mathbb{P}^6$ is birational or the degree of Φ_2 is either 2 or 4, by [MP04, Thm.1.1].

When the degree of Φ_2 is 4 and $K_S^2 = 6$, a complete classification has been given in [MP01]. The classification shows that all these surfaces are Burniat surfaces. It follows that they satisfy Bloch's conjecture, by Theorem 9.1.

Now suppose that the degree of Φ_2 is 2. Then Φ_2 determines an involution τ on S, and Φ_2 is composed with τ. By [MP04, 1.2] there is a fibration $f : S \to \mathbb{P}^1$, whose general fibre F is hyperelliptic of genus 3, such that the bicanonical involution τ on S induces the hyperelliptic involution on F and f has either $r = 4$ or $r = 5$ double fibres. (Both possibilities occur; see [MP04, §4]).

In particular S/τ is rational. Forming the square (2.1), the desingularization Y of S/τ is rational by [MP04, 2.1]. Hence $t_2(Y) = 0$. By Theorem 3.1, $-4 \leq K_Y^2 \leq 0$ and the fixed locus of τ has 10 isolated points. The cases of $r = 4$ or 5 double fibers corresponds to the cases $K_Y^2 = -4$ and $K_Y^2 > -4$.

Although we have been unable to show that Bloch's conjecture holds for every surface with $K_S^2 = 6$ whose bicanonical map has degree 2, we can show this for the examples constructed in [MP04, §4]. We shall use the notation of Section 8, since the construction of these examples are variations of the construction of Inoue's surface described there.

Construction 9.2. In the example in [MP04, 4.1] we assume that the conics F_1, F_2, F_3 all pass through a general point P and that pairwise they intersect transversally at P. In this case the branch locus $D = D_1 + D_2 + D_3$ acquires a singular point of type $(1,1,1)$, and the resulting bidouble cover $X \to \Pi$ has a singularity over the image P' of P. Blowing up P' gives Π'; normalizing $X \times_\Pi \Pi'$ yields a bidouble cover $S \to \Pi'$, and the surface S has $p_g = 0$ and $K_S^2 = 6$. The surface S has a fibration $f : S \to \mathbb{P}^1$, whose general fiber is hyperelliptic of genus 3 and f has 4 double fibers. Moreover the exceptional divisor of $S \to X$ is a smooth rational curve C with $C^2 = -4$.

$$\begin{array}{ccc} S & \xrightarrow{p} & \Pi' \\ \downarrow & & \downarrow \\ X & \xrightarrow{p_0} & \Pi \end{array}$$

Proposition 9.3. *The surface S of 9.2 satisfies Bloch's conjecture.*

Proof. We shall write σ_i for the three nontrivial involutions on X and on S associated to the bidouble covers. We first compute the number k_i' of isolated fixed points on X of σ_i, as we did in Lemma 8.2. Since $P \in D_1 \cap D_2 \cap D_3$ and $F_1 \cdot F_3$ contains only 1 point outside of $P_1, \cdots, P_6$ we get $k_1' = (D_2 \cdot D_3) = 6$. Similarly we get $k_2' = 8$ and $k_3' = 4$. Arguing again as in Lemma 8.2, this implies that the number k_i of isolated fixed points of σ_i on S is: $k_1 = 10$, $k_2 = k_3 = 8$.

By Theorem 3.1, the bicanonical map Φ_2 is composed with σ_1 but not with σ_2 or σ_3. In particular, S/σ_1 is rational. We will show that the desingularizations Y_2 and Y_3 of S/σ_2 and S/σ_3 satisfy Bloch's conjecture, so $t_2(Y_2) = t_2(Y_3) = 0$; Theorem 2.5 will imply that $t_2(S) = 0$, i.e., that S satisfies Bloch's conjecture.

By symmetry it suffices to consider Y_2. Let X_2 denote the blowup of S along the $k_2 = 8$ isolated fixed points of σ_2, and form the square

$$\begin{array}{ccc} X_2 & \xrightarrow{h} & S \\ {\scriptstyle \bar{\sigma}_2}\downarrow & & \downarrow{\scriptstyle \sigma_2} \\ Y_2 & \xrightarrow{g} & S/\sigma_2 \end{array}$$

as in (2.1). The images in Y_2 of the 8 exceptional curves on X_2 are 8 disjoint nodal curves $C_1, \ldots, C_8$. As pointed out after (2.1) the branch locus of $X_2 \to Y_2$ is $B = g^*(D_2) + C + \sum C_j$. We get a minimal resolution W of S/σ_2 with branch locus B by blowing down curves on Y_2. By Theorem 3.5, W cannot have Kodaira dimension 2: B is not the disjoint union of 4 nodal curves, and $k = 8 > K_S^2 = 6$. Hence W and Y_2 are not of general type, so $t_2(Y_2) = 0$ as desired. □

Construction 9.4. The second example of a surface of general type with $p_g = 0$, $K_S^2 = 6$ is constructed in [MP04, 4.2]; the fibration $S \to \mathbb{P}^1$ has a hyperelliptic curve of genus 3 as its general fiber, and 5 double fibers. We start with the same configuration of 6 points $P_1, \ldots, P_6$ as in Section 8, and consider the point $P_7 = \Delta_2 \cap \Delta_3$. Form the blowup Π' of π at P_7, and write e_7 for the corresponding exceptional divisor. Let Δ_2' and Δ_3' denote the strict transform of Δ_2 and Δ_3 on Π' and set $D = D_1 + D_2 + D_3$, where

$$D_1 = C + L_1 + L_2 \ ; \ D_2 = F_3 \ ; \ D_3 = F_1 + F_1' + \Delta_2' + \Delta_3' + L_3 + L_4.$$

Here $F_1, F_1' \in |F_1|$, $F_3 \in |F_3|$ and $C \in |F_2 + F_3 - 2e_7|$ are general curves. As noted in [MP04], C is a smooth irreducible curve of genus 0, and we get a smooth bidouble cover $\pi : X \to \Pi'$ with $p_g(X) = 0$.

For $i = 1, \ldots, 4$ the inverse image of L_i in X is the disjoint union of two (-1)-curves E_{i1}, E_{i2}. Also the inverse image of Δ_2' is the disjoint union of two (-1)-curves E_1 and E_2. The system $|-K_{\Pi'}|$ gives a degree 2 morphism

$\Pi' \to \mathbb{P}^2$. The surface S obtained from X by contracting E_1, E_2 and E_{ij}, with $1 \leq i \leq 4$ and $1 \leq j \leq 2$, is minimal of general type with $K_S^2 = 6$ and $p_g(S) = 0$.

Proposition 9.5. *The surface S of 9.4 satisfies Bloch's conjecture.*

Proof. We shall write σ_i for the three nontrivial involutions on X and on S associated to the bidouble covers. As noted in the proof of Proposition 9.5, the bicanonical map Φ_2 is composed with σ_1 and hence S/σ_1 is rational and $t_2(S/\sigma_1) = 0$.

Next, we compute the number k_2' and k_3' of isolated fixed points on X of σ_2 and σ_3 and the corresponding number k_2 and k_3 of isolated fixed points on S, as we did in Lemma 8.2.

We have $k_2' = (D_1 \cap D_3) = (C \cdot F_1) + (C \cdot F_1') + (C \cdot \Delta_2') + (C \cdot \Delta_3')$. Now $(C \cdot F_1) = (C \cdot F_1') = 4$ and $(C \cdot \Delta_2') = (C \cdot \Delta_3') = 0$, because $e_7 \cdot \Delta_2' = e_7 \cdot \Delta_3' = 1$. Therefore $k_2' = 8$ and $k_2 = k_2' = 8$ because $D_2 = F_3$ does not contain any of the L_i.

Similarly $k_3' = (D_1 \cdot D_2) = (F_2 + F_3) \cdot F_3 = 2$, because $(F_3)^2 = 0$. This last equality follows from adjunction because F_3 has genus 0 and $K_\Pi \cdot F_3 = (-\Delta_1 + \Delta_2 + \Delta_3) \cdot F_3 = -2$. The fixed divisorial part D_3 of σ_3 contains $L_3 + L_4$ and Δ_2' which contract on S to the 6 points corresponding to E_{ij} and to E_1, E_2. Therefore $k_3 = k_3' + 6 = 8$.

Since $k_2 = k_3 = 8$ are less than $K_S^2 + 4 = 10$, Theorem 3.1 implies that Φ_2 is not composed with either σ_2 or σ_3. As in the previous example let X_i be the blowup of S along the fixed points of σ_i, and let $\pi_i : X_i \to Y_i$ be the map to the desingularization Y_i of S/σ_i $(i = 2, 3)$. Then π_i is branched on $B_i = g_i^*(D_i) + \sum_{1 \leq h \leq 8} C_i^h$, with $(C_i^h)^2 = -2$. Because the image $\bar{B}_i$ of the branch locus B_i in a minimal model W_i of Y_i is not the disjoint union of 4 nodal curves, it follows from Theorem 3.5 that the Kodaira dimension of W_i cannot be 2. Therefore both Y_2 and Y_3 satisfy Bloch's conjecture, so $t_2(Y_2) = t_2(Y_3) = 0$. By Theorem 2.5, this shows that $t_2(S) = 0$ and finishes the proof. □

Remark 9.5.1. The same argument applies to the surface in Example 4.3 in [MP04], showing that it also satisfies Bloch's conjecture.

Acknowledgements

The authors would like to thank Rita Pardini and Carlos Rito for their assistence in understanding the details of the constructions of surfaces of general type. We also thank Ingrid Bauer and Fabrizio Catanese for many useful comments during the creation of the paper.

References

[Bar] R. Barlow, *Rational equivalence of zero cycles for some more surfaces with $p_g = 0$*, Invent. math. **79** (1985), 303–308).

[Bau] I. Bauer, *Bloch's conjecture for Inoue surfaces with $p_g = 0, K^2 = 7$*, preprint arXiv:1210.4287v 2 [math.AG], 29 Nov. 2012.

[BC1] I. Bauer and F. Catanese, *Burniat surfaces I: Fundamental groups and moduli of primary Burniat surfaces*, pp. 49–76 in *Classification of algebraic varieties*, EMS Ser. Congr. Rep., Eur. Math. Soc., 2011.

[BC2] I. Bauer and F. Catanese, *Burniat surfaces II: Secondary Burniat surfaces form three connected components of the moduli space*, Invent. Math. 180 (2010), 559–588.

[BC3] I. Bauer and F. Catanese, *The moduli space of Keum-Naie surfaces*, Groups, Geometry, and Dynamics **5** (2011), 231–250.

[BCG] I. Bauer, F. Catanese and F. Grunewald, *The classification of surfaces with $p_g = 0$ isogenous to a product of curves*, Pure Appl. Math Q. **4** (2008), 547–586.

[BCGP] I. Bauer, F. Catanese, F. Grunewald and R. Pignatelli, *Quotients of products of curves, new surfaces with $p_g = 0$*, Amer. J. Math. **134** (2012), 993–1049.

[BCP] I. Bauer, F. Catanese, and R. Pignatelli, *Surfaces of general type with geometric genus 0: A survey*, Complex and Differential Geometry, Springer Proc. in Math. **8** (2011) 1–48.

[Be] A. Beauville, *L'inegalité $p_g \geq 2q - 4$ pour les surfaces de type general*, Appendix to Debarre, Bull. Soc Math. France 110 (1982), 343–346.

[BKL] S. Bloch, A. Kas and D. Lieberman, *Zero cycles on surfaces with $p_g = 0$*, Compositio Math. **33** (1976), 135–145.

[BHPV] W. Barth, K. Hulek, C. Peters and A. Van de Ven, *Compact complex surfaces*, Second edition, Springer, 2004.

[CCM] A. Calabri, C. Ciliberto and M. Mendes Lopes, *Numerical Godeaux surfaces with an involution*, Trans. AMS **359** (2007), 1605–1632.

[CD] F. Catanese and O. Debarre, *Surfaces with $K^2 = 2$, $p_g = 1$, $q = 0$*, J. Reine angew. Math. **395** (1989), 1–55.

[CMP] A. Calabri, M. Mendes Lopes and R. Pardini, *Involutions on numerical Campedelli surfaces*, Tohuku Math. J. **60** (2008), 1–22.

[Ch] Y. Chen, *A new Family of surfaces of general type with $K^2 = 7$ and $p_g = 0$*, preprint arXiv:1210.4633v 3 [math.AG], 1 Nov. 2012.

[CG] H. Clemens and P. Griffiths, *The intermediate Jacobian of a cubic three-fold*, Annals of Math., **95**,(1972), 281–356.

[DMP] I. Dolgachev, M. Mendes Lopes and R. Pardini, *Rational surfaces with many nodes*, Compositio Math. **132** (2002), 349–363.

[Fu] W. Fulton, *Intersection Theory*, Springer-Verlag, Heidelberg-New-York, 1984.

[GP2] V. Guletskii and C. Pedrini, *Finite-dimensional motives and the conjectures of Beilinson and Murre*, *K*-Theory **30** (2003), 243–263.

[GP1] V. Guletskii and C. Pedrini, *The Chow motive of the Godeaux surface*, Algebraic Geometry, a Volume in Memory of Paolo Francia, Walter de Gruyter (2002), 180–195.

[IM] H. Inose and M. Mizukami, *Rational equivalence of 0-cyccles on some surfaces of general type with $p_g = 0$*, Math. Ann. **244** (1979), 205–217.

[I] M. Inoue, *Some new surfaces of general type*, Tokyo J. Math. **17** (1994), 295–319.

[KMP] B. Kahn, J. Murre and C. Pedrini, *On the transcendental part of the motive of a surface*, pp. 143–202 in "Algebraic cycles and Motives Vol II," London Math. Soc. LNS **344**, Cambridge University Press, 2008.

[Ke] J. H. Keum, Some new surfaces of general type with $p_g = 0$', Unpublished preprint, 1988.

[Ki] S. Kimura, *Chow groups are finite dimensional, in some sense*, Math. Ann. **331** (2005), 173–201. Preprint, 1998.

[Ku] V.S. Kulikov, *Old examples and a new example of surfaces of general type with $p_g = 0$*, Izv. Math **68** (2004), 965–1008.

[LS12] Y. Lee and Y. Shin, *Involutions on a surface of general type with $p_g = q = 0$ and $K^2 = 7$*, preprint arXiv:1003.935 v 4 [math.AG], Oct 2012.

[MP01] M. Mendes Lopes and R. Pardini, *A connected component of the moduli space of surfaces with $p_g = 0$*, Topology **40**(2001), 977–991.

[MP01a] M. Mendes Lopes and R. Pardini, *The bicanonical map of surfaces with $p_g = 0$ and $K^2 \geq 7$*, Bull. London Math. Soc. **33** (2001), 265–274.

[MP02] M. Mendes Lopes and R. Pardini, *A survey on the bicanonical map of surfaces with p_g=0 and $K^2 \geq 2$*, pp. 277–287 in Algebraic Geometry, a Volume in Memory of Paolo Francia, Walter de Gruyter (2002).

[MP04] M. Mendez Lopez and R. Pardini, *Surfaces of general type with $p_g = 0$, $K^2 = 6$ and non birational bicanonical map*, Math. Ann. **329** (2004), 535–552.

[Na] D. Naie, *Surfaces d'Enriques et un construction de surfaces de type general avec $p_g = 0$* , Math. Zeit. **215** (1994), 269–280.

[Pa91] R. Pardini, *Abelian covers of algebraic varieties*, J. reine angew. Math. **417** (1991), 191–213.

[Pa03] R. Pardini, *The classification of double planes of general type with $K^2 = 8$ and $p_g = 0$*, J. Algebra **259** (2003), 95–118.

[Ped] C. Pedrini, *On the finite dimensionality of a K3 surface*, Manuscripta Math. **138** (2012), 59–72.

[Pet] C. Peters, *On certain examples of surfaces with $p_g = 0$ due to Burniat*, Nagoya Math J. **66** (1977),109–119.

[Ri1] C. Rito, *Involutions on surfaces with $p_g = q = 0$ and $K_S^2 = 3$*, Geom. Dedicata **157** (2012), 319–330.

[Ri2] C. Rito, *Some bidouble planes with $p_g = q = 0$ and $4 \leq K^2 \leq 7$*, preprint arXiv:1103.2940 v 2 [math.AG], 2012.

[Se] F. Serrano, *Isotrivial fibred surfaces*, Ann. Mat. Pura Appl. (IV) CLXXI (1996), 63–81.

[Vois] C. Voisin, *Bloch's conjecture for Catanese and Barlow surfaces*, preprint arXiv:1203.935 v2 [math.AG], Jan 2013.

[Zh] D.Q. Zhang, *Quotients of K3 surfaces modulo involutions*, Japan J. of Math **24** n.2 (1998), 335–366.

PART III

The arithmetic of periods

III.A Motives, Galois representations, and automorphic forms

13

An introduction to the Langlands correspondence

Wushi Goldring

Contents

W. G. (1) Institut für Mathematik, Mathematisch-naturwissenschaftliche Fakultät Universität Zürich, Winterthurerstrasse 190, CH-8057 Zürich, SWITZERLAND
(2) Department of Mathematics, Washington University in St. Louis, One Brookings Drive, St. Louis, MO 63130, USA
E-mail: `wushijig@gmail.com`

1 Introduction

This is a purely expository article which is based on a series of three lectures given at the conference "Recent Advances in Hodge Theory: Period Domains, Algebraic Cycles, and Arithmetic" held in Vancouver in June 2013.

1.1. The fundamental triangle

Different people use the term "Langlands correspondence" to mean different things. For us the Langlands correspondence is a family of conjectured correspondences which can be summarized by the following fundamental triangle:
(1.1.1)

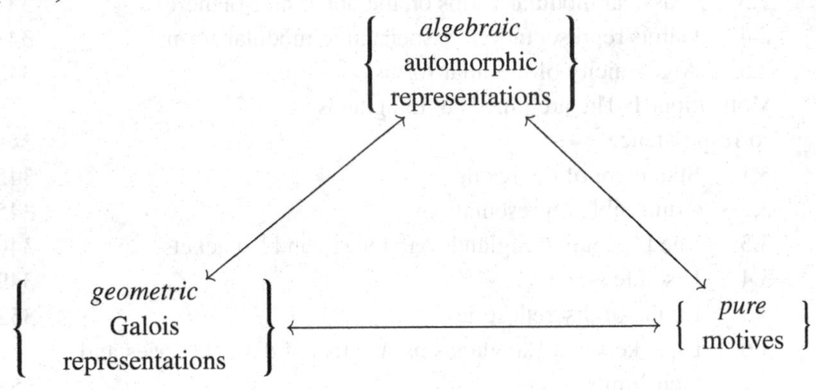

As such, it should be understood from the outset that the Langlands correspondence means something not only due to Langlands, but originating with Langlands and then developed by many people, including Fontaine-Mazur, Grothendieck, Deligne, Serre, Clozel, and Buzzard-Gee.

Most of this article will be concerned only with the arrow in the fundamental triangle (1.1.1) going from algebraic automorphic representations to geometric Galois representations. However, all the objects and arrows in the fundamental triangle (1.1.1) are so inextricably linked to one another that in order to provide a more accurate global conceptual overview we thought it is important to include the full triangle and not just the one arrow.

1.2. The objects

Each of the three collections of objects, algebraic automorphic representations, geometric Galois representations and pure motives, is very complicated and far from understood. It is far beyond the scope of this article to give complete definitions of these objects. Nevertheless, we will try to give some explanation about each collection and to provide references where the interested reader can find the technical details (when such references exist).

Of all the terms in the fundamental triangle (1.1.1), "Galois representation" is easiest to define, see §4.1.4. Intuitively one may think of Galois representations as an efficient means of storing information from Galois theory. The difficulty with defining geometric Galois representations is the "geometric part," which has to do with p-adic Hodge theory.

With automorphic representations the situation is more or less the opposite: It is quite technical to give a full definition of automorphic representations, but explaining the "algebraic" part is relatively elementary, see §4.1.3. Automorphic representations are at the same time generalizations of modular forms and of (infinite-dimensional) representations of real reductive groups.

Finally we come to pure motives. In many ways, "What is a motive?" is an open question. The reader who is uncomfortable with motives can instead restrict attention to algebraic varieties. But more generally, what we intend by "pure motive" is vaguely a piece of the cohomology of an algebraic variety. When we say "the cohomology" we mean any Weil cohomology theory, but especially pertinent to the relationship with Galois representations is étale cohomology.

1.3. A bit of history

At least three sources can be isolated regarding the birth and evolution of the Langlands correspondence into the form of the fundamental triangle (1.1.1). To some extent these three sources can be matched with the three vertices of the triangle.

The first source, which couples with the Galois vertex, is Class Field Theory and its aftermath, particularly Artin's reciprocity law and Artin's conjecture. One point of view is that Langlands developed the Langlands correspondence in order to better understand Artin's conjecture.

A second source, which fits best with the automorphic vertex, is the (infinite-dimensional) representation theory of real reductive groups, as pioneered by Harish-Chandra and then Langlands, culminating in what is now called the Archimedean Local Langlands correspondence.

The third source, which comes closest to the motives vertex, but which really has to do with all three vertices, is the Deligne-Serre theory of modular forms and their associated Galois representations.

The second (resp. third) source will be described in §3 (resp. §2) and used to motivate the Langlands correspondence. Partly due to time constraints, we have chosen not to discuss the first source, but this should not be interpreted as undermining its significance.

1.4. Functoriality

Besides the Langlands correspondence, there is another closely related family of conjectures by Langlands, called Langlands' Principle of Functoriality. As we will explain later, this principle too is motivated by the Archimedean Local Langlands correspondence. Langlands Functoriality predicts that there are many correspondences between automorphic representations of different reductive groups. In this way, Langlands Functoriality may be seen as a family of conjectures that is intrinsic to automorphic representations (and does not mention Galois representations). However, as we will discuss, Langlands Functoriality may also be viewed as a tool for proving cases of the Langlands correspondence.

1.5. Overview

This article is divided into two parts, followed by one appendix. The first part, comprised of §§2, 3 is motivation for the general Langlands correspondence. The second part which consists of §4 is about the Langlands correspondence proper. Appendix A touches briefly on functoriality and its relationship with the Langlands correspondence.

§2 is concerned with the Deligne-Serre theory of modular forms and the Galois representations attached to them. §2 introduces many of the key notions in the base case of modular forms. Many notions which are too technical to describe here in general can be explained more concretely and elementarily in the case of modular forms.

§3 is about the Archimedean Local Langlands correspondence. We give some idea of some of the objects that appear in this correspondence, such as the dual group and L-group (§§3.3.1–3.3.2), and Langlands parameters and L-packets (§3.3.4). We then discuss some special classes of representations, namely discrete series and their limits (§§3.4–3.5). As an illustration of the archimedean Langlands correspondence, we describe some properties of the correspondence for these special classes of representations (§3.6). Finally, §3 ends with a concrete example, that of the symplectic group $Sp(4)$ (§3.7).

§4 is about the general Langlands correspondence itself. The first part §4.1 introduces automorphic representations and Galois representations. Then §4.2 lists the main conjectures. Next, §4.3 records the known results about the Langlands correspondence. Finally, §4.4 states some of the main remaining open problems regarding the correspondence.

Appendix A is concerned with functoriality and its relation to the Langlands correspondence.

2 Motivation I: The Deligne-Serre theory of modular forms

The main goal of §2 is to explain the association of Galois representations to classical modular forms which are Hecke eigenforms, due to Deligne [Del69] and Deligne-Serre [DS74]. The main result is Th. 2.4.1. The results and open problems discussed in §4 are to a large extent attempts to generalize Th. 2.4.1.

2.1. Outline of §2

We begin in §2.2 by noting some important historical developments in the theory of modular forms. Then §2.3 gives the original complex-analytic definition of modular forms and Hecke operators. The main result is stated in §2.4. In §2.5 we list some of the algebro-geometric reinterpretations of aspects of modular forms which form the basis of Th. 2.4.1. We sketch one of the main ideas of the proof of Th. 2.4.1 in §2.5.8.

We have chosen to state Th. 2.4.1 as early as possible, and only then discuss some algebro-geometric notions which make it possible. It is our hope that this ordering of the content will help to elucidate the striking nature of Th. 2.4.1 and the pivotal role played by the algebraic geometry of §2.5.

2.2. A brief history of modular forms

The history of modular forms can be partitioned into four phases. Modular forms first arose in the 19th century (if not earlier) as certain “special” holomorphic functions on the upper half complex plane. In this first phase, the special property of modular forms was expressed as a transformation law (see (2.3.7)) under the action of $SL(2, \mathbf{Z})$ (or a specific subgroup of small index) by fractional linear transformations. During this period, modular forms were considered objects of a complex-analytic nature.

A second phase, dominated by Hecke and Weil saw some fundamental changes in perspective. Hecke introduced the operators which bear his name. The systematic study of Hecke operators led to two related changes in the theory of modular forms which were to have an enduring impact. Hecke operators are linear operators which act on vector spaces of modular forms. Thus, within the idea of Hecke operators lies the idea to focus on *spaces* of modular forms rather than on individual functions. But the discovery of Hecke operators does not stop there. It is also one of the earlier manifestations of a theme which would play a key role in Grothendieck's revolution across mathematics: The "functions theme," which states roughly that, for many spaces X, the functions between X and other spaces reveal deeper structure than the elements of X.

This intermediate stage also saw the birth of a geometric interpretation of modular forms. Developments in algebraic geometry by Weil and others made it possible to think of modular forms as sections of line bundles on algebraic curves. These algebraic curves which became known as modular curves, replaced the upper half-plane in many respects and the awkward transformation law (2.3.7) was reformulated by saying that modular forms are not functions, but rather sections of non-trivial line bundles. It was also understood that Hecke operators could be interpreted geometrically, as correspondences on modular curves.

These new directions reached a completely different level in the third quarter of the 20th century, which is what we call the third phase. The two ideas of (i) spaces rather than individual objects and (ii) functions on spaces rather than spaces were pushed in two initially different directions. On the one hand, Grothendieck's language made it natural to consider modular forms with coefficients in an arbitrary ring. It also made precise the notion that modular forms are invariants of elliptic curves, because Grothendieck made precise– and at the same time conceptually clear– what a moduli space is. As a result, modular curves were seen to be moduli spaces of elliptic curves. In particular, Serre, Deligne and Katz developed a theory of modular forms over $\mathbf{Z}$ and over finite fields.

On the other hand, Langlands developed his theory of automorphic representations. This led to a representation-theoretic point of view concerning modular forms and Hecke operators. The representation theoretic point of view showed that, not only should one consider the space of all modular forms of fixed weight and level at once, but one should in fact allow the level to vary and consider the infinite-dimensional space of all modular forms of fixed weight at once. For modular forms, these developments reached a high point with the association of Galois representations to Hecke eigenforms by Deligne and Deligne-Serre, which is the focus point of §2.

In a fourth stage, pioneered by Wiles in the early 1990's, it was shown that certain motives, including elliptic curves over $\mathbf{Q}$, correspond to Hecke eigenforms. In terms of our fundumental triangle (1.1.1), this concerns the upward arrows in the case of modular forms. We shall almost entirely avoid this aspect, but see §4.3.3.

2.3. Classical modular forms on the upper half-plane

2.3.1. Set-up

Let

$$\mathbf{H} = \{z \in \mathbf{C} | \mathrm{Im}(z) > 0\} \tag{2.3.1}$$

be the upper half-plane. Let $SL(2,\mathbf{R})$ be the group of 2 by 2 real matrices with determinant 1. The group $SL(2,\mathbf{R})$ acts on $\mathbf{H}$ by fractional linear transformations:

$$\begin{pmatrix} a & b \\ c & d \end{pmatrix} z = \frac{az+b}{cz+d} \text{ for all } \begin{pmatrix} a & b \\ c & d \end{pmatrix} \in SL(2,\mathbf{R}), z \in \mathbf{H}. \tag{2.3.2}$$

Let $SL(2,\mathbf{Z})$ be the subgroup of $SL(2,\mathbf{R})$ consisting of matrices with integer entries. Given an integer $N \geq 1$, one has a subgroup $\Gamma(N)$ of $SL(2,\mathbf{Z})$ given by those matrices whose reduction modulo N is the identity matrix i.e.

$$\Gamma(N) = \left\{ \begin{pmatrix} a & b \\ c & d \end{pmatrix} \in SL(2,\mathbf{Z}) \middle| \begin{array}{ll} a \equiv d \equiv 1 & \pmod{N} \\ b \equiv c \equiv 0 & \pmod{N} \end{array} \right\}. \tag{2.3.3}$$

A subgroup of $SL(2,\mathbf{Z})$ is called a congruence subgroup if it contains $\Gamma(N)$ for some N. Two special classes of congruence subgroups are

$$\Gamma_1(N) = \left\{ \begin{pmatrix} a & b \\ c & d \end{pmatrix} \in SL(2,\mathbf{Z}) \middle| \begin{array}{ll} a \equiv d \equiv 1 & \pmod{N} \\ c \equiv 0 & \pmod{N} \end{array} \right\}, \tag{2.3.4}$$

and

$$\Gamma_0(N) = \left\{ \begin{pmatrix} a & b \\ c & d \end{pmatrix} \in SL(2,\mathbf{Z}) \middle| c \equiv 0 \pmod{N} \right\}. \tag{2.3.5}$$

2.3.2. Definition of classical modular forms

Let Γ be a congruence subgroup of $SL(2,\mathbf{Z})$ and let k be a positive integer. A modular form of weight k and level Γ is a holomorphic function

$$f : \mathbf{H} \longrightarrow \mathbf{C} \tag{2.3.6}$$

that satisfies the transformation rule

$$f\left(\frac{az+b}{cz+d}\right) = (cz+d)^k f(z) \text{ for all } \begin{pmatrix} a & b \\ c & d \end{pmatrix} \in \Gamma \tag{2.3.7}$$

and is "holomorphic at infinity" (one also says "holomorphic at the cusps"), an additional condition which we now describe.

Given $\gamma = \begin{pmatrix} a & b \\ c & d \end{pmatrix} \in SL(2,\mathbf{Z})$ and $f : \mathbf{H} \longrightarrow \mathbf{C}$, we record the type of transformation rule occurring in (2.3.7) by defining

$$f|_\gamma(z) = (cz+d)^{-k} f(\gamma z). \tag{2.3.8}$$

In terms of this additional notation, the transformation rule (2.3.7) becomes $f|_\gamma = f$ for all $\gamma \in \Gamma$.

Suppose f is a modular form of weight k and level Γ. Since Γ is assumed to be a congruence subgroup, Γ contains $\Gamma(N)$ for some integer $N \geq 1$. Since $\Gamma(N)$ contains the unipotent matrix $\begin{pmatrix} 1 & N \\ 0 & 1 \end{pmatrix}$, for every $\gamma \in SL(2,\mathbf{Z})$, one has a Fourier-type expansion

$$f|_\gamma(z) = \sum_{\alpha \in \frac{1}{N}\mathbf{Z}} a_{\alpha,\gamma} q^\alpha \text{ with } q = e^{2\pi \mathrm{i} z}. \tag{2.3.9}$$

One says that f is holomorphic at infinity if $a_{\alpha,\gamma} = 0$ for all $\gamma \in SL(2,\mathbf{Z})$ and all $\alpha < 0$ in (2.3.9). If in addition $a_{0,\gamma} = 0$ for all $\gamma \in SL(2,\mathbf{Z})$, then f is called cuspidal, or a cusp form.

In case Γ contains $\Gamma_1(N)$, then Γ contains the unipotent matrix $\begin{pmatrix} 1 & 1 \\ 0 & 1 \end{pmatrix}$. Hence there is an expansion, called the q-expansion,

$$f(z) = \sum_{n \in \mathbf{Z}} a_n q^n \text{ with } q = e^{2\pi \mathrm{i} z}, \tag{2.3.10}$$

where now the sum is over $\mathbf{Z}$ rather than $\frac{1}{N}\mathbf{Z}$.

Note that, in the definitions of holomorphic at infinity and cuspidal, even when $\Gamma \supset \Gamma_1(N)$, it is important to consider all $\gamma \in SL(2,\mathbf{Z})$ and all the corresponding expansions, not just the expansion (2.3.10).

2.3.3. Hecke operators via q-expansions

Much of the richness of modular forms comes from the existence of certain linear operators, called Hecke operators, on spaces of modular forms. We shall give what we believe to be the quickest and most elementary definition of Hecke operators. As mentioned in §2.2, there are other, more sophisticated ways of thinking of Hecke operators, both in terms of algebraic geometry and in terms of representation theory, which shed more light on what they really are.

For concreteness we will only consider level $\Gamma_0(N)$. Let $M_k(N)$ (resp. $S_k(N)$) be the vector space of cusp forms of weight k and level $\Gamma_0(N)$. It is a finite dimensional complex vector space. Let p be a prime not dividing N. Suppose

(2.3.11)
$$f = \sum_{n\geq 0} a_n q^n \in M_k(N).$$

Then define the pth Hecke operator

(2.3.12)
$$T_p : M_k(N) \longrightarrow M_k(N)$$

by

(2.3.13)
$$T_p(f) = \sum_{n\geq 0} a_{pn} q^n + p^{k-1} \sum_{n\geq 0} a_n q^{pn}.$$

The Hecke operator T_p maps cusp forms to cusp forms.

Let p' be another prime not dividing N. Then the Hecke operators T_p and $T_{p'}$ commute. Therefore it makes sense to consider simultaneous eigenvectors of the T_p for all p not dividing N; such a vector is called a Hecke eigenform. Suppose $f = \sum_{n\geq 1} a_n q^n$ is a non-zero, cuspidal Hecke eigenform. The form f is said to be normalized if $a_1 = 1$. If f is normalized, then $T_p f = a_p f$. In words: the coefficients of the q-expansion coincide with the Hecke eigenvalues.

2.4. Galois representations associated to modular forms

Given only the information in §2.3, which seems at first to be analytic, it is striking that there is a close connection between modular forms and Galois theory.

Theorem 2.4.1 (Deligne, Deligne-Serre). *Suppose $f = \sum_{n\geq 1} a_n q^n$ is a normalized, cuspidal Hecke eigenform of weight $k \geq 1$ and level $\Gamma_0(N)$. Then there exists a unique semisimple Galois representation*

(2.4.1)
$$R_\ell(f) : \mathrm{Gal}(\overline{\mathbf{Q}}/\mathbf{Q}) \longrightarrow GL(2, \overline{\mathbf{Q}_\ell})$$

such that if p does not divide $N\ell$, then $R_\ell(f)$ is unramified at p and if Frob_p *denotes a Frobenius element at p, one has*

(2.4.2)
$$\mathrm{tr}(\mathrm{Frob}_p) = a_p.$$

Moreover, if $k = 1$, then the image of $R_\ell(f)$ is finite and therefore $R_\ell(f)$ gives rise to a complex-valued representation

(2.4.3)
$$R_\infty(f) : \mathrm{Gal}(\overline{\mathbf{Q}}/\mathbf{Q}) \longrightarrow GL(2, \mathbf{C}).$$

Remark 2.4.2. Th. 2.4.1 was proved by Deligne [Del69] for $k \geq 2$ and by Deligne-Serre [DS74] for $k = 1$.

Remark 2.4.3. The uniqueness of $R_\ell(f)$ follows from the Cebotarev density theorem, which implies that the set of Frobenius elements Frob_p for p not dividing ℓN are dense in $\mathrm{Gal}(\overline{\mathbf{Q}}/\mathbf{Q})$. See [Ser68, §2.2] for an exposition of the Cebotarev density theorem. The mentioned density of Frobenius elements is Cor.2(a) of §2.2 of *loc. cit.*

Remark 2.4.4. The Galois representations $R_\ell(f)$ of Th. 2.4.1 were proved to in fact be irreducible by Ribet [Rib77, Th.2.3] shortly after the work of Deligne-Serre.

2.5. *Algebraicity of modular forms*

What makes Th. 2.4.1 at all possible is a reinterpretation of modular forms in algebro-geometric terms. The first step is to replace the upper half-plane $\mathbf{H}$ with algebraic curves.

2.5.1. Modular Curves I: Quotients of the upper half-plane

There exists a smooth, affine curve $Y_0(N)$ (resp. $Y_1(N)$), called the (open) modular curve of level $\Gamma_0(N)$ (resp. $\Gamma_1(N)$), whose complex points are given by $Y_0(N)(\mathbf{C}) = \Gamma_0(N)\backslash\mathbf{H}$ (resp. $Y_1(N)(\mathbf{C}) = \Gamma_1(N)\backslash\mathbf{H}$). Let $X_0(N)$ (resp. $X_1(N)$) be the smooth projective curve associated to $Y_0(N)$ (resp. $Y_1(N)$), obtained by adding finitely many points called cusps. The curve $X_0(N)$ (resp. $X_1(N)$) is called the closed modular curve of level $\Gamma_0(N)$ (resp. $\Gamma_1(N)$).

2.5.2. Modular Curves II: Moduli interpretation

The modular curve $Y_0(N)$ (resp. $X_0(N)$) is the moduli stack of pairs (E,C) where E is an elliptic curve (resp. generalized elliptic curve) and C is a subgroup of order N [DR73]. Similarly the modular curve $Y_1(N)$ is the moduli stack of pairs (E,x), with x a point of exact order N on E. One can also give a moduli-theoretic description of $X_1(N)$ using generalized elliptic curves. The curves $Y_0(N)$ and $X_0(N)$ are defined over $\mathbf{Q}$ because this is true of their moduli problems. The curves $Y_1(N)$ and $X_1(N)$ are defined over $\mathbf{Q}(\mu_n)$, the cyclotomic field generated by the nth roots of unity.

2.5.3. Remark about stacks

The modular curves $Y_0(N)$ (resp. $X_0(N)$) are not represented by schemes, since -1 is a non-trivial automorphism of every pair (E,C). For expository purposes, we shall henceforth pretend that $Y_0(N)$ (resp. $X_0(N)$) are schemes. This is justified because it is well known in this situation how to avoid speaking of stacks while remaining perfectly rigorous. For example, one valid option for $Y_0(N)$ is to modify the moduli problem by adding some more level structure. For instance, $Y_1(N)$ is a scheme if $N \geq 5$. However, we believe the notation remains somewhat lighter if we work with $Y_0(N)$.

2.5.4. The Hodge bundle

Let $\pi : \mathcal{E} \longrightarrow Y_0(N)$ be the universal elliptic curve over $Y_0(N)$ and let $\Omega^1_{\mathcal{E}/Y_0(N)}$ denote the sheaf of relative differentials on $\mathcal{E}$. Put $\omega = \pi_*\Omega^1_{\mathcal{E}/Y_0(N)}$. The line bundle ω is called the Hodge bundle. It extends to an ample line bundle on $X_0(N)$ also called ω. The line bundle ω is also defined over $\mathbf{Q}$.

2.5.5. Realization in coherent cohomology

Modular forms of weight at least one may be realized as global sections of powers of the Hodge line bundle. This is the most elementary example of realizing automorphic forms in the coherent cohomology of an algebraic variety. One has

$$M_k(N) \cong H^0(X_0(N), \omega^k) \tag{2.5.1}$$

If D denotes the boundary divisor $X_0(N) - Y_0(N)$, then one can also single out those modular forms which are cuspidal as

$$S_k(N) \cong H^0(X_0(N), \omega^k(-D)). \tag{2.5.2}$$

In case $k \geq 2$, (2.5.2) may also be rewritten as

$$S_k(N) \cong H^0(X_0(N), \omega^{k-2} \otimes K_{X_0(N)}), \tag{2.5.3}$$

where $K_{X_0(N)}$ denotes the canonical bundle of $X_0(N)$. This reflect the fact that the restrictions of ω^2 and $K_{X_0(N)}$ to $Y_0(N)$ are equal, but that the two bundles differ by D on $X_0(N)$.

2.5.6. Hecke operators via the moduli interpretation

Let p be a prime not dividing N and let $f \in H^0(X_0(N), \omega^k)$. Then another way to think of the Hecke operator T_p is

$$T_p(f)(E,C) = \frac{1}{p}\sum \varphi^* f(E',C'), \tag{2.5.4}$$

where the sum ranges over the $p+1$ isogenies $\varphi : E \longrightarrow E'$ of degree p. It follows from this description of T_p that it preserves the rational structure on $H^0(X_0(N), \omega^k)$. Therefore:

Theorem 2.5.1. *Suppose $f \in H^0(X_0(N), \omega^k)$ is a Hecke eigenform. Then the Hecke eigenvalues of f are algebraic numbers.*

2.5.7. Realization in Betti, de Rham and étale cohomology

When the weight k is at least two, one can realize $S_k(N)$ in Betti, de Rham and étale cohomology (and any other Weil cohomology theory). Let $\mathcal{U} = H^1_{dR}(\mathcal{E}/X_0(N))$ be the relative de Rham cohomology sheaf. It may be thought of as a local system on $X_0(N)$. The following result is often attributed to Eichler-Shimura; see [Del69, Th.2.10] for a precise statement.

Theorem 2.5.2 (Eichler-Shimura). *Suppose $k \geq 2$. Then*

$$S_k(N) \oplus \overline{S_k(N)} = \bar{H}^1(X_0(N), \mathrm{Sym}^{k-2}\mathcal{U}), \tag{2.5.5}$$

where $\overline{S_k(N)}$ is the complex-conjugate of $S_k(N)$ and $\bar{H}^1$ means the image of the cohomology with compact support in the cohomology.

The étale cohomology $H^1_{et}(X_0(N) \times \overline{\mathbf{Q}}, \mathrm{Sym}^{k-2}\mathcal{U} \otimes \overline{\mathbf{Q}_\ell})$ has an action of the Galois group $\mathrm{Gal}(\overline{\mathbf{Q}}/\mathbf{Q})$. This is the starting point for proving Th. 2.4.1 in the case $k \geq 2$.

2.5.8. Weight one

There is no analogue of Th. 2.5.2 in weight one. Instead, Deligne and Serre related forms of weight one to forms of higher weight via congruences. Let $E_{\ell-1}$ be the Eisenstein series of weight $\ell - 1$, normalized so that its constant term is 1. Then

$$E_{\ell-1} \equiv 1 \pmod{\ell} \tag{2.5.6}$$

in the sense that if we write $E_{\ell-1} = 1 + \sum_{n \geq 1} a_n q^n$, then ℓ divides a_n for all $n \geq 1$. Therefore

$$fE_{\ell-1} \equiv f \pmod{\ell} \tag{2.5.7}$$

for all modular forms f. In particular, if f is a cusp form of weight 1, then $fE_{\ell-1}$ is a cusp form of weight ℓ. So $fE_{\ell-1}$ is realized in étale cohomology even though f is not. This observation is the beginning of the argument of Deligne-Serre. There is a subtlety in that if f is an eigenform, the product $fE_{\ell-1}$ can fail to be an eigenform.

3 Motivation II: The archimedean Langlands correspondence

We begin by giving a statement of the archimedean Langlands correspondence in §3.1 and then proceed to explain the terms that appear in the statement in §§3.2-3.3. The subsequent §3.4 and §3.5 give a brief exposition of discrete series and their limits. Then §3.6 states how the archimedean Langlands correspondence applies to discrete series and their limits. Finally, in §3.7 we try to make everything as explicit as possible in the case of the rank two symplectic group $Sp(4)$.

3.1. Statement of the result

Theorem 3.1.1 (Langlands [Lan89]). *Let G be a connected, reductive, real algebraic group. Then there is a 'natural' bijection*

$$\frac{\begin{pmatrix} L\text{-packets of} \\ \text{Admissible} \\ \text{representations} \\ \text{of } G \end{pmatrix}}{\begin{pmatrix} \text{Infinitesimal} \\ \text{equivalence} \end{pmatrix}} \longleftrightarrow \frac{\begin{pmatrix} \text{Relevant} \\ \text{Langlands} \\ \text{parameters of } G \end{pmatrix}}{(\text{Conjugation by } G^\vee)} \tag{3.1.1}$$

See [ABV92] for a detailed discussion of the archimedean Langlands correspondence as well as more sophisticated results about the representations of real groups. See [Kna94] for an exposition of the case $G = GL(n)$.

3.2. Admissible representations

Let G be a connected, reductive, real algebraic group. Let K be a maximal compact subgroup. A representation π of G on a (complex) Hilbert space V is called admissible if, for every irreducible representation τ of K, the multiplicity with which τ appears in the restriction of π to K is finite (it may be zero). There is a notion of equivalence for admissible representations, called infinitesimal equivalence. It captures the naive idea that two infinite-dimensional representations might differ only in the topology of the representation space. To every admissible representation there is associated an algebraic object called a $(\mathfrak{g}, K)$ module, or Harish-Chandra module, and then infinitesimal equivalence is defined to be isomorphism of the corresponding Harish-Chandra modules. We shall not define Harish-Chandra modules; rather we state two fundamental results of Harish-Chandra about the relationship between unitarity and admissibility.

Theorem 3.2.1 (Harish-Chandra). *Every irreducible unitary representation of G is admissible. Moreover, two unitary representations of G are unitarily equivalent if and only if they are infinitesimally equivalent.*

One reason to study all admissible representations rather than restricting attention to those which are infinitesimally equivalent to unitary representations is that the former has better closure properties (under certain natural operations) than the latter. Another reason is that it is in general very difficult to tell whether a given admissible representation is unitary or not, see [ABV92].

A brief and relatively elementary exposition of admissible representations, including a statement of Th. 3.2.1 can be found in [Sch05].

3.3. The L-group, Langlands parameters and L-packets

§3.3.1 is based on [Spr79]. §3.3.2 follows [Bor79, §1.2].

3.3.1. The dual group

We now recall how the notion of root datum gives rise to a duality in the collection of connected reductive groups. The theory of root data is much simpler over an algebraically closed field of characteristic zero, and it is already interesting in that case. We note, however, that Grothendieck and his school developed the theory over a general base scheme [ABD+66].

Let $\mathcal{G}$ be a connected, reductive algebraic group over $\mathbf{C}$. Let $\mathcal{T}$ be a maximal torus in $\mathcal{G}$. We write $\mathbf{G}_{m,\mathbf{C}}$ for the base change to $\mathbf{C}$ of the multiplicative group $\mathbf{G}_m$. Since we are working over $\mathbf{C}$, the reader may, at least initially, identify groups over $\mathbf{C}$ with their complex points. In the case of $\mathbf{G}_{m,\mathbf{C}}$, one has $\mathbf{G}_{m,\mathbf{C}}(\mathbf{C}) = \mathbf{C}^\times$.

The root datum associated to the pair $(\mathcal{T},\mathcal{G})$ is the quadruple

$$RD(\mathcal{T},\mathcal{G}) = (\mathbf{X}^*(\mathcal{T}), \mathbf{X}_*(\mathcal{T}), \Delta(\mathcal{T},\mathcal{G}), \Delta^\vee(\mathcal{T},\mathcal{G})), \tag{3.3.1}$$

where $\mathbf{X}^*(\mathcal{T}) = \mathrm{Hom}(\mathcal{T}, \mathbf{G}_{m,\mathbf{C}})$ (resp. $\mathbf{X}_*(\mathcal{T}) = \mathrm{Hom}(\mathbf{G}_{m,\mathbf{C}}, \mathcal{T})$) is the character (resp. cocharacter) group of $\mathcal{T}$ and $\Delta(\mathcal{T},\mathcal{G}) \subset \mathbf{X}^*(\mathcal{T})$ (resp. $\Delta^\vee(\mathcal{T},\mathcal{G}) \subset \mathbf{X}_*(\mathcal{T})$) is the set of roots (resp. coroots) of $\mathcal{T}$ in $\mathcal{G}$.

We briefly recall how to define the roots, coroots and the perfect pairing between characters and cocharacters. The set of roots $\Delta(\mathcal{T},\mathcal{G})$ is the subset of $\mathbf{X}^*(\mathcal{T})$ consisting of those non-trivial characters which appear in the restriction of the adjoint representation $\mathrm{Ad} : \mathcal{G} \longrightarrow GL(\mathfrak{g})$ to the maximal torus $\mathcal{T}$. Since $\mathrm{Hom}(\mathbf{G}_{m,\mathbf{C}}, \mathbf{G}_{m,\mathbf{C}})$ is canonically isomorphic to $\mathbf{Z}$, one has a perfect pairing

$$\begin{array}{rccc} \langle , \rangle : & \mathbf{X}^*(\mathcal{T}) \times \mathbf{X}_*(\mathcal{T}) & \longrightarrow & \mathbf{Z} \\ & \langle \chi, \psi \rangle & \longmapsto & \chi \circ \psi \end{array} . \tag{3.3.2}$$

Finally we arrive at the coroots, which is the piece of the root datum whose definition is most involved. Let α be a root and let $\mathcal{T}_\alpha$ be the connected component of the kernel of α. The coroot $\alpha^\vee$ associated to α is defined to be the unique cocharacter of $\mathcal{T}$ such that (i) the image of $\alpha^\vee$ is contained in the derived subgroup of the centralizer of $\mathcal{T}_\alpha$, (ii) the image of $\alpha^\vee$ and $\mathcal{T}_\alpha$ together generate $\mathcal{T}$, (iii) $\langle \alpha, \alpha^\vee \rangle = 2$.

The properties of the root datum $RD(\mathcal{T},\mathcal{G})$ can be stated without reference to $\mathcal{T}$ and $\mathcal{G}$; in this way one extracts the notion of an abstract root datum $(X^*, X_*, \Delta, \Delta^\vee)$. It was shown by Grothendieck that in fact every abstract root datum arises from a pair $(\mathcal{T},\mathcal{G})$:

Theorem 3.3.1 (Grothendieck, [ABD+66])**.** *There is a bijection between (a) isomorphism classes of pairs $(\mathcal{T},\mathcal{G})$ with $\mathcal{G}$ a connected reductive group over $\mathbf{C}$ and $\mathcal{T}$ a maximal torus and (b) isomorphism classes of abstract root data.*

The dual of a root datum $(X^*, X_*, \Delta, \Delta^\vee)$ is the root datum $(X_*, X^*, \Delta^\vee, \Delta)$. By Th. 3.3.1, there exists a pair $(\mathcal{T}^\vee, \mathcal{G}^\vee)$ whose root datum is the dual of the root datum of $(\mathcal{T}, \mathcal{G})$. The group $\mathcal{G}^\vee$ is called the dual group of $\mathcal{G}$. The isomorphism class of $\mathcal{G}^\vee$ is independent of the choice of maximal torus $\mathcal{T}$.

If $\mathcal{G}$ is of Dynkin type A, D, E, F, G then the dual group $\mathcal{G}^\vee$ is of the same type. On the other hand if $\mathcal{G}$ is of type B (resp. C) then the dual group $\mathcal{G}^\vee$ is of type C (resp. B).

3.3.2. The L-group

Now suppose G is a connected, reductive real algebraic group. Putting $\mathcal{G} = G \otimes \mathbf{C}$ and applying §3.3.1 gives the dual group $\mathcal{G}^\vee$ which we also denote ${}^L G^0$, to stress that it is the connected component of the identity of a bigger group called ${}^L G$ which we now define. Since G is defined over $\mathbf{R}$, there is an action of $\mathrm{Gal}(\mathbf{C}/\mathbf{R}) = \{1, \sigma\}$ on ${}^L G^0$. The L-group ${}^L G$ is then the semidirect product of ${}^L G^0$ with $\mathrm{Gal}(\mathbf{C}/\mathbf{R})$ along this action.

The group G is split over $\mathbf{R}$ if and only if the action of $\mathrm{Gal}(\mathbf{C}/\mathbf{R})$ is trivial, in which case the semidirect product is direct.

3.3.3. The real Weil group $\mathbf{W_R}$

To define Langlands parameters, we must first define the real Weil group, usually denoted $\mathbf{W_R}$. The definition is at the same time elementary and not illuminating. We shall give some remarks following the definition.

The real Weil group $W_\mathbf{R}$ is defined in terms of generators and relations by

$$W_\mathbf{R} = < \mathbf{C}^\times, j | j^2 = -1 \text{ and } jzj^{-1} = \bar{z} \text{ for all } z \in \mathbf{C}^\times >, \tag{3.3.3}$$

where $z \mapsto \bar{z}$ denotes the usual complex conjugation of complex numbers. The group $\mathbf{W_R}$ is a non-split extension of a cyclic group of order two by $\mathbf{C}^\times$. Thinking of this order-two group as $\mathrm{Gal}(\mathbf{C}/\mathbf{R})$, we have a non-split short exact sequence

$$1 \longrightarrow \mathbf{C}^\times \longrightarrow \mathbf{W_R} \longrightarrow \mathrm{Gal}(\mathbf{C}/\mathbf{R}) \longrightarrow 1. \tag{3.3.4}$$

More generally, for every local or global field[1] F, Weil defined a group which we now call the Weil group of F and denote W_F. When $F = \mathbf{R}$, Weil's definition gives the group $\mathbf{W_R}$ defined above. For the non-archimedean local field $\mathbf{Q}_\ell$, one has that $W_{\mathbf{Q}_\ell}$ is the subgroup of $\mathrm{Gal}(\overline{\mathbf{Q}_\ell}/\mathbf{Q}_\ell)$ consisting of elements whose image in $\mathrm{Gal}(\overline{\mathbf{F}_\ell}/\mathbf{F}_\ell)$ is an integer power of the Frobenius automorphism. In view of the close relationship between the Weil and Galois groups in the ℓ-adic case, one may argue that $\mathbf{W_R}$ is a modification of the $\mathrm{Gal}(\mathbf{C}/\mathbf{R})$ which admits a non-trivial connected component (the $\mathbf{C}^\times$).

[1] By global field, we mean either a finite extension of $\mathbf{Q}$ (number field case) or the function field of a connected, smooth projective curve over a finite field (function field case). By a local field, we mean a completion of a global field.

The true nature of $\mathbf{W_R}$ (and even more so of W_F for F a number field) remains mysterious. A number of distinguished mathematicians, starting with Weil himself, have said that understanding these Weil groups would lead to a better understanding of several important open questions.

3.3.4. Langlands parameters and L-packets

A Langlands parameter for G is a continuous homomorphism

$$\varphi : \mathbf{W_R} \longrightarrow {}^L G(\mathbf{C}) \tag{3.3.5}$$

such that $\varphi(j) \in {}^L G(\mathbf{C}) - {}^L G^\circ(\mathbf{C})$ and $\varphi(\mathbf{C}^\times)$ consists of semisimple elements in ${}^L G^\circ(\mathbf{C})$. It is important to note that, in terms of regularity, φ is only required to be continuous; it is not required to be a morphism of algebraic groups in any sense.

We will avoid defining what it means for a Langlands parameter to be relevant; this has to do with Langlands' notion of relevant parabolics. If G is quasi-split over $\mathbf{R}$, which means that G admits a Borel subgroup defined over $\mathbf{R}$, then every Langlands parameter of G is relevant. The groups $U(a,a)$, $U(a,a+1)$, $SL(n)$, $GL(n)$ and $Sp(2n)$ are all quasi-split, the last three actually being split (even over $\mathbf{Q}$).

To each relevant Langlands parameter φ, Langlands associated a non-empty finite set $\mathcal{L}_\varphi$ of infinitesimal equivalence classes of irreducible admissible representations of G. The construction is technically involved, so we shall only note its inductive nature.

First one needs to extend the notion of parabolic subgroup to ${}^L G$. If P is a parabolic subgroup of G, then since P has the same rank as G and can be defined in terms of a root datum for G, one obtains a dual parabolic $P^\vee$, which is naturally a subgroup of $G^\vee$. The action of the Galois group $\mathrm{Gal}(\mathbf{C}/\mathbf{R})$ also makes sense for P and $P^\vee$, so that we may form ${}^L P$ just as was done for ${}^L G$. A parabolic subgroup of ${}^L G$ is then a subgroup which has the form ${}^L P$ for some parabolic P of G. One has a similar definition of a Levi subgroup of ${}^L G$, and if M is the Levi of P in G, then ${}^L M$ will be a Levi of ${}^L P$ in ${}^L G$.

With the definition of parabolics and their Levis at hand, we may next define discrete Langlands parameters, which constitute the base case of the inductive procedure. A Langlands parameter φ is discrete if its image in ${}^L G(\mathbf{C})$ is not contained in any proper parabolic ${}^L P(\mathbf{C})$. Langlands first associates an L-packet to every discrete Langlands parameter of every group G. If φ is a discrete Langlands parameter, the L-packet $\mathcal{L}_\varphi$ is defined to be the set of discrete series representations of G whose infinitesimal character is dictated by the restriction of φ to $\mathbf{C}^\times \subset \mathbf{W_R}$. Discrete series will be discussed in §3.4.

Next suppose φ is a Langlands parameter which is not discrete. Then one observes that there is a smallest parabolic ${}^L P(\mathbf{C})$ containing the image of φ. Since the definition of Langlands parameter requires the image of φ to consist of semisimple elements, it follows that in fact $\varphi(\mathbf{W_R})$ is contained in the Levi subgroup ${}^LM$ of ${}^LP$. Morever, since we chose the smallest P containing the image, the map $\varphi_M : \mathbf{W_R} \longrightarrow {}^LM$ obtained by restricting the codomain of φ to ${}^LM$ is a discrete Langlands parameter for M. Therefore, we have an L-packet of discrete series $\mathcal{L}_{\varphi,M}$ attached to φ_M. Now one induces these discrete series of M to G. Great care must be taken in how this induction is defined; one must use an appropriate version/normalization of parabolic induction.

The resulting representations of G need not be irreducible. However, each of the induced representations will have a unique irreducible quotient, which is sometimes called the Langlands quotient. The set of Langlands quotients obtained from $\mathcal{L}_{\varphi,M}$ is a good approximation of the sought after L-packet $\mathcal{L}_\varphi$. In many cases, it will precisely give $\mathcal{L}_\varphi$, but in general, a further step needs to be taken, involving the so-called R-group.

The Langlands parameter of an admissible representation determines its infinitesimal character (in fact the restriction of the Langlands parameter to $\mathbf{C}^\times$ already determines it). There are many examples of representations with the same infinitesimal character that do not have the same Langlands parameter. So the Langlands parameter is really a finer invariant than the infinitesimal character.

3.4. *Discrete series*

We shall now discuss a special class of infinite-dimensional, irreducible unitary representations called the discrete series. Such representations are arguably both the most important admissible, infinite-dimensional representations and also the closest analogue (for non-compact groups) of finite-dimensional representations of compact Lie groups. A good place to learn about discrete series is [Sch97].

3.4.1. Remark about semisimple versus reductive and connectedness

We would like to simplify the ensuing discussion of discrete series and later in §3.5 also limits of discrete series by assuming that G is semisimple and that G is connected in the classical topology of the real line. However, we would like to alert the reader to the fact that there are serious issues lurking in the background here. Many groups which play an important role in the Langlands correspondence satisfy neither of these conditions. A key example of this is $GL(n,\mathbf{R})$. Moreover, it is often difficult (or not possible) to reduce statements to groups satisfying these conditions from ones that do not. The collection of groups satisfying the more stringent properties has very bad closure properties. For example, a Levi subgroup of a semisimple group which is connected in the classical topology may satisfy neither of these properties (consider $GL(n,\mathbf{R})$ as a Levi of the rank n split symplectic group $Sp(2n,\mathbf{R})$). We have seen that

Levi subgroups play a crucial role in the definition of L-packets in §3.3.4. In general, it has proven to be simplest and most effective to develop theories which apply directly to a class of groups at least as large as that for which Th. 3.3.1 holds.

3.4.2. Definition

Let π be a representation of G on a (complex) Hilbert space V, with inner product denoted $\langle , \rangle$. As a preliminary definition, we recall that, for every pair of vectors $u, v \in V$, the (u,v)-matrix coefficient of π is the function $\mathfrak{m}_{(u,v)} : G \longrightarrow \mathbf{C}$ defined by $m_{(u,v)}(g) = \langle gu, v \rangle$. A non-trivial matrix coefficient is a matrix coefficient $\mathfrak{m}_{(u,v)}$ such that $u \neq 0$ and $v \neq 0$.

An irreducible representation π of G is said to be a discrete series representation if all of its matrix coefficients are square integrable i.e. if $\mathfrak{m}_{(u,v)} \in L^2(G)$ for all $u, v \in V$. Equivalently, for π to be discrete series it is necessary and sufficient that one non-trivial matrix coefficient is square integrable.[2]

Every discrete series representation is tempered (see §3.5) which implies it is unitary (or, to be more precise, unitarizable).

3.4.3. Existence

Harish-Chandra gave a simple criterion classifying those groups G which admit discrete series in terms of basic structure theory of G.

Theorem 3.4.1 (Harish-Chandra). *The group G admits discrete series representations if and only if it admits a maximal torus which is compact.*

Three important families of groups which admit discrete series representations are: The unitary groups $SU(a,b)$ (associated to a Hermitian form of signature (a,b)), the split symplectic groups $Sp(2n)$ (preserving an non-degenerate alternating form in $\mathbf{R}^{2n}$) and the orthogonal groups $SO(c,d)$ with $c+d$ odd (preserving a symmetric bilinear form of signature (c,d)). An important family of groups which do not admit discrete series is that of the (special) linear groups $SL(n)$ for $n > 2$.

3.4.4. Harish-Chandra parametrization

Suppose G admits a compact maximal torus T. Let K be a maximal compact subgroup of G containing T. Let $T_{\mathbf{C}}$, $K_{\mathbf{C}}$ and $G_{\mathbf{C}}$ be the complexifications and $\mathfrak{t}$, $\mathfrak{k}$, $\mathfrak{g}$ the complexified Lie algebras of T, K, G, respectively. Let $\mathbf{X}^*(T_{\mathbf{C}})$ be the character group of $T_{\mathbf{C}}$.

[2] When G is reductive but not necessarily semisimple, one asks instead that the matrix coefficients be square-integrable modulo the center of G. This "modulo center" condition is crucial to ensure that the definition of discrete Langlands parameters disussed in §3.3.4 works properly.

Let $\Delta = \Delta(\mathfrak{t},\mathfrak{g})$ be the roots of $G_{\mathbf{C}}$ with respect to $T_{\mathbf{C}}$ and $\Delta_c = \Delta(\mathfrak{t},\mathfrak{k})$ the roots of $K_{\mathbf{C}}$ with respect to $T_{\mathbf{C}}$. Put $\Delta_n = \Delta - \Delta_c$. A root of G is called compact (resp. non-compact) if it belongs to Δ_c (resp. Δ_n). Let $W = W(\mathfrak{t},\mathfrak{g})$ (resp. $W_c = W(\mathfrak{t},\mathfrak{k})$) be the Weyl group of $G_{\mathbf{C}}$ (resp. $K_{\mathbf{C}}$) relative to $T_{\mathbf{C}}$. We call W the full Weyl group and W_c the compact Weyl group.

Let Δ^+ be a choice of positive roots. Let ρ be one-half the sum of the positive roots. The coset $\mathbf{X}^*(T_{\mathbf{C}})+\rho$ of $\mathbf{X}^*(T_{\mathbf{C}})$ in $\mathbf{X}^*(T_{\mathbf{C}})\otimes\mathbf{R}$ is independent of the choice of positive roots. An element $\lambda \in \mathbf{X}^*(T_{\mathbf{C}})\otimes\mathbf{R}$ is called regular if no coroot is orthogonal to λ. Otherwise λ is termed singular, or irregular. Let $(\mathbf{X}^*(T_{\mathbf{C}})\otimes\mathbf{R})^{\mathrm{reg}}$ be the subset of $\mathbf{X}^*(T_{\mathbf{C}})\otimes\mathbf{R}$ consisting of regular elements and put

$$(\mathbf{X}^*(T_{\mathbf{C}})+\rho)^{\mathrm{reg}} = (\mathbf{X}^*(T_{\mathbf{C}})+\rho)\cap(\mathbf{X}^*(T_{\mathbf{C}})\otimes\mathbf{R})^{\mathrm{reg}}. \tag{3.4.1}$$

Then one has the following fundamental parametrization:

Theorem 3.4.2 (Harish-Chandra). *There is a 'natural' bijection*

$$(\mathbf{X}^*(T_{\mathbf{C}})+\rho)^{\mathrm{reg}}/W_c \longleftrightarrow \left\{\begin{matrix}\text{Discrete series}\\ \text{representations of } G\end{matrix}\right\}\Big/\left(\begin{matrix}\text{Unitary}\\ \text{equivalence}\end{matrix}\right) \tag{3.4.2}$$

such that, in particular, the infinitesimal character of the discrete series representation corresponding to the orbit $W_c\lambda$ is identified with the orbit $W\lambda$ via the Harish-Chandra isomorphism.

In view of Th. 3.4.2, it makes sense to call the elements of $(\mathbf{X}^*(T_{\mathbf{C}})+\rho)^{\mathrm{reg}}/W_c$ regular Harish-Chandra parameters. The Harish-Chandra parameter of a discrete series representation is then defined to be the regular Harish-Chandra parameter corresponding to it by Th. 3.4.2. Given a discrete series representation π, one often chooses a representative λ of the corresponding orbit $W_c\lambda$ and calls λ "the" Harish-Chandra parameter of π. There is often an at least somewhat natural choice of representative, so this terminology is not too abusive.

3.4.5. Holomorphic discrete series

The existence of a compact maximal torus T in G implies that the quotient G/T admits a complex structure, because given a choice Δ^+ of positive roots, the direct sum of the positive root spaces yields an integrable almost complex structure on G/T. We remark that in fact the quotient G/T admits $|W/W_c|$ inequivalent complex structures. In contrast, the quotient G/K may or may not admit a complex structure (and if it does, it admits exactly two). For example, returning to our three families $SU(a,b)$, $Sp(2n)$ and $SO(c,d)$ (with $c+d$ odd) which admit discrete series, the first two have the property that G/K admits a complex structure, while for the third G/K admits a complex structure if and only if $c = 2$ or $d = 2$. In general the structure theory of G reveals whether G/K admits a complex structure.

Lemma 3.4.3. *Suppose $G_{\mathbf{C}}$ is simple. Then the quotient G/K admits a complex structure if and only if the center of K contains a circle.*

Suppose G/K admits a complex structure and fix one. Let Δ_c^+ be a system of positive roots for $K_{\mathbf{C}}$ relative to $T_{\mathbf{C}}$. The holomorphic tangent space of G/K is a direct sum of noncompact root spaces; let Δ_n^+ be the set of those noncompact roots that appear in the direct sum. Then putting $\Delta^+ = \Delta_n^+ \cup \Delta_c^+$ yields a system of positive roots for $G_{\mathbf{C}}$ relative to $T_{\mathbf{C}}$.

We say that a discrete series π with Harish-Chandra parameter λ is holomorphic if $\langle \lambda, \alpha^\vee \rangle > 0$ for all $\alpha \in \Delta_n^+$ (this is in fact independent of the choice of representative λ in the orbit $W_c\lambda$ because W_c maps compact roots to compact roots and noncompact roots to noncompact roots). There are several reasons for the terminology "holomorphic." We will come back to one of them when we discuss the realization of automorphic representations in the coherent cohomology of Shimura varieties.

3.5. *Limits of discrete series*

There is an extension of Th. 3.4.2 which allows for some irregular elements in $(\mathbf{X}^*(T_{\mathbf{C}}) + \rho)$. The corresponding representations are called limits of discrete series.

3.5.1. The problem of defining limits of discrete series

Unfortunately, unlike the case for discrete series (§3.4.2), there is no simple intrinsic definition of limits of discrete series. In fact we don't know if there is an intrinsic definition that avoids the archimedean Langlands correspondence. At least most, if not all, definitions of limits of discrete series reference discrete series in a complicated way. Two examples of this are via Harish-Chandra's character formula and via Zuckerman tensoring [Zuc77].

For this reason, we shall not give a definition of limits of discrete series. Rather, we will give a definition of being tempered and mention that, like discrete series, limits of discrete series are tempered. The property of being tempered can be defined in a way similar to how we defined discrete series. Namely, a representation π of G on a Hilbert space V is tempered if, for every real number $\epsilon > 0$ the matrix coefficients of π are in $L^{2+\epsilon}(G)$. Admittedly this definition is not very enlightening, but at least it is simple to state.

3.5.2. Harish-Chandra parametrization

Second, let us explain the generalization of the Harish-Chandra parametrization to limits of discrete series. By a Weyl chamber of G, we mean the closure of a connected component of $(\mathbf{X}^*(T_\mathbf{C}) \otimes \mathbf{R})^{\mathrm{reg}}$. Recall that the Weyl group W acts simply transitively on the set of Weyl chambers and that the set of Weyl chambers is in one-to-one correspondence with systems of positive roots for $G_\mathbf{C}$ relative to $T_\mathbf{C}$. An irregular Harish-Chandra parameter is a pair (λ, C) where C is a Weyl chamber of G and $\lambda \in (\mathbf{X}^*(T_\mathbf{C}) + \rho) \cap C$ such that λ is not orthogonal to any C-simple compact coroot. Note that a fixed λ may pertain to several irregular Harish-Chandra parameters.

Theorem 3.5.1 (Harish-Chandra-Knapp-Zuckerman). *There is a bijection between W_c-orbits of irregular Harish-Chandra parameters and limits of discrete series up to unitary equivalence. Moreover, the infinitesimal character of a limit of discrete series corresponding to the orbit $W_c(\lambda, C)$ is identified with $W\lambda$ under the Harish-Chandra isomorphism.*

3.5.3. Different kinds of limits

Given an irregular Harish-Chandra parameter (λ, C) we denote the corresponding limit of discrete series by $\pi(\lambda, C)$. Following Knapp-Zuckerman, we say that a limit of discrete series $\pi(\lambda, C)$ is non-degenerate if λ is not orthogonal to any compact coroot; otherwise $\pi(\lambda, C)$ is called degenerate.

As in §3.4.5, suppose G/K admits a complex structure and fix one. Then we explained how to define a set of noncompact roots that we called Δ_n^+. We define a Weyl chamber C to be holomorphic if for any λ' in the interior of C we have $\langle \lambda', \alpha^\vee \rangle > 0$ for all $\alpha \in \Delta_n^+$. There are $|W_c|$ holomorphic Weyl chambers; they form a single orbit for the action of W_c.

A limit of discrete series $\pi(\lambda, C)$ is said to be holomorphic if C is. A holomorphic limit of discrete series is necessarily non-degenerate. When $G_\mathbf{C}$ is simple, this follows since every holomorphic chamber C admits a unique C-simple root which is noncompact.

3.6. L-packets and Langlands parameters of discrete series and their limits

We note here some facts about the L-packets and Langlands parameters of discrete series and their limits. An L-packet which contains a discrete series (resp. limit of discrete series) consists entirely of discrete series (resp. limits of discrete series). For discrete series this follows from the definition (see §3.3.4), but it is a non-trivial fact for limits. Furthermore, an L-packet which contains a non-degenerate (resp. degenerate) limit of discrete series consists entirely of non-degenerate (resp. degenerate) limits of discrete series [Gol11]. Two discrete series (resp. limits) are in the same L-packet if and only if they have the same infinitesimal characters. If G is connected in the classical topology and semisimple then the cardinality of an L-packet of discrete series is the order of the quotients of Weyl groups $|W/W_c|$. Under the same assumptions, the cardinality of an L-packet of limits of discrete series is less than or equal to $|W/W_c|$, but it may be strictly smaller, see §3.7.3.

3.7. *An example: The group Sp(4)*

We exhibit some of the notions discussed earlier in §3 in the special case of the rank two, split symplectic group $Sp(4)$. This is one of the first non-trivial examples "beyond $GL(2)$." Several features which are common in general but not visible for $GL(2)$ (or $SL(2)$) are noticeable in $Sp(4)$.

3.7.1. Structure theory

Let

$$J = \begin{pmatrix} 0 & 0 & -1 & 0 \\ 0 & 0 & 0 & -1 \\ 1 & 0 & 0 & 0 \\ 0 & 1 & 0 & 0 \end{pmatrix}. \tag{3.7.1}$$

Then J is the matrix of a non-degenerate alternating form on $\mathbf{R}^4$. Let $G = G(J)$ be the group preserving J i.e.

$$G = \{g \in GL(4,\mathbf{R}) | gJg^t = J\} \tag{3.7.2}$$

Writing

$$g = \begin{pmatrix} A & B \\ C & D \end{pmatrix} \tag{3.7.3}$$

with A, B, C, D 2-by-2 matrices, (3.7.2) translates to the three equations

$$\begin{cases} AD^t - BC^t = I \\ AB^t - BA^t = 0 \\ CD^t - DC^t = 0 \end{cases} \tag{3.7.4}$$

Let $K = \{g \in G | C = -B, D = A\}$. Then K is a maximal compact subgroup of G and $g \longmapsto A + Bi$ is an isomorphism between K and $U(2)$. Let

$$T = \{g \in K | A = \mathrm{diag}(x_1, x_2) \text{ and } B = \mathrm{diag}(y_1, y_2)\}, \tag{3.7.5}$$

where $\mathrm{diag}(x_1, x_2)$ means the diagonal two-by-two matrix with x_1 followed by x_2 on the diagonal. Then T is a maximal torus in both K and G. We identify $\mathbf{X}^*(T_\mathbf{C})$ with $\mathbf{Z}^2$ by associating to $(n_1, n_2) \in \mathbf{Z}^2$ the character which maps $g \in T$ as above to $(x_1 + iy_1)^{n_1}(x_2 + iy_2)^{n_2}$.

Let $\mathbf{e}_1, \mathbf{e}_2$ be the standard basis of $\mathbf{R}^2$. For $i \in \{1,2\}$, let sgn_i be the linear transformation of $\mathbf{R}^2$ which inverts the sign of the ith coordinate. Let Sgn_2 be the group of order four generated by sgn_1 and sgn_2. The group

S_2 of permutations of $\{1,2\}$ acts on $\mathbf{R}^2$ by permuting the coordinates. The permutation group S_2 acts on the sign group Sgn_2 and we have $W = S_2 \rtimes \mathrm{Sgn}_2$ (so W is also isomorphic to the dihedral group of order 8). Moreover $W_c = S_2$. The set of roots $\Delta = \Delta(T_\mathbf{C}, G_\mathbf{C})$ of $T_\mathbf{C}$ in $G_\mathbf{C}$ is identified with

$$\Delta = \{e_1 - e_2, e_1 + e_2, 2e_1, 2e_2, -e_1 + e_2, -e_1 - e_2, -2e_1, -2e_2\} \tag{3.7.6}$$

and the subset of compact roots is $\Delta_c = \{e_1 - e_2, -e_1 + e_2\}$. We choose as positive roots $\Delta^+ = \{e_1 - e_2, e_1 + e_2, 2e_1, 2e_2\}$ and corresponding compact positive root $e_1 - e_2$.

3.7.2. Harish-Chandra parametrization

There are four Δ_c^+-dominant Weyl chambers C_0, C_1, C_2, C_3 of G. They are given by

$$\begin{cases} C_0 = \{(a_1,a_2) \in \mathbf{R}^2 | a_1 \geq a_2 \geq 0\} \\ C_1 = \{(a_1,a_2) \in \mathbf{R}^2 | a_1 \geq -a_2 \geq 0\} \\ C_2 = \{(a_1,a_2) \in \mathbf{R}^2 | -a_2 \geq a_1 \geq 0\} \\ C_3 = \{(a_1,a_2) \in \mathbf{R}^2 | -a_2 \geq -a_1 \geq 0\} \end{cases} \tag{3.7.7}$$

Of these four, C_0 is the unique holomorphic chamber. The holomorphic limits of discrete series of G are the $\pi((a_1,0), C_0)$ with $a_1 > 0$. The nondegenerate, non-holomorphic limits of discrete series are: $\pi((a_1,0), C_1)$ with $a_1 > 0$, $\pi((a_1,a_2), C_1)$ with $a_1 = -a_2$ and $a_1 > 0$, $\pi((a_1,a_2), C_2)$ with $a_1 = -a_2$ and $a_1 > 0$, $\pi((0,a_2), C_2)$ with $a_2 < 0$. and $\pi((0,a_2), C_3)$ with $a_2 < 0$. Finally there are two degenerate limits given by $\pi((0,0), C_1)$ and $\pi((0,0), C_2)$ and this exhausts all the limits of discrete series of G.

3.7.3. Langlands parameters and L-packets

Combining §3.6 with §3.7.2, we conclude the following: Every discrete series L-packet of $Sp(4)$ contains four discrete series. The L-packet of a holomorphic limit of discrete series also contains four elements: One is a holomorphic limit of discrete series and the other three are non-holomorphic non-degenerate limits of discrete series. On the other hand, the L-packets of the non-degenerate limits $\pi((a_1,a_2), C_1)$ and $\pi((a_1,a_2), C_2)$ with $a_1 = -a_2$ and $a_1 > 0$ have L-packets of size two. Finally, the two degenerate limits $\pi((0,0), C_1)$ and $\pi((0,0), C_2)$ form together a single L packet of size two.

We remark that there exist L-packets for $Sp(4)$ which are tempered and have the same infinitesimal character as a limit of discrete series, but the elements of the L-packet are not limits of discrete series. This even happens with holomorphic limits of discrete series.

The dual group of $Sp(4)$ is the projective group $PSp(4, \mathbf{C})$ (the quotient of $Sp(4, \mathbf{C})$ by its center). Since $Sp(4)$ is split, the action of the Galois group is trivial and the L-group is simply the direct product $PSp(4, \mathbf{C}) \times \{1, \sigma\}$.

3.7.4. Comparison with [Tay91]

We remark that §§3.7.1-3.7.2 are very similar to the first part of §3.1 of *loc. cit.*, with two notable exceptions: First, the group of similitudes $GSp(4)$ is studied in *loc. cit.*, instead of $Sp(4)$. This is the reason the number of Weyl chambers considered in *loc. cit.* is half of the number considered here. Second, there is a mistake in *loc. cit.* when it comes to the Harish-Chandra parametrization: The claim in *loc. cit.* about degenerate limits is false.

4 The Langlands correspondence

We now come to the main topic of this paper, the Langlands correspondence. We shall see that this can be seen as a vast generalization of the topics in both §2 and §3. We begin with a few words about some of the key objects that figure in the Langlands correspondence: Number fields and their adele rings are recalled in §4.1.1. We sketch a definition of automorphic representations in §4.1.2. Several kinds of Galois representations are introduced in §4.1.4–§4.1.6.

The following §4.2 states some of the fundamental conjectures concerning the Langlands correspondence; these may be seen as detailing the arrows of the fundamental triangle (1.1.1). Known results regarding these conjectures are summarized in §4.3. Finally §4.4 discusses some of the open problems concering the conjectures of §4.2.

4.1. Definitions

4.1.1. Number Fields

By a number field we mean a finite extension of the rational numbers $\mathbf{Q}$. Let F be a number field. Let $\mathcal{P}_F$ be the set of places of F. For every place v of F, we have the completion F_v at v. Either $F_v \cong \mathbf{R}$, or $F_v \cong \mathbf{C}$, or F_v is the fraction field of a discrete valuation ring, in which case we say that v is a real, complex or a finite place respectively. In case v is finite denote by $\mathcal{O}_{F,v}$ the discrete valuation ring whose fraction field is F_v. If v is not finite it is called an infinite place. Let $\mathcal{P}_{F,f}$ (resp. $\mathcal{P}_{F,\infty}$) be the set of finite (resp. infinite) places of F. In the case $F = \mathbf{Q}$, the finite places correspond to the prime numbers p with the completions being the p-adic numbers $\mathbf{Q}_p$ and there is a unique infinite place which is real.

The topological ring

$$\mathbf{A}_{F,f} = \{(a_v)_{v \in \mathcal{P}_{F,f}} \in \prod_{v \in \mathcal{P}_F^f} F_v | a_v \in \mathcal{O}_{F,v} \text{ for all but finitely many } v\} \tag{4.1.1}$$

is called the ring of finite adeles of F. Put $F_\infty = F \otimes_{\mathbf{Q}} \mathbf{R}$; we have

$$F_\infty = \prod_{v \in \mathcal{P}_{F,\infty}} F_v \tag{4.1.2}$$

The topological ring $\mathbf{A}_F = \mathbf{A}_{F,f} \times F_\infty$ is called the ring of adeles of F. In case $F = \mathbf{Q}$, we remove the subscript F and write simply $\mathbf{A}_f$ and $\mathbf{A}$. The field F embeds diagonally in both $\mathbf{A}_F$ and $\mathbf{A}_{F,f}$.

4.1.2. Automorphic representations

Let $\mathbf{G}$ be a connected, reductive algebraic group over the number field F. A key example to keep in mind is $\mathbf{G} = GL(n)$ and $F = \mathbf{Q}$. As in this example, it is important to note that while we assume that $\mathbf{G}$ is connected as an algebraic group, its real points $\mathbf{G}(\mathbf{R})$ may be disconnected in the classical topology. Also note that we are no longer assuming that $\mathbf{G}$ is semisimple.

An automorphic representation π is a special kind of representation of the adelic points $\mathbf{G}(\mathbf{A}_F)$. The precise definition is that a representation π of $\mathbf{G}(\mathbf{A}_F)$ is automorphic if π appears in the space of automorphic forms $\mathcal{A}(\mathbf{G})$. The space of automorphic forms $\mathcal{A}(\mathbf{G})$ is the space of smooth functions

$$f : \mathbf{G}(F)\backslash\mathbf{G}(\mathbf{A}_F) \longrightarrow \mathbf{C} \tag{4.1.3}$$

satisfying several growth conditions which we omit. A good approximation for the space of automorphic forms is the space of functions (4.1.3) which are in L^2. In any case, what is more important than either the growth conditions or the L^2 condition is the invariance by $\mathbf{G}(F)$: This is the generalization of the transformation law (2.3.7) under the congruence subgroup Γ for classical modular forms.

Let π be an automorphic representation of $\mathbf{G}$. Then π decomposes as a (restricted) tensor product

$$\pi = {\bigotimes_{v\in\mathcal{P}_F}}' \pi_v \tag{4.1.4}$$

with π_v a representation of the F_v-points $\mathbf{G}(F_v)$. In particular, when v is real or complex, π_v is a representation of the kind studied in §3. Put

$$\pi_\infty = \bigotimes_{v\in\mathcal{P}_{F,\infty}} \pi_v \tag{4.1.5}$$

and

$$\pi_f = \bigotimes_{v\in\mathcal{P}_{F,f}} \pi_v. \tag{4.1.6}$$

We call π_∞ (resp. π_f) the archimedean component (resp. finite part) of π.

There is a property for automorphic representations called being cuspidal which generalizes the notion of cusp forms that we described for modular forms in §2. For those automorphic representations that we know how to relate to algebraic geometry, there is a geometric characterization of being cuspidal which is analogous to the one defined in §2 for modular forms. In case $\mathbf{G}$ is anisotropic over F, which means that $\mathbf{G}$ possesses no non-trivial F-parabolic subgroup, then every automorphic representation of $\mathbf{G}$ is automatically cuspidal.

Put $G = \mathbf{G}(F_\infty)$. We will use the notation established in §3.4.2 with respect to G even though the context is now much more general. So for example T will denote a maximal torus in G, but, unlike in §3.4, T is no longer required to be compact. Let χ_∞ be the infinitesimal character of π_∞. We shall also call χ_∞ the infinitesimal character of π. By the Harish-Chandra isomorphism we identify χ_∞ with an element of $(\mathbf{X}^*(T_{\mathbf{C}}) \otimes C)/W$. We say π is regular if χ_∞ is regular (recall this means χ_∞ is not orthogonal to any coroot); otherwise we say π is irregular.

4.1.3. Algebraic automorphic representations

Let π be an automorphic representation. Following Buzzard-Gee [BG11], we say that π (resp. π_∞) is L-algebraic if $\chi_\infty \in \mathbf{X}^*(T_{\mathbf{C}})$ and that π (resp. π_∞) is C-algebraic if $\chi_\infty \in \mathbf{X}^*(T_{\mathbf{C}}) + \rho$. In particular, the two notions of algebraicity coincide if and only if $\rho \in \mathbf{X}^*(T_{\mathbf{C}})$.

Note that, in view of Ths. 3.4.2 and 3.5.1, when G is semisimple and connected in the classical topology, discrete series and their limits are C-algebraic. In general there is a complication because one may have twists by characters that are not algebraic, and one must allow such twists to get the general statement of the arhimedean local Langlands correspondence (Th. 3.1.1).

4.1.4. Galois representations

Let F be a number field. The absolute Galois group $\mathrm{Gal}(\overline{F}/F)$ of F (or simply Galois group of F for short) is the projective limit of the groups $\mathrm{Gal}(K/F)$, as K ranges over all finite Galois extensions of F. Hence $\mathrm{Gal}(\overline{F}/F)$ is a profinite group, meaning it inherits a natural topology in which it is compact and totally disconnected. Let v be a place of F. Then similarly we have the (local) absolute Galois group $\mathrm{Gal}(\overline{F_v}/F_v)$. Since F embeds into F_v, we have an embedding of $\mathrm{Gal}(\overline{F_v}/F_v)$ into $\mathrm{Gal}(\overline{F}/F)$, which is well-defined up to conjugacy.

Let ℓ be a prime number. A Galois representation ρ is a continuous homomorphism from $\mathrm{Gal}(\overline{F}/F)$ into $GL(n, \overline{\mathbf{Q}_\ell})$ for some $n \geq 1$. An Artin representation is a Galois representation with finite image. An Artin representation may equivalently be viewed as a continuous homomorphism from $\mathrm{Gal}(\overline{F}/F)$ to $GL(n, \mathbf{C})$, since any continuous homomorphism from $\mathrm{Gal}(\overline{F}/F)$ to $GL(n, \mathbf{C})$ necessarily has finite image.

Let

$$\rho : \mathrm{Gal}(\overline{F}/F) \longrightarrow GL(n, \overline{\mathbf{Q}_\ell}) \tag{4.1.7}$$

be a Galois representation. Since the embedding of $\mathrm{Gal}(\overline{F_v}/F_v)$ is well-defined up to conjugacy, it makes sense to restrict ρ to $\mathrm{Gal}(\overline{F_v}/F_v)$. One says that ρ is unramified at v if the restriction of ρ to the inertia subgroup I_w is trivial for some (equivalently all) place(s) w of $\overline{F_v}$.

To learn the basics concerning (ℓ-adic) Galois representations, we suggest consulting [Ser68], especially Chap. 1, §§1–2.

4.1.5. Geometric Galois representations

Let ρ be as in (4.1.7). Using Fontaine's p-adic rings of periods from p-adic Hodge theory, Fontaine-Mazur [FM95] defined what it means for ρ to be potentially semi-stable (equivalently de-Rham) at a place v dividing ℓ. This is a very technical definition, so we omit it, but we stress that being de Rham at a place v dividing ℓ only depends on the restriction of ρ to $\mathrm{Gal}(\overline{F_v}/F_v)$.

Admitting the notion of being de Rham, we say following Fontaine-Mazur that ρ is geometric if ρ is unramified at all but finitely many primes and ρ is de Rham at all places v of F dividing ℓ.

4.1.6. Motivic Galois representations

Suppose X is a smooth projective variety over F. Then the étale cohomology $H^i_{\mathrm{et}}(X \times \overline{\mathbf{Q}}, \overline{\mathbf{Q}_\ell})$ of X is a Galois representation of $\mathrm{Gal}(\overline{F}/F)$, which is known to be geometric (*cf.* [Tay04, p. 77, 79-80]). We will say that ρ is motivic if it is a subquotient of a 'Tate twist' (*cf. loc. cit.*) of $H^i_{\mathrm{et}}(X \times \overline{\mathbf{Q}}, \overline{\mathbf{Q}_\ell})$ for some F, X and i.

4.2. Conjectures

We will omit a careful definition of what it means for a Galois representation ρ to correspond to an automorphic representation π. First one defines what it means, given a place v, that π_v correspond to the restriction of ρ to $\mathrm{Gal}(\overline{F_v}/F_v)$. Then one says that π and ρ correspond if π_v corresponds to the restriction of ρ to $\mathrm{Gal}(\overline{F_v}/F_v)$ for a density one set of places v (in particular if the two correspond for all but finitely many v). At all but finitely many places, the correspondence between π_v and the restriction of ρ to $\mathrm{Gal}(\overline{F_v}/F_v)$ is defined in terms completely analogous to the statement of Th. 2.4.1: Roughly speaking it is an equality of Hecke eigenvalues on the one hand, and traces of Frobenius elements on the other. At the remaining finite set of places things are much more complicated; this is what is called the Local Langlands correspondence, see [HT01] for the case of $GL(n)$.

4.2.1. Automorphic to Galois

Building on the ideas of Langlands and a conjecture of Clozel in the case of $\mathbf{G} = GL(n)$ (see [Clo88]), Buzzard-Gee made the following conjecture:

Conjecture 4.2.1. *Suppose π is an L-algebraic automorphic representation of* **G** *over F. Then for every prime ℓ, there exists a continuous homomorphism*

$$R_\ell(\pi) : \mathrm{Gal}(\overline{F}/F) \longrightarrow {}^L\mathbf{G}(\overline{\mathbf{Q}_\ell}) \tag{4.2.1}$$

such that, for any finite-dimensional algebraic representation

$$r : {}^L\mathbf{G}(\overline{\mathbf{Q}_\ell}) \longrightarrow GL(n, \overline{\mathbf{Q}_\ell}) \tag{4.2.2}$$

the composite $r \circ R_\ell(\pi)$ is a geometric Galois representation associated to π.

There is a similar but more complicated conjecture in the C-algebraic case. It is more complicated because it is not the L-group of **G** which appears in the statement, but rather of a certain extension of **G**, called a z-extension.

4.2.2. Galois to automorphic

The existence of an arrow from the bottom left to the bottom right of the fundamental triangle 1.1.1 was conjectured by Fontaine and Mazur.

Conjecture 4.2.2 (Fontaine-Mazur, [FM95]). *Every geometric Galois representation is motivic.*

Combining this with conjectures of Langlands one obtains:

Conjecture 4.2.3 (Fontaine-Mazur-Langlands). *Every geometric Galois representation arises from an automorphic representation.*

4.3. Known Results

The fundamental marker of difficulty of establishing the Langlands correspondence for an automorphic representation π of a group **G** is the nature of the archimedean component π_∞. A second measure of difficulty is the type of number field F over which the group **G** is considered. There has been much more progress in the Langlands correspondence in the regular case than in the irregular case. Moreover, all of the results to date are restricted to the case that the number field F is either totally real or a CM field.

4.3.1. Automorphic to Galois I: The regular case

The following theorem was proved almost as stated below by Shin [Shi11], building on work of many people, including Kottwitz [Kot92], Clozel [Clo91] and Harris-Taylor [HT01].

Theorem 4.3.1. *Suppose F is a CM field and π is a cuspidal automorphic representation of $GL(n,F)$ which is regular, C-algebraic and satisfies $\pi^\vee \cong \pi \circ c$, where c is complex conjugation. Then for every prime ℓ there exists a* *unique semisimple Galois representation*

$$R_\ell(\pi) : \mathrm{Gal}(\overline{F}/F) \longrightarrow GL(n, \overline{\mathbf{Q}_\ell}) \tag{4.3.1}$$

associated to π. *Moreover* $R_\ell(\pi)$ *is geometric and under a slightly stronger regularity hypothesis* $R_\ell(\pi)$ *is even motivic.*

The proof of Th. 4.3.1 is extremely complicated and uses Arthur's trace formula. However, the initial part of the argument is a generalization of §2.5.7. Namely, one begins by realizing the finite part π_f of π in the étale cohomology of a Shimura variety, which is a generalization of modular curves. While we saw that the modular curves can be viewed as moduli of elliptic curves with extra structure, the Shimura variety in question is a moduli space of abelian varieties with extra structure.

For a long time it seemed very difficult to remove the condition $\pi^\vee \cong \pi \circ c$, since if π does not satisfy this condition, then it does not appear directly in the cohomology of a Shimura variety. A roundabout way was finally discovered using Th. 4.3.1 as a starting block.

Theorem 4.3.2 (Harris-Lan-Taylor-Thorne [HLTT13]). *Suppose, as in* Th. 4.3.1, *that* F *is a CM field and* π *is a cuspidal automorphic representation of* $GL(n,F)$ *which is regular and C-algebraic, but do not assume the relation* $\pi^\vee \cong \pi \circ c$. *Then there still exists a Galois representation* $R_\ell(\pi)$ *as in* (4.3.1) *associated to* π.

Remark 4.3.3. In contrast to Th. 4.3.1, in the context of Th. 4.3.2 it is no longer known whether $R_\ell(\pi)$ is geometric, let alone motivic.

Combining Arthur's work [Art13] on endoscopy (see App. Appendix A) with either Th. 4.3.1 or Th. 4.3.2 gives results about other groups. For example one has:

Theorem 4.3.4. *Suppose* π *is a cuspidal automorphic representation of* $Sp(2n)$ *over a totally real field* F *with* π *regular and C-algebraic. Then there exists a* $2n+1$ *dimensional Galois representation* $R_\ell(\pi)$ *associated to* π *which is geometric and under a slightly stronger regularity hypothesis even motivic.*

4.3.2. Automorphic to Galois II: The irregular case

The only known general result in the irregular case is the following, which generalizes the case $k = 1$ of Th. 2.4.1 (the Deligne-Serre Theorem) and work of Taylor on $GSp(4)$ [Tay91].

Theorem 4.3.5 (Goldring, Goldring-Nicole [Gol14], [GN14]). *Suppose* π *is a cuspidal automorphic representation of a* **Q***-algebraic unitary group such that* π_∞ *is a holomorphic limit of discrete series. Then for all but finitely many primes* ℓ, *there exists a Galois representation* $R_\ell(\pi)$ *associated to* π.

The idea of the proof of Th. 4.3.5 is to reduce to Th. 4.3.1 by using congruences, as was outlined in the case of modular forms in §2.5.8. First one realizes π_f in the coherent cohomology of a Shimura variety, as global sections of vector bundles. Then one wants a form which is "congruent to 1 modulo ℓ." In the case of modular forms, we remarked that the Eisenstein series $E_{\ell-1}$ has this property. In the generality of Th. 4.3.5 we do not know of an Eisenstein series that has the desired properties. Instead we use a form called the Hasse invariant, which is defined using techniques of finite characteristic, namely the Frobenius.

Note added in proof: The preprints of Pilloni-Stroh [PS] and Goldring-Koskivirta [GK15], which appeared after this article was submitted, generalize Th. 4.3.5 to groups admitting a Shimura variety of Hodge type and to arbitrary non-degenerate limit of discrete series archimedean component. A preprint of Boxer [Box15] contains related results in the more restricted context of Shimura varieties of PEL type A and C, but it fails to state or prove any results about Galois representations. The reason this case is within reach is that for such π, the finite part π_f appears in the coherent cohomology of a Shimura variety. The reason the nondegenerate limit case is harder than the holomorphic limit case is that the degree of coherent cohomology in which the realization occurs is no longer zero, but strictly positive.

4.3.3. Galois to automorphic

Starting with the work of Wiles [Wil95] and Taylor-Wiles [TW95], there has also been much progress on the Fontaine-Mazur and Fontaine-Mazur-Langlands conjecture. For two-dimensional Galois representations of $\mathrm{Gal}(\overline{\mathbf{Q}}/\mathbf{Q})$, aside from some technical hypotheses, the conjectures have been established, the last big step being taken by Kisin [Kis09]. For higher-dimensional Galois representations, it is again much more difficult to treat the irregular case. In the regular case, the best results to date are contained in the work of Barnet-Lamb-Gee-Geraghty-Taylor [BLGGT14]. In the irregular case a promising program has been launched by Calegari-Geraghty [CG].

4.4. Open Problems

Let us restrict attention to the problem of associating Galois representations to automorphic representations. Regarding this problem, several outstanding open problems remain.

4.4.1. Degenerate limits of discrete series

Suppose π is a cuspidal automorphic representation of a $\mathbf{Q}$-algebraic group $\mathbf{G}$ such that π_∞ is a degenerate limit of discrete series. Then there is no known realization of π_f in the cohomology (coherent or étale) of an algebraic variety.

It is in fact a theorem of Mirkovic [Mir88] that π_f does not appear (directly) in the cohomology of a Shimura variety. Thus it seems very difficult to relate π to Galois theory.

In a series of papers, Carayol ([Car98], [Car00], [Car05]) has embarked on a program to understand such π by observing that π_f occurs in the cohomology of Griffiths-Schmid manifolds [GS69]. These are homogenous complex manifolds that are known not to be algebraic. This non-algebraicity result was suspected for many years, and finally established by Griffiths-Robles-Toledo [GRT14]. Some of the ideas of Carayol were also further developed and tested in other examples by Kerr [Ker14] and Green-Griffiths-Kerr [GGK13].

4.4.2. Non-limit archimedean component

As was mentioned in §3.7.3, there are C-algebraic automorphic representations whose archimedean component is tempered and has the same infinitesimal character as a limit of discrete series and yet this archimedean component is not a limit of discrete series. Such automorphic representations appear to be even harder to understand, since then there is even no realization in the cohomology of a Griffiths-Schmid manifold. However, one can attempt to use functoriality to reduce to the case of degenerate limits of discrete series, see Appendix A.

4.4.3. Other Number Fields

All the known results mentioned above have the property that the field F is either totally real or a CM field. When F is not of this kind, even if π is regular, there is no known way to directly relate π to the cohomology of an algebraic variety, or even of a Griffiths-Schmid manifold.

4.4.4. Establishing the motivic property in the irregular case

Another problem which seems to require completely new ideas is to prove that the Galois representations constructed in either Th. 4.3.5 or Th. 4.3.2 are motivic.

4.4.5. Image in the L-group

Although Conjecture 4.2.1 predicts a homomorphism whose image is contained in the L-group, aside from some low dimensional coincidences, or cases when the L-group is closely related to $GL(n)$, known results only produce the composite $r \circ R_\ell(\pi)$, for some representation $r : {}^L\mathbf{G} \longrightarrow GL(n)$.

Note added in proof: In the case $\mathbf{G} = GSp(2g)$ and π discrete series, the forthcoming work of Arno Kret and Sug Woo Shin proves the existence of $R_l(\pi)$, valued in the L-group $GSpin(2g+1)$ (these groups are split, so the Galois action is trivial).

Acknowledgements

It is a pleasure to thank the organizers of the conference "Recent Advances in Hodge Theory: Period Domains, Algebraic Cycles, and Arithmetic" for inviting me to speak at the conference and for their encouragement that I write-up my lectures. I am also very grateful to Henri Carayol, Laurent Clozel, Pierre Deligne, David Geraghty, Phillip Griffiths, Matt Kerr, Jean-Stefan Koskivirta, Marc-Hubert Nicole, Jonathan Pottharst, Benoit Stroh and Jacques Tilouine for discussions about the Langlands correspondence. Finally, I would like to thank the anonymous referee for his/her helpful comments.

Appendix A Functoriality

Let H and G be two connected, real algebraic groups. In view of Th. 3.1.1, a homomorphism of L-groups

$$\psi : {}^L H \longrightarrow {}^L G \tag{A.0.1}$$

induces a map from L-packets of H to L-packets of G. Now let $\mathbf{H}$ and $\mathbf{G}$ be $\mathbf{Q}$-algebraic groups (so we assume here for simplicity that $F = \mathbf{Q}$). Langlands introduced a notion of L-packet for automorphic represention and conjectured that given a homomorphism of L-groups, there should similarly be a map from L-packets of $\mathbf{H}$ to L-packets of $\mathbf{G}$. Moreover, such a map from L-packets of $\mathbf{H}$ to those of $\mathbf{G}$ is conjectured to satisfy all kinds of nice properties.

Suppose π and π' are two automorphic representations in the same L-packet. If ρ is a Galois representation correspnding to π, then ρ also corresponds to π'. In other words, it makes sense to talk about a Galois representation corresponding to an L-packet of automorphic representations. Furthermore, suppose that one has an L-packet $\mathcal{L}$ of automorphic representations of $\mathbf{H}$ to which one would like to associate a Galois representation. It is possible that it is easier to associate a Galois representation to the image L-packet $\psi(\mathcal{L})$ of $\mathbf{G}$ and that this also implies the existence of the original sought after Galois representation for $\mathcal{L}$.

The example where this strategy has been most successful is in Arthur's work [Art13]. Arthur has proved the existence of the map from L-packets of $\mathbf{H}$ to those of $\mathbf{G}$ in the case that $\mathbf{H}$ is a symplectic or orthogonal group, $\mathbf{G}$ is $GL(n)$ and ψ is the standard representation. In turn, this functoriality result implies Th. 4.3.4.

In a similar vein, one could hope that, given π with degenerate limit of discrete series archimedean component, there exists ψ such that the image L-packet is non-degenerate. However, we showed this is impossible [Gol11].

To contrast this negative result, we remark that Clozel and Carayol have given an example (unpublished) of a representation which is not limit of discrete series and which transfers via functoriality to one that is.

References

[ABD+66] M. Artin, J. E. Bertin, M. Demazure, P. Gabriel, A. Grothendieck, M. Raynaud, and J.-P. Serre, *Schémas en groupes. Fasc. 7: Exposés 23 à 26*, Séminaire de Géométrie Algébrique de l'Institut des Hautes Études Scientifiques, vol. 1963/64, Institut des Hautes Études Scientifiques, Paris, 1965/1966. MR MR0207710 (34 #7525)

[ABV92] J. Adams, D. Barbasch, and D. Vogan, *The Langlands classification and irreducible characters for real reductive groups*, Progress in Math., vol. 104, Birkhauser, 1992.

[Art13] J. Arthur, *The endoscopic classification of representations: Orthogonal and symplectic groups*, AMS Colloquium publications, vol. 61, AMS, 2013.

[BG11] K. Buzzard and T. Gee, *The conjectural connections between automorphic representations and Galois representations*, Automorphic forms and Galois representations (Durham), LMS Symposia in pure math., LMS, 2011, To appear, available at http://www2.imperial.ac.uk/~buzzard/maths/research/papers/index.html.

[BLGGT14] T. Barnet-Lamb, T. Gee, D. Geraghty, and R. Taylor, *Potential automorphy and change of weight*, Ann. Math. **179** (2014), 501–609.

[Bor79] A. Borel, *Automorphic L-functions*, Automorphic Forms, representations, and L-Functions (A. Borel and W. Casselman, eds.), Proc. Symp. Pure Math., vol. 33, Amer. Math. Soc., 1979, pp. 247–289.

[Box15] G. Boxer, *Torsion in the Coherent Cohomology of Shimura Varieties and Galois Representations*. Preprint, arXiv:1507.05922.

[Car98] H. Carayol, *Limites dégénérées de séries discrètes, formes automorphes et variétés de Griffiths-Schmid*, Comp. Math. **111** (1998), 51–88.

[Car00] ———, *Quelques relations entre les cohomologies des variétés de Shimura et celles de Griffiths-Schmid (cas du groupe $SU(2,1)$)*, Comp. Math. **121** (2000), 305–335.

[Car05] ———, *Cohomologie automorphe et compactifications partielles de certaines variétés de Griffiths-Schmid*, Comp. Math. **141** (2005) 1081–1102.

[CG] F. Calegari and D. Geraghty, *Modularity lifting beyond the Taylor-Wiles method*, preprint.

[Clo88] L. Clozel, *Motifs et formes automorphes: Applications du principe de fonctorialité*, Automorphic Forms, Shimura Varieties, and L-Functions (Ann Arbor, MI) (L. Clozel and J. Milne, eds.), vol. 1, Academic Press, Inc., July 6-16 1988, pp. 77–160.

[Clo91] ———, *Représentations Galoisiennes associees aux représentations automorphes autoduales de $GL(n)$*, Publ. Math. IHES **73** (1991), 97–145.

[Del69] P. Deligne, *Formes modulaires et représentations ℓ-adiques*, Seminaire Bourbaki, 1968-1969, Exposé No. 355, 34p.

[DR73] P. Deligne and M. Rapoport, *Les schémas de modules de courbes elliptiques*, Modular functions of one variable II (P. Deligne and W. Kuyk, eds.), Lect. Notes in Math., vol. 349, Springer, Antwerp, Belgium, 1973, pp. 143–316.

[DS74] P. Deligne and J.-P. Serre, *Formes modulaires de poids 1*, Ann. Sci. ENS **7** (1974), no. 4, 507–530.

[FM95] J.-M. Fontaine and B. Mazur, *Geometric Galois representations*, Elliptic curves, modular forms and Fermat's Last Theorem, Ser. Number Theory, vol. 1, Int. Press, 1995, pp. 41–78.

[GGK13] M. Green, P. Griffiths, and M. Kerr, *Hodge theory, complex geometry and representation theory*, CBMS Regional Conference Series, vol. 118, AMS, 2013.

[GK15] W. Goldring and J.-S. Koskivirta, *Strata Hasse invariants, Hecke algebras and Galois representations*. Preprint, arXiv:1507.05032.

[GN14] W. Goldring and M.-H. Nicole, *The μ-ordinary Hasse invariant of unitary Shimura varieties*, To appear in Crelle's Journal, available at https://sites.google.com/site/wushijig/.

[Gol11] W. Goldring, *Stability of degenerate limits of discrete series under functoriality*, March 2011, Submitted, available at https://sites.google.com/site/wushijig/.

[Gol14] ______, *Galois representations associated to holomorphic limits of discrete series*, Compositio Math. **150** (2014), 191–228, with an appendix by S.-W. Shin.

[GRT14] P. Griffiths, C. Robles, and D. Toledo, *Quotients of non-classical flag domains are not algebraic*, Alg. Geom. **1** (2014).

[GS69] P. Griffiths and W. Schmid, *Locally homogeneous complex manifolds*, Acta. Math. **123** (1969), 253–302.

[HLTT13] M. Harris, K.-W. Lan, R. Taylor, and J. Thorne, *On the rigid cohomology of certain Shimura varieties*, preprint available at https://www.math.ias.edu/~rtaylor.

[HT01] M. Harris and R. Taylor, *The geometry and cohomology of some simple Shimura varieties*, Annals of Math. Studies, vol. **151**, Princeton Univ. Press, 2001.

[Ker14] M. Kerr, *Cup products in automorphic cohomology: The case of $sp(4)$*, Hodge Theory, Complex Geometry, and Representation Theory (Doran, Freidman, and Nollet, eds.), Contemp. Math., vol. 608, AMS, 2014, pp. 199–234.

[Kis09] M. Kisin, *The Fontaine-Mazur conjecture for GL_2*, J. Amer. Math. Soc. **22** (2009), no. 3, 641–690.

[Kna94] A. Knapp, *Local Langlands correspondence: the archimedean case*, Motives (Uwe Jannsen et. al., ed.), Proc. Symp. Pure. Math., vol. 55, 1994, pp. 393–410.

[Kot92] R. Kottwitz, *Points on some Shimura varieties over finite fields.*, J. Amer. Math. Soc. **5** (1992), no. 2, 373–444.

[Lan89] R. Langlands, *On the classification of irreducible representations of real algebraic groups*, Representation theory and harmonic analysis on semisimple Lie groups, Math. Surveys Monographs, vol. 31, Amer. Math. Soc., 1989, pp. 101–170.

[Mir88] I. Mirkovic, *Localization for singular infinitesimal characters*, 1988, unpublished.

[PS] V. Pilloni and B. Stroh, *Cohomologie cohérente et représentations Galoisiennes*. Preprint, available at http://www.math.univ-paris13.fr/stroh/.

[Rib77] K. Ribet, *Galois representations attached to eigenforms with nebentypus*, Modular functions of one variable V (Bonn, Germany) (J.-P. Serre and D. Zagier, eds.), Lecture Notes in Math., no. 601, Springer-Verlag, July 1976 1977, pp. 17–51.

[Sch97] W. Schmid, *Discrete series*, Representation theory and automorphic forms, Proc. Symp. Pure Math., no. 61, AMS, 1997, pp. 83–113.

[Sch05] ———, *Geometric methods in representation theory*, Poisson geometry, deformation quantisation and group representations, London Math. Soc. Lecture Notes, no. 323, Cambridge Univ. Press, 2005, Lecture notes taken by M. Libine, pp. 273–323.

[Ser68] J.-P. Serre, *Abelian l-adic representations and elliptic curves*, W. A. Benjamin, Inc., New York-Amsterdam, 1968, McGill Univ. Lecture Notes written with the collaboration of W. Kuyk and J. Labute.

[Shi11] S. W. Shin, *Galois representations arising from some compact Shimura varieties*, Ann. of Math. **173** (2011), 1645–1741.

[Spr79] T. Springer, *Reductive groups*, Automorphic Forms, representations, and L-Functions (A. Borel and W. Casselman, eds.), Proc. Symp. Pure Math., vol. 33, Amer. Math. Soc., 1979, pp. 3–28.

[Tay91] R. Taylor, *Galois representations associated to Siegel modular forms of low weight*, Duke Math. J. **63** (1991), no. 2, 281–332.

[Tay04] ———, *Galois representations*, Ann. Fac. Sci. Toulouse **13** (2004), 73–119.

[TW95] R. Taylor and A. Wiles, *Ring-theortic properties of certain Hecke algebras*, Ann. Math. **141** (1995), no. 3, 553–572.

[Wil95] A. Wiles, *Modular elliptic curves and Fermat's Last Theorem*, Ann. Math. **141** (1995), no. 3, 443–551.

[Zuc77] G. Zuckerman, *Tensor products of finite and infinite dimensional representations of semisimple groups*, Ann. Math. **106** (1977), no. 2, 295–308.

14

Generalized Kuga-Satake theory and rigid local systems I: the middle convolution

Stefan Patrikis

1 Background and motivation

Let $X/\mathbb{C}$ be a complex K3 surface. The classical construction of Kuga and Satake associates to X a complex abelian variety $KS(X)/\mathbb{C}$ along with a morphism

$$H^2(X,\mathbb{Q}) \hookrightarrow H^1(KS(X),\mathbb{Q}) \otimes H^1(KS(X),\mathbb{Q})$$

of $\mathbb{Q}$-Hodge structures. Here is the Hodge-theoretic description of the construction, as rephrased by Deligne ([Del72]). For later purposes, we take X to be projective, let η be an ample line bundle on X, and replace the full cohomology group $H^2(X,\mathbb{Q})$ by the η-primitive cohomology $\mathrm{Prim}^2_\eta(X,\mathbb{Q})$ (the kernel of $\cup\eta$); for notational convenience we let $V_\mathbb{Q}$ be the weight-zero (Tate-twisted) $\mathbb{Q}$-Hodge structure

$$V_\mathbb{Q} = \mathrm{Prim}^2_\eta(X,\mathbb{Q})(1).$$

The Hodge bi-grading is then given by a representation of the Deligne torus $\mathbb{S} = \mathrm{Res}_{\mathbb{C}/\mathbb{R}}\,\mathbb{G}_m$ on $V_\mathbb{R}$; this preserves the intersection pairing, and so we can set up the 'lifting problem' of whether the dotted arrow in the following diagram can be filled in:

$$\begin{array}{ccc} & & \mathrm{GSpin}(V_\mathbb{R}) \\ & \overset{\tilde{h}}{\nearrow} & \downarrow \\ \mathbb{S} & \xrightarrow{h} & \mathrm{SO}(V_\mathbb{R}). \end{array} \tag{1}$$

It is easily seen that this is possible, and that the ambiguity in such a lift is a choice of Tate twist. Now, associated to the quadratic space $V_\mathbb{Q}$ there is an (even–this is immaterial) Clifford algebra $C^+(V_\mathbb{Q})$, and there is a natural

I am very grateful to the organizers for the opportunity to speak at the conference, and I would like to thank the anonymous referee for some helpful expository suggestions. This work was conceived while unemployed in my parents' basement, but only brought to fruition with the support of NSF grant DMS-1303928.

representation of $\mathrm{GSpin}(V_{\mathbb{Q}})$ on $C^+(V_{\mathbb{Q}})$. In these terms, the Kuga-Satake abelian variety is the complex abelian variety associated to the weight 1 polarizable $\mathbb{Q}$-Hodge structure whose underlying $\mathbb{Q}$-vector space is $C^+(V_{\mathbb{Q}})$, and whose Hodge bi-grading is given by the representation of $\mathbb{S}$ induced by the lift (normalized to weight 1) $\tilde{h}$.

It is a numerical miracle, crucially depending on the fact that $h^{2,0}(X) = 1$, that the Hodge structure $(C^+(V_{\mathbb{Q}}), \tilde{h})$ has Hodge bi-grading of type $\{(1,0),(0,1)\}$. A number of people have studied variants of the original construction (for example, [Mor85], [Gal00], [Voi05], [Moo]), always beginning with some Hodge structure resembling that of a K3 surface or abelian variety, and extracting some other Hodge structure (the 'lift') that can by Riemann's theorem be shown to come from an abelian variety.

But the diagram (1) above suggests a much broader question: abandon K3 surfaces and abelian varieties, and consider *any* orthogonally polarized weight zero Hodge structure $V_{\mathbb{Q}}$ of motivic origin, for instance $\mathrm{Prim}^{2k}(X,\mathbb{Q})(k)$ for *any* smooth projective variety $X/\mathbb{C}$. Again it is easy to find (unique up to Tate twist) a lift $\tilde{h}$, so we again have a (polarizable) $\mathbb{Q}$-Hodge structure $(C^+(V_{\mathbb{Q}}), \tilde{h})$. Is it motivic? Characterizing the essential image of the 'Hodge realization' functor on the category $\mathcal{M}_{\mathbb{C}}$ of pure homological motives over $\mathbb{C}$ is as far as anyone knows a totally intractable problem; and indeed I know of no *geometric* reason for believing $(C^+(V_{\mathbb{Q}}), \tilde{h})$ should come from algebraic geometry. If, however, the theory of motives is to reflect 'l'identité profonde entre la géométrie et l'arithmétique,' we can hope to turn to arithmetic for guidance.

Arithmetic in fact provides two compelling reasons for optimism, one Galois-theoretic and one automorphic. For a full discussion of these matters, see [Pat12]; here I will review the Galois-theoretic aspect. Let F be a number field, with an algebraic closure $\overline{F}$, and let $\Gamma_F = \mathrm{Gal}(\overline{F}/F)$ be its absolute Galois group. In parallel to the Hodge-realization

$$H_B^*\colon \mathcal{M}_{\mathbb{C}} \to \mathbb{Q}\mathrm{HS}^{pol}$$

to polarizable $\mathbb{Q}$-Hodge structures, there are ℓ-adic realizations

$$H_\ell^*\colon \mathcal{M}_F \to \mathrm{Rep}_{\mathbb{Q}_\ell}(\Gamma_F)$$

to the category of continuous representations of Γ_F on finite-dimensional $\mathbb{Q}_\ell$-vector spaces. In contrast to the Hodge-Betti realization, here there is a remarkable conjecture (Conjecture 1.2 below), due to Fontaine and Mazur, which in combination with the Tate conjecture predicts the essential image of H_ℓ^*. To be precise, this prediction takes place after extending H_ℓ^* via an embedding $\iota\colon \overline{\mathbb{Q}} \hookrightarrow \overline{\mathbb{Q}}_\ell$ to a functor

$$H_\iota^*\colon \mathcal{M}_{F,\overline{\mathbb{Q}}} \to \mathrm{Rep}_{\overline{\mathbb{Q}}_\ell}(\Gamma_F)$$

on the category $\mathcal{M}_{F,\overline{\mathbb{Q}}}$ of pure motives over F with $\overline{\mathbb{Q}}$ coefficients.

Definition 1.1 (§1 of [FM95]). A Galois representation $\rho\colon \Gamma_F \to \mathrm{GL}_N(\overline{\mathbb{Q}}_\ell)$ is said to be *geometric*[1] if for all but finitely many places v of F, the restriction of ρ to a decomposition group Γ_{F_v} at v is unramified; and for all $v|\ell$, $\rho|_{\Gamma_{F_v}}$ is de Rham in the sense of Fontaine (see [BC] for a textbook-style introduction to the theory of Fontaine).

It follows from the base-change theorems for ℓ-adic cohomology and Faltings' p-adic de Rham comparison isomorphism ([Fal89]) that for any smooth projective X/F, $H^*_\ell(X)$ is geometric; in fact, the same holds for any X/F separated of finite-type (the assertion at $v \nmid \ell$ follows from [Del77, Théorème 1.9], while at $v|\ell$ it is the main result of [Bei12]). The Fontaine-Mazur conjecture gives a partial converse, in the case of irreducible (or semi-simple) geometric Γ_F-representations:

Conjecture 1.2 (Conjecture 1 of [FM95]). *Any irreducible geometric representation of Γ_F is isomorphic to a sub-quotient of $H^n(X_{\overline{F}}, \overline{\mathbb{Q}}_\ell(r))$ for some (smooth projective) variety X/F and integers n and r.*

Combining this with the Tate conjecture and the Grothendieck-Serre conjecture on semi-simplicity of ℓ-adic cohomology of smooth projective varieties, we can predict that the essential image of H^*_ι on the category of pure motives over F (with $\overline{\mathbb{Q}}$-coefficients) is those Γ_F-representations that are semi-simple and geometric.[2]

Now we can re-draw diagram (1) in its ℓ-adic incarnation, letting V_ℓ be the ℓ-adic realization of an orthogonally polarized motive over F, and replacing the representation h of the Deligne torus $\mathbb{S}$ with the (enormously richer) representation of Γ_F on V_ℓ. Here, again, it is clarifying to strip away unnecessary features of the original context: what turns out to be important about the surjection GSpin $\to$ SO is that its kernel is equal to a *central torus*. We are naturally led, then, to the following question.[3] Let $\widetilde{H} \to H$ be any

[1] Here we follow the original definition of 'geometric' as in [FM95, §1], except we replace their *a priori* stronger condition of 'potential semi-stability' at $v|\ell$ by the 'de Rham' condition; a deep theorem of Berger ([Ber02]) shows the two notions are equivalent.

[2] We will not give the details of this implication, since we do not require it. But the interested reader can deduce it by (1) observing it first with $\mathcal{M}_{F,\overline{\mathbb{Q}}_\ell}$ in place of $\mathcal{M}_{F,\overline{\mathbb{Q}}}$; (2) using the fact that for all $M \in \mathcal{M}_F$, the *semi-simple* algebra $\mathrm{End}_{\mathcal{M}_F}(M)$ splits after scalar extension to $\overline{\mathbb{Q}}$. A cautionary tale against making such a conjecture for the essential image of $H^*_\ell\colon \mathcal{M}_F \to \mathrm{Rep}_{\mathbb{Q}_\ell}(\Gamma_F)$ itself is provided, eg, by algebraic Hecke characters. For suitable CM fields F and E, there are rank one objects $\psi \in \mathcal{M}_{F,E}$ corresponding to algebraic Hecke characters of F that do not, even up to finite-order twist, descend to its totally real subfield. If $\lambda|\ell$ is a place of E split over $\mathbb{Q}$, then the λ-adic realization of ψ is a geometric representation $\Gamma_F \to \mathbb{Q}_\ell^\times = E_\lambda^\times$. This is not isomorphic to the ℓ-adic realization of any object of $\mathcal{M}_F$. But note that all the endomorphism algebras $\mathrm{End}_{\mathcal{M}_F}(M)$ in fact split over the maximal CM extension $\mathbb{Q}^{\mathrm{cm}}$ of $\mathbb{Q}$ (see [Pat12, Lemma 4.1.22]), so in fact we can conjecturally characterize the essential image of H^*_ι on $\mathcal{M}_{F,\mathbb{Q}^{\mathrm{cm}}}$.

[3] This question first arose, with no hint of its connection to the Kuga-Satake construction, in the paper [Con11] of Brian Conrad.

surjection of linear algebraic groups over $\overline{\mathbb{Q}}_\ell$ whose kernel is a central torus in $\widetilde{H}$, and suppose we are given a geometric Galois representation, or even one arising from the ℓ-adic realization of a motive over F, $\Gamma_F \xrightarrow{\rho} H(\overline{\mathbb{Q}}_\ell)$; by an $H(\overline{\mathbb{Q}}_\ell)$-valued geometric representation, we simply mean a homomorphism ρ whose composition with some (equivalently, any) faithful finite-dimensional representation $H \hookrightarrow \mathrm{GL}_N$ is geometric in the sense of Definition 1.1. Does there exist a geometric lift $\tilde{\rho}$ filling in the following diagram?

(2)

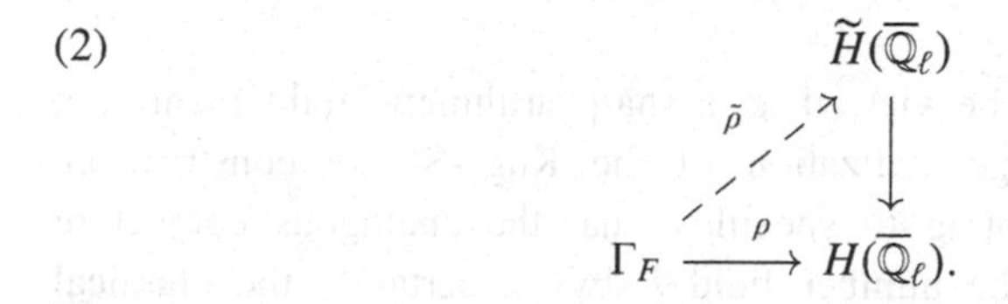

A beautiful theorem of Tate ([Ser77, Theorem 4]), relying on the full force of global class field theory, asserts that the Galois cohomology group $H^2(\Gamma_F, \mathbb{Q}/\mathbb{Z})$ is zero. This implies (see [Con11, Lemma 5.3]) that some, not necessarily geometric, lift $\tilde{\rho}$ exists; and also that when $\overline{\mathbb{Q}}_\ell$ is replaced by $\mathbb{C}$ (i.e., the Galois representations in question have finite image) it is possible to find geometric lifts (of course, here there is no 'geometry,' since the only non-zero Hodge numbers will be $h^{0,0}$). In general, there are some subtleties when F admits real embeddings, but the answer to this geometric lifting question is essentially 'yes': see [Pat12, Theorem 3.2.10] for the positive result over totally imaginary number fields (and [Pat14b, Proposition 5.5] for a complete description of the obstructions when F has a real embedding; the details of this will not concern us).

In combination with the Fontaine-Mazur-Tate conjecture, we are therefore led to the following conjecture. Let $\mathcal{M}_{F,E}$ denote the category of pure motives over F with coefficients in some finite extension $E/\mathbb{Q}$; to have at our disposal an unconditionally Tannakian category of motives, we take this to mean André's category of motives for motivated cycles (see [And96]), but one could alternatively assume the Standard Conjectures and take $\mathcal{M}_{F,E}$ to be the category of Grothendieck motives for any homological (or numerical) equivalence. A choice of fiber functor– say, Betti cohomology, after embedding $F \hookrightarrow \mathbb{C}$– defines by Tannakian theory a motivic Galois group $\mathcal{G}_{F,E}$, a pro-reductive group over E, such that $\mathcal{M}_{F,E}$ is equivalent to the category of algebraic representations of $\mathcal{G}_{F,E}$.

Conjecture 1.3 (See §4.3 of [Pat12]). *Let $\widetilde{H} \to H$ be a surjection of linear algebraic E-groups whose kernel is equal to a central torus in $\widetilde{H}$, and let*

$$\rho \colon \mathcal{G}_{F,E} \to H$$

be a motivic Galois representation. Then if either F is totally imaginary, or the 'Hodge numbers' of ρ satisfy the (necessary) parity condition of [Pat14b,

Proposition 5.5], then there exists a finite extension E'/E and a lifting of motivic Galois representations

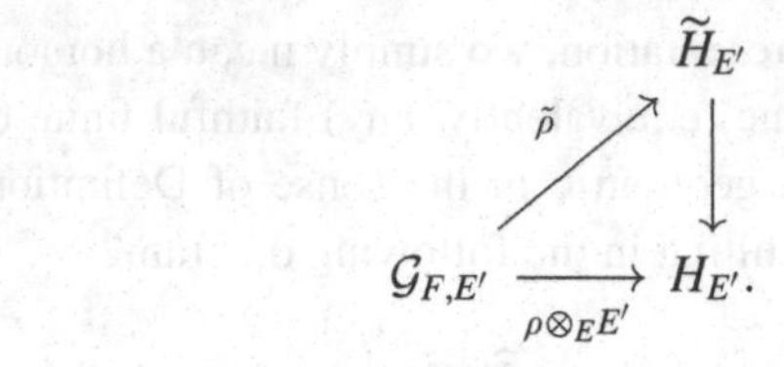

This conjecture should be viewed as a sharp arithmetic refinement (not enlarging F) of a vast generalization of the Kuga-Satake construction. It is of course also tempting to speculate that the analogous conjecture holds when we replace the number field F by $\mathbb{C}$; certainly the classical Kuga-Satake construction provides evidence for this. Along these lines, Serre has asked ([Ser94, 8.3]; this reference unfortunately has no explanation for this speculation) whether for any algebraically closed field k ($\overline{F}$ or $\mathbb{C}$, here) the derived group $\mathcal{G}_k^{der}$ is simply-connected.

The geometric aspect of Conjecture 1.3 (replacing F by $\overline{F}$) already seems to be far out of reach, and the arithmetic refinement would seem to require combining a geometric result (over $\overline{F}$) with the Tate conjecture.[4] The aim of the present paper is to provide examples of (a somewhat weakened version of) Conjecture 1.3, examples in which the motives in question do not lie in the Tannakian sub-category $\mathcal{AV}_F$ of $\mathcal{M}_F$ generated by abelian varieties and Artin motives. If one wants to avoid difficult problems related to the Tate (or in our case 'motivated-Tate') conjecture, it is convenient to phrase a weakened version of the problem as follows:

Question 1.4. Let $\widetilde{H} \to H$ be a surjection of linear algebraic E-groups with kernel equal to a central torus, as in Conjecture 1.3. Given a motivic Galois representation $\rho \colon \mathcal{G}_{F,E} \to H$, does there exist

1. a finite extension E'/E;
2. a finite extension F'/F;
3. for each embedding $E' \hookrightarrow \overline{\mathbb{Q}}_\ell$, inducing a place $\lambda'|\ell$ of E', a geometric lifting

$$\begin{array}{ccc} & & \widetilde{H}(E'_{\lambda'}) \\ & \overset{\tilde{\rho}_{\lambda'}}{\nearrow} & \downarrow \\ \Gamma_F & \xrightarrow[(\rho\otimes E')_{\lambda'}]{} & H(E'_{\lambda'}); \end{array}$$

[4] See [Pat12, Theorem 1.1.12] for an example of how this works in the classical Kuga-Satake setting, where the Tate conjecture for abelian varieties is available (Faltings' theorem).

4. and a faithful finite-dimensional representation $r\colon \widetilde{H}_{E'} \hookrightarrow \mathrm{GL}_{N,E'}$ such that

$$r \circ \tilde{\rho}_{\lambda'}|_{\Gamma_{F'}} : \Gamma_{F'} \to \mathrm{GL}_N(E'_{\lambda'})$$

is isomorphic to the λ'-adic realization of a motive, i.e. an object of $\mathcal{M}_{F',E'}$?

Alternatively, letting $F = \mathbb{C}$, we can ask the same question about the Hodge-Betti realizations, finding a rank N motive whose E'-Hodge structure lifts the one given by the Betti realization of ρ.

The truly tantalizing cases of this question are, in the spirit of the Vancouver conference, those arising from the 'non-classical' context in which the motives (or Hodge structures) do not lie in $\mathcal{AV}_F$. We will content ourselves in this paper with producing many examples in the case (avoiding rationality questions) $\widetilde{H} = \mathrm{GSpin}_5 \to H = \mathrm{SO}_5$, with $r = r_{spin}$ the (4-dimensional) spin representation;[5] these examples provide the first non-trivial examples of 'generalized Kuga-Satake theory' beyond the Tannakian subcategory $\mathcal{AV}_F$ of $\mathcal{M}_F$. We will work with the ℓ-adic version of Question 1.4, and therefore our results as written will only apply to (certain) motives over number fields; but it would also be possible to work purely Hodge-theoretically and deduce consequences for (certain) motives defined over $\mathbb{C}$ but not $\overline{\mathbb{Q}}$. We postpone until §2 a discussion of the precise setting of our motivic lifting theorem, and now conclude this introduction by giving one family of examples.

Consider the well-known family (with parameter t) of Calabi-Yau hypersurfaces

$$\mathbf{X}_t = \{X_0^6 + X_1^6 + X_2^6 + X_3^6 + X_4^6 + X_5^6 = 6tX_0X_1X_2X_3X_4X_5\} \hookrightarrow \mathbb{P}^5.$$

We denote by $\mathbf{X} \to \mathbb{P}^1 \setminus \{\mu_6, \infty\}$ this family of smooth projective sextic four-folds, whose t-fiber is $\mathbf{X}_t$. Each fiber $\mathbf{X}_t$ carries an action of the finite group scheme

$$\mathrm{A} = \{(\zeta_0, \zeta_1, \ldots \zeta_5) \in \mu_6^6 : \prod_i \zeta_i = 1\} \subset \mu_6^6,$$

with of course the diagonal subgroup $(\zeta, \zeta, \ldots, \zeta)$ acting trivially. For each $t \in (\mathbb{P}^1 - \{\mu_6, \infty\})(\overline{\mathbb{Q}})$, we can form the object of A-invariants in primitive cohomology

$$M_t = \mathrm{Prim}^4(\mathbf{X}_t)^{\mathrm{A}}(2) \in \mathcal{M}_{\mathbb{Q}(t)},$$

where $\mathbb{Q}(t)$ is just the field of definition of the point t, and where we have Tate-twisted to weight zero. For any embedding $\mathbb{Q}(t) \hookrightarrow \mathbb{C}$, the associated Hodge numbers are

$$h^{2,-2} = h^{1,-1} = h^{0,0} = h^{-1,1} = h^{-2,2} = 1,$$

[5] A less interesting case also considered – see Theorem 3.6 – is $\mathrm{GSpin}_6 \to \mathrm{SO}_6$, with $r = r_{spin}$ the sum of the two half-spin representations.

and from this it is easy to see that M_t does not lie in $\mathcal{AV}_{\mathbb{Q}(t)}$, nor after base-change does $M_t|_{\mathcal{G}_{\mathbb{C}}}$ lie in $\mathcal{AV}_{\mathbb{C}}$: as noted in [Del72, 7.3 Proposition], it suffices to observe that the resulting Hodge structure on the adjoint representation (of SO_5) is not of type $\{(1,-1),(0,0),(-1,1)\}$. M_t gives rise to an orthogonal motivic Galois representation $\rho_t\colon \mathcal{G}_{\mathbb{Q}(t)} \to O(M_t)$, where the orthogonal pairing is given by Poincaré duality. There is an at most degree 2 extension of $\mathbb{Q}(t)$ such that for all ℓ, the restriction of the ℓ-adic realization $\rho_{t,\ell}$ factors through the special orthogonal group. Here, then, is a sample result:

Theorem 1.5. *For all $t \in \mathbb{P}^1(\overline{\mathbb{Q}}) - \{\mu_6, \infty\}$, there exists a number field F, in fact a quadratic extension (depending on t) of $\mathbb{Q}(\zeta_{24})$, such that:*

1. *for all places λ of $\mathbb{Q}(\zeta_{24})$, there is a lift $\tilde{\rho}_{t,\lambda}$ of $\rho_{t,\lambda}$:*

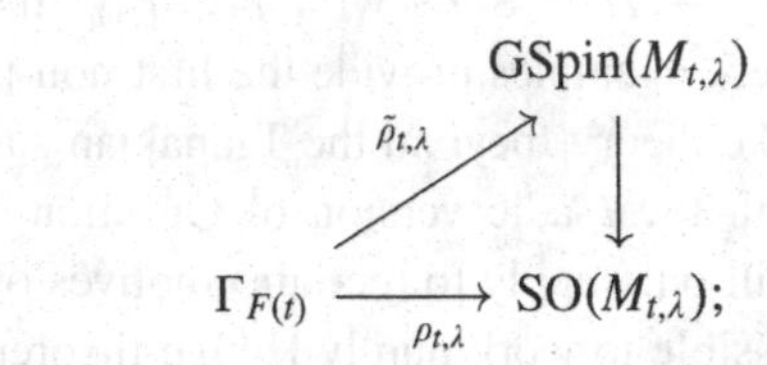

2. *and an object $\widetilde{M}_t$ of $\mathcal{M}_{F(t),\mathbb{Q}(\zeta_{24})}$ such that the λ-adic realization $\widetilde{M}_{t,\lambda}$ is isomorphic (as $\Gamma_{F(t)}$-representation) to $r_{spin} \circ \tilde{\rho}_{t,\lambda}$.*

Informally, the motives M_t each admits a generalized Kuga-Satake lift. Note that the field F has bounded degree, independent of t and λ. This is some consolation for its unfortunate occurence in the statement of the theorem; sometimes with more work sharper results are possible.

2 Rigid local systems and Katz's middle convolution algorithm

What is special about each of the motives M_t arising from the Dwork family of hypersurfaces, substituting for the fact that we no longer have at our disposal the Hodge theory of abelian varieties? Roughly speaking, it is that the ℓ-adic realization $M_{t,\ell}$ is isomorphic to a 'fiber' of a rigid local system over suitably punctured $\mathbb{P}^1$. We will now take some time to describe Katz's beautiful theory of such rigid local systems. Let k be an algebraically closed field, and let ℓ be a prime not equal to the characteristic of k. Let $S \subset \mathbb{P}^1$ be a finite set of closed points, and let $j\colon U = \mathbb{P}^1 - S \hookrightarrow \mathbb{P}^1$ be the open complement. For $s \in S$, we let I_s denote the inertia group at s. To put our results in the proper context, we will introduce not just Katz's notions of rigidity for local systems, but also their (obvious) extensions to H-local systems for any linear algebraic group H over $\overline{\mathbb{Q}}_\ell$ (see too [Yun13] for these definitions). Recall that an H-local system on U

is simply a continuous homomorphism $\rho\colon \pi_1(U,x) \to H(\overline{\mathbb{Q}}_\ell)$, where $\pi_1(U,x)$ denotes the étale (as always, in this paper) fundamental group defined with respect to a fixed geometric point x of U. For $H = \mathrm{GL}_N$, this is is the same (i.e., there is an equivalence of tensor categories, depending on our choice x of geometric point) as a lisse $\overline{\mathbb{Q}}_\ell$-sheaf of rank N on U, and we freely go back and forth between these interpretations. Two H-local systems are isomorphic if they are conjugate by an element of $H(\overline{\mathbb{Q}}_\ell)$. There are at least two relevant notions of rigidity for such an H-local system:

Definition 2.1. An H-local system ρ on U is H-physically rigid if for all H-local systems ρ' on U such that $\rho|_{I_s} \cong \rho'|_{I_s}$ for all $s \in S$, we in fact have $\rho \cong \rho'$.[6] We say that ρ is H-cohomologically rigid if

$$H^1(\mathbb{P}^1, j_* \mathrm{Ad}(\rho)) = 0,$$

where of course $\mathrm{Ad}\colon H \to \mathrm{GL}(\mathrm{Lie}(H))$ denotes the adjoint representation.

If ρ is tamely ramified at all $s \in S$ – this is the only case that will concern us– then the Euler-Poincaré formula simply says

$$\chi(\mathbb{P}^1, j_* \mathrm{Ad}(\rho)) = \dim H \cdot \chi(U) + \sum_{s \in S} \dim \mathrm{Ad}(\rho)^{I_s}.$$

If H is moreover reductive, then $\chi(\mathbb{P}^1, j_* \mathrm{Ad}(\rho)) = 2\dim \mathrm{Ad}(\rho)^{\pi_1(U,x)} - \dim H^1(\mathbb{P}^1, j_* \mathrm{Ad}(\rho))$, so ρ is cohomologically rigid if and only if

$$\dim H \cdot (2 - |S|) + \sum_{s \in S} \dim \mathrm{Ad}(\rho)^{I(s)} = 2\dim \mathrm{Ad}(\rho)^{\pi_1(U,x)}. \tag{3}$$

Note that the analogue of irreducibility (when $H = \mathrm{GL}_N$) is the condition $\mathrm{Cent}_H(\rho) = Z(H)$, in which case the right-hand side of this formula is just $2\dim Z(H)$.

For $H = \mathrm{GL}_N$, Katz has shown ([Kat96, Theorem 5.0.2]) that if an irreducible lisse sheaf $\mathcal{F}$ is cohomologically rigid, then it is physically rigid; and conversely that if $k \subset \mathbb{C}$, and if the analytic sheaf $\mathcal{F}^{an}$ associated to $\mathcal{F}$ is physically rigid on U^{an}, then $\mathcal{F}$ is cohomologically rigid (note that the 'passage to the analytic' functor is fully faithful but not essentially surjective – see [Kat96, Proposition 5.9.2] for a discussion – so that physical rigidity of $\mathcal{F}^{an}$ does not follow immediately from physical rigidity of $\mathcal{F}$; for the proof of this converse, see [Kat96, Theorem 1.1.2]). Note that the implication 'cohomologically rigid implies physically rigid' does not continue to hold for general $H \neq \mathrm{GL}_N$; this has the potential to complicate significantly the kinds of arguments given in this paper, although here (essentially because of low-dimensional coincidences) it does not intervene.

6 We paraphrase this condition as saying that 'if ρ and ρ' are everywhere-locally conjugate, then they are globally conjugate.'

Underlying the approach of this paper, and our hope to exploit rigid local systems for purposes of the Kuga-Satake problem, is a (variant of a) conjecture of Simpson (see [Sim92, p.9]), which we will state as a guiding philosophy rather than as a precise conjecture:

Scholium 2.2. *Let H be a reductive group, and let ρ be an H-rigid local system on U with quasi-unipotent local monodromies. Then ρ is 'motivic,' i.e. arises as a direct factor of the monodromy representation on the cohomology of a family of varieties $Y \to U$.*

Note that we have not specified which kind of rigidity. For the purposes of heuristic argument, let us blur the two. Extrapolating from this principle, if we have a (known to be motivic) Galois representation or Hodge structure ρ_t isomorphic to the t-fiber of some H-rigid local system ρ on U, and we have a surjection $\widetilde{H} \to H$, then to find a Kuga-Satake lift we should lift ρ to $\widetilde{H}$ (this is obviously possible over algebraically closed k; in the Galois case, when we work over a number field, we achieve this after a finite, but *a priori* hard to specify, base-change). The lift $\tilde{\rho}$ will be $\widetilde{H}$-rigid (compare equation 3 for H and $\widetilde{H}$), so it ought to be motivic. The (motivic) t-fiber $\tilde{\rho}_t$ should then provide the Kuga-Satake lift of ρ_t.

Of course, nothing so general as Scholium 2.2 is proven; but, remarkably, Katz (see Theorem 2.3 below) has proven it for tame irreducible ρ, in the case $H = \mathrm{GL}_N$! Bogner and Reiter ([BR13]) have combined Katz's work with a clever trick to establish a similar result for $H = \mathrm{Sp}_4$ as well. These two results are essentially all we will require. For the rest of this section, we proceed more formally and explain Katz's work as needed for our purposes.

As before, let k be an algebraically closed field, and let ℓ be a prime invertible in k. We say that a lisse $\overline{\mathbb{Q}}_\ell$-sheaf $\mathcal{F}$ on U/k is tame with index of quasi-unipotence dividing N if each of the local monodromy representations, i.e. for each $s \in S$ the action of I_s, factors through the tame inertia group $I_s^{\mathrm{tame}} \cong \hat{\mathbb{Z}}$, where a choice of topological generator acts quasi-unipotently with eigenvalues contained in the Nth roots of unity. We restrict to N invertible in k. Rigid local systems over k will admit arithmetic descents, and for this purpose we consider the ring $R_{N,\ell} = \mathbb{Z}[\zeta_N, \frac{1}{N\ell}]$, with a *fixed* choice of embedding $R_{N,\ell} \hookrightarrow k$. Now, we may always apply an automorphism of $\mathbb{P}^1$ to assume that $\infty \in S$, and we choose an ordering of the remaining points of S, writing $U = \mathbb{A}^1 - \{\alpha_1, \dots, \alpha_n\}$. It turns out that a rigid local system on U extends to a 'universal' local system in which the points $\{\alpha_i\}$ are allowed to move in the plane, so we introduce the arithmetic configuration space

$$S_{N,n,\ell} = R_{N,\ell}[T_1, \dots, T_n]\left[\frac{1}{\prod_{i \neq j}(T_i - T_j)}\right].$$

The 'universal' version of a given rigid local system will then live on the relative affine line with n sections deleted,

$$\mathbb{A}^1_{S_{N,n,\ell}} - \{T_1, \dots, T_n\} = \operatorname{Spec} R_{N,n,\ell}[T_1, \dots, T_n, X]\left[\frac{1}{\prod_{i \neq j}(T_i - T_j)} \cdot \frac{1}{\prod_i (X - T_i)}\right].$$

Specialization to a local system on U/k is then achieved by extending the fixed $R_{N,\ell} \hookrightarrow k$ to

$$\phi\colon S_{N,n,\ell} \to k$$
$$T_i \mapsto \alpha_i.$$

Now we can state (a slightly less precise version of) Katz's theorem on the 'motivic' description of rigid local systems:

Theorem 2.3 (Theorem 8.4.1 of [Kat96]). *Let $\mathcal{F}_k$ be a lisse $\overline{\mathbb{Q}}_\ell$-sheaf on U/k that is*

- *tamely ramified;*
- *irreducible;*
- *cohomologically rigid;*
- *quasi-unipotent of index dividing N.*

Fix a faithful character $\chi: \mu_N(R_{N,\ell}) \hookrightarrow \overline{\mathbb{Q}}_\ell^\times$ (equivalently, an embedding $R_{N,\ell} \hookrightarrow \overline{\mathbb{Q}}_\ell$). Then there exists a smooth affine family of $R_{N,\ell}$-schemes

$$\pi : \mathrm{Hyp} \to \mathbb{A}^1_{S_{N,n,\ell}} - \{T_1, \dots, T_n\}$$

such that

- *the fibers of π are geometrically connected of some dimension r;*
- *there is an action of μ_N on* Hyp *for which π is μ_N-equivariant (acting trivially on the base);*
- *after the base-change $\phi\colon S_{N,n,\ell} \to k$, the lisse sheaf*
$$\mathcal{F} = \mathrm{Gr}^W_r\left(\mathrm{R}^r\pi_!\overline{\mathbb{Q}}_\ell\right)^\chi$$
on $\mathbb{A}^1_{S_{N,n,\ell}} - \{T_1, \dots, T_n\}$ becomes isomorphic to $\mathcal{F}_k$. Here $\mathrm{Gr}^W_\bullet$ denotes the weight filtration of [Del80].

For example, if $k = \overline{\mathbb{Q}}$ and the points α_i lie in $\mathbb{A}^1(\mathbb{Q}(\zeta_N))$, this result descends the given rigid local system $\mathcal{F}_{\overline{\mathbb{Q}}}$ to an arithmetic local system on $U/\mathbb{Q}(\zeta_N)$.

To prove Theorem 2.3, Katz introduces a 'geometric' operation on sheaves, the middle convolution, that can be iterated (in combination with a simpler twisting operation) to generate all cohomologically rigid local systems starting from easily-understood rank one local systems. We now describe (one version of) the middle convolution. First, for any $r \geq 0$, let $\mathbb{A}(n,r+1)_{R_{N,\ell}}$ be the space

$$\mathbb{A}(n,r+1)_{R_{N,\ell}} = \operatorname{Spec} R_{N,\ell}[T_1,\dots,T_n,X_1,\dots,X_{r+1}] \times \left[\frac{1}{\prod_{i\neq j}(T_i-T_j)\prod_{a,j}(X_a-T_j)\prod_k(X_{k+1}-X_k)}\right].$$

For $r=0$, we recover $\mathbb{A}(n,1)_{R_{N,\ell}} = \mathbb{A}^1_{S_{N,n,\ell}} - \{T_1,\dots,T_n\}$ (in this case the product over k in the above definition is understood to be 1). To ease notation, we will suppress the $R_{N,\ell}$ subscript and simply write $\mathbb{A}(n,r+1)$. Now, let $\mathrm{Lisse}(N,n,\ell)$ be the category of lisse $\overline{\mathbb{Q}}_\ell$-sheaves on $\mathbb{A}(n,1)$. For any non-trivial character $\chi\colon \mu_N(R_{N,\ell}) \to \overline{\mathbb{Q}}_\ell^\times$, incarnated as a character sheaf $\mathcal{L}_\chi$ on $(\mathbb{G}_m)_{R_{N,\ell}}$, and then as the pull-back $\mathcal{L}_{\chi(X_2-X_1)}$ on $\mathbb{A}(n,2)$ via the difference map

$$d\colon \mathbb{A}(n,2) \xrightarrow{X_2-X_1} (\mathbb{G}_m)_{R_{N,\ell}},$$

Katz defines two operations, naïve and middle convolution by χ,

$$\begin{aligned} \mathrm{NC}_\chi &\colon \mathrm{Lisse}(N,n,\ell) \to \mathrm{Lisse}(N,n,\ell), \\ \mathrm{MC}_\chi &\colon \mathrm{Lisse}(N,n,\ell) \to \mathrm{Lisse}(N,n,\ell), \end{aligned}$$

via the following recipe. For all $r \geq 0$ and $1 \leq i \leq r+1$, there are projections

$$\begin{aligned} \mathrm{pr}_i\colon \mathbb{A}(n,r+1) &\to \mathbb{A}(n,1) \\ (T_1,\dots,T_n,X_1,\dots,X_{r+1}) &\mapsto (T_1,\dots,T_n,X_i). \end{aligned}$$

The naïve convolution of $\mathcal{F}$ by χ is given by ($r=1$ in the above notation)

$$\mathrm{NC}_\chi(\mathcal{F}) = \mathrm{R}^1(\mathrm{pr}_{2,!})\left(\mathrm{pr}_1^*\mathcal{F} \otimes d^*\mathcal{L}_\chi\right). \tag{4}$$

From now on, we will abbreviate $\mathrm{pr}_1^*\mathcal{F} \otimes d^*\mathcal{L}_\chi$ by simply $\mathcal{F} \boxtimes \mathcal{L}_\chi$. Note that if $\mathcal{F}$ is mixed of weights $\leq w$, then $\mathrm{NC}_\chi(\mathcal{F})$ is mixed of weights $\leq w+1$. The middle convolution is then a variant that will take pure sheaves to pure sheaves. To define it, view pr_2 as a relative $\mathbb{A}^1$ (with coordinate X_1) with the sections $T_1,\dots,T_n,X_2$ deleted, and compactify it to a relative $\mathbb{P}^1$:

$$\begin{array}{ccc} \mathbb{A}(n,2) & \xrightarrow{\ j\ } & \mathbb{P}^1 \times_{R_{N,\ell}} \mathbb{A}(n,1) \\ \downarrow{\scriptstyle \mathrm{pr}_1} & \searrow^{\mathrm{pr}_2} & \downarrow{\scriptstyle \overline{\mathrm{pr}}_2} \\ \mathbb{A}(n,1) & & \mathbb{A}(n,1). \end{array}$$

For all $\mathcal{F} \in \mathrm{Lisse}(N,n,\ell)$, set

$$\mathrm{MC}_\chi(\mathcal{F}) = \mathrm{R}^1(\overline{\mathrm{pr}}_{2,*})\left(j_*(\mathcal{F} \boxtimes \mathcal{L}_\chi)\right). \tag{5}$$

See [Kat96, Lemma 8.3.2] for the basic facts about $\mathrm{MC}_\chi(\mathcal{F})$: note for now that it is lisse, and for non-trivial χ and $\mathcal{F}$ pure of weight w, the natural map

$$\mathrm{NC}_\chi(\mathcal{F}) \to \mathrm{MC}_\chi(\mathcal{F})$$

is a surjection identifying $\mathrm{MC}_\chi(\mathcal{F})$ with the top graded piece $\mathrm{Gr}^W_{w+1}(\mathrm{NC}_\chi(\mathcal{F}))$ of the weight filtration on $\mathrm{NC}_\chi(\mathcal{F})$.

The proof of Theorem 2.3 is a combination of Katz's main theorem (see [Kat96, Theorem 5.2.1]) on the structure of tamely ramified quasi-unipotent rigid local systems on U/k (k algebraically closed, recall), which shows that they are all obtained by starting from rank one local systems and iterating some combination of the middle convolution (specialized via $\phi\colon S_{N,n,\ell} \to k$) and the simpler operation of twisting by rank one local systems, and understanding the total geometric effect of iterating the convolution. Our main result will require invoking a variant of Katz's results. We will need to analyze local systems that are not necessarily rigid but that nevertheless admit a description as an iteration of middle convolutions and twists (as in Katz), as well as Schur functors. We will need to check that the natural arithmetic descents of these local systems specialize (over number fields) to the ℓ-adic realizations of motives for motivated cycles (in the sense of [And96]). This will make precise the sense in which the Kuga-Satake lifts we construct are 'motivic.' Our notation here is that for any object M of $\mathcal{M}_{F,E}$, and any embedding $\iota\colon E \hookrightarrow \overline{\mathbb{Q}}_\ell$, the corresponding ℓ-adic realization (a Γ_F-representation) is denoted $M_{\iota,\ell}$. We now explain all of this in detail.

Now we come to the key technical lemma, which is a simple variant of the arguments used by Katz in [Kat96, Chapter 8] to establish Theorem 2.3; Dettweiler has conveniently abstracted many of these arguments in [Det08], and we will be able to reduce to these already-treated cases. To simplify the statement, we first make a definition:

Definition 2.4. An object $\mathcal{F}$ of Lisse(N,n,ℓ) is said to be *geometrically concentrated* if it is of the following form: there exists a smooth morphism $f\colon \mathbf{X} \to \mathbb{A}(n,1)$ such that

- f is equivariant for the action of a finite group G on $\mathbf{X}$, trivial on the base $\mathbb{A}(n,1)$;
- there is an idempotent $e \in \overline{\mathbb{Q}}_\ell[G]$;
- there is an integer $r \geq 0$ such that for all $i \neq r$, $e\mathrm{R}^i f_! \overline{\mathbb{Q}}_\ell = 0$, and:
- $\mathcal{F} \cong \mathrm{Gr}^W_r(e\mathrm{R}^r f_! \overline{\mathbb{Q}}_\ell)$.

We will also use the terminology 'geometrically concentrated' for sheaves on $\mathbb{A}^1 - \{\alpha_1, \ldots, \alpha_n\}$ admitting a similar description, where we have specialized the parameters T_i to some $\alpha_i \in \mathbb{A}^1(k)$.

Here is the result:

Proposition 2.5. *Let $\mathcal{F}$ be any object of* Lisse(N,n,ℓ) *obtained by some iteration, starting from the constant sheaf, of the following three operations:*

- *middle convolution by a non-trivial character $\rho\colon \mu_N(R_{N,\ell}) \to \overline{\mathbb{Q}}_\ell^\times$;*
- *Schur functors;*
- *tensoring by geometrically concentrated sheaves in* Lisse(N,n,ℓ).

Then $\mathcal{F}$ is geometrically concentrated.

Moreover, let F be a field of characteristic zero (eg, a number field) containing $R_{N,\ell}$, and let $t\colon \operatorname{Spec} F \to \mathbb{A}(n,1)$ be any map of $R_{N,\ell}$-schemes. Then there exists a number field E and, for each embedding $\iota\colon E \hookrightarrow \overline{\mathbb{Q}}_\ell$, an object M of $\mathcal{M}_{F,E}$ such that $t^*\mathcal{F}$, as Γ_F-representation, is isomorphic to the ℓ-adic realization $M_{\iota,\ell}$.

Proof. Let $\mathcal{F}$ be geometrically concentrated. Arguing inductively, we have to check the following three statements:

- if $\rho\colon \mu_N(R_{N,\ell}) \to \overline{\mathbb{Q}}_\ell^\times$ is a non-trivial character, then $\mathrm{MC}_\rho(\mathcal{F})$ is geometrically concentrated;
- if λ is a partition of a positive integer m, then the image of the associated Schur functor $\mathbb{S}_\lambda\mathcal{F}$ is geometrically concentrated;
- and if $\mathcal{G}$ is a second geometrically concentrated object of Lisse(N,n,ℓ), then $\mathcal{F}\otimes\mathcal{G}$ is geometrically concentrated.

The first assertion follows from [Det08, 2.6.1 Theorem]. For the second bulleted assertion, let $(\mathcal{F},f,e,r)$ be as in Definition 2.4, and let

$$f^m\colon \mathbf{X}\times_{\mathbb{A}(n,1)}\mathbf{X}\times_{\mathbb{A}(n,1)}\cdots\times_{\mathbb{A}(n,1)}\mathbf{X}\to\mathbb{A}(n,1)$$

denote the projection from the m-fold fiber product. This projection is now equivariant for the natural action of $G^m \rtimes S_m$. By the Künneth formula and the assumption of concentration,

$$(e\times e\times\cdots\times e)\mathrm{R}^{mr}f^m_!\overline{\mathbb{Q}}_\ell \cong (e\times e\times\cdots\times e)\bigoplus_{i_1+\cdots+i_m=mr}(\otimes_{j=1}^m \mathrm{R}^{i_j}f_!\overline{\mathbb{Q}}_\ell)\cong(e\mathrm{R}^rf_!\overline{\mathbb{Q}}_\ell)^{\otimes m}.$$

Applying Gr^W_{mr}, and abbreviating $e^m = e\times\cdots\times e$, we deduce that

$$e^m\,\mathrm{Gr}^W_{mr}\mathrm{R}^{mr}f^m_!\overline{\mathbb{Q}}_\ell \cong \mathcal{F}^{\otimes m}.$$

Let $s_\lambda\in\overline{\mathbb{Q}}_\ell[S_m]$ be the idempotent projector such that for any vector space V, $\mathbb{S}_\lambda(V) = s_\lambda(V^{\otimes m})$. Then the element $s_\lambda\cdot e^m$ of $\overline{\mathbb{Q}}_\ell[G^m\rtimes S_m]$ is still idempotent, since the operators s_λ and e^m commute. Thus

$$\mathbb{S}_\lambda(\mathcal{F}) = s_\lambda e^m\left(\mathrm{Gr}^W_{mr}(\mathrm{R}^{mr}f^m_!\overline{\mathbb{Q}}_\ell)\right)$$

is geometrically concentrated.

For the last bulleted point, again apply the Künneth formula, using concentration of $\mathcal{F}$ and $\mathcal{G}$.

The second half of the proposition – the claim that for all specializations t, $t^*\mathcal{F}$ is the ℓ-adic realization of a motivated motive – is deduced from the first part using the fact that for any smooth (not necessarily projective) variety U/F, and for all $k \geq 0$, the top graded quotient $\mathrm{Gr}_k^W H_c^k(U_{\overline{F}}, \overline{\mathbb{Q}}_\ell)$ can be conveniently described in terms of a smooth compactification of U (with a smooth normal crossings divisor at the boundary). For details, see the discussion in between Remark 2.6 and Corollary 2.7 in [PT15]. □

Remark 2.6. The following rank 1 geometrically concentrated sheaves are, along with the middle convolution, the ingredients in Katz's 'motivic' construction of rigid local systems: fix n characters

$$\chi_i \colon \mu_N(R_{N,\ell}) \to \overline{\mathbb{Q}}_\ell^\times,$$

$i = 1, \ldots, n$, $a = 1, \ldots, r+1$, and set

$$\mathcal{F} = \otimes_{i=1}^n (X - T_i)^*(\mathcal{L}_{\chi_i}),$$

where X denotes the 'relative $\mathbb{A}^1$' coordinate on $\mathbb{A}(n,1)$, whence a map $(X - T_i) \colon \mathbb{A}(n,1) \to \mathbb{G}_m$. All of the 'universal' extensions of rigid local systems to $\mathrm{Lisse}(N,n,\ell)$ are then expressed by iterating middle convolution and tensoring by such rank 1 sheaves $\mathcal{F}$.

3 The main result

We will now demonstrate the existence of 'generalized Kuga-Satake lifts' of motives whose ℓ-adic realizations are isomorphic to some fiber of an SO_5 or SO_6-cohomologically rigid local system. Note that an SO_N-local system that is GL_N-cohomologically rigid is automatically SO_N-cohomologically rigid: $H^1(\mathbb{P}^1, j_*\mathfrak{so}_N)$ is a summand of $H^1(\mathbb{P}^1, j_*\mathfrak{gl}_N)$, so vanishing of the latter group of course implies vanishing of the former group. In light of Scholium 2.2, the essential content of our result is the combination of Katz's theory with the following theorem of Bogner-Reiter; what we will do is reinterpret this theorem and give a couple examples:

Theorem 3.1 (Theorem 3.1 of [BR13]). *Each $\mathrm{Sp}_4(\mathbb{C})$-cohomologically rigid local system with quasi-unipotent local monodromies comes from geometry: it can be constructed by a sequence of 'geometric operations' (direct sum, tensor product, exterior square, symmetric square, rational pull-back, and middle convolution), starting from rank 1 local systems.*

The argument crucially depends on 'low-dimensional coincidences' in the Dynkin diagrams of the classical Lie groups. To give a concrete sense of the difficulties involved in generalizing to higher rank, we will construct (in Examples 3.7 and 3.8 below) motivic, SO_7-rigid local systems whose $Spin_7$ lifts do not seem to be provably motivic using the methods of this paper.

Theorem 3.2. *Let F and E be number fields, and fix an embedding $\iota \colon E \hookrightarrow \overline{\mathbb{Q}}_\ell$. Suppose M is a motivated motive in $\mathcal{M}_{F,E}$ whose (ι,ℓ)-adic realization is orthogonal of rank 5, and is isomorphic to a Γ_F-representation of the following form: there exists*

- *a finite set of points $S \subset \mathbb{P}^1(F)$, with complement $U = \mathbb{P}^1 - S$;*
- *an orthogonal rank 5 local system $\mathcal{F}$ on U over F, such that $\mathcal{F}_{\overline{F}}$ (the corresponding geometric local system) has quasi-unipotent local monodromies and is SO_5-cohomologically rigid;*
- *and a point $t\colon \operatorname{Spec} F \to U$;*
- *such that $M_{\iota,\ell}$ is isomorphic to $t^*\mathcal{F}$ as $\overline{\mathbb{Q}}_\ell[\Gamma_F]$-module.*

For simplicity, assume either that $M_{\iota,\ell}$ is irreducible, or that the geometric monodromy group of $\mathcal{F}_{\overline{F}}$ acts irreducibly in the standard 5-dimensional representation. Then (enlarging F if necessary by a quadratic extension so that $M_{\iota,\ell}$ is special orthogonal) there exists a geometric lift

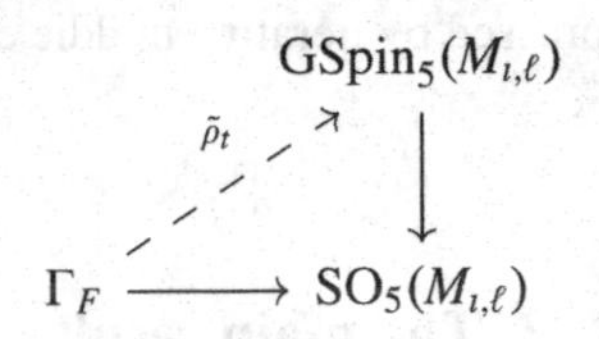

such that for some finite extensions F'/F and E'/E of number fields, $r_{spin} \circ \tilde{\rho}|_{\Gamma_{F'}}$ is isomorphic to the $E' \xrightarrow{\iota'} \overline{\mathbb{Q}}_\ell$-realization of an object of $\mathcal{M}_{F',E'}$, for a suitable embedding ι' extending ι.

Remark 3.3. The hypothesis that either $M_{\iota,\ell}$ or $\mathcal{F}_{\overline{F}}$ has 'big' monodromy is to avoid (more) tedious case-by-case arguments; it will be met in the examples discussed below.

Proof. The first part of the proof is just bookkeeping to reduce to the case where $\mathcal{F}_{\overline{F}}$ has monodromy group containing SO_5; the reader may wish simply to skip to that part of the proof (fourth paragraph). If $M_{\iota,\ell}$ is irreducible, then by [Pat12, Proposition 3.4.1] it can be written in the form $\operatorname{Ind}_{\Gamma_L}^{\Gamma_F}(r \otimes \tau)$ where r is Lie-irreducible (irreducible after every finite base-change) and τ has finite image. Since the rank of $M_{\iota,\ell}$ is 5, this means $M_{\iota,\ell}$ is either the induction of a character from a rank 5 extension L/F, or is Lie-irreducible, or has finite-image. The finite-image case is obvious, since then by Tate's theorem

([Ser77, Theorem 4]) there exists a $\tilde{\rho}$ with finite image, for which $r_{spin} \circ \tilde{\rho}$ is the realization of an Artin motive. The induction case follows since any geometric Galois character $\Gamma_L \to \overline{\mathbb{Q}}_\ell^\times$, at least after a finite base-change L'/L can be cut out as a motivated sub-motive of a CM abelian variety. It is easy to see that the same will then (after a finite base-change) be true of any geometric lift $\tilde{\rho}$ (and such geometric lifts are easy to produce in this case). So we may assume $M_{\iota,\ell}$ is Lie-irreducible. There are then only two possibilities for the connected component of the Zariski closure of the image of this Γ_F-representation: either it is the image of the symmetric fourth power of SL_2, or it is all of SO_5.

Let us denote by $\rho\colon \pi_1(U,\bar{t}) \to O(\mathcal{F}_{\bar{t}}) \cong O_5(\overline{\mathbb{Q}}_\ell)$ the representation associated to the lisse sheaf $\mathcal{F}$ (with $\bar{t}$ the geometric point over t corresponding to our choice of algebraic closure $\overline{F}/F$). If in the last paragraph, $M_{\iota,\ell}$ had (after passing to the connected component) SO_5 as its monodromy group, then of course $\overline{\rho(\pi_1(U,\bar{t}))}^{Zar}$ contains SO_5. In particular, the geometric monodromy group $\overline{\rho(\pi_1(U_{\overline{F}},\bar{t}))}^{Zar}$ has connected component equal to a connected normal subgroup of SO_5, i.e. either SO_5 itself or the trivial group. If trivial, then the finite group $\overline{\rho(\pi_1(U_{\overline{F}},\bar{t}))}$ is normalized by the connected group $\overline{\rho(\pi_1(U,\bar{t}))}^{Zar,0} = SO_5$; but of course any map (conjugation, here) from a connected group to a finite group is trivial, so SO_5 centralizes $\overline{\rho(\pi_1(U_{\overline{F}},\bar{t}))}$, i.e. the latter group is contained in $Z(O_5) = \{\pm 1\}$. But such a local system cannot be rigid.

Next, if $M_{\iota,\ell}$ has monodromy group PGL_2 acting through Sym^4, then the connected component $\overline{\rho(\pi_1(U_{\overline{F}},\bar{t}))}^{Zar,0}$ is either $\{1\}$ or PGL_2 or SO_5. By the argument of the previous paragraph, it cannot be trivial, so let us consider the case of PGL_2. The normalizer of PGL_2 in SO_5 is simply PGL_2,[7] so $\overline{\rho(\pi_1(U,\bar{t}))}^{Zar}$ must equal either PGL_2 or $PGL_2 \times \{\pm 1\}$. In this case a considerably simpler version of the argument we give below will apply, solving an $SL_2 \to PGL_2$ lifting problem rather than a $\mathrm{Spin}_5 \to SO_5$ lifting problem; we omit the details, noting only that when we lift the geometric local system $\mathcal{F}_{\overline{F}}$ to SL_2, we obtain a linearly rigid rank 2 local system, which can be described by Katz's algorithm.

Finally, we treat the basic case – this is the heart of the theorem – in which the geometric monodromy group of $\mathcal{F}_{\overline{F}}$ is irreducible. We make one initial adjustment to ρ – denoting as before the monodromy representation of $\mathcal{F}$ – which will eventually cost us at most a quadratic extension on the field F' in

[7] More generally, one can identify the normalizer in a simple algebraic group of the principal SL_2, i.e. the output of the Jacobson-Morosov theorem for a regular nilpotent element. For example, for $G = SO_{2n+1}$, it is easy to see that the centralizer $Z_G(PGL_2) = \{1\}$, and since PGL_2 has no outer automorphisms, we deduce that $N_G(PGL_2) = PGL_2$.

the conclusion of the theorem. Namely, ρ may have non-trivial determinant

$$\det \rho \colon \pi_1(U, \bar{t}) \to \{\pm 1\}.$$

The underlying geometric (rank 1) local system $\det \rho|_{\pi_1(U_{\overline{F}}, \bar{t})}$, if non-trivial, is isomorphic to a tensor product of translated Kummer sheaves (see [Kat96, §8.1] for this notion) of order 2. Such translated Kummer sheaves descend to $\pi_1(U, \bar{t})$, so there exists a lisse character sheaf $\mathcal{L}$ on U over F of order at most 2 such that $\det(\rho \otimes \mathcal{L})$ is geometrically trivial, i.e. factors through a character $\Gamma_F \to \{\pm 1\}$. Thus, there exists an at most quadratic extension F'/F such that $\det(\rho \otimes \mathcal{L}) = 1$ as $\pi_1(U_{F'}, \bar{t})$-representation. From now on, we replace F by F' and ρ (respectively, $\mathcal{F}$) by $\rho \otimes \mathcal{L}$ (respectively, $\mathcal{F} \otimes \mathcal{L}$), so we can work in the setting of the following diagram:

(6)
$$\begin{array}{ccccc} & & \mathrm{Spin}(\mathcal{F}_{\bar{t}}) & \xrightarrow[\sim]{r_{spin}} \mathrm{Sp}(W_{spin,\bar{t}}) & \longrightarrow \mathrm{GL}(W_{spin,\bar{t}}) \\ & {\scriptstyle ???} \nearrow & \downarrow{\scriptstyle \pi} & & \\ \pi_1(U, \bar{t}) & \xrightarrow[\rho]{} & \mathrm{SO}(\mathcal{F}_{\bar{t}}), & & \end{array}$$

where we let $W_{spin,\bar{t}}$ denote the spin representation, with its natural (in this case symplectic) bilinear pairing – this pairing is well-defined up to $\overline{\mathbb{Q}}_\ell^\times$-scaling. For the time being we set aside the arithmetic local system $\mathcal{F}$ and only work with $\mathcal{F}_{\overline{F}}$.

The representation ρ lands in a profinite subgroup of $\mathrm{SO}(\mathcal{F}_{\bar{t}})$ (namely, a group isomorphic to $\mathrm{SO}_5(\mathcal{O})$ for the ring of integers $\mathcal{O}$ in some finite extension of $\mathbb{Q}_\ell$). We claim that the restriction $\rho|_{\pi_1(U_{\overline{F}}, \bar{t})}$ lifts to $\mathrm{Spin}(\mathcal{F}_{\bar{t}})$; but we can obviously lift the restriction of ρ to the free group (not yet profinitely-completed) on $|S| - 1$ elements to a homomorphism landing in a profinite subgroup of $\mathrm{Spin}(\mathcal{F}_{\bar{t}})$, and so the universal property of profinite completion implies the claim. Let us call such a lift $\tilde{\rho}_{\overline{F}}$.

The Theorem 3.1 of Bogner and Reiter now says that there is a local system $\mathcal{G}_{\overline{F}}$ on $U_{\overline{F}}$, constructed by a series of 'geometric' operations, that is isomorphic as rank 4 local system to $r_{spin} \circ \tilde{\rho}_{\overline{F}}$. Choose an isomorphism, so that we can identify $\rho_{\mathcal{G}_{\overline{F}}}$ and $r_{spin} \circ \tilde{\rho}_{\overline{F}}$ as homomorphisms to $\mathrm{GL}(W_{spin,\bar{t}} \xrightarrow{\sim} \mathcal{G}_{\overline{F},\bar{t}})$. Note that the necessary 'geometric operations' are, starting from rank one sheaves (all of which are products of translated Kummer sheaves): middle convolution by non-trivial characters, application of Schur functors, tensor product of two sheaves obtained through the previous operations – note that these all fall under the umbrella of Proposition 2.5 – and finally pull-backs along rational maps $\mathbb{P}^1 \to \mathbb{P}^1$. We have not taken these rational pull-backs into account in Proposition 2.5, and we do so now in an *ad hoc* manner, working one-by-one through the relevant cases in the proof of [BR13, Theorem

3.1].[8] We will use freely the notation of that argument; the only cases then needing to be considered are $P_3(4,8,10,10)$, $P_4(6,6,10,10)$, $P_4(6,8,8,10)$, and $P_5(8,8,8,10,10)$ (the other cases of their theorem making use of pullbacks all correspond to *reducible* SO_5-local systems, which we have by assumption excluded).

- The cases $P_3(4,8,10,10)$, $P_4(6,6,10,10)$, $P_4(6,8,8,10)$ are all dealt with similarly; for these, [BR13, Theorem 3.1] shows that a series of middle convolution and tensoring by rank 1 sheaves reduces our given local system $r_{spin} \circ \tilde{\rho}_{\overline{F}}$ to a GO_2-local system. Any such sheaf $\mathcal{H}$ is the induction from an index two subgroup of a rank 1 local system $\mathcal{L}$, i.e. there exists a finite étale (degree 2) cover $\pi: U' \to U$ such that $\mathcal{H} \cong \pi_* \mathcal{L}$. This is of course geometrically concentrated, so applying (in reverse) the relevant middle convolutions and tensor products by Kummer sheaves we obtain a geometrically concentrated construction of $r_{spin} \circ \tilde{\rho}_{\overline{F}}$.
- In the remaining case, $P_5(8,8,8,10,10)$, [BR13, Theorem 3.1] yields an identity of the form

$$\mathcal{F}_1 \otimes \mathcal{F}_2 \cong f^* \left(\mathrm{MC}_{-1}(r_{spin} \circ \tilde{\rho}_{\overline{F}}) \right),$$

where $f: \mathbb{P}^1 \to \mathbb{P}^1$ is a degree 2 map, and the $\mathcal{F}_i$ are GL_2-rigid. Restricting f to an étale, necessarily Galois, cover $U' \to U$, and applying Frobenius reciprocity, we find that

$$f_*(\mathcal{F}_1 \otimes \mathcal{F}_2) \cong \mathrm{MC}_{-1}(r_{spin} \circ \tilde{\rho}_{\overline{F}}) \oplus \left(\mathrm{MC}_{-1}(r_{spin} \circ \tilde{\rho}_{\overline{F}}) \otimes \delta_{U'/U} \right),$$

where $\delta_{U'/U}$ denotes the non-trivial quadratic character of $\pi_1(U_{\overline{F}})/\pi_1(U'_{\overline{F}})$. We can then construct an idempotent $e \in \overline{\mathbb{Q}}_\ell[\mathrm{Aut}(U'/U)]$ such that

$$e\left(f_*(\mathcal{F}_1 \otimes \mathcal{F}_2)\right) \cong \mathrm{MC}_{-1}(r_{spin} \circ \tilde{\rho}_{\overline{F}}).$$

It follows that $r_{spin} \circ \tilde{\rho}_{\overline{F}}$ admits a description as a geometrically concentrated local system.

The explicit construction of $\mathcal{G}_{\overline{F}}$ gives us more than just a geometric local system, however. The discussion of §2 shows that for some cyclotomic extension K of F, depending only on the geometric local monodromies of ρ,[9] there exists an arithmetic descent $\mathcal{G}$ of $\mathcal{G}_{\overline{F}}$ to a lisse sheaf on U_K. We denote by $\tilde{\rho}$ the corresponding representation of $\pi_1(U_K, \bar{t})$. It takes values in the normalizer of the geometric monodromy group of $\mathcal{G}_{\overline{F}}$, which is necessarily contained in $\mathrm{GSp}(W_{spin,\bar{t}})$. Since we can extend π (technically, $\pi \circ r_{spin}^{-1}$) to a

[8] The reason anything more needs to be said at this point is that pullback is not, strictly speaking, a 'reversible operation': applying a series of middle convolutions, twists, and pull-backs to our given $r_{spin} \circ \tilde{\rho}_{\overline{F}}$ to produce something visibly geometric does not immediately allow us to express $r_{spin} \circ \tilde{\rho}_{\overline{F}}$ as a 'geometrically concentrated' (Definition 2.4) sheaf.

[9] If all eigenvalues of local monodromy of ρ are N^{th} roots of unity, we can take $F(\mu_{2N})$.

quotient map (killing the center) $\pi : \mathrm{GSp}(W_{spin,\bar{t}}) \to \mathrm{SO}(W_{spin,\bar{t}})$, we can now compare $\pi \circ \tilde{\rho}$ with ρ as $\pi_1(U_K,\bar{t})$-representations. The space

$$\mathrm{Hom}_{\pi_1(U_{\overline{F}},\bar{t})}(\pi \circ \tilde{\rho}, \rho)$$

is a one-dimensional $\overline{\mathbb{Q}}_\ell$-vector space with an action of $\pi_1(U_K,\bar{t})/\pi_1(U_{\overline{K}},\bar{t}) \cong \Gamma_K$. The fact that both $\pi \circ \tilde{\rho}$ and ρ land in SO_5 forces this action to be trivial.[10] Thus in fact $\pi \circ \tilde{\rho} = \rho$ as $\pi_1(U_K,\bar{t})$-representation (they are isomorphic, and we already knew they were equal on $\pi_1(U_{\overline{F}},\bar{t})$). In particular, pulling back along t: $\mathrm{Spec}\, K \to U_K$ (the restriction of the original $t \in U(F)$), we have a commutative diagram of Γ_K-representations

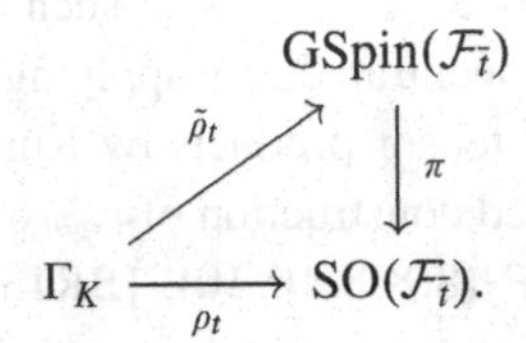

Crucially, by Proposition 2.5, $\tilde{\rho}_t$ (which by definition is $\rho_{\mathcal{G},t}$) is the ℓ-adic realization of a motivated motive. We have therefore produced a generalized Kuga-Satake lift of the original motive M, after some base-change: the ℓ-adic realization (now restricted to Γ_K) of M was assumed to be isomorphic to ρ_t. □

Remark 3.4. Let us record how inefficient the argument was in preserving the original field of definition F of the motive M, in the most important case of the theorem in which the geometric monodromy group of $\mathcal{F}_{\overline{F}}$ was irreducible. We were given a local system $\mathcal{F}$ with eigenvalues of local monodromy generating some cyclotomic field $\mathbb{Q}(\mu_N)$; after replacing $\mathcal{F}$ by a quadratic twist $\mathcal{F} \otimes \mathcal{L}$, we (maybe) had to enlarge F by a quadratic extension to make $\mathcal{F}$ a special orthogonal local system; the lifted arithmetic local system $\widetilde{\mathcal{F}}$ was defined over $F(\mu_{2N})$; and finally, taking into account the original twist $\mathcal{F} \mapsto \mathcal{F} \otimes \mathcal{L}$, we had to replace F by possibly one further quadratic extension. In sum, we have passed from F to $F'(\mu_{2N})$, where F' is some (at most) biquadratic extension, and N is determined by the local monodromies of the local system $\mathcal{F}$.

We will now make this procedure explicit for motivic lifting of fibers of the sextic Dwork family, establishing Theorem 1.5:

Example 3.5. Recall that we are considering motives of the form $M_t = \mathrm{Prim}^4(\mathbf{X}_t)^{\mathrm{A}}(2)$, where

$$\mathbf{X} \xrightarrow{f} \mathbb{P}^1 - \{\infty, \mu_6\}$$

[10] Letting χ denote the resulting character of Γ_K, χ^2 and χ^5 both equal 1, by comparing orthogonal multipliers and determinants.

is the smooth projective family with t-fiber

$$\mathbf{X}_t = \left\{ \sum_{i=0}^{5} X_i^6 = 6t \prod_{i=0}^{5} X_i \right\} \hookrightarrow \mathbb{P}^5,$$

acted on by the finite group scheme A given by the kernel of the multiplication map $\mu_6^6 \xrightarrow{mult} \mu_6$. M_t is a well-defined object of $\mathcal{M}_{\mathbb{Q}(t)}$. For any ℓ, the lisse $\mathbb{Q}_\ell$-sheaf of A-invariants in primitive cohomology,

$$\mathcal{F} = (\mathrm{R}^4 f_* \mathbb{Q}_\ell)^{\mathrm{A}}_{prim},$$

on $\mathbb{P}^1 - \{\infty, \mu_6\}$ over $\mathbb{Q}$ has $t^*\mathcal{F} \cong M_{t,\ell}$ as $\Gamma_{\mathbb{Q}(t)}$-representation. The geometric local system $\mathcal{F}_{\overline{\mathbb{Q}}}$ is not rigid, but it is closely related to a rigid local system on $\mathbb{P}^1 - \{0,1,\infty\}$. Namely, there exists a smooth projective family $\mathbf{X}' \xrightarrow{f'} \mathbb{P}^1 - \{0,1,\infty\}$ with A-action, and a A-equivariant pullback diagram

$$\begin{array}{ccc} \mathbf{X} & \longrightarrow & \mathbf{X}' \\ \downarrow f & & \downarrow f' \\ \mathbb{P}^1 - \{\infty, 0, \mu_6\} & \xrightarrow{[6]} & \mathbb{P}^1 - \{0,1,\infty\}, \end{array}$$

with [6] denoting the map $z \mapsto z^6$.[11] Now the lisse $\overline{\mathbb{Q}}_\ell$-sheaf $\mathcal{F}' = (\mathrm{R}^4 f'_* \overline{\mathbb{Q}}_\ell)^{\mathrm{A}}_{prim}$ satisfies $[6]^*\mathcal{F}' \cong \mathcal{F}$, hence setting $M'_t = \mathrm{Prim}^4(X'_t)^{\mathrm{A}}(2)$, we have

$$M_{t,\ell} \cong M'_{t^6,\ell},$$

so for our purposes we can work with the motives M'_t instead. Now, $\mathcal{F}'_{\overline{\mathbb{Q}}}$ is a GL_5-cohomologically (hence physically) rigid local system satisfying an orthogonal autoduality (Poincaré duality). Its local monodromies are regular unipotent (at ∞), a reflection (at 1), and regular semi-simple (at 0) with eigenvalues $\{\zeta_6^i\}_{i=1,\dots,5}$. As in the proof of Theorem 3.2, we twist $\mathcal{F}'$ by the product of translated Kummer sheaves on $\mathbb{P}^1 - \{0,1,\infty\}$ over $\mathbb{Q}$ having geometric local monodromies -1 at 0 and 1. The result is an SO_5-local system (at least geometrically) whose lift $\widetilde{\mathcal{F}}$ to Sp_4 has local monodromies with Jordan forms

$$U(4) \qquad -1,-1,1,1 \qquad \eta, \eta^3, \eta^9, \eta^{11},$$

where $U(4)$ indicates a 4-by-4 unipotent Jordan block and $\eta^2 = \zeta_6$. Note that the sum of the centralizer dimensions is $4+8+4 = 16$, so this local system

[11] There are various ways of producing such an $\mathbf{X}'$, for instance taking $\mathbf{X}'$ to be the family, with parameter s,

$$s^{-1}Y_0^6 + \sum_{i=1}^{5} Y_i^6 = 6\prod_{i=0}^{5} Y_i.$$

This is observed by Katz in [Kat09].

is not GL_4-cohomologically rigid. The Bogner-Reiter prescription in this case allows us to produce a rank 4 local system with these monodromies as follows. We first compute $MC_{-1}(\widetilde{\mathcal{F}})$, where -1 denotes the non-trivial character $\mu_2 \to \overline{\mathbb{Q}}_\ell^\times$. This has local monodromies, in the above notation,

$$-U(5), -1 \qquad -\eta, -\eta^3, -\eta^9, -\eta^{11}, 1, 1 \qquad U(2), U(2), 1, 1.$$

After twisting (this is not essential) by the rank 1 sheaf with local monodromies -1 at 0 and ∞, we note that this SO_6 local system lifts (via $\wedge^2$) to an SL_4-local system with local monodromies

$$U(4) \qquad \eta^2, \eta^4, \eta^7, \eta^{11} \qquad U(2), 1, 1.$$

The sum of the centralizer dimensions here is $4+4+10 = 18$, so this rank 4 local system is cohomologically rigid, and therefore can be constructed (geometrically) by Katz's algorithm. In summary, we can run this procedure in reverse to construct, via the allowable geometric operations, a rank 4 symplectic local system $\mathcal{G}$ isomorphic to $\widetilde{\mathcal{F}}$. Thus we have made explicit, in the case of the sextic Dwork family, the procedure of Theorem 3.2; for a precise geometric description of these Kuga-Satake lifts, we need only combine Proposition 2.5 with [Kat96, Theorem 8.4.1], where the variety denoted Hyp in Theorem 2.3 is described as an explicit affine hypersurface.

Let us also remark, without lingering over any details, that we similarly (but more easily) obtain a Kuga-Satake lifting result for motives whose ℓ-adic realizations arise as fibers of certain SO_6-local systems:

Theorem 3.6. *Let F and E be number fields, and fix an embedding $\iota \colon E \hookrightarrow \overline{\mathbb{Q}}_\ell$. Suppose M is a motivated motive in $\mathcal{M}_{F,E}$ whose (ι,ℓ)-adic realization is orthogonal of rank 6, and is isomorphic to a Γ_F-representation of the following form: there exists*

- *a finite set of points $S \subset \mathbb{P}^1(F)$, with complement $U = \mathbb{P}^1 - S$;*
- *a local system $\mathcal{F}$ on U over F such that $\mathcal{F}_{\overline{F}}$ has quasi-unipotent local monodromies and is SO_6-cohomologically rigid (for example, GL_6-cohomologically rigid) and irreducible;*
- *and a point $t \colon \operatorname{Spec} F \to U$;*
- *such that $M_{\iota,\ell}$ is isomorphic to $t^*\mathcal{F}$ as $\overline{\mathbb{Q}}_\ell[\Gamma_F]$-module.*

Then (enlarging F if necessary by a quadratic extension so that $M_{\iota,\ell}$ is special orthogonal) there exists a geometric lift

$$\begin{array}{ccc} & & \mathrm{GSpin}_6(M_{\iota,\ell}) \\ & \overset{\tilde{r}}{\nearrow} & \downarrow \\ \Gamma_F & \longrightarrow & SO_6(M_{\iota,\ell}) \end{array}$$

such that for some finite extensions F'/F and E'/E of number fields, $r_{spin} \circ \tilde{r}|_{\Gamma_{F'}}$ is isomorphic to the $E' \xrightarrow{\iota'} \overline{\mathbb{Q}}_\ell$-realization of an object of $\mathcal{M}_{F',E'}$, for a suitable embedding ι' extending ι. (Recall that r_{spin} is now the sum of half-spin representations.)

Proof. (Sketch) Let $\rho\colon \pi_1(U,\bar{t}) \to O_6(\overline{\mathbb{Q}}_\ell)$ denote the monodromy representation of $\mathcal{F}$, with restriction $\rho_{\overline{F}}$ to $\pi_1(U_{\overline{F}},\bar{t})$. The relevant low-dimensional coincidence is $\mathrm{Spin}_6 \cong \mathrm{SL}_4$, so we begin by choosing a lift

$$\begin{array}{ccc} & & \mathrm{SL}_4(\overline{\mathbb{Q}}_\ell) \\ & \overset{\tilde{\rho}_{\overline{F}}}{\nearrow} & \downarrow \pi \\ \pi_1(U_{\overline{F}},\bar{t}) & \xrightarrow[\rho_{\overline{F}}]{} & \mathrm{SO}_6(\overline{\mathbb{Q}}_\ell). \end{array}$$

Since $\rho_{\overline{F}}$ is SO_6-cohomologically rigid, $\tilde{\rho}_{\overline{F}}$ is necessarily (by equation 3) SL_4-cohomologically rigid, hence GL_4-cohomologically rigid. Theorem 2.3 yields a geometric construction of $\tilde{\rho}$ via iterated middle convolution and tensoring with translated Kummer sheaves. This also yields an arithmetic descent $\tilde{\rho}$ of $\tilde{\rho}_{\overline{F}}$, and by irreducibility of $\rho_{\overline{F}}$, we can make the same argument as in Theorem 3.2 to show that $\pi \circ \tilde{\rho} = \rho$ as $\pi_1(U_{F'},\bar{t})$-representation, for some finite extension F'/F. We conclude the proof as in Theorem 3.2. □

We conclude with some examples indicating the difficulty of generalizing the arguments of this note:

Example 3.7. With the Dwork family at hand, it makes sense to ask, for any even n, whether we can understand Kuga-Satake lifts of the orthogonal motives

$$\mathrm{Prim}^{n-2}(X_t)^{\mathrm{A}}\left(\frac{n-2}{2}\right)$$

arising from the family of varieties $\mathbf{X} \xrightarrow{f} \mathbb{P}^1 - \{\infty, \mu_n\}$ with t-fiber

$$X_0^n + X_1^n + \cdots + X_{n-1}^n = ntX_0 \cdots X_{n-1},$$

carrying as before the natural action of $\mathrm{A} = \ker(\mu_n^n \xrightarrow{mult} \mu_n)$. As for $n = 6$, there is a GL_{n-1}-cohomologically rigid local system, on $\mathbb{P}^1 - \{0,1,\infty\}$ whose pullback under $[n]$ is isomorphic to the lisse sheaf $(\mathrm{R}^{n-2}f_*\overline{\mathbb{Q}}_\ell)^{\mathrm{A}}_{prim}$. As before, we can twist to local system $\rho\colon \pi_1(\mathbb{P}^1 - \{0,1,\infty\}) \to \mathrm{SO}_{n-1}(\overline{\mathbb{Q}}_\ell)$, and ask: after lifting ρ to $\tilde{\rho}\colon \pi_1(\mathbb{P}^1 - \{0,1,\infty\}) \to \mathrm{Spin}_{n-1}(\overline{\mathbb{Q}}_\ell)$, can we find a 'geometric' construction of a local system with local monodromies equal to those of $r_{spin} \circ \tilde{\rho}$? I do not know how to do this for $n \geq 8$. An indicator of the difficulty, even for $n = 8$, using the methods of this paper might be the following: the spin representation factors $r_{spin}\colon \mathrm{Spin}_7 \to \mathrm{SO}_8 \subset \mathrm{GL}_8$. The argument for $n =$

6 crucially used the fact that composition with $r_{spin}\colon \mathrm{Spin}_5 \to \mathrm{Sp}_4$ yields a symplectic local system $r_{spin} \circ \tilde{\rho}$ that is (tautologically) Sp_4-cohomologically rigid; so we might ask whether for $n = 8$ $r_{spin} \circ \tilde{\rho}$ is SO_8-cohomologically rigid? (*A priori* it is only Spin_7-rigid.) The answer is no: the Jordan forms of the local monodromies are

$$U(7), 1 \qquad \iota^{\oplus 4}, -\iota^{\oplus 4} \qquad \zeta_8^{\pm 1}, \zeta_8^{\pm 2}, \zeta_8^{\pm 3}, 1, 1,$$

for which the sum of the centralizer dimensions (in SO_8) is $4 + 16 + 4 = 24 < 28 = \dim \mathfrak{so}_8$. Thus we will need in some way to come to terms with Spin_7-rigidity.

Is there any hope at all for tackling such problems? At least in some limited cases, I believe the answer is yes. I will close with one more SO_7 example, which is SO_7-cohomologically but not GL_7-cohomologically rigid, and which admits a geometric construction. I expect as a result of work in progress to be able to show that the Spin_7-lifts of this local system are (in the spin representation) geometric; indeed for every n there will be an analogous example for $\mathrm{Spin}_{2n+1} \to \mathrm{SO}_{2n+1}$ lifts, so this should eventually provide examples of generalized Kuga-Satake theory in arbitrarily large rank.[12]

Example 3.8. There exists an SO_7-cohomologically rigid local system on $U = \mathbb{P}^1 - \{0, 1, \infty\}$ with local monodromies (at, in order, $\infty, 0, 1$):

$$U(7) \qquad U(3), U(2), U(2) \qquad 1, -U(2), -U(2), -1, -1.$$

Here is a construction. For a pair of characters $\alpha, \beta\colon \mu_N(R_{N,\ell}) \to \overline{\mathbb{Q}}_\ell^\times$, with associated Kummer sheaves $\mathcal{L}_\alpha$ and $\mathcal{L}_\beta$, let us introduce the short-hand $\mathcal{L}(\alpha, \beta)$ for the local system on U given by $\mathcal{L}_\alpha \otimes (x-1)^* \mathcal{L}_\beta$ (product of translated Kummer sheaves, so that α is placed at 0 and β is placed at 1). Let π denote the quotient map $\pi : \mathrm{Sp}_4 \to \mathrm{SO}_5$ (for an implicit choice of pairing that will arise below). Then the sheaf

$$\mathcal{L}(1,-1) \otimes \mathrm{MC}_{-1}\left(\mathcal{L}(-1,1) \otimes \mathrm{MC}_{-1}\left(\pi\left(\mathcal{L}(-1,1) \otimes \mathrm{MC}_{-1}(\mathrm{Sym}^2 \mathcal{G})\right)\right)\right)$$

has the desired local monodromies, where $\mathcal{G}$ is the GL_2-rigid sheaf (a well-known classical example) having local monodromies

$$U(2) \qquad -U(2) \qquad U(2).$$

Here too the Spin_7-lifts are Spin_7-rigid but not SO_8-rigid, reflecting the increased difficulty of showing they are motivic. Note too that even were we to find a Spin_7-local system $\tilde{\rho}$ whose local monodromies in the 8-dimensional spin representation were the desired ones, the argument as in Theorem 3.2 would not quite work: the original SO_7-local system is not *a priori* physically

[12] For examples of a similar flavor, see [Pat14a].

rigid, which means that the comparison of local monodromies does not suffice to test global isomorphism of this original local system and the SO_7-reduction of $\tilde{\rho}$. I expect that this difficulty too can be handled.

References

[And96] Yves André, *Pour une théorie inconditionnelle des motifs*, Inst. Hautes Études Sci. Publ. Math. (1996), no. 83, 5–49. MR 1423019 (98m:14022)

[BC] O. Brinon and B. Conrad, *Cmi summer school notes on p-adic hodge theory: preliminary version.*

[Bei12] A. Beilinson, *p-adic periods and derived de Rham cohomology*, J. Amer. Math. Soc. **25** (2012), no. 3, 715–738. MR 2904571

[Ber02] Laurent Berger, *Représentations p-adiques et équations différentielles*, Invent. Math. **148** (2002), no. 2, 219–284. MR 1906150 (2004a:14022)

[BR13] Michael Bogner and Stefan Reiter, *On symplectically rigid local systems of rank four and Calabi-Yau operators*, J. Symbolic Comput. **48** (2013), 64–100. MR 2980467

[Con11] Brian Conrad, *Lifting global representations with local properties*, Preprint available at http://math.stanford.edu/~conrad/ (2011).

[Del72] Pierre Deligne, *La conjecture de Weil pour les surfaces K3*, Invent. Math. **15** (1972), 206–226. MR 0296076 (45 #5137)

[Del77] P. Deligne, *Cohomologie étale*, Lecture Notes in Mathematics, Vol. 569, Springer-Verlag, Berlin-New York, 1977, Séminaire de Géométrie Algébrique du Bois-Marie SGA 41øer2, Avec la collaboration de J. F. Boutot, A. Grothendieck, L. Illusie et J. L. Verdier. MR 0463174 (57 #3132)

[Del80] Pierre Deligne, *La conjecture de Weil. II*, Inst. Hautes Études Sci. Publ. Math. (1980), no. 52, 137–252. MR 601520 (83c:14017)

[Det08] M. Dettweiler, *On the middle convolution of local systems*, preprint, available as arXiv:0810.3334v1 (2008).

[Fal89] Gerd Faltings, *Crystalline cohomology and p-adic Galois-representations*, Algebraic analysis, geometry, and number theory (Baltimore, MD, 1988), Johns Hopkins Univ. Press, Baltimore, MD, 1989, pp. 25–80. MR MR1463696 (98k:14025)

[FM95] Jean-Marc Fontaine and Barry Mazur, *Geometric Galois representations*, Elliptic curves, modular forms, & Fermat's last theorem (Hong Kong, 1993), Ser. Number Theory, I, Int. Press, Cambridge, MA, 1995, pp. 41–78. MR 1363495 (96h:11049)

[Gal00] Federica Galluzzi, *Abelian fourfold of Mumford-type and Kuga-Satake varieties*, Indag. Math. (N.S.) **11** (2000), no. 4, 547–560. MR 1909819 (2003g:14061)

[Kat96] Nicholas M. Katz, *Rigid local systems*, Annals of Mathematics Studies, vol. 139, Princeton University Press, Princeton, NJ, 1996. MR 1366651 (97e:14027)

[Kat09] ———, *Another look at the Dwork family*, Algebra, arithmetic, and geometry: in honor of Yu. I. Manin. Vol. II, Progr. Math., vol. 270, Birkhäuser Boston Inc., Boston, MA, 2009, pp. 89–126. MR 2641188 (2011f:14016)

[Moo] Ben Moonen, *personal communication, paper in preparation.*

[Mor85] David R. Morrison, *The Kuga-Satake variety of an abelian surface*, J. Algebra **92** (1985), no. 2, 454–476. MR 778462 (86j:14041)

[Pat12] S. Patrikis, *Variations on a theorem of Tate*, ArXiv e-prints 1207.6724v4 (2012).

[Pat14a] ______, *Generalized Kuga-Satake theory and rigid local systems, II: rigid Hecke eigensheaves*, ArXiv e-prints 1407.1941 (2014).

[Pat14b] Stefan Patrikis, *On the sign of regular algebraic polarizable automorphic representations*, Mathematische Annalen (2014), 1–25 (English).

[PT15] Stefan Patrikis and Richard Taylor, *Automorphy and irreducibility of some l-adic representations*, Compositio Mathematica **FirstView** (2015), 1–23.

[Ser77] J.-P. Serre, *Modular forms of weight one and Galois representations*, Algebraic number fields: *L*-functions and Galois properties (Proc. Sympos., Univ. Durham, Durham, 1975), Academic Press, London, 1977, pp. 193–268. MR 0450201 (56 #8497)

[Ser94] Jean-Pierre Serre, *Propriétés conjecturales des groupes de Galois motiviques et des représentations l-adiques*, Motives (Seattle, WA, 1991), Proc. Sympos. Pure Math., vol. 55, Amer. Math. Soc., Providence, RI, 1994, pp. 377–400. MR 1265537 (95m:11059)

[Sim92] Carlos T. Simpson, *Higgs bundles and local systems*, Inst. Hautes Études Sci. Publ. Math. (1992), no. 75, 5–95. MR 1179076 (94d:32027)

[Voi05] Claire Voisin, *A generalization of the Kuga-Satake construction*, Pure Appl. Math. Q. **1** (2005), no. 3, part 2, 415–439. MR 2201323 (2006k:14012)

[Yun13] Zhiwei Yun, *Rigidity in the Langlands correspondence and applications*, preprint, to appear in ICCM 2013 proceedings (2013).

Department of Mathematics MIT Cambridge, MA 02139

E-mail address: `patrikis@math.mit.edu`

15

On the fundamental periods of a motive

Hiroyuki Yoshida

ABSTRACT. For a motive over $\mathbf{Q}$, the author introduced the fundamental periods which contain Deligne's periods as a special case. We will generalize the fundamental periods to a motive over an algebraic number field. We will discuss a few applications.

Introduction

Let M be a motive over $\mathbf{Q}$ with coefficients in an algebraic number field. A well-known conjecture of Deligne ([D]) describes the critical values of the L-function $L(M,s)$ of M in terms of the periods $c^{\pm}(M)$. We are interested in investigating the change of Deligne's periods $c^{\pm}(M)$ under various algebraic operations, for example the tensor product of two motives. For this purpose, the author introduced the fundamental periods of a motive M ([Y2]) and showed that the concept can be applied efficiently to study the periods of Siegel modular forms.

The purpose of this paper is twofold. First we will give a few more illustrative applications. These are: (i) To express Harris' quadratic periods ([HA]) by the fundamental periods. (ii) A simple derivation of a formula of Blasius ([B1]) for periods of a tensor product of rank 2 motives. (iii) On periods of a Siegel modular form. Second we will generalize the concept of the fundamental periods to motives over a number field F. This is necessary, for example, to describe the change of periods for the tensor product of motives over F.

Now we will explain the organization of this paper. In section 1, we will review the fundamental periods briefly. In section 2, we will give applications to Blasius' formula and to quadratic periods. In section 3, we will give an application of the fundamental periods to Siegel modular forms. In section 4, we will generalize the fundamental periods to a motive over a number field.

The topic of this paper is different from the author's talk in Vancouver. I wrote down the contents of my talk at the conference on "p-adic modular forms and arithmetic" in UCLA, June 2012, which was not published before.

Notation. An algebraic number field means an algebraic extension of $\mathbf{Q}$ of finite degree contained in $\mathbf{C}$. For an algebraic number field E, J_E denotes the finite set of all isomorphisms of E into $\mathbf{C}$. For a commutative ring A, $M(m,n,A)$ denotes the set of all $m \times n$-matrices with entries in A. $M(m,n)$ stands for the mn-dimensional affine space whose set of A-valued points is $M(m,n,A)$. For a set S, $|S|$ denotes the cardinality of S.

§1 Fundamental periods of a motive over Q

We first review the fundamental periods of a motive over $\mathbf{Q}$. Let M be a motive over $\mathbf{Q}$ with coefficients in E of pure weight w, where E is an algebraic number field.[1] Put $R = E \otimes_{\mathbf{Q}} \mathbf{C}$. It is important to know the change of Deligne's periods under various algebraic operations for motives; for example $M \otimes N$, the tensor product of two motives. The fundamental periods work for this purpose quite efficiently.

Let $H_{\mathrm{DR}}(M)$ be the de Rham realization of M; it is a finite dimensional vector space over E. Let d be its dimension. We call $d = d(M)$ the *rank* of M. There exists a spectral sequence

$$E_1^{p,q} = H^q(M, \Omega^p) \Longrightarrow H_{\mathrm{DR}}^{p+q}(M).$$

The Hodge filtration $\{F^p\}$ is the filtration on $H_{\mathrm{DR}}^{p+q}(M)$ associated to this spectral sequence. Since $H_{\mathrm{DR}}^{w}(M) = H_{\mathrm{DR}}(M)$, we have

$$(1.1) \qquad F^p(H_{\mathrm{DR}}(M))/F^{p+1}(H_{\mathrm{DR}}(M)) \cong E_{\infty}^{p,w-p}, \qquad p \in \mathbf{Z}.$$

Since this spectral sequence degenerates at the E_1-term, we have

$$(1.2) \quad F^p(H_{\mathrm{DR}}(M))/F^{p+1}(H_{\mathrm{DR}}(M)) \cong E_1^{p,w-p} = H^{w-p}(M, \Omega^p), \qquad p \in \mathbf{Z}.$$

We write $F^p(H_{\mathrm{DR}}(M))$ as $F^p(M)$ or simply as F^p.

Let $H_B(M)$ be the Betti realization of M; it is a d-dimensional vector space over E. Complex conjugation F_∞ acts on $H_B(M)$ and we have

$$H_B(M) = H_B^+(M) \oplus H_B^-(M),$$

where $H_B^{\epsilon}(M) = \{v \in H_B(M) \mid F_\infty v = \epsilon v\}$, $\epsilon = \pm$. $H_B(M)$ has the Hodge decomposition

$$H_B(M) \otimes_{\mathbf{Q}} \mathbf{C} = \oplus_{p+q=w} H^{pq}(M), \qquad \overline{H^{pq}(M)} = H^{qp}(M).$$

Let

$$I : H_B(M) \otimes_{\mathbf{Q}} \mathbf{C} \cong H_{\mathrm{DR}}(M) \otimes_{\mathbf{Q}} \mathbf{C}$$

[1] By a motive, I mean a motive for absolute Hodge cycles ([D], [DM], §6). In this paper, we will be concerned with problems on linear algebra and invariant theory related to the Hodge filtration. So I think that many results will apply to other categories of motives as well.

be the comparison isomorphism, which is an isomorphism of R-modules. We have

$$(1.3) \qquad I(\oplus_{p'\geq p} H^{p'q}(M)) = F^p(M) \otimes_{\mathbf{Q}} \mathbf{C}.$$

Now we are going to define a period matrix of M. Let $\{v_1^+, v_2^+, \ldots, v_{d^+}^+\}$ be a basis of $H_B^+(M)$ over E and $\{v_1^-, v_2^-, \ldots, v_{d^-}^-\}$ be a basis of $H_B^-(M)$ over E. We write the Hodge filtration as

$$H_{\mathrm{DR}}(M) = F^{i_1} \supsetneqq F^{i_2} \supsetneqq \cdots \supsetneqq F^{i_m} \supsetneqq F^{i_{m+1}} = \{0\}$$

so that there are no different filtrations between successive members. (The choice of numbers i_μ may not be unique for F^{i_μ}. For the sake of simplicity, we assume that i_μ is chosen, for $1 \leq \mu \leq m$, so that it is the maximum number.) Put

$$s_\mu = \mathrm{rank}(H^{i_\mu, w-i_\mu}(M)), \qquad 1 \leq \mu \leq m$$

where rank means the rank as a free R-module. Then we have

$$i_c + i_{m+1-c} = w, \quad 1 \leq c \leq m, \qquad s_\mu = s_{m+1-\mu}, \quad 1 \leq \mu \leq m,$$

and a partion of d:

$$(1.4) \qquad d = s_1 + s_2 + \cdots + s_m, \qquad s_\mu > 0, \quad 1 \leq \mu \leq m.$$

We have

$$s_\mu = \dim_E F^{i_\mu} - \dim_E F^{i_{\mu+1}},$$

$$\dim_E F^{i_\mu} = s_\mu + s_{\mu+1} + \cdots + s_m, \qquad 1 \leq \mu \leq m.$$

We take a basis $\{w_1, w_2, \ldots, w_d\}$ of $H_{\mathrm{DR}}(M)$ over E so that $\{w_{s_1+s_2+\ldots+s_{\mu-1}+1}, \ldots, w_d\}$ is a basis of F^{i_μ} for $1 \leq \mu \leq m$. Writing

$$I(v_j^\pm) = \sum_{i=1}^{d} x_{ij}^\pm w_i, \qquad x_{ij}^\pm \in R, \quad 1 \leq j \leq d^\pm,$$

we obtain matrices $X^\pm = (x_{ij}^\pm) \in M(d, d^\pm, R)$.

Let P_M be the lower parabolic subgroup of $GL(d)$ which corresponds to the partition (1.4) of d. Then the class of $X^\pm$ in

$$P_M(E) \backslash M(d, d^\pm, R) / GL(d^\pm, E)$$

does not depend on the choices of basis. We put $X = (X^+ X^-) \in M(d, d, R)$ and call it a *period matrix* of M. The class of X in

$$P_M(E) \backslash M(d, d, R) / (GL(d^+, E) \times GL(d^-, E))$$

is well defined. Here $GL(d^+, E) \times GL(d^-, E)$ is embedded in $GL(d)$ as diagonal blocks. Thus we are interested in a polynomial function on $M(d, d)$ rational over $\mathbf{Q}$, which satisfies

(1.5) $f(px\gamma) = \lambda_1(p)\lambda_2(\gamma)f(x)$ for all $p \in P_M$, $\gamma \in GL(d^+) \times GL(d^-)$.

Here λ_1 and λ_2 are characters of P_M and $GL(d^+) \times GL(d^-)$ respectively given by

$$\lambda_1\left(\begin{pmatrix} p_{11} & 0 & \cdots & 0 \\ * & p_{22} & \cdots & 0 \\ * & * & \ddots & \vdots \\ * & * & * & p_{mm} \end{pmatrix}\right) = \det(p_{11})^{a_1}\det(p_{22})^{a_2}\cdots\det(p_{mm})^{a_m},$$

where $p_{ii} \in GL(s_i)$, $a_i \in \mathbf{Z}$,

$$\lambda_2\left(\begin{pmatrix} a & 0 \\ 0 & b \end{pmatrix}\right) = (\det a)^{k^+}(\det b)^{k^-}, \qquad a \in GL(d^+),\ b \in GL(d^-).$$

Here $k^\pm \in \mathbf{Z}$. We call f to be *admissible* of the *type* (λ_1, λ_2) or of the *type* $\{(a_1, a_2, \ldots, a_m); (k^+, k^-)\}$. All admissible f generate a graded algebra over $\mathbf{Q}$. For the following Theorems 1.1 and 1.2, see [Y2], Theorems 1 and 3 respectively.

Theorem 1.1. *The graded algebra of all admissible f is isomorphic to a polynomial ring with explicitly given generators. Each graded component corresponding to (λ_1, λ_2) is at most one dimensional.*

Let us recall the definition of Deligne's periods ([D], subsection 1.7). If w is even, put $p = w/2$ and we assume that F_∞ acts on $H^{pp}(M)$ by a scalar. If $F_\infty = 1$ on $H^{pp}(M)$, we put

$$F^+(M) = F^p(H_{\mathrm{DR}}(M)), \qquad F^-(M) = F^{p+1}(H_{\mathrm{DR}}(M)),$$

and if $F_\infty = -1$ on $H^{pp}(M)$, we put

$$F^+(M) = F^{p+1}(H_{\mathrm{DR}}(M)), \qquad F^-(M) = F^p(H_{\mathrm{DR}}(M)).$$

If w is odd, put $w = 2p+1$ and let $F^\pm(M) = F^{p+1}(H_{\mathrm{DR}}(M))$. We set $H^\pm_{\mathrm{DR}}(M) = H_{\mathrm{DR}}(M)/F^\mp(M)$. Then, from the comparison isomorphism I, we have

$$I^\pm : H_B^\pm(M) \otimes_{\mathbf{Q}} \mathbf{C} \cong H^\pm_{\mathrm{DR}}(M) \otimes_{\mathbf{Q}} \mathbf{C}$$

Deligne's periods $c^\pm(M)$ are defined as the determinants of these maps when we take bases of $H_B^\pm(M)$ and of $H^\pm_{\mathrm{DR}}(M)$ as vector spaces over E. Since they are determinants of the isomorphisms of R-modules, we have $c^\pm(M) \in R^\times$. Similarly the period $\delta(M) \in R^\times$ is defined as the determinant of the isomorphism I.

Let $f(x) = \det(x)$, $x \in M(d,d)$. Then $f(x)$ is of the type $\{(1,1,\ldots,1);(1,1)\}$ and $f(X)$ is the period $\delta(M)$. Assume that $s_1 + s_2 + \cdots + s_{p^+} = d^+$ for some

p^+. Let $f^+(x)$ be the determinant of the upper left $d^+ \times d^+$-submatrix of $x \in M(d,d)$. Then $f^+(x)$ is of the type $\{(\overbrace{1,1,\ldots,1}^{p^+},0,\ldots,0);(1,0)\}$ and $f^+(X)$ is the period $c^+(M)$. Similarly if $s_1+s_2+\cdots+s_{p^-}=d^-$ for some p^-, let $f^-(x)$ be the determinant of the upper right $d^- \times d^-$-submatrix of x. Then $f^-(x)$ is of the type $\{(\overbrace{1,1,\ldots,1}^{p^-},0,\ldots,0);(0,1)\}$ and $f^-(X)$ is the period $c^-(M)$. Either one of the above conditions is equivalent to that $F^{\mp}(M)$, hence also $c^{\pm}(M)$, can be defined. Then we have $F^{\mp}(M)=F^{i_{p^{\pm}+1}}(M)$; $F^{\pm}(M)$ can be defined if M has a critical value.

Let $\mathcal{P}=\mathcal{P}(M)$ denote the set of integers p such that $s_1+s_2+\cdots+s_p<\min(d^+,d^-)$. Fix $p\in\mathcal{P}$ and put $q=m-p$. Then $p<q$ and $s_1+s_2+\cdots+s_q=d-(s_1+\cdots+s_p)$.

Theorem 1.2. *For every $p\in\mathcal{P}$, there exists a non-zero $\mathbf{Q}$-rational polynomial f_p of the type $\{(\overbrace{2,\ldots,2}^{p},\overbrace{1,\ldots,1}^{m-2p},\overbrace{0,\ldots,0}^{p});(1,1)\}$. Every admissible polynomial can be written uniquely as a monomial of $\det(x), f^+(x), f^-(x), f_p(x)$, $p\in\mathcal{P}$.*

We put

$$c_p(M)=f_p(X),\qquad p\in\mathcal{P}. \tag{1.6}$$

We call $\delta(M), c^{\pm}(M), c_p(M), p\in\mathcal{P}$ the *fundamental periods* of M. By Theorem 1.2, any period invariant of M can be written as a monomial of the fundamental periods.

The L-function $L(M,s)$ of a motive M takes values in $R=E\otimes_{\mathbf{Q}}\mathbf{C}\cong\prod_{\sigma\in J_E}\mathbf{C}$. We can naturally embed E into R by $E\ni x\mapsto(\sigma(x))_{\sigma\in J_E}\in\prod_{\sigma\in J_E}\mathbf{C}$. Deligne's conjecture states that $L(M,0)/c^+(M)\in E$ if 0 is critical for M. We may interpret Deligne's conjecture as saying that the mathematical nature has chosen the simplest fundamental period for the critical value. Other period invariants are hidden in the relation to $L(M,0)$ but will manifest themselves when we make various algebraic operations on M (tensor products with other motives, exterior powers, etc.). By the construction of Deligne, we have $c^{\pm}(M)\in R^{\times}$. We can prove that other fundamental periods are also invertible elements of R (cf. [Y2], Cor. 2 to Prop. 12). Hereafter we understand the equality between period invariants mod $E^{\times}$.

Let us explain a general principle to study the variation of fundamental periods of motives under algebraic operations. As an example, we take the case $M=M_1\otimes M_2\otimes\cdots\otimes M_n$. Here all motives are defined over $\mathbf{Q}$, and with coefficients in E. Let X_i be a period matrix of M_i for $1\le i\le n$ and X be a period matrix of M. We see that

(i) Every entry of X can be written as a polynomial of entries of X_i with coefficients in $\mathbf{Q}$.

(ii) If we replace X_i by a matrix in $P_{M_i}(E)X_i(GL(d^+(M_i),E)\times GL(d^-(M_i),E))$, then X is replaced by a matrix in $P_M(E)X(GL(d^+(M),E)\times GL(d^-(M),E))$.

Now let $p\in\mathcal{P}(M)$ and let $c_p(M)=f_p(X)$ be a fundamental period of M. Here f_p is an admissible polynomial on $M(d(M),d(M))$. We put $d_i=d(M_i)$, $1\le i\le n$. By (i), we have $c_p(M)=f_p(g(X_1,\ldots,X_n))$ with a polynomial g. We are going to prove the equality of the form

$$f_p(g(X_1,\ldots,X_n))=ch_1(X_1)\cdots h_n(X_n),\qquad c\in\mathbf{Q}^\times \tag{1.7}$$

where h_i is a polynomial on $M(d_i,d_i)$ of some admissible type. To this end, we may assume that X_i is a variable matrix on $M(d_i,d_i)$. (If we establish (1.7) when X_i are variables, then a substitution will give the desired result. The reader may imagine a standard calculation of the Vandermonde determinant.) We see that (i) and (ii) hold even when X_i is a variable matrix. By (ii), we see that $f_p(g(X_1,\ldots,X_n))$ is, as a polynomial of X_i, a polynomial h_i of some admissible type. By the one dimensionality (Theorem 1.1), we conclude that $c_p(M)=ch_1(X_1)\cdots h_n(X_n)$ with $c\in\mathbf{Q}^\times$. (Thus we can determine h_i by just looking at the type of $f_p(g(X_1,\ldots,X_n))$ as a polynomial of X_i.) Here we have tacitly used the next Lemma, whose proof may be left to the reader.

Lemma 1.3. *Let X_i be a variable on $M(d_i,d_i)$, $1\le i\le n$ and let $U(X_1,X_2,\ldots,X_n)$ be a $\mathbf{Q}$-rational polynomial of X_1, …, X_n. Suppose that there exists a $\mathbf{Q}$-rational polynomial $u(X_1)$ of X_1 such that $U(X_1,Y_2,\ldots,Y_n)$ is a rational multiple of $u(X_1)$ for every substitution $X_2\mapsto Y_2\in M(d_2,d_2,\mathbf{Q})$, …, $X_n\mapsto Y_n\in M(d_n,d_n,\mathbf{Q})$. Then there exists a $\mathbf{Q}$-rational polynomial $V(X_2,\ldots,X_n)$ of X_2, …, X_n such that $U(X_1,X_2,\ldots,X_n)=u(X_1)V(X_2,\ldots,X_n)$.*

The same argument can be used also for $c^\pm(M)$, $\delta(M)$. Therefore every fundamental period of M is a monomial of fundamental periods of M_i, $1\le i\le n$.

Reflecting on the above argument, we may say that Theorem 1.1 could be foreseen without any calculation. In fact, if it were false, the calculation of the periods of $M_1\otimes M_2\otimes\cdots\otimes M_n$ would be a chaos.

§2 Applications

As an example, we first reproduce Blasius' famous formula for periods of a tensor product of motives [B1]. Suppose that there are given motives M_i, $1\le i\le n$ of rank 2 over $\mathbf{Q}$ with coefficients in E. We write the Hodge decomposition as

$$H_B(M_i)\otimes\mathbf{C}=H^{a_i(+),a_i(-)}\oplus H^{a_i(-),a_i(+)}$$

for $1 \le i \le n$. We assume that $a_i(+) > a_i(-)$, $1 \le i \le n$. Then M_i has three fundamental periods: $c^+(M_i)$ which is of type $\{(1,0);(1,0)\}$; $c^-(M_i)$ which is of type $\{(1,0);(0,1)\}$; $\delta(M_i)$ which is of type $\{(1,1);(1,1)\}$.

Let $M = M_1 \otimes M_2 \otimes \cdots \otimes M_n$. Let Λ be the set of all mappings from $\{1,2,\ldots,n\}$ to $\{\pm 1\}$ and put $\Lambda_{\pm} = \{\lambda \in \Lambda \mid \prod_{i=1}^n \lambda(i) = \pm 1\}$. We assume

$$(2.1) \qquad \sum_{i=1}^n a_i(\lambda(i)) \neq \sum_{i=1}^n a_i(-\lambda(i)), \qquad \forall \lambda \in \Lambda$$

which is a necessary condition for $c^{\pm}(M)$ to be defined. We put

$$\Lambda^+ = \left\{ \lambda \in \Lambda \,\middle|\, \sum_{i=1}^n a_i(\lambda(i)) > \sum_{i=1}^n a_i(-\lambda(i)) \right\}.$$

By (2.1), we have $|\Lambda^+| = 2^{n-1}$. We set

$$n_i = |\{\lambda \in \Lambda^+ \mid \lambda(i) = 1\}|, \qquad m_i = |\{\lambda \in \Lambda^+ \mid \lambda(i) = -1\}|.$$

Then we have

$$n_i + m_i = |\Lambda^+| = 2^{n-1},$$
$$H_B^+(M) = \oplus_{\lambda \in \Lambda_+} (\otimes_{i=1}^n H_B^{\lambda(i)}(M_i)).$$

Let

$$X_i^{\pm} = \begin{pmatrix} x_{i,1}^{\pm} \\ x_{i,2}^{\pm} \end{pmatrix}, \qquad X_i = (X_i^+ X_i^-)$$

be the period matrix of M_i. Put

$$x_{i,1}^{\pm} = x_{i,+}^{\pm}, \qquad x_{i,2}^{\pm} = x_{i,-}^{\pm}.$$

Then we have

$$c^+(M) = \det\Bigl(\prod_{i=1}^n x_{i,\mu(i)}^{\lambda(i)}\Bigr), \qquad \lambda \in \Lambda_+,\ \mu \in \Lambda^+.$$

(The determinant of a $2^{n-1} \times 2^{n-1}$-matrix.) As explained in the end of the last section, we have

$$c^+(M) = c \prod_{i=1}^n h_i(X_i), \qquad c \in \mathbf{Q}^{\times},$$

where h_i is an admissible polynomial on $M(2,2)$. Let $\{(a,b);(\alpha,\beta)\}$ be the type of h_i. To determine it, we examine the change of $c^+(M)$ for

$$X_i = \begin{pmatrix} x_{i,+}^+ & x_{i,+}^- \\ x_{i,-}^+ & x_{i,-}^- \end{pmatrix} \mapsto \begin{pmatrix} t & 0 \\ 0 & 1 \end{pmatrix} \begin{pmatrix} x_{i,+}^+ & x_{i,+}^- \\ x_{i,-}^+ & x_{i,-}^- \end{pmatrix}.$$

Then we see $a = n_i$. Similarly we have $b = m_i$, $\alpha = \beta = 2^{n-2}$. Therefore we obtain

$$(2.2) \qquad c^+(M) = \prod_{i=1}^{n} (c^+(M_i)c^-(M_i))^{(n_i-m_i)/2}\delta(M_i)^{m_i}$$

and similarly

$$c^-(M) = c^+(M).$$

As the next example, we will express Harris' quadratic periods in terms of fundamental periods.

Lemma 2.1. *Let f be a rational function on $M(n,n)$ of the type (λ_1,λ_2), i.e., it is a rational function which satisfies (1.5). Then there exist polynomial functions g and h on $M(n,n)$ of types (λ_1',λ_2'), $(\lambda_1'',\lambda_2'')$ respectively such that $f = g/h$, $\lambda_1 = \lambda_1'(\lambda_1'')^{-1}$, $\lambda_2 = \lambda_2'(\lambda_2'')^{-1}$.*

Lemma 2.2. *When (λ_1,λ_2) is given, a rational function on $M(n,n)$ of the type (λ_1,λ_2) is unique up to the constant multiple.*

Lemma 2.1 is easily proved by writing $f = g/h$ with relatively prime polynomials g and h. Lemma 2.2 is then obvious in view of Theorem 1.1.

Let M be a motive over $\mathbf{Q}$ with coefficients in E of pure weight w, rank d. We assume that M is *regular*, which means that the Hodge filtration can be written as

$$H_{\mathrm{DR}}(M) = F^{i_1} \supsetneqq F^{i_2} \supsetneqq \cdots \supsetneqq F^{i_d} \supsetneqq \{0\}, \qquad \dim_E F^{i_\mu} = d+1-\mu.$$

We also assume that E is totally real. Let $X = (X^+ X^-) \in M(d,d,R)$ be a period matrix of M. Put $Y = (y_{ij}) = X^{-1}$. Then we have

$$(2.3) \qquad I^{-1}(w_j) = \sum_{i=1}^{d^+} y_{ij}v_i^+ + \sum_{i=d^++1}^{d} y_{ij}v_{i-d^+}^-, \quad 1 \le j \le d,$$

$$(2.4) \qquad F_\infty I^{-1}(w_j) = \sum_{i=1}^{d^+} y_{ij}v_i^+ - \sum_{i=d^++1}^{d} y_{ij}v_{i-d^+}^-, \quad 1 \le j \le d.$$

Now assume that M is polarizable, that is, there exists a non-degenerate morphism of motives

$$\phi:\ M \otimes M \longrightarrow T.$$

Here T is a motive of rank 1, of weight $2w$ over $\mathbf{Q}$ with coefficients in E (cf. Deligne-Milne [DM], §4). The polarization ϕ gives an E-linear map

$$\phi_B : H_B(M) \otimes H_B(M) \longrightarrow H_B(T).$$

Since $H_B(T) \cong E$, ϕ_B corresponds to a non-degenerate E-bilinear form $(\ ,\ ):$ $H_B(M) \times H_B(M) \longrightarrow E$. It extends to an R-bilinear form $(H_B(M) \otimes_{\mathbf{Q}} \mathbf{C}) \times (H_B(M) \otimes_{\mathbf{Q}} \mathbf{C}) \longrightarrow R$, for which we use the same symbol $(\ ,\ )$. Let ω_j be the projection of $I^{-1}(w_j) \in H_B(M) \otimes \mathbf{C} = \oplus H^{pq}(M)$ to $H^{i_j, w-i_j}(M)$. Put

$$(2.5) \qquad Q_j = (\omega_j, F_\infty \omega_j), \qquad 1 \leq j \leq d.$$

Then Q_j is well defined modulo $(E^\times)^2$; Q_j is the *quadratic period* defined by Harris [HA]. (Actually Q_j of Harris is our Q_{d+1-j}.) In view of the pairing $H_{\mathrm{DR}}(M) \times H_{\mathrm{DR}}(M) \longrightarrow H_{\mathrm{DR}}(T) \cong E$ and the functoriality of I, we see that Q_j may be defined by

$$(2.6) \qquad F_\infty \omega_j = Q_j \omega_{d+1-j}, \qquad 1 \leq j \leq d,$$

when considered modulo $E^\times$. Note that $H^{i_j, w-i_j}(M)$ is a free R-module of rank 1.

Lemma 2.3. *There exist* **Q***-rational polynomials g_j and h_j on $M(d,d)$ such that $Q_j = h_j(X)/g_j(X)$, $g_j(X) \neq 0$, where X is a period matrix of M.*

We admit this lemma for a moment. To simplify the notation, we understand the equality between period invariants mod $E^\times$. Obviously Q_j does not depend on the choice of $\{v_i^\pm\}$. By (2.6), we immediately see that the type of Q_j is

$$\{0, \ldots, 0, -1, 0, \ldots, 0, 1, 0, \ldots, 0); (0,0)\}, \qquad 1 \leq j \leq d, \quad 2j \neq d+1.$$

Here -1 (resp. 1) is placed at the jth (resp. $d+1-j$th) place. In other words, $f = h_j/g_j$ satisfies (1.5) with λ_2 is trivial, $a_j = -1$, $a_{d+1-j} = 1$, other a_i's are zero. Hence the type of $(\prod_{j=1}^{[d/2]} Q_{d+1-j})\delta(M)$ is

$$\{(2,2,\ldots,2,0,0,\ldots,0);(1,1)\} \qquad \text{if } d \text{ is even},$$
$$\{(2,2,\ldots,2,1,0,0,\ldots,0);(1,1)\} \quad \text{if } d \text{ is odd}.$$

By Lemma 2.2, we obtain

$$c^+(M)c^-(M) = \delta(M) \prod_{j=1}^{[d/2]} Q_{d+1-j}.$$

Thus we essentially recover a result of Harris ([HA], Proposition 1.5.7). We note that

$$(2.7) \qquad Q_d \delta(M) = c_1(M),$$

$$(2.8) \qquad Q_{d-j} = c_j(M)/c_{j+1}(M), \qquad 1 \leq j < \min(d^+, d^-) - 1,$$

where $c_j(M)$ is the fundamental period (cf. (1.6)).

Proof of Lemma 2.3. Fix j. We can write (cf. (1.3))

$$\omega_j = I^{-1}(w_j) + \sum_{k=j+1}^{d} c_k I^{-1}(w_k), \qquad c_k \in R,$$

$$\omega_{d+1-j} = I^{-1}(w_{d+1-j}) + \sum_{l=d+2-j}^{d} d_l I^{-1}(w_l), \qquad d_l \in R.$$

By (2.6), we get

$$\begin{aligned} &F_\infty I^{-1}(w_j) + \sum_{k=j+1}^{d} c_k F_\infty I^{-1}(w_k) \\ &= Q_j\left(I^{-1}(w_{d+1-j}) + \sum_{l=d+2-j}^{d} d_l I^{-1}(w_l)\right). \end{aligned} \tag{2.9}$$

Since

$$F_\infty I^{-1}(F^{i_{j+1}} \otimes \mathbf{C}) \cap I^{-1}(F^{i_{d+1-j}} \otimes \mathbf{C}) = \{0\},$$

we see that d quantities Q_j, c_k, d_l are uniquely determined by (2.9). Using (2.3), (2.4), we see that (2.9) is equivalent to

$$\begin{aligned} &Q_j y_{id+1-j} + \sum_{l=d+2-j}^{d} Q_j d_l y_{il} - \sum_{k=j+1}^{d} c_k y_{ik} = y_{ij} \quad 1 \le i \le d^+, \\ &Q_j y_{id+1-j} + \sum_{l=d+2-j}^{d} Q_j d_l y_{il} + \sum_{k=j+1}^{d} c_k y_{ik} = -y_{ij} \quad d^+ + 1 \le i \le d. \end{aligned} \tag{2.10}$$

(2.10) is simultaneous d inhomogeneous linear equations in d unknowns Q_j, c_k, d_l. By the uniqueness of the solution of (2.10), we see that there exist $\mathbf{Q}$-rational polynomials p_j and q_j on $M(d,d)$ such that $Q_j = p_j(Y)/q_j(Y)$. Since $Y = X^{-1}$, the assertion of Lemma 2.3 follows.

Remark 2.4. We can define Q_j whenever $F^{i_j}/F^{i_{j+1}}$ is one dimensional, without assuming the regularity of M. Then Lemma 2.3 and the assertion on the type of Q_j remain true.

§3 Quadratic periods and Petersson norms of Siegel modular forms

We are going to relate the Petersson norm of a Siegel modular form to a quadratic period of a motive associated with it.

Let X be a projective smooth algebraic variety of dimension n defined over a field k of characteristic 0. We have the spectral sequence

$$E_1^{p,q} = H^q(X,\Omega^p) \Longrightarrow H_{\mathrm{DR}}^{p+q}(X),$$

which degenerates at the E_1-term. Hence we have

$$F^p(H_{\mathrm{DR}}^w(X))/F^{p+1}(H_{\mathrm{DR}}^w(X)) \cong E_1^{p,w-p} = H^{w-p}(X,\Omega^p). \tag{3.1}$$

The polarization on $H_B^w(X)$ can be given by the Hodge-Riemann form as follows. We assume that E is totally real and that $k \subset E$. Let L be an E-rational $(1,1)$-form which represents a polarization of X. Take $c \in \mathbf{C}$ so that $c\int_{X(\mathbf{C})} L^n = 1$. For $\omega \in H^{p,q}(X)$, $\omega' \in H^{p',q'}(X)$, $p+q = p'+q' = w$, we have

$$(\omega,\omega') = c\int_{X(\mathbf{C})} \omega \wedge \omega' \wedge L^{n-w}. \tag{3.2}$$

This choice of c is made to satisfy the demand on the form $(\ ,\ )$ on $H_B(M)$ stated in §2. The pairing is non-zero only for $p = q'$, $q = p'$. For a motive M which is a factor of $H^w(X)$, quadratic periods can be calculated as follows. Assume that $\dim F^p(M)/F^{p+1}(M) = 1$. Take $\alpha \in F^p(M)$, $\alpha \notin F^{p+1}(M)$ and take $\tilde{\alpha} \in H^{w-p}(X,\Omega^p)$ which corresponds to α by (3.1). Take the $(p,w-p)$ form ω on X corresponding to $\tilde{\alpha}$ by the isomorphism $\mathbf{C} \otimes H^{w-p}(X,\Omega^p) \cong H^{p,w-p}(X)$. Then the quadratic period which corresponds to $F^p(M)/F^{p+1}(M)$ (i.e., Q_j if $i_j = p$) can be given by $(\omega,\bar{\omega})$ (cf. (2.5)).

Let Γ be a congruence subgroup of $Sp(m,\mathbf{Z})$. Let $S_k(\Gamma)$ denote the space of Siegel modular cusp forms of degree m, of weight k with respect to Γ. Now let $f \in S_k(\Gamma)$. We denote the normalized Petersson norm of f by $\langle f,f \rangle$, i.e.,

$$\langle f,f \rangle = \mathrm{vol}(\Gamma\backslash\mathfrak{H}_m)^{-1} \int_{\Gamma\backslash\mathfrak{H}_m} |f(z)|^2 (\det y)^{k-m-1} dx\, dy,$$

where $\mathfrak{H}_m$ denotes the Siegel upper half space of degree m and $z = x+iy$ with real symmetric matrices x and y. We assume that f is a Hecke eigenform and that the Fourier coefficients of f belong to an algebraic number field E. We also assume that E is totally real. In this case, we can conjecture the existence of motives $M_{\mathrm{st}}(f)$ and $M_{\mathrm{sp}}(f)$, which are motives over $\mathbf{Q}$ with coefficients in E, associated with f as in [Y2], §5; only (5.2) and (5.10) have to be suitably modified. We note that

$$L(M_{\mathrm{st}}(f),s) = (L_{\mathrm{st}}(s,f^\sigma))_{\sigma\in J_E}, \qquad L(M_{\mathrm{sp}}(f),s) = (L_{\mathrm{sp}}(s,f^\sigma))_{\sigma\in J_E},$$

where $L_{\mathrm{st}}(s,f^\sigma)$ (resp. $L_{\mathrm{sp}}(s,f^\sigma)$) denotes the standard (resp. spinor) L-function attached to f^σ.

First assume $m = 2$, $k = 3$. Put $M = M_{\mathrm{sp}}(f)$. The Hodge filtrations is

$$H_{\mathrm{DR}}(M) = F^0 \supsetneqq F^1 \supsetneqq F^2 \supsetneqq F^3 \supsetneqq F^4 = \{0\}, \qquad \dim_E F^i = 4-i.$$

By [Y2], (3.3), we have

$$c^+(\wedge^2 M) = c_1(M).$$

Assume [Y2], (4.9) for $m = n = 2$ (as far as the author knows, this is an exceptional case for the critical values of $L_{\mathrm{st}}(s,f)$ which is not established yet), then Deligne's conjecture gives

$$c^+(\wedge^2 M) = (\langle f^\sigma, f^\sigma \rangle)_{\sigma \in J_E}.$$

By (2.7) and [Y2], (4.7), we obtain

$$Q_4 = c_1(M)\delta(M)^{-1} = \pi^6 (\langle f^\sigma, f^\sigma \rangle)_{\sigma \in J_E}. \tag{3.3}$$

This is what we expected[2], since f defines an element of $H^0(M,\Omega^3)$. (Note that $p = 3$, $i_4 = 3$.) For Q_3, we can expect that if a non-holomorphic automorphic form f', which is generic and in the same L-packet as f, gives an E-rational element of $H^1(M,\Omega^2)$, then we would have

$$Q_3 = \pi^6 (\langle f'^\sigma, f'^\sigma \rangle)_{\sigma \in J_E}.$$

This is related to a conjecture of Furusawa-Shalika [FS].[3]

Next let us consider general m. It is widely believed that the main part of the zeta function of the Shimura variety of Siegel modular type can be described by the spinor L-functions of (not necessarily holomorphic) Siegel modular forms.[4] Thus we conjecture that the quadratic period Q attached to the deepest filtration of $H_{\mathrm{DR}}(M_{\mathrm{sp}}(f))$ can be given by the normalized Petersson norm of f if $k > m$. (In view of (3.1), the deepest filtration corresponds to the holomorphic part. Also note that the motive $M_{\mathrm{sp}}(f)$ is not regular in general even for $k > 2m$. cf. Remark 2.4.) In fact, we can prove:

Proposition 3.1. *Assume that $\Gamma = Sp(m,\mathbf{Z})$, $k > 2m$ and that E is totally real. We also assume Tate's conjecture as in [Y2], Theorem 14. Put $Q = c_1(M_{sp}(f))\delta(M_{sp}(f))^{-1}$. Then we have*

$$Q^2 = \pi^{2mk} (\langle f^\sigma, f^\sigma \rangle^2)_{\sigma \in J_E}.$$

Proof. First assume $k > m$. By [Y2], Proposition 7, the type of $c_1(\mathrm{Sym}^2 M_{\mathrm{sp}}(f))$ is

$$\{(2^m+3, 2^m+1, \ldots, 2^m+1, 2^m-1); (2^m+1, 2^m+1)\}.$$

2 For the exponent of π, we note that $(2\pi i)^3 f$ is E-rational in $H^0(M,\Omega^3)$. In general, $(2\pi i)^{m+1} f$ is E-rational for $f \in S_{m+1}(\Gamma)$. This can be seen by considering the differentiation of E-rational automorphic functions.

3 Furusawa told the author that the quantity on the right-hand side is essentially the constant (period) in a conjecture of Böcherer, which is discussed in the introduction of [FS]. Apparently this speculation was never in print.

4 For this, see the discussion in the introduction of Langlands [L]. We see that "pois minuscule" in p. 1124 corresponds to the spinor representation of the L-group by a table of Bourbaki ([Bo], p. 198). For the conjectural form of the zeta functions of Shimura varieties, see also Kottwitz [K]. As far as the author knows, a complete proof including non-compact case is not available yet, even for Siegel modular varieties.

Hence

$$c_1(\mathrm{Sym}^2 M_{\mathrm{sp}}(f)) = c_1(M_{\mathrm{sp}}(f))^2 \delta(M_{\mathrm{sp}}(f))^{2^m-1}.$$

Put $w = mk - m(m+1)/2$. We get

$$c_1(\mathrm{Sym}^2 M_{\mathrm{sp}}(f)(w)) = c_1(M_{\mathrm{sp}}(f))^2 \delta(M_{\mathrm{sp}}(f))^{2^m-1}(1\otimes 2\pi i)^{2^{m-1}(2^m+1)w},$$

$$\delta(M_{\mathrm{sp}}(f)) = (1\otimes 2\pi i)^{-2^{m-1}w} \qquad \text{by [Y2], (5.10),}$$

(3.4) $$c_1(\mathrm{Sym}^2 M_{\mathrm{sp}}(f)(w)) = c_1(M_{\mathrm{sp}}(f))^2 \delta(M_{\mathrm{sp}}(f))^{-2}.$$

Here (w) denotes the Tate twist. Comparing the Euler p-factors of L-functions, we have[5]

$$L(\mathrm{Sym}^2 M_{\mathrm{sp}}(f), s+w) = L(\wedge^m M_{\mathrm{st}}(f), s)\Big(\prod_{j=0}^{m-1} L(\wedge^j M_{\mathrm{st}}(f), s)^{\delta_j}.$$

Here we understand that $\wedge^0 M_{\mathrm{st}}(f) = T(0)$, the Tate motive of weight 0. Then by the Tate conjecture, we have the crucial isomorphism ([Y2], (5.16))

(3.5) $$\mathrm{Sym}^2 M_{\mathrm{sp}}(f)(w) \cong \wedge^m M_{\mathrm{st}}(f) \oplus (\oplus_{j=0}^{m-1} \delta_j \wedge^j M_{\mathrm{st}}(f)).$$

Let N be the motive on the right-hand side of (3.5). We see that $c_1(N)\delta(N)^{-1}$ is the period invariant of N of the type $\{(1,0,\ldots,0,-1);(0,0)\}$; it depends also on a period matrix of $M_{\mathrm{st}}(f)$, and regarded as a period invariant of $M_{\mathrm{st}}(f)$ defined by its Hodge data, we see that it is of the type[6]

$$\{(\overbrace{1,\ldots,1}^{m},0,\overbrace{-1,\ldots,-1}^{m});(0,0)\}.$$

Hence we get

$$c_1(N)\delta(N)^{-1} = c^+(M_{\mathrm{st}}(f))c^-(M_{\mathrm{st}}(f))\delta(M_{\mathrm{st}}(f))^{-1}.$$

Since $\delta(N) = \delta(M_{\mathrm{st}}(f)) = 1$, we obtain

(3.6) $$c_1(M_{\mathrm{sp}}(f))^2 \delta(M_{\mathrm{sp}}(f))^{-2} = c^+(M_{\mathrm{st}}(f))c^-(M_{\mathrm{st}}(f)).$$

Assume $k > 2m$. Then by [Y2], (5.8), we have

$$c^{\pm}(M_{\mathrm{st}}(f)) = \pi^{mk}(\langle f^\sigma, f^\sigma\rangle)_{\sigma\in J_E}.$$

The proof is now complete by this formula and by (3.6).

Remark 3.2. When $m = 2$, $Q = \pi^{2k}(\langle f^\sigma, f^\sigma\rangle)_{\sigma\in J_E}$ holds ([Y2], (4.12), (3.3)). In general, it may be difficult to take off the squares in the formula of Proposition 3.1.

[5] $\delta_j = 1$ if $j \equiv m$ or $m+1 \mod 4$, $\delta_j = 0$ otherwise; see [Y2], p. 1196.

[6] Write period matrices of $M_{\mathrm{st}}(f)$ and N as in [Y2], p. 1181 and look at the vector of the smallest weight of a period matrix of N.

The results in this section and the results on quadratic periods in §2 were exposed in my letter to Shalika [Y3].

§4. Fundamental periods of a motive over a number field

Let E and F be algebraic number fields. Let M be a motive over F with coefficients in E of pure weight w. We have

$$L(M,s) = L(\mathrm{Res}_{F/\mathbf{Q}}(M),s),$$

where $\mathrm{Res}_{F/\mathbf{Q}}$ denotes the restriction of scalars functor. So apparently, we need only consider motives over $\mathbf{Q}$ to study critical values of L-functions. However to study the behavior of the periods $c^{\pm}(\mathrm{Res}_{F/\mathbf{Q}}(M))$ under the tensor product of two motives over F, it is necessary to introduce the notion of τ-period ([Y1]. §2), where τ is an isomorphism of F into $\mathbf{C}$. To understand the τ-period intuitively, imagine that M corresponds to a cohomology of a projective smooth algebraic variety defined over F. Then the periods of M depend on an embedding $\tau : F \hookrightarrow \mathbf{C}$. The problem is also discussed in Hida [HI] and in Blasius [B2]. In this paper, I will discuss the problem in the context of the fundamental periods of a motive explained in section 1.

Let J_F^0 be the set of all infinite places of F. Let c be complex conjugation of $\mathbf{C}$. If $\tau \in J_F$ and $\tau(F) \not\subset \mathbf{R}$, then τ and $c\tau$ define the same imaginary place (cf. [W], Chapter III). We regard J_F^0 as a subset of J_F, choosing one isomophism in J_F corresponding to an imaginary place.

Let $H_{\mathrm{DR}}(M)$ be the de Rham realization of M. It is a free $E \otimes_{\mathbf{Q}} F$-module of rank d. We call $d = d(M)$ the *rank* of M. We write the Hodge filtration as

$$H_{\mathrm{DR}}(M) = \mathcal{F}^{j_1} \supsetneqq \mathcal{F}^{j_2} \supsetneqq \cdots \supsetneqq \mathcal{F}^{j_n} \supsetneqq \mathcal{F}^{j_{n+1}} = \{0\}. \tag{4.1}$$

Here $\mathcal{F}^{j_\nu}$ is an $E \otimes_{\mathbf{Q}} F$-module, which is not necessarily free. In (4.1), we assume that there are no different filtrations between successive members. Let $\tau \in J_F$. We have the Betti realization $H_{\tau,B}(M)$; it is a d-dimensional vector space over E. We have the comparison isomorphism

$$I_\tau : H_{\tau,B}(M) \otimes_{\mathbf{Q}} \mathbf{C} \cong H_{\mathrm{DR}}(M) \otimes_{F,\tau} \mathbf{C}. \tag{4.2}$$

(I_τ is an isomorphism of R-modules, $R = E \otimes_{\mathbf{Q}} \mathbf{C}$.) $H_{\tau,B}(M)$ has an E-rational Hodge structure, that is we have the Hodge decomposition

$$H_{\tau,B}(M) \otimes_{\mathbf{Q}} \mathbf{C} = \oplus_{p+q=w} H^{pq}(\tau,M).$$

Here $H^{pq}(\tau,M)$ is an R-module, which is not necessarily free. We have

$$I_\tau(\oplus_{p' \geqq p} H^{p'q}(\tau,M)) = \mathcal{F}^p \otimes_{F,\tau} \mathbf{C}. \tag{4.3}$$

If τ is real, complex conjugation $F_{\infty,\tau}$ at τ acts on $H_{\tau,B}(M)$ and we have the decomposition

$$H_{\tau,B}(M) = H^+_{\tau,B}(M) \oplus H^-_{\tau,B}(M).$$

If τ is imaginary, complex conjugation $F_{\infty,\tau}$ at τ permutes $H_{\tau,B}(M)$ and $H_{c\tau,B}(M)$. We have

$$\begin{aligned}&H_{\tau,B}(M)\oplus H_{c\tau,B}(M)\\&=(H_{\tau,B}(M)\oplus H_{c\tau,B}(M))^{+}\oplus(H_{\tau,B}(M)\oplus H_{c\tau,B}(M))^{-},\end{aligned}$$

the notation being self-explanatory.

Let us first briefly review a generalization of Deligne's periods $c^{\pm}(M)$, $\delta(M)$ to the τ-periods following [Y1]. If w is odd, we put $w = 2p+1$, $\mathcal{F}^{\pm}(M) = \mathcal{F}^{p+1}$. If w is even, we put $w = 2p$. We assume that F_∞ acts on $H^{pp}(\mathrm{Res}_{F/\mathbf{Q}}(M))$ by a scalar and define

$$\begin{aligned}&\mathcal{F}^{+}(M)=\mathcal{F}^{p}, \quad F^{-}(M)=\mathcal{F}^{p+1} \quad \text{if } F_\infty = 1 \text{ on } H^{pp}(\mathrm{Res}_{F/\mathbf{Q}}(M))\\&\mathcal{F}^{+}(M)=\mathcal{F}^{p+1}, \quad F^{-}(M)=\mathcal{F}^{p} \quad \text{if } F_\infty = -1 \text{ on } H^{pp}(\mathrm{Res}_{F/\mathbf{Q}}(M)).\end{aligned}$$

In both cases, we set $H^{\pm}_{\mathrm{DR}}(M) = H_{\mathrm{DR}}(M)/\mathcal{F}^{\mp}(M)$. If F has a real infinite place, then $H^{\pm}_{\mathrm{DR}}(M)$ is a free $E\otimes_{\mathbf{Q}} F$-module. If F is totally imaginary, then $H^{\pm}_{\mathrm{DR}}(M)\oplus cH^{\pm}_{\mathrm{DR}}(M)$ is a free $E\otimes_{\mathbf{Q}} F$-module. Here, for an $E\otimes_{\mathbf{Q}} F$-module V, cV denotes the module whose $E\otimes_{\mathbf{Q}} F$-module structure is given changing the action of F by $f\cdot v = c(f)\cdot v$, $f\in F$, $v\in V$. Let $\tau\in J^0_F$. If τ is real, we have a canonical homomorphism

$$I^{\pm}_{\tau} : H^{\pm}_{\tau,B}(M)\otimes_{\mathbf{Q}}\mathbf{C}\cong H^{\pm}_{\mathrm{DR}}(M)\otimes_{F,\tau}\mathbf{C}$$

and if τ is imaginary, we have

$$I^{\pm}_{\tau} : (H_{\tau,B}(M)\oplus H_{c\tau,B}(M))^{\pm}\otimes_{\mathbf{Q}}\mathbf{C}\cong (H^{\pm}_{\mathrm{DR}}(M)\otimes_{F,\tau}\mathbf{C})\oplus(H^{\pm}_{\mathrm{DR}}(M)\otimes_{F,c\tau}\mathbf{C}).$$

We set

$$c^{\pm}_{\tau}(M) = \det(I^{\pm}_{\tau}),$$

$$\delta_\tau(M) = \begin{cases}\det(I_\tau) & \text{if } \tau \text{ is real},\\ \det(I_\tau)\det(I_{c\tau}) & \text{if } \tau \text{ is imaginary}.\end{cases}$$

Here the determinant is calculated by an E-rational basis of the left-hand side and by a basis over $E\otimes_{\mathbf{Q}} F$ as a free $E\otimes_{\mathbf{Q}} F$-module on the right-hand side. $c^{\pm}_{\tau}(M)$ and $\delta_\tau(M)$ are called τ-periods. Let $\widetilde{F}$ be the normal closure of F over $\mathbf{Q}$. Each τ-period is an element of $(E\otimes_{\mathbf{Q}}\mathbf{C})^{\times}$ and determined up to multiplication by an element of $E\cdot\widetilde{F}$. Suppose that $H^{\pm}_{\mathrm{DR}}(M)$ is a free $E\otimes_{\mathbf{Q}} F$-module and a fixed basis of it is used to calculate τ-periods for $\tau\in J^0_F$. Then the product $\prod_{\tau\in J^0_F} c^{+}_{\tau}(M)$, $\prod_{\tau\in J^0_F} c^{-}_{\tau}(M)$ and $\prod_{\tau\in J^0_F}\delta_\tau(M)$ are determined up to multiplication by an element of E. Now we have ([Y1], Proposition 2.2):

Proposition 4.1. *If F has a real infinite place (resp. is totally imaginary), let $d^{\pm}(M)$ (resp. $2d^{\pm}(M)$) denote the rank of free $E\otimes_{\mathbf{Q}} F$-modules $H^{\pm}_{\mathrm{DR}}(M)$ (resp.*

$H^{\pm}_{\mathrm{DR}}(M) \oplus cH^{\pm}_{\mathrm{DR}}(M)$). *Then we have*

$$c^{+}(\mathrm{Res}_{F/\mathbf{Q}}(M)) = D_F^{d^{+}(M)/2} \prod_{\tau \in J_F^0} c_\tau^{+}(M),$$

$$c^{-}(\mathrm{Res}_{F/\mathbf{Q}}(M)) = D_F^{d^{-}(M)/2} \prod_{\tau \in J_F^0} c_\tau^{-}(M),$$

$$\delta(\mathrm{Res}_{F/\mathbf{Q}}(M)) = D_F^{d(M)/2} \prod_{\tau \in J_F^0} \delta_\tau^{+}(M),$$

modulo $E \cdot \widetilde{F}$. *If* $H^{\pm}_{\mathrm{DR}}(M)$ *are free* $E \otimes_{\mathbf{Q}} F$*-modules and* τ*-periods are defined using a fixed free basis, the equalities hold modulo* E. *Here* D_F *is the discriminant of* F *and we put* $D_F^{1/2} = 1 \otimes D_F^{1/2} \in E \otimes \mathbf{C}$.

Now we are going to define τ-fundamental periods. To this end, we first define a period matrix at τ. For $\sigma \in J_E$, $\tau \in J_F$,

$$E \otimes_{\mathbf{Q}} F \ni x \otimes y \mapsto \sigma(x)\tau(y) \in \mathbf{C}$$

defines a $\mathbf{Q}$-algebra homomorphism. Therefore $\mathbf{C}$ is an $E \otimes_{\mathbf{Q}} F$-module with respect to σ and τ. We have

$$\begin{aligned} H_{\mathrm{DR}}(M) \otimes_{F,\tau} \mathbf{C} &= H_{\mathrm{DR}}(M) \otimes_{E \otimes_{\mathbf{Q}} F} ((E \otimes_{\mathbf{Q}} F) \otimes_{F,\tau} \mathbf{C}) \\ &= H_{\mathrm{DR}}(M) \otimes_{E \otimes_{\mathbf{Q}} F} (E \otimes_{\mathbf{Q}} \mathbf{C}), \end{aligned}$$

where $E \otimes_{\mathbf{Q}} \mathbf{C} = \prod_{\sigma \in J_E} \mathbf{C}$ is an F-module by τ. Hence we have

$$H_{\mathrm{DR}}(M) \otimes_{F,\tau} \mathbf{C} = \prod_{\sigma \in J_E} H_{\mathrm{DR}}(M) \otimes_{(E \otimes_{\mathbf{Q}} F, \sigma \otimes \tau)} \mathbf{C},$$

$$\begin{aligned} H_{\tau,B}(M) \otimes_{\mathbf{Q}} \mathbf{C} &= H_{\tau,B}(M) \otimes_E (E \otimes_{\mathbf{Q}} \mathbf{C}) \\ &= \prod_{\sigma \in J_E} (H_{\tau,B}(M) \otimes_{E,\sigma} \mathbf{C}). \end{aligned}$$

Restricting I_τ to the identity component $\mathrm{id} \in J_E$, we get

$$I_\tau^0 : H_{\tau,B}(M) \otimes_{E,\mathrm{id}} \mathbf{C} \cong H_{\mathrm{DR}}(M) \otimes_{(E \otimes_{\mathbf{Q}} F, \mathrm{id} \otimes \tau)} \mathbf{C},$$

which is an isomorphism of vector spaces over $\mathbf{C}$. Let $E^{(\tau)}$ be the composite field $E \cdot \tau(F)$. Since

$$E \otimes_{\mathbf{Q}} F \ni x \otimes y \mapsto x\tau(y) \in E^{(\tau)}$$

is a $\mathbf{Q}$-algebra homomorphism, $E^{(\tau)}$ is an $E \otimes_{\mathbf{Q}} F$-submodule of $\mathbf{C}$. From (4.1), by taking the tensor product and omitting possible duplications, we obtain the filtration by vector spaces over $E^{(\tau)}$:

$$\begin{aligned} H_{\mathrm{DR}}(M) \otimes_{(E\otimes_{\mathbf{Q}}F,\mathrm{id}\otimes\tau)} E^{(\tau)} &= \mathcal{F}^{i_1} \otimes_{(E\otimes_{\mathbf{Q}}F,\mathrm{id}\otimes\tau)} E^{(\tau)} \\ &\supsetneqq \mathcal{F}^{i_2} \otimes_{(E\otimes_{\mathbf{Q}}F,\mathrm{id}\otimes\tau)} E^{(\tau)} \supsetneqq \cdots \supsetneqq \mathcal{F}^{i_m} \otimes_{(E\otimes_{\mathbf{Q}}F,\mathrm{id}\otimes\tau)} E^{(\tau)} \\ &\supsetneqq \mathcal{F}^{i_{m+1}} \otimes_{(E\otimes_{\mathbf{Q}}F,\mathrm{id}\otimes\tau)} E^{(\tau)} = \{0\}. \end{aligned}$$

Suppose that τ is real. Put $\mathcal{T}^{i_\mu} = \mathcal{F}^{i_\mu} \otimes_{(E\otimes_{\mathbf{Q}}F,\mathrm{id}\otimes\tau)} E^{(\tau)}$,

$$s_\mu = \dim_{E^{(\tau)}} \mathcal{T}^{i_\mu} - \dim_{E^{(\tau)}} \mathcal{T}^{i_{\mu+1}}, \qquad 1 \le \mu \le m.$$

Then

$$d = \dim_{E^{(\tau)}} \mathcal{T}^{i_1} = s_1 + s_2 + \cdots + s_m.$$

Let P_M be the lower parabolic subgroup of $\mathrm{GL}(d)$ corresponding to this partition of d. We choose a basis $\{w_1,\ldots,w_d\}$ of $H_{\mathrm{DR}}(M)\otimes_{(E\otimes_{\mathbf{Q}}F,\mathrm{id}\otimes\tau)} E^{(\tau)}$ over $E^{(\tau)}$ so that $\{w_{s_1+s_2+\ldots+s_{\mu-1}+1},\ldots,w_d\}$ is a basis of $\mathcal{T}^{i_\mu}$ over $E^{(\tau)}$, for $1 \le \mu \le m$. Let $\{v_1^+, v_2^+,\ldots,v_{d^+}^+\}$ be a basis of $H^+_{\tau,B}(M)$ over E and $\{v_1^-, v_2^-,\ldots,v_{d^-}^-\}$ be a basis of $H^-_{\tau,B}(M)$ over E. ($d^\pm = d^\pm_\tau$ may depend on τ in general.) Let

$$I^0_\tau(v_j^\pm) = \sum_{i=1}^{d} x^\pm_{ij} w_i, \quad x^\pm_{ij} \in \mathbf{C}, \quad 1 \le j \le d^\pm_\tau,$$

$$X^\pm_\tau = (x^\pm_{ij}) \in M(d, d^\pm_\tau, \mathbf{C}).$$

Then the class of $X^\pm_\tau$ in

$$P_M(E^{(\tau)})\backslash M(d, d^\pm_\tau, \mathbf{C})/\mathrm{GL}(d^\pm_\tau, E)$$

does not depend on the choice of the basis $\{v_j^\pm\}$, $\{w_j\}$. Let

$$X_\tau = (X^+_\tau X^-_\tau) \in M(d,d,\mathbf{C})$$

and call it a *period matrix at* τ. The class of X_τ in

$$P_M(E^{(\tau)})\backslash M(d,d,\mathbf{C})/(\mathrm{GL}(d^+_\tau, E)\times \mathrm{GL}(d^-_\tau, E))$$

does not depend on the choice of the basis $\{v_j^\pm\}$, $\{w_j\}$. Now the τ-fundamental periods are defined similarly to the case $F = \mathbf{Q}$. Let $f(x)$ be an admissible $\mathbf{Q}$-rational polynomial function on $M(d,d)$ of the type $\{(a_1,\ldots,a_m);(k^+,k^-)\}$. Then $f(X_\tau) \in \mathbf{C}$ is well defined modulo $(E^{(\tau)})^\times$.

$f(x) = \det(x)$ is of the type $\{(1,1,\ldots,1);(1,1)\}$. We have $\delta_\tau(M) = f(X_\tau)$. If $s_1 + s_2 + \cdots + s_{p^+} = d^+$ for some p^+ (we fix τ and abbreviate $d^\pm_\tau$ to $d^\pm$), then let $f^+(x)$ be the upper left minor of size d^+ of x. $f^+(x)$ is of the type $\{(\overbrace{1,1,\ldots,1}^{p^+},0,\ldots,0);(1,0)\}$. We have $c^+_\tau(M) = f^+(X_\tau)$. If $s_1 + s_2 + \cdots + s_{p^-} = d^-$ for some p^-, then let $f^-(x)$ be the upper right minor of size d^- of x. $f^-(x)$ is of the type $\{(\overbrace{1,1,\ldots,1}^{p^-},0,\ldots,0);(0,1)\}$. We have $c^-_\tau(M) = f^-(X_\tau)$.

Let p be a positive integer such that for some integer q, $p < q \le m$, two conditions

$$s_1 + s_2 + \cdots + s_p < \min(d^+, d^-),$$
$$s_1 + s_2 + \cdots + s_q = d - (s_1 + s_2 + \cdots + s_p)$$

hold. Then there exists a non-zero $\mathbf{Q}$-rational admissible polynomial f_p of the type $\{(\overbrace{2,\ldots,2}^{p},\overbrace{1,\ldots,1}^{m-2p},\overbrace{0,\ldots,0}^{p});(1,1)\}$. We define a τ-fundamental period by $c_{p,\tau}(M) = f_p(X_\tau)$.

Remark 4.2. We can show that $c_{p,\tau}(M) \in \mathbf{C}^\times$ in a similar manner to Cor. 2 to Prop. 12 of [Y2], using a result of Blasius-Rogawski [BR].

Next suppose that $\tau \in J_F^0$ is imaginary. We have

$$I_\tau \oplus I_{c\tau} : (H_{\tau,B}(M) \otimes_{\mathbf{Q}} \mathbf{C}) \oplus (H_{c\tau,B}(M) \otimes_{\mathbf{Q}} \mathbf{C})$$
$$\cong (H_{\mathrm{DR}}(M) \otimes_{F,\tau} \mathbf{C}) \oplus (H_{\mathrm{DR}}(M) \otimes_{F,c\tau} \mathbf{C}),$$
$$I_\tau^0 \oplus I_{c\tau}^0 : (H_{\tau,B}(M) \otimes_{E,\mathrm{id}} \mathbf{C}) \oplus (H_{c\tau,B}(M) \otimes_{E,\mathrm{id}} \mathbf{C})$$
$$\cong (H_{\mathrm{DR}}(M) \otimes_{(E\otimes_{\mathbf{Q}}F,\mathrm{id}\otimes\tau)} \mathbf{C}) \oplus (H_{\mathrm{DR}}(M) \otimes_{(E\otimes_{\mathbf{Q}}F,\mathrm{id}\otimes c\tau)} \mathbf{C}).$$

Let $\{v_1^+, v_2^+, \ldots, v_d^+\}$ be a basis of $(H_{\tau,B}(M) \oplus H_{c\tau,B}(M))^+$ over E, and let $\{v_1^-, v_2^-, \ldots, v_d^-\}$ be a basis of $(H_{\tau,B}(M) \oplus H_{c\tau,B}(M))^-$ over E. Let $E^{(\tau)}$ be the composite field of E, $\tau(F)$ and $c\tau(F)$. We put

$$\mathcal{T}^{j_\nu} = (\mathcal{F}^{j_\nu} \otimes_{(E\otimes_{\mathbf{Q}}F,\mathrm{id}\otimes\tau)} E^{(\tau)}) \oplus (\mathcal{F}^{j_\nu} \otimes_{(E\otimes_{\mathbf{Q}}F,\mathrm{id}\otimes c\tau)} E^{(\tau)}).$$

From (4.1), by taking the tensor product and omitting possible duplications, we obtain the filtration by vector spaces over $E^{(\tau)}$:

$$(H_{\mathrm{DR}}(M) \otimes_{(E\otimes_{\mathbf{Q}}F,\mathrm{id}\otimes\tau)} E^{(\tau)}) \oplus (H_{\mathrm{DR}}(M) \otimes_{(E\otimes_{\mathbf{Q}}F,\mathrm{id}\otimes c\tau)} E^{(\tau)}) = \mathcal{T}^{i_1}$$
$$\supsetneqq \mathcal{T}^{i_2} \supsetneqq \cdots \supsetneqq \mathcal{T}^{i_m} \supsetneqq \mathcal{T}^{i_{m+1}} = \{0\}.$$

Put

$$s_\mu = \dim_{E^{(\tau)}} \mathcal{T}^{i_\mu} - \dim_{E^{(\tau)}} \mathcal{T}^{i_{\mu+1}}, \qquad 1 \le \mu \le m.$$

Then

$$2d = \dim_{E^{(\tau)}} \mathcal{T}^{i_1} = s_1 + s_2 + \cdots + s_m.$$

Let P_M be the lower parabolic subgroup of $\mathrm{GL}(2d)$ corresponding to this partition of $2d$. We choose a basis $\{w_1, \ldots, w_{2d}\}$ of $(H_{\mathrm{DR}}(M) \otimes_{(E\otimes_{\mathbf{Q}}F,\mathrm{id}\otimes\tau)} E^{(\tau)}) \oplus (H_{\mathrm{DR}}(M) \otimes_{(E\otimes_{\mathbf{Q}}F,\mathrm{id}\otimes c\tau)} E^{(\tau)})$ over $E^{(\tau)}$ so that $\{w_{s_1+s_2+\ldots+s_{\mu-1}+1}, \ldots, w_{2d}\}$ is a basis of $\mathcal{T}^{i_\mu}$ over $E^{(\tau)}$ for $1 \le \mu \le m$. Let

$$(I_\tau^0 \oplus I_{c\tau}^0)(v_j^\pm) = \sum_{i=1}^{2d} x_{ij}^\pm w_i, \quad x_{ij}^\pm \in \mathbf{C}, \quad 1 \le j \le d,$$

$$X_\tau^\pm = (x_{ij}^\pm) \in M(2d, d, \mathbf{C}).$$

Then the class of $X_\tau^\pm$ in

$$P_M(E^{(\tau)})\backslash M(2d,d,\mathbf{C})/\mathrm{GL}(d,E)$$

does not depend on the choice of the basis $\{v_j^\pm\}$, $\{w_j\}$. Let

$$X_\tau = (X_\tau^+ X_\tau^-) \in M(2d,2d,\mathbf{C})$$

The class of $X = X_\tau$ in

$$P_M(E^{(\tau)})\backslash M(2d,2d,\mathbf{C})/(\mathrm{GL}(d,E)\times \mathrm{GL}(d,E))$$

does not depend on the choice of the basis $\{v_j^\pm\}$, $\{w_j\}$. Now we can define τ-fundamental periods using admissible polynomials on $M(2d,2d)$ similarly to the case where τ is real. Let $f(x)$ be an admissible $\mathbf{Q}$-rational polynomial function on $M(2d,2d)$ of the type $\{(a_1,\ldots,a_m);(k^+,k^-)\}$. Then $f(X_\tau)\in\mathbf{C}$ is well defined modulo $(E^{(\tau)})^\times$.

References

[B1] D. Blasius, Appendix to Orloff: Critical values of certain tensor product *L*-functions, Invent. Math. 90 (1987), 181–188.

[B2] D. Blasius, Period relations and critical values of *L*-functions, Pacific J. Math. 181 (1997), 53–83.

[BR] D. Blasius and J. D. Rogawski, Motives for Hilbert modular forms, Invent. Math. 114 (1993), 55–87.

[Bo] N. Bourbaki, Groupes et algèbres de Lie, Chapitres 7 et 8, Diffusion C. C. L. S. , Paris, 1975.

[D] P. Deligne, Valeurs de fonctions *L* et périodes des intégrales, Proc. Sympos. Pure Math. 33 (1979), part 2, 313–346.

[DM] P. Deligne and J. S. Milne, Tannakian categories, Lecture notes in Math. 900 (1982), 101-228, Springer-Verlag.

[FS] M. Furusawa and J. Shalika, On central critical values of the degree four *L*-functions for GSp(4): The fundamental lemma, Memoirs of Amer. Math. Soc. Volume 164, Number 782, 2003.

[HA] M. Harris, *L*-functions and periods of polarized regular motives, J. reine angew. Math. 483 (1997), 75–161.

[HI] H. Hida, On the critical values of *L*-functions of GL(2) and GL(2) × GL(2), Duke Math. J. 74 (1994), 431–529.

[JKS] U. Jannsen, S. Keiman and J. -P. Serre (eds.), Motives, Proc. Symposia Pure Math. 55 (1994), part 1.

[K] R. E. Kottwitz, Shimura varieties and λ-adic representations, in Automorphic forms, Shimura varieties and *L*-functions I, Perspect. Math. 10, Academic Press, Boston, 1990, 161–209.

[L] R. P. Langlands, On the zeta-functions of some simple Shimura varieties, Can. J. Math. Vol. XXXI (1976), 1121-1216.

[S] G. Shimura, Arithmeticity in the Theory of Automorphic Forms, Math. Surveys Monogr., vol. 82, Amer. Math. Soc., 2000.

[W] A. Weil, Basic Number Theory, Die Grundlehren der mathematischen Wissenschaften, Band 144, Springer Verlag, 1967.

[Y1] H. Yoshida, On the zeta functions of Shimura varieties and periods of Hilbert modular forms, Duke Math. J. 75 (1994), 121–191.

[Y2] H. Yoshida, Motives and Siegel modular forms, Amer. J. Math. 123 (2001), 1171–1197.

[Y3] H. Yoshida, A letter to Shalika, April 16, 2001.

[Y4] H. Yoshida, Motivic Galois groups and L-groups, Clay Mathematics Proceedings 13 (2011), "On certain L-functions," 603–647.

Department of Mathematics, Kyoto University, Kyoto 606-8502, Japan
E-mail: hyoshida111@gmail.com

III.B Modular forms and iterated integrals

16

Geometric Hodge structures with prescribed Hodge numbers

Donu Arapura

Let us call a pure, and necessarily effective polarizable, Hodge structure geometric if it is contained in the cohomology of a smooth complex projective variety. A natural question is which polarizable Hodge structures are geometric? Our goals in this note are twofold. We first of all show that there are no numerical restrictions on geometric Hodge structures. That is for any set of Hodge numbers (subject to the obvious constraints imposed by Hodge symmetry and effectivity), there exists a geometric Hodge structure with precisely these Hodge numbers. The argument is fairly easy, but we hope that there is some pedagogical value in writing it out. Our example is contained in the cohomology of a power E^N of a CM elliptic curve. Since the Hodge conjecture is known for such varieties, we get a slightly stronger statement, that our example is the Hodge realization of a Grothendieck motive. The second part of this note is less elementary and more speculative. We try to address the question about where all the two dimensional irreducible geometric Hodge structures come from. The answer, assuming the Hodge conjecture, is that they come from either from elliptic curves or motives over $\bar{\mathbb{Q}}$. Examples of the latter type include the Hodge structures constructed in the first section, and Hodge structures associated to modular forms discussed in the second. Although we do not have a good conjecture about how to describe all the two dimensional $\bar{\mathbb{Q}}$-motivic Hodge structures, the ones arising from motives over $\mathbb{Q}$ should be exactly the modular examples.

Coming back to part one, we should mention that Schreieder [Scr] has solved the related, but more difficult, problem of finding a smooth projective variety with a prescribed set of Hodge numbers in a given degree k (under suitable hypotheses when k is even). This does imply the main result here for odd k. Nevertheless, the construction is different and somewhat more involved. We thank Greg Pearlstein for first bringing Schreieder's work to our attention, and the referee for other useful remarks.

Partially supported by the NSF and at the IAS by the Ralph E. and Doris M. Hansmann fund.

1 Main theorem

We will work exclusively with rational Hodge structures below. It is worth observing that since the category of polarizable Hodge structures is semisimple [M], we have:

Proposition 1.1 *A geometric Hodge structure of weight k is a direct summand of $H^k(X,\mathbb{Q})$ for some smooth complex projective variety X.*

Theorem 1.1 *Given a set of nonnegative integers $g^{k,0}, g^{k-1,1}, \dots g^{0,k}$, satisfying $k \geq 0$ and $g^{p,q} = g^{q,p}$, there exists a geometric rational Hodge structure H with $\dim H^{p,q} = g^{p,q}$.*

Lemma 1.1 *The Tate structure $\mathbb{Q}(-n)$ is geometric for $n \geq 0$. Let V_1 and V_2 be geometric Hodge structures of weight n_1 and n_2 respectively, then $V_1 \otimes V_2$ is geometric. If $n_1 = n_2$, then $V_1 \oplus V_2$ is also geometric.*

Proof. We have $\mathbb{Q}(-n) = H^{2n}(\mathbb{P}^n, \mathbb{Q})$. By assumption, there exists smooth projective varieties X_i such that V_i is a summand of $H^{n_i}(X_i)$. Then $V_1 \otimes V_2$ and $V_1 \oplus V_2$ are summands of $H^k(X_1 \times X_2)$ where $k = n_1 n_2$ and $k = n_1 = n_2$ respectively. □

Lemma 1.2 *For any $n > 0$, there exists a geometric Hodge structure G with $\dim G^{n,0} = \dim G^{0,n} = 1$ and the remaining Hodge numbers equal to zero.*

Proof. Although, it is not difficult to prove by hand, it is a bit cleaner if we make use of basic facts about Mumford-Tate groups, cf [L, appendix B] or [M]. In order to spell things out as explicitly as possible, we work with the specific elliptic curve $E = \mathbb{C}/\mathbb{Z}i \oplus \mathbb{Z}$. This has complex multiplication by $K = \mathbb{Q}(i)$. Let $V = H^1(E,\mathbb{Q})$. We choose a basis v_1, v_2 for V dual to the basis $i, 1$ of the lattice $\mathbb{Z}i \oplus \mathbb{Z}$. Then $dz = iv_1 + v_2$ and $d\bar{z} = -iv_1 + v_2$ determines the Hodge decomposition on $V \otimes \mathbb{C}$. We define a homomorphism h of the unit circle into $GL(V \otimes \mathbb{R})$ by rotations

$$h(\theta) = \begin{pmatrix} \cos\theta & -\sin\theta \\ \sin\theta & \cos\theta \end{pmatrix}$$

Then $h(\theta)dz = e^{i\theta}dz$, and $h(\theta)d\bar{z} = e^{-i\theta}d\bar{z}$. Let $SMT(V)$ denote the Weil restriction

$$Res_{K/\mathbb{Q}}\mathbb{G}_m = \left\{ \begin{pmatrix} a & -b \\ b & a \end{pmatrix} \mid (a,b) \in \mathbb{Q}^2, (a,b) \neq (0,0) \right\}$$

This is clearly the smallest $\mathbb{Q}$-algebraic group whose real points contain $\operatorname{im} h$. In other words, $SMT(V)$ is the special Mumford-Tate group or Hodge group of V. The Mumford-Tate group $MT(V)$ is defined similarly, as the smallest

$\mathbb{Q}$-algebraic group whose real points contain the image of Deligne's torus $\mathbb{S}(\mathbb{R}) = \mathbb{C}^*$. It works out to the product of $SMT(V)$ and the group of nonzero scalar matrices. The significance of this group comes from the Tannakian interpretation: $MT(V)$ is the group whose category of representations is equivalent to the tensor category generated by V. Since $MT(V)$ was described as a matrix group, it comes with an obvious representation $\rho : MT(V) \to GL(V)$. For an integer n, let V_n be the representation of this group given by composing ρ with the nth power homomorphism $MT(V) \to MT(V)$. Now let us suppose that $n > 0$. Then V_n is irreducible, and therefore simple as a Hodge structure. The elements $dz, d\bar{z}$ still give a basis of $V_n \otimes \mathbb{C}$, but now the circle acts by $dz \mapsto e^{in\theta} dz$ and $d\bar{z} \mapsto e^{-in\theta} dz$. Thus these vectors span $V_n^{n,0}$ and $V_n^{0,n}$. We have an embedding $V_n \subseteq S^n V \subset V^{\otimes n}$ given by identifying it with the span of v_1^n and v_2^n. Therefore the previous lemma implies that V_n is geometric. Thus $G = V_n$ is the desired Hodge structure. □

Lemma 1.3 *Given integers $p > q \geq 0$, there exists a geometric Hodge structure $H(p,q)$ with $\dim H(p,q)^{p,q} = \dim H(p,q)^{q,p} = 1$ and the remaining Hodge numbers equal to zero.*

Proof. Let G be the Hodge structure constructed in the previous lemma with $n = p - q$. Then $H(p,q) = G \otimes \mathbb{Q}(-q)$ will satisfy the above conditions. □

Remark 1.2 *We can replace E by any CM elliptic curve E', and the same construction works. We denote the corresponding Hodge structure by $H_{E'}(p,q)$. However, the method will fail for non CM curves, because the Mumford-Tate group will no longer be abelian.*

Proof of theorem. We get the desired Hodge structure by taking sums of the Hodge structures constructed in lemma 1.3, and the appropriate number of Tate structures when k is even. More explicitly

$$H = \bigoplus_{p>q} H(p,q)^{\oplus g^{pq}} \underbrace{\oplus \mathbb{Q}(-k/2)^{\oplus g^{k/2,k/2}}}_{k \text{ even}}$$

□

Remark 1.3 *As the referee points out, this can also proved by invoking [A] but we prefer to keep the argument elementary and self-contained.*

The proof actually shows that $H \subset H^k(E^N, \mathbb{Q})$ for some N. We can use this fact to get the stronger conclusion stated in the introduction; it says roughly that the inclusion is also defined algebro-geometrically. In order to make a precise statement, we recall that a motive, or more precisely an effective pure motive with respect to homological equivalence, consists of a smooth projective variety X together with an algebraic cycle $p \in H^*(X \times X, \mathbb{Q})$ such that

$p \circ p = p$ [K]. The composition $\circ$ is the usual one for correspondences [F2, chap 16]. It follows that p is an idempotent of the ring of Hodge endomorphisms $\prod_i End_{HS}(H^i(X))$. Thus we get a mixed Hodge structure $p(H^*(X,\mathbb{Q}))$ which is by definition the Hodge realization of the motive (X,p). We quickly run up against fundamental difficulties. For instance, it is unknown, whether $H^k(X)$ is the realization of a motive for arbitrary X and k. For this, we would need to know that the Künneth components of the diagonal $\Delta \subset X \times X$ are algebraic, and this is one Grothendieck's standard conjectures [G]. Fortunately, it is not an issue in our example, because the Hodge conjecture holds for E^{2N} [L, appendix B, §3] and this implies Grothendieck's conjecture. Thus $H^k(E^N)$ is the realization of a motive. Since H is a summand of $H^k(E^N)$, it is given by the image of an idempotent in $End_{HS}(H^k(E^N))$. By the Hodge conjecture, this is algebraic. Thus to summarize:

Proposition 1.2 *The geometric Hodge structure H, given in the theorem, is the Hodge realization of a motive.*

2 Two dimensional geometric Hodge structures

Our goal now is to obtain an understanding of where two dimensional geometric Hodge structures come from. We may as well assume that the structures are irreducible because otherwise they are just sums of Tate structures. Given a subfield $K \subset \mathbb{C}$, by a motive over K we will mean a pair (X,p), as described earlier, such that both X and the algebraic cycle representing p are defined over K.

Theorem 2.1 *Let H be an irreducible two dimensional geometric Hodge structure. Then, assuming the Hodge conjecture, there are two nonexclusive possibilities.*

(a) *$H = H^1(E)(-m)$, where E is an elliptic curve and $m \geq 0$, or*
(b) *H comes from a motive over $\bar{\mathbb{Q}}$.*

Proof. We recall that the level of H is the largest value of $|p-q|$ such that $H^{pq} \neq 0$. If the level is 1 then after Tate twisting, we can assume that H is of type $(1,0),(0,1)$ and of course still 2 dimensional. So it must coincide with $H^1(E)$ where $E = H/(H^{10} + H_{\mathbb{Z}})$ for some lattice $H_{\mathbb{Z}} \subset H_{\mathbb{Q}}$. If the level of H is greater than 1, we obtain (b) from the next proposition. □

Proposition 2.1 *Assuming the Hodge conjecture, any geometric Hodge structure with no consecutive nonzero Hodge numbers comes from a motive over $\bar{\mathbb{Q}}$. In particular, this conclusion applies to two dimensional geometric Hodge structures of level at least two.*

Proof. Fix a period domain D parametrizing polarized Hodge structures with fixed Hodge numbers $h^{m,0} = *, h^{m-1,1} = 0, h^{m-2,2} = *, \ldots$ [CMP]. Pick a point $t_0 \in D$ corresponding to a Hodge structure H which lies in $H^m(X,\mathbb{Q})$, where X is a complex smooth projective variety. Since the category of polarizable Hodge structures is semisimple, H is the image of an idempotent $p \in End_{HS}(\bigoplus_i H^i(X,\mathbb{Q}))$. By the Hodge conjecture, this can be lifted to an algebraic cycle that we also denote by $p = \sum n_i p_i$ on $X \times X$. The pair (X,p) is defined over a finitely generated extension of $K/\bar{\mathbb{Q}}$. We can regard K as the function field of a variety $S/\bar{\mathbb{Q}}$. After shrinking S if necessary, we can assume it is smooth and that X is the generic fibre (based changed to $\mathbb{C}$) of a smooth projective family $f : \mathcal{X} \to S$. We can also assume, after further shrinking, that the relative cycles $\mathcal{P}_i \subset \mathcal{X} \times_S \mathcal{X}$ given by the scheme theoretic closure of the components p_i are flat over S. Let $\mathcal{P}$ denote the relative correspondence $\sum n_i \mathcal{P}_i$. Then the rank of the image of the action of the fibre $\mathcal{P}_s$ on $H^*(\mathcal{X}_s(\mathbb{C}))$ is constant as s varies. It follows that $im[\mathcal{P} : R^* f_* \mathbb{Q} \to R^* f_* \mathbb{Q}]$ defines a variation of Hodge structure over $S(\mathbb{C})$. Let $\pi : \tilde{S}(\mathbb{C}) \to D$ be the associated period map on the universal cover. Now comes the key point. With our choice of Hodge numbers, Griffiths transversality forces π to be constant. Therefore we can realize t_0 by the motive determined by any pair $(\mathcal{X}_s, \mathcal{P}_s)$ with $s \in S(\mathbb{C})$. In particular, we can choose $s \in S(\bar{\mathbb{Q}})$. □

We describe a few examples of Hodge structures arising from motives over $\bar{\mathbb{Q}}$. Suppose that E is a CM elliptic curve. Then as is well known [Si, chap II, §2], E is defined over $\bar{\mathbb{Q}}$. Thus the examples $H_E(p,q)$ constructed earlier all arise from motives over $\bar{\mathbb{Q}}$. These examples have CM, and so are rather special. Recall that a Hodge structure H has CM if $SMT(H)$ is abelian. It turns out that any irreducible two dimensional polarizable Hodge structure of even weight has CM. To see this, observe that the special Mumford-Tate group would be abelian since it would have to contain SO_2 as a maximal compact. (Our thanks to the referee for this remark.) It is not difficult to conclude, as a consequence, that two dimensional irreducible Hodge structures of even weight are of the form $H_E(p,q)$, for some CM elliptic curve E.

In addition to the $H_E(p,q)$, there is one very natural source of examples: modular forms. Let us explain the basic set up. For simplicity, we work with the principal congruence group $\Gamma(n) = \ker[SL_2(\mathbb{Z}) \to SL_2(\mathbb{Z}/n)]$ with $n \geq 3$, but there are several other natural choices. Recall that a weight k cusp form $f \in S_k(\Gamma(n))$ is given by a holomorphic function on the upper half plane $\mathbb{H}$ satisfying

$$f\left(\frac{az+b}{cz+d}\right) = (cz+d)^k f(z), \quad \begin{pmatrix} a & b \\ c & d \end{pmatrix} \in \Gamma(n)$$

and such that the Fourier expansion $f = \sum a_j q^j$, $q = e^{2\pi iz/n}$, has only positive terms [DS]. The moduli space $Y(n) = \mathbb{H}/\Gamma(n)$ of elliptic curves with a full level n structure is a fine moduli space, so it comes with a universal family $\pi_n : \mathcal{E}(n) \to Y(n)$. This can be completed to a minimal elliptic surface $\bar{\mathcal{E}}(n) \to X(n)$ over the nonsingular compactification $j : Y(n) \hookrightarrow X(n)$. By a theorem of Zucker [Z], the intersection cohomology of the symmetric power

$$H = IH^1(X(n), S^{k-1}R^1\pi_{n*}\mathbb{Q}) = H^1(X(n), j_*S^{k-1}R^1\pi_{n*}\mathbb{Q})$$

carries a pure Hodge structure of weight k. This Hodge structure turns out to be isomorphic to one constructed by Shimura [Z, §12]. It has only $(k,0)$ and $(0,k)$ parts, and the $(k,0)$ part is isomorphic to the space $S_{k+1}(\Gamma(n))$ of cusp forms of weight $k+1$. We note that H is geometric, since it can be shown to lie the cohomology of a desingularization of the $(k-1)$-fold fibre product $\bar{\mathcal{E}}(n) \times_{X(n)} \times \ldots \bar{\mathcal{E}}(n)$. The quickest way to see this is by applying the decomposition theorem for Hodge modules [Sa], although we won't actually need this. The space $X(n)$ has a large collection of commuting self-correspondences called Hecke operators [DS]. These operators act on H [Sc], and they can be used to break it up into pieces. Suppose that $f = \sum a_j q^j \in S_{k+1}(\Gamma(n))$ is a suitably normalized nonzero simultaneous eigenvector for the Hecke operators, and further assume that the coefficients a_j are all rational. Then f and $\bar{f}$ span a two dimensional Hodge structure $H(f) \subset H$. These examples arise from motives over $\mathbb{Q}$ constructed by Scholl [Sc]. We want to argue that the converse is also true.

Given a smooth complex projective variety X, we can always choose a model X_k over a finitely generated field k. Then in addition to singular cohomology $H^*(X,\mathbb{Q})$ with its Hodge structure, we can consider étale cohomology $H^*_{et}(X_{\bar{k}}, \mathbb{Q}_\ell) \cong H^*(X,\mathbb{Q}) \otimes \mathbb{Q}_\ell$ with its canonical $Gal(\bar{k}/k)$-action [Mi]. Given a modular form f of the above type, we can associate a Galois representation by replacing ordinary cohomology by étale cohomlogy in the discussion of the previous paragraph (c.f. [D]). Serre [Se] conjectured, and Khare and Wintenberger [KW] proved, that a certain general class of two dimensional Galois representations agrees "modulo ℓ" with those coming from modular forms. Although we won't try to make this precise, we will spell out one key consequence [Se, thm 6]:

Theorem 2.2 (Serre, Khare-Wintenberger) *Given a smooth projective variety defined over $\mathbb{Q}$ such that $H^m(X(\mathbb{C}))$ is two dimensional and of type $(m,0),(0,m)$, then the Galois representation on $H^m_{et}(X_{\bar{\mathbb{Q}}}, \mathbb{Q}_\ell)$ comes from a modular form of weight $m+1$.*

If we assume Tate's conjectures [Ta], then this correspondence would have to come from an isomorphism of the underlying motives. In particular, it should be compatible with the Hodge structures. From here it seems a short step to conjecture the following:

Conjecture 2.3 *Any two dimensional Hodge structure of odd weight associated to a motive over $\mathbb{Q}$ should be isomorphic to one given by a modular form.*

This is hardly an original idea; a version of this is stated in the introduction to [FM], where it is referred to as a "well known conjecture." Having stated this conjecture, it seems a good idea to actually try to check it in some cases. A natural source of two dimensional Hodge structures, beyond those already considered, comes from the world of Calabi-Yau (or CY) threefolds. These are three dimensional smooth projective varieties X with trivial canonical bundle and hence $h^{30} = 1$. Usually one also requires $h^{10} = h^{20} = 0$ but this plays no role here. We say that X is rigid if it has no infinitesimal deformations, i.e. $H^1(X,T) = 0$. By Serre duality, this is equivalent to the vanishing of $h^{12} = h^{21}$. Thus $H^3(X,\mathbb{C}) = H^{30} \oplus H^{03}$ is two dimensional. There is by now a large collection of known examples of rigid CY's defined over $\mathbb{Q}$, see [Me]. In this note we will be content to consider just one example. We recall that to a fan, by which we mean subdivision of a Euclidean space into rational polyhedral cones, we can associate a special kind of algebraic variety called a toric variety [F1]. This construction can be applied to a root lattice, where the Weyl chambers form a fan. Verrill [V] constructed a rigid CY 3-fold V by first starting with the toric variety associated to the root lattice A_3, taking a double cover branched over an appropriate divisor in twice the anti-canonical system, and then choosing a suitable desingularization. It can be described more directly as a desingularization of the hypersurface

$$(x+y+z+w)(yzw+zxw+xyw+xyz)t - (t+1)^2xyzw = 0$$

in $\mathbb{P}^3 \times \mathbb{P}^1$ where t an inhomogeneous coordinate on the second factor. Using this equation, Saito and Yui [SY, thm 5.1] gives an explicit birational equivalence between Verrill's variety V and a desingularization Y of the fibre product of the elliptic modular surface associated to the congruence group $\Gamma_1(6) \subset SL_2(\mathbb{Z})$ with itself. Since h^{30} is a birational invariant, $h^{30}(Y) = 1$. So under this birational equivalence $H^3(V)$ will correspond to the necessarily unique irreducible sub Hodge structure of $H^3(Y)$ of type $(3,0),(0,3)$. On the other hand $IH^1(X_1(6), S^2R^1\pi_*\mathbb{Q}) \subseteq H^3(Y)$ is also of this type. So these must match, and the conjecture is verified in this example.

References

[A] S. Abdulali, *Hodge structures of CM-type.* J. Ramanujan Math. Soc. 20 (2005), no. 2, 155-162.

[CMP] J. Carlson, S. Müller-Stach, C. Peters, *Period mappings and period domains.* Cambridge Studies in Advanced Mathematics, 85. Cambridge University Press, (2003)

[D] P. Deligne, *Formes modulaires et representations ℓ-adic* , Sem. Bourbaki 355 (1968-69)

[DS] F. Diamond, J. Shurman, *A first course in modular forms*, Graduate Texts in Mathematics, 228. Springer-Verlag, New York, 2005.

[FM] J-M Fontaine, B. Mazur, *Geometric Galois representations*, Elliptic curves, modular forms, Fermat's last theorem (Hong Kong, 1993), 41 - 78, Ser. Number Theory, I, Int. Press, Cambridge, MA, 1995.

[F1] W. Fulton, *Introduction to toric varieties.* Annals of Mathematics Studies, 131. Princeton University Press, Princeton, NJ, 1993

[F2] W. Fulton, *Intersection theory*, Second edition. Ergebnisse der Mathematik und ihrer Grenzgebiete. Springer-Verlag, Berlin, 1998.

[G] A. Grothendieck, *Standard conjectures on algebraic cycles*, 1969 Algebraic Geometry (Internat. Colloq., Tata Inst. Fund. Res., Bombay, 1968) pp. 193-199 Oxford Univ. Press

[K] S. Kleiman, *Motives*, Algebraic geometry, Oslo 1970 (Proc. Fifth Nordic Summer-School in Math., Oslo, 1970), pp. 5382. Wolters-Noordhoff, 1972

[KW] C. Khare, J-P, Wintenberger, *Serre's modularity conjecture*, Proceedings of the International Congress of Mathematicians. Volume II, 280-293, Hindustan Book Agency, (2010)

[L] J. Lewis, *A survey of the Hodge conjecture. Second edition. Appendix B by B. Brent Gordon.* CRM Monograph Series, 10. American Mathematical Society, Providence, RI, 1999

[Me] C. Meyer, *Modular Calabi-Yau threefolds* . Fields Institute Monographs, 22. AMS (2005)

[Mi] J. Milne, *Etale cohomology*, Princeton University Press (1980)

[M] B. Moonen, *Notes on Mumford-Tate groups*, http://www.math.ru.nl/personal/bmoonen/

[SY] M-H. Saito, N. Yui, *The modularity conjecture for rigid Calabi-Yau threefolds over Q.* J. Math. Kyoto Univ. 41 (2001), no. 2, 403-419.

[Sa] M. Saito, *Modules de Hodge polarisables.* Publ. Res. Inst. Math. Sci. 24 (1988), no. 6, 849 - 995 (1989).

[Sc] A. Scholl, *Motives for modular forms*, Invent. Math. 100 (1990), no. 2, 419 - 430

[Scr] S. Schreieder, *On the construction problem for Hodge numbers*, ArXiv 1301.0478v3.pdf

[Se] J-P. Serre, *Sur les représentations modulaires de degré 2 de* $Gal(\bar{Q}/Q)$. Duke Math. J. 54 (1987), no. 1, 179-230.

[Si] J. Silverman, *Advanced topics in the arithmetic of elliptic curves.* Graduate Texts in Mathematics, 151. Springer-Verlag, New York, 1994

[Ta] J. Tate, *Algebraic cycles and poles of zeta functions.* Arithmetical Algebraic Geometry (Proc. Conf. Purdue Univ., 1963) pp. 93-110 Harper & Row, New York (1965)

[T] B. Totaro, *Hodge structures of type* $(n,0,\ldots,0,n)$, IMRN (2014)

[V] H Verrill, *The L-series of certain rigid Calabi-Yau threefolds*, J Numb. Theory, 81, (2000)

[Z] S. Zucker, *Hodge theory with degenerating coefficients.* L^2 *cohomology in the Poincaré metric.* Ann. of Math. (2) 109 (1979), no. 3, 415 - 476

Department of Mathematics, Purdue University, West Lafayette, IN 47907, U.S.A.

17

The Hodge-de Rham Theory of Modular Groups

Richard Hain

Contents

Supported in part by grant DMS-1005675 from the National Science Foundation

1 Introduction

The completion $\mathcal{G}_\Gamma$ of a finite index subgroup Γ of $\mathrm{SL}_2(\mathbb{Z})$ with respect to the inclusion $\rho : \Gamma \hookrightarrow \mathrm{SL}_2(\mathbb{Q})$ is a proalgebraic group, defined over $\mathbb{Q}$, which is an extension

$$1 \to \mathcal{U}_\Gamma \to \mathcal{G}_\Gamma \to \mathrm{SL}_2 \to 1$$

of SL_2 by a prounipotent group $\mathcal{U}_\Gamma$, and a Zariski dense homomorphism $\Gamma \to \mathcal{G}_\Gamma(\mathbb{Q})$. The main result of [15] implies that, for each choice of a base point of the associated orbi-curve $X_\Gamma = \Gamma\backslash\backslash\mathfrak{h}$, the coordinate ring $\mathcal{O}(\mathcal{G}_\Gamma)$ has a canonical mixed Hodge structure (MHS) that is compatible with its product, coproduct and antipode. This MHS induces one on the Lie algebra $\mathfrak{g}_\Gamma$ of $\mathcal{G}_\Gamma$.[1]

In this paper we give a detailed exposition of the construction and basic properties of the natural MHS on (the coordinate ring and Lie algebra of) relative completions of modular groups. Part 1 is an exposition of the basic properties of relative completion. It also contains a direct construction of the MHS on the relative completion of the fundamental group (and, more generally, of path torsors) of a smooth affine (orbi) curve with respect to the monodromy representation of a polarized variation of Hodge structure (PVHS). Part 2 is an exploration of the MHS on relative completions of modular groups and their associated path torsors, especially in the case of the full modular group $\mathrm{SL}_2(\mathbb{Z})$.

Completions of modular groups are interesting because of their relationship to modular forms and to categories of admissible variations of MHS over modular curves. Because the inclusion $\Gamma \to \mathrm{SL}_2(\mathbb{Q})$ is injective, one might expect the prounipotent radical $\mathcal{U}_\Gamma$ of $\mathcal{G}_\Gamma$ to be trivial. However, this is not the case. Its Lie algebra $\mathfrak{u}_\Gamma$ is a pronilpotent Lie algebra freely topologically generated (though not canonically) by

$$\prod_{m\geq 0} H^1(\Gamma, S^m H)^* \otimes S^m H \tag{1.1}$$

where H denotes the defining representation of SL_2 and $S^m H$ its mth symmetric power. Because $H^1(\Gamma, S^m H)$ is a space of modular forms of Γ of weight $m+2$ (Eichler-Shimura), there should be a close relationship between the MHS on $\mathcal{G}_\Gamma$ and the geometry and arithmetic of elliptic curves.

To explain the connection with admissible variations of MHS, consider the category $\mathsf{MHS}(X_\Gamma, \mathbb{H})$ of admissible variations of MHS $\mathbb{V}$ over X_Γ whose weight graded quotients have the property that the monodromy representation

$$\Gamma \to \operatorname{Aut} \operatorname{Gr}^W_m \mathbb{V}$$

[1] There is more structure: if X_Γ is defined over the number field K and if $x \in X_\Gamma$ is a K-rational point, then one also has a Galois action on $\mathcal{G} \otimes \mathbb{Q}_\ell$. This and the canonical MHS on $\mathcal{G}_\Gamma$ should be the Hodge and ℓ-adic étale realizations of a motivic structure on $\mathcal{G}$ that depends on x.

factors through an action of the algebraic group SL_2 for all m. The monodromy representation

$$\Gamma \cong \pi_1(X_\Gamma, x) \to \mathrm{Aut}\, V_x$$

of such a variation $\mathbb{V}$ factors through the canonical homomorphism $\Gamma \to \mathcal{G}_\Gamma(\mathbb{Q})$, so that one has a natural coaction

$$V_x \to V_x \otimes \mathcal{O}(\mathcal{G}_\Gamma). \tag{1.2}$$

In Section 8 we show that there is an equivalence of categories between $\mathsf{MHS}(X_\Gamma, \mathbb{H})$ and the category of "Hodge representations" of $\mathcal{G}_\Gamma$ — that is the category of representations of Γ on a MHS V that induce a homomorphism $\mathcal{G}_\Gamma \to \mathrm{Aut}\, V$ for which the coaction (1.2) is a morphism of MHS. The prounipotent radical $\mathcal{U}_\Gamma$ of $\mathcal{G}_\Gamma$, and hence modular forms via (1.1), control extensions in $\mathsf{MHS}(X_\Gamma, \mathbb{H})$. This is a special case of a more general result which is proved in [24].

Modular forms give simple extensions in $\mathsf{MHS}(X_\Gamma, \mathbb{H})$. The fundamental representation H of SL_2 corresponds to the polarized variation of Hodge structure $\mathbb{H}$ of weight 1 over X_Γ whose fiber over the point $x \in X_\Gamma$ is the first cohomology group of the corresponding elliptic curve. The classification of admissible variations of MHS over X_Γ in the previous paragraph and the computation (1.1) imply that there are extensions of variations

$$0 \to H^1(X_\Gamma, S^m\mathbb{H})^* \otimes S^m\mathbb{H} \to \mathbb{E} \to \mathbb{Q} \to 0.$$

When Γ is a congruence subgroup, this variation splits as the sum of extensions

$$0 \to \check{M}_f \otimes S^m\mathbb{H} \to \mathbb{E}_f \to \mathbb{Q} \to 0,$$

where f is a normalized Hecke eigenform of weight $m+2$, M_f is the corresponding Hodge structure, and $\check{M}_f = M_f(m+1)$ is its dual. When f is a cusp form, $\check{M}_f \otimes S^m\mathbb{H}$ has weight -1. In the case where $\Gamma = \mathrm{SL}_2(\mathbb{Z})$, we give an explicit construction of these extensions and the corresponding normal functions in Section 13.4. When f is an Eisenstein series, $\check{M}_f = \mathbb{Q}(m+1)$ and the extension is of the form

$$0 \to S^m\mathbb{H}(m+1) \to \mathbb{E}_f \to \mathbb{Q} \to 0.$$

These extensions are constructed explicitly in Section 13.3 when $\Gamma = \mathrm{SL}_2(\mathbb{Z})$. They correspond to the elliptic polylogarithms of Beilinson and Levin [1].

This work also generalizes and clarifies Manin's work on "iterated Shimura integrals" [29, 30]. The exact relationship is discussed in Section 13.2. The periods of the MHS on $\mathcal{O}(\mathcal{G}_\Gamma)$ are iterated integrals (of the type defined in [15]) of the logarithmic forms in Zucker's mixed Hodge complex that computes the MHS on the cohomology groups $H^1(X_\Gamma, S^m\mathbb{H})$, whose definition is recalled in Section 6. Manin's iterated Shimura integrals are iterated integrals of elements of the subcomplex of holomorphic forms in Zucker's complex. They form a Hopf algebra whose spectrum is a quotient $\mathcal{U}_A$ of $\mathcal{U}_\Gamma$ by the normal subgroup generated by $F^0\mathcal{U}_\Gamma$. This quotient is not motivic as it does not support a MHS for which the quotient mapping $\mathcal{U}_\Gamma \to \mathcal{U}_A$ is a morphism of MHS. There is a further quotient $\mathcal{U}_B$ of $\mathcal{U}_A$ that is dual to the Hopf algebra generated by iterated integrals of Eisenstein series. In Section 19 we show that it is not motivic by relating it to the *Eisenstein quotient* of $\mathcal{U}_\Gamma$, described below.

Fix a base point $x \in X_\Gamma$, so that $\mathcal{G}_\Gamma$ denotes the completion of $\pi_1(X_\Gamma, x) \cong \Gamma$ with its natural MHS. The "Eisenstein quotient" $\mathcal{G}_\Gamma^{\mathrm{eis}}$ of $\mathcal{G}_\Gamma$, defined in Section 16, is the maximal quotient of $\mathcal{G}_\Gamma$ whose Lie algebra $\mathfrak{g}_\Gamma^{\mathrm{eis}}$ has a MHS whose weight graded quotients are sums of Tate twists of the natural Hodge structure on $S^n H_x$. Its isomorphism type does not depend on the base point x. As x varies in X_Γ, the coordinate rings of the Eisenstein quotients form an admissible VMHS over X_Γ.

Denote the Lie algebra of $\mathcal{U}_B$ by $\mathfrak{u}_B$ and of the prounipotent radical $\mathcal{U}_\Gamma^{\mathrm{eis}}$ of $\mathcal{G}_\Gamma^{\mathrm{eis}}$ by $\mathfrak{u}_\Gamma^{\mathrm{eis}}$. Since the Hodge structure $\check{M}_f \otimes S^n H_x$ associated to an eigencuspform f is not of this type, such Hodge structures will lie in the kernel of

$$H_1(\mathfrak{u}_\Gamma) \to H_1(\mathfrak{u}_\Gamma^{\mathrm{eis}}),$$

which implies that $H_1(\mathfrak{u}_\Gamma^{\mathrm{eis}})$ is generated by one copy of $S^m H_x(m+1)$ for each normalized Eisenstein series of weight $m+2$. In particular, when $\Gamma = \mathrm{SL}_2(\mathbb{Z})$,

$$H_1(\mathfrak{u}_\Gamma^{\mathrm{eis}}) \cong \prod_{n \geq 1} S^{2n} H_x(2n+1).$$

There is a natural projection $\mathcal{U}_B \to \mathcal{U}_\Gamma^{\mathrm{eis}}$ from Manin's quotient of $\mathcal{U}_\Gamma$ to $\mathcal{U}_\Gamma^{\mathrm{eis}}$ that induces an isomorphism

$$H_1(\mathfrak{u}_B) \cong H_1(\mathfrak{u}_\Gamma^{\mathrm{eis}}).$$

But, as we show in Section 19, the cuspidal generators $\check{M}_f \otimes S^{2n} H_x$ of $\mathfrak{u}_\Gamma$ become non-trivial relations in $\mathfrak{u}_\Gamma^{\mathrm{eis}}$. Such relations were suggested by computations in the ℓ-adic étale version with Makoto Matsumoto (cf. [23]). Evidence for them was provided by Aaron Pollack's undergraduate thesis [33] in which he found quadratic relations between the generators of the image of the representation $\mathrm{Gr}^W_\bullet \mathfrak{u}_\Gamma^{\mathrm{eis}} \to \mathrm{Der}\,\mathbb{L}(H)$ induced by the natural action of $\mathcal{U}_\Gamma^{\mathrm{eis}}$ on the unipotent fundamental group of a once punctured elliptic curve, which we construct in Section 15.[2] The arguments in Section 19 and the computations of Brown [3] and Terasoma [43] (Thm. 19.3) imply that Pollack's quadratic relations also hold in $\mathfrak{u}_\Gamma^{\mathrm{eis}}$. Since $\mathfrak{u}_B$ is free and since $\mathfrak{u}^{\mathrm{eis}}$ is not, Manin's quotient $\mathfrak{u}_B$ does not support a natural MHS.

[2] He also found, for each cusp form, relations of all degrees ≥ 3 that hold in a certain quotient of the image of this representation.

The starting point of much of this work is the theory of "universal mixed elliptic motives" [23] developed with Makoto Matsumoto. The origin of that project was a computation in 2007 of the ℓ-adic weighted completion of $\pi_1(\mathcal{M}_{1,1/\mathbb{Z}[1/\ell]})$ in which we observed that cuspidal generators of the relative completion of the geometric fundamental group of $\mathcal{M}_{1,1/\mathbb{Z}[1/\ell]}$ appeared to become relations in the weighted completion of its arithmetic fundamental group. Pollack's thesis [33] added evidence that these cuspidal generators had indeed become relations in the weighted completion.

Finally, we mention related work by Levin and Racinet [28], Brown and Levin [4], and Calaque, Enriquez and Etingof [6], and subsequent work of Enriquez.

Although the paper contains many new results, it is expository. The intended audience is somebody who is familiar with modern Hodge theory. Several standard topics, such as a discussion of modular symbols, are included to fix notation and point of view, and also to make the paper more accessible. The reader is assumed to be familiar with the basics of mixed Hodge structures, their construction and their variations, including the basics of computing limit mixed Hodge structures.

Acknowledgments: It is a pleasure to acknowledge the mathematicians with whom I have had fruitful discussions, which helped shape my view of the subject of these notes. In particular, I would like to thank my long term collaborator, Makoto Matsumoto, as well as Aaron Pollack and Francis Brown. I am indebted to Francis Brown and Tomohide Terasoma, each of whom communicated their computation of the cup product, Theorem 19.3. I am also grateful to Francis Brown for his interest in the project and for his numerous constructive comments and corrections.

1.1. Notation and Conventions

1.1.1. Path multiplication and iterated integrals

In this paper we use the topologist's convention (which is the opposite of the algebraist's convention) for path multiplication. Two paths $\alpha, \beta : [0,1] \to X$ in a topological space X are composable when $\alpha(1) = \beta(0)$. The product $\alpha * \beta$ of two composable paths first traverses α and then β.

Denote the complex of smooth $\mathbb{C}$-valued forms on a smooth manifold M by $E^\bullet(M)$. Iterated integrals are defined using Chen's original definition: if $\omega_1, \ldots, \omega_r \in E^1(M) \otimes A$ are 1-forms on a manifold M that take values in an associative $\mathbb{C}$-algebra A and $\alpha : [0,1] \to M$ is a piecewise smooth path, then

$$\int_\alpha \omega_1 \omega_2 \ldots \omega_r = \int_{\Delta^r} f_1(t_1) \ldots f_r(t_r) dt_1 dt_2 \ldots dt_r.$$

where $f_j(t)dt = \alpha^*\omega_j$ and Δ^r is the "time ordered" r-simplex

$$\Delta^r = \{(t_1,\dots,t_r) \in \mathbb{R}^n : 0 \le t_1 \le t_2 \le \cdots \le t_r \le 1\}.$$

An exposition of the basic properties of iterated integrals can be found in [13, 17].

1.1.2. Filtrations

The lower central series (LCS) $L^\bullet G$ of a group G is defined by

$$G = L^1 G \supseteq L^2 G \supseteq L^3 G \supseteq \cdots$$

where $L^{m+1}G = [G, L^m G]$. Its associated graded $\mathrm{Gr}^\bullet_{\mathrm{LCS}} G$ is a graded Lie algebra over $\mathbb{Z}$ whose mth graded quotient is $\mathrm{Gr}^m_{\mathrm{LCS}} G := L^m G / L^{m+1} G$.

The lower central series $L^\bullet\mathfrak{g}$ of a Lie algebra $\mathfrak{g}$ is defined similarly. A Lie algebra $\mathfrak{g}$ is *nilpotent* if $L^N\mathfrak{g} = 0$ for some $N \ge 0$.

1.1.3. Hodge theory

All mixed Hodge structures will be $\mathbb{Q}$ mixed Hodge structures unless otherwise stated. The category of $\mathbb{Q}$-mixed Hodge structures will be denoted by MHS. The category of $\mathbb{R}$-mixed Hodge structures will be denoted by $\mathsf{MHS}_{\mathbb{R}}$.

Often we will abbreviate mixed Hodge structure by MHS, variation of MHS by VMHS, mixed Hodge complex by MHC, cohomological MHC by CMHC. The category of admissible VMHS over a smooth variety X will be denoted by $\mathsf{MHS}(X)$.

2 Preliminaries

2.1. Proalgebraic groups

In this paper, the term *algebraic group* will refer to a linear algebraic group. Suppose that F is a field of characteristic zero. A proalgebraic group G over F is an inverse limit of algebraic F-groups G_α. The coordinate ring $\mathcal{O}(G)$ of G is the direct limit of the coordinate rings of the G_α. The Lie algebra $\mathfrak{g}$ of G is the inverse limit of the Lie algebras $\mathfrak{g}_\alpha$ of the G_α. It is a Hausdorff topological Lie algebra. The neighbourhoods of 0 are the kernels of the canonical projections $\mathfrak{g} \to \mathfrak{g}_\alpha$.

The continuous cohomology of $\mathfrak{g} = \varprojlim \mathfrak{g}_\alpha$ is defined by

$$H^\bullet(\mathfrak{g}) := \varinjlim_\alpha H^\bullet(\mathfrak{g}_\alpha).$$

Its homology is the full dual:

$$H_\bullet(\mathfrak{g}) := \mathrm{Hom}_F(H^\bullet(\mathfrak{g}), F) \cong \varprojlim H_\bullet(\mathfrak{g}_\alpha)$$

Each homology group is a Hausdorff topological vector space; the neighbourhoods of 0 are the kernels of the natural maps $H_\bullet(\mathfrak{g}) \to H_\bullet(\mathfrak{g}_\alpha)$.

Continuous cohomology can be computed using continuous Chevalley-Eilenberg cochains:

$$C^\bullet(\mathfrak{g}) = \mathrm{Hom}_F^{\mathrm{cts}}(\Lambda^\bullet \mathfrak{g}, F) := \varinjlim_\alpha \mathrm{Hom}_F(\Lambda^\bullet \mathfrak{g}_\alpha, F)$$

with the usual differential.

If, instead, $\mathfrak{g} = \bigoplus_m \mathfrak{g}_m$ is a graded Lie algebra, then the homology and cohomology of $\mathfrak{g}$ are also graded. This follows from the fact that the grading of $\mathfrak{g}$ induces a grading of the Chevalley-Eilenberg chains and cochains of $\mathfrak{g}$.

2.2. Prounipotent groups and pronilpotent Lie algebras

A prounipotent F-group is a proalgebraic group that is an inverse limit of unipotent F-groups.

A pronilpotent Lie algebra over a F is an inverse limit of finite dimensional nilpotent Lie algebras. The Lie algebra of a prounipotent group is a pronilpotent Lie algebra. The functor that takes a prounipotent group to its Lie algebra is an equivalence of categories between the category of unipotent F-groups and the category of pronilpotent Lie algebras over F.

The following useful result is an analogue for pronilpotent Lie algebras of a classical result of Stallings [40]. A proof can be found in [18, §3].

Proposition 2.1. *For a homomorphism $\varphi : \mathfrak{n}_1 \to \mathfrak{n}_2$ of pronilpotent Lie algebras, the following are equivalent:*

(i) *φ is an isomorphism,*
(ii) *$\varphi^* : H^\bullet(\mathfrak{n}_2) \to H^\bullet(\mathfrak{n}_1)$ is an isomorphism,*
(iii) *$\varphi^* : H^j(\mathfrak{n}_2) \to H^j(\mathfrak{n}_1)$ is an isomorphism when $j = 1$ and injective when $j = 2$.*

□

Another useful fact that we shall need is the following exact sequence, which is essentially due to Sullivan [42].

Proposition 2.2. *If $\mathfrak{n}$ is a pronilpotent Lie algebra over F, then the sequence*

$$0 \longrightarrow (\mathrm{Gr}^2_{\mathrm{LCS}}\, \mathfrak{n})^* \xrightarrow{[\ ,\ ]^*} \Lambda^2 H^1(\mathfrak{n}) \xrightarrow{cup} H^2(\mathfrak{n})$$

is exact, where $(\ )^* = \mathrm{Hom}^{\mathrm{cts}}(\ , F)$ and $[\ ,\ ]^*$ denotes the continuous dual of the bracket $\Lambda^2 H_1(\mathfrak{n}) \to \mathrm{Gr}^2_{\mathrm{LCS}}\, \mathfrak{n}$.

Remark 2.3. When $H_1(\mathfrak{n})$ is finite dimensional, one can dualize to obtain the exact sequence

$$H_2(\mathfrak{n}) \xrightarrow{\text{cup}^*} \Lambda^2 H_1(\mathfrak{n}) \xrightarrow{[\,,\,]} \mathrm{Gr}^2_{\mathrm{LCS}}\,\mathfrak{n} \longrightarrow 0.$$

However, when $H_1(\mathfrak{n})$ is infinite dimensional, this sequence is not exact. Instead, one needs to replace $\Lambda^2 H_1(\mathfrak{n})$ by the alternating part of the completed tensor product $H_1(\mathfrak{n})\hat{\otimes} H_1(\mathfrak{n})$.

2.3. Free Lie algebras

Suppose that F is a field of characteristic 0 and that V is a vector space over F. Here we are not assuming V to be finite dimensional. The free Lie algebra generated by V will be denoted by $\mathbb{L}(V)$. It is characterized by the property that a linear map $V \to \mathfrak{g}$ into a Lie algebra over F induces a unique Lie algebra homomorphism $\mathbb{L}(V) \to \mathfrak{g}$. The Poincaré-Birkhoff-Witt Theorem implies [37] that $\mathbb{L}(V)$ is the Lie subalgebra of the tensor algebra $T(V)$ (with bracket $[A,B] = AB - BA$) generated by V and that the inclusion $\mathbb{L}(V) \to T(V)$ induces an isomorphism $U\mathbb{L}(V) \to T(V)$ from the enveloping algebra of $\mathbb{L}(V)$ to $T(V)$. The cohomology of $\mathbb{L}(V)$ with trivial coefficients vanishes in degrees > 1.

If $\mathfrak{f}$ is a Lie algebra, then any splitting of the projection $\mathfrak{f} \to H_1(\mathfrak{f})$ induces a homomorphism $\mathbb{L}(H_1(\mathfrak{f})) \to \mathfrak{f}$. If $\mathfrak{f}$ is free, then this homomorphism is an isomorphism [37]. It induces a *canonical* isomorphism

$$\mathrm{Gr}^\bullet_{\mathrm{LCS}}\,\mathfrak{f} \cong \mathbb{L}(H_1(\mathfrak{f})) \tag{2.4}$$

of the graded Lie algebra associated to the lower central series (LCS) of $\mathfrak{f}$ with the free Lie algebra generated by its first graded quotient $H_1(\mathfrak{f}) = \mathfrak{f}/L^2\mathfrak{f}$.

The *free completed Lie algebra* $\mathbb{L}(V)^\wedge$ generated by V is defined to be

$$\mathbb{L}(V)^\wedge = \varprojlim \mathfrak{n},$$

where $\mathfrak{n}$ ranges over all finite dimensional nilpotent quotients of $\mathbb{L}(V)$. It is viewed as a topological Lie algebra. It is useful to note that there is a canonical isomorphism

$$\mathbb{L}(V)^\wedge = \varprojlim_{W,n} \mathbb{L}(W)/L^n\mathbb{L}(W)$$

of topological Lie algebras, where W ranges over all finite dimensional quotients of V and n over all positive integers.

We can regard V as a topological vector space: the neighbourhoods of 0 are the subspaces of V of finite codimension. Every continuous linear mapping $V \to \mathfrak{u}$ from V into a pronilpotent Lie algebra induces a unique continuous homomorphism $\mathbb{L}(V)^\wedge \to \mathfrak{u}$. The continuous cohomology of $\mathbb{L}(V)^\wedge$ vanishes in degrees ≥ 2.

If $\mathfrak{n}$ is a pronilpotent Lie algebra, then any continuous section of the quotient mapping $\mathfrak{n} \to H_1(\mathfrak{n})$ induces a continuous surjective homomorphism $\mathbb{L}(H_1(\mathfrak{n}))^\wedge \to \mathfrak{n}$. Applying Proposition 2.1 to this homomorphism, we obtain:

Proposition 2.5. *A pronilpotent Lie algebra is free if and only if* $H^2(\mathfrak{n}) = 0$. □

Part 1. Completed Path Torsors of Affine Curves

3 Relative Completion in the Abstract

Suppose that Γ is a discrete group and that R is a reductive algebraic group over a field F of characteristic zero. The completion of Γ relative to a Zariski dense representation $\rho : \Gamma \to R(F)$ is a proalgebraic F-group $\mathcal{G}$ which is an extension

$$1 \to \mathcal{U} \to \mathcal{G} \to R \to 1$$

of R by a prounipotent group $\mathcal{U}$, and a homomorphism $\hat{\rho} : \Gamma \to \mathcal{G}(F)$ such that the composite

$$\Gamma \xrightarrow{\hat{\rho}} \mathcal{G}(F) \longrightarrow R(F)$$

is ρ. It is universal for such groups: if G is a proalgebraic F group that is an extension of R by a prounipotent group, and if $\phi : \Gamma \to G(F)$ is a homomorphism whose composition with $G \to R$ is ρ, then there is a homomorphism $\hat{\phi} : \mathcal{G} \to G$ of proalgebraic F-groups such that the diagram

$$\begin{array}{ccc} \Gamma & \xrightarrow{\hat{\rho}} & \mathcal{G}(F) \\ {\scriptstyle\phi}\downarrow & \swarrow{\scriptstyle\hat{\phi}} & \downarrow \\ G(F) & \longrightarrow & R(F) \end{array}$$

commutes.

When R is trivial, ρ is trivial and $\mathcal{G} = \mathcal{U}$ is the unipotent completion of Γ over F.

Relative completion can be defined as follows: Let $\mathcal{L}(\Gamma, R)$ denote the category of finite dimensional F-linear representations V of Γ that admit a filtration

$$0 = V_0 \subset V_1 \subset \cdots \subset V_N = V$$

by Γ-submodules with the property that each graded quotient V_j/V_{j-1} is an R-module and the action of Γ on it factors through ρ. It is a neutral tannakian category. The completion of Γ relative to ρ is the fundamental group of this category with respect to the fiber functor that takes a representation to its underlying vector space.

We will generally be sloppy and not distinguish between a proalgebraic group G and its group $G(F)$ of F-rational points. For example, in the context of relative completion, ρ will be a homomorphism $\Gamma \to R$.

3.1. Levi splittings

The following generalization of Levi's Theorem implies that the relative completion $\mathcal{G}$ of a finitely generated group Γ can be expressed (non-canonically) as a semi-direct product $\mathcal{G} \cong R \ltimes \mathcal{U}$. The Lie algebra $\mathfrak{u}$ of $\mathcal{U}$ is then a pronilpotent Lie algebra in the category of R-modules. The isomorphism type of $\mathcal{G}$ is determined by $\mathfrak{u}$ with its R-action.

Suppose that F is a field of characteristic 0 and that R is a reductive F-group. Call an extension

$$1 \to \mathcal{U} \to \mathcal{G} \to R \to 1$$

of R by a prounipotent group in the category of affine F-groups *quasi-finite* if for all finite dimensional R-modules V, $\mathrm{Hom}_R(V, H_1(\mathcal{U}))$ is finite dimensional. The results in the following section imply that the completion of a finitely generated group Γ relative to a homomorphism $\rho : \Gamma \to R(F)$ is a quasi finite extension of R.

Proposition 3.1. *Every quasi-finite extension of R by a prounipotent group $\mathcal{U}$ is split. Moreover, any two splittings are conjugate by an element of $\mathcal{U}(F)$.*

Sketch of Proof. The classical case where $\mathcal{U}$ is an abelian unipotent group (i.e., a finite dimensional vector space) was proved by Mostow in [32]. (See also, [2, Prop. 5.1].)

First consider the case where $\mathcal{U}$ is an abelian proalgebraic group. The quasi-finiteness assumption implies that there are (finite dimensional) abelian unipotent groups $\mathcal{U}_\alpha$ with R-action and an R-equivariant isomorphism

$$\mathcal{U} \cong \prod_\alpha \mathcal{U}_\alpha.$$

The extension of $\mathcal{G}$ by $\mathcal{U}$ can be pushed out along the projection $\mathcal{U} \to \mathcal{U}_\alpha$ to obtain extensions

$$1 \to \mathcal{U}_\alpha \to \mathcal{G}_\alpha \to R \to 1.$$

The classical case, stated above, implies that each of these has a splitting s_α and that this splitting is unique up to conjugation by an element of $\mathcal{U}_\alpha$. These sections assemble to give a section $s = (s_\alpha)$ of $\mathcal{G} \to R$ that is defined over F.

Every section of $\mathcal{G} \to R$ is of this form. Any two are conjugate by an element of $\mathcal{U}(F)$.

To prove the general case, consider the extensions

$$1 \to \mathcal{U}_n \to \mathcal{G}_n \to R \to 1 \tag{3.2}$$

where $\mathcal{G}_n = \mathcal{G}/L^{n+1}\mathcal{U}$, $\mathcal{U}_n = \mathcal{U}/L^{n+1}\mathcal{U}$, and where $L^n\mathcal{U}$ denotes the nth term of the LCS of $\mathcal{U}$. The result is proved by constructing a compatible sequence of sections of these extensions. We have already established the $n = 1$ case. Suppose that $n > 1$ and that we have constructed a splitting of s_{n-1} of $\mathcal{G}_{n-1} \to R$ and shown that any two such splittings are conjugate by an element of $\mathcal{U}_{n-1}$.

Pulling back the extension

$$1 \to \mathrm{Gr}^n_{\mathrm{LCS}}\mathcal{U} \to \mathcal{G}_n \to \mathcal{G}_{n-1} \to 1$$

along s_{n-1} gives an extension

$$1 \to \mathrm{Gr}^n_{\mathrm{LCS}}\mathcal{U} \to G \to R \to 1.$$

The quasi-finite assumption implies that the R-module $\mathrm{Gr}^n_{\mathrm{LCS}}\mathcal{U}$ is a product of finite dimensional R-modules. The $n = 1$ case implies that this extension is split and that any two splittings are conjugate by an element of $\mathrm{Gr}^n_{\mathrm{LCS}}\mathcal{U}$. If s is a section of $G \to R$, then the composition of s with the inclusion $G \hookrightarrow \mathcal{G}_n$ is a section s_n of (3.2) that is compatible with s_{n-1}:

$$\begin{array}{ccccccccc}
1 & \longrightarrow & \mathrm{Gr}^n_{\mathrm{LCS}}\mathcal{U} & \longrightarrow & G & \overset{s}{\underset{\longrightarrow}{\curvearrowleft}} & R & \longrightarrow & 1 \\
 & & \| & & \downarrow & \swarrow_{s_n} & \downarrow s_{n-1} & & \\
1 & \longrightarrow & \mathrm{Gr}^n_{\mathrm{LCS}}\mathcal{U} & \longrightarrow & \mathcal{G}_n & \longrightarrow & \mathcal{G}_{n-1} & \longrightarrow & 1
\end{array}$$

The uniqueness of s implies that any two such lifts of s_{n-1} are conjugate by an element of $\mathrm{Gr}^n_{\mathrm{LCS}}\mathcal{U}(F)$. This and the fact that s_{n-1} is unique up to conjugation by an element of $\mathcal{U}(F)$ implies that s_n is as well. □

3.2. Cohomology

We continue with the notation above, where $\mathcal{G}$ is the relative completion of Γ. When R is reductive, the structure of $\mathfrak{g}$ and $\mathfrak{u}$ are closely related to the cohomology of Γ with coefficients in rational representations of R. We will assume also that $H^j(\Gamma, V)$ is finite dimensional when $j \leq 2$ for all rational representations V of R. This condition is satisfied when Γ is finitely presented and thus by fundamental groups of all complex algebraic varieties.

For each rational representation V of R there are natural isomorphisms

$$\mathrm{Hom}^{\mathrm{cts}}_R(H_\bullet(\mathfrak{u}), V) \cong [H^\bullet(\mathfrak{u}) \otimes V]^R \cong H^\bullet(\mathcal{G}, V).$$

The homomorphism $\Gamma \to \mathcal{G}(F)$ induces a homomorphism

$$(3.3) \qquad H^\bullet(\mathfrak{u}, V)^R \cong H^\bullet(\mathcal{G}, V) \to H^\bullet(\Gamma, V)$$

It is an isomorphism in degrees ≤ 1 and an injection in degree 2.

Denote the set of isomorphism classes of finite dimensional irreducible representations of R by $\check{R}$. Fix an R-module V_λ in each isomorphism class $\lambda \in \check{R}$. If each irreducible representation of R is absolutely irreducible[3] and if $H^j(\Gamma, V)$ is finite dimensional for all rational representations V of R when $j = 1, 2$, then (3.3) implies that there is an isomorphism

$$(3.4) \qquad \prod_{\lambda \in \check{R}} [H^1(\Gamma, V_\lambda)]^* \otimes_F V_\lambda \cong H_1(\mathfrak{u})$$

of topological modules, and that there is a continuous R-invariant surjection

$$\prod_{\lambda \in \check{R}} [H^2(\Gamma, V_\lambda)]^* \otimes_F V_\lambda \to H_2(\mathfrak{u}).$$

In both cases, the LHS has the product topology.

3.3. Base change

When discussing the mixed Hodge structure on a relative completion of the fundamental group of a complex algebraic manifold X, we need to be able to compare the completion of $\pi_1(X, x)$ over $\mathbb{R}$ (or $\mathbb{Q}$) with its completion over $\mathbb{C}$. For this reason we need to discuss the behaviour of relative completion under base change.

The cohomological properties of relative completion stated above imply that it behaves well under base change. To explain this, suppose that K is an extension field of F. Then $\rho_K : \Gamma \to R(K)$ is Zariski dense in $R \times_F K$, so one has the completion $\mathcal{G}_K$ of Γ relative to ρ_K. It is an extension of $R \times_F K$ by a prounipotent group. The universal mapping property of $\mathcal{G}_K$ implies that the homomorphism $\Gamma \to \mathcal{G}(K)$ induces a homomorphism $\mathcal{G}_K \to \mathcal{G} \times_F K$ of proalgebraic K-groups. The fact that (3.3) is an isomorphism in degree 1 and injective in degree 2 implies that this homomorphism is an isomorphism.

3.4. Examples

Here the coefficient field F will be $\mathbb{Q}$. But because of base change, the discussion is equally valid when F is any field of characteristic 0.

[3] This is the case when $R = \mathrm{Sp}_g$ over any field of characteristic zero.

3.4.1. Free groups

Suppose that Γ is a finitely generated free group and that $\rho : \Gamma \to R(F)$ is a Zariski dense reductive representation. Denote the completion of Γ with respect to ρ by $\mathcal{G}$ and its unipotent radical by $\mathcal{U}$. Denote their Lie algebras by $\mathfrak{g}$ and $\mathfrak{u}$. Since $H^j(\Gamma, V)$ vanishes for all R-modules V for all $j \geq 2$, $\mathfrak{u}$ is free. Consequently, the homomorphism (3.3) is an isomorphism in all degrees.

3.4.2. Modular groups

Suppose that Γ is a modular group — that is, a finite index subgroup of $\mathrm{SL}_2(\mathbb{Z})$. Let $R = \mathrm{SL}_2$ and $\rho : \Gamma \to \mathrm{SL}_2(\mathbb{Q})$ be the inclusion. This has Zariski dense image. Denote the completion of Γ with respect to ρ by $\mathcal{G}$ and its unipotent radical by $\mathcal{U}$.

Every torsion free subgroup Γ' of $\mathrm{SL}_2(\mathbb{Z})$ is the fundamental group of the quotient $\Gamma' \backslash \mathfrak{h}$ of the upper half plane by Γ'. Since this is a non-compact Riemann surface, Γ' is free. Since $\mathrm{SL}_2(\mathbb{Z})$ has finite index torsion free subgroups (e.g., the matrices congruent to the identity mod m for any $m \geq 3$), every modular group is virtually free. This implies that $H^j(\Gamma, V) = 0$ whenever $j \geq 2$ and V is a rational vector space. The results of Section 3.2 imply that the Lie algebra $\mathfrak{u}$ of $\mathcal{U}$ is a free pronilpotent Lie algebra. As in the case of a free group, this implies that the homomorphism (3.3) is an isomorphism in all degrees.

The set $\check{R}$ of isomorphism classes of irreducible R-modules is $\mathbb{N}$. The natural number n corresponds to the nth symmetric power $S^n H$ of the defining representation H of SL_2. The results of Section 3.2 imply that there is a non-canonical isomorphism

$$\mathfrak{u} \cong \mathbb{L}\Big(\bigoplus_{n \geq 0} H^1(\Gamma, S^n H)^* \otimes S^n H\Big)^{\wedge}$$

of pronilpotent Lie algebras in the category of SL_2 modules. (Cf. Remarks 3.9 and 7.2 in [16].) So we have a complete description of $\mathcal{G}$ as a proalgebraic group:

$$\mathcal{G} \cong \mathrm{SL}_2 \ltimes \exp \mathfrak{u}.$$

In Section 4 we give a method for constructing a homomorphism $\Gamma \to \mathrm{SL}_2(\mathbb{C}) \ltimes \exp \mathfrak{u}$ that induces an isomorphism $\mathcal{G} \to \mathrm{SL}_2(\mathbb{C}) \ltimes \exp \mathfrak{u}$.

3.4.3. Unipotent completion of fundamental groups of punctured elliptic curves

Here E is a smooth elliptic curve over $\mathbb{C}$ and $\Gamma = \pi_1(E', x)$ where $E' = E - \{0\}$ and $x \in E'$. In this case we take R to be trivial. The corresponding completion of Γ is the unipotent completion of $\pi_1(E', x)$. Since $H^2(E') = 0$, the results of Section 3.2 imply that the Lie algebra $\mathfrak{p}$ of the unipotent completion of

$\pi_1(E',x)$ is (non-canonically isomorphic to) the completion of the free Lie algebra generated by $H_1(E,\mathbb{Q})$:

$$\mathfrak{p} \cong \mathbb{L}(H_1(E))^{\wedge}.$$

This induces a canonical isomorphism $\operatorname{Gr}^{\bullet}_{\mathrm{LCS}} \mathfrak{p} \cong \mathbb{L}(H_1(E))$ of the associated graded Lie algebra of the lower central series (LCS) of $\mathfrak{p}$ with the free Lie algebra generated by $H_1(E)$.

3.5. Naturality and Right exactness

The following naturality property is easily proved using either the universal mapping property of relative completion or its tannakian description.

Proposition 3.5. *Suppose that Γ and Γ' are discrete groups and that R and R' are reductive F-groups. If one has a commutative diagram*

$$\begin{array}{ccc} \Gamma' & \xrightarrow{\rho'} & R' \\ \downarrow & & \downarrow \\ \Gamma & \xrightarrow{\rho} & R \end{array}$$

in which ρ and ρ' are Zariski dense, then one has a commutative diagram

$$\begin{array}{ccccc} \Gamma' & \longrightarrow & \mathcal{G}' & \longrightarrow & R' \\ \downarrow & & \downarrow & & \downarrow \\ \Gamma & \longrightarrow & \mathcal{G} & \longrightarrow & R \end{array}$$

where $\mathcal{G}$ and $\mathcal{G}'$ denote the completions of Γ and Γ' with respect to ρ and ρ'. □

Relative completion is not, in general, an exact functor. However, it is right exact. The following is a special case of this right exactness. It can be proved using the universal mapping property of relative completion. (A similar argument can be found in [22, §4.5].)

Proposition 3.6. *Suppose that Γ, Γ' and Γ'' are discrete groups and that R, R' and R'' are reductive F-groups. Suppose that one has a diagram*

$$\begin{array}{ccccccccc} 1 & \longrightarrow & \Gamma' & \longrightarrow & \Gamma & \longrightarrow & \Gamma'' & \longrightarrow & 1 \\ & & \downarrow{\scriptstyle \rho'} & & \downarrow{\scriptstyle \rho} & & \downarrow{\scriptstyle \rho''} & & \\ 1 & \longrightarrow & R' & \longrightarrow & R & \longrightarrow & R'' & \longrightarrow & 1 \end{array}$$

with exact rows in which ρ, ρ' and ρ'' are Zariski dense, then the corresponding diagram

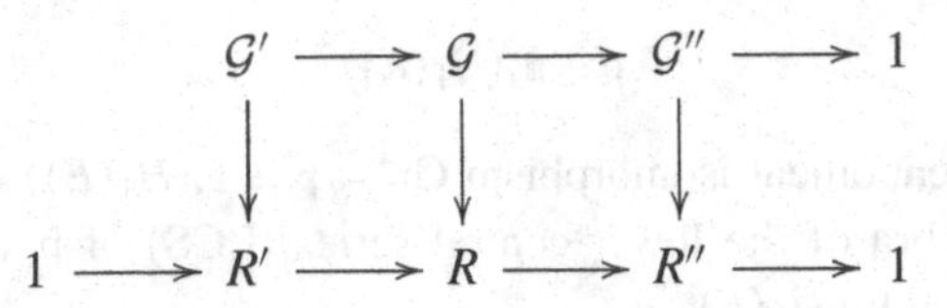

of relative completions has right exact top row. □

Example 3.7. The moduli space of $n \geq 1$ pointed genus 1 curves will be denoted by $\mathcal{M}_{1,n}$. It will be regarded as an orbifold. It is isomorphic to (and will be regarded as) the moduli space of elliptic curves $(E,0)$ with $n-1$ distinct labelled points $(x_1,\ldots,x_{n-1})$. The point of $\mathcal{M}_{1,n}$ that corresponds to $(E,0,x_1,\ldots,x_{n-1})$ will be denoted by $[E,x_1,\ldots,x_{n-1}]$.

The fiber of the projection $\mathcal{M}_{1,2} \to \mathcal{M}_{1,1}$ that takes $[E,x]$ to $[E]$ is $E' := E - \{0\}$. Fix a base point $x_o = [E,x]$ of $\mathcal{M}_{1,2}$ and $t_o = [E] \in \mathcal{M}_{1,1}$. The (orbifold) fundamental group of $\mathcal{M}_{1,2}$ is an extension

$$1 \to \pi_1(E',x) \to \pi_1(\mathcal{M}_{1,2},x_o) \to \pi_1(\mathcal{M}_{1,1},t_o) \to 1.$$

Denote the completion of $\pi_1(\mathcal{M}_{1,2},x_o)$ with respect to the natural homomorphism to $\mathrm{SL}(H_1(E)) \cong \mathrm{SL}_2(\mathbb{Q})$ by $\widetilde{\mathcal{G}}$. Functoriality and right exactness of relative completion implies that we have an exact sequence

$$\pi_1(E',x)^{\mathrm{un}} \to \widetilde{\mathcal{G}} \to \mathcal{G} \to 1.$$

In this case, we can prove exactness on the left as well.

This is proved using the conjugation action of $\pi_1(\mathcal{M}_{1,2},x_o)$ on $\pi_1(E',x)$, which induces an action of $\pi_1(\mathcal{M}_{1,2},x_o)$ on the Lie algebra $\mathfrak{p}$ of $\pi_1(E',x)^{\mathrm{un}}$. This action preserves the lower central series filtration of $\mathfrak{p}$ and therefore induces an action on $\mathrm{Gr}^\bullet_{\mathrm{LCS}}\mathfrak{p} \cong \mathbb{L}(H_1(E))$. This action is determined by its action on $H_1(E)$, and therefore factors through the homomorphism $\pi_1(\mathcal{M}_{1,2},x_o) \to \mathrm{SL}_2(\mathbb{Q})$. The universal mapping property of relative completion implies that this induces an action $\widetilde{\mathcal{G}} \to \operatorname{Aut}\mathfrak{p}$ and the corresponding Lie algebra homomorphism $\tilde{\mathfrak{g}} \to \operatorname{Der}\mathfrak{p}$. The composite $\mathfrak{p} \to \tilde{\mathfrak{g}} \to \operatorname{Der}\mathfrak{p}$ is the homomorphism induced by the conjugation action of $\pi_1(E',x)^{\mathrm{un}}$ on itself and is therefore the adjoint action. Since $\mathfrak{p}$ is free of rank > 1, it has trivial center, which implies that the adjoint action is faithful and that $\mathfrak{p} \to \tilde{\mathfrak{g}}$ is injective.

3.6. Hodge Theory

Suppose that X is the complement of a normal crossings divisor in a compact Kähler manifold. Suppose that $F = \mathbb{Q}$ or $\mathbb{R}$ and that $\mathbb{V}$ is a polarized variation of F-Hodge structure over X. Pick a base point $x_o \in X$.[4] Denote the fiber over $\mathbb{V}$ over x_o by V_o. The Zariski closure of the image of the monodromy representation

$$\rho : \pi_1(X, x_o) \to \mathrm{Aut}(V_o)$$

is a reductive F-group, [11, 4.2.6]. Denote it by R. Then one has the relative completion $\mathcal{G}$ of $\pi_1(X, x_o)$ with respect to $\rho : \pi_1(X, x_o) \to R(F)$.

Theorem 3.8 ([15]). *The coordinate ring $\mathcal{O}(\mathcal{G})$ is a Hopf algebra in the category of Ind-mixed Hodge structures over F. It has the property that $W_{-1}\mathcal{O}(\mathcal{G}) = 0$ and $W_0\mathcal{O}(\mathcal{G}) = \mathcal{O}(R)$.*

A slightly weaker version of the theorem is stated in terms of Lie algebras. Denote the prounipotent radical of $\mathcal{G}$ by $\mathcal{U}$. Denote their Lie algebras by $\mathfrak{g}$ and $\mathfrak{u}$, and the Lie algebra of R by $\mathfrak{r}$.

Corollary 3.9 ([15]). *The Lie algebra $\mathfrak{g}$ is a Lie algebra in the category of pro-mixed Hodge structures over F. It has the property that*

$$\mathfrak{g} = W_0\mathfrak{g},\ \mathfrak{u} = W_{-1}\mathfrak{g},\ \text{and } \mathrm{Gr}_0^W \mathfrak{g} \cong \mathfrak{r}.$$

If $\mathbb{V}$ is a PVHS over X with fiber V_o over the base point x_o, then the composite

$$H^\bullet(\mathfrak{u}, V_o)^R \to H^\bullet(\Gamma, V_o) \to H^\bullet(X, \mathbb{V})$$

of (3.3) with the canonical homomorphism is a morphism of MHS. It is an isomorphism in degrees ≤ 1 and injective in degree 2. When X is an (orbi) curve, it is an isomorphism in all degrees.

The existence of the mixed Hodge structure on $\mathfrak{u}$ in the unipotent case and when X is not necessarily compact is due to Morgan [31] and Hain [15]. The results in this section also hold in the orbifold case.

Example 3.10. The local system $R^1 f_* \mathbb{Q}$ over $\mathcal{M}_{1,1}$ associated to the universal elliptic curve $f : \mathcal{E} \to \mathcal{M}_{1,1}$ is a polarized variation of Hodge structure of weight 1. This variation and its pullback to $\mathcal{M}_{1,n}$ will be denoted by $\mathbb{H}$. It has fiber $H^1(E)$ over $[E]$. The Zariski closure of the monodromy representation $\pi_1(\mathcal{M}_{1,1}, [E]) \to \mathrm{Aut}\, H^1(E)$ is $\mathrm{SL}(H_1(E))$, which is isomorphic to SL_2. Poincaré duality $H_1(E) \cong H^1(E)(1)$ induces an isomorphism of $\mathbb{H}(1)$ with the local system over $\mathcal{M}_{1,1}$ whose fiber over $[E]$ is $H_1(E)$.

The choice of an elliptic curve E and a non-zero point x of E determines compatible base points of E', $\mathcal{M}_{1,1}$ and $\mathcal{M}_{1,2}$. Denote the Lie algebras of the relative completions of $\pi_1(\mathcal{M}_{1,1}, [E])$ and $\pi_1(\mathcal{M}_{1,2}, [E, x])$ by $\mathfrak{g}$ and $\tilde{\mathfrak{g}}$, respectively. Denote the Lie algebra of the unipotent completion of $\pi_1(E', x)$

[4] We also allow tangential base points.

by $\mathfrak{p}$. The results of this section imply that each has a natural MHS and that the sequence

$$0 \to \mathfrak{p} \to \tilde{\mathfrak{g}} \to \mathfrak{g} \to 0$$

is exact in the category of MHS. The adjoint action of $\tilde{\mathfrak{g}}$ on $\mathfrak{p}$ induces an action

$$\tilde{\mathfrak{g}} \to \operatorname{Der} \mathfrak{p}$$

Since the inclusion $\mathfrak{p} \to \tilde{\mathfrak{g}}$ is a morphism of MHS, this homomorphism is a morphism of MHS.

Since the functor $\operatorname{Gr}^\bullet_W$ is exact on the category of MHS, one can study this action by passing to its associated graded action

$$\operatorname{Gr}^W_\bullet \tilde{\mathfrak{g}} \to \operatorname{Der} \mathbb{L}(H_1(E)).$$

4 A Concrete Approach to Relative Completion

Suppose that M is the orbifold quotient[5] $\Gamma \backslash X$ of a simply connected manifold X by a discrete group Γ. We suppose that Γ acts properly discontinuously and virtually freely[6] on X. Our main example will be when $X = \mathfrak{h}$ and Γ is a finite index subgroup of $\mathrm{SL}_2(\mathbb{Z})$.

Suppose that R is a complex (or real) Lie group and that $(\mathfrak{u}_\alpha)_\alpha$ is an inverse system of finite dimensional nilpotent Lie algebras in the category of left R-modules. Its limit

$$\mathfrak{u} = \varprojlim_\alpha \mathfrak{u}_\alpha$$

is a pronilpotent Lie algebra in the category of R-modules. Denote the unipotent Lie group corresponding to $\mathfrak{u}_\alpha$ by U_α. The prounipotent Lie group corresponding to $\mathfrak{u}$ is the inverse limit of the U_α.

The action of R on $\mathfrak{u}$ induces an action of R on $\mathcal{U}$, so we can form the semi-direct product[7] $R \ltimes \mathcal{U}$. This is the inverse limit

$$R \ltimes \mathcal{U} = \varprojlim_\alpha (R \ltimes U_\alpha)$$

If R is an algebraic group, then $R \ltimes \mathcal{U}$ is a proalgebraic group.

Suppose that $\rho : \Gamma \to R$ is a representation. At this stage, we do not assume that ρ has Zariski dense image. The following assertion is easily proved.

5 For a detailed and elementary description of what this means, see [19, §3].

6 That is, Γ has a finite index subgroup that acts freely on X.

7 To be clear, the group $R \ltimes U$ is the set $U \times R$ with multiplication

$$(u_1, r_1)(u_2, r_2) = (u_1(r_1 \cdot u_2), r_1 r_2),$$

where $r \cdot u$ denotes the action of R on U. We will omit the dot when it is clear from the context that ru means the action of R on U.

Lemma 4.1. *Homomorphisms $\hat{\rho} : \Gamma \to R \ltimes \mathcal{U}$ that lift ρ correspond to functions $F : \Gamma \to \mathcal{U}$ that satisfy the 1-cocycle condition*

$$F(\gamma_1\gamma_2) = F(\gamma_1)(\gamma_1 \cdot F(\gamma_2)).$$

The homomorphism $\hat{\rho}$ corresponds to the function $\Gamma \to \mathcal{U} \times R$

$$\gamma \mapsto \big(F(\gamma), \rho(\gamma)\big)$$

under the identification of $R \ltimes \mathcal{U}$ with $\mathcal{U} \times R$. □

Cocycles can be constructed from Γ-invariant 1-forms on X with values in $\mathfrak{u}$. Define

$$E^\bullet(X)\hat{\otimes}\mathfrak{u} := \varprojlim_\alpha E^\bullet(X) \otimes \mathfrak{u}_\alpha,$$

where $E^\bullet(X)$ denotes the complex of smooth $\mathbb{C}$-valued forms on X. The group Γ acts on $E^\bullet(X)\hat{\otimes}\mathfrak{u}$ by

$$\gamma \cdot \omega = \big((\gamma^*)^{-1} \otimes \gamma\big)\omega.$$

Such a form ω is invariant if

$$(\gamma^* \otimes 1)\omega = (1 \otimes \gamma)\omega$$

for all $\gamma \in \Gamma$.

Let Γ act on $X \times \mathcal{U}$ diagonally: $\gamma : (u,x) \to (\gamma u, \gamma x)$. The projection

$$X \times \mathcal{U} \to X$$

is a Γ-equivariant principal right $\mathcal{U}$-bundle. Its sections correspond to functions $f : X \to \mathcal{U}$. Each $\omega \in E^1(X)\hat{\otimes}\mathfrak{u}$ defines a connection on this bundle invariant under the right $\mathcal{U}$ action via the formula

$$\nabla f = df + \omega f,$$

where f is a $\mathcal{U}$-valued function defined locally on X. The connection is Γ-invariant if and only if ω is Γ-invariant. It is flat if and only if ω is integrable:

$$d\omega + \frac{1}{2}[\omega, \omega] = 0 \quad \text{in } E^2(X)\hat{\otimes}\mathfrak{u}.$$

In this case, parallel transport defines a function

$$T : PX \to \mathcal{U}$$

from the path space of X into $\mathcal{U}$. With our conventions, this satisfies $T(\alpha * \beta) = T(\beta)T(\alpha)$. When ω is integrable, $T(\alpha)$ depends only on the homotopy class of γ relative to its endpoints. Chen's transport formula implies that the inverse transport function is given by the formula

$$T(\alpha)^{-1} = 1 + \int_\alpha \omega + \int_\alpha \omega\omega + \int_\alpha \omega\omega\omega + \cdots \tag{4.2}$$

Cf. [20, Cor. 5.6].

Fix a point $x_o \in X$. Since X is simply connected, for each $\gamma \in \Gamma$ there is a unique homotopy class c_γ of paths from x_o to $\gamma \cdot x_o$.

Proposition 4.3. *If $\omega \in E^1(X)\hat{\otimes}\mathfrak{u}$ is Γ-invariant and integrable, then the function $\Theta_{x_o} : \Gamma \to \mathcal{U}$ defined by*

$$\Theta_{x_o}(\gamma) = T(c_\gamma)^{-1}$$

is a well-defined (left) 1-cocycle with values in $\mathcal{U}$:

$$\Theta_{x_o}(\gamma\mu) = \Theta_{x_o}(\gamma)\big(\gamma \cdot \Theta_{x_o}(\mu)\big).$$

Consequently, the function $\tilde{\rho}_{x_o} : \Gamma \to R \ltimes \mathcal{U}$ defined by $\gamma \mapsto \big(\Theta_{x_o}(\gamma), \rho(\gamma)\big)$ is a homomorphism.

Proof. This follows directly from the fact that $c_{\gamma\mu} = c_\gamma * (\gamma \cdot c_\mu)$ and the transport formula above. □

Remark 4.4. The dependence of Θ_{x_o} and $\tilde{\rho}_{x_o}$ on x_o is easily determined. Suppose that x' is a second base point. If e is the unique homotopy class of paths in X from x_o to x', then $c'_\gamma := e^{-1} * c_\gamma * (\gamma \cdot e)$ is the unique homotopy class of paths in X from x' to $\gamma \cdot x'$. Thus

$$T(c'_\gamma)^{-1} = T(e)T(c_\gamma)^{-1}(\gamma \cdot T(e)^{-1}).$$

Since, for $u, v \in \mathcal{U}$ and $r \in R$,

$$(v,1)(u,r)(v^{-1},1) = (v(r \cdot v^{-1}), \gamma)$$

in $R \ltimes \mathcal{U}$, the previous formula implies that $\tilde{\rho}_{x'}$ is obtained from $\tilde{\rho}_{x_o}$ by conjugation by $T(e) \in \mathcal{U}$ and that

$$\Theta_{x'}(\gamma) = T(e)\Theta_{x_o}(\gamma \cdot T(e)^{-1}).$$

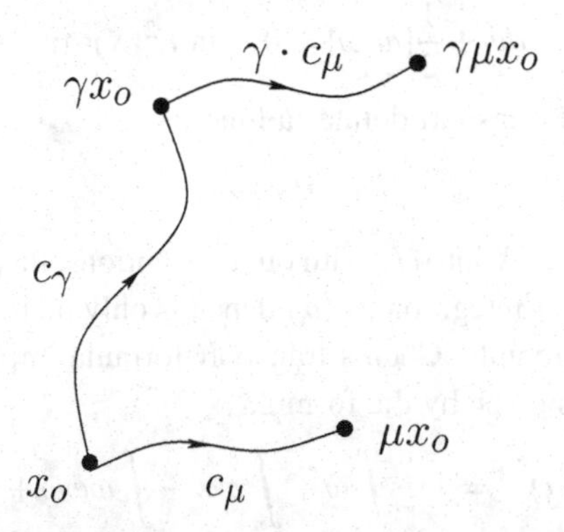

Figure 1. The cocycle relation $c_{\gamma\mu} = c_\gamma * (\gamma \cdot c_\mu)$.

4.1. Variant

Later we will sometimes need a slightly more general setup. This material is standard. More details can be found in [20, §5]. As above, R acts on $\mathcal{U}$ on the left.

Recall that a *factor of automorphy* is a smooth function

$$M : \Gamma \times X \to R, \quad (\gamma, x) \mapsto M_\gamma(x)$$

that satisfies $M_{\gamma\mu}(x) = M_\gamma(\mu x) M_\mu(x)$ for all $x \in X$ and $\gamma, \mu \in \Gamma$. One checks easily that the function

$$\gamma : (x,u) \mapsto (\gamma x, M_\gamma(x) u)$$

defines a left action of Γ on the right principal $\mathcal{U}$ bundle $X \times \mathcal{U}$.

Suppose that ∇_0 is a connection on the bundle $X \times \mathcal{U} \to X$. A 1-form $\omega \in E^1(X)\hat{\otimes}\mathfrak{u}$ defines a right $\mathcal{U}$-invariant connection ∇ on this bundle by

$$\nabla f = \nabla_0 f + \omega f,$$

where $f : X \to \mathcal{U}$ is a section. This connection is Γ-invariant if and only if

$$(\gamma^* \otimes 1)\omega = (1 \otimes M_\gamma)\omega - (\nabla_0 M_\gamma) M_\gamma^{-1} \text{ all } \gamma \in \Gamma$$

and flat if and only if $\nabla_0 \omega + [\omega, \omega]/2 = 0$.

When ∇ is a flat Γ-invariant connection, the monodromy representation

$$\Theta_{x_o} : \Gamma \to R \ltimes \mathcal{U}$$

is given by

$$\Theta_{x_o} : \gamma \mapsto \big(T(c_\gamma)^{-1}, M_\gamma(x_o)\big),$$

where for a path $\alpha \in X$, $T(\alpha)$ is given by the formula (4.2).

The setup of the previous section is a special case where $M_\gamma(x) = \rho(\gamma)$ and where the connection ∇_0 is trivial; that is, $\nabla_0 = d$.

4.2. A characterization of relative unipotent completion

To give a concrete construction of the completion of Γ relative to $\rho : \Gamma \to R$, we need a useful criterion for when a homomorphism $\Gamma \to R \ltimes \mathcal{U}$ induces an isomorphism $\mathcal{G} \to R \ltimes \mathcal{U}$ with the relative completion. In the present situation, the criterion is cohomological.

Suppose that $\omega \in E^1(X)\hat{\otimes}\mathfrak{u}$ is an integrable Γ-invariant 1-form on X, as above. Suppose that V is a finite dimensional R-module and that $\mathcal{U} \to \operatorname{Aut} V$ is an R-invariant homomorphism.[8] Denote the action of $u \in \mathcal{U}$ on $v \in V$ by $u \cdot v$.

[8] Equivalently, V is an $R \ltimes \mathcal{U}$-module.

Let $\mathbb{V}$ be the orbifold quotient of $X \times V$ by the diagonal Γ-action: $\gamma : (x,v) \mapsto (\gamma x, \rho(\gamma)v)$. The formula

$$\nabla v = dv + \omega \cdot v$$

defines a flat Γ-invariant connection on the vector bundle $X \times V \to X$. The complex

$$(E^\bullet(X) \otimes V)^\Gamma := \{\omega \in E^\bullet(X) \otimes V : (\gamma^* \otimes 1)\omega = (1 \otimes \gamma)\omega\}$$

of Γ-invariant V-valued forms on X has differential defined by

$$\nabla(\eta \otimes v) = d\eta \otimes v + \eta \wedge (\omega \cdot v),$$

where $\eta \in E^\bullet(X)$ and $v \in V$. It computes the cohomology $H^\bullet(M, \mathbb{V})$ of the orbifold M with coefficients in the orbifold local system $\mathbb{V}$ of locally constant sections of the flat vector bundle $\Gamma \backslash (X \times V)$ over M.

Lemma 4.5. *If R is reductive, each integrable Γ-invariant 1-form $\omega \in E^1(X) \hat{\otimes} \mathfrak{u}$ induces a ring homomorphism*

$$H^\bullet(\mathfrak{u}, V)^R \to H^\bullet(M, \mathbb{V}).$$

Proof. Note that Γ acts on $\mathfrak{u}$ on the left by the formula $\gamma : u \mapsto \rho(\gamma) \cdot u$. A 1-form $\omega \in E^1(X) \hat{\otimes} \mathfrak{u}$ can be regarded as a function

$$\theta_\omega : \mathrm{Hom}^{\mathrm{cts}}(\mathfrak{u}, \mathbb{C}) \to E^1(X), \quad \varphi \mapsto (1 \otimes \varphi)\omega.$$

It is Γ-equivariant if and only if ω is. The induced algebra homomorphism

$$C^\bullet(\mathfrak{u}) = \Lambda^\bullet \mathrm{Hom}^{\mathrm{cts}}(\mathfrak{u}, \mathbb{C}) \to E^\bullet(X)$$

commutes with differentials if and only if ω is integrable. So a Γ-invariant and integrable 1-form $\omega \in E^1(X) \hat{\otimes} \mathfrak{u}$ induces a dga homomorphism

$$C^\bullet(\mathfrak{u}, V)^R = [(\Lambda^\bullet \mathrm{Hom}^{\mathrm{cts}}(\mathfrak{u}, \mathbb{C})) \otimes V]^R \to (E^\bullet(X) \otimes V)^\Gamma.$$

Since R is reductive, the natural map

$$H^\bullet(C^\bullet(\mathfrak{u}, V)^R) \to H^\bullet(C^\bullet(\mathfrak{u}, V))^R$$

is an isomorphism. The result follows. □

The desired characterization of relative completion is:

Proposition 4.6. *If R is reductive and $\rho : \Gamma \to R$ has Zariski dense image, then the homomorphism $\tilde{\rho}_{x_o} : \Gamma \to R \ltimes \mathcal{U}$ constructed from $\omega \in (E^1(X) \hat{\otimes} \mathfrak{u})^\Gamma$ above is the completion of Γ with respect to ρ if and only if the homomorphism*

$$\theta_\omega^* : [H^j(\mathfrak{u}) \otimes V]^R \to H^j(M, \mathbb{V})$$

induced by θ_ω is an isomorphism when $j = 0, 1$ and injective when $j = 2$ for all R-modules V.

When X is a universal covering of a manifold, such universal 1-forms can be constructed using a suitable modification of Chen's method of power series connections [7].

Proof. Denote the completion of Γ relative to ρ by $\mathcal{G}$. The universal mapping property of relative completion implies that the homomorphism $\tilde{\rho}_{x_o} : \Gamma \to R \ltimes \mathcal{U}$ induces a homomorphism $\Psi : \mathcal{G} \to R \ltimes \mathcal{U}$ that commutes with the projections to R. Denote the prounipotent radical of $\mathcal{G}$ by $\mathcal{N}$ and its Lie algebra by $\mathfrak{n}$. Then Ψ induces a homomorphism $\mathcal{N} \to \mathcal{U}$ and an R-invariant homomorphism

$$\Psi^* : H^\bullet(\mathfrak{u}) \to H^\bullet(\mathfrak{n}).$$

Then for each finite dimensional R-module V, one has the commutative diagram

$$\begin{array}{ccc} [H^\bullet(\mathfrak{n}) \otimes V]^R & \longrightarrow & H^\bullet(\Gamma, V) \\ \Psi^* \uparrow & & \downarrow \\ [H^\bullet(\mathfrak{u}) \otimes V]^R & \underset{\theta_\omega^*}{\longrightarrow} & H^\bullet(M, \mathbb{V}) \end{array}$$

where the right-hand vertical mapping is induced by the orbifold morphism $M \to B\Gamma$ of M into the classifying space of Γ. Standard topology implies that this is an isomorphism in degrees 0 and 1 and injective in degree 2. Results in Section 3.2 imply that the top row is an isomorphism in degrees 0 and 1, and injective in degree 2. The assumption implies that the left hand vertical map is an isomorphism in degrees 0 and 1 and injective in degree 2. Since $H^\bullet(\mathfrak{u})$ and $H^\bullet(\mathfrak{n})$ are direct limits of finite dimensional R-modules, by letting V run through all finite dimensional irreducible R-modules, we see that $H^\bullet(\mathfrak{u}) \to H^\bullet(\mathfrak{n})$ is an isomorphism in degrees 0 and 1 and an injective in degree 2. Proposition 2.1 implies that $\mathfrak{n} \to \mathfrak{u}$ is an isomorphism. This implies that Ψ is an isomorphism. □

4.3. Rational structure

To construct a MHS on the completion of (say) a modular group Γ, we will first construct its complex form together with its Hodge and weight filtrations using an integrable, Γ-invariant 1-form, as above. An easy formal argument, given below, implies that this relative completion has a natural $\mathbb{Q}$ structure provided that R and ρ are defined over $\mathbb{Q}$. To understand the MHS on $\mathcal{O}(\mathcal{G})$, we will need a concrete description of this $\mathbb{Q}$-structure on $R \ltimes \mathcal{U}$ in terms of periods. Explaining this is the goal of this section.

In general, we are not distinguishing between a proalgebraic F-group and its group of F rational points. Here, since we are discussing Hodge theory, we will distinguish between a $\mathbb{Q}$-group, and its groups of $\mathbb{Q}$ and $\mathbb{C}$ rational points.

Suppose that R is a reductive group that is defined over $\mathbb{Q}$ and that $\rho : \Gamma \to R$ takes values in the $\mathbb{Q}$-rational points $R(\mathbb{Q})$ of R. Denote the completion of Γ with respect to ρ over $\mathbb{Q}$ by $\mathcal{G}$ and its prounipotent radical by $\mathcal{N}$. These are proalgebraic $\mathbb{Q}$-groups. We also have the completion $\mathcal{G}_{\mathbb{C}}$ of Γ over $\mathbb{C}$ relative to ρ, where the coefficient field is $\mathbb{C}$. Base change (cf. Section 3.3) implies that the natural homomorphism $\mathcal{G} \otimes_{\mathbb{Q}} \mathbb{C} \to \mathcal{G}_{\mathbb{C}}$ is an isomorphism.

When the hypotheses of Proposition 4.6 are satisfied we obtain a canonical $\mathbb{Q}$-structure on $R \ltimes \mathcal{U}$ from the isomorphism

$$\psi : \mathcal{G} \otimes_{\mathbb{Q}} \mathbb{C} \xrightarrow{\simeq} R \ltimes \mathcal{U}.$$

induced by $\tilde{\rho}_{x_o}$. The $\mathbb{Q}$-structure on $\mathcal{U}$ is the image of the restriction $\mathcal{N} \otimes_{\mathbb{Q}} \mathbb{C} \to \mathcal{U}$ of this isomorphism to $\mathcal{N}$. This induces a canonical $\mathbb{Q}$-structure on $\mathfrak{u}$ via the isomorphism $\mathfrak{u} \cong \mathfrak{n} \otimes_{\mathbb{Q}} \mathbb{C}$.

Proposition 4.7. *The canonical $\mathbb{Q}$-structure on $\mathfrak{u}$ is the $\mathbb{Q}$-Lie subalgebra of $\mathfrak{u}$ generated by the set* $\{\log \Theta_{x_o}(\gamma) : \gamma \in \Gamma\}$.

Proof. Fix a $\mathbb{Q}$-splitting s of the surjection $\mathcal{G} \to R_{\mathbb{Q}}$. This gives an identification of $\mathcal{G}$ with $R_{\mathbb{Q}} \ltimes \mathcal{N}$. Levi's Theorem (Prop. 3.1) implies that there is a $u \in \mathcal{U}$ such that the composition of

$$R_{\mathbb{Q}} \xrightarrow{s} \mathcal{G} \xrightarrow{\psi} R \ltimes \mathcal{U}$$

is conjugated to the section $R_{\mathbb{Q}} \to R \ltimes \mathcal{U}$ that takes r to $(1, r)$ by u. That is, the first section takes $r \in R$ to $(u(r \cdot u)^{-1}, r) \in R \ltimes \mathcal{U}$. The composite

$$R_{\mathbb{Q}} \ltimes \mathcal{N} \xrightarrow{\simeq} \mathcal{G} \longrightarrow \mathcal{G}_{\mathbb{C}} \xrightarrow[\psi]{\simeq} R \ltimes \mathcal{U}$$

is thus given by the formula

$$(n, r) \mapsto (\psi(n), 1)(u(r \cdot u)^{-1}, r) = (\psi(n)u(r \cdot u)^{-1}, r) \in R \ltimes \mathcal{U}. \tag{4.8}$$

The composite

$$\Gamma \longrightarrow \mathcal{G}(\mathbb{Q}) \xrightarrow{\simeq} R(\mathbb{Q}) \ltimes \mathcal{N}(\mathbb{Q})$$

takes $\gamma \in \Gamma$ to $(F(\gamma), \gamma) \in \mathcal{N}(\mathbb{Q}) \times R(\mathbb{Q})$ for some 1-cocycle $F : \Gamma \to \mathcal{N}(\mathbb{Q})$. Note that every divisible subgroup of the rational points of a prounipotent $\mathbb{Q}$-group N is the set of $\mathbb{Q}$-rational points of a $\mathbb{Q}$-subgroup of N. This implies that the smallest subgroup of $\mathcal{N}(\mathbb{Q})$ that contains $\{F(\gamma) : \gamma \in \Gamma\}$ is the set of $\mathbb{Q}$-points of a $\mathbb{Q}$-subgroup $\mathcal{S}$ of $\mathcal{N}$. The subgroup $\mathcal{S}$ must be $\mathcal{N}$. This is because the cocycle condition implies that it is a Γ-invariant subgroup:

$$\gamma_1 \cdot F(\gamma_2) = F(\gamma_1)^{-1} F(\gamma_1 \gamma_2) \in \mathcal{S}(\mathbb{Q}).$$

Since Γ is Zariski dense in $\mathcal{G}$, $\mathcal{S}$ is a normal subgroup of $\mathcal{G}$. This implies that $R_{\mathbb{Q}} \ltimes \mathcal{S}$ is a subgroup of $\mathcal{G} = R_{\mathbb{Q}} \ltimes \mathcal{N}$ that contains the image of the canonical homomorphism $\tilde{\rho} : \Gamma \to \mathcal{G}(\mathbb{Q})$. But since $\tilde{\rho}$ is Zariski dense, we must have $\mathcal{G} = R_{\mathbb{Q}} \ltimes \mathcal{S}$.

Formula (4.8) now implies that the image of $\mathcal{N}(\mathbb{Q})$ in $\mathcal{U}(\mathbb{C})$ is the smallest divisible subgroup of $\mathcal{U}(\mathbb{C})$ that contains the set $\{\psi \circ F(\gamma) : \gamma \in \Gamma\}$. But the formula (4.8) and the commutativity of the diagram

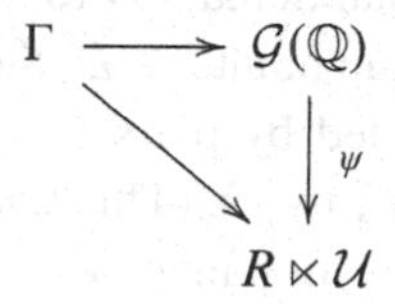

imply that $\psi \circ F = \Theta_{x_o}$, so that the image of $\mathcal{N}(\mathbb{Q})$ in $\mathcal{U}$ is the smallest divisible subgroup of $\mathcal{U}$ that contains the set $\{\Theta_{x_o}(\gamma) : \gamma \in \Gamma\}$. The result now follows from the Baker-Campbell-Hausdorff formula. □

Remark 4.9. One might think that this $\mathbb{Q}$ structure on $\mathcal{U}$ can be constructed as the image of the unipotent completion over $\mathbb{Q}$ of $\ker \rho$ in $\mathcal{U}$. This often works, but it does not when Γ is a modular group as $\rho : \Gamma \to R$ is injective and $\mathcal{U}$ is non-trivial in this case.

4.3.1. Complements

The coordinate ring $\mathcal{O}(\mathcal{G})$ of $\mathcal{G}$ is isomorphic, as a ring, to $\mathcal{O}(R) \otimes \mathcal{O}(\mathcal{U})$. Its coproduct is twisted by the action of R on $\mathcal{U}$. The $\mathbb{Q}$-form of $\mathcal{O}(\mathcal{G})$ consists of those elements of $\mathcal{O}(R) \otimes \mathcal{O}(\mathcal{U})$ that take rational values on the image of $\Gamma \to R \ltimes \mathcal{U}$. Since the exponential map $\mathfrak{u} \to \mathcal{U}$ is an isomorphism of affine schemes, the coordinate ring of $\mathcal{U}$ consists of the polynomial functions on $\mathfrak{u}$ that are continuous functions $\mathfrak{u} \to \mathbb{C}$. Since the coefficients of the logarithm and exponential functions are rational numbers, $\mathcal{O}(\mathcal{U}_{\mathbb{Q}})$ is the ring of continuous polynomials on $\mathfrak{u}_{\mathbb{Q}}$.

5 Relative Completion of Path Torsors

This section can be omitted on a first reading. Here we consider the relative completion $\mathcal{G}_{x,y}$ of the torsor of paths from x to y in a manifold M with respect to a reductive local system $\mathbb{H}$. This can be described using tannakian formalism.[9] Here we outline a direct approach partly because it is more concrete and better suits our needs.

[9] The category of local systems of finite dimensional F-vector spaces over M that admit a filtration whose graded quotients are local systems that correspond to representations of the Zariski closure of $\Gamma \to \operatorname{Aut} H_x$ is tannakian. The completion $\mathcal{G}_{x,y}$ of the torsor $\Pi(M;x,y)$ of paths in

We use the setup of the previous section. So $M = \Gamma\backslash X$ where X is a simply connected manifold, Γ is a discrete group that acts properly discontinuously and virtually freely on X, and $\rho : \Gamma \to R$ is a representation of Γ into an affine F-group ($F = \mathbb{R}$ or $\mathbb{C}$), not yet assumed to be reductive or Zariski dense.

If M is a manifold and $x, y \in M$, then $\Pi(M;x,y)$ denotes the set of homotopy classes of paths from x to y. We need to define what we mean by $\Pi(M;x,y)$ when the action of Γ on X is not free, in which case, M is an orbifold, but not a manifold. Choose a fundamental domain D for the action of Γ on X. The orbit of each $x \in M$ contains a unique point $\tilde{x} \in D$. Suppose that $x, y \in M$. Elements of $\Pi(M;x,y)$ can be represented by pairs (γ, c_γ), where $\gamma \in \Gamma$ and c_γ is a homotopy class of paths from $\tilde{x}$ to $\gamma\tilde{y}$. (This homotopy class is unique as X is simply connected.) The composition map

$$\Pi(M;x,y) \times \Pi(M;y,z) \to \Pi(M;x,z)$$

is given by

$$\big((\gamma, c_\gamma), (\mu, c_\mu)\big) \mapsto (\gamma\mu, c_{\gamma\mu}) := \big(\gamma\mu, c_\gamma * (\gamma \cdot c_\mu)\big).$$

Note that $\Pi(M;x,y)$ is a torsor under the left action of $\pi_1(M,x) := \Pi(M,x,x)$ and a torsor under the right action of $\pi_1(M,y)$. This definition of $\Pi(M;x,y)$ agrees with the standard definition when the action of Γ is fixed point free.

Now suppose that H is a finite dimensional vector space over $F = \mathbb{C}$ (or $\mathbb{R}$), that R is a reductive subgroup of $\mathrm{GL}(H)$, and $\rho : \Gamma \to R(F) = R$ is a Zariski dense representation. Let $\mathbb{H}$ be the corresponding (orbifold) local system over the orbifold M.

As in the previous section, we suppose that $\omega \in E^1(X)\hat{\otimes}\mathfrak{u}$ is an integrable, Γ-invariant 1-form on X. Define

$$\Theta_{x,y} : \Pi(M;x,y) \to R \ltimes \mathcal{U} \text{ by } (\gamma, c_\gamma) \mapsto \big(\rho(\gamma), T(c_\gamma)^{-1}\big).$$

Note that, unless $x = y$, this is not a group homomorphism. The universal mapping property of $\mathcal{G}_{x,y}$ implies that $\Theta_{x,y}$ induces a morphism $\mathcal{G}_{x,y} \to R \ltimes \mathcal{U}$ of affine schemes such that the diagram

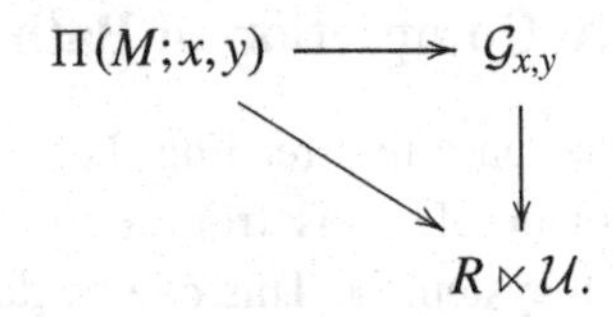

If ω satisfies the assumptions of Proposition 4.6, then the vertical morphism is an isomorphism. This follows as, in this case, $\mathcal{G}_{x,y}$ and $R \ltimes \mathcal{U}$ are both torsors

M from x to y is the torsor of isomorphisms between the fiber functors at x and y. It is an affine scheme over F.

under the left action of the relative completion $\mathcal{G}_{x,x} \cong R \ltimes \mathcal{U}$ of $\Gamma \cong \pi_1(M,x)$ with respect to ρ.

If R and ρ are defined over $\mathbb{Q}$, then $\mathcal{G}_{x,y}$ has a natural $\mathbb{Q}$ structure consisting of those elements of $\mathcal{O}(\mathcal{G}_{x,y})$ that take rational values on the image of $\Pi(M;x,y)$ in $\mathcal{G}_{x,y}$.

6 Zucker's Mixed Hodge Complex

In this section we recall the construction of the natural MHS on the cohomology of a smooth curve with coefficients in a polarized variation of Hodge structure.

Suppose that C is a compact Riemann surface and that D is a finite subset, which we assume to be non-empty. Then $C' := C - D$ is a smooth affine curve. Suppose that $\mathbb{V}$ is a polarized variation of Hodge structure over C' of weight m. For simplicity, we assume that the local monodromy about each $x \in D$ is unipotent. Zucker [45, §13] constructs a cohomological Hodge complex $\mathbf{K}(\mathbb{V})$ that computes the MHS on $H^\bullet(C', \mathbb{V})$. In this section we recall the definition of its complex component $\mathbf{K}_\mathbb{C}(\mathbb{V})$, together with its Hodge and weight filtrations. We first recall a few basic facts about mixed Hodge complexes.

6.1. Review of mixed Hodge complexes

This is a very brief outline of how one constructs a mixed Hodge structure on a graded vector space using a mixed Hodge complex. Full details can be found in [11].

The standard method for constructing a mixed Hodge structure on a graded invariant $M^\bullet$ of a complex algebraic variety is to express the invariant as the cohomology of a mixed Hodge complex (MHC). Very briefly, a MHC $\mathbf{K}$ consists of:

(i) two complexes $K_\mathbb{Q}^\bullet$ and $K_\mathbb{C}^\bullet$, each endowed with a weight filtration $W_\bullet$ by subcomplexes,
(ii) a $W_\bullet$ filtered quasi-isomorphism between $K_\mathbb{Q} \otimes \mathbb{C}$ and $K_\mathbb{C}$.
(iii) a Hodge filtration $F^\bullet$ of $K_\mathbb{C}^\bullet$ by subcomplexes.

These are required to satisfy several technical conditions, which we shall omit, although the lemma below encodes some of them. The complexes $K_\mathbb{Q}^\bullet$ and $K_\mathbb{C}^\bullet$ compute the $\mathbb{Q}$- and $\mathbb{C}$-forms of the invariant M:

$$M_\mathbb{Q}^\bullet \cong H^\bullet(K_\mathbb{Q}^\bullet) \quad \text{and} \quad M_\mathbb{C}^\bullet \cong H^\bullet(K_\mathbb{C}^\bullet)$$

The quasi-isomorphism between them is compatible with these isomorphisms. The weight filtration of $K_\mathbb{Q}^\bullet$ induces a weight filtration of $M_\mathbb{Q}$ by

$$W_m M_\mathbb{Q}^j = \operatorname{im}\{H^j(W_{m-j}K_\mathbb{Q}^\bullet) \to M_\mathbb{Q}^j\}.$$

The assumption that the quasi-isomorphism between $K^\bullet_\mathbb{Q} \otimes \mathbb{C}$ and $K^\bullet_\mathbb{C}$ be $W_\bullet$-filtered implies that the weight filtrations of $K^\bullet_\mathbb{Q}$ and $\mathcal{K}^\bullet_\mathbb{C}$ induce the same weight filtration on $M^\bullet$. That is,

$$(W_m M^j_\mathbb{Q}) \otimes \mathbb{C} = \operatorname{im}\{H^j(W_{m-j}K^\bullet_\mathbb{C}) \to M^j_\mathbb{C}\}$$

Finally, the Hodge filtration of $K^\bullet_\mathbb{C}$ induces the Hodge filtration of $M^\bullet_\mathbb{C}$ via

$$F^p M^\bullet_\mathbb{C} := \operatorname{im}\{H^\bullet(F^p K^\bullet_\mathbb{C}) \to M^\bullet_\mathbb{C}\}.$$

If $\mathbf{K}$ is a MHC, then $M^\bullet$ is a MHS with these Hodge and weight filtrations.

We shall need the following technical statement. As pointed out above, this is equivalent to several of the technical conditions satisfied by a MHC.

Lemma 6.1 ([14, 3.2.8]). *Suppose that $(K^\bullet_\mathbb{C}, W_\bullet, F^\bullet)$ is the complex part of a MHC. If $u \in F^p W_m K^j_\mathbb{C}$ is exact in $K^\bullet_\mathbb{C}$, then there exists $v \in F^p W_{m+1} K^{j-1}_\mathbb{C}$ such that $dv = u$.*

Finally, a *cohomological mixed Hodge complex* (CMHC) is a collection of filtered complexes of sheaves on a variety (or a topological space) with the property that the global sections of a collection of acyclic resolutions of its components is a MHC. For details, see [11].

6.2. *Zucker's cohomological MHC*

We'll denote Zucker's cohomological MHC for computing the MHS on $H^\bullet(C', \mathbb{V})$ by $\mathbf{K}(\mathbb{V})$. We describe only its complex part $\mathbf{K}_\mathbb{C}(\mathbb{V})$. When trying to understand the definition of its weight filtration, the reader may find it useful to read the following section on the limit MHS of $\mathbb{V}$ at a tangent vector to a cusp $P \in D$.

Set $\mathcal{V} = \mathbb{V} \otimes \mathcal{O}_{C'}$. This has a canonical flat connection ∇. This extends to a meromorphic connection

$$\nabla : \overline{\mathcal{V}} \to \Omega^1_C(\log D) \otimes \overline{\mathcal{V}}$$

on Deligne's canonical extension $\overline{\mathcal{V}}$ of it to C. Schmid's Nilpotent Orbit Theorem [36] implies that the Hodge sub-bundles of $\mathcal{V}$ extend to holomorphic sub-bundles $F^p\overline{\mathcal{V}}$ of $\overline{\mathcal{V}}$.

As a sheaf, $\mathbf{K}_\mathbb{C}(\mathbb{V})$ is simply $\Omega^\bullet_C(\log D) \otimes \overline{\mathcal{V}}$ with the differential ∇. Standard arguments imply that $H^\bullet(C, \mathbf{K}_\mathbb{C}(\mathbb{V}))$ is isomorphic to $H^\bullet(C', \mathbb{V})$. Its Hodge filtration is defined in the obvious way:

$$F^p \mathbf{K}_\mathbb{C}(\mathbb{V}) := \sum_{s+t=p} \left(F^s \Omega^\bullet_C(\log D)\right) \otimes F^t \overline{\mathcal{V}}.$$

In degree 0 the weight filtration is simply

$$0 = W_{m-1}\mathbf{K}^0_\mathbb{C}(\mathbb{V}) \subseteq W_m \mathbf{K}^0_\mathbb{C}(\mathbb{V}) = \mathbf{K}^0_\mathbb{C}(\mathbb{V}).$$

In degree 1, $W_r \mathbf{K}^1_{\mathbb{C}}(\mathbb{V})$ vanishes when $r < m$. To define the remaining terms in degree 1, consider the reside mapping

$$\operatorname{Res}_P : \Omega^1_C(\log D) \otimes \overline{\mathcal{V}} \to V_P$$

at $P \in C$, which takes values in the fiber V_P of $\overline{\mathcal{V}}$ over P. The residue N_P of the connection ∇ on $\overline{\mathcal{V}}$ at $P \in C$ is the local monodromy logarithm divided by $2\pi i$. It acts on V_P. When $r \geq 0$, the stalk of $W_{m+r}\mathbf{K}^1_{\mathbb{C}}(\mathbb{V})$ at P is

$$W_{m+r}\mathbf{K}^1_{\mathbb{C}}(\mathbb{V})_P := \operatorname{Res}_P^{-1}(\operatorname{im} N_P + \ker N_P^r).$$

Note that, when $P \notin D$, the reside map vanishes (and so does N_P), so that the stalk of $W_m \mathbf{K}^1_{\mathbb{C}}(\mathbb{V})$ when $P \notin D$ is $\mathbf{K}_{\mathbb{C}}(\mathbb{V})_P$.

The Hodge and weight filtrations on $H^j(C', \mathbb{V})$ are defined by

$$F^p H^j(C', \mathbb{V}) = \operatorname{im}\{H^j(C, F^p \mathbf{K}_{\mathbb{C}}(\mathbb{V})) \to H^j(C', \mathbb{V})\}$$

and

$$W_m H^j(C', \mathbb{V}) = \operatorname{im}\{H^j(C, W_{m-j}\mathbf{K}_{\mathbb{C}}(\mathbb{V})) \to H^j(C', \mathbb{V})\}.$$

The definition of the weight filtration implies that $H^0(C', \mathbb{V})$ has weight m and that the weights on $H^1(C', \mathbb{V})$ are $\geq 1 + m$.

Remark 6.2. Let $j : C' \to C$ denote the inclusion. The complex of sheaves $W_m \mathbf{K}_{\mathbb{C}}(\mathbb{V})$ on C is a cohomological Hodge complex that is easily seen to be quasi-isomorphic to the sheaf $j_* \mathbb{V}$ on C. It therefore computes the intersection cohomology $IH^j(C, \mathbb{V})$ and shows that it has a canonical pure Hodge structure of weight $m + j$. For more details, see Zucker's paper [45].

6.3. The limit MHS on V_P

Suppose that $P \in D$. For each choice of a non-zero tangent vector $\vec{v} \in T_P C$ there is a limit MHS, denoted $V_{\vec{v}}$ on V_P. The pth term of the Hodge filtration is the fiber of $F^p \overline{\mathcal{V}}$ over P. The weight filtration is the monodromy weight filtration shifted so that its average weight is m, the weight of $\mathbb{V}$. The $\mathbb{Q}$ (or $\mathbb{Z}$ structure, if that makes sense) is constructed by first choosing a local holomorphic coordinate t defined on a disk Δ containing P where $t(P) = 0$. Assume that $\Delta \cap D = \{P\}$. Standard ODE theory (cf. [44, Chapt. II]) implies that there is a trivialization $\Delta \times V_P$ of the restriction of $\overline{\mathcal{V}}$ to Δ such that the connection ∇ is given by

$$\nabla f = df + N_P(f)\frac{dt}{t}$$

with respect to this trivialization, where $f : \Delta \to V_P$. Suppose that $Q \in \Delta - \{P\}$. The $\mathbb{Q}$-structure on V_P corresponding to the tangent vector $\vec{v} = t(Q)\partial/\partial t$ is obtained from the $\mathbb{Q}$ structure $V_{Q,\mathbb{Q}}$ on V_Q by identifying V_P with V_Q via the trivialization. This MHS depends only on the tangent vector and not on the choice of the holomorphic coordinate t.

For all non-zero $\vec{v} \in T_P C$, the monodromy logarithm $N_P : V_{\vec{v}} \to V_{\vec{v}}$ acts as a morphism of type $(-1,-1)$. This implies that $V_{\vec{v}}/\operatorname{im} N_P$ has a natural MHS for all $\vec{v} \neq 0$. Since N_P acts trivially on this, the MHS on $V_P/\operatorname{im} N_P$ has a natural MHS that is independent of the choice of the tangent vector $\vec{v} \in T_P C$. The definition of the weight filtration on $V_{\vec{v}}$ implies that the weight filtration on $V_P/\operatorname{im} N_P$ is

$$W_{m+r}(V_P/\operatorname{im} N_P) = (\operatorname{im} N_P + \ker N_P^{r-1})/\operatorname{im} N_P$$

when $r \geq 0$ and $W_{m+r}(V_P/\operatorname{im} N_P) = 0$ when $r < 0$.

There is a canonical isomorphism

$$H^1(\Delta - \{P\}, \mathbb{V}) \cong V_P/\operatorname{im} N_P.$$

from which it follows that this cohomology group has a canonical MHS for each $P \in D$.

6.4. An exact sequence

Observe that

$$\mathbf{K}_{\mathbb{C}}(\mathbb{V})/W_m \mathbf{K}_{\mathbb{C}}(\mathbb{V}) = \bigoplus_{P \in D} i_{P*}(V_P/N_P V_P)(-1)[-1].$$

This and the exact sequence of sheaves

$$0 \to W_m \mathbf{K}(\mathbb{V}) \to \mathbf{K}(\mathbb{V}) \to \mathbf{K}(\mathbb{V})/W_m \to 0$$

on C gives the exact sequence of MHS

$$\begin{aligned} 0 \to W_{m+1} H^1(C', \mathbb{V}) &\to H^1(C', \mathbb{V}) \\ &\to \bigoplus_{P \in D} (V_P/\operatorname{im} N_P)(-1) \to IH^2(C, \mathbb{V}) \to 0. \end{aligned} \tag{6.3}$$

Here we are assuming that we are in the "interesting case" where D is non-empty.

Since $H^0(C', \mathbb{V}) = IH^0(C, \mathbb{V})$, and since this is dual to $IH^2(C, \mathbb{V})$, we see that the sequence

$$0 \to W_{m+1} H^1(C', \mathbb{V}) \to H^1(C', \mathbb{V}) \to \bigoplus_{P \in D} (V_P/\operatorname{im} N_P)(-1) \to 0$$

is exact when $H^0(C', \mathbb{V}) = 0$.

6.5. A MHC of smooth forms

To extend this MHS from the cohomology groups $H^1(C',\mathbb{V})$ to one on relative completion of its fundamental group, we will need the complex part of a global MHC of smooth forms. The construction of this from $\mathbf{K}_\mathbb{C}(\mathbb{V})$ is standard. We recall the construction.

The resolution of $\mathbf{K}_\mathbb{C}(\mathbb{V})$ by smooth forms is the total complex of the double complex

$$\mathcal{K}_\mathbb{C}^{\bullet\bullet}(\mathbb{V}) := \mathbf{K}_\mathbb{C}(\mathbb{V}) \otimes_{\mathcal{O}_C} \mathcal{E}_C^{0,\bullet}, \tag{6.4}$$

where $\mathcal{E}_C^{0,\bullet}$ denotes the sheaf of smooth forms on C of type $(0,\bullet)$. The Hodge and weight filtrations extend as

$$F^p\mathcal{K}_\mathbb{C}^{\bullet\bullet}(\mathbb{V}) := \mathcal{K}_\mathbb{C}^{\geq p,\bullet}(\mathbb{V}) = (F^p\mathbf{K}_\mathbb{C}(\mathbb{V})) \otimes_{\mathcal{O}_C} \mathcal{E}_C^{0,\bullet}$$

and

$$W_r\mathcal{K}_\mathbb{C}^{\bullet,\bullet}(\mathbb{V}) := (W_r\mathbf{K}_\mathbb{C}(\mathbb{V})) \otimes_{\mathcal{O}_C} \mathcal{E}_C^{0,\bullet}.$$

The global sections

$$K^\bullet(C,D;\mathbb{V}) := H^0(C, \operatorname{tot}\mathcal{K}_\mathbb{C}^{\bullet,\bullet}(\mathbb{V})).$$

of (6.4) is a sub dga of $E^\bullet(C',\mathbb{V})$. It has Hodge and weight filtrations defined by taking the global sections of the Hodge and weight filtrations of (6.4). It is the complex part of a mixed Hodge complex.

The Hodge and weight filtrations on $H^j(C',\mathbb{V})$ are

$$F^pH^j(C',\mathbb{V}) = \operatorname{im}\{H^j(F^pK^\bullet(C,D;\mathbb{V})) \to H^j(C',\mathbb{V})\}$$

and

$$W_mH^j(C',\mathbb{V}) = \operatorname{im}\{H^j(W_{m-j}K^\bullet(C,D;\mathbb{V})) \to H^j(C',\mathbb{V})\}.$$

Zucker's MHC and its resolution by smooth forms are natural in the local system $\mathbb{V}$ and are compatible with tensor products: if $\mathbb{V}_1$, $\mathbb{V}_2$ and $\mathbb{V}_3$ are PVHS over C', then a morphism $\mathbb{V}_1 \otimes \mathbb{V}_2 \to \mathbb{V}_3$ of PVHS induces morphism

$$\mathbf{K}(\mathbb{V}_1) \otimes \mathbf{K}(\mathbb{V}_2) \to \mathbf{K}(\mathbb{V}_1 \otimes \mathbb{V}_2) \to \mathbf{K}(\mathbb{V}_3).$$

of CMHCs and dga homomorphism

$$K^\bullet(C,D;\mathbb{V}_1) \otimes K^\bullet(C,D;\mathbb{V}_2) \to K^\bullet(C,D;\mathbb{V}_1 \otimes \mathbb{V}_2) \to K^\bullet(C,D;\mathbb{V}_3)$$

that preserve the Hodge and weight filtrations.

6.6. Remarks about the orbifold case

Zucker's work extends formally to the orbifold case. For us, an orbi-curve $C' = C - D$ is the orbifold quotient of a smooth curve $X' = X - E$ by a finite group G. This action does not have to be effective. (That is, $G \to \operatorname{Aut} X$ does not have to be injective.) An orbifold variation of MHS $\mathbb{V}$ over C' is an admissible variation of MHS $\mathbb{V}_X$ over X' together with a G-action such that the projection $\mathbb{V} \to X'$ is G-equivariant. For each $g \in G$, we require that the map $g : \mathbb{V}_X \to \mathbb{V}_X$ induce an isomorphism of variations of MHS $g^*\mathbb{V}_X \cong \mathbb{V}_X$.

With these assumptions, G acts on $\mathbf{K}^\bullet(X,E;\mathbb{V}_X)$ and $\mathbf{K}^\bullet(X,E;\mathbb{V}_X)^G$ is a sub MHC. Define

$$\mathbf{K}^\bullet(C,D;\mathbb{V}) = \mathbf{K}^\bullet(X,E;\mathbb{V}_X)^G.$$

This computes the cohomology $H^\bullet(C',\mathbb{V})$ and implies that it has a MHS such that the canonical isomorphism $H^\bullet(C',\mathbb{V}) \cong H^\bullet(X',\mathbb{V}_X)^G$ is an isomorphism of MHS.

7 Relative Completion of Fundamental Groups of Affine Curves

In this section we use the results of the last two sections to construct, under suitable hypotheses, a MHS on the relative completion of the fundamental group of an affine curve. As in the previous section, we suppose that C is a compact Riemann surface and that D is a finite subset of C. Here we suppose, in addition, that D is non-empty, so that $C' = C - D$ is an affine curve. Suppose that $\mathbb{V}$ is a polarized variation of $\mathbb{Q}$-HS over C' with unipotent monodromy about each $P \in D$. Denote the fiber of $\mathbb{V}$ over x by V_x. The Zariski closure of the monodromy representation

$$\rho : \pi_1(C',x) \to \operatorname{Aut}(V_x)$$

is a reductive $\mathbb{Q}$-group [39, Lem. 2.10], which we will denote by R_x. Fix a base point $x_o \in C$. Set $R = R_{x_o}$. Each monodromy group R_x is isomorphic to R; the isomorphism is unique mod inner automorphisms. We thus have Zariski dense monodromy representations

$$\rho_x : \pi_1(C',x) \to R_x(\mathbb{Q})$$

for each $x \in X$. Denote the completion of $\pi_1(C',x)$ with respect to ρ_x by $\mathcal{G}_x$, and its Lie algebra by $\mathfrak{g}_x$. We will construct natural MHSs on $\mathcal{O}(\mathcal{G}_x)$ and on $\mathfrak{g}_x$ that are compatible with their algebraic structures (Hopf algebra, Lie algebra). Before doing this we need to show that the connection form Ω can be chosen to have coefficients in Zucker's MHC and be compatible with the various Hodge and weight filtrations.

For simplicity, we make the following assumptions:

1. Every irreducible representation of R is absolutely irreducible. That is, it remains irreducible when we extend scalars from $\mathbb{Q}$ to $\mathbb{C}$.

2. For each irreducible representation V_λ of R, the corresponding local system $\mathbb{V}_\lambda$ over C' underlies a PVHS over C'. The Theorem of the Fixed Part (stated below) implies that this PVHS is unique up to Tate twist.

These hold in our primary example, where C' is a quotient X_Γ of the upper half plane by a finite index subgroup of $\mathrm{SL}_2(\mathbb{Z})$ and $R \cong \mathrm{SL}_2$.

7.1. The bundle $\mathfrak{u}$ of Lie algebras

Denote the set of isomorphism classes of irreducible R-modules by $\check{R}$. Fix a representative V_λ of each $\lambda \in \check{R}$ and the structure of a PVHS on the corresponding local system $\mathbb{V}_\lambda$ over C'. Filter

$$\check{R}_1 \subset \check{R}_2 \subset \check{R}_3 \subset \cdots \subset \bigcup_n \check{R}_n = \check{R}$$

$\check{R}$ by finite subsets such that if $\lambda \in \check{R}_n$ and $\mu \in \check{R}_m$, then the isomorphism class of $V_\lambda \otimes V_\mu$ is in $\check{R}_{m+n}$. For example, when $R = \mathrm{SL}_2$, one can take $\check{R}_n$ to be the set of all symmetric powers $S^m H$ with $m \leq n$ of the defining representation H of SL_2.

For each $\lambda \in \check{R}$, the variation MHS $H^1(C',\mathbb{V}_\lambda)^* \otimes \mathbb{V}_\lambda$, being the tensor product of a constant MHS with a PVHS, is an admissible variation of MHS over C'. Note that the VMHS structure on it does not change when $\mathbb{V}_\lambda$ is replaced by $\mathbb{V}_\lambda(n)$, so that the VMHS $H^1(C',\mathbb{V}_\lambda)^* \otimes \mathbb{V}_\lambda$ is independent of the choice of the PVHS structure on $\mathbb{V}_\lambda$. Since the weights of $H^1(C,\mathbb{V}_\lambda)$ are at least 1+ the weight of $\mathbb{V}_\lambda$, the weights of $H^1(C',\mathbb{V}_\lambda)^* \otimes \mathbb{V}_\lambda$ are ≤ -1.

The inverse limit

$$\mathbf{u}_1 := \varprojlim_n \bigoplus_{\lambda \in \check{R}_n} H^1(C',\mathbb{V}_\lambda)^* \otimes \mathbb{V}_\lambda$$

is pro-variation of MHS over $\mathbb{C}$ of negative weight.[10] That is, $\mathbf{u}_1 = W_{-1}\mathbf{u}_1$. Observe that its fiber

$$\varprojlim_n \bigoplus_{\lambda \in \check{R}_n} H^1(C',\mathbb{V}_\lambda)^* \otimes V_{\lambda,o} = \prod_{\lambda \in \check{R}} H^1(C',\mathbb{V}_\lambda)^* \otimes V_{\lambda,o}$$

over x_o is the abelianization of the prounipotent radical of the completion of $\pi_1(C',x_o)$ with respect to the homomorphism to R.

The degree n part $V \mapsto \mathbb{L}_n(V)$ of the free Lie algebra is a Schur functor, so that it makes sense to apply it to bundles. Set

$$\mathbf{u}_n = \mathbb{L}_n(\mathbf{u}_1) := \varprojlim_m \mathbb{L}_n\big(\bigoplus_{\lambda \in \check{R}_m} H^1(C',\mathbb{V}_\lambda)^* \otimes \mathbb{V}_\lambda\big)$$

[10] Note that this is a very special kind of variation of MHS — it is a direct sum of variations that are the tensor product of a constant MHS with a PVHS. Their asymptotic behaviour is determined by that of the PVHS that occur in the summands.

and

$$\mathbf{u} := \varprojlim_{n} \bigoplus_{j=1}^{n} \mathbf{u}_j \quad \text{and} \quad \mathbf{u}^N := \varprojlim_{n \geq N} \bigoplus_{j=N}^{n} \mathbf{u}_j.$$

These are admissible pro-variations of MHS over C'. Denote the fiber of $\mathbf{u}$ over x by $\mathfrak{u}_x$. It is abstractly isomorphic to the prounipotent radical of the completion of $\pi_1(C',x)$ relative to the monodromy representation $\pi_1(C',x) \to R$. The fiber of $\mathbf{u}^N$ over x is the nth term $L^n\mathfrak{u}_x$ of the lower central series (LCS) of $\mathfrak{u}_x$.

7.2. Some technicalities

The Theorem of the Fixed Part states that if $\mathbb{A}$ is an admissible VMHS over a smooth variety X, then $H^0(X,\mathbb{A})$ has a natural MHS with the property that for each $x \in X$, the natural inclusion $H^0(X,\mathbb{A}) \to A_x$ is a morphism of MHS. In the algebraic case, it is enough to prove this when X is a curve. When $\mathbb{A}$ is pure, this follows from Zucker's MHS [45] on $H^\bullet(X,\mathbb{V})$. The general case follows from Saito's theory of Hodge and mixed Hodge modules [34, 35].

The following is a direct consequence of the Theorem of the Fixed Part. Its proof is left as an exercise.

Lemma 7.1. *Assume that $\mathbb{A}$ is an admissible VMHS over C' whose monodromy representation $\pi_1(C',x_o) \to \operatorname{Aut} A_{x_o}$ factors through $\pi_1(C',x_o) \to R$. With the assumptions above, each R isotypical component of $\mathbb{A}$ over C' is an admissible VMHS. If $\mathbb{V}_\lambda$ is a PVHS that corresponds to the irreducible R-module V_λ, then the natural mapping*

$$\bigoplus_{\lambda \in \check{R}} H^0\big(C', \operatorname{Hom}(\mathbb{V}_\lambda, \mathbb{A})\big) \otimes \mathbb{V}_\lambda \to \mathbb{A}$$

is an isomorphism of admissible VMHS. In particular, the structure of a PVHS on $\mathbb{V}_\lambda$ is unique up to Tate twist. □

Every pro object of the category of admissible VMHS $\mathbb{A}$ over C' is thus of the form

$$\mathbb{A} = \prod_{\lambda \in \check{R}} \mathbb{V}_\lambda \otimes A_\lambda,$$

where each A_λ is a MHS. Define

$$\mathbf{K}(\mathbb{A}) = \prod_{\lambda \in \check{R}} \mathbf{K}(\mathbb{V}_\lambda) \otimes A_\lambda.$$

This is a pro-CMHC. In particular, its complex part

$$K^\bullet(C,D;\mathbb{A}) = \prod_{\lambda \in \check{R}} K^\bullet(C,D;\mathbb{V}_\lambda) \otimes A_\lambda$$

has naturally defined Hodge and weight filtrations; its differential is strict with respect to the Hodge and weight filtrations (cf. Lemma 6.1). In particular, for all $n \geq 1$, the complexes $K^\bullet(C,D;\mathbf{u}_n)$ have this strictness property.

7.3. The connection form Ω

Observe that $H^1(C',H^1(C',\mathbb{V}_\lambda)^* \otimes \mathbb{V}_\lambda)$ is naturally isomorphic (as a MHS) to

$$H^1(C',\mathbb{V}_\lambda) \otimes H^1(C',\mathbb{V}_\lambda)^* \cong \mathrm{Hom}(H^1(C',\mathbb{V}_\lambda), H^1(C',\mathbb{V}_\lambda)).$$

So, for each $\lambda \in \check{R}$, there is an element $\xi_\lambda \in F^0W_0H^1(C',H^1(C',\mathbb{V}_\lambda)^* \otimes \mathbb{V})$ that corresponds to the identity mapping $H^1(C',\mathbb{V}_\lambda) \to H^1(C',\mathbb{V}_\lambda)$. Lemma 6.1 implies that this is represented by a 1-form

$$\omega_\lambda \in F^0W_{-1}K^1(C,D;H^1(C';\mathbb{V}_\lambda)^* \otimes \mathbb{V}_\lambda).$$

Set

$$\Omega_1 := \prod_{\lambda \in \check{R}} \omega_\lambda \in K^1(C,D;\mathbf{u}_1).$$

Note that $d\Omega_1 = 0$ and that

$$\frac{1}{2}[\Omega_1,\Omega_1] \in F^0W_{-2}K^2(C,D;\mathbf{u}_2).$$

Since C is a surface, $[\Omega_1,\Omega_1]$ is closed. Since C' is not compact, it is exact. Lemma 6.1 implies that we can find Ξ_2 in $F^0W_{-1}K^1(C,D;\mathbf{u}_2)$ such that $d\Xi_2 = \frac{1}{2}[\Omega_1,\Omega_1]$. Set

$$\Omega_2 = \Omega_1 - \Xi_2 \in F^0W_{-1}K^1(C,D;\mathbf{u}_1 \oplus \mathbf{u}_2).$$

Then

$$d\Omega_2 + \frac{1}{2}[\Omega_2,\Omega_2] \in F^0W_{-2}K^2(C,D;\mathbf{u}^3).$$

Its component that lies in $K^2(C,D;\mathbf{u}_3)$ is closed and thus exact. So it is the exterior derivative of some $\Xi_3 \in F^0W_{-2}K^1(C,D;\mathbf{u}_3)$. Set

$$\Omega_3 = \Omega_2 - \Xi_3 \in F^0W_{-1}K^1(C,D;\mathbf{u}_1 \oplus \mathbf{u}_2 \oplus \mathbf{u}_3).$$

Then

$$d\Omega_3 + \frac{1}{2}[\Omega_3,\Omega_3] \in F^0W_{-2}K^2(C,D;\mathbf{u}^4).$$

Continuing this way, we obtain a sequence of elements $\Xi_n \in F^0W_{-1}K^1(C,D;\mathbf{u}_n)$ such that for all $N \geq 2$

$$\Omega_N := \Omega_1 - (\Xi_2 + \cdots + \Xi_N) \in F^0W_{-1}K^1(C,D;\oplus_{n=1}^N \mathbf{u}_n)$$

satisfies

$$d\Omega_N + \frac{1}{2}[\Omega_N,\Omega_N] \in F^0W_{-2}K^2(C,D;\mathbf{u}^{N+1}).$$

Then the $\mathbf{u}$-valued 1-form

$$\Omega := \varprojlim_N \Omega_N \in F^0 W_{-1} K^1(C,D;\mathbf{u}) \tag{7.2}$$

is integrable:

$$d\Omega + \frac{1}{2}[\Omega,\Omega] = 0.$$

To understand the significance of the form Ω, note that the bundle $\mathbf{u}_1$ over C', and hence each $\mathbf{u}_n = \mathbb{L}_n(\mathbf{u}_1)$, is a flat bundle over C'. The monodromy of each factors through the representation $\rho_x : \pi_1(C',x) \to R$. Summing these, we see that for each $N \geq 1$, the bundle

$$\mathbf{u}/\mathbf{u}^{N+1} \cong \mathbf{u}_1 \oplus \cdots \oplus \mathbf{u}_N$$

is flat with monodromy that factors through ρ_x. Denote the limit of these flat connections by ∇_0. Then

$$\nabla := \nabla_0 + \Omega$$

defines a new connection on the bundle $\mathbf{u}$ over C' which is flat as Ω is integrable. Here we view $\mathbf{u}$ (and hence Ω) as acting on each fiber by inner derivations.

The restriction of the filtration

$$\mathbf{u} = \mathbf{u}^1 \supset \mathbf{u}^2 \supset \mathbf{u}^3 \supset \cdots \tag{7.3}$$

of $\mathbf{u}$ to each fiber is the lower central series. Note that

$$\mathrm{Gr}^{\bullet}_{\mathrm{LCS}}\,\mathbf{u} := \mathbf{u}^n/\mathbf{u}^{n+1}$$

is naturally isomorphic to $\mathbf{u}_n$.

Lemma 7.4. *Each term $\mathbf{u}^n$ of the lower central series filtration of $\mathbf{u}$ is a flat sub-bundle of $(\mathbf{u},\nabla)$. The induced connection on $\mathrm{Gr}^n_{\mathrm{LCS}}\,\mathbf{u} \cong \mathbf{u}_n$ is ∇_0.*

Proof. This follows from the fact that Ω takes values in $\mathbf{u}$ and that the inner derivations act trivially on $\mathrm{Gr}^{\bullet}_{\mathrm{LCS}}\,\mathbf{u}$. □

7.4. Hodge and weight bundles and their extensions to C

The flat connection ∇ on $\mathbf{u}$ defines a new complex structure as a (pro) holomorphic vector bundle over C'. To understand it, write $\Omega = \Omega' + \Omega''$, where Ω' is of type $(1,0)$ and Ω'' is of type $(0,1)$. Set

$$D' = \nabla_0 + \Omega' \text{ and } D'' = \bar{\partial} + \Omega''$$

so that $\nabla = D' + D''$. Then D'' is a $(0,1)$-valued form taking values in $\mathbf{u}$. Note that $(D'')^2 = 0$.

Lemma 7.5. *A section s of $\mathfrak{u}$ is holomorphic with respect to the complex structure on $\mathfrak{u}$ defined by the flat connection ∇ if and only if $D''s = 0$.*

Proof. Since D'' is $\mathcal{O}_{C'}$ linear, it suffices to show that $D''s = 0$ when s is a flat local section of $\mathfrak{u}$. But this follows as $D''s$ is the $(0,1)$ component of ∇s, which vanishes as s is flat. □

Denote $\mathfrak{u}$ with this complex structure by $(\mathfrak{u}, D'')$. Since all holomorphic sections of $(\mathfrak{u}, D'')$ are $\mathcal{O}$-linear combinations of flat sections, Lemma 7.4 implies:[11]

Lemma 7.6. *The lower central series filtration (7.3) of $\mathfrak{u}$ is a filtration by holomorphic sub-bundles. The isomorphism*

$$\mathrm{Gr}^n_{\mathrm{LCS}}(\mathfrak{u}, D'') \cong \mathfrak{u}_n$$

is an isomorphism of holomorphic vector bundles. □

Denote the canonical extension of $\mathfrak{u}_1$ to C by $\overline{\mathfrak{u}}_1$. Then, since the local monodromy operators are unipotent, $\overline{\mathfrak{u}}_n := \mathbb{L}_n(\overline{\mathfrak{u}}_1)$ is the canonical extension of $\mathfrak{u}_n$ to C. Define

$$\overline{\mathfrak{u}} := \varprojlim_n \bigoplus_{j=1}^{n} \overline{\mathfrak{u}}_j \text{ and } \overline{\mathfrak{u}}^N := \varprojlim_{n \geq N} \bigoplus_{j=N}^{n} \overline{\mathfrak{u}}_j.$$

Then $(\overline{\mathfrak{u}}, \nabla_0)$ is the canonical extension of $(\mathfrak{u}, \nabla_0)$ to C.

Our next task is to show that $(\overline{\mathfrak{u}}, \nabla)$ is the canonical extension of $(\mathfrak{u}, \nabla)$ to C. Since smooth logarithmic $(0,1)$-forms on C with poles on D are smooth on C, it follows that Ω'' is a smooth, $\overline{\mathfrak{u}}$-valued $(0,1)$-form on C. It follows that D'' extends to a $(0,1)$ form-valued operator on smooth sections of $\overline{\mathfrak{u}}$. Since $(D'')^2 = 0$, it defines a complex structure on $\overline{\mathfrak{u}}$. A smooth locally defined section s of $\overline{\mathfrak{u}}$ is holomorphic if and only if $D''s = 0$. We'll denote this complex structure by $(\overline{\mathfrak{u}}, D'')$.

Suppose that $P \in D$ and that t is a local holomorphic coordinate on C centered at P. Since $t\Omega$ is a smooth $\overline{\mathfrak{u}}$-valued form on C in a neighbourhood Δ of P, and since $t\nabla_0$ takes smooth sections of $\overline{\mathfrak{u}}$ defined on Δ to smooth 1-forms with values in $\overline{\mathfrak{u}}$, it follows that $t\nabla$ is a differential operator on sections of $\overline{\mathfrak{u}}$ over Δ. This implies that ∇ is a meromorphic connection on $(\overline{\mathfrak{u}}, D'')$ with regular singular points along D.

Proposition 7.7. *The bundle $(\overline{\mathfrak{u}}, D'')$ with the connection D' is Deligne's canonical extension of $(\mathfrak{u}, \nabla)$ to C.*

[11] This also follows from the fact that Ω'' is $\mathfrak{u}$-valued thus acts trivially on the graded quotients of the lower central series filtration of $\mathfrak{u}$.

Proof. Since all singularities of ∇ are regular singular points, it suffices to check that the residue of ∇ at each $P \in D$ is pronilpotent endomorphism of $\mathfrak{u}_P$, the fiber of $\overline{\mathfrak{u}}$ over P. This is an immediate consequence of the fact that the residue of $(\mathrm{Gr}^\bullet_{\mathrm{LCS}}\mathfrak{u},\nabla) \cong (\mathfrak{u},\nabla_0)$ at P is pronilpotent by assumption and that the residue of Ω at P is an element of $\mathfrak{u}_P$, which acts trivially on $\mathrm{Gr}^\bullet_{\mathrm{LCS}}\mathfrak{u}_P$. □

We now turn our attention to the behaviour of the Hodge bundles. Since each $\mathfrak{u}_n$ is a sum of variations of MHS that are tensor products of a constant MHS with a PVHS, they behave well near each cusp $P \in D$. In particular, the Hodge bundles $F^p\mathfrak{u}_n$ extend to sub-bundles of $\overline{\mathfrak{u}}_n$. This implies that the Hodge bundles $F^p\mathfrak{u}$ extend to holomorphic sub-bundles of $\overline{\mathfrak{u}}$. Consequently, they extend as C^∞ sub-bundles of $(\overline{\mathfrak{u}},D'')$.

Lemma 7.8. *The Hodge sub-bundles of $\overline{\mathfrak{u}}$ are holomorphic and the connection ∇ satisfies Griffiths transversality: if s is a local holomorphic section of $F^p\overline{\mathfrak{u}}$, then ∇s is a local section $\Omega^1_C(\log D)\otimes F^{p-1}\overline{\mathfrak{u}}$.*

Proof. To prove that $F^p\overline{\mathfrak{u}}$ is a holomorphic sub-bundle with respect to the complex structure D'', it suffices to show that if s is a local C^∞ section of $F^p\overline{\mathfrak{u}}$, then $D''s$ is a $(0,1)$-form with values in $F^p\overline{\mathfrak{u}}$. Since $F^p\mathfrak{u}$ is a holomorphic sub-bundle of $(\mathfrak{u},\bar\partial)$, it follows that $\bar\partial s$ is a $(0,1)$-form with values in $F^p\overline{\mathfrak{u}}$. And since

$$\Omega'' \in F^0K^{0,1}(C,D;\mathfrak{u}) = E^{0,1}(C)\hat\otimes F^0\overline{\mathfrak{u}}$$

it follows that $\Omega''(s) \in E^{0,1}(C)\hat\otimes F^p\overline{\mathfrak{u}}$, which implies that $D''s \in E^{0,1}(C)\hat\otimes F^p\overline{\mathfrak{u}}$, as required.

Griffiths transversality follows for similar reasons. Suppose that s is a local C^∞ section of $F^p\mathfrak{u}$. Since $(\mathfrak{u},\nabla_0)$ satisfies Griffiths transversality, $\nabla_0 s$ is a 1-form valued section of $F^{p-1}\overline{\mathfrak{u}}$. Since

$$\Omega \in F^0K^1(C,D;\mathfrak{u}) \subseteq E^1(C\log D)\hat\otimes F^{p-1}\overline{\mathfrak{u}},$$

$\Omega(s)$ is a 1-form valued section of $F^{p-1}\overline{\mathfrak{u}}$. It follows that $\nabla(s)$ takes values in $\Omega^1_C(\log D)\otimes F^{p-1}\overline{\mathfrak{u}}$. □

Lemma 7.9. *The weight sub-bundles $W_m\mathfrak{u}$ are flat sub-bundles of $(\overline{\mathfrak{u}},\nabla)$. Moreover, the identity induces an isomorphism of PVHS*

$$(\mathrm{Gr}^W_m\overline{\mathfrak{u}},\nabla) \cong (\mathrm{Gr}^W_m\overline{\mathfrak{u}},\nabla_0).$$

Proof. Both assertions follow from the fact that Ω (and hence Ω'' as well) takes values in $\mathfrak{u}$ and that $\mathfrak{u} = W_{-1}\mathfrak{u}$. This implies that the adjoint action of Ω and Ω'' on $\mathrm{Gr}^W_\bullet\mathfrak{u}$ is trivial. It follows that ∇ respects the weight filtration of $\mathfrak{u}$ and that the induced connection on $\mathrm{Gr}^W_\bullet\mathfrak{u}$ is ∇_0. □

Since Ω acts trivially on $\mathrm{Gr}^\bullet_{\mathrm{LCS}}\mathfrak{u}$, we have:

Lemma 7.10. *For all $n \geq 1$ there is a natural isomorphism of*

$$(\mathrm{Gr}^n_{\mathrm{LCS}}\mathbf{u}, \nabla) \cong (\mathrm{Gr}^n_{\mathrm{LCS}}\mathbf{u}, \nabla_0) \cong \mathbf{u}_n$$

local systems. □

7.5. The ind-variation $\mathbb{O}(P_x)$

Here we dispense of a few technicalities, in preparation for the construction of a MHS on $\mathcal{O}(\mathcal{G}_x)$ in the next section.

Denote the fibers of $\mathbb{V}$ over $x, y \in C'$ by V_x and V_y, respectively. Denote the Zariski closure of the image of the parallel transport mapping

$$\rho_{x,y} : \Pi(C'; x, y) \to \mathrm{Hom}(H_x, H_y)$$

by $R_{x,y}$. Set $R_x = R_{x,x}$. Then $R_{x,y}$ is a left R_x torsor and a right R_y torsor.

The theorem of the fixed part (Sec. 7.2) implies that the coordinate ring of $R_{x,y}$ is an algebra in the category of ind-Hodge structures of weight 0. This HS is characterized by the property that the monodromy coaction

$$E_x \to \mathcal{O}(R_{x,y}) \otimes E_y \tag{7.11}$$

is a morphism of HS for all polarized variations of Hodge structure $\mathbb{E}$ over C' whose monodromy representation factors through $\rho_x : \pi_1(C', x) \to R_x$. (Here E_z denotes the fiber of $\mathbb{E}$ over $z \in C'$.) Under our assumption that each irreducible representation V_λ of R is absolutely irreducible and that the corresponding local system $\mathbb{V}_\lambda$ underlies a PVHS over C', there is an isomorphism of HS

$$\mathcal{O}(R_{x,y}) \cong \bigoplus_\lambda \mathrm{Hom}(V_{\lambda,x}, V_{\lambda,y})^*.$$

When $C' = \Gamma\backslash\mathfrak{h}$, Γ a finite index subgroup of $\mathrm{SL}_2(\mathbb{Z})$, and $\mathbb{H}$ is the standard variation of HS,

$$\mathcal{O}(R_{x,y}) = \bigoplus_{n \geq 0} \mathrm{Hom}(S^n H_x, S^n H_y)^*.$$

Denote the local system over C' whose fiber over y is $\mathcal{O}(R_{y,x})$ by $\mathbb{O}_x$. In concrete terms

$$\mathbb{O}_x = \bigoplus_\lambda \mathrm{Hom}(\mathbb{V}_\lambda, V_{\lambda,x})^* \cong \bigoplus_\lambda \mathbb{V}_\lambda \otimes V^*_{\lambda,x}.$$

Note that this is local system of algebras and that there is a natural left R_x action that preserves the algebra structure.

For every local system $\mathbb{E}$ whose monodromy representation factors through $\rho : \pi_1(X, x) \to R_x$, there is a natural isomorphism

$$\mathbb{E} \cong [\mathbb{O}_x \otimes E_x]^R. \tag{7.12}$$

The R-finite vectors in the de Rham complex of C' with coefficients in $\mathbb{O}_x$ form a complex $E^\bullet_{\mathrm{fin}}(C',\mathbb{O}_x)$. In concrete terms, this is

$$E^\bullet_{\mathrm{fin}}(C',\mathbb{O}_x) = \bigoplus_\lambda E^\bullet(C',\mathbb{V}_\lambda)\otimes V^*_{\lambda,x}.$$

It is a (graded commutative) differential graded algebra. Similarly, one defines the ind-MHC

$$\mathbf{K}(\mathbb{O}_x) = \bigoplus_\lambda \mathbf{K}(\mathbb{V}_\lambda)\otimes V^*_{\lambda,x}$$

whose complex part is

$$K^\bullet(C,D;\mathbb{O}_x) = \bigoplus_\lambda K^\bullet(C,D;\mathbb{V}_\lambda)\otimes V^*_{\lambda,x}.$$

The relevance of these complexes is that the iterated integrals of their elements are elements of the coordinate ring of $\mathcal{G}_x$. (Cf. [15].)

Lemma 7.13. *If $\mathbb{E}$ is a PVHS over C', then the isomorphism (7.12) induces an isomorphism of bifiltered complexes:*

$$K^\bullet(C,D;\mathbb{E}) \cong (K^\bullet(C,D;\mathbb{O}_x)\otimes\mathbb{E})^R. \quad \square$$

The relevance of the preceding discussion is that iterated integrals of elements of

$$E^\bullet(C',\mathbb{V}_\lambda)\otimes V^*_{\lambda,x}$$

that occur in Section 4 are iterated integrals of elements of $E^{\mathrm{fin}}(C',\mathbb{O}_x)$ considered in [15]. This implies that the iterated Shimura integrals considered by Manin [29, 30] are examples of the iterated integrals constructed in [15].

7.6. The MHS on the relative completion of $\pi_1(C',x)$

Denote the fiber of $\mathbf{u}$ over $x\in C'$ by $\mathfrak{u}_x$. Denote the corresponding prounipotent group by $\mathcal{U}_x$. The Hodge and weight bundles of $\mathbf{u}$ restrict to Hodge and weight filtrations on $\mathfrak{u}_x$. Set $\Gamma=\pi_1(C',x)$ and write C' as the quotient $\Gamma\backslash X$ of a simply connected Riemann surface by Γ. Implicit here is that we have chosen a point $\tilde{x}\in X$ that lies over $x\in C'$.

Trivialize the pullback of each local system $\mathbf{u}_n$ to X so that the flat sections are constant.[12] This determines a trivialization

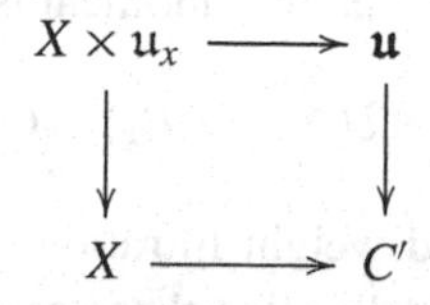

of the pullback of $\mathbf{u}$ to X as the product of the pullbacks of the $\mathbf{u}_n$. The pullback of the connection $\nabla = \nabla_0 + \Omega$ on $\mathbf{u}$ is Γ-invariant with respect to the diagonal Γ-action on $X \times \mathfrak{u}_x$ and is of the form $d + \widetilde{\Omega}$, where

$$\widetilde{\Omega} \in (E^1(X) \hat{\otimes} \mathfrak{u}_x)^\Gamma.$$

Proposition 4.6 and the fact that Ω_1 represents the product of the identity maps $H^1(C', \mathbb{V}_\lambda) \to H^1(C', \mathbb{V}_\lambda)$ for all $\lambda \in \check{R}$ imply that the transport of $\widetilde{\Omega}$ induces an isomorphism

$$\Theta_x : \mathcal{G}_x \to R_x \ltimes \mathcal{U}_x.$$

The MHS on $\mathcal{O}(\mathcal{G}_x)$ is constructed by pulling back the natural Hodge and weight filtrations on $\mathcal{O}(R_x) \otimes \mathcal{O}(\mathcal{U}_x)$, which we now recall.

To describe $\mathcal{O}(\mathcal{U}_x)$ we recall some basic facts about prounipotent groups. Suppose that $\mathcal{N}$ is a prounipotent group over F with Lie algebra $\mathfrak{n}$. The enveloping algebra $U\mathfrak{n}$ of $\mathfrak{n}$ is a Hopf algebra. The exponential mapping $\exp : \mathfrak{n} \to \mathcal{N}$ is a bijection, so we can identify $\mathcal{N}$ with the subspace $\mathfrak{n}$ of $U\mathfrak{n}$. The Poincaré-Birkhoff-Witt Theorem implies that the restriction mapping induces a Hopf algebra isomorphism

$$\mathcal{O}(\mathcal{N}) \cong \operatorname{Hom}^{\text{cts}}(U\mathfrak{n}, F) \cong \operatorname{Sym}^{\text{cts}}(\mathfrak{n}) := \varinjlim_{\alpha} \operatorname{Sym} \mathfrak{n}_\alpha.$$

In particular, $\mathcal{O}(\mathcal{U}_x) \cong \operatorname{Hom}^{\text{cts}}(U\mathfrak{u}_x, \mathbb{C})$. The construction of the Hodge and weight filtrations of $\mathfrak{u}_x$ and (7.11) imply that the coaction

$$U\mathfrak{u}_x \to \mathcal{O}(R_x) \otimes U\mathfrak{u}_x \tag{7.14}$$

that defines the semi direct product $R_x \ltimes \mathcal{U}_x$ preserves the Hodge and weight filtrations.[13]

[12] It is also natural to trivialize $\mathbf{u}_1$ (and hence all $\mathbf{u}_n$) so that the Hodge bundles are trivialized. In this case, we are in the setup of Section 4.1. The monodromy representation will be the same as it is with this "constant trivialization." Trivializing the Hodge bundles is better when computing the MHS on completed path torsors. For example, in Section 9.1, we can trivialize $\mathcal{H} := \mathbb{H} \otimes \mathcal{O}_\mathfrak{h}$ over the upper half plane $\mathfrak{h}$ using the flat sections $\mathbf{a}, \mathbf{b}$, as we do here, or we can use the sections $\mathbf{a}, \mathbf{w}$ that are adapted to the Hodge filtration. The second trivialization is better suited to studying asymptotic properties of VMHS.

[13] In fact, this is a morphism of MHS — and thus strict with respect with respect to the Hodge and weight filtrations — if we give $\mathfrak{u}_x$ MHS via the identification $\mathfrak{u}_x \cong \prod_n (\operatorname{Gr}_n^{\text{LCS}} \mathfrak{u})_x$. The canonical MHS on $\mathfrak{u}_x$ has the same underlying complex vector space, the same Hodge and weight filtrations, but its $\mathbb{Q}$-structure is deformed using the deformed connection $\nabla_0 + \Omega$.

The Hodge and weight filtrations of $\mathfrak{u}_x$ induce Hodge and weight filtrations on $\mathrm{Hom}^{\mathrm{cts}}(U\mathfrak{u}_x,\mathbb{C})$, and thus on $\mathcal{O}(\mathcal{U}_x)$. Define Hodge and weight filtrations on the coordinate ring of $R_x \ltimes \mathcal{U}_x$ via the canonical isomorphism

$$\mathcal{O}(R_x \ltimes \mathcal{U}_x) \cong \mathcal{O}(R_x) \otimes \mathcal{O}(\mathcal{U}_x).$$

These pullback to Hodge and weight filtrations on $\mathcal{O}(\mathcal{G}_x)$ along the isomorphism Θ_x. Equation (7.14) implies that these are compatible with the product and coproduct of $\mathcal{O}(R_x \times \mathcal{U}_x)$.

A filtration $W_\bullet$ of a Lie algebra $\mathfrak{u}$ is said to be *compatible with the bracket* if $[W_i\mathfrak{u}, W_j\mathfrak{u}] \subseteq W_{i+j}\mathfrak{u}$ for all i,j. Similarly, a filtration $W_\bullet$ of a Hopf algebra A is *compatible with its multiplication* μ *and comultiplication* Δ if $\mu(W_iA \otimes W_jA) \subseteq W_{i+j}A$ and $\Delta(W_mA) \subset \sum_{i+j} W_iA \otimes W_jA$. The Hodge and weight filtrations on $\mathcal{O}(R_x \ltimes \mathcal{U}_x)$ defined above are compatible with its product and coproduct.

Denote the maximal ideal of $\mathcal{O}(\mathcal{G}_x)$ of functions that vanish at the identity by $\mathfrak{m}_e$. The Lie algebra $\mathfrak{g}_x$ of $\mathcal{G}_x$ is isomorphic to $\mathrm{Hom}(\mathfrak{m}_e/\mathfrak{m}_e^2, F)$ and its bracket is induced by the Lie cobracket of $\mathcal{O}(\mathcal{G}_x)$. The Hodge and weight filtrations of $\mathcal{O}(\mathcal{G}_x)$ thus induce Hodge and weight filtrations on $\mathfrak{g}_x$ that are compatible with its bracket.

Theorem 7.15. *These Hodge and weight filtrations define a MHS on* $\mathcal{O}(\mathcal{G}_x)$ *for which the multiplication and comultiplication are morphisms. This MHS agrees with the one constructed in [15].*

Proof. The natural isomorphism $\Theta_x^* : \mathcal{O}(R_x \ltimes \mathcal{U}_x) \to \mathcal{O}(\mathcal{G}_x)$ respects $\mathbb{Q}$-structures (essentially by definition). To prove the result it suffices to show that it takes the Hodge and weight filtrations on $\mathcal{O}(R_x \ltimes \mathcal{U}_x)$ constructed above onto the Hodge and weight filtrations of the canonical MHS on $\mathcal{O}(\mathcal{G}_x)$ constructed in [15]. This will imply that the weight filtration defined on $\mathcal{O}(R_x \ltimes \mathcal{U}_x)$ above is defined over $\mathbb{Q}$ and that $\mathcal{O}(R_x \otimes \mathcal{U}_x)$ has a MHS and that this MHS is isomorphic to the canonical MHS on $\mathcal{O}(\mathcal{G}_x)$ via Θ_x^*.

The first point is that Saito's MHC for computing the MHS on $H^\bullet(C',\mathbb{V})$, which is used in [15], is a generalization of Zucker's MHC used here and agrees with it in the curve case. The iterated integrals of elements of $K^\bullet(C,D;\mathbb{V}) \otimes V_x^*$ in this paper are a special case of the iterated integrals defined in [15, §5] by Lemma 7.13.

The next point is that, by Equation (7.2) and Lemma 7.13,

$$\Omega \in F^0W_{-1}K^1(C,D;\mathbf{u}) \cong F^0W_{-1}\big(K^1(C,D;\mathbb{O}_x)\hat{\otimes}\mathfrak{u}_x\big).$$

This implies, and this is the key point, that — with the Hodge and weight filtrations on the bar construction defined in [15, §13] and the Hodge and weight filtrations on $\mathfrak{u}_x$ defined above — the $U\mathfrak{u}_x$-valued *formal transport*

$$T := 1 + [\widetilde{\Omega}] + [\widetilde{\Omega}|\widetilde{\Omega}] + [\widetilde{\Omega}|\widetilde{\Omega}|\widetilde{\Omega}] + \cdots \in B\big(\mathbb{C}, K^\bullet_{\text{fin}}(C,D;\mathbb{O}_x),\mathbb{C}\big)\hat{\otimes}\, U\mathfrak{u}_x$$

of $\widetilde{\Omega}$, which takes values in the completed enveloping algebra of $\mathfrak{u}_x$, satisfies

$$T \in F^0 W_0 H^0\Big(B\big(\mathbb{C}, K^\bullet_{\text{fin}}(C,D;\mathbb{O}_x),\mathbb{C}\big)\hat{\otimes}\, U\mathfrak{u}_x\Big).$$

This implies that the induced Hopf algebra homomorphism

$$f : \mathcal{O}(\mathcal{U}_x) \cong \text{Hom}^{\text{cts}}(U\mathfrak{u}_x,\mathbb{C}) \to H^0\Big(B\big(\mathbb{C}, K^\bullet_{\text{fin}}(C,D;\mathbb{O}_x),\mathcal{O}(R_x)\big)\Big) \cong \mathcal{O}(\mathcal{G}_x)$$

that corresponds to the function

$$\mathcal{G}_x \xrightarrow{\Theta_x} R_x \ltimes \mathcal{U}_x \xrightarrow{\text{projn}} \mathcal{U}_x$$

preserves the Hodge and weight filtrations.

The constructions in [15] imply that the homomorphism $\pi^* : \mathcal{O}(R_x) \to \mathcal{O}(\mathcal{G}_x)$ induced by the projection $\pi : \mathcal{G}_x \to R_x$ is a morphism of MHS. Since the Hodge and weight filtrations of $\mathcal{O}(\mathcal{G}_x)$ are compatible with multiplication and since both f and π^* preserve the Hodge and weight filtrations, the homomorphism

$$\mathcal{O}(R_x) \otimes \mathcal{O}(\mathcal{U}_x) \xrightarrow{\pi^* \otimes f} \mathcal{O}(\mathcal{G}_x)$$

does too. This homomorphism is Θ_x^* when $\mathcal{O}(R_x) \otimes \mathcal{O}(\mathcal{U}_x)$ is identified with $\mathcal{O}(R_x \ltimes \mathcal{U}_x)$.

It remains to prove that this isomorphism is an isomorphism of bifiltered vector spaces. Since π^* is a morphism of MHS, it suffices to show that the isomorphism $j^* \circ f$

$$\mathcal{O}(\mathcal{U}_x) \xrightarrow{f} \mathcal{O}(\mathcal{G}_x) \xrightarrow{j^*} \mathcal{O}(\mathcal{N}_x),$$

where $j : \mathcal{N}_x \to \mathcal{G}_x$ is the inclusion of the prounipotent radical of $\mathcal{G}_x$, is a bifiltered isomorphism. To prove this, it suffices to show that $\mathfrak{n}_x \to \mathfrak{u}_x$ is a bifiltered isomorphism, where $\mathfrak{n}_x$ denotes the Lie algebra of $\mathcal{N}_x$. But this follows from Lemma 7.10, which implies that $H_1(\mathfrak{u}_x)$ has a MHS and that the induced homomorphism

$$\text{Gr}^{\text{LCS}}_\bullet \mathfrak{n}_x \to \text{Gr}^{\text{LCS}}_\bullet \mathfrak{u}_x$$

is an isomorphism of graded MHS. □

The result also gives an explicit description of the MHS on $\mathfrak{u}_x$.

Corollary 7.16. *The Hodge and weight filtrations of the natural MHS on* $\mathfrak{u}_x$ *are those induced on it from* $\mathbf{u}$*; its* $\mathbb{Q}$*-structure is the one described in Proposition 4.7.* □

To complete the story, we show that the $\mathbf{u}$ is a pro-admissible variation of MHS.

Theorem 7.17. *With the Hodge and weight filtrations and* $\mathbb{Q}$*-structure defined above,* $(\mathbf{u}, \nabla)$ *is a pro-object of the category of admissible variation of MHS over* C'*. Its lower central series is a filtration of* $\mathbf{u}$ *by pro-admissible variations of MHS. The natural isomorphism*

$$(\mathrm{Gr}^n_{\mathrm{LCS}}\, \mathbf{u}, \nabla_0) \cong (\mathrm{Gr}^n_{\mathrm{LCS}}\, \mathbf{u}, \nabla)$$

is an isomorphism of admissible variations of MHS. In particular, there are natural MHS isomorphisms

$$H_1(\mathfrak{u}_x) \cong \prod_{\lambda \in \check{R}} H^1(C', \mathbb{V}_\lambda)^* \otimes V_{\lambda,x}$$

for all base points x of C'.[14] *Finally, for all PVHS* $\mathbb{V}$ *over* C' *whose monodromy representation factors through* ρ*, the natural homomorphism*

$$H^\bullet(\mathcal{G}_x, V_x) = [H^\bullet(\mathfrak{u}_x) \otimes V_x]^R \to H^\bullet(C', \mathbb{V})$$

is an isomorphism of MHS.

With a little more work, one can show that the local system with fiber $\mathcal{O}(\mathcal{G}_x)$ over $x \in C'$ is an admissible VMHS.

Proof. Proposition 7.7 and Lemma 7.8 imply that the Hodge bundles are holomorphic sub-bundles of $\mathbf{u}$ and extend to holomorphic sub-bundles of $\overline{\mathbf{u}}$. Lemma 7.9 implies that the weight bundle's are flat and extend to $\overline{\mathbf{u}}$. Theorem 7.15 implies that $\mathbf{u}$ has a natural $\mathbb{Q}$-form and that, with respect to these structures, each fiber of $\mathbf{u}$ has a natural MHS.

To complete the proof, we need to show that at each $P \in D$ there is a relative weight filtration of the fiber $\mathfrak{u}_P$ of $\overline{\mathbf{u}}$ over P. First an easily verified fact. Suppose that $\mathbb{V}$ is a PVHS over C' of weight m. Let $M_\bullet$ be the monodromy weight filtration of its fiber V_P over $P \in D$ shifted so that it is centered at m. Then if A is a constant MHS, then the filtration

$$M_r(A \otimes V_P) := \sum_{i+j=r} W_i A \otimes M_j V_P$$

[14] This statement holds, even when x is a tangential base point.

defines a relative weight filtration of the fiber over $P \in D$ of the admissible variation of MHS $A \otimes \mathbb{V}$. From this it follows that the fiber over $P \in D$ of each

$$H^1(C',\mathbb{V}_\lambda)^* \otimes \mathbb{V}_\lambda$$

has a relative weight filtration. Adding these implies that the fiber over $P \in D$ of each of the pro-variations $\mathbf{u}_n$ has a relative weight filtration. Write the residue at $P \in D$ of ∇ as the sum

$$N_P = N_0 + N_{\mathrm{u}}$$

of the residues of ∇_0 and Ω. The discussion above implies that the product of the weight filtrations on the fibers over P of the $\mathbf{u}_n$ defines a relative weight filtration for N_0 on the fiber $\mathfrak{u}_P$ of $\overline{\mathbf{u}}$ over P. We have to show that this is also a relative weight filtration for N. To prove this, it suffices to show that $N_{\mathrm{u}} \in W_{-2}\mathfrak{u}_P$. (See the definition of the relative weight filtration in [41].)

Let t be a local holomorphic coordinate on C centered at P. Then, near P, we can write

$$\Omega = N_{\mathrm{u}} \otimes \frac{dt}{t} + \text{ a smooth 1-form with values in } \overline{\mathbf{u}}$$

Since $\Omega^1 \in W_{-1}K^1(C,D;\mathbf{u})$, and since dt/t has weight 1, we see that $N_{\mathrm{u}} \in W_{-2}\mathfrak{u}_P$, as required.

The last statement is a direct consequence of (3.4.2), the de Rham construction of the homomorphism in Section 4.2, and the fact that $\Omega \in F^0W_{-1}K^1(C,D;\mathbf{u})$, which implies that $\theta_\Omega : H^0(C',\mathrm{Hom}(\Lambda^\bullet\mathbf{u},\mathbb{V})) \to K^\bullet(C,D,\mathbb{V})$ preserves the Hodge filtration and satisfies

$$\theta_\Omega\Big(W_m H^0(C',\mathrm{Hom}(\Lambda^j\mathbf{u},\mathbb{V}))\Big) \subseteq W_{m-j}K^j(C,D,\mathbb{V})$$

in degree j.[15] The last ingredient is the Theorem of the Fixed Part, which implies that for each $x \in C'$, the restriction mapping

$$H^0\big(C',\mathrm{Hom}(\Lambda^\bullet\mathbf{u},\mathbb{V})\big) \to C^\bullet(\mathfrak{u}_x, V_x)^R$$

is an isomorphism of MHS for all $x \in C'$. □

7.7. A MHS on completed path torsors

Here we give a brief description of how to extend the methods of the previous section to construct the canonical MHS on the coordinate ring $\mathcal{O}(\mathcal{G}_{x,y})$ of the relative completion of the path torsor $\Pi(C';x,y)$ of C' with respect to a polarized VHS $\mathbb{V}$. Here C' may be an orbifold of the form $\Gamma\backslash X$.

[15] This is equivalent to the statement that θ_Ω is a filtration preserving dga homomorphism to $\mathrm{Dec}_W K^j(C,D,\mathbb{V})$, where Dec_W is Deligne's shifting functor (with respect to the weight filtration), [11].

As in Section 7.6, write C' as the quotient of a simply connected Riemann surface X by a discrete group Γ isomorphic to $\pi_1(C',x)$. Trivialize the pullback of each $\mathbf{u}_n$ to X using the flat sections and use this to trivialize the pullback of $\mathbf{u}$ to X. Here, unlike in the previous section, it is useful to denote the common fiber of the trivialization by $\mathfrak{u}$ so that the pullback of $\mathbf{u}$ to X is $X \times \mathfrak{u}$. The fiber over $t \in X$ will be regarded as $\mathfrak{u}$ with a Hodge filtration $F_t^\bullet$, which depends on t, and a weight filtration $W_\bullet$, which does not. We can therefore identify $\mathrm{Hom}^{\mathrm{cts}}(\mathfrak{u}_s, \mathfrak{u}_t)$ with $\mathrm{End}^{\mathrm{cts}}\,\mathfrak{u}$. The Hodge and weight filtrations of $\mathfrak{u}_s$ and $\mathfrak{u}_t$ induce Hodge and weight filtrations on $\mathrm{Hom}^{\mathrm{cts}}(\mathfrak{u}_s, \mathfrak{u}_t)$.

For $s,t \in X$, set $\mathcal{U}_{s,t}$ be the subscheme of $\mathrm{Isom}^{\mathrm{cts}}(\mathfrak{u}_s, \mathfrak{u}_t)$ that corresponds to the subgroup $\mathcal{U} := \exp \mathfrak{u}$ of $\mathrm{Aut}\,\mathfrak{u}$. It is a proalgebraic variety. Denote the subspace of $\mathrm{Hom}^{\mathrm{cts}}(\mathfrak{u}_s, \mathfrak{u}_t)$ that corresponds to the image of

$$U\mathfrak{u} \in \mathrm{End}(\mathfrak{u}), \quad u \mapsto \{v \mapsto uv\}$$

by $U\mathfrak{u}_{s,t}$. The coordinate ring of $\mathcal{U}_{s,t}$ is $\mathcal{O}(\mathcal{U}_{s,t}) \cong \mathrm{Hom}^{\mathrm{cts}}(U\mathfrak{u}_{s,t}, \mathbb{C})$. It has natural Hodge and weight filtrations induced from those on $\mathrm{Hom}^{\mathrm{cts}}(\mathfrak{u}_s, \mathfrak{u}_t)$.

The pullback connection is $d + \widetilde{\Omega}$, where $\widetilde{\Omega} \in E^1(X) \hat{\otimes} \mathfrak{u}$. Since the structure group of this connection is $\mathcal{U}$, the parallel transport map $T_{s,t} : \mathfrak{u}_s \to \mathfrak{u}_t$ lies in $\mathcal{U}_{s,t}$.

As in Section 5, we choose a fundamental domain D of the action of Γ on X. Denote the unique lift of $z \in C'$ to D by $\tilde{z}$. For each homotopy class of paths in C' from x to y one has $\gamma \in \Gamma$ and a homotopy class c_γ of paths from $\tilde{x}$ to $\gamma\tilde{y}$. Parallel transport defines a function

$$\Theta_{x,y} : \Pi(C';x,y) \to \mathcal{U}_{x,y} \times R_{x,y} \cong \mathcal{G}_{x,y}$$

by $(\gamma, c_\gamma) \mapsto \big(T(c_\gamma)^{-1}, \rho_{x,y}(\gamma)\big)$. This is the relative completion of $\Pi(C';x,y)$.

The isomorphism

$$\mathcal{O}(\mathcal{G}_{x,y}) \cong \mathcal{O}(\mathcal{U}_{x,y}) \otimes \mathcal{O}(R_{x,y})$$

induces Hodge and weight filtrations on the coordinate ring of $\mathcal{G}_{x,y}$. It also has a natural $\mathbb{Q}$-structure as relative completion is defined over $\mathbb{Q}$ and behaves well under base change from $\mathbb{Q}$ to $\mathbb{C}$.

The following theorem generalizes Theorems 7.15 and 7.17. It is proved using a similar arguments. A more general version of all but the last statement is proved in [15, §12–13].

Theorem 7.18. *These Hodge and weight filtrations define a MHS on $\mathcal{O}(\mathcal{G}_{x,y})$, making it a ring in the category of ind-mixed Hodge structures. This MHS coincides with the one constructed in [15]. It has the property that if $x,y,z \in X$, then the maps $\mathcal{G}_{x,y} \to \mathcal{G}_{y,x}$ and $\mathcal{G}_{x,y} \times \mathcal{G}_{y,z} \to \mathcal{G}_{x,z}$ induced by inverse and path multiplication, respectively, induce morphisms of MHS*

$$\mathcal{O}(\mathcal{G}_{y,x}) \to \mathcal{O}(\mathcal{G}_{x,y}) \text{ and } \mathcal{O}(\mathcal{G}_{x,z}) \to \mathcal{O}(\mathcal{G}_{x,y}) \otimes \mathcal{O}(\mathcal{G}_{y,z}).$$

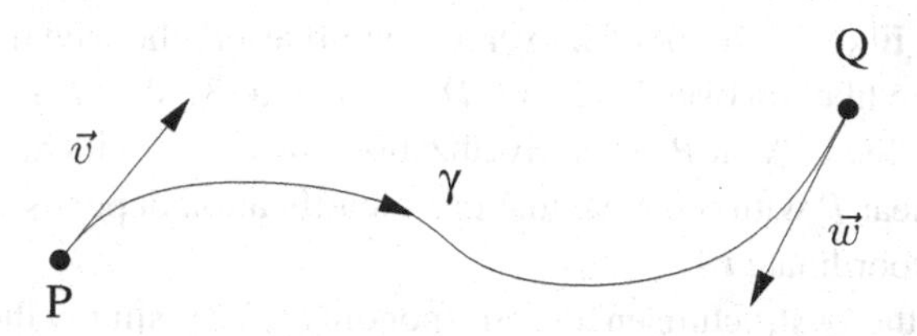

Figure 2. A path from $\vec{v}$ to $\vec{w}$ in C'.

In addition the local system $\mathcal{G}_{x,\_}$ over C' whose fiber over x,y is $\mathcal{O}(\mathcal{G}_{x,y})$ is an Ind object of the category of admissible variations of MHS over C'.

7.8. Tangential base points and limit MHSs

Theorem 7.17 implies that for each choice of a non-zero tangent vector $\vec{v}$ of C at $P \in D$, there is a limit MHS on the fiber $\mathfrak{u}_P$ of $\overline{\mathfrak{u}}$ at P. We will denote this MHS by $\mathfrak{u}_{\vec{v}}$. It is natural to think of it as a MHS on the unipotent radical of the relative completion of the fundamental group $\pi_1(C',\vec{v})$ of C' with (tangential) base point $\vec{v}$.

More generally, we consider path torsors between tangential base points,. We first recall the definition from [12] of the torsor of paths $\Pi(C';\vec{v},\vec{w})$. Here $P,Q \in D$ and $\vec{v} \in T_P C$, $\vec{w} \in T_Q C$ are non-zero tangent vectors. Elements of $\Pi(C;\vec{v},\vec{w})$ are homotopy classes of piecewise smooth paths $\gamma : I \to C$ satisfying

(i) $\gamma(0) = P$, $\gamma(1) = Q$,
(ii) $\gamma(t) \in C'$, when $0 < t < 1$,
(iii) $\gamma'(0) = \vec{v}$ and $\gamma'(1) = -\vec{w}$.

This definition can be modified to define $\Pi(C;\vec{v},x)$ and $\Pi(C;x,\vec{w})$ when $x \in C'$. One defines $\pi_1(C',\vec{v}) = \Pi(C',\vec{v},\vec{v})$. For any 3 base points (tangential or regular) b, b', b'', there are composition maps

$$\Pi(C';b,b') \times \Pi(C';b',b'') \to \Pi(C';b,b'').$$

7.8.1. The MHS on $\mathfrak{u}_{\vec{v}}$

Suppose that $P \in D$ and that $\vec{v}$ is a non-zero tangent vector of C at P. The complex vector space underlying the limit MHS on $\mathfrak{u}_{\vec{v}}$ is the fiber $\mathfrak{u}_P$ of $\overline{\mathfrak{u}}$ over P. Its Hodge and weight filtrations $F^\bullet$ and $W_\bullet$ are the restrictions to $\mathfrak{u}_P$ of the Hodge and weight filtrations of $\overline{\mathfrak{u}}$ that were constructed above. There is also the relative weight filtration $M_\bullet$ of $\mathfrak{u}_P$, which was constructed in the proof of Theorem 7.17. These data depend only on P and not on $\vec{v}$.

To construct the $\mathbb{Q}$-structure, choose a local holomorphic coordinate $t : \Delta \to \mathbb{C}$ on C centered at P with the property that $\vec{v} = \partial/\partial t$. Then there is a unique trivialization $\overline{\mathbf{u}}|_\Delta \cong \Delta \times u_P$ of $\overline{\mathbf{u}}$ over Δ such that (1) the trivialization is the identity on the fiber u_P over P, (2) $\Delta \cap D = \{P\}$, and (3) $\nabla = d + N_P dt/t$, where N_P is the residue of ∇ at P. This trivialization allows us to identify fibers of $\mathbf{u}$ over points near P with u_P. Note that this identification depends on the choice of the local coordinate t.

For $t \in \Delta$, the $\mathbb{Q}$-structure on u_P corresponding to $t\vec{v}$ is simply the $\mathbb{Q}$ structure on u_P obtained by identifying u_P with the fiber u_t of $\mathbf{u}$ over $t \in \Delta$ and taking the $\mathbb{Q}$-structure to be that of u_t. The $\mathbb{Q}$-structure corresponding to $\vec{v}$ is

$$u_{\vec{v},\mathbb{Q}} := t^{N_P} u_{t\vec{v},\mathbb{Q}}.$$

That is, it is the unique $\mathbb{Q}$-structure on u_P such that

$$u_{t\vec{v},\mathbb{Q}} := t^{-N_P} u_{\vec{v},\mathbb{Q}}.$$

for all $t \in \mathbb{C}^*$. Although the trivialization above depends on the choice of the local coordinate t, the $\mathbb{Q}$-structure on u_P corresponding to $\vec{v}$ depends only on $\vec{v}$ the tangent vector $\vec{v}$.

With a little more effort, one can construct the limit MHS on $\mathcal{O}(\mathcal{G}_{\vec{v}})$. Full details will appear in [24]. As in the case of cohomology, where periods of limit MHSs can be computed by regularizing integrals, the periods of the limit MHS on $u_{\vec{v}}$ are regularized iterated integrals.

7.8.2. Limit MHS on completed path torsors

Similarly, one can construct the limit MHS on $\mathcal{O}(\mathcal{G}_{x,\vec{w}})$ and $\mathcal{O}(\mathcal{G}_{\vec{v},\vec{w}})$, etc. Full details will appear in [24].

8 Completed Path Torsors and Admissible Variations of MHS

Here we state two results that relate the Hodge theory of relative completion of fundamental groups and path torsors to admissible variations of MHS. These are special cases of results in [24].

Let C, D and $C' = C - D$ be as above. Let $\mathbb{V}$ be a PVHS over C' and R_x the Zariski closure of the monodromy representation $\pi_1(C',x) \to \operatorname{Aut} V_x$. Let ρ_x be the homomorphism $\pi_1(X,x) \to R_x$. Let $\mathcal{G}_x$ be the completion of $\pi_1(C',x)$ with respect to ρ_x. For base points (regular or tangential) b, b' of C', let $\mathcal{G}_{x,y}$ denote the relative completion of $\Pi(C';b,b')$.

Denote by $\mathsf{MHS}(C',\mathbb{V})$ the category of admissible VMHS $\mathbb{A}$ over C' with the property that the monodromy representation of $\operatorname{Gr}^W_\bullet \mathbb{A}$ factors through ρ_x. This condition implies that the monodromy representation

$$\pi_1(C',x) \to \operatorname{Aut}(A_x)$$

factors through $\pi_1(C',x) \to \mathcal{G}_x$.

Theorem 8.1. *For all variations* $\mathbb{A}$ *in* $\mathsf{MHS}(C',\mathbb{V})$ *and all base points* b,b' *(possibly tangential) of* C' *the parallel transport mapping*

$$A_b \to A_{b'} \otimes \mathcal{O}(\mathcal{G}_{b,b'})$$

induced by $A_b \times \Pi(C';b,b') \to A_{b'}$ *is a morphism of MHS. When* b *or* b' *is tangential, then the monodromy preserves both the weight filtration* $W_\bullet$ *and the relative weight filtration* $M_\bullet$*.*

Definition 8.2. Suppose that $\mathcal{G}$ is a proalgebraic group whose coordinate ring $\mathcal{O}(\mathcal{G})$ is a Hopf algebra in the category of ind mixed Hodge structures. A *Hodge representation* of $\mathcal{G}$ on a MHS A is a homomorphism for which the action

$$A \to A \otimes \mathcal{O}(\mathcal{G})$$

is a morphism of MHS. The category of Hodge representations of $\mathcal{G}$ will be denoted by $\mathsf{HRep}(\mathcal{G})$.

The previous result implies that taking the fiber at b defines a functor from $\mathsf{MHS}(C',\mathbb{V})$ to $\mathsf{HRep}(\mathcal{G}_b)$. The next theorem follows from Theorem 7.15 by a tannakian argument. Full details will be given in [24].

Theorem 8.3. *For all base points* b *of* C'*, the "fiber at* b*" functor* $\mathsf{MHS}(C',\mathbb{V}) \to \mathsf{HRep}(\mathcal{G}_b)$ *is an equivalence of categories.*

Part 2. Completed Path Torsors of Modular Curves

In this part, we apply the general constructions of the first part to explore the relative completions of modular groups, mainly in the case of the full modular group $\mathrm{SL}_2(\mathbb{Z})$. Throughout we use the following notation.

Suppose that Γ is a finite index subgroup of $\mathrm{SL}_2(\mathbb{Z})$. The associated curve X_Γ is the quotient $\Gamma\backslash\mathfrak{h}$ of the upper half plane by Γ. It is a smooth affine curve when Γ is torsion free. When Γ has torsion, it will be regarded as an orbifold as follows: Choose a finite index, torsion free normal subgroup Γ' of Γ. Set $G = \Gamma/\Gamma'$. Then X_Γ is the orbifold quotient of $X_{\Gamma'}$ by G. To work on the orbifold X_Γ, one can work either G-equivariantly on $X_{\Gamma'}$ or Γ-equivariantly on $\mathfrak{h}$.

The (orbi) curve X_Γ can be completed to a smooth (orbi) curve by adding the finite set $D := \Gamma\backslash\mathbb{P}^1(\mathbb{Q})$ of "cusps." Denote the compactified curve by $\overline{X}_\Gamma$. When $\Gamma = \mathrm{SL}_2(\mathbb{Z})$, the modular curve X_Γ is the moduli space $\mathcal{M}_{1,1}$ of elliptic curves and $\overline{X}_\Gamma$ is $\overline{\mathcal{M}}_{1,1}$, it's the Deligne-Mumford compactification, which is obtained by adding a single cusp.

If $P \in D$ is in the orbit of $\infty \in \mathbb{P}^1(\mathbb{Q})$, then

$$\Gamma \cap \begin{pmatrix} 1 & \mathbb{Z} \\ 0 & 1 \end{pmatrix} = \begin{pmatrix} 1 & n\mathbb{Z} \\ 0 & 1 \end{pmatrix}.$$

for some $n \geq 1$. A punctured neighbourhood of P in X_Γ is the quotient of $\{\tau \in \mathbb{C} : \mathrm{Im}(\tau) > 1\}$ by this group. This is a punctured disk with coordinate $e^{2\pi i n\tau}$. In particular, when $\Gamma = \mathrm{SL}_2(\mathbb{Z})$, the coordinate about the cusp is $q := e^{2\pi i\tau}$.

9 The Variation of Hodge Structure $\mathbb{H}$

9.1. The Local System $\mathbb{H}$

The universal elliptic curve $f : \mathcal{E} \to X_\Gamma$ over X_Γ is the quotient of $\mathbb{C} \times \mathfrak{h}$ by $\Gamma \ltimes \mathbb{Z}^2$, which acts via

$$(m,n) : (z,\tau) \mapsto \left(z + \begin{pmatrix} m & n \end{pmatrix}\begin{pmatrix} \tau \\ 1 \end{pmatrix}, \tau\right)$$

and $\gamma : (z,\tau) \mapsto \left((c\tau+d)^{-1}z, \gamma\tau\right)$, where

$$\gamma = \begin{pmatrix} a & b \\ c & d \end{pmatrix} \in \Gamma.$$

When Γ is not torsion free, it should be regarded as an orbifold family of elliptic curves.

The local system $\mathbb{H}$ over X_Γ is $R^1 f_* \mathbb{Q}$. When Γ is torsion free, this is the local system over X_Γ with fiber $H^1(f^{-1}(x), \mathbb{Q})$ over x. It is a polarized variation of HS of weight 1. Since Poincaré duality induces an isomorphism

$$H_1(E) \cong H^1(E)(1)$$

for all elliptic curves E, the polarized variation $\mathbb{H}(1)$ of weight -1 is the local system over X_Γ whose fiber over x is $H_1(f^{-1}(x), \mathbb{Q})$. The polarization is the intersection pairing. Denote the corresponding holomorphic vector bundle $\mathbb{H} \otimes \mathcal{O}_{X_\Gamma}$ by $\mathcal{H}$. Its Hodge filtration satisfies

$$\mathcal{H} = F^0\mathcal{H} \supset F^1\mathcal{H} \supset F^2\mathcal{H} = 0.$$

The only interesting part is $F^1\mathcal{H}$.

In general we will work with the pullback $\mathbb{H}_\mathfrak{h}$ of $\mathbb{H}$ to $\mathfrak{h}$. Its fiber over $\tau \in \mathfrak{h}$ is $H^1(E_\tau, \mathbb{Z})$, where

$$E_\tau := \mathbb{C}/\Lambda_\tau \text{ and } \Lambda_\tau := \mathbb{Z} \oplus \mathbb{Z}\tau.$$

Denote the basis of $H_1(E_\tau, \mathbb{Z})$ corresponding to the elements 1 and τ of Λ_τ by $\mathbf{a}, \mathbf{b}$. Denote the dual basis of $H^1(E_\tau)$ by $\check{\mathbf{a}}, \check{\mathbf{b}}$. These trivialize $\mathbb{H}_\mathfrak{h}$.

If we identify $H^1(E_\tau)$ with $H_1(E_\tau)$ via Poincaré duality, then

$$\check{\mathbf{a}} = -\mathbf{b} \text{ and } \check{\mathbf{b}} = \mathbf{a}.$$

We regard these as sections of $\mathbb{H}_\mathfrak{h}$.

For each $\tau \in \mathfrak{h}$, let $\omega_\tau \in H^0(E_\tau, \Omega^1)$ be the unique holomorphic differential that takes the value 1 on $\mathbf{a}$. It spans $F^1H^1(E_\tau)$. In terms of the framing, it is given by

$$\omega_\tau = \check{\mathbf{a}} + \tau\check{\mathbf{b}} = \tau\mathbf{a} - \mathbf{b} = \begin{pmatrix}\mathbf{a} & -\mathbf{b}\end{pmatrix}\begin{pmatrix}\tau \\ 1\end{pmatrix}. \tag{9.1}$$

The map $\omega : \tau \mapsto \omega_\tau$ is thus a holomorphic section of $\mathcal{H}_\mathfrak{h} := \mathbb{H}_\mathfrak{h} \otimes \mathcal{O}_\mathfrak{h}$ whose image spans $F^1\mathcal{H}_\mathfrak{h}$.

Since $\mathfrak{h}$ is contractible, $H_1(\mathcal{E},\mathbb{Z}) \cong \mathbb{Z}\mathbf{a} \oplus \mathbb{Z}\mathbf{b}$. The left action of $\mathrm{SL}_2(\mathbb{Z})$ on $\mathcal{E}$ induces a *left* action on $H_1(\mathcal{E})$, and therefore a *right* action on frames. The following result gives a formula for this action on frames.

Lemma 9.2. *For all $\gamma \in \Gamma$,*

(i) $\gamma : \begin{pmatrix}\mathbf{a} & -\mathbf{b}\end{pmatrix} \mapsto \begin{pmatrix}\mathbf{a} & -\mathbf{b}\end{pmatrix}\gamma$,
(ii) *the section ω of $\mathcal{H}_\mathfrak{h}$ satisfies $(1 \otimes \gamma)\omega = (c\tau + d)(\gamma^* \otimes 1)\omega$.*

Proof. Let $\gamma = \begin{pmatrix} a & b \\ c & d\end{pmatrix}$. Regard $\mathbf{a}$ and $\mathbf{b}$ as sections of $\mathbb{H}_\mathfrak{h}$. Denote the values of $\mathbf{a}$ and $\mathbf{b}$ at τ by $\mathbf{a}, \mathbf{b}$ and at $\gamma(\tau)$ by $\mathbf{a}', \mathbf{b}'$. Then (cf. Figure 3)

$$\begin{pmatrix}\mathbf{a}' & -\mathbf{b}'\end{pmatrix} = \begin{pmatrix}\mathbf{a} & -\mathbf{b}\end{pmatrix}\begin{pmatrix} d & -b \\ -c & a\end{pmatrix} = \begin{pmatrix}\mathbf{a} & -\mathbf{b}\end{pmatrix}\gamma^{-1}.$$

Thus $\gamma_*\begin{pmatrix}\mathbf{a}(\tau) & -\mathbf{b}(\tau)\end{pmatrix} = \begin{pmatrix}\mathbf{a}(\gamma\tau) & -\mathbf{b}(\gamma\tau)\end{pmatrix}\gamma$, from which the first assertion follows.

The second assertion now follows:

$$(1\otimes\gamma)\omega = \begin{pmatrix}\mathbf{a} & -\mathbf{b}\end{pmatrix}\gamma\begin{pmatrix}\tau \\ 1\end{pmatrix} = (c\tau+d)\begin{pmatrix}\mathbf{a} & -\mathbf{b}\end{pmatrix}\begin{pmatrix}\gamma\tau \\ 1\end{pmatrix} = (c\tau+d)(\gamma^*\otimes 1)\omega.$$

□

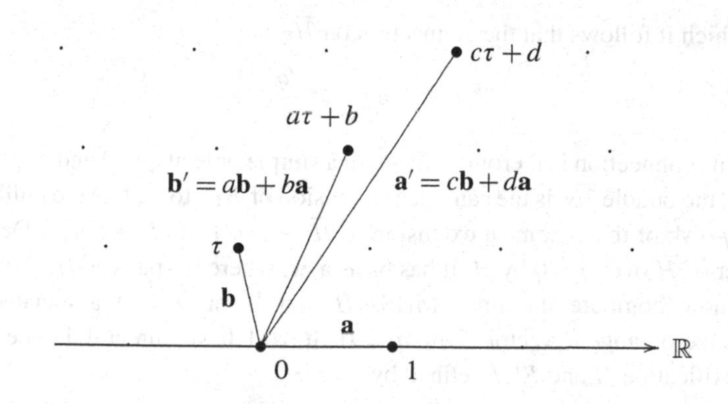

Figure 3. The $\mathrm{SL}_2(\mathbb{Z})$ action on frames.

Define $\mathbf{w}$ to be the section

$$\mathbf{w} : \tau \mapsto 2\pi i \omega_\tau = 2\pi i(\tau \mathbf{a} - \mathbf{b}). \tag{9.3}$$

of $\mathcal{H}_\mathfrak{h}$. Since $\langle \mathbf{w}, \mathbf{a} \rangle = 2\pi i \langle \tau \mathbf{a} - \mathbf{b}, \mathbf{a} \rangle = 2\pi i$. $\mathbf{w}, \mathbf{a}$ is a framing of $\mathcal{H}_\mathfrak{h}$. The sections $\mathbf{a}$ and $\mathbf{w}$ trivialize $\mathcal{H}_\mathfrak{h}$ over $\mathfrak{h}$. As we shall see below, this trivialization is better suited to computing limit MHSs. The following computation is immediate. The proof of an equivalent formula can be found in [20, Ex. 3.4].

Corollary 9.4. *The factor of automorphy associated to the trivialization*

$$\mathcal{H}_\mathfrak{h} \cong (\mathbb{C}\mathbf{a} \oplus \mathbb{C}\mathbf{w}) \times \mathfrak{h}$$

of the pullback of $\mathcal{H}$ to $\mathfrak{h}$ is

$$M_\gamma(\tau) : \begin{pmatrix} \mathbf{a} & \mathbf{w} \end{pmatrix} = \begin{pmatrix} \mathbf{a} & \mathbf{w} \end{pmatrix} \begin{pmatrix} (c\tau + d)^{-1} & 0 \\ c/2\pi i & c\tau + d \end{pmatrix}.$$

That is, the action of $\mathrm{SL}_2(\mathbb{Z})$ *on* $(\mathbb{C}\mathbf{a} \oplus \mathbb{C}\mathbf{w}) \times \mathfrak{h}$ *is*

$$\gamma : (\mathbf{a}, \tau) \mapsto \big((c\tau + d)^{-1}\mathbf{a} + (c/2\pi i)\mathbf{w}, \gamma\tau\big) \textit{ and } \gamma : (\mathbf{w}, \tau) \mapsto \big((c\tau + d)\mathbf{w}, \gamma\tau\big). \quad \square$$

The sections $\mathbf{a}$ and $\mathbf{w}$ of $\mathcal{H}_\mathfrak{h}$ are both invariant under $\tau \mapsto \tau + 1$. They therefore descend to sections of the quotient $\mathcal{H}_{\mathbb{D}^*} \to \mathbb{D}^*$ of $\mathcal{H}_\mathfrak{h} \to \mathfrak{h}$ by $\begin{pmatrix} 1 & \mathbb{Z} \\ 0 & 1 \end{pmatrix}$. They trivialize $\mathcal{H}_{\mathbb{D}^*}$. The vector bundle

$$\overline{\mathcal{H}}_\mathbb{D} := \mathcal{O}_\mathbb{D}\mathbf{a} \oplus \mathcal{O}_\mathbb{D}\mathbf{w}$$

is thus an extension of $\mathcal{H}_{\mathbb{D}^*}$ to the q-disk. The Hodge bundle $F^1\mathcal{H}_{\mathbb{D}^*}$ extends to the sub-bundle $\mathcal{O}_\mathbb{D}\mathbf{w}$ of $\overline{\mathcal{H}}_\mathbb{D}$.

Denote the natural flat connection on $\mathcal{H}$ by ∇_0. Equation (9.1) implies that

$$\nabla_0 \begin{pmatrix} \mathbf{a} & \mathbf{w} \end{pmatrix} = \begin{pmatrix} \mathbf{a} & \mathbf{w} \end{pmatrix} \begin{pmatrix} 0 & 2\pi i \\ 0 & 0 \end{pmatrix} d\tau = \begin{pmatrix} \mathbf{a} & \mathbf{w} \end{pmatrix} \begin{pmatrix} 0 & 1 \\ 0 & 0 \end{pmatrix} \frac{dq}{q}$$

from which it follows that the connection on $\overline{\mathcal{H}}_\mathbb{D}$ is

$$\nabla_0 = d + \mathbf{a}\frac{\partial}{\partial \mathbf{w}} \otimes \frac{dq}{q}. \tag{9.5}$$

Since this connection is meromorphic with a simple pole at $q = 0$ and nilpotent residue, the bundle $\overline{\mathcal{H}}_\mathbb{D}$ is the canonical extension of $\mathcal{H}_{\mathbb{D}^*}$ to $\mathbb{D}$. It is the pullback to the q-disk of the canonical extension of $\overline{\mathcal{H}} \to \overline{\mathcal{M}}_{1,1}$ of $\mathcal{H} \to \mathcal{M}_{1,1}$. Denote the fiber of $\overline{\mathcal{H}}$ over $q = 0$ by H. It has basis $\mathbf{a}, \mathbf{w}$, where $\mathbf{w}$ spans $F^1 H$.

We now compute the limit MHSs $H_{\vec{v}}$ on $\mathbb{H}$ at $q = 0$ associated to the non-zero tangent vector $\vec{v}$ of $q = 0$. It will have integral lattice $H_\mathbb{Z}$, complexification H and $F^1 H$ defined by

$$H_\mathbb{Z} = \mathbb{Z}\mathbf{a} \oplus \mathbb{Z}\mathbf{b},\ H = \mathbb{C}\mathbf{a} \oplus \mathbb{C}\mathbf{w} \text{ and } F^1 H = \mathbb{C}\mathbf{w} \subset H.$$

To specify the MHS, we give an isomorphism $(H_{\mathbb{Z}}) \otimes \mathbb{C} \cong H$, which will depend on the tangent vector $\vec{v}$.[16]

Proposition 9.6. *The $\mathbb{Z}$-MHS $H_{\vec{v}}(1)$ on H that corresponds to the non-zero tangent vector $\vec{v} = z\partial/\partial q$ is the MHS determined by the linear isomorphism $(H_{\mathbb{Z}}) \otimes \mathbb{C} \cong H_{\mathbb{C}}$ given by*

$$\begin{pmatrix}\mathbf{a} & \mathbf{w}\end{pmatrix} = \begin{pmatrix}\mathbf{a} & -\mathbf{b}\end{pmatrix}\begin{pmatrix}1 & \log z \\ 0 & 2\pi i\end{pmatrix}.$$

It is the extension of $\mathbb{Z}(-1)$ by $\mathbb{Z}(0)$ that corresponds to $z \in \mathbb{C}^*$ under the standard isomorphism $\operatorname{Ext}^1_{\mathsf{MHS}}(\mathbb{Z}(-1), \mathbb{Z}(0)) \cong \mathbb{C}^*$. It splits if and only if $\vec{v} = \partial/\partial q$.

Proof. Most of this is proved above. The integral lattice of $H_{z\partial/\partial q}$ is computed using the standard prescription and the fact that the value of $\mathbf{w}$ at $z \in \mathbb{D}^*$ is

$$\mathbf{w}(z) = \log z\, \mathbf{a} + 2\pi i(-\mathbf{b}),$$

which follows from (9.1). The weight filtration $M_\bullet$ of the limit MHS is the monodromy filtration of the nilpotent endomorphism $\mathbf{a}\partial/\partial\mathbf{w}$ shifted by the weight 1 of H:

$$M_{-1}H = 0, M_0H = M_1H = \mathbb{Q}\mathbf{a},\ M_2H = H. \tag{9.7}$$

□

This result can also be stated by saying that the limit MHS on $H(1)$ of the variation $\mathbb{H}(1)$ associated to $z\partial/\partial q$ is the extension of $\mathbb{Z}(0)$ by $\mathbb{Z}(1)$ corresponding to $z \in \mathbb{C}^* \cong \operatorname{Ext}^1_{\mathsf{MHS}}(\mathbb{Z}(0), \mathbb{Z}(1))$.

Remark 9.8. Note that if $\vec{v} \in \mathbb{Q}^\times \partial/\partial q$, then $H_{\vec{v}}$ splits as an extension of $\mathbb{Q}(-1)$ by $\mathbb{Q}(0)$ if and only if $\vec{v} = \pm\partial/\partial q$. These are also the only two tangent vectors of the cusp $q = 0$ of $\overline{\mathcal{M}}_{1,1}$ that are defined over $\mathbb{Z}$ and remain non-zero at every prime p. For this reason it is natural to identify the fiber H of $\overline{\mathcal{H}}$ over $q = 0$ with $H_{\partial/\partial q}$. In particular, $\mathbf{w} = -2\pi i\mathbf{b}$.

These considerations suggest a natural choice of Cartan subalgebra of $\mathfrak{sl}(H)$ and of the positive root vectors. Namely, the Cartan is the one that acts diagonally with respect to the isomorphism

$$H_{\pm\partial/\partial q} = \mathbb{Q}(0) \oplus \mathbb{Q}(-1) = \mathbb{Q}\mathbf{a} \oplus \mathbb{Q}\mathbf{b}.$$

The natural choice of a positive weight vector is the one with positive Hodge theoretic weight. Since $\mathbf{b}$ has Hodge weight 2 and $\mathbf{a}$ has weight 0 with respect

[16] For the more arithmetically inclined, $\mathbf{a}, \mathbf{b}$ is a basis of the Betti component H^B of the MHS H, and $\mathbf{a}, \mathbf{w}$ is a basis of the $\mathbb{Q}$-de Rham component H^{DR} of H.

to the weight filtration (9.7), the natural choice of positive root vectors in $\mathfrak{sl}(H)$ is $\mathbb{Q}\mathbf{b}\partial/\partial\mathbf{a}$. This choice determines a Borel subalgebra of $\mathfrak{sl}(H)$.

With this choice of Cartan subalgebra, there are two natural choices of symplectic bases of $H_{\mathbb{Z}}$. Namely $\mathbf{a},\mathbf{b}$ and $-\mathbf{b},\mathbf{a}$. Because of formula (i) in Lemma 9.2, which is dictated by the standard formula for the action of $\mathrm{SL}_2(\mathbb{Z})$ on $\mathfrak{h}$, we will use the basis $-\mathbf{b},\mathbf{a}$. This choice determines corresponding isomorphisms $\mathfrak{sl}(H) \cong \mathfrak{sl}_2$ and $\mathrm{SL}(H) \cong \mathrm{SL}_2$. With respect to the above choice of Cartan subalgebra, $\mathbf{b}$ has $\mathfrak{sl}_2$-weight 1 and $\mathbf{a}$ has $\mathfrak{sl}_2$-weight -1.

10 Representation Theory of SL_2

This is a quick review of the representation theory of $\mathfrak{sl}_2$ and SL_2. Much of the time, SL_2 will be $\mathrm{SL}(H)$, where $H = H_{\partial/\partial q}$ is the fiber of $\mathbb{H}$ over $\partial/\partial q$. As pointed out above, this has a natural basis, which leads us to a natural choice of Cartan and Borel subalgebras, which we make explicit below.

Let F denote $\mathbb{Q}$, $\mathbb{R}$ or $\mathbb{C}$. Let V be a two dimensional vector space over F endowed with a symplectic form (i.e., a non-degenerate, skew symmetric bilinear form)

$$\langle\ ,\ \rangle : V \otimes V \to F.$$

The choice of a symplectic basis $\mathbf{v}_1, \mathbf{v}_2$ of V determines actions of $\mathrm{SL}_2(F)$ and $\mathfrak{sl}_2(F)$ on V via the formula

$$\begin{pmatrix} a & b \\ c & d \end{pmatrix} : (\mathbf{v}_2 \quad \mathbf{v}_1)\begin{pmatrix} x_1 \\ x_2 \end{pmatrix} \mapsto (\mathbf{v}_2 \quad \mathbf{v}_1)\begin{pmatrix} a & b \\ c & d \end{pmatrix}\begin{pmatrix} x_1 \\ x_2 \end{pmatrix} \tag{10.1}$$

where $x_1, x_2 \in F$, and isomorphisms $\mathrm{SL}_2 \cong \mathrm{SL}(V)$ and $\mathfrak{sl}_2 \cong \mathfrak{sl}(H)$.

We will fix the choice of Cartan subalgebra of $\mathfrak{sl}_2$ to be the diagonal matrices. This fixes a choice of Cartan subalgebra of $\mathfrak{sl}(V)$. We will take $\mathbf{v}_1$ to have $\mathfrak{sl}_2$ weight $+1$ and $\mathbf{v}_2$ to have $\mathfrak{sl}_2$-weight -1. The element

$$\begin{pmatrix} 0 & 1 \\ 0 & 0 \end{pmatrix}$$

of $\mathfrak{sl}_2$ corresponds to $\mathbf{v}_2\partial/\partial\mathbf{v}_1 \in \mathfrak{sl}(V)$ and has weight -2. Denote it by $\mathbf{e}_0$.

Isomorphism classes of irreducible representations of $\mathrm{SL}(V)$ correspond to non-negative integers. The integer $n \in \mathbb{N}$ corresponds to the nth symmetric power S^nV of the defining representation V. To distinguish distinct but isomorphic representations of SL_2, we use the notation

$$S^n(\mathbf{e}) := \mathrm{span}\{\mathbf{e}_0^j \cdot \mathbf{e} : \mathbf{e}_0^{n+1} \cdot \mathbf{e} = 0\}$$

to denote the SL_2-module with highest weight vector $\mathbf{e}$ of weigh n.

For example, motivated by the discussion in Remark 9.8, we will typically work in the following situation:

1. $V = H$ and $\langle\ ,\ \rangle$ is the intersection pairing,
2. $\mathbf{v}_1 = -\mathbf{b}$ and $\mathbf{v}_2 = \mathbf{a}$, so that $\mathbf{b}$ has $\mathfrak{sl}_2$ weight $+1$ and $\mathbf{a}$ has $\mathfrak{sl}_2$ weight -1,
3. $S^n H = S^n(\mathbf{b}^n)$,
4. $\mathbf{e}_0 = -\mathbf{a}\frac{\partial}{\partial \mathbf{b}}$, which has $\mathfrak{sl}_2$ weight -2. It is also $-2\pi i$ times the residue at $q = 0$ of the connection ∇_0 on $\overline{\mathcal{H}}$.

Note that the weight filtration

$$0 = M_{-1}S^n H \subset M_0 S^n H \subset \cdots \subset M_{2n-1}S^n H \subset M_{2n}S^n H = S^n H$$

associated to the nilpotent endomorphism $\mathbf{e}_0$ of $S^n H$, shifted by the weight n of $S^n H$, is the filtration obtained by giving $\mathbf{b}$ weight 2 and $\mathbf{a}$ weight 0:

$$M_m S^n H = \operatorname{span}\{\mathbf{a}^{n-j}\mathbf{b}^j : 2j \le m\},$$

Observe that

$$\operatorname{Gr}^M_{2n} S^n H = S^n H / \operatorname{im} \mathbf{e}_0 \cong \mathbb{Q}(-n). \tag{10.2}$$

It is generated by the highest weight vector $\mathbf{b}^n$.

11 Modular Forms and Eichler-Shimura

Suppose that Γ is a finite index subgroup of $\mathrm{SL}_2(\mathbb{Z})$. Recall that a holomorphic function $f : \mathfrak{h} \to \mathbb{C}$ is a *modular form of weight* w for Γ if

(i) $f(\gamma\tau) = (c\tau + d)^w f(\tau)$ for all $\gamma \in \Gamma$. This implies that f has a Fourier expansion $\sum_{k=-\infty}^{\infty} a_k q^{k/n}$ for some $n \ge 1$.
(ii) f is holomorphic at each cusp $P \in \Gamma\backslash\mathbb{P}^1(\mathbb{Q})$. If, for example, $P = \infty$, this means that the coefficients a_k of the Fourier expansion of f at P vanish when $k < 0$.

A modular form f is a *cusp form* if it vanishes at each cusp — that is, its Fourier coefficients a_k vanish for all $k \le 0$.

Assume the notation of Section 10. For an indeterminate $\mathbf{e}$ we have the SL_2-module $S^m(\mathbf{e})$ that is isomorphic to $S^m H$. Denote the corresponding local system over X_Γ by $\mathbb{S}^m(\mathbf{e})$. As a local system, it is isomorphic to $S^m\mathbb{H}$. Lemma 7.1 implies that if $\mathbb{S}^m(\mathbf{e})$ has the structure of a PVHS, then it is isomorphic to $S^m\mathbb{H}(r)$ for some $r \in \mathbb{Z}$.

For the time being, we will suppose that the PVHS $\mathbb{S}^m(\mathbf{e})$ over X_Γ has weight m, so that it is an isomorphic copy of $S^m\mathbb{H}$. Define a function $v : \mathfrak{h} \to S^m(\mathbf{e})$ by

$$v(\tau, \mathbf{e}) := \exp(\tau \mathbf{e}_0)\mathbf{c}.$$

The discussion preceding Lemma 9.2 implies that $v(\tau, \mathbf{e})$ is a trivializing section of $F^m(\mathbb{S}^m(\mathbf{e}) \otimes \mathcal{O}_\mathfrak{h})$.

Lemma 11.1. *For all* $\gamma \in \mathrm{SL}_2(\mathbb{Z})$ *we have* $(c\tau+d)^m\gamma^*v = \gamma_* v$.

Proof. Write $\mathbf{e} = \mathbf{b}^m$, where $\mathbf{b} \in H$. Then $\exp(\tau\mathbf{e}_0)\mathbf{e} = \big(\exp(\tau\mathbf{e}_0)\mathbf{b}\big)^m$. So it suffices to consider the case $m = 1$. In this case $v(\tau,\mathbf{b}) = (-\mathbf{a},\mathbf{b})(\tau,1)^T$ and

$$(\gamma_* v)(\tau,\mathbf{b}) = -\begin{pmatrix}\mathbf{a} & -\mathbf{b}\end{pmatrix}\begin{pmatrix}a & b\\ c & d\end{pmatrix}\begin{pmatrix}\tau\\ 1\end{pmatrix}$$

$$= -(c\tau+d)\begin{pmatrix}\mathbf{a} & -\mathbf{b}\end{pmatrix}\begin{pmatrix}\gamma\tau\\ 1\end{pmatrix} = (\gamma^* v)(\tau,\mathbf{b}).$$

□

For a modular form f of Γ weight $w = m+2$ and an indeterminate $\mathbf{e}$, define[17]

$$\omega_f(\mathbf{e}) = 2\pi i f(\tau)v(\tau,\mathbf{e})d\tau \in E^1(\mathfrak{h}) \otimes S^m(\mathbf{e}).$$

A routine calculation shows that $\omega_f(\mathbf{e})$ is Γ-invariant in the sense that

$$(\gamma^* \otimes 1)\omega_f(\mathbf{e}) = (1 \otimes \gamma)\omega_f(\mathbf{e}).$$

It follows that

$$\omega_f(\mathbf{e}) \in (E^1(\mathfrak{h}) \otimes S^m(\mathbf{e}))^\Gamma \cong E^1(X_\Gamma, \mathbb{S}^m(\mathbf{e})).$$

Since $\omega_f(\mathbf{e})$ is closed, it determines a class in $H^1(X_\Gamma, \mathbb{S}^m(\mathbf{e}))$. Its complex conjugate

$$\overline{\omega}_f(\mathbf{e}) = \overline{f(\tau)}v(\overline{\tau},\mathbf{e})^m d\overline{\tau}$$

also defines a class in $H^1(X_\Gamma, \mathbb{S}^m(\mathbf{e}))$.

Recall that $X_\Gamma = \overline{X}_\Gamma - D$, where $D = \Gamma\backslash\mathbb{P}^1(\mathbb{Q})$. And recall from Section 6 that $K^\bullet(\overline{X}, D; \mathbb{S}^n(\mathbf{e}))$ is Zucker's mixed Hodge complex that computes the MHS on $H^\bullet(X_\Gamma, \mathbb{S}^n(\mathbf{e}))$. It is straightforward to check that:

Proposition 11.2 (Zucker [45]). *If f is a modular form of Γ of weight $w = m+2$, then*

$$\omega_f(\mathbf{e}) \in F^{m+1}K^1(\overline{X}_\Gamma, D; \mathbb{S}^m(\mathbf{e})).$$

If f is cusp form, then $\omega_f(\mathbf{e}) \in F^{m+1}W_m K^1(\overline{X}_\Gamma, D; \mathbb{S}^m(\mathbf{e}))$.

When $m > 0$ the exact sequence (6.3) from Section 6.4 becomes

(11.3)
$$0 \to W_{m+1}H^1(X_\Gamma, \mathbb{S}^m(\mathbf{e})) \to H^1(X_\Gamma, \mathbb{S}^m(\mathbf{e})) \to \bigoplus_{P\in D}\big((S^m V_P)/\operatorname{im} N_P\big)(-1) \to 0,$$

where V_P denotes the fiber of the canonical extension of $\mathbb{S}^m(\mathbf{e}) \otimes \mathcal{O}_{X_\Gamma}$ to C and N_P the associated local monodromy operator. (Cf. Section 6.2.) It is an

[17] The particular scaling by $2\pi i$ is chosen so that, when $\Gamma = \mathrm{SL}_2(\mathbb{Z})$, if $v(\tau,\mathbf{e})$ corresponds (locally) to a $\mathbb{Q}$-rational section of the canonical extension $S^m\mathcal{H}(r)$ of $S^m\mathbb{H}$ to $\overline{\mathcal{M}}_{1,1}$, then ω_f is $\mathbb{Q}$-rational in the sense that $\omega_f(\mathbf{e}) \in H^1_{\mathrm{DR}}(\mathcal{M}_{1,1/\mathbb{Q}}, S^m\mathcal{H})$. See, e.g., [20, §21].

exact sequence of MHS. Note that each $S^m V_P/\operatorname{im} N_P$ is one dimensional and is isomorphic to $\mathbb{Q}(-m)$ by (10.2).

The following result is equivalent to Eichler-Shimura combined with the observations that

$$K^1(\overline{X}_\Gamma, D; \mathbb{S}^m(\mathbf{e})) = W_{2m+1}K^1(\overline{X}_\Gamma, D; \mathbb{S}^m(\mathbf{e})) \text{ and } F^{m+2}K^1(\overline{X}_\Gamma, D; \mathbb{S}^m(\mathbf{e})) = 0.$$

This version was proved by Zucker in [45].

Denote the space of modular forms of Γ of weight w by $\mathfrak{M}_w(\Gamma)$ and the subspace of cusp forms by $\mathfrak{M}_w^o(\Gamma)$. When $\Gamma = \mathrm{SL}_2(\mathbb{Z})$, we will omit Γ and simply write $\mathfrak{M}_w$ and $\mathfrak{M}_w^o$.

Theorem 11.4 (Shimura, Zucker). *If Γ is a finite index subgroup of $\mathrm{SL}_2(\mathbb{Z})$, then $H^1(X_\Gamma, \mathbb{S}^m(\mathbf{e}))$ is spanned by the classes of modular forms of weight $w = m+2$ and their complex conjugates. The only non-vanishing Hodge numbers $h^{p,q}$ occur when $(p,q) = (m+1,0)$, $(0,m+1)$ and $(m+1,m+1)$. The weight $m+1$ part is spanned by the classes of cusp forms and their complex conjugates. Moreover, the function that takes a modular form f to the class of $\omega_f(\mathbf{e})$ induces an isomorphism*

$$\mathfrak{M}_{m+2}(\Gamma) \cong F^{m+1}H^1(X_\Gamma, \mathbb{S}^m(\mathbf{e})).$$

The map that takes a cusp form f to the class of its complex conjugate $\overline{\omega}_f(\mathbf{e})$ induces a conjugate linear isomorphism of $\mathfrak{M}_{k+2}^o(\Gamma)$ with the $(0,m+1)$ part of the Hodge structure $W_{m+1}H^1(X_\Gamma, \mathbb{S}^m(\mathbf{e}))$.

When $\Gamma = \mathrm{SL}_2(\mathbb{Z})$, there is only one cusp. It is routine to show that the cohomology with coefficients in $\mathbb{S}^m(\mathbf{e})$ vanishes when m is odd. So for each $n > 0$ we have the exact sequence

$$0 \to W_{2n+1}H^1(\mathcal{M}_{1,1}, \mathbb{S}^{2n}(\mathbf{e})) \to H^1(\mathcal{M}_{1,1}, \mathbb{S}^{2n}(\mathbf{e})) \to \mathbb{Q}(-2n-1) \to 0$$

The class of the 1-form $\omega_f(\mathbf{e})$ associated to the Eisenstein series $f = G_{2n+2}$ of weight $2n+2$ projects to a generator of $\mathbb{Q}(-2n-1)$. The MHS on $H^1(X_\Gamma, \mathbb{S}^{2n}(\mathbf{e}))$ can be described in terms of modular symbols. We will return to this in Section 17.2.

11.1. Cohomology of congruence subgroups

Recall that a congruence subgroup of $\mathrm{SL}_2(\mathbb{Z})$ is one that contains a principal congruence subgroup

$$\mathrm{SL}_2(\mathbb{Z})[N] := \{\gamma \in \mathrm{SL}_2(\mathbb{Z}) : \gamma \equiv \text{identity} \mod N\}.$$

When Γ is a congruence subgroup, one has Hecke operators

$$T_p \in \operatorname{End}_{\mathsf{MHS}} H^1(X_\Gamma, S^m\mathbb{H})$$

for each prime number p. Since the Hecke algebra (the algebra generated by the T_p) is semi-simple, $H^1(X_\Gamma, S^m\mathbb{H}_\mathbb{Q})$ decomposes into simple factors. Each is a MHS.

Let $\mathfrak{B}_w(\Gamma)$ be the set of normalized Hecke eigen cusp forms of Γ of weight $w \geq 2$. This is a basis of $\mathfrak{M}^o_w(\Gamma)$, the weight w cusp forms. For each $f \in \mathfrak{B}_w(\Gamma)$, let K_f be the subfield of $\mathbb{C}$ generated by its Fourier coefficients. Since the restriction of T_p to the cusp forms is self adjoint with respect to the Petersson inner product, its eigenvalues are real. Since this holds for all p, it implies that $K_f \subset \mathbb{R}$. Consequently, the smallest subspace V_f of $H^1(X_\Gamma, S^m\mathbb{H}_{K_f})$ whose complexification contains $\omega_f(\mathbf{e})$ is a K_f-sub HS of the MHS $H^1(X_\Gamma, S^m\mathbb{H})$.

Denote by M_f the smallest $\mathbb{Q}$-Hodge sub-structure of $H^1(X_\Gamma, S^m\mathbb{H}_\mathbb{Q})$ with the property that $M_f \otimes K_f$ contains V_f. It is a sum of the Hodge structures V_h of the eigenforms conjugate to f. Call two eigenforms f and h *equivalent* if $M_f = M_h$.

When f is a normalized Eisenstein series the smallest sub-MHS of $H^1(X_\Gamma, S^m\mathbb{H}_\mathbb{Q})$ that contains the corresponding cohomology class is one dimensional and spans a Tate MHS $\mathbb{Q}(-m-1)$.

Theorem 11.5. *If Γ is a congruence subgroup of $\mathrm{SL}_2(\mathbb{Z})$, then the MHS Hodge structure on $H^1(X_\Gamma, S^m\mathbb{H})$ splits. In particular, when $m > 0$, there is a canonical isomorphism*

$$H^1(X_\Gamma, S^m\mathbb{H}) \cong \bigoplus_f M_f \oplus \bigoplus_{P \in D} \mathbb{Q}(-m-1),$$

where f ranges over the equivalence classes of eigen cusp forms of weight $m+2$. As a real mixed Hodge structure

$$H^1(X_\Gamma, S^m\mathbb{H}) \cong \bigoplus_{f \in \mathfrak{B}_{m+2}(\Gamma)} V_f \oplus \bigoplus_{P \in D} \mathbb{R}(-m-1).$$

Sketch of Proof. The splitting of the MHS on $H^1(X_\Gamma, S^m\mathbb{H})$ follows from the fact that each Hecke operator T_p is a morphism of MHS. The weight filtration splits because T_p acts on Gr^W_{m+1} with eigenvalues of modulus bounded by $Cp^{1+m/2}$ and on Gr^W_{2m+2} with eigenvalues of size $O(p^{m+1})$. (For $\mathrm{SL}_2(\mathbb{Z})$ this is proved in [38]. See p. 94 and p. 106. For general Γ see [27, Lem. 2, p. 13].) The V_f are the common eigenspaces of the T_p acting on $W_{m+1}H^1(X_\Gamma, S^m\mathbb{H})$. The M_f are their Galois orbits and are $\mathbb{Q}$-HS. □

12 Hodge Theory of the Relative Completion of Modular Groups

Here we make the construction of the mixed Hodge structure on the unipotent radical of the completion of a modular group with respect to its inclusion into $\mathrm{SL}_2(\mathbb{Q})$ explicit.

We retain the notation of previous sections: Γ is a finite index subgroup of $\mathrm{SL}_2(\mathbb{Z})$, $X_\Gamma = \Gamma\backslash\mathfrak{h}$ is the associated curve, $D = \Gamma\backslash\mathbb{P}^1(\mathbb{Q})$ is the set of cusps, and $\overline{X}_\Gamma = X_\Gamma \cup D$ is its smooth completion. As in Section 9, H denotes the fiber over the unique cusp $q = 0$ of $\overline{\mathcal{M}}_{1,1}$ of the canonical extension $\overline{\mathcal{H}}$ of the local system $\mathbb{H}$. The pullback of $\overline{\mathcal{H}}$ along the quotient morphism $\overline{X}_\Gamma \to \overline{\mathcal{M}}_{1,1}$ is the canonical extension of $\mathbb{H}\otimes\mathcal{O}_{X_\Gamma}$ to $\overline{X}_\Gamma$, so that the fiber of $\overline{\mathcal{H}}_{\overline{X}_\Gamma}$ over each $P \in D$ is naturally isomorphic to H.

Fix a base point x_o of X_Γ. We allow x_o to be a non-zero tangent vector at a cusp $P \in D$. Denote the completion of $\pi_1(X_\Gamma, x_o)$ with respect to its inclusion into $\mathrm{SL}_2(\mathbb{Q})$ by $\mathcal{G}_{x_o}$ and its prounipotent radical by $\mathcal{U}_{x_o}$. Their Lie algebras (and coordinate rings) have natural mixed Hodge structures. Recall that the polarized variation $\mathbb{H}$ over X_Γ has weight 1. Denote its fiber over x_o by H_o.

We also fix a lift τ_o of x_o to $\mathfrak{h}$. This determines an isomorphism $\pi_1(X_\Gamma, x_o) \cong \Gamma$ and isomorphisms of $\mathcal{G}_{x_o}$ and $\mathcal{U}_{x_o}$ with the completion $\mathcal{G}$ of Γ with respect to the inclusion $\Gamma \hookrightarrow \mathrm{SL}_2(\mathbb{Q})$, and $\mathcal{U}$, its prounipotent radical.

12.1. General considerations

As pointed out in Section 3.4.2, $\mathfrak{u}$ is free. So, up to a non-canonical isomorphism, it is determined by its abelianization $H_1(\mathfrak{u})$. Theorem 7.15, implies that $\mathfrak{u}$ has negative weights, so there is an exact sequence

$$0 \to W_{-2}H_1(\mathfrak{u})_{\mathrm{eis}} \to H_1(\mathfrak{u}) \to \mathrm{Gr}^W_{-1}\mathfrak{u} \to 0.$$

of pro-MHS with SL_2 action. Eichler-Shimura (Thm. 11.4) and the computation (3.4) imply that the weight -1 quotient comes from cusp forms:

$$\mathrm{Gr}^W_{-1}\mathfrak{u} = \prod_{m\geq 0}(W_{m+1}H^1(X_\Gamma, S^m\mathbb{H}))^*\otimes S^m H_o = \prod_{m\geq 0} IH^1(X_\Gamma, S^m\mathbb{H})^*\otimes S^m H_o.$$

The exact sequence (6.3) implies that the weight < -1 part

$$W_{-2}H_1(\mathfrak{u}) = \widetilde{H}_0(D;\mathbb{Q}(1)) \oplus \bigoplus_{P\in D}\prod_{m\geq 0} S^m H_o(m+1)$$

is a direct product of Hodge structures. Note that when x_o is a finite base point (i.e., $x_o \in X_\Gamma$), then $S^m H_o(m+1)$ has weight $-m-2$. If x_o is a tangent vector at a cusp, then $S^m H_o(m+1)$ has weight graded quotients $\mathbb{Q}(1)$, $\mathbb{Q}(2), \ldots, \mathbb{Q}(m+1)$. In this case, $W_{-2}H_1(\mathfrak{u})$ is mixed Tate.

The Manin-Drinfeld Theorem (Thm. 11.5) implies:

Proposition 12.1. *If Γ is a congruence subgroup of $\mathrm{SL}_2(\mathbb{Z})$, then $H_1(\mathfrak{u})$ is the product*

$$H_1(\mathfrak{u}) \cong \prod_{r<0}\mathrm{Gr}^W_r H_1(\mathfrak{u})$$

of its weight graded quotients in the category of pro-MHS with SL_2 action. □

12.2. Hodge theory of congruence subgroups

Now assume that Γ is a congruence subgroup of $SL_2(\mathbb{Z})$. The first step in writing down a formal connection $\Omega \in K^1(\overline{X}_\Gamma, D; \mathbf{u})$ is to write down, for each $m > 0$, a form

$$\Omega_{1,m} \in F^0 W_{-1}\big(K^1(\overline{X}_\Gamma, D; S^m\mathbb{H}) \otimes H^1(X_\Gamma, S^m\mathbb{H})^*\big)$$

that represents the identity $H^1(X_\Gamma, S^m\mathbb{H}) \to H^1(X_\Gamma, S^m\mathbb{H})$.

Since Γ is a congruence subgroup, the Hecke algebra acts on the modular forms $\mathfrak{M}_w(\Gamma)$ of Γ of weight w. When $w > 2$, for each cusp $P \in D$, there is a normalized Eisenstein series $E_{w,P}(\tau)$ that vanishes at the other cusps. When $w = 2$, Eisenstein series give elements of $H^1(X_\Gamma, \mathbb{C})$ with non-zero residues at least two cusps. Fix a cusp $P_o \in D$. For each $P \in D' := D - \{P_o\}$, choose an Eisenstein series $E_{2,P}$ that is non-zero at P and vanishes at all other points of D'.

Now suppose that $m \geq 0$. Identify $S^m\mathbb{H}$ with $\mathbb{S}^m(\mathbf{b}^m)$. For each $f \in \mathfrak{B}_{m+2}(\Gamma)$ we have the 1-forms $\omega_f(\mathbf{b}^m)$ and $\overline{\omega}_f(\mathbf{b}^m)$. When $m > 0$ (resp. $m = 0$) and $P \in D$ (resp. $P \in D'$) set

$$\psi_{m,P}(\mathbf{b}^m) := \omega_{E_{m,P}}(\mathbf{b}^m).$$

This will be viewed as an element of $K^1(\overline{X}_\Gamma, D; S^m\mathbb{H})$ and of $E^1(\mathfrak{h}, S^n H)^\Gamma$. Then

$$(12.2)\quad \{\omega_f(\mathbf{b}^m), \overline{\omega}_f(\mathbf{b}^m) : f \in \mathfrak{B}_{m+2}(\Gamma)\} \cup \{\psi_{m+2,P}(\mathbf{b}^m) : (P \in D \text{ and } m > 0) \text{ or } (P \in D' \text{ and } m = 0)\}$$

is a subset of $K^1(\overline{X}_\Gamma, D; S^m\mathbb{H})$ that represents a basis of $H^1(X_\Gamma, S^m\mathbb{H})$. Let

$$\{u'_f, u''_f, u_{m+2,P} : f \in \mathfrak{B}_{m+2}(\Gamma),\ P \in D\}$$

be a basis of $H^1(X_\Gamma, S^m\mathbb{H})$ dual to the cohomology classes of the closed forms (12.2). The Hodge types of u'_f, u''_f and $u_{m+2,P}$ are $(-m-1,0)$, $(0,-m-1)$ and $(-m-1,-m-1)$, respectively

Set

$$(12.3)\qquad \mathbf{e}'_f := \mathbf{b}^m \otimes u_f,\ \mathbf{e}''_f := \mathbf{b}^m \otimes \overline{u}_f,\ \mathbf{e}_{m+2,P} = \mathbf{b}^m \otimes u_{m+2,P}.$$

These are elements of $S^m H \otimes H^1(X_\Gamma, S^m\mathbb{H})^*$. Then

$$\omega_f(\mathbf{b}^m) \otimes u'_f = \omega_f(\mathbf{e}'_f),\ \overline{\omega}_f(\mathbf{b}^m) \otimes u''_f = \overline{\omega}_f(\mathbf{e}''_f),$$
$$\text{and } \psi_{m+2,P}(\mathbf{b}^m) \otimes u_{m+2,P} = \psi_{m+2,P}(\mathbf{e}_{m+2,P}).$$

All are elements of $K^1(\overline{X}_\Gamma, D; S^m\mathbb{H}) \otimes H^1(X_\Gamma, S^m\mathbb{H})^*$ and

$$(12.4)\qquad \Omega_{1,m} = \sum_{f \in \mathfrak{B}_{m+2}(\Gamma)} \big(\omega_f(\mathbf{e}'_f) + \overline{\omega}_f(\mathbf{e}''_f)\big) + \sum_P \psi_{m+2,P}(\mathbf{e}_{m+2,P})$$

is a closed 1-form that represents the identity $H^1(X_\Gamma, S^m\mathbb{H}) \to H^1(X_\Gamma, S^m\mathbb{H})$. Here the second sum is over $P \in D$ when $m > 0$ and $P \in D'$ when $m = 0$.

Lemma 12.5. *For each $m \geq 0$*

$$\Omega_{1,m} \in F^0 W_{-1}\big(K^1(\overline{X}_\Gamma, D; S^m\mathbb{H}) \otimes H^1(X_\Gamma, S^m\mathbb{H})^*\big).$$

Proof. Since the Hodge types of u_f', u_f'' and $u_{m+2,P}$ are $(-m-1,0)$, $(0,-m-1)$ and $(-m-1,-m-1)$, the definitions of the Hodge and weight filtrations of Zucker's mixed Hodge complex $K^\bullet(\overline{X}_\Gamma, D; S^m\mathbb{H})$ imply that

$$\omega_f(\mathbf{b}^m) \in F^{m+1} W_m K^1(\overline{X}_\Gamma, D; S^m\mathbb{H}) \quad \text{and} \quad \overline{\omega}_f(\mathbf{b}^m) \in F^0 W_m K^1(\overline{X}_\Gamma, D; S^m\mathbb{H})$$

when $f \in \mathfrak{B}_{m+2}(\Gamma)$ and $\psi_{m+2,P}(\mathbf{b}^m) \in F^{m+1} W_{2m+1} K^1(\overline{X}_\Gamma, D; S^m\mathbb{H})$ for each P. The result follows as u_f', u_f'' and $u_{m+2,P}$ have Hodge types $(-m-1,0)$, $(0,-m-1)$ and $(-m-1,-m-1)$, respectively. □

The Lie algebra $\mathfrak{u}$ of the prounipotent radical $\mathcal{U}$ of the relative completion of Γ is the free pronilpotent Lie algebra $\mathfrak{u} = \mathbb{L}(V)^\wedge$, where $V = \bigoplus_{m \geq 0} V_m$ and

$$\begin{aligned} V_m &:= H^1(X_\Gamma, S^m\mathbb{H})^* \otimes S^m H \\ &= \begin{cases} H_1(X_\Gamma, \mathbb{C}) & m = 0, \\ \bigoplus_{f \in \mathfrak{B}_{m+2}(\Gamma)} \big(S^m(\mathbf{e}_f') \oplus S^m(\mathbf{e}_f'')\big) \oplus \bigoplus_{P \in D} S^m(\mathbf{e}_{m+2,P}) & m > 0. \end{cases} \end{aligned}$$

It is a Lie algebra in the category of pro-representations of SL_2.

The 1-form

$$\Omega_1 := \sum_{m \geq 0} \Omega_{1,m} \in E^1(\mathfrak{h}) \hat{\otimes} \mathfrak{u}$$

is Γ-invariant and represents the identity. It can thus be completed to a power series connection

$$\Omega \in F^0 W_{-1} K^1(\overline{X}_\Gamma, D; \mathfrak{u})$$

using the method described in Section 7, which determines the MHS on $\mathfrak{u}_{x_o}$.

Before discussing the case $\Gamma = \mathrm{SL}_2(\mathbb{Z})$, note that since

$$\begin{aligned} S^m(\mathbf{e}_f') &= \mathrm{span}\{\mathbf{e}_0^j \cdot \mathbf{e}_f' : \mathbf{e}_0^{m+1} \cdot \mathbf{e}_f' = 0\}, & f &\in \mathfrak{B}_{m+2}(\Gamma) \\ S^m(\mathbf{e}_f'') &= \mathrm{span}\{\mathbf{e}_0^j \cdot \mathbf{e}_f'' : \mathbf{e}_0^{m+1} \cdot \mathbf{e}_f'' = 0\}, & f &\in \mathfrak{B}_{m+2}(\Gamma) \\ S^m(\mathbf{e}_{m+2,P}) &= \mathrm{span}\{\mathbf{e}_0^j \cdot \mathbf{e}_{m+2,P} : \mathbf{e}_0^{m+1} \cdot \mathbf{e}_{m+2,P} = 0\}, & P &\in D, \end{aligned}$$

$\mathfrak{u}$ is the free Lie algebra topologically generated by

$$\{\mathbf{e}_{2,P} : P \in D'\} \cup \bigcup_{m > 0} \{\mathbf{e}_0^j \cdot \mathbf{e}_f',\ \mathbf{e}_0^j \cdot \mathbf{e}_f'',\ \mathbf{e}_0^j \cdot \mathbf{e}_{P,m+2} : 0 \leq j \leq m, f \in \mathfrak{B}_{m+2}, P \in D\}.$$

The Hodge and weight filtrations of $\mathfrak{u}$ are defined by giving $\mathbf{b} \in H$ type $(1,0)$. The generators (12.3) thus have types given in Figure 4. So, for example, $\mathbf{e}_0^j \cdot \mathbf{e}_f'$

		Hodge type	W-weight	M-weight
$\mathbf{e}_0$		$(-1,1)$	0	-2
$\mathbf{e}'_f$	$f \in \mathfrak{B}_{m+2}(\Gamma)$	$(-1,0)$	-1	$m-1$
$\mathbf{e}''_f$	$f \in \mathfrak{B}_{m+2}(\Gamma)$	$(m,-m-1)$	-1	$m-1$
$\mathbf{e}_{m+2,P}$	$P \in D$	$(-1,-m-1)$	$-m-2$	-2

Figure 4. Hodge types of the generators of $\mathfrak{u}$.

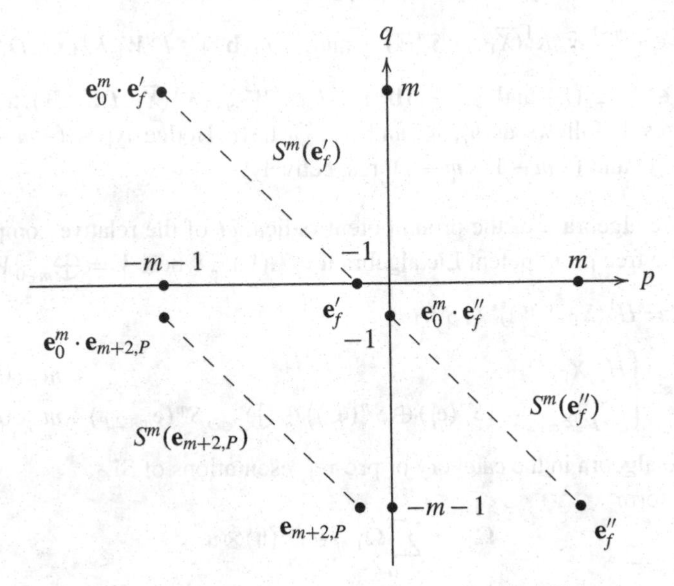

Figure 5. Hodge numbers of $S^m(\mathbf{e}'_f)$, $S^m(\mathbf{e}''_f)$ and $S^m(\mathbf{e}_{m+2,P})$.

has type $(-1-j,j)$. The Hodge and weight filtrations on the generators extend naturally to Hodge and weight filtrations on $\mathfrak{u}$.

The Hodge types on the $\mathfrak{sl}_2$ module with highest weight vectors $\mathbf{e}'_f, \mathbf{e}''_f, \mathbf{e}_{m+2,P}$ are illustrated in Figure 5.

12.3. *The case of* $\mathrm{SL}_2(\mathbb{Z})$

In this case, $X_\Gamma = \mathcal{M}_{1,1}$, $C = \overline{\mathcal{M}}_{1,1}$ and D consists of a single point, which we shall denote by P. The modular parameter $q = e^{2\pi i\tau}$ is a local holomorphic coordinate on the orbifold $\overline{\mathcal{M}}_{1,1}$ centered at P.

There are no modular forms for $\mathrm{SL}_2(\mathbb{Z})$ of odd weight. Since there is a single cusp, there is a 1-dimensional space of Eisenstein series for each weight

$2n \geq 4$. The normalized Eisenstein series of even weight $2n$ is

$$G_{2n}(\tau) = \frac{1}{2}\frac{(2n-1)!}{(2\pi i)^{2n}} \sum_{\substack{\lambda \in \mathbb{Z} \oplus \mathbb{Z}\tau \\ \lambda \neq 0}} \frac{1}{\lambda^{2n}} = -\frac{B_{2n}}{4n} + \sum_{k=1}^{\infty} \sigma_{2n-1}(k) q^k. \tag{12.6}$$

This has value $(2n-1)!\zeta(2n)/(2\pi i)^{2n}$ at the cusp P.[18] The dual homology class $\mathbf{e}_{2n,P}$ will be denoted by $\mathbf{e}_{2n}$ and the form $\psi_{2n,P}$ by ψ_{2n}.

Later, we will use the tangent vector $\vec{v} := \partial/\partial q$ of P as a base point. We will also write $\mathfrak{B}_w$ instead of $\mathfrak{B}_w(\mathrm{SL}_2(\mathbb{Z}))$.

13 VMHS associated to Modular Forms and their Period Maps

This section considers three related topics: relative higher albanese maps (which are related to period mappings of VMHS), Manin's iterated Shimura integrals, and the existence of extensions of VMHS coming from Hecke eigenforms. The construction of these extensions is a special case of the general technique for constructing admissible VMHS sketched in the proof of Theorem 8.3. Such extensions correspond to normal functions, so every Hecke eigenform produces a normal function.

13.1. Relative Albanese maps

This construction generalizes the non-abelian albanese manifolds of [25] from the unipotent case to the relative case. Although this discussion applies more generally, here we restrict to the case of modular curves.

Fix a base point τ_o of $\mathfrak{h}$. Let Γ be a finite index subgroup of $\mathrm{SL}_2(\mathbb{Z})$. Denote the image of τ_o in X_Γ by x_o. The choice of τ_o determines an isomorphism $\Gamma \cong \pi_1(X_\Gamma, x_o)$. Let $\mathcal{G}_o$ be the complex form of the relative completion of $\pi_1(X_\Gamma, x_o)$. Let $\mathcal{U}_o$ be its prounipotent radical, and let $\mathfrak{g}_o$, $\mathfrak{u}_o$ be their Lie algebras. Since the bracket of $\mathfrak{g}_o$ respects the Hodge filtration, $F^0\mathfrak{g}_o$ is a subalgebra of $\mathfrak{g}_o$. Denote the corresponding subgroup of $\mathcal{G}_o$ by $F^0\mathcal{G}_o$.

Let

$$\Omega \in F^0 W_{-1} K^1(\overline{X}_\Gamma, D; \mathbf{u})$$

be as above. Trivialize the pullback of $\mathbf{u}$ to $\mathfrak{h}$ using the sections $\{\mathbf{e}_0^j \cdot \mathbf{e}'_f, \mathbf{e}_0^j \cdot \mathbf{e}''_f, \mathbf{e}_0^j \cdot \mathbf{e}_{2n}\}$:

$$\mathfrak{u} \times \mathfrak{h} \to \mathfrak{h}.$$

[18] Here B_{2n} is the $2n$th Bernoulli number, $\zeta(s)$ is the Riemann zeta function and $\sigma_k(n)$ is the sum of the kth powers of the divisors of n.

This trivializes both the Hodge and weight filtrations. It also fixes an isomorphism $\mathcal{G}_x \cong \mathrm{SL}(H_x) \ltimes \mathcal{U}_x$ for all $x \in X_\Gamma$. Denote the pullback of Ω to $\mathfrak{h}$ by

$$\widetilde{\Omega} \in (E^1(\mathfrak{h}) \hat{\otimes} \mathfrak{u})^\Gamma.$$

For convenience we now take the base point τ_o to be i. Since $\widetilde{\Omega}$ is integrable, the function $F : \mathfrak{h} \to \mathcal{U}_o \times \mathrm{SL}(H_o) \cong \mathcal{G}_o$ defined by

$$\widetilde{F}(\tau) := \left(1 + \int_{\tau_o}^{\tau} \widetilde{\Omega} + \int_{\tau_o}^{\tau} \widetilde{\Omega}\widetilde{\Omega} + \int_{\tau_o}^{\tau} \widetilde{\Omega}\widetilde{\Omega}\widetilde{\Omega} + \cdots, \begin{pmatrix} v & u/v \\ 0 & v^{-1} \end{pmatrix}\right)$$

where $\tau = u + iv^2$ (u, v real), is well defined and smooth. It induces a function $F : \mathfrak{h} \to \mathcal{G}_o/F^0\mathcal{G}_o$ which is equivariant with respect to the natural left Γ-actions on each.

Proposition 13.1. *The function $F : \mathfrak{h} \to \mathcal{G}_o/F^0\mathcal{G}_o$ is holomorphic. (That is, it is an inverse limit of holomorphic functions.)*

Proof. Set $H = H_o$. The map $\mathrm{SL}(H_\mathbb{R}) \to \mathfrak{h}$ that takes g to $g\tau_o = gi$ induces an isomorphism $\mathrm{SL}(H_\mathbb{R})/\mathrm{SO}(2) \to \mathfrak{h}$. The inclusion $\mathrm{SL}(H_\mathbb{R}) \hookrightarrow \mathrm{SL}(H_\mathbb{C})$ induces an inclusion

$$\mathfrak{h} = \mathrm{SL}(H_\mathbb{R})/\mathrm{SO}(2) \hookrightarrow \mathrm{SL}(H_\mathbb{C})/F^0 \cong \mathbb{P}^1.$$

Since the right hand matrix in $\widetilde{F}$ is in $\mathrm{SL}(H_\mathbb{R})$ and takes i to τ, the composition of F with the projection $\mathcal{G}_o/F^0\mathcal{G}_o \to \mathrm{SL}(H_\mathbb{C})/F^0$ is the inclusion $\mathfrak{h} \hookrightarrow \mathrm{SL}(H_\mathbb{C})/F^0$. Consequently, to show that F is holomorphic, we need only check that its first factor is holomorphic. Write $\widetilde{\Omega} = \widetilde{\Omega}' + \widetilde{\Omega}''$, where $\widetilde{\Omega}'$ has type $(1,0)$ and $\widetilde{\Omega}''$ has type $(0,1)$. Since $\Omega \in F^0K^1(\overline{X}_\Gamma, D; \mathbf{u})$,

$$\widetilde{\Omega}'' \in E^{0,1}(\mathfrak{h}) \hat{\otimes} F^0\mathfrak{u}.$$

The fundamental theorem of calculus implies that F satisfies the differential equation $d\widetilde{F} = \widetilde{F}\widetilde{\Omega}$ from which it follows that $\bar{\partial}\widetilde{F} = \widetilde{F}\Omega''$. This implies the vanishing of $\bar{\partial}F$. □

The projection $\mathcal{G}_o \to \mathrm{SL}(H_{o,\mathbb{C}})$ preserves the Hodge filtration and induces a holomorphic, Γ-invariant projection $\mathcal{G}_o/F^0\mathcal{G}_o \to \mathrm{SL}(H_o)/F^0 \cong \mathbb{P}^1$. Then one has the diagram

$$\begin{array}{ccccc} (\mathcal{G}_o/F^0\mathcal{G}_o)|_\mathfrak{h} & \longrightarrow & \mathcal{G}_o/F^0\mathcal{G}_o & & \\ \downarrow & \overset{F}{\nearrow} & \downarrow & & \\ \mathfrak{h} & \hookrightarrow & \mathrm{SL}(H_{x,\mathbb{C}})/F^0 & \xrightarrow{\cong} & \mathbb{P}^1. \end{array}$$

Recall that our choices have fixed an isomorphism $\mathcal{G}_o \cong \mathrm{SL}(H_o) \ltimes \mathcal{U}_o$. The natural homomorphism $\Gamma \to \mathcal{G}_o$ corresponds to a non-abelian 1-cocycle $\Theta_o : \Gamma \to \mathcal{U}_o(\mathbb{Q})$.

Lemma 13.2. *The function* $F : \mathfrak{h} \to \mathcal{G}_o/F^0$ *satisfies* $F(\gamma\tau) = (\Theta_o(\gamma), \gamma)F(\tau)$ *for all* $\gamma \in \Gamma$.

Proof. The identification $\mathcal{G}_o \cong \mathrm{SL}(H_o) \ltimes \mathcal{U}_o$ induces an identification of $\mathcal{G}_o/F^0$ with $\mathcal{U}_o \times \mathbb{P}^1$. If c_γ is a path from τ_o to $\gamma\tau_o$ and c is a path from τ_o to τ, then $c_\gamma * (\gamma \cdot c)$ is a path from τ_o to $\gamma\tau$. So

$$F(\gamma\tau) = \big(\Theta_o(\gamma)(\gamma \cdot T(c)^{-1}), \gamma\tau\big) = (\Theta_o(\gamma), \gamma)F(\tau),$$

where T^{-1} denotes $1 + \int\widetilde{\Omega} + \int\widetilde{\Omega}\widetilde{\Omega} + \cdots$. □

Let $\mathcal{U}_{o,\mathbb{Z}}$ be the subgroup of $\mathcal{U}_o$ that is generated by $\{\Theta_o(\gamma) : \gamma \in \Gamma\}$. Let $\mathcal{G}_{o,\mathbb{Z}}$ be the subgroup of $\mathcal{G}_o \cong \mathrm{SL}(H_o) \cong \mathcal{U}_o$ that corresponds to $\mathrm{SL}(H_{o,\mathbb{Z}}) \ltimes \mathcal{U}_{o,\mathbb{Z}}$. The previous result implies that $F(\gamma\tau)$ and $\gamma F(\tau)$ lie in the same left $\mathcal{U}_{o,\mathbb{Z}}$ orbit.

The *universal relative albanese manifold* $\mathcal{A}_\Gamma$ of X_Γ is defined by

$$\mathcal{A}_\Gamma = \mathcal{G}_{o,\mathbb{Z}} \backslash \big((\mathcal{G}_o/F^0\mathcal{G}_o)|_{\mathfrak{h}}\big).$$

There is a natural quotient mapping to X_Γ. Taking the quotient of the left-hand map of the previous diagram by Γ gives the *universal non-abelian Albanese map*

$$\mathcal{A}_\Gamma \underset{\xleftarrow{F}}{\longrightarrow} X_\Gamma.$$

The fiber over $x_o \in X_\Gamma$ is isomorphic to $\mathcal{U}_{o,\mathbb{Z}} \backslash \mathcal{U}_o / F^0$.

Remark 13.3. For each finite dimensional quotient G_α of $\mathcal{G}_o$ in the category of groups with a MHS, one can define

$$\mathcal{A}_\alpha = G_{\alpha,\mathbb{Z}} \backslash \big((G_\alpha/F^0G_\alpha)|_{\mathfrak{h}}\big).$$

Since $\mathcal{G}_o$ is the inverse limit of the G_α, $\mathcal{A}_\Gamma = \varprojlim \mathcal{A}_\alpha$. The reduction of the section F is a holomorphic section F_α of this bundle. In particular, when Γ is a congruence subgroup, by taking G_α to be the quotient

$$0 \to \check{M}_f \otimes S^n H_o \to G_\alpha \to \mathrm{SL}(H_o) \to 1$$

of $\mathcal{G}_o$, where M_f is the smallest $\mathbb{Q}$-sub HS of $H^1(X_\Gamma, S^m\mathbb{H})$ that contains $\omega_f(\mathbf{e})$ and $\check{M}_f$ its dual, we see that for each $f \in \mathfrak{B}_{n+2}(\Gamma)$, there is a bundle over X_Γ with fiber over $x \in X_\Gamma$ the intermediate jacobian

$$\mathrm{Ext}^1_{\mathsf{MHS}}(\mathbb{Z}, \check{M}_{f,\mathbb{Z}} \otimes S^n H_{x,\mathbb{Z}})$$

with a holomorphic section induced by F. We will see below that this is a normal function which is the period mapping of an extension of $\mathbb{Q}$ by $\check{M}_f \otimes S^n\mathbb{H}$. When Γ is a congruence subgroup, each Eisenstein series determines a normal function that corresponds to an extension of $\mathbb{Z}$ by $S^n\mathbb{H}(n+1)$.

The coefficients of F are holomorphic functions on $\mathfrak{h}$ which can be realized as periods of admissible variations of MHS. These include iterated integrals of (holomorphic) modular forms, but there are many more. Below is an example of such a holomorphic iterated integral that is not of the type considered by Manin. It is, in some sense, a generalization of the Riemann theta function. (Cf. [17, Ex. 4.4].)

Example 13.4. Suppose that $f \in M^o_{2n+2}$ and $g \in M^o_{2m+2}$ are cusp forms, not necessarily of the same weight. Since $\mathrm{SL}_2(\mathbb{Z})$ has cohomological dimension 1, the form $\overline{\omega}_f(\mathbf{e}''_f) \wedge \omega_g(\mathbf{e}'_g)$ is exact. Lemma 6.1 implies that there is

$$\xi \in F^0 W_{-1} K^1\big(\overline{X}_\Gamma, D; \mathbb{S}^{2n}(\mathbf{e}''_f) \otimes \mathbb{S}^m(\mathbf{e}'_g)\big)$$

such that

$$\bar{\partial}\xi + \overline{\omega}_f(\mathbf{e}''_f) \wedge \omega_g(\mathbf{e}'_g) = 0.$$

Then

$$\tau \mapsto \int_{\tau_o}^{\tau} \overline{\omega}_f(\mathbf{e}''_f)\,\omega_g(\mathbf{e}'_g) + \xi$$

is a well defined function from $\mathfrak{h}$ to $S^{2n}H_o \otimes S^{2m}H_o$. An elementary argument (cf. [17, Prop. 4.3]) implies that it is holomorphic.

Perhaps the most interesting version of this example is where $f = g$ and one composes it with an invariant bilinear form $S^{2n}H_o \otimes S^{2n}H_o \to \mathbb{C}$. Such iterated integrals occur as periods of biextensions.

13.2. Iterated Shimura integrals

In [29, 30] Manin considered iterated integrals of holomorphic modular forms and non-abelian generalizations of modular symbols. Here we briefly discuss the relationship of his work to the Hodge theory of modular groups. Recall that $\mathfrak{B}_w = \mathfrak{B}_w(\mathrm{SL}_2(\mathbb{Z}))$.

Manin considers iterated integrals with values in the algebras

$$A = \mathbb{C}\langle\langle \mathbf{e}_0^j \cdot \mathbf{e}_{2n},\ \mathbf{e}_0^j \cdot \mathbf{e}'_f : f \in \mathfrak{B}_{2n},\ 0 \le j \le 2n-2,\ n \ge 2 \rangle\rangle$$

and

$$B = \mathbb{C}\langle\langle \mathbf{e}_0^j \cdot \mathbf{e}_{2n} : 0 \le j \le 2n-2,\ n \ge 2 \rangle\rangle.$$

These iterated integrals are of the form

$$1 + \int_{\tau_o}^{\tau} \Omega + \int_{\tau_o}^{\tau} \Omega\Omega + \int_{\tau_o}^{\tau} \Omega\Omega\Omega + \cdots$$

where

$$\Omega = \Omega_A := \sum_{n \ge 2} \Big(\psi_{2n}(\mathbf{e}_{2n}) + \sum_{f \in \mathfrak{B}_{2n}} \omega_f(\mathbf{e}'_f) \Big)$$

in the first case and

(13.5) $$\Omega = \Omega_B := \sum_{n\geq 2} \psi_{2n}(\mathbf{e}_{2n})$$

in the second case. Both of these forms are Γ-invariant.

Let $\mathfrak{u}_A$ (resp. $\mathfrak{u}_B$) be the set of primitive elements of A (resp. $\mathfrak{u}_B$). Then Ω_A (resp. Ω_B) takes values in $\mathfrak{u}_A$ (resp. $\mathfrak{u}_B$). Set

$$\mathfrak{u} = \mathbb{L}(\mathbf{e}_0^j \cdot \mathbf{e}_{2n},\ \mathbf{e}_0^j \cdot \mathbf{e}_f',\ \mathbf{e}^0 \cdot \mathbf{e}_f'' : f \in \mathfrak{B}_{2n},\ 0 \leq j \leq 2n-2,\ n \geq 2)^\wedge.$$

It follows from Figure 5 that $F^0\mathfrak{u}$ is generated by $\{\mathbf{e}_0^j \cdot \mathbf{e}_f'' : f \in \mathfrak{B}_{2n}, n \geq 2\}$, so that $\mathfrak{u}_A$ is the quotient of $\mathfrak{u}$ by the ideal $(F^0\mathfrak{u})$ *generated by* $F^0\mathfrak{u}$, and $\mathfrak{u}_B$ is the quotient of $\mathfrak{u}$ by the ideal generated by all $\mathbf{e}_0^j \cdot \mathbf{e}_f'$ and $\mathbf{e}_0 \cdot \mathbf{e}_f''$. His iterated integral is the reduction of the one in the previous section (the first argument of $\widetilde{F}$) mod these ideals. This implies that there are many interesting holomorphic iterated integrals which do not occur as iterated Shimura integrals.

One can ask whether the $\mathrm{SL}_2(\mathbb{Z})$ connection on the local system

$$\mathfrak{u}_B \times \mathfrak{h} \to \mathfrak{h}$$

defined by Ω_B descends to an admissible VMHS over $\mathcal{M}_{1,1}$. It will follow from Theorem 19.4 that it does not as we explain in Remark 19.5.

13.3. Extensions of variations of MHS associated to Eisenstein series

Here we suppose, for simplicity, that $\Gamma = \mathrm{SL}_2(\mathbb{Z})$. In this section, we sketch an explicit construction of an extension

$$0 \to S^{2n}\mathbb{H}(2n+1) \to \mathbb{E} \to \mathbb{Q} \to 0$$

for each Eisenstein series G_{2n+2}.[19]

Let $H = \mathbb{C}\mathbf{a} \oplus \mathbb{C}\mathbf{w}$. Define the Hodge filtration on H by $F^0 H = H$ and $F^1 H = \mathbb{C}\mathbf{w}$. This induces a Hodge filtration on $S^{2n}H$. Trivialize the bundle $\mathcal{H}_\mathfrak{h} \to \mathfrak{h}$ with the sections $\mathbf{a}$ and $\mathbf{w}$:

$$\mathcal{H}_\mathfrak{h} \cong H \times \mathfrak{h}.$$

Trivialize $S^{2n}\mathcal{H}_\mathfrak{h}$ using monomials in $\mathbf{a}$ and $\mathbf{w}$. Then $F^p S^{2n}\mathcal{H}_\mathfrak{h}$ is trivialized by the sections $\{\mathbf{a}^{2n-j}\mathbf{w}^j : j \geq p\}$.

Set $V = \mathbb{C}\mathbf{e} \oplus S^{2n}H(2n+1)$ and $\mathcal{V}_\mathfrak{h} = V \times \mathfrak{h}$. Define Hodge and weight filtrations on V by giving $\mathbf{e}$ type $(0,0)$ and $\mathbf{a}^{2n-j}\mathbf{w}^j$ type $(j-2n-1, -j-1)$. Let $\mathrm{SL}_2(\mathbb{Z})$ act in this bundle by acting trivially on $\mathbf{e}$, and on $\mathbf{a}$ and $\mathbf{w}$ by the factor of automorphy given in Corollary 9.4. The Hodge and weight filtrations are

[19] This construction works equally well when Γ is a congruence subgroup. The construction in the more general case is sketched at the end of the next section.

invariant under this action, so that they descend to Hodge and weight filtrations on the (orbifold) quotient bundle

$$\mathcal{V} := \mathrm{SL}_2(\mathbb{Z})\backslash \mathcal{V}_{\mathfrak{h}} \to \mathcal{M}_{1,1}.$$

This bundle is trivial over the punctured q-disk Δ^*. Extend it to a bundle $\overline{\mathcal{V}}$ over $\overline{\mathcal{M}}_{1,1}$ by defining its sections over the q-disk Δ to be $V \otimes \mathcal{O}_\Delta$. The Hodge and weight bundles clearly extend to sub-bundles of $\overline{\mathcal{V}}$.

Define a connection on $\mathcal{V}_{\mathfrak{h}}$ by $d + \Omega$, where

$$\Omega = \begin{pmatrix} 0 & 0 \\ \psi_{2n+2}(\mathbf{w}^{2n}) & \mathbf{a}\frac{\partial}{\partial \mathbf{w}}\frac{dq}{q} \end{pmatrix} \in \begin{pmatrix} \mathbb{C} & 0 \\ S^{2n}H & \mathrm{End}\, S^{2n}H \end{pmatrix} \frac{dq}{q}.$$

It is holomorphic, flat and Γ-invariant. It therefore descends to a flat connection ∇ on $\mathcal{V}$ which has a regular singular point at the cusp when viewed as a connection on $\overline{\mathcal{V}}$. This implies that the extended bundle is Deligne's canonical extension of $(\mathcal{V}, \nabla)$ to $\overline{\mathcal{M}}_{1,1}$. Since

$$\Omega \in (F^{-1} W_{-1} \mathrm{End}\, V) \otimes \mathcal{O}(\Delta) \frac{dq}{q},$$

the weight filtration is flat and the connection satisfies Griffiths transversality. Since $\psi_{2n+2}(\mathbf{w}^{2n})$ has rational periods, it follows that the local system $\mathbb{V}$ associated to $(\mathcal{V}, \nabla)$ has a natural $\mathbb{Q}$-form. The associated weight graded local system is

$$\mathrm{Gr}^W_\bullet \mathbb{V} = \mathbb{Q}(0) \oplus S^{2n}\mathbb{H}(2n+1).$$

The existence of a relative weight filtration at $q = 0$ follows from the argument in the proof of Theorem 7.17 It follows that $\mathbb{V}$ is an admissible variation of MHS over $\mathcal{M}_{1,1}$. The results of Section 18.4 imply that every extension of $\mathbb{Q}$ by $S^{2n}\mathbb{H}(m)$ over $\mathcal{M}_{1,1}$ is a multiple of this extension when $m = 2n+1$ and trivial otherwise.

13.4. Extensions of variations of MHS associated to cusp forms

The construction of the extension corresponding to an eigen cusp form is similar, but a little more elaborate. Suppose that Γ is a congruence subgroup of $\mathrm{SL}_2(\mathbb{Z})$ and that $m \geq 0$. The first step is to construct an extension

$$0 \to H^1(X_\Gamma, S^m\mathbb{H}_\mathbb{Z})^* \otimes S^m\mathbb{H}_\mathbb{Z} \to \mathbb{V} \to \mathbb{Z} \to 0 \tag{13.6}$$

in the category of $\mathbb{Z}$-MHS over X_Γ.

Denote the completion of X_Γ by $\overline{X}_\Gamma$. Let $\overline{\mathcal{V}}$ be the C^∞ vector bundle over $\overline{X}_\Gamma$ associated to the canonical extension

$$\mathcal{O}_{\overline{X}_\Gamma} \oplus H^1(X_\Gamma, S^m\mathbb{H})^* \otimes S^m\overline{\mathcal{H}}$$

of the admissible variation $\mathbb{Q}\oplus H^1(X_\Gamma,S^m\mathbb{H})^*\otimes S^m\mathbb{H}$ over X_Γ. This has natural Hodge and weight sub-bundles. Denote the restriction of $\overline{\mathcal{V}}$ to X_Γ by $\mathcal{V}$ and the direct sum connection on it by ∇_0.

Define a C^∞ connection on $\mathcal{V}$ by $\nabla = \nabla_0 + \Omega_{1,m}$, where $\Omega_{1,m}$ is the form defined in Equation (12.4). This connection is flat, and thus defines a new holomorphic structure on the bundle $\mathcal{V}$. Arguments almost identical to those in Section 7.4 show that $(\overline{\mathcal{V}},\nabla)$ is Deligne's canonical extension of $(\mathcal{V},\nabla)$, that the Hodge bundles are holomorphic sub-bundles of $\overline{\mathcal{V}}$ with respect to this new complex structure, and that the connection ∇ satisfies Griffiths transversality. The existence of a relative weight filtration at each cusp is established as in the proof of Theorem 7.17. The fact that $\Omega_{1,m}$ represents the identity $H^1(X_\Gamma,S^m\mathbb{H}) \to H^1(X_\Gamma,S^m\mathbb{H})$ implies that the local system $\mathbb{V}$ underlying the flat bundle $(\mathcal{V},\nabla)$ has a natural $\mathbb{Z}$-form. It follows that there is an admissible $\mathbb{Z}$-VMHS $\mathbb{V}$ over X_Γ whose corresponding C^∞ vector bundle is $\mathcal{V}$ and whose weight graded quotients are $\mathbb{Q}(0)$ and $H^1(X_\Gamma,S^m\mathbb{H}_\mathbb{Z})^*\otimes S^m\mathbb{H}$. (Cf. Lemma 7.9.)

Having constructed the extension (13.6), we can now construct the extension corresponding to a Hecke eigen cusp form $f\in\mathfrak{B}_{m+2}(\Gamma)$. The smallest sub $\mathbb{Q}$-HS M_f of $H^1(X_\Gamma,S^m\mathbb{H}_\mathbb{Q})$ whose complexification contains $\omega_f(\mathbf{e})$ is pure of weight $m+1$. So $\check{M}_f\otimes S^m\mathbb{H}$ is pure of weight -1. The corresponding extension

$$0\to \check{M}_f\otimes S^m\mathbb{H}_\mathbb{Z}\to \mathbb{E}_f\to\mathbb{Q}\to 0$$

is obtained by pushing out the extension (13.6) along the dual of the inclusion

$$M_f\hookrightarrow H^1(X_\Gamma,S^m\mathbb{H}_\mathbb{Z}).$$

This extension has a natural $\mathbb{Z}$-form, which we denote by $E_{f,\mathbb{Z}}$.

The extension $\mathbb{E}_{f,\mathbb{Z}}$ corresponds to a holomorphic section of the associated bundle of intermediate jacobians, which has fiber

$$J(H^1(X_\Gamma,S^m\mathbb{H})^*\otimes S^m H_x)$$

over $x\in X_\Gamma$, where for a $\mathbb{Z}$-MHS V with negative weights

$$J(V):=V_\mathbb{C}/\big(V_\mathbb{Z}+F^0V_\mathbb{C}\big)\cong \operatorname{Ext}^1_{\mathsf{MHS}}(\mathbb{Z},V).$$

The section is obtained by integrating the invariant 1-form $\omega_f(\mathbf{e}_f')+\omega_f(\mathbf{e}_f'')$. More sections can be obtained by applying elements of $\operatorname{Aut}M_{f,\mathbb{Z}}$.

A similar construction can be used to construct the extension of a normalized Eisenstein series f. When $\Gamma=\mathrm{SL}_2(\mathbb{Z})$ this reduces to the construction in the previous section. In this case, the smallest $\mathbb{Q}$-Hodge sub structure M_f of $H^1(X_\Gamma,S^m\mathbb{H})$ that contains $\psi_f(\mathbf{e})$ is $M_f=\mathbb{Q}(-m-1)$. Pushing out the extension (13.6) along the inclusion $M_f\to H^1(X_\Gamma,S^m\mathbb{H})$ gives the extension

$$0\to S^m\mathbb{H}(m+1)\to\mathbb{E}_f\to\mathbb{Q}\to 0.$$

corresponding to f.

14 The Relative Completion of $\pi_1(\mathcal{M}_{1,\vec{1}}, x)$

By a sleight of hand, can deduce the MHS on the unipotent radical of the relative completion of the fundamental group of $\mathcal{M}_{1,\vec{1}}$ from the MHS on the unipotent radical of the relative completion of $\mathrm{SL}_2(\mathbb{Z})$. The MHS on this completion is of interest as it acts on the unipotent completion of the fundamental group of a once punctured elliptic curve.

First recall some classical facts. (Detailed proofs can be found, for example, in [19].) The moduli space $\mathcal{M}_{1,\vec{1}}$ of elliptic curves with a non-zero tangent vector at the identity is the complement of the discriminant locus $u^3 - 27v^2 = 0$ in $\mathbb{C}^2$. For us, it is more useful to write it as the quotient of $\mathbb{C}^* \times \mathfrak{h}$ by the action

$$\gamma : (\xi, \tau) \mapsto \left((c\tau + d)^{-1}\xi, \gamma\tau\right),$$

where $\gamma = \begin{pmatrix} a & b \\ c & d \end{pmatrix}$. This action is fixed point free, so that $\mathcal{M}_{1,\vec{1}}$ is an analytic variety. The projection $\mathbb{C} \times \mathfrak{h} \to \mathfrak{h}$ induces a projection $\pi : \mathcal{M}_{1,\vec{1}} \to \mathcal{M}_{1,1}$ that is the $\mathbb{C}^*$ bundle associated to the orbifold line bundle $\mathcal{L} \to \mathcal{M}_{1,1}$ with factor of automorphy $c\tau + d$. Modular forms of $\mathrm{SL}_2(\mathbb{Z})$ of weight m are sections of $\mathcal{L}^{\otimes m}$. (Cf. [19, §4].) The cusp form Δ of $\mathrm{SL}_2(\mathbb{Z})$ of weight 12 trivializes $\mathcal{L}^{\otimes 12}$.

The $\mathrm{SL}_2(\mathbb{Z})$ action lifts to an action of a central extension

$$0 \to \mathbb{Z} \to \widehat{\Gamma} \to \mathrm{SL}_2(\mathbb{Z}) \to 1 \tag{14.1}$$

on $\mathbb{C} \times \mathfrak{h}$.[20] The group $\widehat{\Gamma}$ is the mapping class group of a genus 1 surface with one boundary component.[21] This extension corresponds to the orbifold $\mathbb{C}^*$-bundle $\mathcal{M}_{1,\vec{1}} \to \mathcal{M}_{1,1}$.

Denote the completion of $\widehat{\Gamma}$ with respect to the homomorphism

$$\widehat{\Gamma} \to \mathrm{SL}_2(\mathbb{Z}) \hookrightarrow \mathrm{SL}_2(\mathbb{Q})$$

by $\widehat{\mathcal{G}}$ and its prounipotent radical by $\widehat{\mathcal{U}}$. Denote the completion of $\mathrm{SL}_2(\mathbb{Z})$ with respect to its inclusion into $\mathrm{SL}_2(\mathbb{Q})$ by $\mathcal{G}$ and its prounipotent radical by $\mathcal{U}$. Denote the Lie algebras of $\mathcal{U}$ and $\widehat{\mathcal{U}}$ by $\mathfrak{u}$ and $\widehat{\mathfrak{u}}$, respectively. The projection $\widehat{\Gamma} \to \mathrm{SL}_2(\mathbb{Z})$ induces a homomorphism $\widehat{\mathcal{G}} \to \mathcal{G}$ that commutes with the projections to SL_2.

Proposition 14.2. *For each choice of a base point $x \in \mathcal{M}_{1,1}$ and each lift $\widehat{x}$ of x to $\mathcal{M}_{1,\vec{1}}$, there is a natural isomorphism*

$$\widehat{\mathcal{G}}_{\widehat{x}} \cong \mathcal{G}_x \times \mathbb{G}_a(1),$$

[20] This action can be understood as follows: The quotient of $\mathbb{C} \times \mathfrak{h}$ by the central $\mathbb{Z}$ in $\widehat{\Gamma}$ is $\mathbb{C}^* \times \mathfrak{h}$. The quotient mapping is the exponential mapping on the first factor; $\mathrm{SL}_2(\mathbb{Z})$ acts on $\mathbb{C}^* \times \mathfrak{h}$ with factor of automorphy $(c\tau + d)^2$.

[21] The group $\widehat{\Gamma}$ is isomorphic to the 3-string braid group B_3 and also to the inverse image of $\mathrm{SL}_2(\mathbb{Z})$ in the universal covering group of $\mathrm{SL}_2(\mathbb{R})$. (Cf. [19, §8].)

where $\mathbb{G}_a(1)$ denotes the copy of $\mathbb{G}_a$ with the MHS $\mathbb{Q}(1)$. This induces an isomorphism of MHS

$$\widehat{\mathfrak{g}}_{\vec{x}} \cong \mathfrak{g}_x \oplus \mathbb{Q}(1).$$

where $\widehat{\mathfrak{g}}_{\vec{x}}$ is given the natural MHS constructed in [15].

Proof. Since the weight 12 cusp form Δ trivializes $\mathcal{L}^{\otimes 12}$ and since $\mathcal{M}_{1,\vec{1}}$ is $\mathcal{L}^*$, there is a 12-fold covering

$$\mathcal{M}_{1,\vec{1}} \to \mathcal{M}_{1,1} \times \mathbb{C}^*$$

that commutes with the projections to $\mathcal{M}_{1,1}$. It induces an inclusion

$$\begin{array}{ccccccc} 0 \longrightarrow & \mathbb{Z} & \longrightarrow & \widehat{\Gamma} & \longrightarrow & \mathrm{SL}_2(\mathbb{Z}) & \longrightarrow 1 \\ & \downarrow \times 12 & & \downarrow \phi & & \| & \\ 0 \longrightarrow & \mathbb{Z} & \longrightarrow & \mathrm{SL}_2(\mathbb{Z}) \times \mathbb{Z} & \longrightarrow & \mathrm{SL}_2(\mathbb{Z}) & \longrightarrow 1 \end{array}$$

of extensions. The completion of $\mathrm{SL}_2(\mathbb{Z}) \times \mathbb{Z}$ with respect to the obvious homomorphism to $\mathrm{SL}_2(\mathbb{Q})$ is $\mathcal{G} \times \mathbb{G}_a$. This and the right exactness (Prop. 3.6) of relative completion imply that the commutative diagram

$$\begin{array}{ccccccc} & \mathbb{G}_a & \longrightarrow & \widehat{\mathcal{G}} & \longrightarrow & \mathcal{G} & \longrightarrow 1 \\ & \downarrow \times 12 & & \downarrow \phi_* & & \| & \\ 0 \longrightarrow & \mathbb{G}_a & \longrightarrow & \mathcal{G} \times \mathbb{G}_a & \longrightarrow & \mathcal{G} & \longrightarrow 1 \end{array}$$

has exact rows. It follows that $\phi_* : \widehat{\mathcal{G}} \to \mathcal{G} \times \mathbb{G}_a$ is an isomorphism. The Hodge theoretic statements follow from the functoriality of the MHS on relative completion. □

There is therefore an isomorphism $\widehat{\mathfrak{u}} \cong \mathfrak{u} \oplus \mathbb{C}\mathbf{e}_2$, in the category of pronilpotent Lie algebras with an SL_2 action. The new generator $\mathbf{e}_2$ spans a copy of the trivial representation of SL_2 and commutes with the remaining generators

$$\bigcup_{n>0} \{\mathbf{e}_0^j \cdot \mathbf{e}_f', \mathbf{e}_0^j \cdot \mathbf{e}_f'', \mathbf{e}_0^j \cdot \mathbf{e}_{2n} : 0 \le j \le 2n-2, f \in \mathfrak{B}_{2n}\}.$$

The Hodge type of $\mathbf{e}_2$ is $(-1,-1)$, which is consistent with the Hodge types of the other $\mathbf{e}_{2m+2,P}$ given in Figure 4.

Remark 14.3. One can lift the power series connection Ω whose monodromy representation $\mathrm{SL}_2(\mathbb{Z}) \to \mathrm{SL}_2 \ltimes \mathcal{U}$ is the relative completion of $\mathrm{SL}_2(\mathbb{Z})$ to a power series connection $\widehat{\Omega}$ whose monodromy homomorphism $\widehat{\Gamma} \to \mathrm{SL}_2 \ltimes \widehat{\mathcal{U}}$ is the relative completion of $\widehat{\Gamma}$.

The normalized Eisenstein series $G_2(\tau)$ is also defined by the series (12.6), suitably summed. Although G_2 is not a modular form, it satisfies (cf. [38, pp. 95–96])

$$G_2(\gamma\tau) = (c\tau+d)^2 G_2(\tau) + ic(c\tau+d)/4\pi .$$

This implies that

$$\psi_2 := 2\pi i G_2(\tau)d\tau - \frac{1}{2}\frac{d\xi}{\xi} \in E^1(\mathbb{C}^*\times\mathfrak{h})$$

is $\mathrm{SL}_2(\mathbb{Z})$-invariant, and thus a closed 1-form on $\mathcal{M}_{1,\vec{1}}$.[22]

If $\Omega \in K^1(\overline{\mathcal{M}}_{1,1}, P; \mathbf{u})$ is a power series connection (as constructed above), then

$$\widehat{\Omega} = \Omega + \psi_2 \mathbf{e}_2$$

is an integrable $\mathrm{SL}_2(\mathbb{Z})$-invariant power series connection with values in $\widehat{\mathfrak{u}} := \mathfrak{u} \oplus \mathbb{C}\mathbf{e}_2$. For each choice of a lift $\widehat{x} \in \mathbb{C}\times\mathfrak{h}$ of a base point $x \in \mathcal{M}_{1,\vec{1}}$, the monodromy representation

$$\widehat{\Gamma} \to \mathrm{SL}_2(\mathbb{C}) \ltimes \widehat{\mathcal{U}}$$

induces isomorphisms $\widehat{\mathcal{G}}_x \cong \mathrm{SL}_2 \ltimes \widehat{\mathcal{U}} \cong (\mathrm{SL}_2 \ltimes \mathcal{U}) \times \mathbb{G}_a$.

15 The Monodromy Representation

Let E be an elliptic curve with identity 0. Set $E' = E - \{0\}$ and let $\vec{v} \in T_0 E$ be a non-zero tangent vector. Denote the Lie algebra of the unipotent completion of $\pi_1(E', \vec{v})$ by $\mathfrak{p}(E, \vec{v})$. Recall from Section 3.4.3 that this is a completed free Lie algebra with abelianization $H_1(E)$.

Denote by $\mathfrak{p}$ the local system over $\mathcal{M}_{1,\vec{1}}$ whose fiber over $[E, \vec{v}]$ by $\mathfrak{p}(E, \vec{v})$. Fix a base point $x_o = [E_o, \vec{v}_o]$ of $\mathcal{M}_{1,\vec{1}}$. Set $H_o = H_1(E_o)$ and $\mathfrak{p}_o = \mathfrak{p}(E_o, \vec{v}_o)$. Denote the completion of $\pi_1(\mathcal{M}_{1,\vec{1}}, x_o)$ relative to the standard homomorphism to $\mathrm{SL}(H_o)$ by $\widehat{\mathcal{G}}_o$ and its prounipotent radical by $\widehat{\mathcal{U}}_o$. Denote their Lie algebras by $\widehat{\mathfrak{g}}_o$ and $\widehat{\mathfrak{u}}_o$, respectively.

The monodromy action $\pi_1(\mathcal{M}_{1,\vec{1}}, x_o) \to \operatorname{Aut}\mathfrak{p}_o$ respects the lower central series of $\mathfrak{p}_o$ and acts on each graded quotient through an action of $\mathrm{SL}(H_o)$. The universal mapping property of relative completion implies that the monodromy representation above induces a homomorphism

$$\widehat{\mathcal{G}}_o \to \operatorname{Aut}\mathfrak{p}_o .$$

22 It is useful to note that $\psi_2 = -\frac{1}{24}\frac{dD}{D}$, where, where $D = u^3 - 27v^2$ denotes the discriminant function on $\mathcal{M}_{1,\vec{1}}$. This is because there is a unique logarithmic 1-form on $\mathcal{M}_{1,\vec{1}}$ with given residue along the divisor of nodal cubics. Cf. [20, Eqn. 19.1], where D is denoted Δ.

This induces a homomorphism $\widehat{\mathfrak{g}}_o \to \operatorname{Der}\mathfrak{p}_o$ that we shall call the *infinitesimal monodromy action.*

Proposition 15.1. *The infinitesimal monodromy action* $\widehat{\mathfrak{g}}_o \to \operatorname{Der}\mathfrak{p}_o$ *is a morphism of MHS.*

Sketch of Proof. The universal punctured elliptic curve $\mathcal{E}' \to \mathcal{M}_{1,\vec{1}}$ has fiber E' over $[E,\vec{v}] \in \mathcal{M}_{1,\vec{1}}$. The tangent vector $\vec{v}_o$ of E at 0 can be regarded as a tangential base point of $\mathcal{E}'$. The diagram

$$\begin{array}{ccccccccc} 1 & \longrightarrow & \pi_1(E'_o,\vec{v}_o) & \longrightarrow & \pi_1(\mathcal{E}',\vec{v}_o) & \longrightarrow & \pi_1(\mathcal{M}_{1,\vec{1}},x_o) & \longrightarrow & 1 \\ & & \downarrow & & \downarrow & & \downarrow & & \\ & & 1 & \longrightarrow & \mathrm{SL}(H_o) & = & \mathrm{SL}(H_o) & \longrightarrow & 1 \end{array}$$

gives rise to an exact sequence

$$1 \to \mathcal{P}_o \to \mathcal{G}_{\mathcal{E},o} \to \widehat{\mathcal{G}}_o \to 1.$$

of completions that is compatible with mixed Hodge structures. Here $\mathcal{G}_{\mathcal{E},o}$ denotes the completion of $\pi_1(\mathcal{E}',\vec{v}_o)$ with respect to the natural homomorphism to $\mathrm{SL}(H_o)$ and $\mathcal{P}_o$ the unipotent completion of $\pi_1(E'_o,\vec{v}_o)$. One has exactness on the left as $\mathcal{P}_o$ has trivial center. The conjugation action of $\mathcal{G}_{\mathcal{E},o}$ on $\mathcal{P}_o$ induces a homomorphism $\mathfrak{g}_{\mathcal{E},o} \to \operatorname{Der}\mathfrak{p}_o$ of their Lie algebras (cf. 3.10) that is a morphism of MHS.

The tangent vectors $\vec{v}$ induce a section of $\pi_1(\mathcal{E}',\vec{v}_o) \to \pi_1(\mathcal{M}_{1,\vec{1}},x_o)$. It induces a section of $\widehat{\mathcal{G}}_o \to \mathcal{G}_{\mathcal{E},o}$ that is compatible with mixed Hodge structures. The natural action of $\widehat{\mathfrak{g}}_o$ on $\mathfrak{p}_o$ is the composite $\widehat{\mathfrak{g}}_o \to \mathfrak{g}_{\mathcal{E},o} \to \operatorname{Der}\mathfrak{p}_o$ and is therefore a morphism of MHS. $\square$

Since $L^m\mathfrak{p}_o = W_{-m}\mathfrak{p}_o$, there is a canonical isomorphism (cf. (2.4))

$$\mathrm{Gr}^W_\bullet\,\mathfrak{p}_o \cong \mathbb{L}(H_o).$$

of graded Lie algebras in the category of $\mathrm{SL}(H_o)$ modules. The map on each graded quotient is an isomorphism of mixed Hodge structures.

The element σ of $\pi_1(E'_o,\vec{v}_o)$ obtained by rotating the tangent vector once around the identity is trivial in homology and thus lives in the commutator subgroup. The image of its logarithm in

$$\mathrm{Gr}^W_{-2}\,\mathfrak{p}_o \cong \Lambda^2 H_o$$

is $[\mathbf{a},\mathbf{b}]$, where $\mathbf{a},\mathbf{b}$ is any symplectic basis of $H_1(E)$. It spans a copy of the trivial representation. Let

$$\mathrm{Der}^0\,\mathbb{L}(H_o) = \{\delta \in \operatorname{Der}\mathbb{L}(H_o) : \delta([\mathbf{a},\mathbf{b}]) = 0\}.$$

Since the natural action of $\pi_1(\mathcal{M}_{1,\vec{1}},x_o)$ on $\pi_1(E'_o,\vec{v}_o)$ fixes σ, we have:

Corollary 15.2. *The image of the infinitesimal monodromy representation*

$$\operatorname{Gr}^W_\bullet \widehat{\mathfrak{u}}_o \to \operatorname{Der}\mathbb{L}(H_o) \tag{15.3}$$

lies in $\operatorname{Der}^0\mathbb{L}(H_o)$. □

The Lie algebra $\operatorname{Gr}^W_\bullet \widehat{\mathfrak{u}}_o$ is freely generated by the image of any $\mathrm{SL}(H_o)$-invariant Hodge section of $\operatorname{Gr}^W_\bullet \widehat{\mathfrak{u}}_o \to \operatorname{Gr}^W_\bullet H_1(\widehat{\mathfrak{u}}_o)$. Since this projection is an isomorphism in weight -1, each cuspidal generator $\mathbf{e}'_f$ and $\mathbf{e}''_f$ has a canonical lift to $\operatorname{Gr}^W_{-1}\widehat{\mathfrak{u}}_o$. Fix a lift $\tilde{\mathbf{e}}_{2n}$ of each Eisenstein generator $\mathbf{e}_{2n}$ to $\operatorname{Gr}^W_\bullet \widehat{\mathfrak{u}}_o$.

Theorem 15.4. *The image of the graded monodromy representation (15.3) is generated as an* $\mathrm{SL}(H_o)$*-module by the images of the* $\tilde{\mathbf{e}}_{2n}$, $n \geq 1$.

Proof. First observe that $\operatorname{Gr}^W_{-1}\operatorname{Der}^0\mathbb{L}(H_o) = 0$. This is because

$$\operatorname{Gr}^W_{-1}\mathbb{L}(H_o) = H_o \text{ and } \operatorname{Gr}^W_{-2}\mathbb{L}(H_o) = \mathbb{Q}[\mathbf{a},\mathbf{b}].$$

The element $u \in H_o$ corresponds to the derivation ad_u. Since $\operatorname{ad}_u([\mathbf{a},\mathbf{b}]) = [u,[\mathbf{a},\mathbf{b}]] \neq 0$ for all non-zero $u \in H_o$, $\operatorname{Gr}^W_{-1}\operatorname{Der}^0\mathbb{L}(H) = 0$.

Since each $\mathbf{e}'_f$ and $\mathbf{e}''_f$ in $\operatorname{Gr}^W_{-1}\widehat{\mathfrak{u}}_o$ has weight -1, this vanishing implies that $\mathbf{e}_0^j \cdot \mathbf{e}'_f$ and $\mathbf{e}_0^j \cdot \mathbf{e}''_f$ are in the kernel of the graded monodromy representation. It follows that the image is generated by the images of the remaining generators – the Eisenstein generators $\mathbf{e}_0^j \cdot \tilde{\mathbf{e}}_{2n}$. □

The next task is to identify the images of the $\tilde{\mathbf{e}}_{2n}$ in $\operatorname{Der}^0\mathbb{L}(H_o)$. For each $n \geq 0$ a basis $\mathbf{v}_1, \mathbf{v}_2$ of H define derivations $\epsilon_{2n}(\mathbf{v}_1,\mathbf{v}_2)$ by

$$\epsilon_{2n}(\mathbf{v}_1,\mathbf{v}_2) := \begin{cases} -\mathbf{v}_2\frac{\partial}{\partial \mathbf{v}_1} & n = 0; \\ \operatorname{ad}_{\mathbf{v}_1}^{2n-1}(\mathbf{v}_2) - \sum_{\substack{j+k=2n-1 \\ j>k>0}}(-1)^j[\operatorname{ad}^j_{\mathbf{v}_1}(\mathbf{v}_2),\operatorname{ad}^k_{\mathbf{v}_1}(\mathbf{v}_2)]\frac{\partial}{\partial \mathbf{v}_2} & n > 0. \end{cases} \tag{15.5}$$

Here we are identifying $\mathbb{L}(H)$ with its image in $\operatorname{Der}\mathbb{L}(H)$ under the inclusion $\operatorname{ad} : \mathbb{L}(H) \hookrightarrow \operatorname{Der}\mathbb{L}(H)$.

The following result implies that the image of $\tilde{\mathbf{e}}_{2n}$ in $\operatorname{Der}^0\mathbb{L}(H)$ depends only on $\mathbf{e}_{2n}$ and not on the choice of the lift $\tilde{\mathbf{e}}_{2n}$.

Proposition 15.6 (Hain-Matsumoto). *For each* $n \geq 1$ *there is a unique copy of* $S^{2n}H(2n+1)$ *in* $\operatorname{Gr}^W_{-2n-2}\operatorname{Der}^0\mathbb{L}(H)$. *It has highest weight vector the derivation* $\epsilon_{2n}(\mathbf{v}_1,\mathbf{v}_2)$, *where* $\mathbf{v}_1,\mathbf{v}_2 \in H$ *are non-zero vectors of* $\mathfrak{sl}_2$*-weight* 1 *and* -1, *respectively.* □

It follows that the image of $\tilde{\mathbf{e}}_{2n}$ in $\operatorname{Der}\mathbb{L}(H_o)$ is a multiple (possibly zero) of $\epsilon_{2n}(\mathbf{b},\mathbf{a})$. We compute this multiple using the universal elliptic KZB-connection [6, 28, 20], which provides an explicit formula for the connection on the bundle $\mathfrak{p}$ over $\mathcal{M}_{1,\vec{1}}$.

Theorem 15.7. *For all choices of the lift* $\tilde{\mathbf{e}}_{2n}$*, the image of* $\tilde{\mathbf{e}}_{2n}^B = 2\pi i \tilde{\mathbf{e}}_{2n}$ *under the graded monodromy representation (15.3) is* $2\epsilon_{2n}(\mathbf{b},\mathbf{a})/(2n-2)!$ *when* $n>0$ *and* ϵ_0 *when* $n=0$.

Proof. Trivialize the pullback of $\mathcal{H}$ to $\mathbb{C}^* \times \mathfrak{h}$ by the sections

$$T := \tau\mathbf{a} - \mathbf{b} = \exp(\tau\mathbf{e}_0)(-\mathbf{b}) \text{ and } A := (2\pi i)^{-1}\mathbf{a}.$$

In [20, §13–14] it is shown that the pullback of $\mathfrak{p}$ to $\mathbb{C}^* \times \mathfrak{h}$ may be identified with the trivial bundle

$$\mathbb{L}(T,A)^\wedge \times \mathbb{C}^* \times \mathfrak{h} \to \mathbb{C}^* \times \mathfrak{h}$$

with the connection $\nabla = d + \omega'$, where

$$\omega' = -2\pi i\Big(d\tau \otimes \epsilon_0(T,A) + \sum_{m\geq 1}\frac{2}{(2n-2)!}G_{2n}(\tau)d\tau \otimes \epsilon_{2n}(T,A)\Big).$$

To prove the result, we need to rewrite this in terms of the frame $-\mathbf{b},\mathbf{a}$ of $\mathcal{H}$. First note that $\epsilon_{2n}(c_1\mathbf{v}_1, c_2\mathbf{v}_2) = c_1^{2n-1}c_2\epsilon_{2n}(\mathbf{v}_1,\mathbf{v}_2)$ and that if $g \in \mathrm{SL}(H)$, then $\epsilon_{2n}(g\mathbf{v}_1, g\mathbf{v}_2) = g\cdot\epsilon_{2n}(\mathbf{v}_1,\mathbf{v}_2)$, where $g\in\mathrm{SL}(H)$ acts on a derivation δ by $g\cdot\delta := g\delta g^{-1}$. Since $\mathbf{e}_0\cdot\mathbf{a} = 0$, these imply that

$$\begin{aligned}2\pi i\epsilon_{2n}(T,A) &= \epsilon_{2n}(T,\mathbf{a}) = \epsilon_{2n}\big(\exp(\tau\mathbf{e}_0)(-\mathbf{b}), \exp(\tau\mathbf{e}_0)\mathbf{a}\big)\\ &= -\exp(\tau\mathbf{e}_0)\cdot\epsilon_{2n}(\mathbf{b},\mathbf{a}).\end{aligned}$$

It follows that $2\pi i G_{2n}(\tau)d\tau \otimes \epsilon_{2n}(T,A) = -\psi_{2n}(\epsilon_{2n}(\mathbf{b},\mathbf{a}))/2\pi i$.

Since the natural connection ∇_0 on $\mathcal{H}$ is given by

$$\nabla_0 = d - 2\pi i\epsilon_0(T,A)\otimes d\tau,$$

the pullback connection may be written

$$\begin{aligned}\nabla &= \nabla_0 - 2\pi i\sum_{m\geq 1}\frac{2}{(2n-2)!}G_{2n}(\tau)d\tau\otimes\epsilon_{2n}(T,A)\\ &= \nabla_0 + \sum_{m\geq 1}\frac{2}{(2n-2)!}\psi_{2n}\big(\epsilon_{2m}(\mathbf{b},\mathbf{a})/2\pi i\big).\end{aligned}$$

It follows that, regardless of the choice of the lifts of the $\tilde{\mathbf{e}}_{2n}$, the de Rham generator $\tilde{\mathbf{e}}_{2n}$ goes to $2\epsilon_{2n}(\mathbf{b},\mathbf{a})/2\pi i(2n-2)!$ under the graded monodromy representation. Since $\mathbf{e}_{2n}$ spans a copy of $\mathbb{Q}(1)$, the Betti generator is $\mathbf{e}_{2n}^B = \mathbf{e}_{2n}^{\mathrm{DR}}/2\pi i$. □

Remark 15.8. Since $\widehat{\mathfrak{g}} = \mathfrak{g}\oplus\mathbb{Q}(1)$, there are natural representations

$$\mathfrak{g}\to\operatorname{Der}\mathfrak{p} \quad\text{and}\quad \operatorname{Gr}_\bullet^W\mathfrak{g}\to\operatorname{Der}^0\mathbb{L}(H).$$

There are also the outer actions

$$\mathfrak{g} \to \operatorname{OutDer} \mathfrak{p} \quad \text{and} \quad \operatorname{Gr}^W_\bullet \mathfrak{g} \to \operatorname{OutDer} \mathbb{L}(H).$$

The representations $\operatorname{Gr}^W_\bullet \mathfrak{g} \to \operatorname{Der} \mathbb{L}(H)$ and $\operatorname{Gr}^W_\bullet \mathfrak{g} \to \operatorname{OutDer} \mathbb{L}(H)$ have the same kernel as $\mathbf{e}_2 \notin \mathfrak{g}$ and as

$$\operatorname{InnDer} \mathbb{L}(H) \cap \operatorname{Der}^0 \mathbb{L}(H) = \mathbb{Q}\epsilon_2(\mathbf{b}, \mathbf{a}).$$

Exactness of $\operatorname{Gr}^W_\bullet$ implies that $\mathfrak{g} \to \operatorname{Der} \mathfrak{p}$ and $\mathfrak{g} \to \operatorname{OutDer} \mathfrak{p}$ have the same kernel. Since it is generally easier to work with derivations than with outer derivations, we will work with $\mathfrak{g} \to \operatorname{Der} \mathfrak{p}$.

16 The Eisenstein Quotient of a Completed Modular Group

The results of the previous section imply that for all $x \in \mathcal{M}_{1,1}$, each weight graded quotient of the image of $\mathfrak{g}_x$ in $\operatorname{Der} \mathfrak{p}_x$ is a sum of Tate twists $S^m H_x(r)$ of symmetric powers of H_x. Any such Lie algebra quotient of $\mathfrak{g}_x$ has the property that its weight associated graded is generated by the images of the Eisenstein generators $\mathbf{e}_{2n}$.

Suppose that Γ is a finite index subgroup of $\mathrm{SL}_2(\mathbb{Z})$. Denote the completion of $\pi_1(X_\Gamma, x)$ with respect to the inclusion $\Gamma \to \mathrm{SL}_2(H_x)$ by $\mathcal{G}_x$ and its pronilpotent radical by $\mathcal{U}_x$. Denote their Lie algebras by $\mathfrak{g}_x$ and $\mathfrak{u}_x$, respectively.

Proposition 16.1. *For each $x \in X_\Gamma$ there is a unique maximal quotient $\mathfrak{g}_x^{\mathrm{eis}}$ of $\mathfrak{g}_x$ in the category of Lie algebras with a mixed Hodge structure with the property that each weight graded quotient of $\mathfrak{g}_x^{\mathrm{eis}}$ is a sum of Tate twists of symmetric powers of H_x. Moreover, the Lie algebra isomorphism $\mathfrak{g}_x \to \mathfrak{g}_y$ corresponding to a path from x to y in X_Γ induces a Lie algebra (but not a Hodge) isomorphism $\mathfrak{g}_x^{\mathrm{eis}} \to \mathfrak{g}_y^{\mathrm{eis}}$. The corresponding local system $\boldsymbol{\mathfrak{g}}^{\mathrm{eis}} := (\mathfrak{g}_x^{\mathrm{eis}})_{x \in X_\Gamma}$ underlies an admissible VMHS.*

The quotient $\mathfrak{g}_x^{\mathrm{eis}}$ will be called the *Eisenstein quotient* of $\mathfrak{g}_x$. The corresponding quotient of $\mathcal{G}_x$ will be denoted by $\mathcal{G}_x^{\mathrm{eis}}$.

Remark 16.2. Note that if we instead use a tangential base point $\vec{v}$, then $H_{\vec{v}}$ is an extension of $\mathbb{Q}$ by $\mathbb{Q}(1)$, so that $\mathfrak{g}_{\vec{v}}^{\mathrm{eis}}$ is a mixed Hodge-Tate structure. (That is, all of its $M_\bullet$ weight graded quotients are of type (p,p).) In this case, $\mathfrak{g}_{\vec{v}}^{\mathrm{eis}}$ is the "maximal Tate quotient" of $\mathfrak{g}_{\vec{v}}$ in the category pro-Lie algebras with MHS.

Corollary 16.3. *When $\Gamma = \mathrm{SL}_2(\mathbb{Z})$, the monodromy homomorphism $\mathfrak{g}_x \to \operatorname{Der} \mathfrak{p}_x$ factors through $\mathfrak{g}_x^{\mathrm{eis}}$:*

$$\mathfrak{g}_x \longrightarrow \mathfrak{g}_x^{\mathrm{eis}} \longrightarrow \operatorname{Der} \mathfrak{p}_x$$

□

Note that if Γ has finite index in $\mathrm{SL}_2(\mathbb{Z})$, then the Eisenstein quotient of its relative completion surjects onto the Eisenstein quotient of the relative completion of $\mathrm{SL}_2(\mathbb{Z})$.

Proof of Proposition 16.1. If $\mathfrak{h}_1$ and $\mathfrak{h}_2$ are quotients of $\mathfrak{g}_x^{\mathrm{eis}}$ whose weight graded quotients are sums of Tate twists of symmetric powers of H_x, then the image $\mathfrak{h}$ of $\mathfrak{g}_x^{\mathrm{eis}} \to \mathfrak{h}_1 \oplus \mathfrak{h}_2$ is a quotient of $\mathfrak{g}_x^{\mathrm{eis}}$ whose weight graded quotients are sums of twists of symmetric powers of H_x. It also surjects onto $\mathfrak{h}_1$ and $\mathfrak{h}_2$. This implies that the "Eisenstein quotients" of $\mathfrak{g}_x$ form an inverse system from which it follows that the Eisenstein quotient is unique. Note that since $\mathfrak{sl}(H_x) \cong S^2 H_x$, $\mathfrak{g}_x$ surjects onto $\mathfrak{sl}(H_x)$.

We begin the proof of the second part with an observation. Suppose that V is a MHS and that K is a subspace of V that is defined over $\mathbb{Q}$. Give it the induced weight filtration. Then there is a natural isomorphism

$$\mathrm{Gr}_m^W (V/K) \cong (\mathrm{Gr}_m^W V)/(\mathrm{Gr}_m^W K).$$

This isomorphism respects the Hodge filtration on each induced from the Hodge filtration $F^\bullet \cap (W_m V)$ of $W_m V$. These are defined as the images of the maps

$$F^p W_m V \to (W_m V)/(W_m K) \to \mathrm{Gr}_m^W (V/K)$$
$$\text{and } F^p W_m V \to \mathrm{Gr}_m^W V \to (\mathrm{Gr}_m^W V)/(\mathrm{Gr}_m^W K).$$

The observation is that if this Hodge filtration defines a Hodge structure on each $\mathrm{Gr}_m^W (V/K)$, then K is a sub MHS of V and V/K is a quotient MHS of V.

Now apply this with $V = \mathfrak{g}_y$ and K the ideal of $\mathfrak{g}_y$ that corresponds to the kernel of $\mathfrak{g}_x \to \mathfrak{g}_x^{\mathrm{eis}}$ under the isomorphism $\mathfrak{g}_x \cong \mathfrak{g}_y$ given by parallel transport. This implies that $\mathfrak{g}_y/K$ is a quotient of $\mathfrak{g}_y$ in the category of MHS whose weight graded quotients are sums of twists of symmetric powers of H_y. It is therefore an Eisenstein quotient of $\mathfrak{g}_y$.

The last statement follows from the fact that if $\mathbb{W}$ is a quotient of the local system underlying an admissible variation of MHS $\mathbb{V}$ with the property that each fiber of W_x has a MHS that is the quotient of the MHS on V_x, then $\mathbb{W}$ is an admissible variation of MHS. □

The significance of $\mathfrak{g}^{\mathrm{eis}}$ lies in the following result, which follows directly from Theorem 8.3.

Corollary 16.4. *For all base points (finite, tangential), the category* $\mathsf{HRep}(\mathfrak{g}_x^{\mathrm{eis}})$ *of Hodge representations (Def. 8.2) of* $\mathfrak{g}_x^{\mathrm{eis}}$ *is equivalent to the category of admissible VMHS over* X_Γ *whose weight graded quotients are sums of Tate twists* $S^n\mathbb{H}(r)$ *of symmetric powers of* $\mathbb{H}$. □

Remark 16.5. We will call such variations over a modular curve *Eisenstein variations of MHS*. Like Tate VMHS in the unipotent case (cf. [26]), Eisenstein variations over a modular curve can be written down reasonably explicitly. This follows from the constructions of Section 8. The explanation we give below is somewhat technical. Suppose that $\mathbb{A}$ is an Eisenstein variation over X_Γ. Set

$$A_{m,n} = H^0\big(X_\Gamma, \mathrm{Hom}(S^n\mathbb{H}, \mathrm{Gr}^W_m \mathbb{A})\big).$$

This is a Tate Hodge structure of weight $m-n$. There is a natural isomorphism

$$\mathrm{Gr}^W_m \mathbb{A} \cong \bigoplus_n A_{m,n} \otimes S^n\mathbb{H}$$

of MHS. For each r satisfying $0 \le r \le \min(n,\ell)$, fix a highest weight vector $\mathbf{h}^{(r)}_{n,\ell}$ of $\mathfrak{sl}_2$-weight $n+\ell-2r$ in $\mathrm{Hom}(S^nH, S^\ell H)$, so that

$$\mathrm{Hom}(S^n\mathbb{H}, S^\ell\mathbb{H}) \cong \bigoplus_{r=0}^{\min(n,\ell)} \mathbb{S}^{n+\ell-2r}(\mathbf{h}^{(r)}_{n,\ell}).$$

Implicit here is that $\mathbf{h}^{(r)}_{n,\ell}$ has Hodge weight $\ell - n$. Set $\mathcal{A} = \mathbb{A} \otimes \mathcal{O}_{X_\Gamma}$. Denote the natural connection on it by ∇. Set

$$\mathcal{A}_m = (\mathrm{Gr}^W_m \mathbb{A}) \otimes \mathcal{O}_{X_\Gamma} \cong \bigoplus_n A_{m,n} \otimes S^n\mathcal{H}.$$

The standard connection on $\mathcal{H}$ induces a connection on each of these that we denote by ∇_0. The construction of Section 8 implies that for each cusp $P \in D$ of X_Γ, there are linear maps

$$\varphi^{m,n}_{k,\ell,P} \in F^{\ell-r+1} W_{2\ell-2r+2} \mathrm{Hom}_{\mathbb{C}}(A_{m,n}, A_{k,\ell}) \tag{16.6}$$

such that

$$(\mathcal{A}, \nabla) \cong \Big(\bigoplus_m \mathcal{A}_m, \nabla_0 + \Omega\Big) \tag{16.7}$$

where

$$\Omega = \sum_{P\in D} \sum_{n,\ell \ge 0} \sum_{r=0}^{\min(n,\ell)} \sum_{m,k} \varphi^{m,n}_{k,\ell,P} \otimes \psi_{n+\ell-2r+2,P}(\mathbf{h}^{(r)}_{n,\ell}).$$

Implicit in this statement is that the canonical extension $\overline{\mathcal{A}}$ of $(\mathcal{A}, \nabla)$ is isomorphic to

$$\bigoplus_{m,n} A_{m,n} \otimes (S^n\overline{\mathcal{H}}, \nabla_0).$$

The isomorphism (16.7) is bifiltered with respect to the Hodge and weight filtrations. The condition (16.6) implies that $\Omega \in F^0 W_{-1} K^1\big(\overline{X}_\Gamma, D; \mathrm{End}(\mathrm{Gr}^W_\bullet \mathbb{A})\big)$.

Caution: not every such 1-form Ω defines the structure of an admissible variation of MHS over X_Γ. The issue is that one needs the monodromy of

$(\mathcal{A}, \nabla_0 + \Omega)$ to be defined over $\mathbb{Q}$. Determining which Ω give rise to Eisenstein variations is closely related to the problem of determining the relations in $\mathfrak{u}^{\mathrm{eis}}$.

17 Modular Symbols and Pollack's Quadratic Relations

Motivic arguments (cf. [23]) suggested that $\mathfrak{u}_x^{\mathrm{eis}}$ may not be freely generated by the $\mathbf{e}_0^j \cdot \mathbf{e}_{2m}$ and predict that the relations that hold between the $\mathbf{e}_0^j \cdot \mathbf{e}_m$ arise from cusp forms. In other words, cusp forms go from being generators of $\mathfrak{u}_x$ to relations in $\mathfrak{u}_x^{\mathrm{eis}}$. The goal of the rest of this paper is to sketch a Hodge theoretic explanation for these relations. For this we will need to recall some basic facts about modular symbols, which record the periods of cusp forms and determine the Hodge structure on $H^1_{\mathrm{cusp}}(\mathcal{M}_{1,1}, S^{2n}\mathbb{H})$. In this and subsequent sections, $\mathfrak{B}_w = \mathfrak{B}_w(\mathrm{SL}_2(\mathbb{Z}))$, the set of normalized Hecke eigen cusp forms of $\mathrm{SL}_2(\mathbb{Z})$.

17.1. Modular Symbols

Modular symbols are homogeneous polynomials attached to cusp forms of $\mathrm{SL}_2(\mathbb{Z})$. They play two roles: they give a concrete representation of the cohomology class associated to a cusp form; secondly, modular symbols of degree m record the periods of the MHS on $H^1(\mathcal{M}_{1,1}, S^m\mathbb{H}) \cong H^1(\mathrm{SL}_2(\mathbb{Z}), S^m H)$. A standard reference is [27, Ch. IV].

Recall that $\mathrm{SL}_2(\mathbb{Z})$ has presentation

$$\mathrm{SL}_2(\mathbb{Z}) = \langle S, U : S^2 = U^3,\ S^4 = U^6 = I \rangle.$$

where

$$S = \begin{pmatrix} 0 & -1 \\ 1 & 0 \end{pmatrix}, \quad T = \begin{pmatrix} 1 & 1 \\ 0 & 1 \end{pmatrix}, \quad U := ST = \begin{pmatrix} 0 & -1 \\ 1 & 1 \end{pmatrix}$$

The action of $\mathrm{SL}_2(\mathbb{Z})$ on $\mathfrak{h}$ factors through

$$\mathrm{PSL}_2(\mathbb{Z}) := \mathrm{SL}_2(\mathbb{Z})/(\pm I) = \langle S, U : S^2 = U^3 = I \rangle.$$

17.1.1. Group cohomology

Suppose that Γ is a group and that V is a left Γ-module. Then one has the complex

$$0 \longrightarrow C^0(\Gamma, V) \xrightarrow{\delta} C^1(\Gamma, V) \xrightarrow{\delta} C^2(\Gamma, V) \xrightarrow{\delta} \cdots$$

of standard cochains, where $C^j(\Gamma, V) = \{\text{functions } \phi : \Gamma^j \to V\}$. The differential takes $v \in V = C^0(\Gamma, V)$ to the function $\delta v : \gamma \mapsto (\gamma - 1)v$ and $\delta : C^0(\Gamma, V) \to$

$C^1(\Gamma, V)$ takes $\phi : \Gamma \to V$ to the function

$$(\delta\phi)(\gamma_1, \gamma_2) \mapsto \phi(\gamma_1) - \phi(\gamma_1\gamma_2) + \gamma_1 \cdot \phi(\gamma_2).$$

So ϕ is a 1-cocycle if and only if $\phi(\gamma_1\gamma_2) = \phi(\gamma_1) + \gamma_1 \cdot \phi(\gamma_2)$. The cohomology $H^\bullet(\Gamma, V)$ of Γ with coefficients in V is defined to be the homology of this complex.

Now suppose that V is a real or complex vector space and that Γ acts on a simply connected manifold X. Fix a base point $x_o \in X$. As in Proposition 4.3, to each $\gamma \in \Gamma$ we can associate the unique homotopy class c_γ of paths in X from x_o to $\gamma \cdot x_o$. If $\omega \in E^1(X) \otimes V$ is Γ invariant, then the function

$$\phi : \gamma \mapsto \int_{c_\gamma} \omega$$

is a 1-cocycle. Changing the base point from x_o to x' changes ϕ by the coboundary of $\int_{x_o}^{x'} \omega \in V$. (Cf. Remark 4.4.) This construction induces a map

$$H^1(E^1(X \times V)^\Gamma) \to H^1(\Gamma, V),$$

which is an isomorphism when Γ acts properly discontinuously and virtually freely on X. This is the case when Γ is a modular group and X is the upper half plane.

Suppose that V is divisible as an abelian group. When $-I$ acts trivially on V, V is the pullback of a $\mathrm{PSL}_2(\mathbb{Z})$-module and

$$H^1(\mathrm{PSL}_2(\mathbb{Z}), V) \to H^1(\mathrm{SL}_2(\mathbb{Z}), V)$$

is an isomorphism.

17.1.2. Cuspidal cohomology

Suppose that F is a field of characteristic zero. Set $H_F = F\mathbf{a} \oplus F\mathbf{b}$. Define $C^\bullet_{\mathrm{cusp}}(\mathrm{SL}_2(\mathbb{Z}), S^{2n}H_F)$ to be the kernel of the restriction mapping

$$C^\bullet(\mathrm{SL}_2(\mathbb{Z}), S^{2n}H_F) \to \widetilde{C}^\bullet(\langle T \rangle, S^{2n}H_F)$$

where the right hand complex is the quotient of $C^\bullet(\langle T \rangle, S^{2n}H_F)$ by $\mathbf{a}^{2n}$ in degree 0. Set

$$H^\bullet_{\mathrm{cusp}}(\mathrm{SL}_2(\mathbb{Z}), S^{2n}H_F) := H^\bullet(C^\bullet_{\mathrm{cusp}}(\mathrm{SL}_2(\mathbb{Z}), S^{2n}H_F)).$$

The corresponding long exact sequence gives the exact sequence

$$\begin{aligned} 0 \to H^1_{\mathrm{cusp}}(\mathrm{SL}_2(\mathbb{Z}), S^{2n}H_F) &\to H^1(\mathrm{SL}_2(\mathbb{Z}), S^{2n}H_F) \\ &\to H^1(\langle T \rangle, S^{2n}H_F) \to H^2_{\mathrm{cusp}}(\mathrm{SL}_2(\mathbb{Z}), S^{2n}H_F) \to 0. \end{aligned}$$

This is an instance of the exact sequences (6.3) and (11.3) where $C' = \mathcal{M}_{1,1}$.

The cuspidal cohomology group $H^1_{\text{cusp}}(\mathrm{SL}_2(\mathbb{Z}), S^{2n}H_F)$ has a nice description. Recall that $\mathrm{SL}_2(\mathbb{Z})$ acts on H via the formula (10.1) with $\mathbf{v}_1 = -\mathbf{b}$ and $\mathbf{v}_2 = \mathbf{a}$.

Proposition 17.1. *Suppose that F is a field of characteristic zero. For all $n \geq 1$, there is an isomorphism*

$$H^1_{\text{cusp}}(\mathrm{SL}_2(\mathbb{Z}), S^{2n}H_F)$$

with the vector space of $\mathsf{r}(\mathbf{a},\mathbf{b}) \in S^{2n}H_F$ that satisfy

$$(I+S)\mathsf{r}(\mathbf{a},\mathbf{b}) = 0 \text{ and } (I+U+U^2)\mathsf{r}(\mathbf{a},\mathbf{b}) = 0 \tag{17.2}$$

modulo $\mathbf{a}^{2n} - \mathbf{b}^{2n}$.

Proof. Suppose that $\phi : \mathrm{SL}_2(\mathbb{Z}) \to S^{2n}H_F$ is a cuspidal 1-cocycle. Since $-I$ acts trivially on $S^{2n}H$, the cocycle condition and the equation $(-I)^2 = 1$ imply that $\phi(-I) = 0$. Since S and U generate $\mathrm{SL}_2(\mathbb{Z})$, ϕ is determined by $\phi(S)$. Since $U = ST$,

$$\phi(U) = \phi(ST) = \phi(S) + S\phi(T) = \phi(S).$$

Thus ϕ is determined by $\phi(S)$. Denote this element of $S^{2n}H_F$ by $\mathsf{r}_\phi(\mathbf{a},\mathbf{b})$. Since $S^2 = U^3 = I$, the cocycle condition implies that $\mathsf{r}_\phi(\mathbf{a},\mathbf{b})$ satisfies the equation (17.2). Conversely, if $\mathsf{r}(\mathbf{a},\mathbf{b}) \in S^{2n}H_F$ satisfies these equations, it determines a well-defined cuspidal cocycle ϕ by $\phi(S) = \phi(U) = \mathsf{r}(\mathbf{a},\mathbf{b})$.

The last statement follows as the only cuspidal coboundaries are scalar multiplies of $\delta\mathbf{a}^{2n}$. This has value $\mathsf{r}(\mathbf{a},\mathbf{b}) = \mathbf{b}^{2n} - \mathbf{a}^{2n}$ on S. □

Remark 17.3. Since $(I+S)\sum_j c_j \mathbf{a}^j \mathbf{b}^{2n-j}$ vanishes if and only if $c_{2n-j} = (-1)^{j+1}c_j$ for all j, the terms of a cocycle $r(\mathbf{a},\mathbf{b})$ of top degree in $\mathbf{a}$ and $\mathbf{b}$ is a multiple of $\mathbf{a}^{2n} - \mathbf{b}^{2n}$. Since this corresponds to the coboundary of $\mathbf{a}^{2n}$, we can identify $H^1_{\text{cusp}}(\mathrm{SL}_2(\mathbb{Z}), S^{2n}H_F)$ with those $\mathsf{r}(\mathbf{a},\mathbf{b})$ that satisfy the cocycle conditions (17.2) and have no terms of degree $2n$ in $\mathbf{a}$ or $\mathbf{b}$.

17.1.3. Modular symbols

If f is a cusp form of weight $2n+2$, the $S^{2n}H$-valued 1-form $\omega_f(\mathbf{b}^{2n})$ is $\mathrm{SL}_2(\mathbb{Z})$-invariant. Since f is a cusp form, it is holomorphic on the q-disk. We can therefore take the base point x_o above to be the cusp $q = 0$. Since T fixes $q = 0$, the function

$$\gamma \mapsto \int_{x_o}^{\gamma x_o} \omega_f(\mathbf{w}^{2n}) = (2\pi i)^{2n} \int_{x_o}^{\gamma x_o} \omega_f(\mathbf{b}^{2n}) \in S^{2n}H_{\mathbb{C}}$$

is a well defined cuspidal 1-cocycle. The *modular symbol* of f is its value[23]

$$\mathsf{r}_f(\mathbf{a},\mathbf{b}) := \int_0^1 f(q)\mathbf{w}^{2n}\frac{dq}{q} = -(2\pi i)^{2n+1}\int_0^\infty f(iy)(\mathbf{b}-iy\mathbf{a})^{2n}d(iy) \in S^{2n}H_\mathbb{C}$$

on S. It satisfies the cocycle condition (17.2) and represents the class

$$(2\pi i)^{2n}\omega_f(\mathbf{b}^{2n}) \in H^1_{\mathrm{cusp}}(\mathcal{M}_{1,1}, S^{2n}\mathbb{H}).$$

It determines f.

17.2. Hodge theory

The Hodge structure

$$H^1_{\mathrm{cusp}}(\mathcal{M}_{1,1}, S^{2n}\mathbb{H}) = H^{2n+1,0} \oplus H^{0,2n+1}.$$

has a description in terms of modular symbols. The underlying $\mathbb{Q}$ vector space is the set of $\mathsf{r}(\mathbf{a},\mathbf{b}) \in S^{2n}H_\mathbb{Q}$ that satisfy (17.2), modulo $\mathbf{a}^{2n}-\mathbf{b}^{2n}$; the Hodge filtration is given by

$$H^{2n+1,0} = F^{2n+1}H^1_{\mathrm{cusp}}(\mathcal{M}_{1,1}, S^{2n}\mathbb{H}) = \{\mathsf{r}_f(\mathbf{a},\mathbf{b}) : f \in M^o_{2n+2}\}/\mathbb{C}(\mathbf{a}^{2n}-\mathbf{b}^{2n}).$$

There is more structure. Each element $r(\mathbf{a},\mathbf{b})$ of $S^{2n}H$ can be written in the form

$$\mathsf{r}(\mathbf{a},\mathbf{b}) = \mathsf{r}^+(\mathbf{a},\mathbf{b}) + \mathsf{r}^-(\mathbf{a},\mathbf{b}),$$

where $\mathsf{r}^+(\mathbf{a},\mathbf{b})$ is the sum of the terms involving only even powers of $\mathbf{a}$ and $\mathbf{b}$ and $\mathsf{r}^-(\mathbf{a},\mathbf{b})$ is the sum of the terms involving only odd powers.

If f has real Fourier coefficients (e.g., $f \in \mathfrak{B}_{2n+2}$), then

$$\mathsf{r}_f(\mathbf{a},\mathbf{b}) = \mathsf{r}_f^+(\mathbf{a},\mathbf{b}) + i\mathsf{r}_f^-(\mathbf{a},\mathbf{b}) \tag{17.4}$$

where $\mathsf{r}_f^\pm(\mathbf{a},\mathbf{b})$ are real. Since the cocycle corresponding to $\bar{f}$ is $\mathsf{r}_f^+(\mathbf{a},\mathbf{b}) - i\mathsf{r}_f^-(\mathbf{a},\mathbf{b})$, $\mathsf{r}_f^+(\mathbf{a},\mathbf{b}), \mathsf{r}_f^-(\mathbf{a},\mathbf{b}) \in S^{2n}H_\mathbb{R}$ also satisfy the cocycle condition. Since the classes $\omega_f(\mathbf{b}^{2n}), f \in \mathfrak{B}_{2n+2}$ span the cuspidal cohomology, we deduce:

Proposition 17.5. *If* $\mathsf{r}(\mathbf{a},\mathbf{b}) \in S^{2n}H_F$ *satisfies the cocycle condition (17.2), then so do* $\mathsf{r}^+(\mathbf{a},\mathbf{b})$ *and* $\mathsf{r}^-(\mathbf{a},\mathbf{b})$. □

This gives a decomposition $V_F = V_F^+ \oplus V_F^-$ of $V_F := H^1_{\mathrm{cusp}}(\mathcal{M}_{1,1}, S^{2n}\mathbb{H}_F)$. Since

$$H^{2n+1,0}_{\mathrm{cusp}}(\mathcal{M}_{1,1}, S^{2n}\mathbb{H}) \cap H^1(\mathcal{M}_{1,1}, S^{2n}\mathbb{H}_\mathbb{R}) = 0$$

both parts $\mathsf{r}_f^\pm(\mathbf{a},\mathbf{b})$ of the modular symbol of a cusp form with real Fourier coefficients are non-zero. In particular, for each $f \in \mathfrak{B}_{2n+2}$ we can write

$$V_{f,\mathbb{R}} = V_{f,\mathbb{R}}^+ \oplus V_{f,\mathbb{R}}^- := \mathbb{R}\,\mathsf{r}_f^+(\mathbf{a},\mathbf{b}) \oplus \mathbb{R}\,\mathsf{r}_f^-(\mathbf{a},\mathbf{b}).$$

Note that V_f^+ and V_f^- are real sub Hodge structures of V_f.

[23] We use this normalization because, if $f \in \mathfrak{B}_{2n+2}$, then $f(q)\mathbf{w}^{2n}dq/q \in H^1_{\mathrm{DR}}(\mathcal{M}_{1,1/\mathbb{Q}}, S^{2n}\mathcal{H})$. Cf. [20, §21].

17.2.1. The action of real Frobenius

Complex conjugation (aka "real Frobenius") $\mathcal{F}_\infty \in \mathrm{Gal}(\mathbb{C}/\mathbb{R})$ acts on $\mathcal{M}_{1,1}$ and on the local system $\mathbb{H}_\mathbb{R}$ as we shall explain below. It therefore acts on $V_\mathbb{R} = H^1_{\mathrm{cusp}}(\mathcal{M}_{1,1}, S^{2n}\mathbb{H}_\mathbb{R})$. In this section we show that its eigenspaces are $V^\pm_\mathbb{R}$.

The stack $\mathcal{M}_{1,1/\mathbb{C}}$ has a natural real (even $\mathbb{Q}$) form, viz, $\mathbb{G}_m\backslash\backslash(\mathbb{A}^2_\mathbb{R} - D)$, where D is the discriminant locus $u^3 - 27v^2 = 0$ and where $t\cdot(u,v) = (t^4u, t^6v)$. The involution $\mathcal{F}_\infty : (u,v) \mapsto (\bar{u},\bar{v})$ of $\mathcal{M}_{1,1}$ is covered by the involution of $\tau \mapsto -\overline{\tau}$ of $\mathfrak{h}$.

The universal curve $\mathcal{E}$ over it also has a natural real form as it is defined over $\mathbb{R}$. The projection $\mathcal{E} \to \mathcal{M}_{1,1}$ is invariant under complex conjugation. This implies that $\mathcal{F}_\infty$ acts on $\mathbb{H}_\mathbb{R}$. This action is determined by the action of $\mathcal{F}_\infty$ on H, the fiber over the tangent vector $\partial/\partial q$, which is real and therefore fixed by $\mathcal{F}_\infty$. This induced map is easily seen to be the involution $\sigma : H \to H$ defined by[24]

$$\mathbf{b} \mapsto -\mathbf{b},\ \mathbf{a} \mapsto \mathbf{a}. \tag{17.6}$$

The monodromy representation $\mathrm{SL}_2(\mathbb{Z}) \to \mathrm{Aut} H$ of $\mathcal{F}^*_\infty\mathbb{H}$ is the standard representation conjugated by σ. There is therefore a natural action

$$\mathcal{F}_\infty : H^1_{\mathrm{cusp}}(\mathcal{M}_{1,1}, S^{2n}\mathbb{H}_\mathbb{R}) \to H^1_{\mathrm{cusp}}(\mathcal{M}_{1,1}, S^{2n}\mathbb{H}_\mathbb{R}).$$

Let $\overline{\mathcal{F}}_\infty : H^1_{\mathrm{cusp}}(\mathcal{M}_{1,1}, S^{2n}\mathbb{H}_\mathbb{C}) \to H^1_{\mathrm{cusp}}(\mathcal{M}_{1,1}, S^{2n}\mathbb{H}_\mathbb{C})$ be its composition with complex conjugation $H^1_{\mathrm{cusp}}(\mathcal{M}_{1,1}, S^{2n}\mathbb{H}_\mathbb{C}) \to H^1_{\mathrm{cusp}}(\mathcal{M}_{1,1}, S^{2n}\mathbb{H}_\mathbb{C})$. This is the de Rham involution.

Lemma 17.7. *Real Frobenius $\mathcal{F}_\infty$ acts on $V_\mathbb{R} = H^1_{\mathrm{cusp}}(\mathcal{M}_{1,1}, S^{2n}\mathbb{H}_\mathbb{R})$ by multiplication by $+1$ on $V^+_\mathbb{R}$ and -1 on $V^-_\mathbb{R}$.*

Proof. The result follows from equations (17.4) and (17.6) and the fact that the image of $F^{2n+1}H^1_{\mathrm{DR}}(\mathcal{M}_{1,1/\mathbb{R}}, S^{2n}\mathcal{H})$ in $H^1_{\mathrm{cusp}}(\mathcal{M}_{1,1}, S^{2n}\mathbb{H}_\mathbb{C})$ is invariant under $\overline{\mathcal{F}}_\infty$. □

17.2.2. Extensions of MHS associated to cusp forms

Since $V_{f,\mathbb{C}} = V_{f,\mathbb{R}} \oplus F^rV_f$ when $r < 2n+2$ and $F^0V_f(r) = 0$ when $r \geq 2n+2$, we have

$$\mathrm{Ext}^1_{\mathsf{MHS}}\left(\mathbb{R}, V_f(r)\right) \cong \begin{cases} 0 & r < 2n+2, \\ iV_{f,\mathbb{R}} & r \geq 2n+2 \text{ even}, \\ V_{f,\mathbb{R}} & r \geq 2n+2 \text{ odd}. \end{cases} \tag{17.8}$$

We can now compute the extensions invariant under the de Rham involution.

[24] Here H is identified with $H^1(E_{\partial/\partial q})$ which is isomorphic to $H_1(E_{\partial/\partial q})(-1)$. So the actions of $\mathcal{F}_\infty$ on $H_1(E_{\partial/\partial q})$ and $H^1(E_{\partial/\partial q})$ differ by -1.

Proposition 17.9. *If $f \in \mathfrak{B}_{2n+2}$, then*

$$\operatorname{Ext}^1_{\mathsf{MHS}}\big(\mathbb{R}, V_f(r)\big)^{\overline{\mathcal{F}}_\infty} \cong \begin{cases} 0 & r < 2n+2, \\ iV^-_{f,\mathbb{R}} & r \geq 2n+2 \text{ even}, \\ V^+_{f,\mathbb{R}} & r \geq 2n+2 \text{ odd}. \end{cases}$$

In particular, it is one dimensional for all $r \geq 2n+2$.

Proof. Since $\mathcal{F}_\infty : V_{f,\mathbb{C}} \to V_{f,\mathbb{C}}$ is $\mathbb{C}$-linear and since twisting by $\mathbb{R}(r)$ multiplies this action by $(-1)^r$, Lemma 17.7 implies that $\overline{\mathcal{F}}_\infty$ acts on $iV^{\pm}_{\mathbb{R}}(r)$ by multiplication by $(-1)^{r+1}$ and on $iV^-_{\mathbb{R}}$ by $(-1)^r$. The result follows from the $\overline{\mathcal{F}}_\infty$-equivariant isomorphism $\operatorname{Ext}^1_{\mathsf{MHS}}(\mathbb{R}, V_f(r)) \cong iV_{f,\mathbb{R}}(r)$ which holds for all $r \geq 2n+2$. □

So each normalized Hecke eigen cusp form $f \in \mathfrak{B}_{2n+2}$ determines an element of $\operatorname{Ext}^1_{\mathsf{MHS}}\big(\mathbb{R}, V_f(r)\big)^{\overline{\mathcal{F}}_\infty}$ for each $r \geq 2n+2$. Namely, the extension corresponding to $ir^+_f(\mathbf{a},\mathbf{b}) \in iV^+_{f,\mathbb{R}}$ when r is even, and $r^-_f(\mathbf{a},\mathbf{b}) \in V^-_{f,\mathbb{R}}$ when r is odd.

17.3. Pollack's quadratic relations

Motivic considerations [23] suggest that $\mathfrak{u}^{\mathrm{eis}}_x$ is not free and that a minimal set of relations that hold between the $\mathbf{e}_0^j \cdot \mathbf{e}_{2n}$ are parametrized by cusp forms. In fact, each cusp form should determine relations between the $\mathbf{e}_0^j \cdot \mathbf{e}_{2n}$ of every degree ≥ 2. (For this purpose, $\mathbf{e}_0$ is considered to have degree 1.) One way to guess such relations is to find relations that hold between their images $\mathbf{e}_0^j \cdot \epsilon_{2n}(\mathbf{b},\mathbf{a})$ in $\operatorname{Der}\operatorname{Gr}^W_\bullet \mathfrak{p} \cong \operatorname{Der}\mathbb{L}(H)$.[25] For convenience, set $\epsilon_{2n} = \epsilon_{2n}(\mathbf{b},\mathbf{a})$.

In his undergraduate thesis [33], Pollack found a complete set of quadratic relations that hold between the ϵ_{2n}, and found relations of all degrees > 2 that hold between the $\mathbf{e}_0^j \cdot \epsilon_{2n}$ modulo a certain filtration of $\operatorname{Der}\mathbb{L}(H)$. Here we state his quadratic relations.

Theorem 17.10 (Pollack [33, Thm. 2]). *The relation*

$$\sum_{\substack{j+k=n \\ j,k>0}} c_j[\epsilon_{2j+2}, \epsilon_{2k+2}] = 0$$

holds in $\operatorname{Der}\mathbb{L}(H)$ *if and only if there is a cusp form f of weight $2n+2$ whose modular symbol is*

$$r^+_f(\mathbf{a},\mathbf{b}) = \sum_{\substack{j+k=n \\ j,k\geq 0}} c_j \mathbf{a}^{2j}\mathbf{b}^{2n-2j}.$$

[25] Since $\operatorname{Gr}^W_\bullet$ is exact, one neither gains nor loses relations in the associated graded.

In view of Theorem 15.7, this suggests each cusp form of weight $2n+2$ with $r_f^+(\mathbf{a},\mathbf{b}) = \sum c_j \mathbf{a}^{2j}\mathbf{b}^{2n-2j}$ might determine a relation

$$\sum_{\substack{j+k=n \\ j,k>0}} c_j (2j)!(2k)![\mathbf{e}_{2j+2}, \mathbf{e}_{2k+2}] = 0 \tag{17.11}$$

in $\mathrm{Gr}^W_\bullet \mathfrak{u}_x^{\mathrm{eis}}$ and that these relations are connected with $\mathrm{Ext}^1_{\mathsf{MHS}}(\mathbb{R}, V_f(2r))$ for appropriate $r>0$. In the remaining sections, we show that Pollack's relations do indeed lift to relations in $\mathrm{Gr}^W_\bullet \mathfrak{u}_x^{\mathrm{eis}}$ and explain the connection to extensions of MHS related to cusp forms. In preparation for proving this, we restate Pollack's result in cohomological terms.

We let the base point be $\vec{v} = \partial/\partial q$, although any base point will do. Denote the image of the monodromy homomorphism $\mathcal{G} \to \mathrm{Aut}\,\mathfrak{p}$ by $\mathcal{D}$ and the Lie algebra of its pronilpotent radical by $\mathfrak{n}$. Since the monodromy homomorphism is a morphism of MHS, $\mathcal{O}(\mathcal{D})$ and $\mathfrak{n}$ have natural MHSs. This implies that $H^\bullet(\mathcal{D}, S^m H(r))$ has a natural MHS for each $m \geq 0$ and $r \in \mathbb{Z}$ and that the natural isomorphism

$$H^\bullet(\mathcal{D}, S^m H(r)) \cong \left[H^\bullet(\mathfrak{n}) \otimes S^m H(r)\right]^{\mathrm{SL}(H)}$$

is an isomorphism of MHS. (Cf. [21].)

Since the monodromy representation factors through $\mathcal{G}^{\mathrm{eis}}$, there is an MHS isomorphism

$$H_1(\mathfrak{n}) \cong \bigoplus_{n>0} S^{2n}(\epsilon_{2n+2}) \cong \bigoplus_{n>0} S^{2n} H(2n+1).$$

This implies that

$$H^1\big(\mathcal{D}, S^{2n} H(2n+1)\big) = \mathrm{Hom}_{\mathrm{SL}(H)}(H_1(\mathfrak{n}), S^n H)(2n+1) \cong \mathbb{Q}(0).$$

Regard $S^{2m}H = S^{2m}(\mathbf{b}^{2m})$. Let $\check{\epsilon}_{2m}$ be the element of

$$H^1(\mathcal{D}, S^{2m} H) \cong \mathrm{Hom}_{\mathrm{SL}(H)}(H_1(\mathfrak{u}), S^{2m} H)$$

that takes the class of ϵ_{2m+2} to $\mathbf{b}^{2m}$.

The standard duality (Prop. 2.2) between quadratic relations in $\mathfrak{n}$ and the cup product $H^1(\mathfrak{n}) \otimes H^1(\mathfrak{n}) \to H^2(\mathfrak{n})$ gives the following dual version of Pollack's quadratic relations, Theorem 17.10.

Proposition 17.12. *There is a surjection*

$$H^2(\mathcal{D}, S^{2n} H) \to H^1_{\mathrm{cusp}}(\mathcal{M}_{1,1}, S^{2n}\mathbb{H}_{\mathbb{Q}})^{\overline{\mathcal{F}}_\infty} \otimes \mathbb{Q}(-2n-2)$$

which is a morphism of MHS when $H^1_{\mathrm{cusp}}(\mathcal{M}_{1,1}, S^{2n}\mathbb{H})^{\overline{\mathcal{F}}_\infty}$ is regarded as a HS of type $(0,0)$. *After tensoring with* $\mathbb{R}$, *it gives a surjection*

$$H^2(\mathcal{D}, S^{2n} H_{\mathbb{R}}) \to \bigoplus_{f \in \mathfrak{B}_{2n+2}} \mathbb{R}(-2n-2). \tag{17.13}$$

The generator of $\mathbb{R}(-2n-2)$ corresponding to $f \in \mathfrak{B}_{2n+2}$ will be denoted by z_f. When $j+k=n$, the composition of the cup product

$$H^1\big(\mathcal{D}, S^{2j}H(2j+1)\big) \otimes H^1\big(\mathcal{D}, S^{2k}H(2k+1)\big) \to H^2\big(\mathcal{D}, S^{2n}H(2n+2)\big)$$

with the projection (17.13) is given by

$$\check{e}_{2j} \otimes \check{e}_{2k} \to \sum_{f \in \mathfrak{B}_{2n+2}} c_{j,k}(f)\, z_f \tag{17.14}$$

where $\mathrm{r}_f^+(\mathbf{a},\mathbf{b}) = \sum_{j+k=n} c_{j,k}(f)\, \mathbf{a}^{2j}\mathbf{b}^{2k}$.

The reason that V_f^- appears in H^2 and the coefficients of elements of V_f^+ occur in the relations is explained by noting that the Petersson inner product induces an isomorphism $V_f^- (V_f^+)^*$, while relations give subspaces of H_2.

Remark 17.15. The occurrence of cusp forms here (without their associated Hodge structure) should be related to, and may help explain, the appearance of modular and cusp forms in the work of Conant, Kassabov and Vogtmann [9, 10, 8] on the $\mathrm{Sp}(H)$-representation theory of the derivation algebra of a once punctured Riemann surface S of genus $g \gg 0$.

18 Deligne Cohomology and Extensions of VMHS

The relations that hold in $\mathfrak{u}_x^{\mathrm{eis}}$ are controlled by relations in the Yoneda ext groups of the category $\mathsf{MHS}(\mathcal{M}_{1,1}, \mathbb{H})$. In this section, we sketch the relationship between Deligne cohomology of the relative completion of the fundamental group of an affine curve $C' = C - D$ and Yoneda ext groups in the categories $\mathsf{MHS}(C', \mathbb{H})$. This generalizes the results of [5] that hold in the unipotent case. We will work in the category of $\mathbb{Q}$-MHS, although the discussion is equally valid in the category of $\mathbb{R}$-MHS. This section is a summary of results from [21] where full details can be found.

18.1. Deligne–Beilinson cohomology of a curve

Let $\mathbb{V}$ be a PVHS over the affine (orbi) curve $C' = C - D$. The Deligne–Beilinson cohomology $H_{\mathcal{D}}^{\bullet}(C', \mathbb{V})$ is the cohomology of the complex

$$\mathrm{cone}\big(F^0 W_0 \mathrm{Dec}_W K_{\mathbb{C}}^{\bullet}(C,D;\mathbb{V}) \oplus W_0 \mathrm{Dec}_W K_{\mathbb{Q}}^{\bullet}(C,D;\mathbb{V}) \\ \to W_0 \mathrm{Dec}_W K_{\mathbb{C}}^{\bullet}(C,D;\mathbb{V})\big)[-1].$$

Here Dec_W is Deligne's *filtration decalée* functor (with respect to $W_\bullet$), which is defined in [11, §1.3]. The DB-cohomology fits in an exact sequence

$$0 \to \mathrm{Ext}^1_{\mathsf{MHS}}\big(\mathbb{Q}, H^{j-1}(C',\mathbb{V})\big) \to H^j_{\mathcal{D}}(C',\mathbb{V}) \to \mathrm{Hom}_{\mathsf{MHS}}\big(\mathbb{Q}, H^j(C',\mathbb{V})\big) \to 0. \tag{18.1}$$

Deligne–Beilinson cohomology of a higher dimensional variety $X = \overline{X} - D$ with coefficients in a PVHS $\mathbb{V}$ can be defined using Saito's mixed Hodge complex that generalizes Zucker's.

The next result follows directly from the Manin-Drinfeld Theorem (Thm. 11.5) using the exact sequence (18.1).

Proposition 18.2. *The DB-cohomology $H^j_{\mathcal{D}}(\mathcal{M}_{1,1}, S^m\mathbb{H}(r))$ vanishes when m is odd and when $j > 2$. When $j = 1$ it vanishes except when $m = 2n$ and $r = 2n+1$, in which case it is isomorphic to $\mathbb{Q}$. When $j = 2$*

$$\begin{aligned} H^2_{\mathcal{D}}(\mathcal{M}_{1,1}, S^{2n}\mathbb{H}(r)) &= \operatorname{Ext}^1_{\mathsf{MHS}}\big(\mathbb{Q}, H^1(\mathcal{M}_{1,1}, S^{2n}\mathbb{H})(r)\big) \\ &\cong \operatorname{Ext}^1_{\mathsf{MHS}}\big(\mathbb{Q}, \mathbb{Q}(r-2n-1)\big) \oplus \bigoplus_f \operatorname{Ext}^1_{\mathsf{MHS}}\big(\mathbb{Q}, M_f(r)\big) \end{aligned}$$

where f ranges over the equivalence classes of $f \in \mathfrak{B}_{2n+2}$. □

As explained in Section 17.2.1, the real Frobenius $\mathcal{F}_\infty$ acts on $\mathcal{M}_{1,1}$ and on the local system $\mathbb{H}$. Since $\overline{\mathcal{F}}_\infty$ preserves the Hodge filtration, it acts on the complex used to compute $H^\bullet_{\mathcal{D}}(\mathcal{M}_{1,1}, S^{2n}\mathbb{H}_\mathbb{R}(r))$ and thus on the Deligne cohomology itself.

Corollary 18.3. *For all $n > 0$ and $r \in \mathbb{Z}$, there are natural $\overline{\mathcal{F}}_\infty$-equivariant isomorphisms*

$$H^2_{\mathcal{D}}(\mathcal{M}_{1,1}, S^{2n}\mathbb{H}_\mathbb{R}(r)) \cong \operatorname{Ext}^1_{\mathsf{MHS}}\big(\mathbb{R}, \mathbb{R}(r-2n-1)\big) \oplus \bigoplus_{f \in \mathfrak{B}_{2n+2}} \operatorname{Ext}^1_{\mathsf{MHS}}\big(\mathbb{R}, V_f(r)\big).$$

These vanish when $r < 2n+2$. The first term on the right hand side corresponds to the Eisenstein series G_{2n+2}. □

18.2. Deligne–Beilinson cohomology of affine groups

Here we recall the definition and basic properties of the Deligne–Beilinson cohomology of an affine group from [21]. Suppose that G is an affine $\mathbb{Q}$ group that is an extension of a reductive $\mathbb{Q}$ group R by a prounipotent group U. Suppose that the coordinate ring $\mathcal{O}(G)$ and its sub algebra $\mathcal{O}(R)$ are Hopf algebras in the category ind-MHS. Then the coordinate ring $\mathcal{O}(U)$ of U is also a Hopf algebra in ind-MHS. This implies that its Lie algebra $\mathfrak{u}$ is a pronilpotent Lie algebra in pro-MHS. Suppose that V is a Hodge representation (Defn. 8.2) of G. The Deligne–Beilinson cohomology of G is defined by

$$H^\bullet_{\mathcal{D}}(G, V) := \operatorname{Ext}^\bullet_{\mathsf{HRep}(G)}(\mathbb{Q}, V).$$

This has a natural product: if V_1, V_2 are in $\mathsf{HRep}(G)$, there are natural multiplication maps

$$H^\bullet_{\mathcal{D}}(G,V_1)\otimes H^\bullet_{\mathcal{D}}(G,V_2)\to H^\bullet_{\mathcal{D}}(G,V_1\otimes V_2).$$

The following result is proved in [21].

Proposition 18.4. *For all $j \geq 0$, there is a short exact sequence*

$$0\to \mathrm{Ext}^1_{\mathsf{MHS}}\big(\mathbb{Q},H^{j-1}(G,V)\big)\to H^j_{\mathcal{D}}(G,V)\to \mathrm{Hom}_{\mathsf{MHS}}\big(\mathbb{Q},H^j(G,V)\big)\to 0. \quad \square$$

18.3. Extensions of VMHS over curves

Suppose that $\mathbb{H}$ is a PVHS over C'. Fix a base point $x \in C'$ and let R_x be the Zariski closure of the monodromy action $\pi_1(C',x) \to \mathrm{Aut}\, H_x$. Denote the completion of $\pi_1(X,x)$ with respect to $\pi_1(C',x) \to R_x$ by $\mathcal{G}_x$. Theorem 8.3 implies that the category $\mathsf{MHS}(C',\mathbb{H})$ is equivalent to $\mathsf{HRep}(\mathcal{G}_x)$, so that there is a natural isomorphism

$$\mathrm{Ext}^\bullet_{\mathsf{MHS}(C',\mathbb{H})}(\mathbb{Q},\mathbb{V})\cong H^\bullet_{\mathcal{D}}(\mathcal{G}_x,V_x).$$

Even more is true:

Theorem 18.5 ([21]). *If $\mathbb{V}$ is an object of $\mathsf{MHS}(C',\mathbb{H})$, then there are natural isomorphisms*

$$\mathrm{Ext}^\bullet_{\mathsf{MHS}(C',\mathbb{H})}(\mathbb{Q},\mathbb{V})\cong H^\bullet_{\mathcal{D}}(\mathcal{G}_x,V_x)\xrightarrow{\simeq} H^\bullet_{\mathcal{D}}(C',\mathbb{V})$$

where V_x denotes the fiber of $\mathbb{V}$ over the base point x. These isomorphisms are compatible with the natural products.[26] $\square$

Denote the category of admissible variations of MHS over C' by $\mathsf{MHS}(C')$. Since the Deligne cohomology $H^\bullet_{\mathcal{D}}(C',\mathbb{V})$ does not depend on the choice of the basic variation $\mathbb{H}$ with $\mathbb{V}\in\mathsf{MHS}(C',\mathbb{H})$, we have the following useful result.

Corollary 18.6 ([21]). *For all admissible variations of MHS over C', there is a natural isomorphism*

$$\mathrm{Ext}^\bullet_{\mathsf{MHS}(C')}(\mathbb{Q},\mathbb{V})\cong H^\bullet_{\mathcal{D}}(C',\mathbb{V})$$

that is compatible with products.

[26] There is a more general result which applies when C' is replaced by a smooth variety X of arbitrary dimension. In that case, there is a natural homomorphism $H^\bullet_{\mathcal{D}}(\mathcal{G}_x,V_x)\to H^\bullet_{\mathcal{D}}(X,\mathbb{V})$ that is an isomorphism in degrees ≤ 1 and injective in degree 2. This is proved in [21].

18.4. Extensions of variations of MHS over modular curves

Suppose that Γ is a congruence subgroup of $\mathrm{SL}_2(\mathbb{Z})$. The following result follows from Corollary 18.6 and the Manin-Drinfeld Theorem (Thm. 11.5).

Proposition 18.7. *If $m > 0$ and A is a Hodge structure, then*

$$\mathrm{Ext}^1_{\mathsf{MHS}(X_\Gamma)}(\mathbb{Q}, A \otimes S^m\mathbb{H}) \cong \mathrm{Hom}_{\mathsf{MHS}}\big(\mathbb{Q}, A \otimes H^1(X_\Gamma, S^m\mathbb{H})\big).$$

When $m = 0$, $\mathrm{Ext}^1_{\mathsf{MHS}(X_\Gamma)}(\mathbb{Q}, A) = \mathrm{Ext}^1_{\mathsf{MHS}}(\mathbb{Q}, A)$. If A is a simple $\mathbb{Q}$-HS, then $\mathrm{Ext}^1_{\mathsf{MHS}(X_\Gamma)}(\mathbb{Q}, A \otimes S^m\mathbb{H})$ is non-zero if and only if either $A = \mathbb{Q}(m+1)$ or $A \cong M_f(m+1)$ for some Hecke eigen cusp form $f \in \mathfrak{B}_{m+2}(\Gamma)$. □

This result can be interpreted as a computation of the group of normal functions (tensored with $\mathbb{Q}$) over X_Γ associated to a PVHS of the form $\mathbb{V} = A \otimes S^m\mathbb{H}$. These are holomorphic sections of the bundle of intermediate jacobians associated to $\mathbb{V}$. The group of normal functions (essentially by definition) is isomorphic to $\mathrm{Ext}^1_{\mathsf{MHS}(X_\Gamma,\mathbb{H})}(\mathbb{Z}, \mathbb{V})$. The normal functions constructed in Section 13.4 generate all simple extensions and normal functions in $\mathsf{MHS}(X_\Gamma, \mathbb{H})$.

19 Cup Products and Relations in $\mathfrak{u}^{\mathrm{eis}}$

In this section we show that Pollack's quadratic relations lift to relations in $\mathfrak{u}^{\mathrm{eis}}$. In particular, we show that $\mathfrak{u}^{\mathrm{eis}}$ is not free. Throughout, the base point is $\vec{v} = \partial/\partial q$, although most of the arguments are valid with any base point. As before, H denotes the fiber of $\overline{\mathcal{H}}$ over $\vec{v}$. In this setup, $\mathbf{w} = 2\pi i\mathbf{b}$. We will omit the base point from the notation.

Recall from Section 17.3 that $\mathcal{D}$ denotes the image of the monodromy homomorphism $\mathcal{G} \to \mathrm{Aut}\,\mathfrak{p}$. As before, we take $S^{2n}H = S^{2n}(\mathbf{b}^{2n})$. Let $\check{\mathbf{e}}_{2n}$ be the element of

$$H^1(\mathcal{G}, S^{2n}H) \cong \mathrm{Hom}_{\mathrm{SL}(H)}\big(H_1(\mathfrak{u}), S^{2n}H\big)$$

that takes the class of $\mathbf{e}_{2n+2}$ to $\mathbf{b}^{2n}$. Recall that the real Frobenius $\overline{\mathcal{F}}_\infty$ acts on $H^\bullet_{\mathcal{D}}(\mathcal{M}_{1,1}, S^{2n}\mathbb{H}_\mathbb{R}(r))$.

Lemma 19.1. *For all $m > 0$, the homomorphisms $\mathcal{G} \to \mathcal{G}^{\mathrm{eis}} \to \mathcal{D}$ induce isomorphisms*

$$H^1_{\mathcal{D}}(\mathcal{D}, S^{2n}\mathbb{H}(2n+1)) \xrightarrow{\simeq} H^1_{\mathcal{D}}(\mathcal{G}^{\mathrm{eis}}, S^{2n}\mathbb{H}(2n+1))$$

$$\xrightarrow{\simeq} H^1_{\mathcal{D}}(\mathcal{G}, S^{2n}\mathbb{H}(2n+1)) \xrightarrow{\simeq} H^1_{\mathcal{D}}(\mathcal{M}_{1,1}, S^{2n}\mathbb{H}(2n+1))^{\overline{\mathcal{F}}_\infty}.$$

Each of these groups is a 1-dimensional $\mathbb{Q}$ *vector space. The first is spanned by* $\check{\epsilon}_{2n+2}$*, the last by* $\check{\mathbf{e}}_{2n+2}/2\pi i$*. The isomorphism identifies* $\check{\epsilon}_{2n+2}$ *with*

$$\frac{2}{(2n)!}\frac{\check{\mathbf{e}}_{2n+2}}{2\pi i}.$$

Proof. For all $n > 0$, each of the groups

$$H^1(\mathcal{D}, S^{2n}H),\ H^1(\mathcal{G}^{\mathrm{eis}}, S^{2n}H),\ H^1(\mathcal{G}^{\mathrm{eis}}, S^{2n}H)$$

is isomorphic to $\mathbb{Q}(-2n-1)$. The left-hand group is generated by $\check{\epsilon}_{2n+2}$ and the right two groups by $\check{\mathbf{e}}_{2n+2}$. Theorem 15.7 implies that the projections $\mathcal{G} \to \mathcal{G}^{\mathrm{eis}} \to \mathcal{D}$ take $\check{\epsilon}_{2n+2}$ to $2\check{\mathbf{e}}_{2n+2}/2\pi i(2n)!$, so that the homomorphisms induced by the projections are isomorphisms.

Proposition 18.4 implies that the projections $\mathcal{G} \to \mathcal{G}^{\mathrm{eis}} \to \mathcal{D}$ induce isomorphisms

$$H^1_{\mathcal{D}}\big(\mathcal{G}^{\mathrm{eis}}, S^{2n}\mathbb{H}(2n+1)\big) \xrightarrow{\simeq} H^1_{\mathcal{D}}\big(\mathcal{G}, S^{2n}\mathbb{H}(2n+1)\big) \cong \mathbb{Q}$$

for all $n > 0$. The corresponding class in Deligne cohomology is easily seen to be $\overline{\mathcal{F}}_\infty$ invariant. □

Lemma 19.2. *There is a natural inclusion*

$$iH^1_{\mathrm{cusp}}(\mathcal{M}_{1,1}, S^{2m}\mathbb{H}_{\mathbb{R}})^{\overline{\mathcal{F}}_\infty} \hookrightarrow H^2_{\mathcal{D}}\big(\mathcal{D}, S^{2m}H_{\mathbb{R}}(2m+2)\big)$$

Proof. Since $\mathfrak{u}$ is free, $H^2(\mathcal{G}, S^m H) \cong H^2(\mathfrak{u}, S^m H)^{\mathrm{SL}(H)}$ vanishes for all $m > 0$. The result follows from Proposition 18.4, the computation (Thm. 7.17) of $H^1(\mathfrak{u})$ and the isomorphism

$$\begin{aligned}&\mathrm{Ext}^1_{\mathsf{MHS}}\big(\mathbb{R}, H^1_{\mathrm{cusp}}(\mathcal{M}_{1,1}, S^{2m}\mathbb{H}(2m+2))\big)^{\overline{\mathcal{F}}_\infty}\\ &\cong iH^1_{\mathrm{cusp}}(\mathcal{M}_{1,1}, S^{2m}\mathbb{H}_{\mathbb{R}})^{\overline{\mathcal{F}}_\infty} = \bigoplus_{f\in\mathfrak{B}_{2m+2}} iV_f^-.\end{aligned}$$

which is well defined up to an even power of $2\pi i$ that depends upon the choice of the first isomorphism. □

By Corollary 18.3, there is an $\overline{\mathcal{F}}_\infty$ invariant projection

$$H^2_{\mathcal{D}}\big(\mathcal{M}_{1,1}, S^{2n}\mathbb{H}_{\mathbb{R}}(2n+2)\big) \to \mathrm{Ext}^1_{\mathsf{MHS}}\big(\mathbb{R}, V_f(2n+2)\big).$$

The following computation is the key to proving that Pollack's quadratic relations are motivic.

Theorem 19.3 (Brown [3, Thm. 11.1], Terasoma [43, Thm. 7.3]). *If $j,k>0$ and $n=j+k$ and if there is a cup form of weight $2n+2$, then the image of the cup product*

$$H^1_{\mathcal{D}}(\mathcal{M}_{1,1},S^{2j}\mathbb{H}_{\mathbb{R}}(2j+1))\otimes H^1_{\mathcal{D}}(\mathcal{M}_{1,1},S^{2k}\mathbb{H}_{\mathbb{R}}(2k+1))$$
$$\to H^2_{\mathcal{D}}(\mathcal{M}_{1,1},S^{2n}\mathbb{H}_{\mathbb{R}}(2n+2))$$

is non-zero. More precisely, the composition of the cup product with the projection

$$H^2_{\mathcal{D}}(\mathcal{M}_{1,1},S^{2n}\mathbb{H}_{\mathbb{R}}(2n+2))\to \operatorname{Ext}^1_{\mathsf{MHS}}\big(\mathbb{R},V_f(2n+2)\big)^{\overline{\mathcal{F}}_\infty}\cong V^-_{f,\mathbb{R}}$$

is non-trivial for all $f\in\mathfrak{B}_{2n+2}$. □

As we shall show below, a direct consequence is that Pollack's quadratic relations hold in $\mathfrak{u}^{\mathrm{eis}}$. Brown's period computations [3] are more detailed and imply that all of Pollack's relations lift from a quotient of $\mathrm{Der}^0\,\mathfrak{p}$ to relations in $\mathfrak{u}^{\mathrm{eis}}$ and are therefore motivic. Full details will appear in [23].

Theorem 19.4. *Pollack's quadratic relations (17.11) hold in $\mathrm{Gr}^W_\bullet\,\mathfrak{u}^{\mathrm{eis}}$. In particular, the pronilpotent radical $\mathfrak{u}^{\mathrm{eis}}$ of $\mathfrak{g}^{\mathrm{eis}}$ is not free.*

We use the notation $\Gamma V := \mathrm{Hom}_{\mathsf{MHS}}(\mathbb{Q},V)$ to denote the set of Hodge classes of type $(0,0)$ of a MHS V.

Proof. Suppose that $n>0$. Since the Hodge structure $H^1_{\mathrm{cusp}}(\mathcal{M}_{1,1},S^{2n}\mathbb{H}(2n+2))$ does not occur in $H^1(\mathfrak{u}^{\mathrm{eis}})$, Proposition 18.4 and the Manin-Drinfeld Theorem imply that the homomorphism

$$\begin{aligned} H^2_{\mathcal{D}}(\mathcal{G}^{\mathrm{eis}},S^{2n}\mathbb{H}(2n+2)) &\to H^2_{\mathcal{D}}(\mathcal{G},S^{2n}\mathbb{H}(2n+2))\\ &\cong \operatorname{Ext}^1_{\mathsf{MHS}}\big(\mathbb{Q},H^1(\mathcal{M}_{1,1},S^{2n}\mathbb{H}(2n+2))\big)\\ &\to \operatorname{Ext}^1_{\mathsf{MHS}}\big(\mathbb{Q},H^1_{\mathrm{cusp}}(\mathcal{M}_{1,1},S^{2n}\mathbb{H}(2n+2))\big)\end{aligned}$$

induces a well-defined map

$$r:\Gamma H^2(\mathcal{G}^{\mathrm{eis}},S^{2n}\mathbb{H}(2n+2))\to \operatorname{Ext}^1_{\mathsf{MHS}}\big(\mathbb{Q},H^1_{\mathrm{cusp}}(\mathcal{M}_{1,1},S^{2n}\mathbb{H}(2n+2))\big).$$

Compose this with the projection to

$$\operatorname{Ext}^1_{\mathsf{MHS}}\big(\mathbb{R},H^1_{\mathrm{cusp}}(\mathcal{M}_{1,1},S^{2n}\mathbb{H}(2n+2))\big)\cong\bigoplus_{f\in\mathfrak{B}_{2n+2}}V_{f,\mathbb{R}}$$

to obtain a projection

$$p:\Gamma H^2(\mathcal{G}^{\mathrm{eis}},S^{2n}\mathbb{H}(2n+2))\to\bigoplus_{f\in\mathfrak{B}_{2n+2}}V_{f,\mathbb{R}}.$$

Consider the diagram

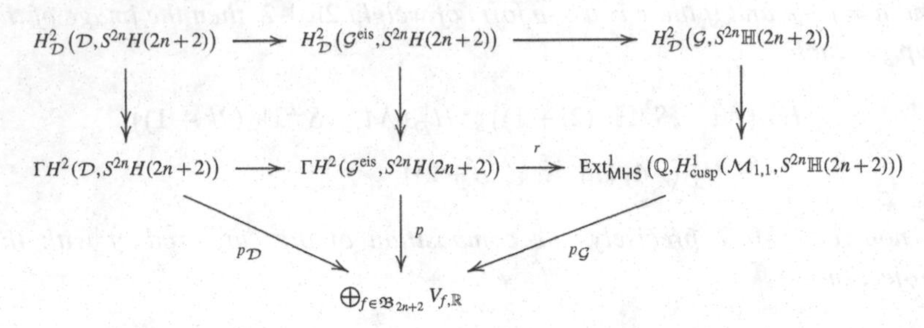

where $p_{\mathcal{D}}$ is the projection dual to the Pollack relations (Prop. 17.12), p is the projection constructed above and $p_{\mathcal{G}}$ is the standard projection. Naturality implies that the top left square commutes; the right hand square commutes by the construction of r; the bottom right triangle commutes by the definition of p. The cup product computation (Thm. 19.3) implies that the sum of the two triangles commute. It follows that the diagram commutes.

Suppose that $j,k > 0$ satisfy $n = j+k$. Lemma 19.1 and Proposition 17.12 now imply that the composite of

$$H^1\big(\mathcal{G}^{\mathrm{eis}}, S^{2j}H(2j+1)\big) \otimes H^1\big(\mathcal{G}^{\mathrm{eis}}, S^{2k}, S^{2k}H(2j+1)\big) \to \Gamma H^2\big(\mathcal{G}^{\mathrm{eis}}, S^{2n}H(2n+2)\big)$$

with the projection p is given by the formula (17.14). The result now follows from the duality between cup product and quadratic relations. □

Much of the discussion in this section can be generalized to modular curves X_Γ where Γ is a congruence subgroup of $\mathrm{SL}_2(\mathbb{Z})$. In particular, the prounipotent radical $\mathfrak{u}_\Gamma^{\mathrm{eis}}$ is not free for all congruence subgroups.

Remark 19.5. This result implies that, when $\Gamma = \mathrm{SL}_2(\mathbb{Z})$, Manin's quotient $\mathfrak{u}_B$ of $\mathfrak{u}_\Gamma$ (cf. 13.2) is not a quotient of $\mathfrak{u}_\Gamma$ in the category of Lie algebras with a MHS. If it were, it would $\mathfrak{u}_\Gamma^{\mathrm{eis}}$. But since $\mathfrak{u}_B$ is free, and since $\mathfrak{u}_\Gamma^{\mathrm{eis}}$ is not, $\mathcal{U}_B \to \mathcal{U}_\Gamma^{\mathrm{eis}}$ cannot be an isomorphism.

References

[1] A. Beilinson, A. Levin: *The elliptic polylogarithm*, in Motives (Seattle, WA, 1991), 123–190, Proc. Sympos. Pure Math., 55, Part 2, Amer. Math. Soc., 1994.

[2] A. Borel, J.-P. Serre: *Théorèmes de finitude en cohomologie galoisienne*, Comment. Math. Helv. 39 (1964), 111–164.

[3] F. Brown: *Multiple modular values for* $\mathrm{SL}_2(\mathbb{Z})$, [arXiv:1407.5167]

[4] F. Brown, A. Levin: *Multiple Elliptic Polylogarithms*, [arXiv:1110.6917]

[5] J. Carlson, R. Hain: *Extensions of variations of mixed Hodge structure*, Theorie de Hodge, Luminy, Juin, 1987, Asterisque no. 179-180, 39–65.

[6] D. Calaque, B. Enriquez, P. Etingof: *Universal KZB equations: the elliptic case*, in Algebra, arithmetic, and geometry: in honor of Yu. I. Manin. Vol. I, 165–266, Progr. Math., 269, Birkhäuser, Boston, 2009, [arXiv:math/0702670]

[7] K.-T. Chen: *Extension of C^∞ function algebra by integrals and Malcev completion of π_1*, Advances in Math. 23 (1977), 181–210.

[8] J. Conant: *The Johnson cokernel and the Enomoto-Satoh invariant*, Algebr. Geom. Topol. 15 (2015), 801–821. [arXiv:1306.3698]

[9] J. Conant, M. Kassabov, K. Vogtmann: *Hairy graphs and the unstable homology of* Mod(g,s), Out(F_n) *and* Aut(F_n), J. Topol. 6 (2013), 119–153. [arXiv:1107.4839]

[10] J. Conant, M. Kassabov, K. Vogtmann: *Higher hairy graph homology*, Geom. Dedicata 176 (2015), 345–374. [arXiv:1308.3825]

[11] P. Deligne: *Théorie de Hodge, II*, Inst. Hautes Études Sci. Publ. Math. No. 40 (1971), 5–57.

[12] P. Deligne: *Le groupe fondamental de la droite projective moins trois points*, Galois groups over $\mathbb{Q}$ (Berkeley, CA, 1987), 79–297, Math. Sci. Res. Inst. Publ., 16, Springer, 1989.

[13] R. Hain: *The geometry of the mixed Hodge structure on the fundamental group*, Algebraic geometry, Bowdoin, 1985, 247–282, Proc. Sympos. Pure Math., 46, Part 2, Amer. Math. Soc., 1987.

[14] R. Hain: *The de Rham homotopy theory of complex algebraic varieties, I*, K-Theory 1 (1987), 271–324.

[15] R. Hain: *Hodge-de Rham theory of relative Malcev completion*, Ann. Sci. École Norm. Sup., t. 31 (1998), 47–92.

[16] R. Hain: *Infinitesimal presentations of the Torelli groups*, J. Amer. Math. Soc. 10 (1997), 597–651.

[17] R. Hain: *Iterated integrals and algebraic cycles: examples and prospects*, Contemporary trends in algebraic geometry and algebraic topology (Tianjin, 2000), 55–118, Nankai Tracts Math., 5, World Sci. Publ., 2002.

[18] R. Hain: *Relative weight filtrations on completions of mapping class groups*, in Groups of Diffeomorphisms, Advanced Studies in Pure Mathematics, vol. 52 (May, 2008), 309–368, Mathematical Society of Japan. [arXiv:0802.0814]

[19] R. Hain: *Lectures on Moduli Spaces of Elliptic Curves*, Transformation Groups and Moduli Spaces of Curves, pp. 95–166, Adv. Lect. Math. (ALM), 16, Int. Press, Somerville, MA, 2011. [arXiv:0812.1803]

[20] R. Hain: *Notes on the Universal Elliptic KZB Equation*, preprint, 2013. [arXiv:1309.0580]

[21] R. Hain: *Deligne–Beilinson cohomology of affine groups*, preprint, 2015. [arXiv:1507.03144]

[22] R. Hain, M. Matsumoto: *Weighted completion of Galois groups and Galois actions on the fundamental group of $\mathbb{P}^1 - \{0,1,\infty\}$*, Compositio Math. 139 (2003), 119–167.

[23] R. Hain, M. Matsumoto: *Universal mixed elliptic motives*, preprint, 2015.

[24] R. Hain, G. Pearlstein: *Tannakian fundamental groups of categories of variations of mixed Hodge structure*, in preparation, 2015.

[25] R. Hain, S. Zucker: *Unipotent variations of mixed Hodge structure*, Invent. Math. 88 (1987), 83–124.

[26] R. Hain, S. Zucker: *A guide to unipotent variations of mixed Hodge structure*, Hodge theory (Sant Cugat, 1985), 92 106, Lecture Notes in Math., 1246, Springer, 1987.

[27] S. Lang: *Introduction to Modular Forms, with appendixes by D. Zagier and Walter Feit.*Corrected reprint of the 1976 original. Grundlehren der Mathematischen Wissenschaften, 222. Springer-Verlag, Berlin, 1995.

[28] A. Levin, G. Racinet: *Towards multiple elliptic polylogarithms*, unpublished preprint, 2007, [arXiv:math/0703237]

[29] Y. Manin: *Iterated integrals of modular forms and noncommutative modular symbols*, Algebraic geometry and number theory, 565–597, Progr. Math., 253, Birkhäuser, 2006.

[30] Y. Manin: *Iterated Shimura integrals*, Mosc. Math. J. 5 (2005), 869–881.

[31] J. Morgan: *The algebraic topology of smooth algebraic varieties*, Inst. Hautes Études Sci. Publ. Math. No. 48 (1978), 137–204.

[32] G. Mostow: *Fully reducible subgroups of algebraic groups* Amer. J. Math. 78 (1956), 200–221.

[33] A. Pollack: *Relations between derivations arising from modular forms*, undergraduate thesis, Duke University, 2009.[27]

[34] M. Saito: *Modules de Hodge polarisables*, Publ. Res. Inst. Math. Sci. 24 (1988), 849–995.

[35] M. Saito: *Mixed Hodge modules*, Publ. Res. Inst. Math. Sci. 26 (1990), 221–333.

[36] W. Schmid: *Variation of Hodge structure: the singularities of the period mapping*, Invent. Math. 22 (1973), 211–319.

[37] J.-P. Serre: *Lie algebras and Lie groups, 1964 lectures given at Harvard University*, Second edition. LNM 1500. Springer-Verlag, 1992.

[38] J.-P. Serre: *A course in arithmetic*, Translated from the French. GTM 7, Springer-Verlag, 1973.

[39] C. Simpson: *Higgs bundles and local systems*, Inst. Hautes Études Sci. Publ. Math. No. 75 (1992), 5–95.

[40] J. Stallings: *Homology and central series of groups*, J. Algebra 2 (1965), 170–181.

[41] J. Steenbrink, S. Zucker: *Variation of mixed Hodge structure, I*, Invent. Math. 80 (1985), 489–542.

[42] D. Sullivan: *On the intersection ring of compact three manifolds*, Topology 14 (1975), 275–277.

[43] T. Terasoma: *Relative Deligne cohomologies and higher regulators for Kuga-Sato fiber spaces*, manuscript dated January, 2011.

[44] W. Wasow: *Asymptotic expansions for ordinary differential equations*. Pure and Applied Mathematics, Vol. XIV Interscience Publishers John Wiley & Sons, 1965.

[45] S. Zucker: *Hodge theory with degenerating coefficients, L_2 cohomology in the Poincaré metric*, Ann. of Math. (2) 109 (1979), 415–476.

Department of Mathematics Duke University Durham, NC 27708-0320
E-mail address: hain@math.duke.edu

[27] Available at: http://dukespace.lib.duke.edu/dspace/handle/10161/1281

Printed in the United States
by Baker & Taylor Publisher Services